# STRUCTURE AND DEVELOPMENT OF
# SOLAR ACTIVE REGIONS

INTERNATIONAL ASTRONOMICAL UNION
UNION ASTRONOMIQUE INTERNATIONALE

SYMPOSIUM No. 35
HELD IN BUDAPEST, HUNGARY, 4–8 SEPTEMBER 1967

# STRUCTURE AND DEVELOPMENT OF SOLAR ACTIVE REGIONS

EDITED BY

K. O. KIEPENHEUER

*(Fraunhofer Institut, Freiburg im Br.)*

D. REIDEL PUBLISHING COMPANY

DORDRECHT-HOLLAND

1968

*Published on behalf of*
*the International Astronomical Union*
*by*
*D. Reidel Publishing Company, Dordrecht, Holland*


© *International Astronomical Union, 1968*

*Softcover reprint of the hardcover 1st edition 1968*

ISBN-13: 978-94-011-6817-5      e-ISBN-13: 978-94-011-6815-1
DOI: 10.1007/978-94-011-6815-1

# CONTENTS

List of participants           XIII

L. Detre     Welcome by the Hungarian Academy of Sciences  1

K.O. Kiepenheuer  Preface and introduction  3

PART I

## GENERAL DEVELOPMENT OF AN ACTIVE REGION

V. Bumba, R. Howard,  Patterns of active region magnetic field development  13
M.J. Martres, and
I. Soru-Iscovici

M.J. Martres   Origine des régions actives solaires 'anormales'  25

Sara F. Smith and  Magnetic classification of active regions  33
Robert Howard

E. Tandberg-Hanssen  Development of magnetic fields in active regions  43
and C. Porter

Jan Olof Stenflo   The balance of magnetic fluxes in active regions  47

T. Fortini and   On the birth of some proton-flare regions  50
M. Torelli

H.W. Dodson and  Some patterns in the development of centers of solar
E.R. Hedeman   activity, 1962–66  56

V. Bumba, J. Kleczek,  Last phases of development of active regions  64
J. Olmr, B. Růžičková-
Topolova, and J. Sýkora

S.I. Gopasyuk   On some properties of the velocity field in a developed active region  68

L. Dezső, O. Gerlei, and The migration of sunspot activity along solar me-
Ágnes Kovács   ridians and parallels  70

David M. Rust   Chromospheric explosions and satellite sunspots  77

C.J. Macris and   Study of an active region of the Sun during three
T.J. Prokakis   rotation periods  85

PART II

## THEORETICAL ASPECTS

H. U. Schmidt — Magnetohydrodynamics of an active region — 95

G. W. Simon and N. O. Weiss — Concentration of magnetic fields in the deep convection zone — 108

R. E. Danielson and B. D. Savage — Oscillatory modes of energy transport in solar magnetic regions — 112

F. A. Ermakov, E. I. Mogilevsky, and B. D. Shelting — On the magnetic structure of an active region — 126

S. I. Syrovatsky and Y. D. Zhugzhda — Oscillatory convection in strong magnetic fields and origin of active regions — 127

M. Kopecký and G. V. Kuklin — Quantitative estimations of the anomalous plasma diffusion in an active region — 131

J. Rayrole et M. Semel — Étude critique d'un champ 'current-free' dans l'atmosphère solaire — 134

G. F. Anderson and D. H. Menzel — Sunspots and magnetohydrodynamic flows — 142

PART III

## OPTICAL STRUCTURE OF AN ACTIVE REGION

Z. B. Korobova, A. K. Tchandaev, and G. Y. Vassilyeva — On the state of the photosphere before the appearance of sunspots — 151

G. A. Chapman and N. R. Sheeley, Jr. — Correlations between brightness fields and magnetic fields on the Sun — 161

M. G. Dmitrieva, M. Kopecký, and G. V. Kuklin — The supergranular pattern and the stable stages of sunspot groups — 174

J. M. Beckers and E. H. Schröter — The intensity, velocity and magnetic structure in and around a sunspot — 178

W. Mattig and J. P. Mehltretter — Fine structure of brightness, velocity and magnetic field in the penumbra — 187

CONTENTS VII

M. D. Altschuler, Y. Nakagawa, and C. G. Lilliequist — Concerning the development of the Evershed motion in sunspots — 193

N. V. Steshenko — The connection of fine-structure photospheric features in active regions with magnetic fields — 201

O. Kjeldseth Moe — On the magnetic-field configuration in sunspots — 202

G. V. Kuklin — The proper motions of sunspots and the magnetic field of active regions — 211

H. Künzel — Fluctuations of the magnetic-field strength of sunspots within one day — 214

E. I. Mogilevsky, L. B. Demkina, B. A. Ioshpa, and V. N. Obridko — On the structure of the magnetic field of sunspots — 215

F. L. Deubner — Magnetic and Doppler oscillations in active regions — 230

A. B. Severny — Preliminary communication on the short-period oscillations of solar magnetic fields — 233

G. Y. Vassilyeva and A. K. Tchandaev — Some comments about correlations between magnetic field and velocity, magnetic field and line intensity in the undisturbed photosphere — 236

Yngve Öhman — On some spectrographic observations related to the structure with height of active regions and particularly solar flares — 240

M. K. V. Bappu and K. R. Sivaraman — Chromospheric heights in active regions — 247

E. Dubov — The structure of the lower solar chromosphere in undisturbed and active regions — 255

Eberhard Wiehr — Problems in the interpretation of polarization measurements in active regions — 259

B. A. Ioshpa — On the magnetic-field structure around filaments — 261

Sara F. Smith — The formation, structure and changes in filaments in active regions — 267

J. Kleczek — Prominences in active regions — 280

Boris Valníček — The 'detwisted' prominence of September 12, 1966 — 282

Z. Švestka                Loop-prominence systems and proton-flare active
                          regions                                                287

A. Bruzek                 Bright points (moustaches) and arch filaments in
                          young active regions                                  293

Kerstin Fredga            Solar active regions in Mg II light                   299

## PART IV

# COOPERATIVE STUDY OF SOLAR ACTIVE REGIONS (CSSAR)

R. Michard                Introductory report                                  307

V. Bumba, L. Křivský,     Flare activity and spotgroup development             311
M.J. Martres, and
I. Soru-Iscovici

M.J. Martres,             A study of the localization of flares in selected active
R. Michard,               regions                                              318
I. Soru-Iscovici, and
T. Tsap

G. Godoli and             Evolution of Ca plages of the CSSAR active regions   326
B.C. Monsignori Fossi

V. Bumba and              Correlation between Ca plages and longitudinal mag-
G. Godoli                 netic fields of the CSSAR active regions             338

J.L. Leroy, J. Rösch, and Évolution des émissions coronales au cours de la vie
M. Trellis                d'un centre actif                                    346

## PART V

# CORONAL AND INTERPLANETARY STRUCTURE
# OF AN ACTIVE REGION

Audouin Dollfus           Observation des jets et concentrations de la couronne
                          au-dessus des régions actives                        359

G. Newkirk,               Influence of magnetic fields on the structure of the
M.D. Altschuler, and      solar corona                                         379
J. Harvey

R. Tousey, G.D. Sandlin,  Photographs of coronal streamers from a rocket on
and M.J. Koomen           May 9, 1967                                          385

B. Bednářová-Nováková and J. Halenka — New aspects of the role of development and structure of solar active regions in the arrangements of the corona based on its geomagnetic displays — 389

J. M. Wilcox, N. F. Ness, and K. H. Schatten — Active regions and the interplanetary magnetic field — 390

W. M. Burton — Extreme ultraviolet observations of active regions in the solar corona — 395

C. Y. Fan, M. Pick, R. Ryle, J. A. Simpson, and D. R. Smith — Protons associated with centres of solar activity and their propagation in interplanetary magnetic-field regions co-rotating with the Sun — 403

Werner M. Neupert — The solar corona above active regions: a comparison of extreme ultraviolet line emission with radio emission — 404

R. Tousey, G. D. Sandlin, and J. D. Purcell — On some aspects of XUV spectroheliograms — 411

R. Michard and Mme E. Ribes — La composante lentement variable des rayons X solaires en relation avec la structure des centres d'activité — 420

K. A. Pounds, K. Evans and P. C. Russell — X-radiation studies of the corona — 431

L. W. Acton and Philip C. Fisher — Observations of energetic X-rays from quiescent solar active regions — 432

G. Elwert — X-ray picture of the Sun taken with Fresnel zone plates — 439

G. Elwert — The significance of the polarization of solar short-wavelength X-rays — 444

PART VI

TRANSIENT PHENOMENA

R. Falciani, M. Landini, A. Righini, and M. Rigutti — Analysis of some solar flares from optical, X-ray, and radio observations — 451

L. Křivský — Interaction of magnetic fields and the origin of proton flares — 465

| P. A. Sturrock | A model of solar flares | 471 |
| C. de Jager | The high-energy flare plasma | 480 |
| J. Houtgast | The occurrence and possible meaning of the 'nimbus' | 483 |
| Friedrich Meyer | Flare-produced coronal waves | 485 |
| R. L. Arnoldy, S. R. Kane, and J. R. Winckler | The observation of 10–50 KeV solar flare X-rays by the OGO satellites and their correlation with solar radio and energetic particle emission | 490 |

PART VII

## PROTON FLARE PROJECT (PFP)

| Z. Švestka | Introduction and summary | 513 |
| J. H. Kinsey and F. B. McDonald | Observations of the solar proton event of August 28, 1966 | 536 |
| Constance Sawyer | Sunspot changes following proton flares | 543 |

PART VIII

## RADIO STRUCTURE OF AN ACTIVE REGION

| A. D. Fokker | Homology of solar radio events | 553 |
| B. Clavelier | Some results on solar activity at 408 MHz | 556 |
| A. Boischot and B. Clavelier | Conditions of acceleration of solar electrons, and determination of the magnetic field in the high corona from the characteristics of a type-IV burst | 565 |
| A. Böhme | Tendencies to repeating of type-IVm bursts and their relations to the stage of development of the sunspot group | 570 |
| V. G. Nagnibeda | Properties of sources of the slowly varying component of 2 cm solar radio emission | 575 |
| G. Swarup, M. R. Kundu, V. K. Kapahi, and J. D. Isloor | Some properties of the sources of slowly varying component and of bursts at 612 Mc/s | 581 |

R.G. Stone,                          Satellite observations of solar radio bursts              585
H.H. Malitson,
J.K. Alexander, and
C.R. Somerlock

V. Efanov, I. Moiseev,               Comparison of 8-mm solar radio features with local
and A. Severny                       magnetic fields and chromospheric features               588

J. Kleczek, J. Olmr, and             Radio emission of spotgroups                              594
A. Krüger

A. Koeckelenbergh                    Quelques relations entre sursauts radioélectriques
                                     solaires sur ondes décimétriques et caractères mor-
                                     phologiques des éruptions chromosphériques as-
                                     sociées                                                   598

S.J. Gopasyuk,                       The effect of compression and expansion of plasma
N.N. Erushev, and                    on the generation of synchrotron radiation               600
Y.I. Neshpor

J.P. Castelli, J. Aarons,            The great burst of May 23, 1967                           601
and G.A. Michael

## SUMMARIZING REVIEW

C. de Jager                          The development and structure of an active region   602

# LIST OF PARTICIPANTS

Acton, Loren W., Lockheed Research Laboratory, Palo Alto, U.S.A.
Altschuler, Martin, High Altitude Observatory, Boulder, U.S.A.
Ambroz, P., Observatory, Ondřejov, C.S.S.R.
Anderson, Gerald, Harvard Observatory, Cambridge, Mass., U.S.A.
Banin, Valery, Sibizmir, Irkutsk, U.S.S.R.
Bappu, M.K., Vainu, Observatory, Kodaikanal, India
Beckers, Jaques M., Sacramento Peak Observatory, Sunspot, U.S.A.
Bednárová-Nováková, Bohumila, Geophysical Institute, Praha, C.S.S.R.
Bhavilai, Rawi, University, Bangkok, Thailand
Boischot, André, Observatory, Meudon, France
Brandt, John C., Goddard Space Flight Center, Greenbelt, Md., U.S.A.
Bruzek, Anton, Fraunhofer Institute, Freiburg, Germany
Bumba, Vaclav, Observatory, Ondřejov, C.S.S.R.
Burton, William, Culham Laboratory, Abingdon, England
Castelli, John P., A.F. Cambridge Research Laboratory, Bedford, Mass., U.S.A.
Chapman, Sydney, High Altitude Observatory, Boulder, Colo., U.S.A.
Cimino, Massimo, Observatory Monte Mario, Roma, Italy
Clavelier, Bernard, Observatory, Meudon, France
Culhane, John L., Physics Dept., University College, London, England
Danielson, Robert E., Observatory, Princeton, N.J., U.S.A.
Davis, Leverett, California Institute of Technology, Pasadena, Calif., U.S.A.
De Feiter, Leendert D., Observatory, Utrecht, Holland
De Graaff, Willem, Space Research Laboratory, Utrecht, Holland
De Jager, Cornelis, Observatory, Utrecht, Holland
Demkina, Ljudmila, Izmiran, Moscow, U.S.S.R.
Deubner, Franz Ludwig, Fraunhofer Institute, Freiburg, Germany
Dizer, Muammer, Kandilli Observatory, Istanbul, Turkey
Dodson-Prince, Helen, McMath-Hulbert Obs., Pontiac, U.S.A.
Dolginova, Julia, Izmiran, Moscow, U.S.S.R.
Dollfus, Audouin, Observatory, Meudon, France
Dubov, Emil, Crimean Astrophysical Observatory, Nauchny, U.S.S.R.
Dunn, Richard, Sacramento Peak Observatory, Sunspot, U.S.A.
Elgarøy, Øistein, Institute Theoretical Astrophysics, Oslo, Norway
Elwert, Gerhard, Institute Theoretical Physics, University, Tübingen, Germany
Evans, John W., Sacramento Peak Observatory, Sunspot, U.S.A.
Evans, Kenton, University, Leicester, England

Falciani, Roberto, Observatory Arcetri, Firenze, Italy
Fokker, Adriaan D., Observatory, Utrecht, Holland
Fortini, Teresa, Observatory Monte Mario, Roma, Italy
Fredga, Kerstin, Dept. Plasma Physics, Stockholm, Sweden
Glaser, Harold, NASA, Washington, D.C., U.S.A.
Glencross, William, University College, London, Dorking, England
Godoli, Giovanni, Observatory, Catania, Italy
Goldberg, Nonna, Pulkovo Observatory, Leningrad, U.S.S.R.
Gopasyuk, Stepan, Crimean Observatory, Nauchny, U.S.S.R.
Grossmann-Doerth, Ulrich, European Space Research Organization, Paris, France
Gulbrandsen, Arild, Institute Cosmic Physics, Oslo, Norway
Halenka, Jaroslav, Geophysical Institute, Praha, C.S.S.R.
Hearn, Antony G., Culham Laboratory, Abingdon, England
Hedeman, E. Ruth, McMath-Hulbert Observatory, Pontiac, U.S.A.
Houtgast, Jacob, Observatory, Utrecht, Holland
Howard, Robert, Mt. Wilson Observatory, Pasadena, Calif., U.S.A.
Huber, Martin, Harvard Observatory, Cambridge, Mass., U.S.A.
Jäger, Friedrich W., Sonnenobservatorium, Potsdam, Germany
Jefferies, John T., University Hawaii, Honolulu, U.S.A.
Jensen, Eberhart, Institute Theoretical Astrophysics, Oslo, Norway
Jordan, Carole, University College, London, England
Khromova, Tatiana, Crimean Astrophysical Observatory, Nauchny, U.S.S.R.
Kiepenheuer, Karl-Otto, Fraunhofer Institute, Freiburg, Germany
Kinsey, James H., Dept. of Astronomy, Maryland University, College Park, U.S.A.
Kleczek, Josip, Observatory, Ondřejov, C.S.S.R.
Knuth, Robert, Ionospheric Observatory, Kühlungsborn, Germany
Koeckelenbergh, André, Observatoire Royal de Belgique, Bruxelles, Belgium
Kopecky, Miloslav, Observatory, Ondřejov, C.S.S.R.
Krat, Vladimir, Pulkovo Observatory, Leningrad, U.S.S.R.
Krawiecka, Jadwiga, Astronomical Institute, Wrocław, Poland
Křivský, Ladislav, Observatory, Ondřejov, C.S.S.R.
Krüger, Albrecht, Heinrich-Hertz Institut, Berlin, Germany
Kuklin, Georgy, Sibizmir, Irkutsk, U.S.S.R.
Kundu, Mukul R., Tata Institute, Bombay, India
Künzel, Horst, Sonnenobservatorium, Potsdam, Germany
Laffineur, Marius, Institut Astrophysique, Paris, France
Landini, Massimo, Observatory Arcetri, Firenze, Italy
Lantos, Pierre, Observatory, Meudon, France
Leighton, Robert B., California Institute of Technology, Pasadena, Calif., U.S.A.
Letfus, Vojteck, Observatory, Ondřejov, C.S.S.R.
Lundbak, Asger, Meteorological Institute, Copenhagen, Denmark

Maltby, Per, Institute Theoretical Astrophysics, Oslo, Norway
Malville, John McKim, High Altitude Observatory, Boulder, Colo., U.S.A.
Mamedov, Mutallim A.O., Shemakha Observatory, Baku, U.S.S.R.
Martres, Marie J., Observatory, Meudon, France
Mattig, Wolfgang, Fraunhofer Institute, Freiburg, Germany
Mavridis, Lyssimachos, Dept. Geodesy and Astronomy, University, Thessaloniki, Greece
McIntosh, Patrick, E.S.S.A., Boulder, Colo., U.S.A.
Mergentaler, Jan, Astronomical Institute, Wrocław, Poland
Meyer, Friedrich, Max-Planck-Institute, München, Germany
Miroshnichenko, Leonty, Izmiran, Moscow, U.S.S.R.
Mogilevsky, Emanuil, Izmiran, Moscow, U.S.S.R.
Monin, Georgij Aleksej, Crimean Astrophysical Observatory, Nauchny, U.S.S.R.
Nagnibeda, Valerij, University Observatory, Leningrad, U.S.S.R.
Nešpor, Juri, Crimean Astrophysical Observatory, Nauchny, U.S.S.R.
Ness, Norman, Goddard Space Flight Center, Greenbelt, Md., U.S.A.
Neupert, Werner, Goddard Space Flight Center, Greenbelt, Md., U.S.A.
Neven, Luc G.J., Observatoire Royal de Belgique, Bruxelles, Belgium
Newkirk, Gordon, High Altitude Observatory, Boulder, Colo., U.S.A.
Noyes, John, Boeing Scientific Research Laboratory, Seattle, Wash., U.S.A.
Nussbaumer, Harry, Dept. Physics, University College, London, England
Öhman, Yngve, Stockholm Observatory, Saltsjöbaden, Sweden
Orrall, Frank, Hawaii Geophysical Institute, Honolulu, U.S.A.
Paciorek, Jana, Astronomical Institute, Wrocław, Poland
Parkinson, William H., Harvard University, Cambridge, Mass., U.S.A.
Pasachoff, Jay M., Harvard University Observatory, Cambridge, Mass., U.S.A.
Pick, Monique, Observatory, Meudon, France
Pinter, Stepan, Geomagnetical Observatory, Hurbanovo, C.S.S.R.
Pounds, Kenneth, University, Leicester, England
Rayrole, Jean, Observatory, Meudon, France
Reeves, Edmond M., Harvard University, Cambridge, Mass., U.S.A.
Righini, Alberto, Observatory Arcetri, Firenze, Italy
Righini, Guglielmo, Observatory Arcetri, Firenze, Italy
Rigutti, Mario, Observatory Arcetri, Firenze, Italy
Ringnes, Truls S., Institute for Theoretical Astrophysics, Oslo, Norway
Rompolt, Bogdan, Astronomical Institute, Wrocław, Poland
Rösch, Jean, Observatoire du Pic-du-Midi, Bagnères-de-Bigorre, France
Rossi, Bruno, M.I.T., Cambridge, Mass., U.S.A.
Sawyer, Constance, E.S.S.A. Boulder, Colo., U.S.A.
Schanda, Erwin, Institut Angew. Phys., University, Bern, Switzerland
Schmidt, Hermann U., Max-Planck-Institute, München, Germany

Schröter, Egon H., Observatory, Göttingen, Germany
Semel, Meir, Observatory, Meudon, France
Severny, Andrei, Crimean Astrophysical Observatory, Nauchny, U.S.S.R.
Sheeley, Neil, Kitt Peak National Observatory, Tucson, Ariz., U.S.A.
Shelting, Bertha, Izmiran, Moscow, U.S.S.R.
Shilova, Natalia, Izmiran, Moscow, U.S.S.R.
Shpitalnaja, Alexandra, Pulkovo Observatory, Leningrad, U.S.S.R.
Simon, George, W., Sacramento Peak Observatory, Sunspot, U.S.A.
Simon, Paul, Observatory, Meudon, France
Sitnik, Gregory, Sternberg Institute, Moscow, U.S.S.R.
Smith, Elske, Dept. Astronomy, Maryland University, College Park, Md., U.S.A.
Smith, Henry, NASA Headquarters, Washington, D.C., U.S.A.
Smith, Sara, Lockheed Observatory, Burbank, U.S.A.
Stenflo, Jan Olof, Observatory, Lund, Sweden
Stepanov, Vladimir, Sibizmir, Irkutsk, U.S.S.R.
Stepanyan, Arnold, Crimean Astrophysical Observatory, Nauchny, U.S.S.R.
Stepanyan, Natalia, Crimean Astrophysical Observatory, Nauchny, U.S.S.R.
Steshenko, Natalia, Crimean Astrophysical Observatory, Nauchny, U.S.S.R.
Steshenko, Nikolaj, Crimean Astrophysical Observatory, Nauchny, U.S.S.R.
Sturrock, Peter, Institute of Plasma Research, University, Stanford, Calif., U.S.A.
Švestka, Zdeněk, Observatory, Ondřejov, C.S.S.R.
Sýkora, Julius, Observatory, Skalnate Pleso, C.S.S.R.
Tavares, José C. T. L., University, Coimbra, Portugal
Thomas, Richard N., J.I.L.A., Boulder, Colo., U.S.A.
Tifrea, Emilia, Observatory, Bucharest, Romania
Tlamicha, Antonin, Observatory, Ondřejov, C.S.S.R.
Topolová-Ruzicková, Blazena, Observatory, Ondřejov, C.S.S.R.
Torelli-Piccione, Maria, Observatory Monte Mario, Roma, Italy
Torrisi, Salvatore, Observatory, Catania, Italy
Tousey, Richard, Naval Research Laboratory, Washington, D.C., U.S.A.
Tuominen, Jaakko, Astrophysical Laboratory, Helsinki, Finland
Urbarz, Hans, Astronomical Institute, University of Tübingen, Weissenau, Germany
Vaiana, Giuseppe S., American Sci. Engin., Cambridge, Mass., U.S.A.
Valnicek, Boris, Observatory, Ondřejov, C.S.S.R.
Vasilyeva, Galina, Pulkovo Observatory, Leningrad, U.S.S.R.
Vyelshin, Gennadij, Pulkovo Observatory, Leningrad, U.S.S.R.
Weiss, Nigel, Dept. Applied Mathematics, University, Cambridge, England
Wiehr, Eberhard, Observatory, Göttingen, Germany
Wilcox, John, Space Sciences Laboratory, Berkely, Calif., U.S.A.
Winckler, John R., School of Astronomy, University of Minnesota, Minneapolis, Minn., U.S.A.

Withbroe, George L., Harvard Observatory, Cambridge, Mass., U.S.A.
Xanthakis, John, Research Center of Astronomy, Athens, Greece
Zhugzhda, Josif, Izmiran, Moscow, U.S.S.R.
Ziganov, Arkadi, University Observatory, Leningrad, U.S.S.R.

### LOCAL COMMITTEE AND SECRETARIAT

Dezső, Lorant ⎫
Gerlei, Ottó ⎪
Guman, István ⎬ Observatory Debrecen
Horváth, Erzsébet ⎪
Kovács, Ágnes ⎪
Ökrös, István ⎭

Neuzilova, Hana, Observatory Ondřejov
Siebler-Ferry, Herta, Fraunhofer Inst., Freiburg

Ill, Márton, Baja Station of Budapest Observatory, Baja
Marik, Miklós, Dept. Astronomy, Eötvös University, Budapest
Kálmán, Béla, Student of Astronomy (from Debrecen Observatory), Moscow,
Dezső, Anna ⎫
Lindenfeld, Erzsébet ⎬ Students (of English), Debrecen
Szeleczky, Beatrix ⎭

# WELCOME BY THE HUNGARIAN ACADEMY OF SCIENCES

L. DETRE

*(Konkoly Observatory, Budapest, Hungary)*

Dear Colleagues,
Ladies and Gentlemen:

On behalf of the Hungarian Academy of Sciences and of the Hungarian National Committee of the IAU I extend a most cordial welcome to you.

Symposium 35 is the first IAU meeting which is held in Hungary. I express sincere thanks to Commissions 10 and 12 for having given our Academy the honour to house it.

I thank all of you for having come to Budapest to attend the symposium. We never had before in Hungary a gathering of so many distinguished astronomers and geophysicists.

Special thanks are due to Professor Kiepenheuer for his continuous help to the Local Organizing Committee with his great experience and valuable advices. In January he made an extra travel to Budapest to talk over with us every detail of the organization, and even now nothing escapes his attention.

In recent years by the help of our Academy we made efforts to modernize our astronomical observatories, and as a first step we have built a new mountain-station 120 km northeast of Budapest. This was especially necessary because the Konkoly Observatory at its old place on the Szabadsághegy near Budapest suffered very much from the smoke and illumination of the city. Till now we have installed a 90/60-cm Schmidt telescope with objective prisms and a 50-cm Cassegrain telescope with a two-channel polarimeter. I hope to obtain later a telescope with an aperture of $1\frac{1}{2}$ m. The next step will be the modernization of the Solar Physics Observatory at Debrecen.

In the last issue of the Irish Astronomical Journal one can read a poem which is an ingenious summary of the problems of solar physics, having the title 'A Solar Physicist's Lament'. I wish you to get rid of some laments at this symposium. But naturally, with the last words of the poem

> It looks as if many a sunny day
> Will pass along on its way
> Before we solve it all.

I very much hope you will have a pleasant stay in Hungary and an exciting symposium.

Now, I declare IAU Symposium 35 on the Structure and Development of Solar Active Regions open and I call on Dr. Leighton to take the chair.

# PREFACE AND INTRODUCTION

K. O. Kɪᴇᴘᴇɴʜᴇᴜᴇʀ
*(Fraunhofer Institut, Freiburg i. Br., Germany)*

The present symposium, to my knowledge the largest ever held in the field of solar research (170 astronomers from 21 countries) was held in the building of the Hungarian Academy of Sciences in Budapest from September 4 to 8, 1967. It was the 35th symposium organized and sponsored by the International Astronomical Union. The majority of participants were financed from national sources. The Organizing Committee consisted of K. O. Kiepenheuer (Chairman), L. Davis, L. Dezső (Local Organizer), A. D. Fokker, R. Michard, A. B. Severny, H. J. Smith, Z. Švestka, and H. Tanaka.

In order to ensure prompt publication, the manuscripts had to be supplied by the authors 1 month after the meeting. The discussions have been recorded on tape. Their reproduction in this book, however, is based almost completely on the contributors' writing down their comments and questions on the spot.

Two special projects have been reported and discussed shortly during the symposium:

The world wide project 'Cooperative Study of Solar Active Regions' (CSSAR) organized by Dr. R. Michard, under the auspices of the IAU, which has put at the disposal of our solar community a precious observing material on Active Regions over a period of 6 months.

The 'Proton Flare Project' (PFP), organized by P. Simon and Z. Švestka, which concentrates on single Active Regions having produced a proton flare, and which tries to supply as much information as possible on the structure and development of these special regions.

Both projects are model cases and exceedingly valuable also for the general understanding of Active Regions.

The subjects of the sessions were the following: General Development of an Active Region (Chairman: R. B. Leighton); Theoretical Aspects (L. Davis); Optical Structure of an Active Region (J. W. Evans and A. B. Severny); Cooperative Study of Solar Active Regions, CSSAR (Z. Švestka); Coronal and Interplanetary Structure of an Active Region, Transient Phenomena (G. Righini); Proton Flare Project, PFP (V. A. Krat); Radio Structure of an Active Region (H. Dodson-Prince).

Summaries for each session were given on the morning of the next day by: R. Howard, H. U. Schmidt, N. Sheeley, G. Newkirk, A. Bruzek, and A. D. Fokker. They are not included in the proceedings because of their provisional character. The diffi-

*Kiepenheuer (ed.), Structure and Development of Solar Active Regions,* 3–9. © *I.A.U.*

cult and delicate task of presenting an overall summary at the end of the symposium was brilliantly solved by C. de Jager (see p. 602).

In addition to the scientific sessions there was a full-day excursion to the Balaton Lake. A banquet in the Citadella was offered to all participants by the Academy. Excursions, meeting rooms, ladies' program, and other activities were splendidly arranged by the local committee which was in the hands of the inexhaustible Dr. L. Dezső, Debrecen. We have to thank him and his staff very much.

I further want to express my sincere gratitude to the members of the Organizing Committee, especially to Dr. Z. Švestka for all his kind assistance from the beginning to the end; to Mrs. H. Siebler-Ferry and Miss M. Ilan for valuable help editing the manuscripts and the discussions, as well as to Mrs. H. Neuzilova and Dr. Bruzek for their kind help during the symposium.

And now let me turn right away to the subject of our symposium: As you will agree, an Active Region (AR) on the Sun is a beautiful one with all its spectacular events in photosphere, chromosphere, and corona. A challenge to the optical observer, a unique and inexhaustible paradise for the investigator of the interplay of magnetic fields, turbulence, hydromagnetics, and radiation in a stellar plasma! An Active Region on the Sun is also a very complex and intriguing phenomenon of solar meteorology, which shows a different aspect almost in every wavelength we look at it, X, EUV, visible, infrared or radio. Apart from the many people working on certain features occurring within an Active Region – as spots, flares faculae, prominences – and their terrestrial effects, there are really rather few, strictly investigating the structure and the development of an Active Region as a whole. The more we are happily surprised about the great number of attendants and contributors to this symposium.

What is an Active Region or a Center of Activity? According to D'Azambuja (1953) it is "the totality of all visible phenomena accompanying the birth of sunspots". Today we should extend this definition a little bit: "The totality of all observable phenomena preceding, accompanying and following the birth of sunspots, incl. radio-, X-, EUV- and particle emission." The Active Region comprises photospheric, chromospheric and coronal features and reaches practically into interplanetary space.

Physically speaking, the observable part of a single Active Region represents only a very small fraction of a much larger machinery which produces the solar cycle and which has been built together so convincingly by Babcock (1963). We know nowadays that this model does not cover all observational aspects of a solar cycle. It is to be expected that at least the field-amplification mechanism and its dependence upon latitude needs adaptation to the new observations of Bumba, Dodson, Hedeman, and Howard, presented during this symposium. It will be indispensable during this symposium – from time to time – to refer to this large-scale mechanism, in order to penetrate with our imagination into the non-observable parts of the solar interior, to reach the roots of the Active Regions.

The most primary *observable* components of an Active Region lie down in the photosphere. For this reason our most primary information will be optical and comes from the visible part of the spectrum. Only in this region a reasonable angular resolution can be obtained and magnetic fields can be measured with some accuracy.

The synopsis of an Active Region is impeded moreover by the great diversity of the involved phenomena as well as by the great number of instruments needed to observe them in the different layers of the solar atmosphere. Last not least – also because of this diversity – the understanding between observers and theoreticians, in this case mostly hydromagneticians, needs definitely to be improved. Let us hope that this symposium will work in this direction.

During this meeting we will probably be tempted again and again to get caught too much by a single interesting phenomenon as perhaps flares, instead of doing our best to find out about the large and simple laws governing the development of an Active Region as a unit. On the other hand we must be aware of the fact that, in spite of an Active Region developing to sizes of considerably more than 100000 km, small-scale events with almost subtelescopic dimensions are of crucial importance. I think of the initial formation of a magnetic region, of a magnetic pore, or of the random-walk mechanism. The limits which are set today to the angular and time resolution of sensitive magnetographs are indeed a serious handicap for the study of the basic processes forming an Active Region. However, good chromospheric filtergrams and spectroheliograms in Hα have proved already helpful, supplementing magnetograms with a kind of high-resolution magnetic vectorgrams (filamentary structure of disturbed chromosphere), in K (Ca II) they represent sensitive indicators for the very first occurrence of field concentration. Also the optical observation of large filaments (on the disk and on the solar limb) as well as of moustaches will tell us about high-resolution details in the magnetic field, which are not in the magnetograms. I am convinced – even knowing that this is a complicated project – that it would pay to bring a magnetograph in the stratosphere or even in space. Studying Active Regions, one is somehow in the difficult position of a man who has to investigate the development of an elephant with a microscope!

Fortunately the whole phenomenon of an Active Region – thanks to a number of important new observations and a significant progress in their hydromagnetic interpretation – has become more transparent, more intelligible during the last decade. I draw your attention to the concept of supergranules, which together with the observations of Leighton (1964), Bumba and Howard (1965) and others that magnetic fields and spots initially appear along the boundaries of adjacent supergranules, has changed drastically the model of an Active Region. Moreover, Leighton's random-walk diffusion, the systematic analysis of magnetic metamorphosis of an Active Region by the Babcocks, Bumba and Howard (1965), Severny (1964) and others using especially the wealth of full-disk magnetograms obtained at Mt. Wilson Observatory have become the most important part of today's model of an Active Region. Just as important, however, are the many theoretical attempts by Sweet (1956), Dungey (1958), Gold and Hoyle (1960), Parker, Petschek (1963) and others, which have been made in the domain of hydromagnetics somehow with the aim to lower the barriers built up by the concept of the frozen-in magnetic field with its much too long time scale. This type of work has mostly been carried out in connection with the magnetic interpretation of flares. We are now allowed again, as in good old times and as in our laboratories, to reconnect field lines, to concentrate or to dilute magnetic flux in the solar atmosphere somewhat quicker again. And last not least, I would like to stress again how strongly Babcock's large-scale magnetic mechanism of the solar cycle has stimulated and influenced last years' work on Active Regions, even if it might turn out necessary to modify it in some respects. I think we are quite lucky to start our symposium with such fertile boundary conditions.

I will not try to give the story or a description of an Active Region here, but let me just recall a few important facts to which we should not forget to direct our attention. We know today for sure that the process producing and structuring an Active Region is of magnetic nature and that all the accompanying features like spots, faculae, flares and associated effects, filaments etc., are byproducts of this magnetic process. Therefore our responsibility must be first of all to analyse and to interpret this process. The first sessions will be mostly devoted to this problem and Messrs. Bumba, Howard and Schmidt have kindly accepted to review shortly the actual situation in this field of research.

An important part of this magnetic process – which is not just easy to observe – goes on under our eyes. Another part, which we will never be able to follow, happens inside the Sun. Concentrating on the observable part, we can at least hope to be able to define clear boundary conditions for the unvisible interior.

What we observe is that – preferably in a region with a weak residual magnetic field, left over from old Active Regions – a very sudden local increase of magnetic flux of the order of $10^{21}$ Maxwell within a few tens of hours occurs. The way of growing, the structure and the configuration of this field, which is reflected quite in detail by the brightening in the K line (Ca II), as well as the initial spot formations are related so closely and in such a detail to the granules and supergranules, that it is very difficult to think that the spot field or the initial field of a BM-region is not being produced or concentrated by the supergranules themselves. Very remarkably also are the close analogy between the birth of a pore in closest interaction with one or a few granules (Schröter, 1962; Bray and Loughhead, 1964) and the formation of a spot group or a BM-region in a just as intimate contact with one or a few supergranules (Bumba and Howard, 1965).

One has to keep in mind, however, that observation shows quite clearly that – at least during the active phase of the birth of an Active Region – the old magnetic field even in the very surrounding of the Active Region's birth place does not change. This means, that the magnetic flux of a young Active Region cannot just simply be the flux of the visible photospheric surrounding. It must come from some depths below the photosphere.

All these observations in combination with some recent results by Parker (1963), Weiss (1966) and others about the concentration of magnetic flux by turbulence in the deep convective zone as well as in higher levels along the boundaries of super-granules make it appear quite promising to discuss the formation of sunspots as occurring *almost in situ* in the photosphere. Sheeley's (1967) observation of strong local magnetic fields outside the spot belts points in the same direction. Mr. Schmidt and Mr. Weiss will talk about this important subject. In any case more reliable informations have to be collected to get a better idea of the height or the depth in which these flux concentrations originate. A study of the metamorphosis of the supergranules in the course of hours and days will probably be of great help in this connection.

Even if the total magnetic flux constituting an Active Region is being concentrated from a quasi-superficial flux by the supergranular convection, still the cooperation of a well-organized deep-seated large scale field is indispensable. We all know that the 11-year variation in the sunspot numbers, the Spörer law of variation of mean latitude of sunspots, the reversal of the Sun's poloidal field, Hale's law of sunspot polarity, the initial orientation of spot groups and BM-regions, and the dominance of the preceding spot to the following *must* be parts of a large-scale global magnetic process. It will be very important for us, however, to find out of what form and what strength this 'magnetic message' from the solar interior has to be, in order to release the observable phenomena. We are on the other side at least qualitatively sure, that the more than 1000 BM-regions, produced during a solar cycle, after expansion, migration and splitting apart cause the large-scale polar reversal and the reversal of the magnetic configuration of spot groups and BM-regions in the coming cycle. Active Regions are therefore not only products of the 11-year cycle. They are indispensable to keep the solar cycle running.

In this connection I would like also to stress the importance of the so-called 'Giant Cells', discussed first by Bumba (1966). They are fairly regular features with a diameter of about 400000 km, formed by the background field pattern. In analogy to the supergranules these cells are bordered by K emission. New Active Regions seem to form preferably at the junctions of these giant structures. Also the long-lived family configurations of Active Regions, as discussed by Helen Dodson during this symposium need special attention. Both phenomena show that Active Regions are not independent events; their interaction plays an important part in their history. It will also be an urgent and interesting task to confront these results with Babcock's model.

In spite of the fact that an Active Region has a lifetime of a number of solar rotations, its intrinsic active phase lasts only a few days, whereas the violent secondary effects cover a period 3 to 4 times longer. As soon as the full magnetic flux of $10^{21}$ to $10^{22}$ Maxwells is being concentrated, the decay sets in. All what is following then is the result of the occurrence of this magnetic flux, its amount, its expansion and its deformation, which obviously does not seem to be seriously affected by the display of optical phenomena produced by it in the solar atmosphere.

The expansion and the deformation of this concentrated flux will be carried out – strange enough to say – by the same type of convection, if not by the same super-granules, which just before took such an active part in concentrating it. Altogether, the expanding and change of form of a BM-region is certainly not just equivalent to the combined action of the Sun's differential rotation and the random-walk diffusion. There is an increasing evidence that the geometry resp. the magnetic structure of a young Active Region will be affected by the structure of the weak remanent fields, while the form of a well-developed BM-region can be strongly influenced by the collision with other magnetic regions.

As compared to the basic active phase of an Active Region, the decay is a very slow process which is difficult to follow. But just this late period in the evolution of an Active Region with all its problems connected with the nature of the background fields, the role of unipolar and 'ghost' unipolar magnetic regions is of significant importance for understanding the solar cycle as a whole.

The optical and all the other features which accompany the concentration and the

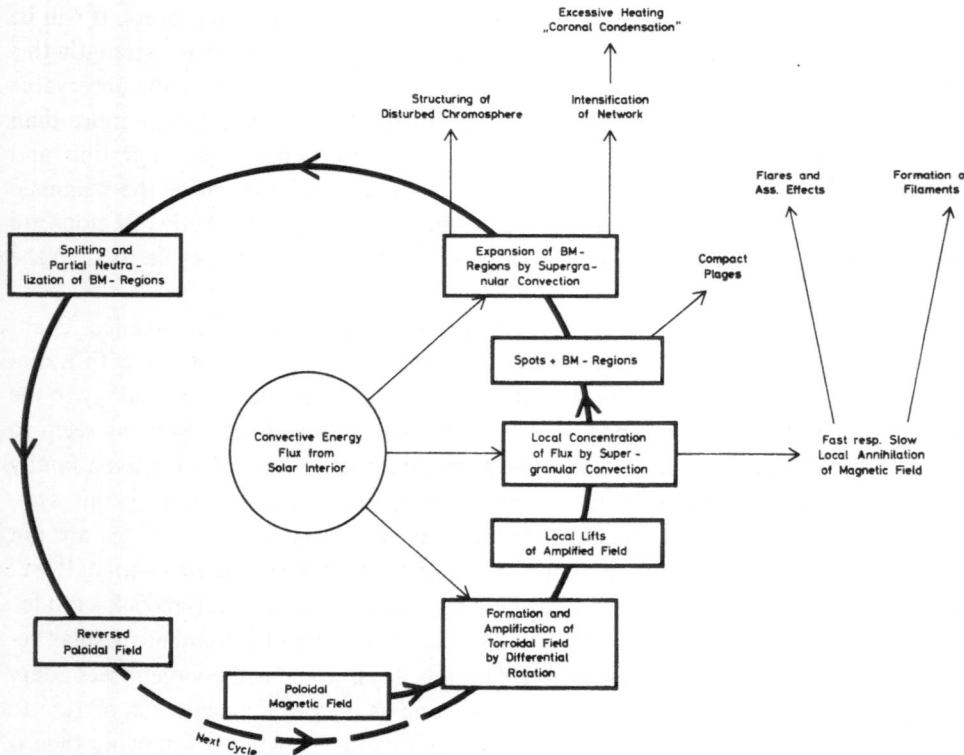

FIG. 1.   *The Magnetic Cycle and the Optical Phenomena of Solar Activity.*

expansion of the magnetic flux, as this becomes clearer every day, are not only influenced or shaped by the magnetic field. They somehow seem to be direct products of it. I am speaking of spots, plages, the structures in the disturbed chromosphere, flares and associated effects, all types of prominences, etc. Flares and prominences, both occurring in regions of horizontal magnetic field (and usually strong gradients of longitudinal field) seem to form as a consequence of a local *annihilation* of magnetic field, very fast in the case of flares, somewhat more slowly for prominences (according to the recent work of Kuperus and Tandberg-Hanssen, 1967). The striking change from the undisturbed to the disturbed structure in the chromosphere, when hit by the expanding field – mainly the formation of thread- and loop-shaped structures – which cover the magnetized region like a curled wig, is not yet quite understood. The fact that these long threads have the same diameter in the photosphere, chromosphere and in the prominences high up in the corona, is probably of some importance.

Altogether, we get the impression, that the formation, the structure and the development of an Active Region is the result of a strange interplay of solar convection, differential rotation, and deeper seated weak magnetic fields. Concentration and expansion of these fields is being done seemingly by the same cellular convection. The more important *observable* phenomena in an Active Region seem to be connected with local concentration as well as annihilation of magnetic fields. The diagram of Figure 1 tries to give a simplified description of this situation.

## References

Babcock, H.W. (1963)    *Annual Rev. Astronomy, Astrophysics*, **1**, 41.
Bumba, V. (1966)    39. Course Varenna (Plasma Astrophysics). Private Circulation.
Bumba, V., Howard, R. (1965)    *Astrophys. J.*, **141**, 2, 1492 and 1502.
D'Azambuja, L. (1953)    *L'Astronomie*, **67**, 430.
Dungey, J.W. (1958)    *Cosmic Electrodynamics*, Cambridge.
Gold, T., Hoyle, F. (1960)    *Mon. Not. R. astr. Soc.*, **120**, 89.
Kuperus, M., Tandberg-Hanssen, E. (1967)    *Solar Phys.*, **2**, 39.
Leighton, R.B. (1964)    *Astrophys. J.*, **140**, 1547.
Parker, N.E. (1963)    *Astrophys. J.* Supplement No. 77, p. 177.
Parker, N.E. (1963)    *Astrophys. J.*, **138**, 552.
Severny, A.B. (1964)    *Space Sci. Rev.*, **3**, 451.
Sheeley, N.R. (1967)    *Solar Phys.*, **1**, 171.
Sweet, P.A. (1956)    *IAU Symposium No. 6, Stockholm 1956*, p. 123.
Weiss, N.O. (1966)    *Proc. RAS*, **A2093**, 310.

Part I

# GENERAL DEVELOPMENT OF AN ACTIVE REGION

# PATTERNS OF ACTIVE REGION
# MAGNETIC FIELD DEVELOPMENT*

V. Bumba, R. Howard,** and M. J. Martres, I. Soru-Iscovici
(Astronomical Institute, (Observatoire de Paris-Meudon,
Ondřejov, Czechoslovakia) (92) Meudon, France)

ABSTRACT

We discuss some characteristics of the appearance and development of magnetic fields within active regions as well as the large-scale ordering of activity into complexes of activity. It is not possible to separate a study of the evolution of active regions from a study of the model of the activity cycle. Many of the results obtained in the last few years concerning the development of active regions and large-scale activity have not been easily explained within any of the solar activity models. A chronological scheme of the development of a 'typical' C- or D-type active region is presented. We point out that the appearance of magnetic flux at the solar surface seems always to be a relatively rapid event, occurring during the course of a day or so. If the region does not receive more magnetic flux to make it a large region or if there is not a resurgence of activity later in its lifetime, the rest of the development is a gradual expansion and mixing of magnetic flux with the surrounding background field pattern.

## 1. The Magnetic Cycle

A theoretical model of the birth and development of an active region cannot be constructed without an understanding of the physical mechanisms involved in the solar cycle. We cannot separate the study of the solar cycle from the study of the development of active regions. In recent years there have been only a few models of the solar activity cycle suggested. The enormous complexity of the problem has so far made it impossible to construct a rigorous theory of the activity cycle. The models suggested have been phenomenological models which help us to visualize a possible solution.

Parker (1955) suggested a model which supposed toroidal magnetic fields – part of a dynamo wave – which move toward the equator under the solar surface. Magnetic buoyancy brings loops of these magnetic-flux ropes to the surface to form active regions. The equatorward drift of these magnetic fields provides the latitude drift of Spoerer's law. This model gives an explanation of some of the phenomena of solar activity, but it is not as complete in its description of the phenomena as is Babcock's model. Moreover, the recent observations of the latitude drift of aging active region

---

\* Presented by R. Howard.
\*\* On leave from the Mount Wilson and Palomar Observatories.

magnetic fields – especially the different behavior of the leading and following polarities – does not fit this model well.

Babcock (1961) proposed a model which starts with an initial bipolar field which in low latitudes consists of subsurface flux tubes. Differential rotation draws these flux tubes out and twists them until, first at high latitudes, magnetic buoyancy brings loops of lines of force to the surface. The active latitudes move to the equator as the cycle progresses. As the old regions expand and their magnetic fields weaken, the following polarity fields drift preferentially to the pole and the leading polarity fields drift to the equator and merge with those from the opposite hemisphere. Thus the polar fields reverse polarity and the stage is set for the next half of the 22-year cycle.

Although Babcock's model is the most successful one suggested so far, it cannot explain all the observed phenomena. One observation seems to us to be particularly difficult to explain. During the first few months of the present cycle, the first new cycle regions were forming in one broad longitude zone, in an area where weak fields from the old cycle were still visible and presumably were moving toward the polar

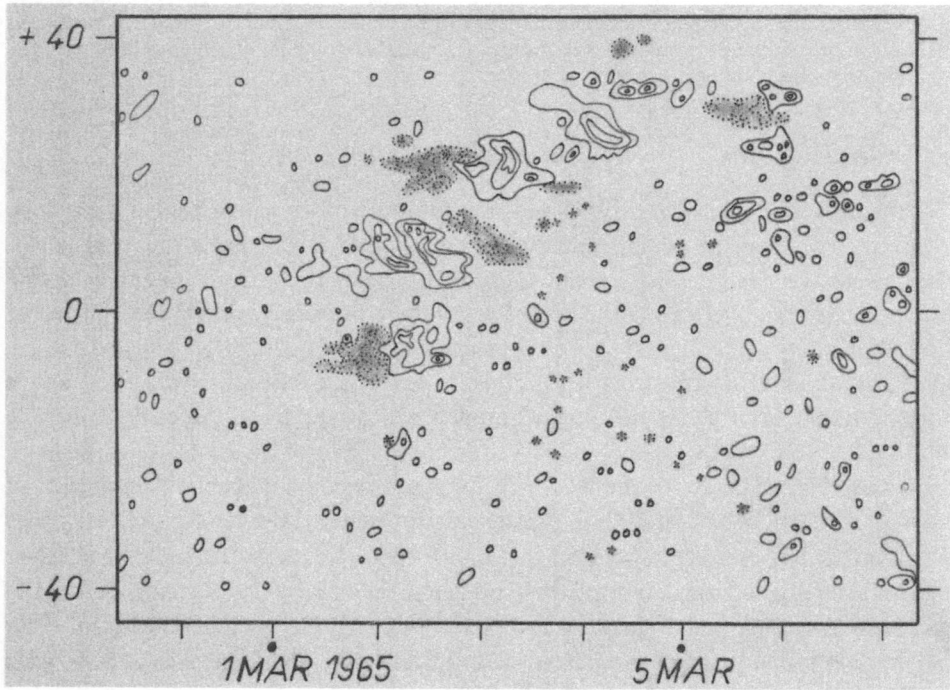

FIG. 1.   *A portion of the magnetic-field synoptic chart for rotation no. 1491. East is to the right. Positive magnetic polarity is indicated by solid lines, and negative polarity by dotted lines and shading. The gauss levels are 6, 12, and 20 gauss. The new-cycle regions in the North have negative magnetic fields in their leading portions.*

regions. An example of this is shown in Figure 1. It is difficult to visualize how this can happen in Babcock's model.

Recent results concerning the development of active region magnetic fields (Bumba and Howard, 1965a) have added to our knowledge of the phenomena, but so far these results and others have not added to our understanding of the activity-cycle model. From the observations we cannot at this time see direct evidence that an active region forms as a result of the emergence to the surface of a tube of magnetic lines of force.

## 2. Large-Scale Activity

Recent results suggesting the large-scale ordering of activity (Bumba and Howard, 1965b; Warwick, 1965; Švestka, 1967) have not found an explanation in any of the models suggested so far.

A complex of activity (see Figure 2) is the grouping together in one large longitude zone of a number of active regions. When the level of activity is sufficiently low, the appearance of the active region within the complex may be seen to follow a characteristic pattern. The active regions during the first few rotations are large and concentrated near the center of the complex. In subsequent rotations the boundaries of the complex expand, and the active regions which form are smaller. The expanding fields of the old active regions may fill a very large area in both hemispheres. A unipolar magnetic region (seen with 23 arc-sec resolution) may form at high latitudes. At lower latitudes large-scale fields can be seen which appear to extend into interplanetary space.

A complex of activity is intermediate in importance between a single active region and the activity cycle itself, and it shows some of the characteristics of both. For example, the development of its activity is rapid compared to its decay, and its area grows steadily at first then decreases irregularly.

## 3. The Birth and Development of an Active Region

We present some characteristics of the birth and later development of an active region which have been noted so far. These concern principally the magnetic fields and the calcium plages. The results are a combination of earlier work (Bumba and Howard, 1965a) and the present study of the Meudon and Crimean magnetic-field data obtained during the period of the Cooperative Study of Solar Active Regions. This material comes from the period of the last solar minimum.

The following characteristics are commonly observed in developing regions:

(1) On a large scale, new regions form at the location of old weak fields. Generally regions form in old fields of leading polarity, and the appearance is that of islands of following polarity fields in extended areas of leading polarity fields which are predominant at low latitudes. In general in a complex of activity there are several regions

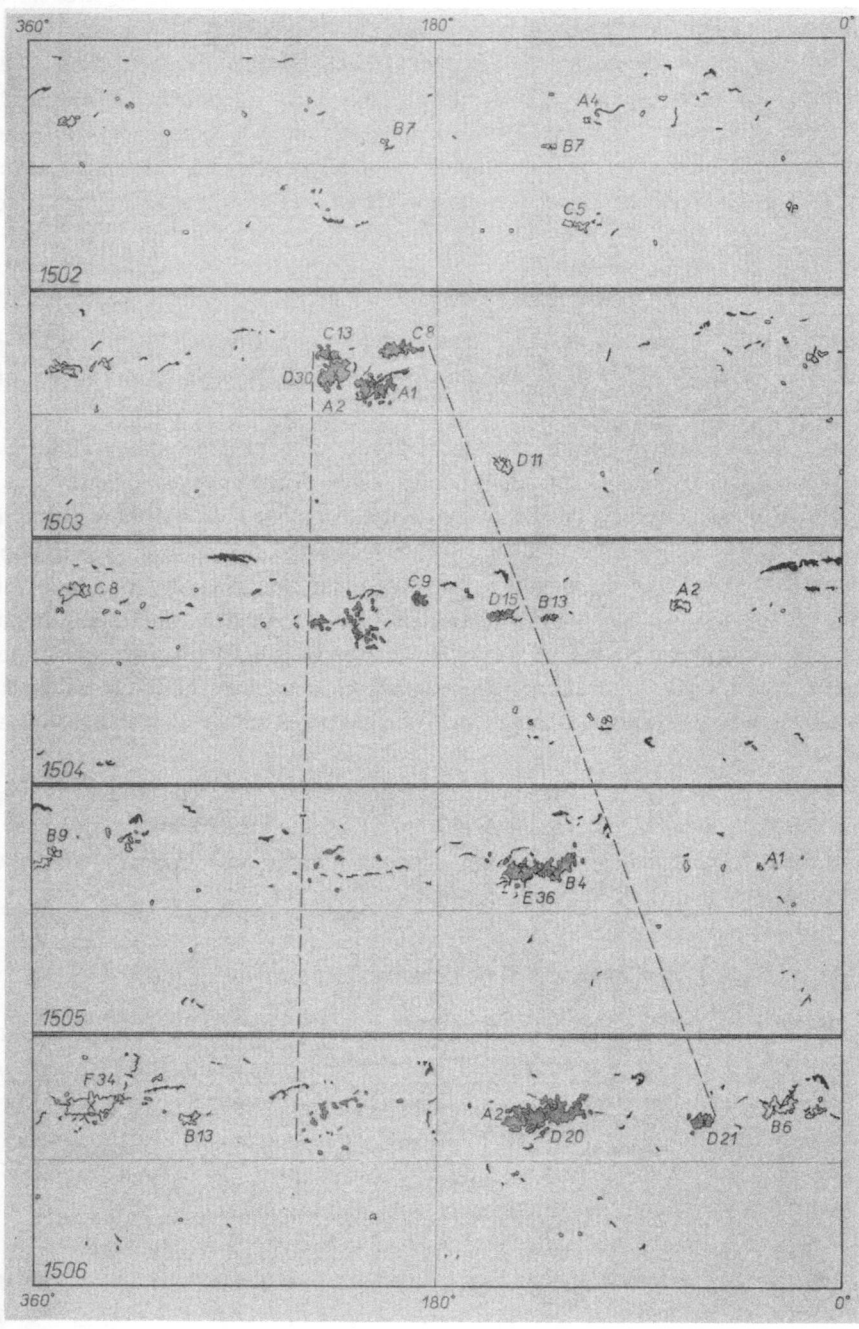

Fig. 2a.  *A series of synoptic charts drawn from Fraunhofer Institute Daily Maps of the Sun. The plages associated with the complex of activity are shaded. All spot groups are indicated with their Zürich types. We have indicated the expanding boundaries of the complex by dashed lines.*

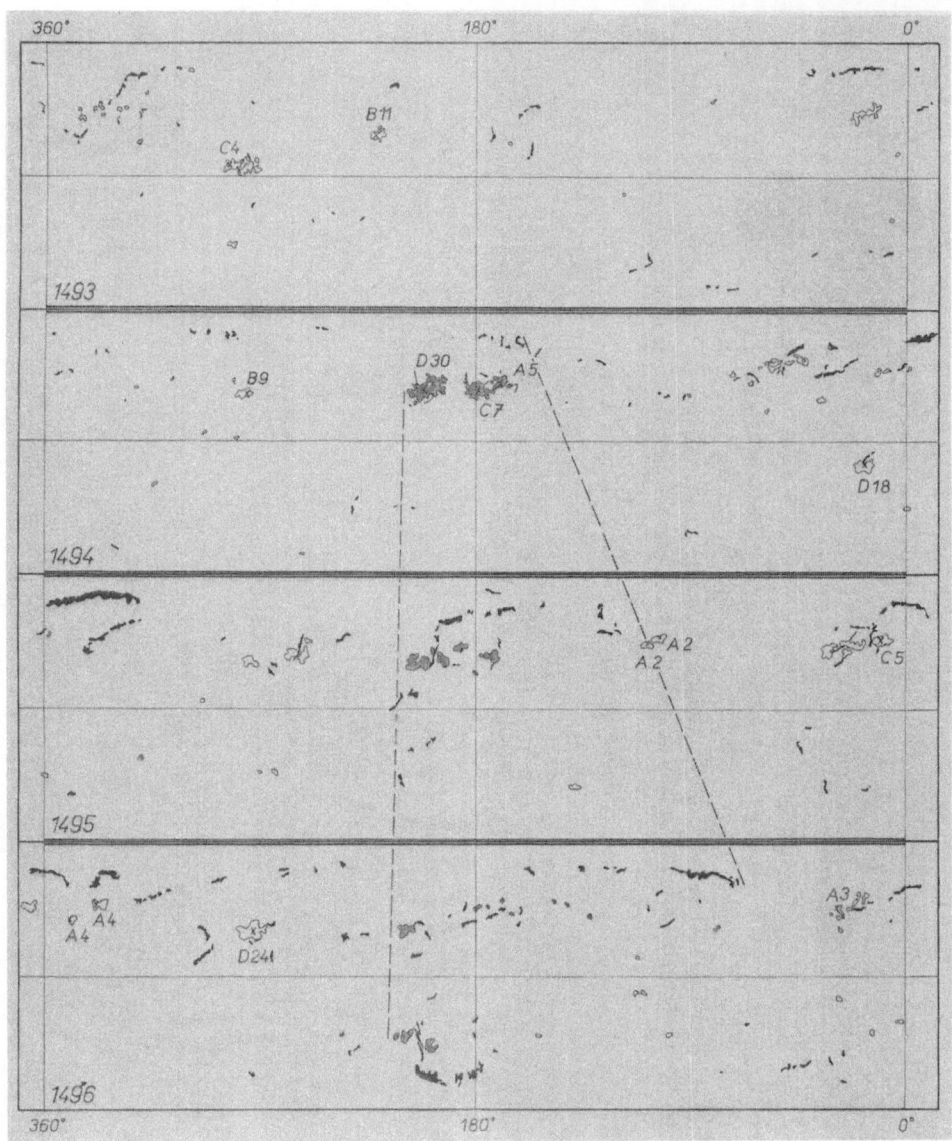

FIG. 2b. *Similar to Figure 2a for a different complex.*

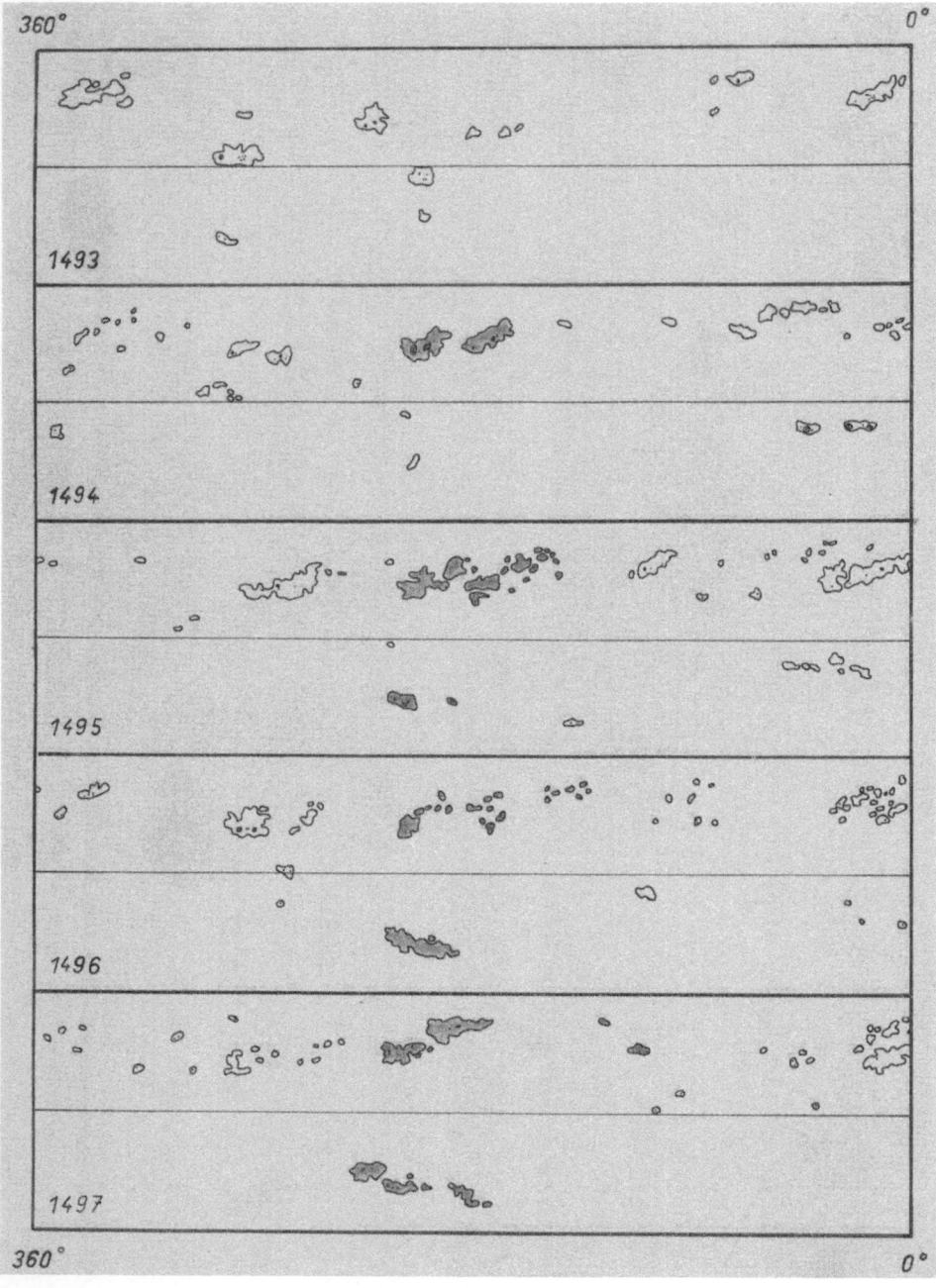

FIG. 2c. *The same complex as Figure 2b, but illustrated by means of the Heliographische Karten der Photosphere from Zürich.*

close together. Often the following polarity fields of several regions appear to form one large feature as do the leading polarity fields. On a small scale the first brightenings of a new region appear at the boundaries of several supergranular cells – in particular at a point where old fields of both polarities meet.

(2) The first appearance of magnetic fields and plage within the young region follows the outline of the pre-existing calcium network. In the case of large regions, whole cells are eventually filled in with plage and field.

(3) The development of a region appears to start from the center of the region, move to the following portion and then to the leading portion. That is to say, the center of gravity of the region moves from what is later the center of the fully developed group to the following portion, then back to the center. When the leading part develops, the leading sunspot also develops. There is now no observational evidence one way or the other about whether the progress of the development is a motion of lines of force or an appearance of new fields.

(4) In many cases the magnetic fluxes of the two polarities appear not to be balanced during the first stages of the development of the group. These results were obtained with relatively crude angular resolution. It will be of interest to obtain more observations of the early stages of development of active regions using high resolution.

(5) As new regions form, the old fields within which they appear remain unchanged for several days. After this period the old and new fields begin to merge.

(6) In the case of groups which reach a maximum importance of C or greater in the Zürich Classification, the development in general proceeds in the following manner: for the first day or perhaps two days the development is identical with that of a small simple region. Then near the boundary of the two polarities new fields appear. Within these fields are the spots which give the region a C or greater classification. Often there is a 'gulf' of one polarity in the other. In such places there are large velocities observed (Bumba, 1963) and most of the flare activity takes place (Bumba *et al.*, 1968). At this time a complex configuration may originate. At this point in the development of the region the addition of new magnetic flux has ended (unless there is a later renewal of activity caused by the addition of new magnetic fields, and the decay of the region sets in. (See Figure 3.)

(7) The early post-maximum stage of the development of an active region is characterized by the cessation of magnetic changes which tend to increase the magnetic complexity of the region or the magnetic gradients within the region. These changes take place near the center of the region at the boundary of the polarities. At this post-maximum stage the plage often acquires a characteristic triangular appearance, and the boundary between the polarities begins to appear less complicated. Also at this stage a small filament often forms pointing to the center of the group between the polarities. This filament may later develop into a large quiescent filament.

(8) As is well known, the later stages of the development of the region are character-

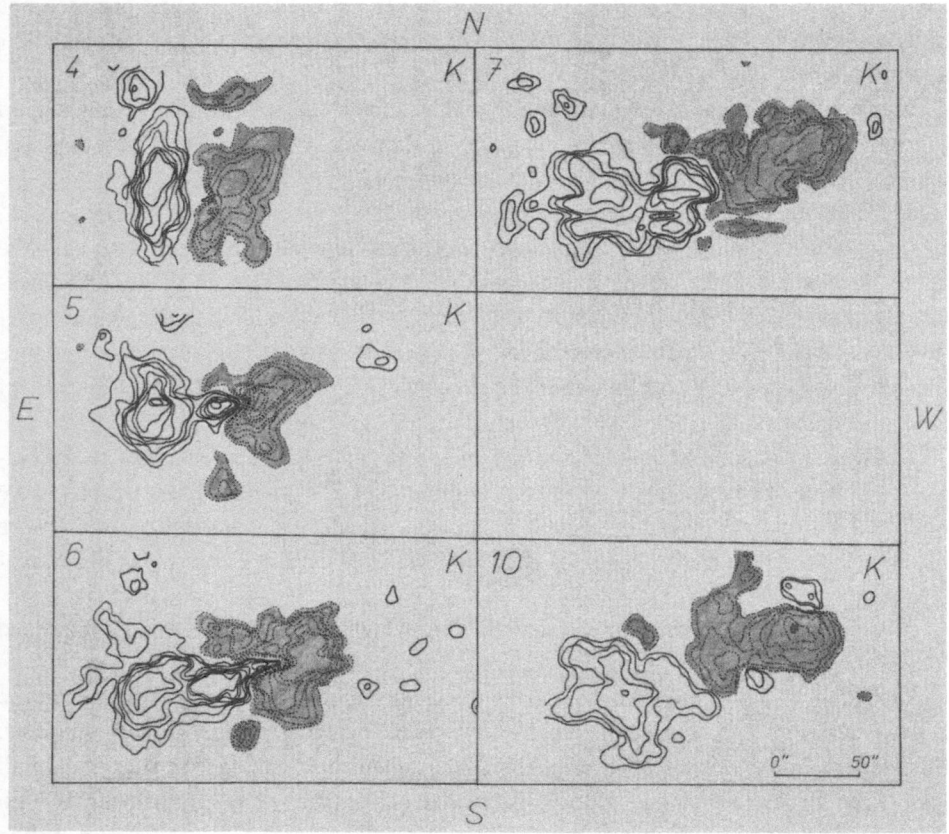

FIG. 3.    *Magnetic maps of a region during September 1965. The data are from the CSSAR magnetic material from the Crimean Observatory and Meudon Observatory. The day of the month is given in the upper left corner of each map. The gauss levels are usually 10, 20, 40, 60, 100, 150, 200, 300, 400. Note the slightly complex situation that develops near the center of the group and then disappears.*

ized by the expansion of the plage, by the continuous growth of the distance between the centers of gravity of the two polarities, and by the characteristic curved teardrop shape of both the leading and the following portions. At this point, depending upon the 'density' of activity in the neighborhood of the region, the magnetic fields will either expand and weaken until they are no longer detectable, or they will merge with and become a part of the background-field pattern.

We present here an idealized chronological scheme of the shape of the magnetic fields associated with the development of a large but magnetically simple active region. (See Figure 4.) These results come mostly from the Meudon magnetic maps. This scheme refers to the stronger magnetic fields ($> \sim 50$ gauss) and the brighter portions of the plage.

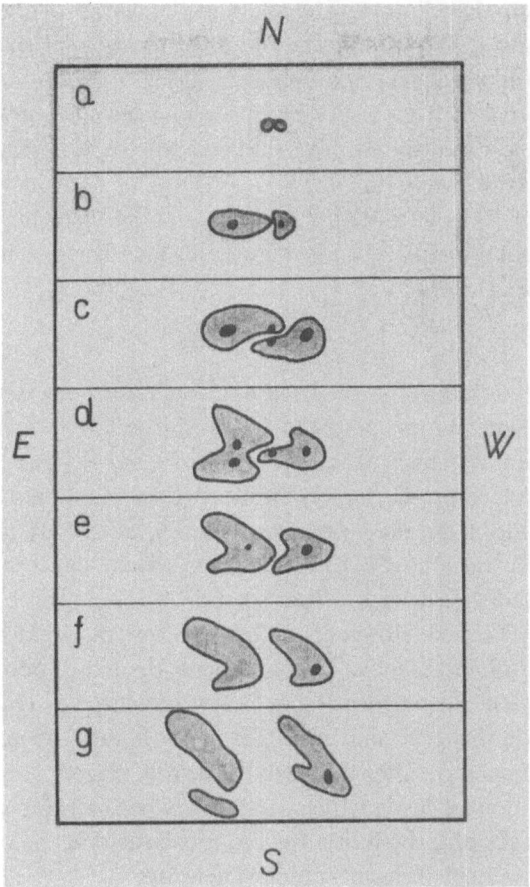

FIG. 4.   *An idealized sketch of the development of an active region. The shaded portions represent the plage and the two magnetic polarities. Spots are indicated by black dots. See the text for details.*

Stage a: Age from several hours to 1 day. The first brightenings at the boundaries of supergranules.

Stage b: Age about 2 days. The development is proceeding most rapidly in the following portion of the region. The first spots have formed.

Stage c: Age from 3 to 4 days. Penumbras are formed and the group takes on a more complicated appearance. The whole field occupies an elliptical area, and the leading polarity fields lie predominantly equatorward of the following polarity fields.

Stage d: Age from 4 to 6 days. This is the start of the post-maximum phase and the boundary between the polarities is still relatively complicated. The following polarity has formed two tails to the East of the center of the region, and the plage has taken on a characteristic triangular appearance.

Stage e: Age from 5 to 7 days. The spots in the center of the group have disappeared. There may still be spots in the following portion of the plage. The 'arrow' shape of both polarities is characteristic of this phase.

Stage f: Age from 6 to 9 days. The two polarities have begun to separate, and the following spots have disappeared. The influence of the differential rotation on the shape of the fields is evident.

Stage g: Age usually greater than 8 days. This is the final stage of the spot group. Only the leading spot remains. The magnetic fields may begin to merge with the background field pattern.

## 4. Discussion

We have presented here some observations concerning the development of active region magnetic fields. We propose the following points as being important in an understanding of the mechanism of the formation of active regions and of the cycle itself.

(1) The solar cycle must be largely influenced by some subsurface phenomena. This conclusion comes from the large-scale ordering of activity in complexes of activity (Bumba and Howard, 1965b) and the recent results concerning the distribution of activity in longitudes (Warwick, 1965; Švestka, 1967). In this connection the 'giant supergranular cells' (Bumba, Howard, and Smith, 1964) may play a role.

(2) In the formation of an active region, at least the major portion of the magnetic flux of the new region must appear from below the surface. This conclusion comes from the observation that the magnetic fields in the immediate neighborhood remain undisturbed for several days after the birth of the new region. A region does not form from the redistribution of fields which are already at the solar surface. Even in the case of the later stages of growth of a region, it is clear that the older magnetic fields are undisturbed initially by the appearance of new flux.

(3) The complexities in the magnetic configuration of an active region develop within the first few days of the birth of the region. Unless there is a later renewal of activity (which may also produce a magnetically complex situation), the tendency after the maximum phase of the development of the region is for the magnetic configuration to become simple. The analogy in the large scale is the formation of regular features in the background-field pattern.

The following questions concerning solar activity present themselves at this point and appear to be of interest. More observations are needed to clear up these matters.

(1) What happens to the large-scale field formed by the expansion of old active regions? A large amount of flux appears to head toward the poles, but this flux is not observed to collect at the poles, and we know very little of what happens at the polar regions.

(2) Does the characteristic shape seen in the evolution of active regions imply an angular velocity of rotation which is greater for stronger fields than for weaker fields – and thus perhaps imply an influence from deeper layers?

(3) Is the fine scale structure of the magnetic field giving us on the large scale of the

resolution of our observations, false data on net fluxes? It seems unlikely that this can be a very great effect (Howard, 1966); however, we must await observations of very high resolution to resolve this problem. Naturally, because of the fine-scale structure of magnetic fields over much of the solar surface, the appearance of a magnetogram depends markedly upon the angular resolution used to obtain the observation. While in general one wishes to use the highest angular resolution possible, for problems involving large-scale features it is sometimes desirable to use low resolution in order to see the 'forest' instead of the 'trees'.

## References

Babcock, H.W. (1961)      Astrophys. J., **133**, 572.
Bumba, V. (1963)      BAC, **14**, 137.
Bumba, V., Howard, R. (1965a)      Astrophys. J., **141**, 1492.
Bumba, V., Howard, R. (1965b)      Astrophys. J., **141**, 1502.
Bumba, V., Howard, R., Smith, S. F. (1964)      Carnegie Institution of Washington Year Book **63**, 7.
Bumba, V., Křivský, L., Martres, M. J., Soru-Iscovici, I. (1968)      the present volume, p. 311 .
Howard, R. (1966)      Observatory, **86**, 160.
Parker, E. N. (1955)      Astrophys. J., **121**, 491.
Švestka, Z. (1968)      the present volume, p. 287.
Warwick, C. S. (1965)      Astrophys. J., **141**, 500.

## DISCUSSION

*Winckler:* Is the emergence of flux loops by magnetic buoyancy thought to apply to BMR's or to the spots themselves? This point is not clear in Babcock's papers.

*Howard:* Both the BMR's and the spots are thought to develop by the emergence of flux tubes.

*Acton:* What predictions did the Babcock theory make which provided observational tests of the theory?

*Howard:* The migration toward the polar latitudes of the following magnetic flux of old active regions was predicted by the model and subsequently observed.

*Bumba:* The observers are in great difficulties to say something more about the observation of emerging flux ropes. If we observe with a lower resolution, we may see very nicely the distribution of magnetic field in two polarities and this distribution may be represented by a model of outcoming tube of magnetic field. But if we observe with high enough resolution we do not see such well-organized picture any more, we then have a complicated situation which we cannot represent by a simple model of buoyant flux rope.

*Sturrock:* If sunspots always form adjacent to neutral lines, they may represent the emergence of new flux. If sunspots may form far from a neutral line, these spots must be formed by the gathering together of field lines which emerged earlier in the development of the active region. Do sunspots always form adjacent to neutral lines?

*Howard:* No, they do not, so it appears that the spots must form from the new plage magnetic fields.

*Severny:* How is your feeling about unipolar magnetic regions? We have got convinced from our observations with high resolution on Crimean magnetograph that these regions disappear when we use high resolution ($2''5 \times 4''$) and they consist of a number of small elements of different polarity (I mean the first part of your talk relating to the local weak magnetic fields).

*Howard:* Do you think that even magnetic flux over these regions can be zero?

*Severny:* Yes. Because the signal of magnetograph is proportional to the field strength and the area of an element and this signal must be higher than noise level to be recorded. If the area is small enough the signal can drop below noise level even for strong field and this element can be omitted at the records. So the excessive flux of one sign could in principle be partly compensated by small elements of opposite sign.

*Howard:* Our observations at high resolution do not indicate that the UM flux disappears. Such an effect could well get us out of the problem of what happens to the flux that moves toward the poles, but even if there were no net flux, the UM's would be obviously of some physical significance since they are clearly observed to have shape and motion.

*Bumba* (to Severny's question): We are using the terms BMR and UMR in the same sense as they were used for the first time by Babcock. If we were to take into account the fine structure of the magnetic field, we must speak only about the complex magnetic fields.

*Nussbaumer:* Where do you put the division between high and low resolution and how good is the best resolution you can obtain?

*Howard:* Our best resolution is 2 arc-sec, and I would put the dividing line between high and low resolution for magnetographs at about 10 arc-sec.

*Wilcox:* Professor Severny has correctly pointed out that the so-called unipolar photospheric magnetic fields contain a complicated structure having fields of both polarities when examined at high resolution. However, it seems that the concept of a 'unipolar' region is still valid if this is defined to mean a large region throughout which the net magnetic flux (as observed with a slit size of perhaps 23″) is of a single polarity. Such 'unipolar' regions seem often to be the sources of interplanetary magnetic-field sector structures within each of which the field polarity is unidirectional.

*Bappu:* Am I correct in assuming that your magnetic-field representation refers only to the longitudinal case? Have you examined the transverse fields in such developing regions?

*Howard:* Our observations are only of the longitudinal component of the field. We have no means at present of observing the transverse component.

*Jäger:* If one observes an active region on the one hemisphere, often one observes a similar region on the other at nearly the same latitude and longitude. Can you derive from your observations how often this occurs in the average?

*Howard:* I do not know, and I also do not know whether the occurrence of two AR's at about the same time at the same longitude in two hemispheres is due to chance or implies some deep-rooted connection.

*Jensen:* I have a question related to the prediction one can make from the idea of magnetic buoyancy. If the field is formed in this way, the direction of the field should change from horizontal to vertical during the development. Is there any observational evidence for such a change?

*Howard:* I am not aware of the existence of such observations, but perhaps we shall hear of some recent observations during this symposium.

*McIntosh:* White-light observations show linear dark features, lanes in the photosphere, during the growth of bipolar sunspot groups, and emergence of new spots in large groups. If we interpret these dark linear features as analogous to the dark penumbral structures then it appears that the emergence of a strong transverse field accompanies the growth of a sunspot.

*Sturrock:* My colleagues at Stanford have carried out an analysis of sunspot data to determine whether there is any correlation between the appearance of another AR at about the same longitude in the hemisphere. We find that there is no statistically significant correlation.

# ORIGINE DES RÉGIONS ACTIVES SOLAIRES 'ANORMALES'

M. J. MARTRES

*(Observatoire de Meudon, S. & O., France)*

ABSTRACT

Solar active regions are considered 'anomalous' when they belong to magnetic classes $\gamma, \beta\gamma$ and $\beta$f-$\alpha$f. The study of the solar activity of the region where, later on, these groups are born shows an evident correlation between the presence of an old active center and the complexity of the new active region.

It is found that the complexity is greater if the old active center is younger, and the superposition better. We also observe that the birth of anomalous sunspots groups occurs much more frequently on the western side of the magnetic inversion line of the old center.

When the birth of an active center occurs *outside* and *on the West* of the faculae, we observe the weakly anomalous groups $\beta$f-$\alpha$f. The 'perturbation' decreases with distance and is extended at least to 10 heliographic degrees of the boundaries of the old faculae.

La classe des groupes de taches $\beta$p-$\alpha$p de la classification magnétique du Mount-Wilson réunit environ 80% de l'ensemble des groupes de taches qui apparaissent sur le soleil, et il est facile de penser que les groupes ainsi définis correspondent à un type *'normal'* de centre actif solaire.

Dans un ordre croissant de malformations, nous trouvons les $\beta$f-$\alpha$f, les $\beta\gamma$ et les $\gamma$. Par opposition nous pouvons donc penser que ce sont là des centres d'activité de type *'anormal'*.

Utilisant d'une part les publications des données magnétiques et le film des observations journalières de Mount-Wilson, d'autre part la collection des spectrohéliogrammes de Meudon, nous avons cherché à associer la formation des centres actifs appartenant aux différentes classes magnétiques, à l'activité solaire préexistante de la région dans laquelle ils apparaissent. Cette étude est parfois gênée par les lacunes d'observations de l'un et l'autre des deux observatoires.

**1.** Au cours de la période 1920–35 nous avons revelé 307 groupes de taches d'importance supérieure à 5 dans l'échelle à 10 échelons des *Cartes Synoptiques de la Chromosphère Solaire* et de classe magnétique quelconque.

Ces groupes se trouvent répartis comme suit:

49 ont été signalés $\gamma$ au moins 1 jour,
119 ont été signalés $\beta\gamma$ au moins 1 jour,
139 n'ont jamais été classés $\gamma$ ou $\beta\gamma$.

Certains de ces groupes sont nés dans des régions solaires où aucune trace faculaire n'est observable sur les clichés $K_1$, d'autres à l'emplacement ou très près de régions actives préexistantes.

Leur répartition en fonction de leur classe magnétique est indiquée dans l'histogramme de la Figure 1.

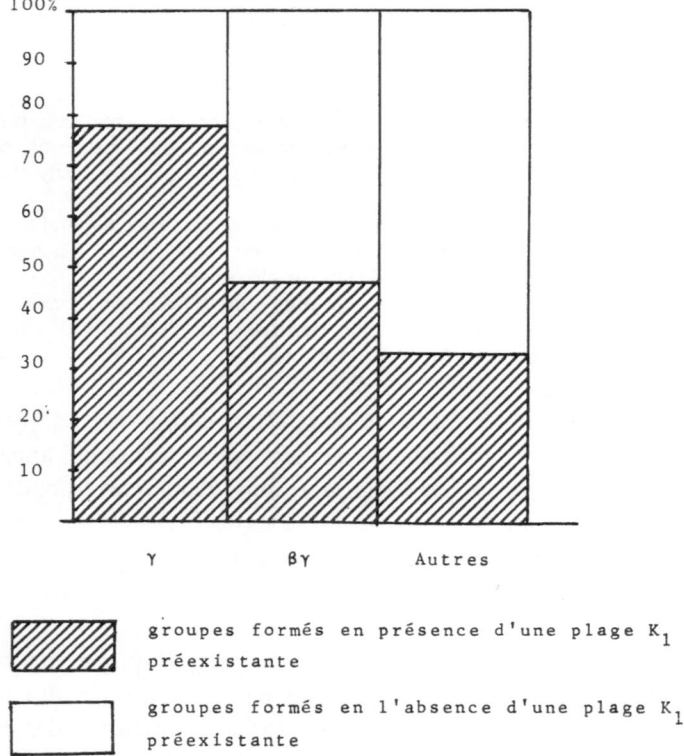

FIG. 1. *Histogramme de la distribution des groupes en fonction de leur classe magnétique et de l'activité de la région où ils apparaissent.*

*Les pourcentages,* malgré la sévérité des critères (complexité éphémère) et les lacunes des observations (un groupe peut naître et mourir pendant les 13 jours de passage d'une région dans l'hémisphère invisible du Soleil) *semblent indiquer que la présence de C.A. préexistant influe sur la complexité des structures nouvelles.*

**2.** Nous avons cherché à préciser certaines lois de l'interaction entre C.A. nouveau et C.A. ancien.

Beaucoup de variables entrent en jeu. Parmi celles-ci on trouve, d'une part, le degré de complexité du C.A. nouveau, d'autre part les importances respectives des deux centres actifs en présence, l'âge ou le stade de l'évolution du C.A. préexistant au moment de la naissance du second et aussi leur disposition respective.

Nous nous sommes bornés à étudier pour les groupes de l'échantillon précédent nés dans des régions où l'existence d'un centre actif était observable, la corrélation avec la complexité magnétique:

(1) de l'âge de la formation préexistante: à savoir si celle-ci possède ou ne possède plus de taches visibles sur $K_{1v}$;

(2) de leurs distances respectives: à savoir s'il y a superposition de leurs facules ou seulement juxtaposition.

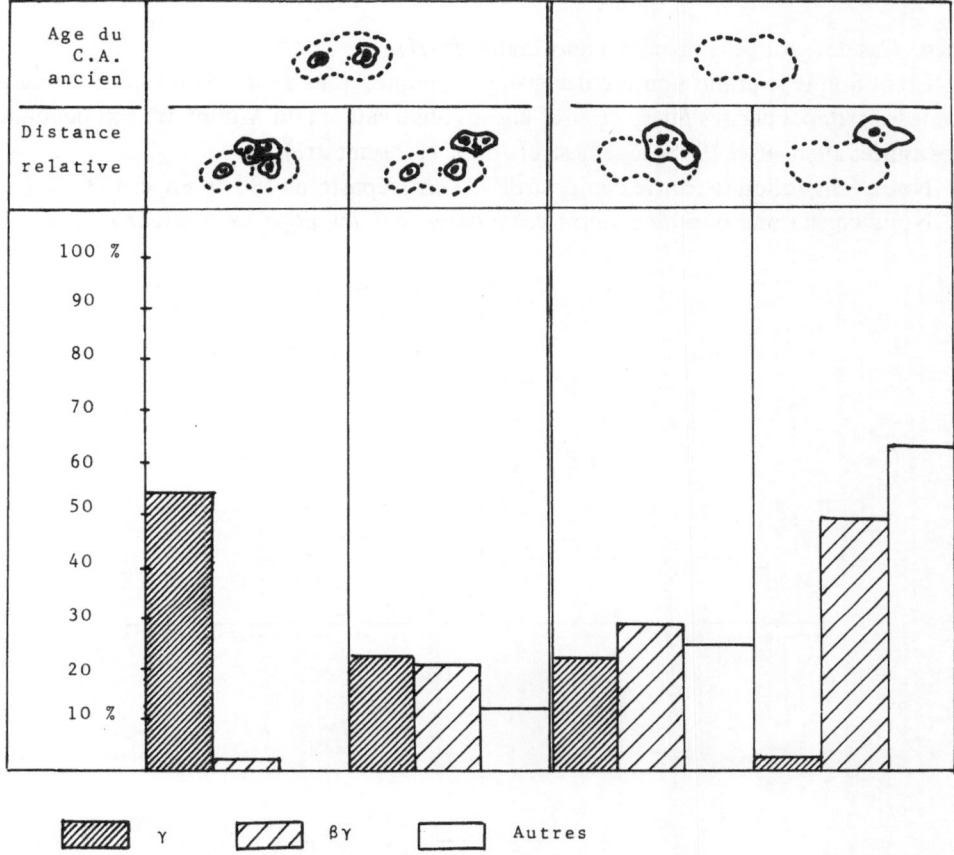

FIG. 2. *Distribution des C.A. formés en présence d'un C.A. ancien en fonction de leur classe magnétique, de l'âge du C.A. ancien et de la distance relative.*

Comme il est naturel *on trouve* (Figure 2) *que la perturbation est d'autant plus forte que le C.A. ancien a des champs magnétiques plus forts (groupe jeune, encore taché) et que la superposition des deux formations est plus complète.*

**3.** Influence de la disposition relative.

Nous avons recherché si le fait qu'un C.A. nouveau se forme à l'Est ou à l'Ouest d'une formation préexistante joue un rôle dans la complexité magnétique et l'évolution du C.A. résultant.

Les points d'apparition des groupes $\gamma$ et $\beta\gamma$ (160 cas) ont été comparés à une droite fictive représentant grossièrement la ligne d'inversion magnétique du C.A. ancien.

*On reconnaît que plus de 70% de ces centres complexes apparaissent à l'ouest de cette droite (quelques uns sur cette droite). En revanche, d'une manière générale, tous les groupes apparus à l'est comptent parmi les moins perturbés.*

**4.** Cas des groupes faiblement anormaux ($\beta f$-$\alpha f$).

Etant donné le grand nombre des groupes simples nous avons réduit notre étude à tous les groupes classés $\beta p$-$\alpha p$ et $\beta f$-$\alpha f$ par les observateurs du Mount-Wilson pendant les années 1936–46 et 1955 (2è semestre) – 1956 (1er semestre).

Notre échantillon se trouve composé de 733 cas, répartis en 580 $\beta p$-$\alpha p$, et 153 $\beta f$-$\alpha f$.

Nous constatons que *dans notre échantillon tous les $\beta f$-$\alpha f$ se trouvent disposés à*

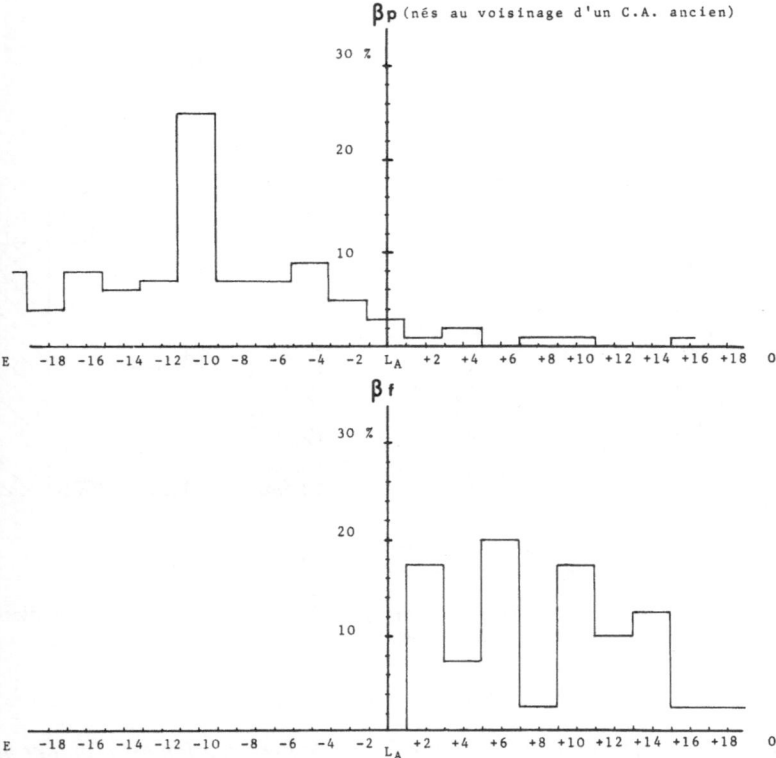

FIG. 3. *Répartition des écarts en longitude $\Delta L$. $\Delta L = L_N - L_A$ des centres de figure des C.A. nouveau et ancien.*

*l'ouest d'une R.A. préexistante\** et en général à une distance égale ou inférieure à 10°
du bord de la facule de l'ancienne formation alors que les *βp-αp apparaissent soit dans
des régions dépourvus de C.A. soit à l'est d'une région active préexistante* (quelques cas
sont juxtaposés mais en pile ordonnée avec celle-ci) (Figure 3).
  La Figure 4 schématise l'ensemble des résultats.

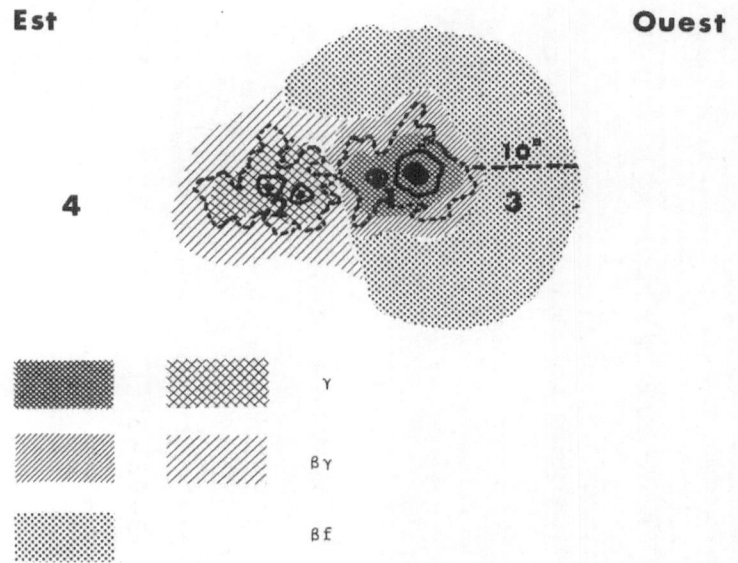

Fig. 4.  *Schéma d'un C.A. de l'hémisphère N. et disposition des aires où peuvent se développer les
groupes anormaux.*

  Autour d'un centre actif solaire taché, supposé unique sur le Soleil, nous avons
délimité les zones favorables à l'apparition des groupes anormaux.
  Sachant que
  72% des groupes $\gamma$ apparaissent dans la zone 1,
  23% des groupes $\gamma$ apparaissent dans la zone 2,
  72% des groupes $\beta\gamma$ apparaissent dans la région limitrophe des zones 1–3,
  21% des groupes $\beta\gamma$ apparaissent dans la région limitrophe des zones 2–4,
  90% des groupes $\beta$f apparaissent dans la zone 3,
nous pouvons délimiter ainsi une aire grossièrement centrée sur la tache de tête du
C.A. existant où, une naissance de C.A. venant à se produire on aura toutes les chances
de voir se développer des groupes anormaux, d'autant plus perturbés que le centre
préexistant sera jeune et qu'ils seront situés près de cette tache de tête.

* Ont été retirés de l'échantillon les groupes situés à moins de 10° de plusieurs C.A. anciens.

M. J. MARTRES

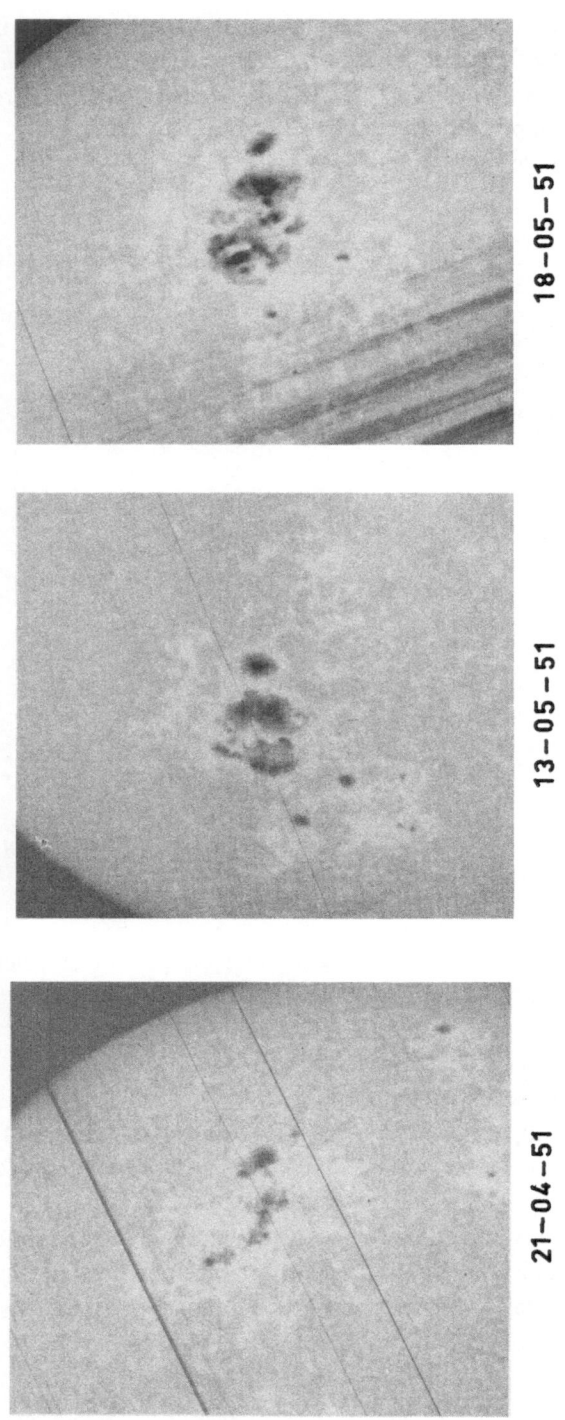

21–04–51    13–05–51    18–05–51

FIG. 5.    *Apparition à l'intérieur d'une Région active tachée d'un centre actif évoluant en gros complexe. Spectrohéliogrammes $K_{1V}$ de l'Observatoire de Meudon.*

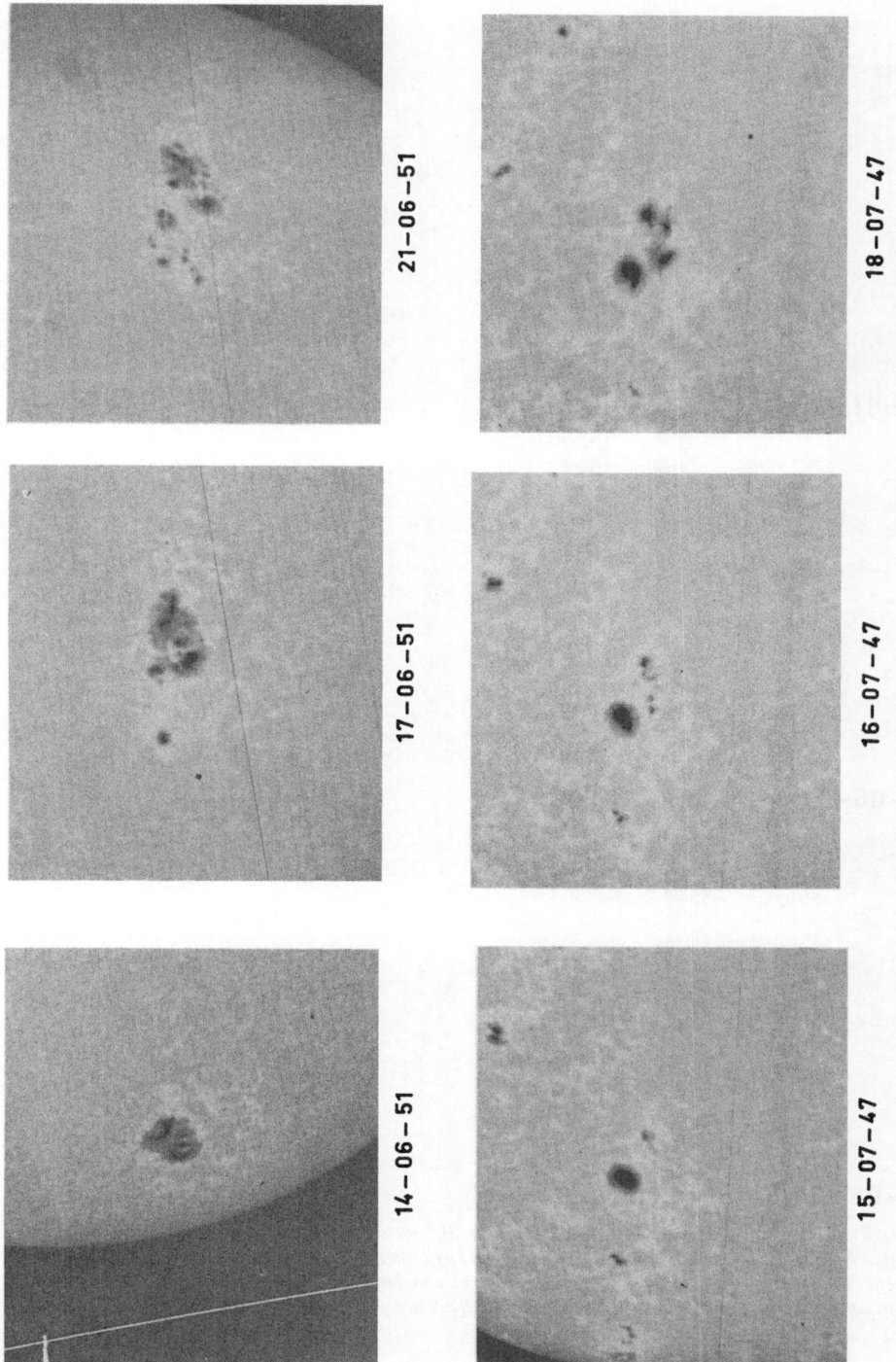

FIG. 6. *Apparition, en juxtaposition avec un C.A. taché de C.A. importants: en haut, à l'Ouest du premier, évolution en βγ; en bas, à l'Est du premier, évolution en γ. Spectrohéliogrammes K₁ᵥ de l'Observatoire de Meudon.*

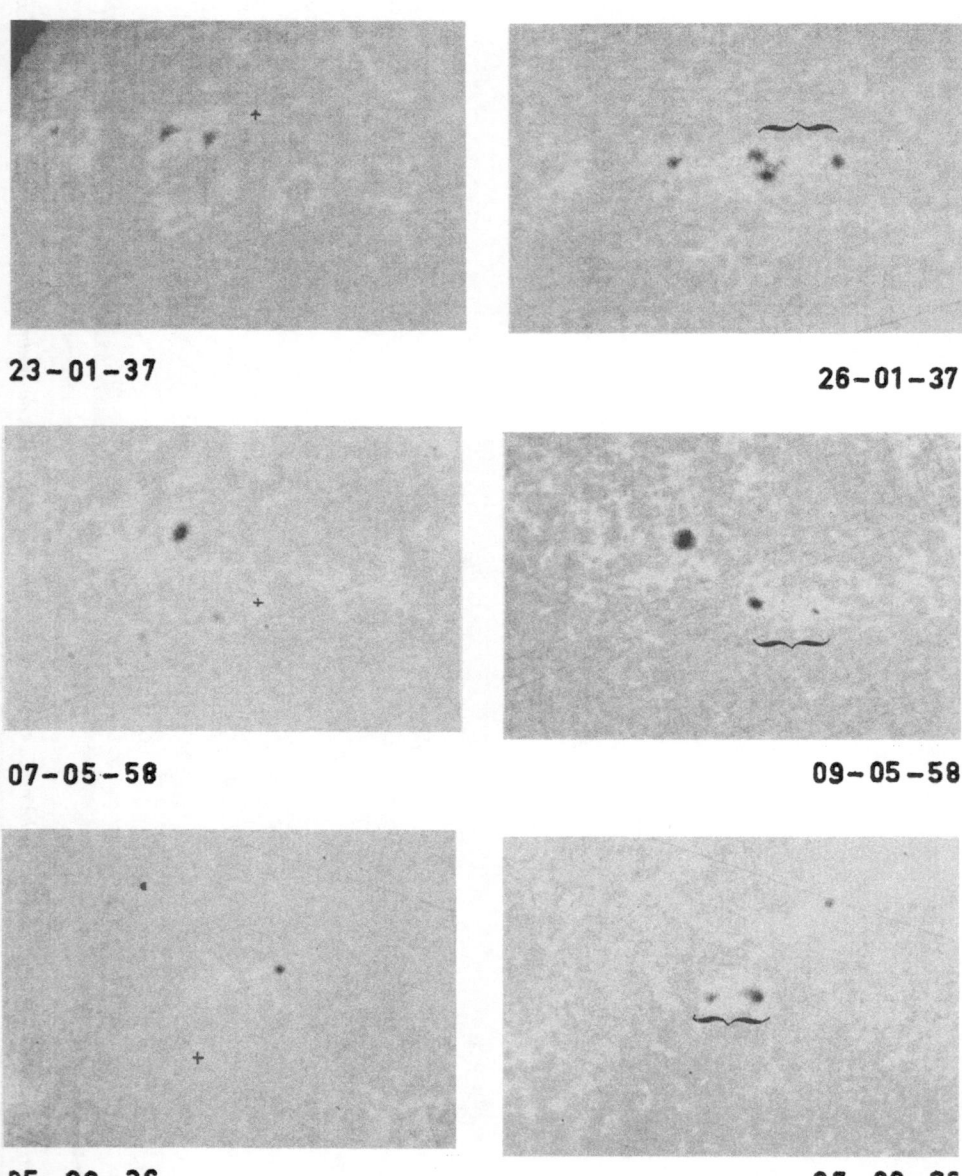

23-01-37                                       26-01-37

07-05-58                                         09-05-58

25-08-36                                         27-08-36

FIG. 7. *Apparition de C.A. simples à moins de 10° du centre préexistant. (1) en haut, centres jointifs. Formation nouvelle à l'Ouest, évolue en β f. (2) au centre, centres non jointifs. Formation nouvelle à l'Ouest, évolue en β f. (3) en bas, centres non jointifs. Formation nouvelle à l'Est, évolue en βp. Spectrohéliogrammes de l'Observatoire de Meudon.*

# MAGNETIC CLASSIFICATION OF ACTIVE REGIONS*

SARA F. SMITH                and                ROBERT HOWARD
(Lockheed Solar Observatory,                    (Mount Wilson
Burbank, Calif., U.S.A.)                         and Palomar Observatories;
                                                 California Institute of Technology,
                                                 Pasadena, Calif., U.S.A.)

ABSTRACT

All the active regions observed on the Sun with the Mount-Wilson magnetograph between August 1959 and December 1962 have been given magnetic classifications in a system similar to the Mount-Wilson sunspot-classification scheme. The flare productivity of regions classified as unipolar, bipolar, and complex bipolar, as well as regions composed of multiple bipolar components has been studied. It has not been necessary to provide a classification corresponding to the $\gamma$ class of sunspots. Although the relatively poor angular resolution employed in the magnetograms limits somewhat the accuracy of the data, it is clear that both complex bipolar regions and regions with multiple bipolar components produce more than three times the number of flares than the simple bipolar regions produce. The most flare-productive class of regions is the reversed polarity complex classification.

The statistical relation of the spot magnetic-field classification to the classification of the corresponding plage fields has been studied and found to be poor for these data.

The distribution of the magnetic regions in latitude shows that many of the regions with polarities reversed from the usual orientation are confined to the equatorial zone.

## 1. Introduction

In 1963, we began a project of studying and classifying magnetic regions and comparing these magnetic regions with sunspots, plages, and flares. The Mount-Wilson magnetograms used in this study were the whole disk daily maps made during the period August 1959 to December 1962. These magnetograms contain the longitudinal field component below about 100 gauss, and have a resolution of 23 sec of arc.

A system for classifying magnetic regions similar to the one we adopted for this work was devised independently by Martres, Michard, and Soru-Iscovici in 1966 based on observations made at Meudon. The Martres, Michard, and Soru-Iscovici system is surprisingly similar to ours, even though they were working primarily with magnetic-field measures of sunspots, and bright plage while we were working with magnetograms of plage and surrounding weaker fields.

Initially, we tried to adapt the Mount-Wilson classification system for sunspot groups to the weak fields observed on the magnetograms. After applying this system to

* Presented by S. Smith.

*Kiepenheuer (ed.), Structure and Development of Solar Active Regions, 33–42. © I.A.U.*

several months of data, it became evident that the system did not adequately distinguish the situations observed in weak fields of low resolution. Therefore, a modified system employing new definitions was developed. All of the regions observed between August 1959 and December 1962 were studied using the modified sunspot group-classification system. After gaining additional familiarity with the characteristics of magnetic regions and their correspondence with plages, still other modifications to the classification scheme seemed to be desirable.

## 2. The Classification of Regions

The system which we discuss in this paper, we have found to be a practical one, although it is likely that we will make more modifications to it in the future. This system is described in Table 1 and illustrated in Figures 1 and 2. In this system all

### Table 1

| Designation | Magnetic Region Classifications<br>Definition |
|---|---|
| B | Simple bipolar region |
| BC | Simple bipolar region with one polarity partially encircling the opposite polarity |
| BS | A simple bipolar region with an area of opposite polarity embedded in one or both of the main bipolar components of the region |
| BY | Bipolar region with a peninsula of one polarity extending into the opposite polarity |
| BCS | Bipolar region with the characteristics of both a BC and a BS region. |
| BYS | Bipolar region with the characteristics of both a BY and a BS region. |
| BB | Two adjacent simple bipolar regions *not* clearly distinguishable as separate plages |
| BBC | A BB region with any of the C, Y or S characteristics |
| B–B | Two adjacent simple bipolar regions distinguished as separate plages |
| BCB | Two adjacent bipolar regions which were distinguished as separate plages but including any of the C, Y, or S characteristics |
| BBB | Three closely spaced B regions of any complexity |
| X | No classification given – poor data |

Greek and lower-case letter designations have been replaced with Roman capital letters which are compatible with computer languages. In Figure 1, solid lines represent positive polarity and dashed lines represent negative polarity. In the upper left corner of the figure is an example of the most common class of magnetic region associated with plages. The letter designation is B meaning simple bipolar region. This class corresponds exactly to the Sn class in the Martres, Michard, Soru-Iscovici system (MMS). In the upper right of Figure 1, BC represents a bipolar region in which one polarity is partially encircled by the opposite polarity. This situation often occurs when a new region develops in a much older region and, therefore, corresponds most

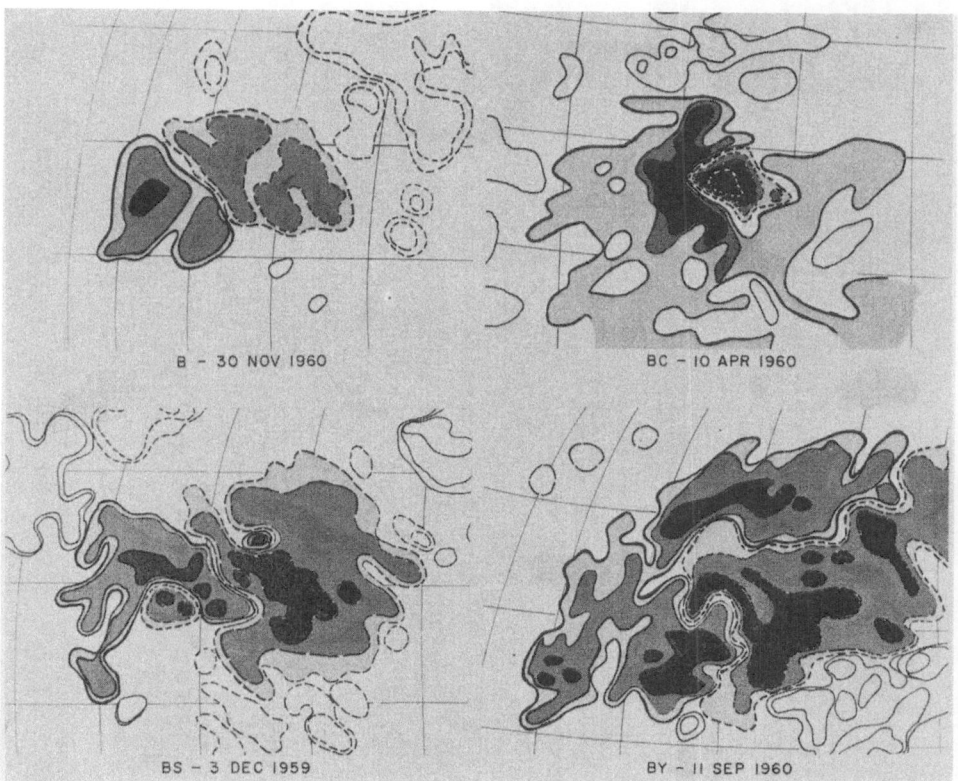

Fig. 1. *B = bipolar region; C = one polarity partially encircling opposite polarity; S = island of one polarity amidst opposite polarity; Y = a peninsula of one polarity extending into opposite polarity.*

closely to the Ca class in the MMS system. In the lower left of Figure 1, BS represents a bipolar region with one or more islands of opposite polarity in either the leading or following polarity of the region. This class is exactly analogous to the Cp class in the MMS system. The example of the BY class in the lower right of Figure 1 is a region in which a peninsula of one polarity extends into the opposite polarity. There is no exact equivalent for the BY class in the MMS system. We also use BCS and BYS designations to refer to a more complex situation. There are no counterparts for these in the MMS system.

Figure 2 shows three examples of 'multiple bipolar regions'. These are two or more bipolar regions which are sufficiently close together that at least at some time during the disk passage, they are considered to be one plage. In the upper left BB represents two simple bipolar regions. In the upper right BBB is used to designate three or more regions of any degree of complexity found in B, BC, BS, and BY regions. In the lower left BBC refers to a double bipolar complex region, also with any of the degrees of

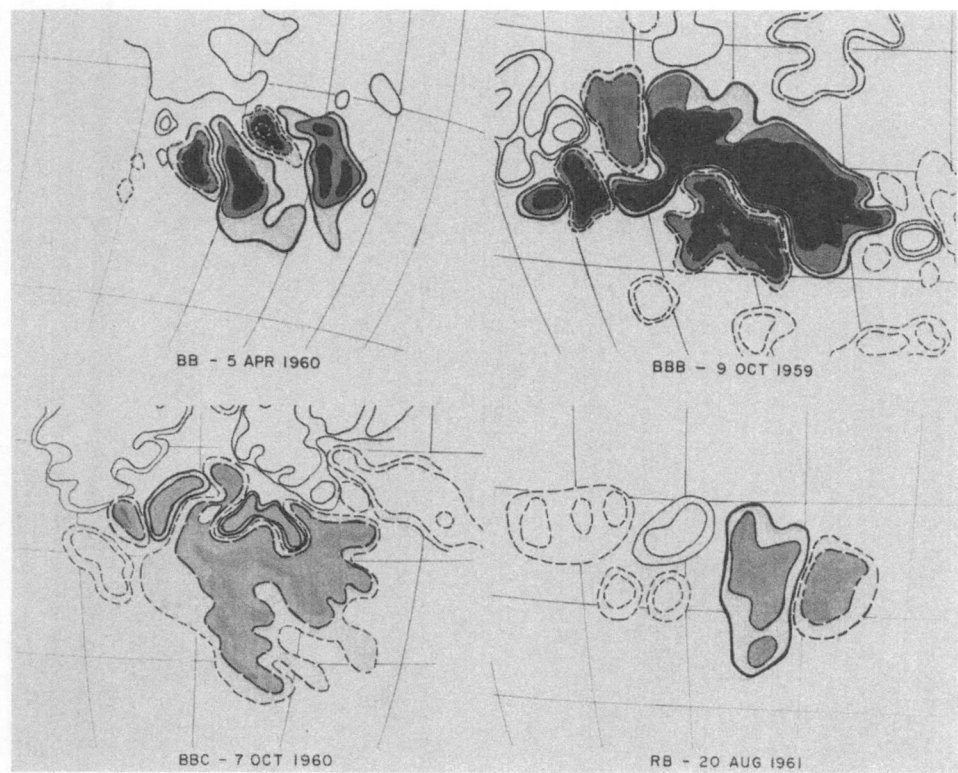

FIG. 2. *BB = two adjacent bipolar regions; BBB = three adjacent bipolar regions; BBC = two adjacent bipolar regions more complex than a BB region; RB = reversed polarity simple bipolar region. All of these regions appeared in the Southern Hemisphere.*

complexity as in the examples in Figure 1. Lastly, in the lower right RB stands for a simple reversed polarity magnetic region. All of the regions in Figure 2 are Southern-Hemisphere regions so that the interchanged polarities of this region are shown in comparison with the normal configuration for this hemisphere and solar cycle. Reversed polarity regions appear to be similar to all other regions except for this polarity reversal. Reversed polarity regions may be of any degree of complexity. They were isolated in this classification system by placing an R in front of the regular classification.

In addition to these classes of magnetic regions there are also positive and negative 'A' regions or areas of single polarity. In all known cases these regions are simply old remnants of bipolar regions. An 'A' region as defined here is an isolated feature of one polarity and bears no relation to the UM regions observed on synoptic charts. Also, a large number of regions were classified as X to indicate the classification was unknown because of poor data.

The MMS system contains two classes, Si and Cγ, which are not equivalent to any of the classes in our system. Si represents regions whose polarities are abnormally inclined such as those in which the leading polarity is North of the following polarity instead of West of the following polarity. This would be a good addition to our system since these regions would consist of an intermediate class between B and RB regions. The Cγ class in the MMS system is a region, like the Mount-Wilson γ class of sunspots, having the polarities very mixed. Since we have several more classes for describing complex configurations we find no need for such complex γ class.

### 3. Statistical Characteristics of Magnetic Regions

In Figure 3 is shown the distribution of magnetic regions among the various classes. All the classifications discussed in this paper are the most complex class attained by a region during one disk passage. Positive and negative A regions are the most numer-

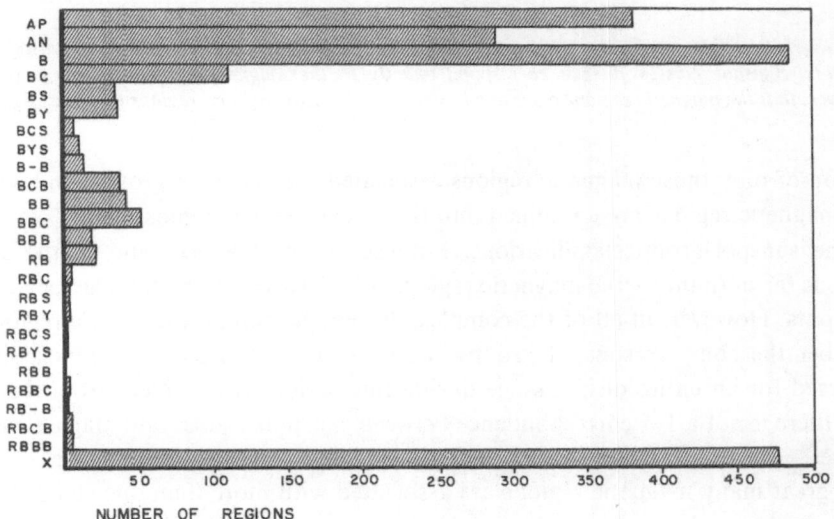

FIG. 3. *Distribution of magnetic regions by classification. Simple bipolar regions, B and BC, are the most common classes of young regions. AP and AN regions are remnants of old bipolar regions. Reversed polarity regions begin with R. X refers to regions not classified because of poor data.*

ous. Next in frequency are simple bipolar regions, B and BC. The complex regions BS, BY, BYC, and BCS comprise only 10% of the entire sample of over 1600 regions classified other than X or A, and are outnumbered by regions composed of multiple bipolar units which comprise 19% of the sample. Least frequent are reversed polarity regions which altogether comprise less than 5% of the sample.

In Figure 4, we compare the relative distribution of sunspot groups with the distri-

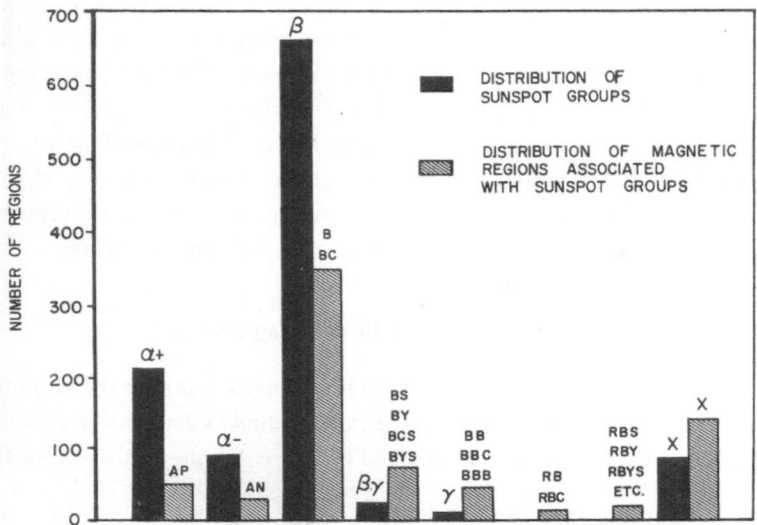

Fig. 4. *Distribution of magnetic regions associated with sunspots compared to the distribution of sunspots. Sunspot classes do not correspond exactly to the magnetic-region classes. It is evident, however, that the magnetic regions contain a larger percentage of complex regions than the sunspot data.*

bution of only those magnetic regions associated with sunspot groups. In this graph the magnetic regions are combined into the categories which most nearly correspond to the sunspot-group classifications. For the classes $\alpha+$, $\alpha-$ and $\beta$, the sunspot regions far outnumber the magnetic regions of similar class which are associated with sunspots. However, in all of the complex classes, the complex magnetic regions outnumber the complex sunspot groups. There were no sunspots groups classified as reversed for an entire disk passage during this period. It is evident from this graph that there is not a 1:1 correspondence between sunspots classes and magnetic regions classes.

A great many magnetic regions are associated with more than one sunspot group. The actual correlation between sunspots and magnetic regions is depicted in Figures 5 and 6. First, in Figure 5, we consider those cases in which there was only one sunspot group per magnetic region. On the horizontal axis in Figure 5, the sunspot classes are shown in five groups. $\beta$, $\beta$p, and $\beta$f are all simple bipolar regions. Because there are so few $\beta\gamma$ and $\gamma$ spot groups we consider them together as complex sunspot groups. $\alpha+$ and $\alpha-$ are considered together as unipolar sunspots. There were no reversed polarity spots in this sample even though there were 37 reversed polarity magnetic regions. Lastly, there are some classified as X for unknown because of lack of data.

On the vertical axis is the distribution of sunspot classes among similar groupings of the magnetic region classes. The primary reason for this arrangement is to show that the correlation between sunspot classification and magnetic-region classifi-

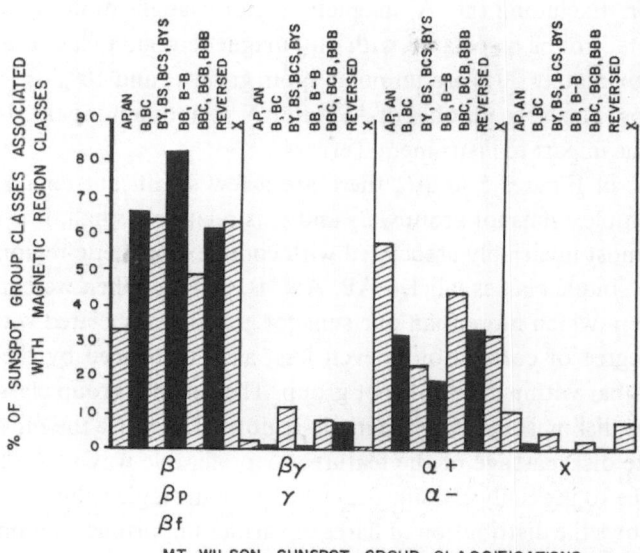

FIG. 5. *Distribution of sunspots groups among groups of magnetic-region classes for only those cases in which one sunspot group corresponded to one magnetic region. Complex sunspot groups tend to be associated with complex magnetic regions but the converse is not true.*

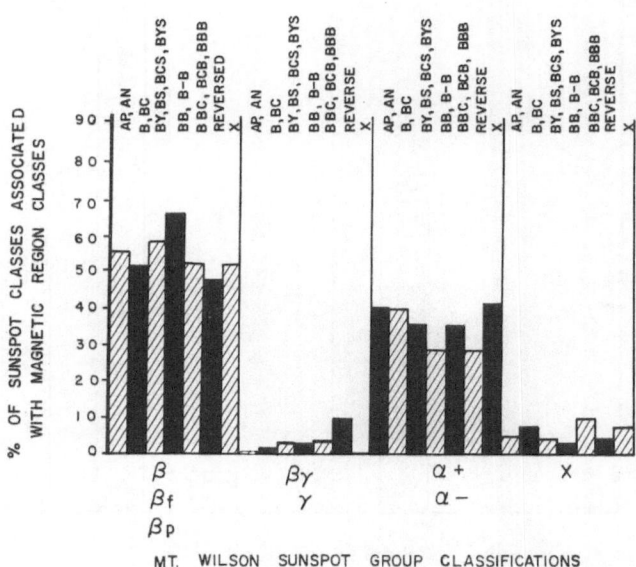

FIG. 6. *Distribution of sunspot groups among groups of magnetic regions for those cases in which more than one sunspot group corresponded to one magnetic region. The correlation between sunspot classification and magnetic-region classification is seen to be poor.*

cation is poor. Excluding the 'A' magnetic region classification, approximately 60% of the sunspots groups associated with any magnetic region class are simple bipolar groups; approximately 30% are unipolar spots groups; and 10% or less are complex sunspot groups. The large number of AP and AN regions among the $\beta$-sunspot classes may also be due in part to instrumental errors.

In addition, in Figures 5 and 6, there are a few significant details. Although the number of complex sunspot groups, $\beta\gamma$ and $\gamma$, is relatively small, it is clear that these groups are almost invariably associated with complex magnetic-region classes, rather than with the simple classes labeled AP, AN, B, or BC. When we consider the cases as in Figure 6 in which more than one sunspot group is associated with one magnetic region, the degree of correlation is even less, as is evidenced by the more uniform height of each bar within each sunspot group. The sunspot group classes are averages for a complete disk passage. The magnetic-regions classes are the most complex class for a complete disk passage of the features. A preferable way to make such a correlation would be to use daily classifications rather than single values for a disk passage.

Figure 7 shows the distribution of flares of various importances among the magnetic-regions classifications. In this figure, flare productivity represents the total number of flares of each importance per region, where it is understood that one region represents

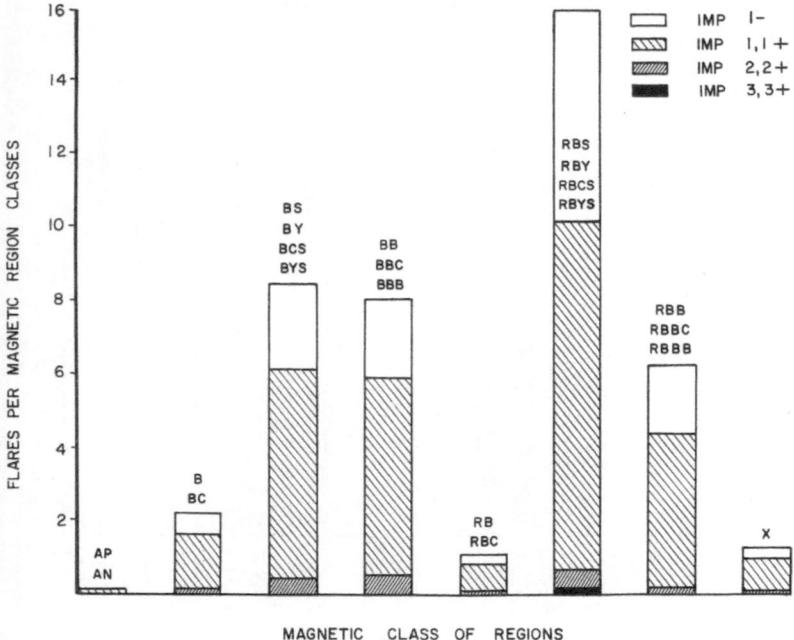

FIG. 7.  *Complex reversed polarity magnetic regions were the most flare-productive regions during this period, Aug. 1959 to Dec. 1962. All complex regions were three times as flare productive as simple bipolar regions B, BC, and RB, RBC.*

one disk passage of a magnetic feature. Only the importance-I flares can be considered to be a reliable statistical sample. Most important to note in this figure is that the reversed polarity complex regions are by far the most flare-productive region classification. This gives us another important parameter in the search for tools for flare prediction. Also important are the complex bipolar regions and multiple bipolar regions which are seen to be approximately three times as flare productive as simple bipolar regions.

We investigated the latitude dependence of regions of various classifications. Some of these results are shown in Figure 8. A feature to note is the distribution of reversed polarity regions relative to B regions. The mean latitude in the North or the South for the reversed regions is definitely shifted toward the equator. The preponderance of Northern-Hemisphere regions is reflected in all of the magnetic classes except for the

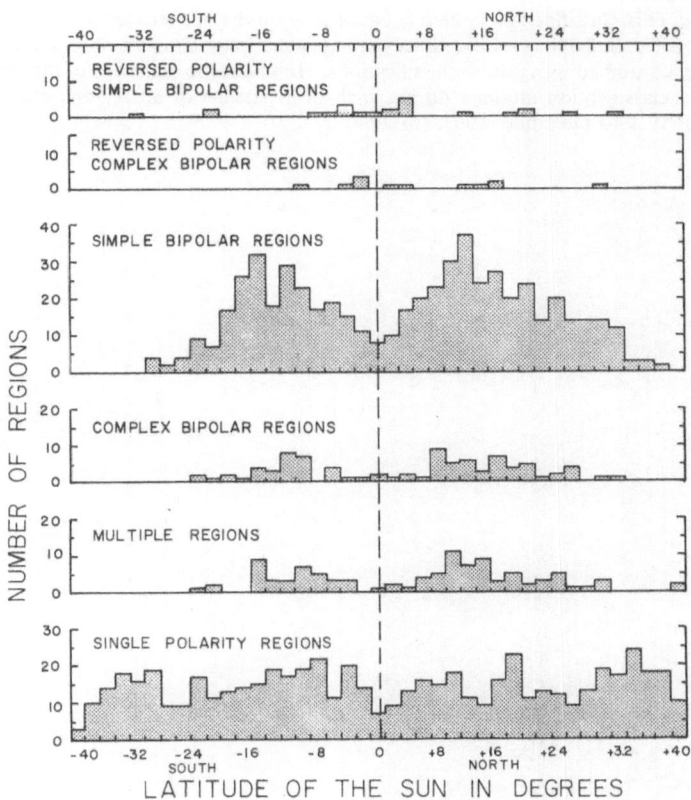

FIG. 8. *Distribution of magnetic regions on the sun, Aug. 1959 – Dec. 1962. Reversed polarity magnetic regions are the only class of magnetic regions which tend to occur most frequently within 10° of the solar equator.*

A (unipolar) regions. 'A' regions have a uniform distribution which extends to higher latitudes than the distribution of sunspots.

It should be noted again that the classifications given in this paper are maximum-complexity classifications for the disk passage of the regions. It would be better to study flare productivity and sunspot correlations using *daily* magnetic-region classifications. We plan further studies using more recent magnetic-field data which are of better quality.

## References

Martres, M.J., Michard, R., Soru-Iscovici, I. (1966)      *Ann. Astrophys.* **29**, 245.

## DISCUSSION

*Bumba:* Does your classification system contain the stage of evolution of active regions, for example the maximum stage?

*Sara Smith:* This classification system is based only on the appearance of regions and not on the way regions evolve. Since the classifications were assigned for each day's observations, the evolution of regions can be studied using these classifications. However, the diagrams in this paper employ the most complex classification attained during each disk passage of all regions observed during the period August 1959 to December 1962.

# DEVELOPMENT OF MAGNETIC FIELDS IN ACTIVE REGIONS*

E. TANDBERG-HANSSEN and C. PORTER

*(High Altitude Observatory, Boulder, Colo., U.S.A.)*

## 1. Introduction

The magnetograph of the High Altitude Observatory station at Climax has been used to study changes in the longitudinal component of the magnetic field in a number of active regions. For a description of the instrument see Lee, Rust, and Zirin (1965). By securing simultaneous Hα pictures of the regions we compare the magnetic data with the optical appearance of the plages. From such studies we have selected the results of two active regions to present here.

## 2. The Observations

The observations were made with a 10″ aperture scanning the active regions to build up a map (magnetogram) of the distribution of the longitudinal component of the magnetic field. Figures 1 and 2 pertain to two different regions, A and B, both observed on 28 February 1967. Figure 1 shows the magnetic field in region A (PA 334, RV 0·75) at 1730 UT and at 2000 UT. There is no great change in the structure of the magnetic field, but the field strength of the Southern bipolar structure (dominated by a bipolar sunspot group) has decreased significantly. For instance, the Southern positive plage area decreased in intensity from a maximum value of about 380 gauss (±15 gauss) to 190 gauss (±11 gauss). To within the accuracy of our measurements, the Northern positive plage area remained unchanged during this same period at 95 gauss (±12 gauss).

Figure 2 shows the development of region B (PA 360, RV 0·60) between 2140 UT and 2304 UT. The main change is an invasion of a positive magnetic field into a previously mainly negative magnetic plage in the Northern part of the observed region. This was not accompanied by any visible change in the Hα plage. The structure of the Southern bipolar area remained in essence unchanged during this time. In addition to the Hα picture we show in Figure 2 an Aerospace white light photograph of region B.

---

* Presented by J. M. Malville.

*Kiepenheuer (ed.), Structure and Development of Solar Active Regions, 43–46. © I.A.U.*

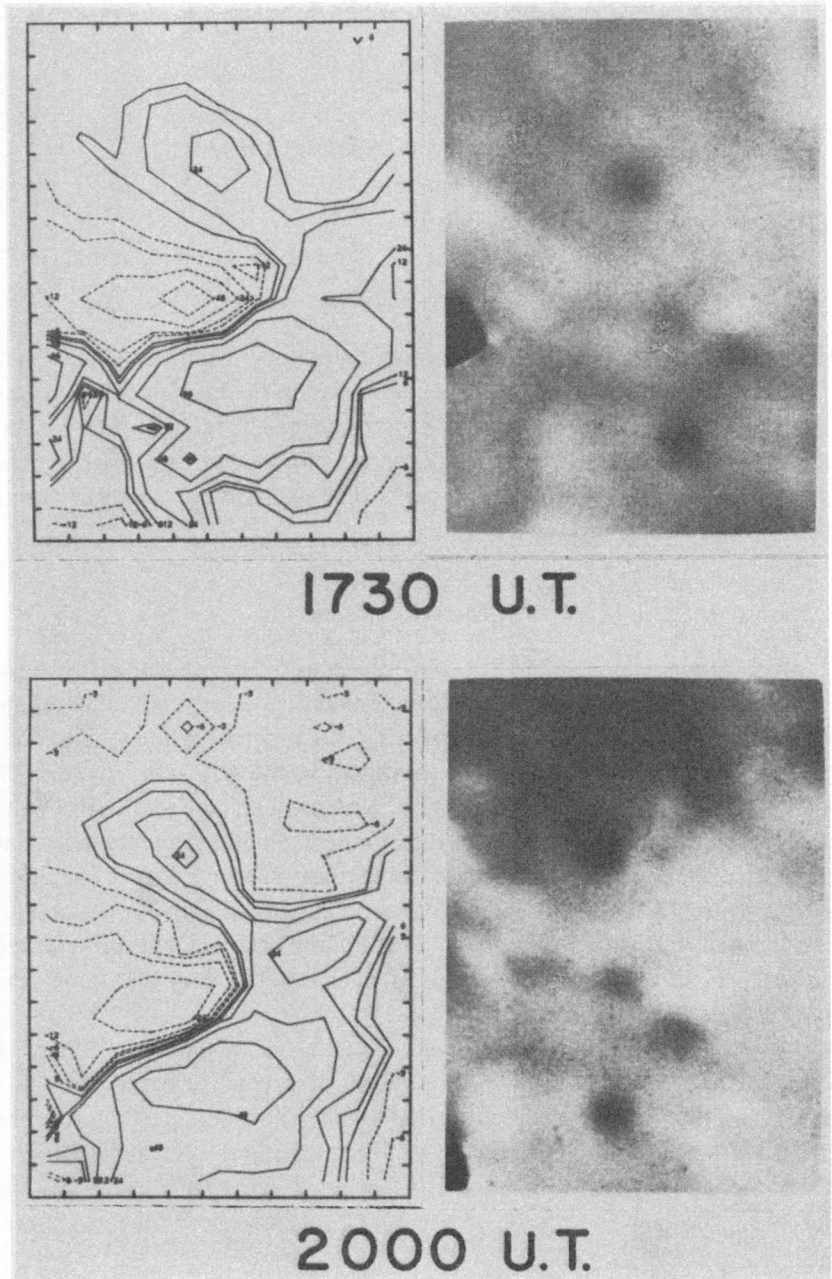

Fig. 1. *Magnetograms (isogauss plots) and Hα filtergrams of active region A (PA 334, RV 0·75) at 1730 UT and at 2000 UT.*

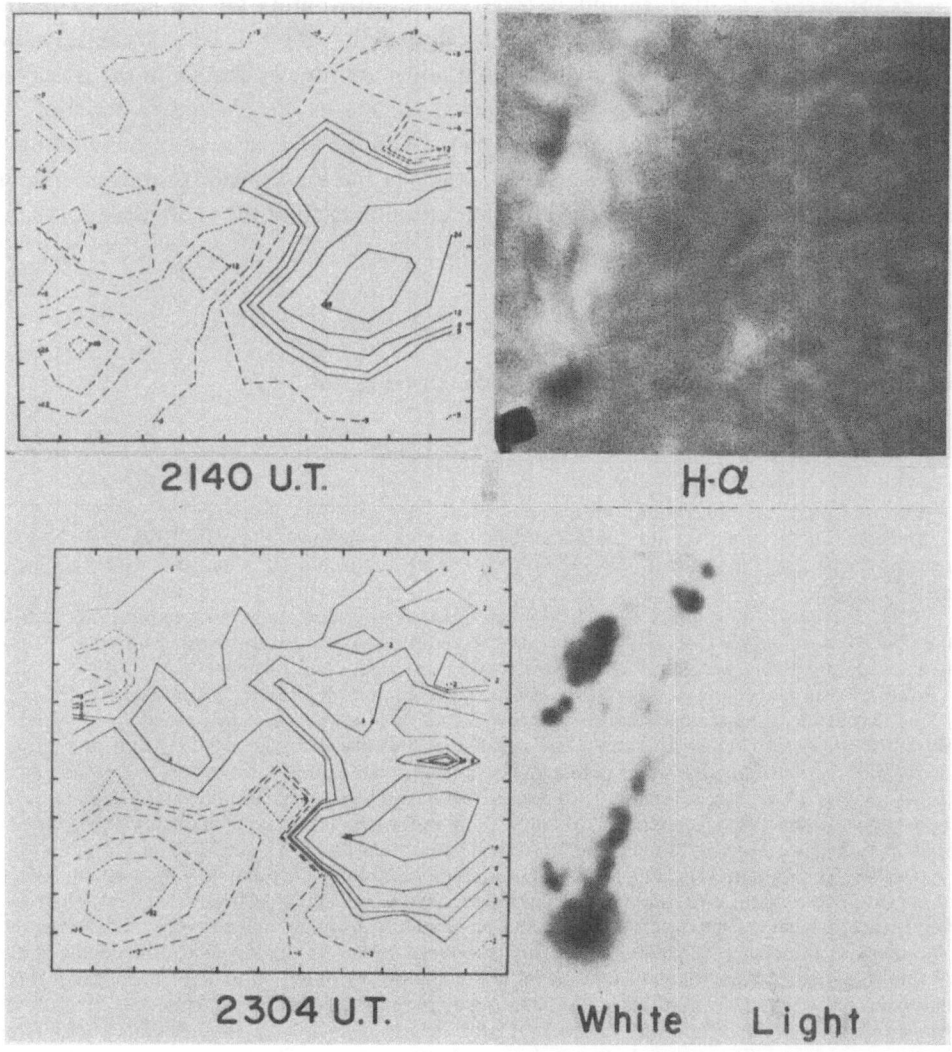

FIG. 2.   *Top; magnetogram and Hα filtergram of active region B (PA 360, RV 0·60) at 2140 UT.*
*Bottom: magnetogram at 2304 UT and white-light picture (Aerospace Corporation) of active region B.*

## 3. Discussion

A straightforward interpretation of the above-mentioned observations indicates
that – even in plages dominated by strong sunspot fields – there are at times significant changes in periods of several hours. The changes pertain either to variations in
the magnetic-field intensity, or to changes in the distribution of positive and negative

fields. However, caution should be exercised in attributing all the changes in the observed longitudinal component to real changes in the magnetic field. We observe the magnetic field using the Hα line and an unknown amount of change in the observed longitudinal field may be due to a temporal change in the Hα source function. A certain change in temperature or density of the plage plasma may occur without a change in the magnetic field. A thorough study of the Hα-transfer problems in magnetic plages must therefore precede any final interpretation of the changes observed in the longitudinal component.

## Reference

Lee, R. H., Rust, D. M., Zirin, H. (1965)     *Applied Optics*, **4**, 1081.

## DISCUSSION

*Maltby:* Could you comment on the accuracy of the magnetic-field observations?

*Malville:* The accuracy of these measurements is approximately ±15 gauss. For a number of reasons the accuracy is less than the usual error of ±2 gauss in our measurements with the Climax magnetograph.

*H. U. Schmidt:* Is the change of the field intensity in one of your examples more pronounced than the change in topology? If so, it would be hard to understand the corresponding migration of flux. It would imply a very peculiar correlation between the velocity and magnetic field.

*Malville:* The slides showed examples of changes of topology and also changes of magnetic intensity.

*Severny:* It seems to me dangerous to ascribe the effect of a change of magnetographic signal in Hα entirely to magnetic field, because the signal is proportional to the intensity and to the field strength. Hα in plage regions appears in emission and the change from absorption to emission leads to the reversal of the signal from plus to minus. We have been recording magnetic fields in Hα for 3 years and found that it is extremely difficult to eliminate the influence of intensity in the case of plage regions even if we have independent simultaneous records of Hα intensity.

*Malville:* The Climax magnetograph is continuously calibrated in order to minimize the effects of varying line profile in Hα due to plage emission. Two signals are synchronously detected at two frequencies, one signal due to switching of the phase of the λ/4 plate and the other due to the displacement of the center of gravity of the line produced by an oscillating relay lens ahead of the double slits of the magnetograph. One signal is divided by the other, so that for equal wavelength shifts of the two systems variations in line shape, even to the extent of emission reversals, are entirely removed. However, the observed variations in the magnetic field, may I believe, also arise from fluctuations of the 'effective' height at which the radiation is formed. Such an alternate explanation for the observed fluctuations cannot be eliminated.

*Beckers:* To elaborate on a remark made by Drs. Krat and Severny about the interpretation of Hα-magnetograph measurements: With a height reversal of the source function as may occur in plages one has indeed a superposition of a background Hα-absorption line on the plage-emission line. Were only the plage-emission line present one would measure an opposite signal from the case in which only the Hα-absorption line is present under the same magnetic-field conditions. It is therefore very hard to estimate either field sign or magnitude for a mixture of the two, if the field is different in both. This is so even if you make a calibration as done at HAO. The changes in the magnetic field in the paper may be real changes but may also be changes in the Hα optical properties in plages (see *Solar Physics* **3** (1968), Section 6).

# THE BALANCE OF MAGNETIC FLUXES
# IN ACTIVE REGIONS

Jan Olof Stenflo

*(Astronomical Observatory, Lund, Sweden)*

According to modern theories of the solar cycle, active regions on the Sun are caused by a magnetic disturbance penetrating the solar surface from below. Sunspots, filaments, flares and other conspicuous events in an active region seem to be only secondary phenomena, the basic feature being the magnetic field itself.

One important property of an active region as a whole is that it is generally asymmetric in the East–West direction. From white-light photographs of sunspot groups we know that the preceding spots generally dominate over the following spots, although there are many exceptions. The same fact is known from the measurements of sunspot magnetic fields. Grotrian and Künzel (1950) found that the magnetic flux through the preceding spots is in the mean 3–4 times greater than the flux through the following spots. Hence, if only the magnetic field of sunspots is taken into account, there is a very strong disbalance between the preceding and following fluxes in active regions.

Sunspots develop when the magnetic field is strong enough to inhibit convection. By measuring the field of sunspot pores at the best available seeing, Steshenko (1967) concluded that sunspots can develop only when the field strength exceeds approximately 1100 gauss. Accordingly, the result of Grotrian and Künzel means that if only fields stronger than about 1100 gauss are considered, there is a strong disbalance between the fluxes in the region. It is therefore of great interest to study how the balance of fluxes varies with the field strength.

Isogauss maps of sunspot groups recorded in 1963 and 1965 with the magnetograph at the Crimean Astrophysical Observatory were used for the study of the flux balance. Only maps which appeared to contain nearly all the magnetic fields belonging to the spot group were studied. 17 maps of 10 spot groups were thus selected for examination. The magnetograph was equipped with a brightness compensator when the records were made, eliminating the influence of the brightness variations within the spot group. Visual determinations of the maximal field strengths in the spots were also available.

When interpreting the isogauss maps and measuring the fluxes we must know the calibration curve of the magnetograph, i.e. the relation between the magnetographic signal and the magnetic-field strength in the line of sight. Severny (1967) has determined the calibration curve of the Crimean magnetograph by observations. The

*Kiepenheuer (ed.), Structure and Development of Solar Active Regions*, 47–49. © *I.A.U.*

observationally determined curve deviates very much from the curve based on the theory for the Zeeman effect in a homogeneous solar atmosphere. It has been considered to be more appropriate to use the observational curve in the study of the flux balance, but we should be aware of systematic errors introduced by errors in the calibration curve.

The magnetic fluxes in different intervals of the longitudinal field strength (i.e. between different isogauss lines on the maps) were measured on the isogauss maps using a planimeter. The measured fluxes in a spot group were then represented in the form of histograms, giving the flux $\Phi_i$ per unit interval of the longitudinal field strength $H$, i.e. $\Phi_i/\Delta_i H$, as a function of the field strength $H$. Some of the common properties of the histograms may best be studied by determining a distribution, which is the

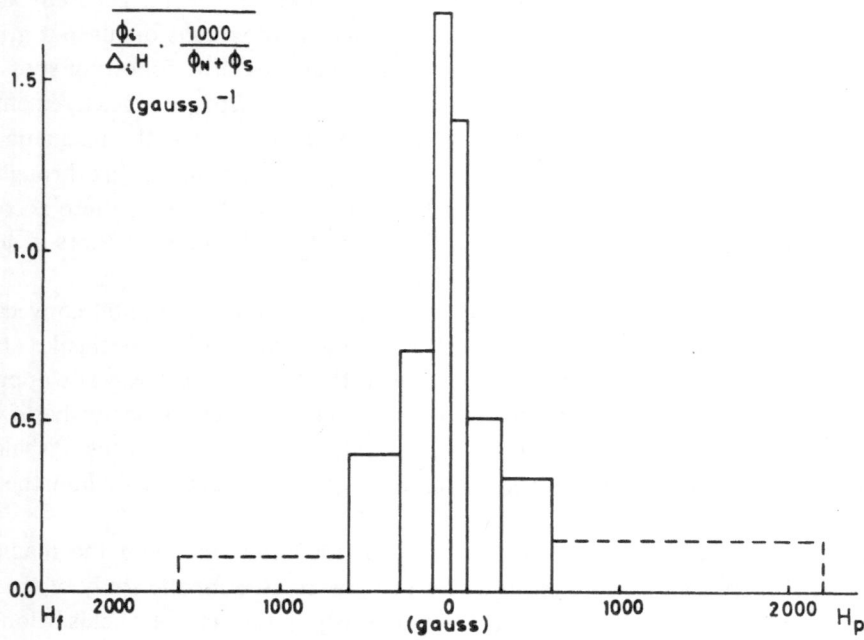

FIG. 1. *Distribution of magnetic fluxes. $H_p$ ($H_t$) means longitudinal field strength of preceding (following) polarity.*

mean of all the individual histograms. The result is shown in Figure 1. As a kind of normalization, the fluxes $\Phi_i$ are divided by the sum of the total fluxes of N- and S-polarity in the region, $\Phi_N + \Phi_S$, before the mean is taken. The histogram is dashed in the strong field intervals, because the upper interval limit is somewhat uncertain. It is seen that the weaker fields are responsible for a considerable part of the total flux. We also notice the asymmetry of the distribution, which is very pronounced for the stronger fields (sunspot fields). Taking the difference between the $p$- and $f$-fluxes in the same $H$-intervals, Figure 2 is obtained, which describes the flux balance. We find that the following part of the region contains more flux, if only longitudinal fields $H$ weaker than about 600 gauss are considered. For stronger fields the situation is reversed. Hence we obtain very nearly a total balance of fluxes in the region.

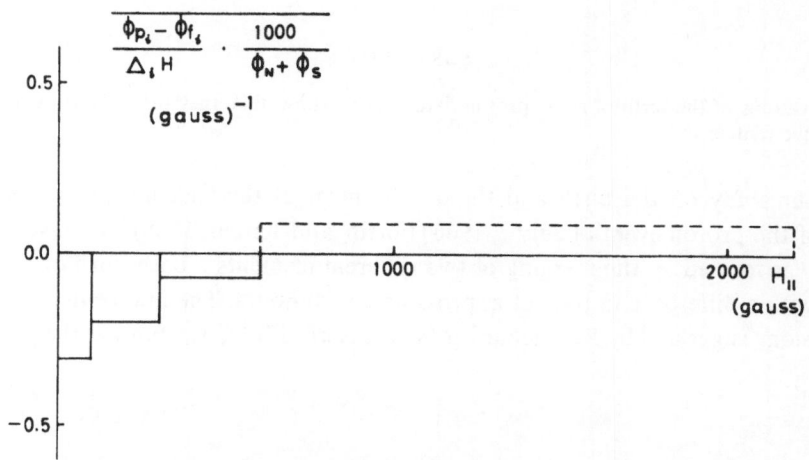

FIG. 2.   *Flux-balance diagram.*

We can now better understand the result obtained by Grotrian and Künzel. Let us study the ratio between the $p$- and $f$-fluxes, $\Phi_p/\Phi_f$, for different $H$-intervals. In the interval 0–600 gauss we find this ratio to be about 0·8. For $H > 600$ gauss it is 2·5. Hence the result of Grotrian and Künzel that $\Phi_p/\Phi_f \approx 3·5$ for sunspots, i.e. for $H \gtrsim 1100$ gauss, agrees very well with the present results, although our statistical material (the number of isogauss maps) is rather limited.

On the basis of what has been said, the general magnetic character of an active region as a whole could very briefly be described as follows: The magnetic field is bipolar, and the preceding and following fluxes balance each other. There is an East–West asymmetry. The magnetic lines of force are more concentrated in the preceding part of the region than in the following part, where they are spread out over a larger area.

## References

Grotrian, W., Künzel, H. (1950)    *Z. Astrophys.*, **28**, 28.
Severny, A. B. (1967)    *Izv. Krym. astrofiz. Obs.*, **36**, 22.
Steshenko, N. V. (1967)    *Izv. Krym. astrofiz. Obs.*, **37**, 21.

# ON THE BIRTH OF SOME PROTON-FLARE REGIONS*

T. FORTINI and M. TORELLI

*(Osservatorio Astronomico Monte Mario, Roma, Italy)*

ABSTRACT

Observations of the birth of some proton-flare regions show that they originate in two or more close active centers.

On our study on the birth and the development of the Calcium plage (McMath 8362) of the proton event of July 7, 1966 (Fortini and Torelli, 1968), we have realized that it was formed by the merging of two different elements – both born on the solar disk – with a difference in time of approximately 80 hours. The first center, using the terminology suggested by R. Michard (Martres *et al.*, 1966), was born at the periphery

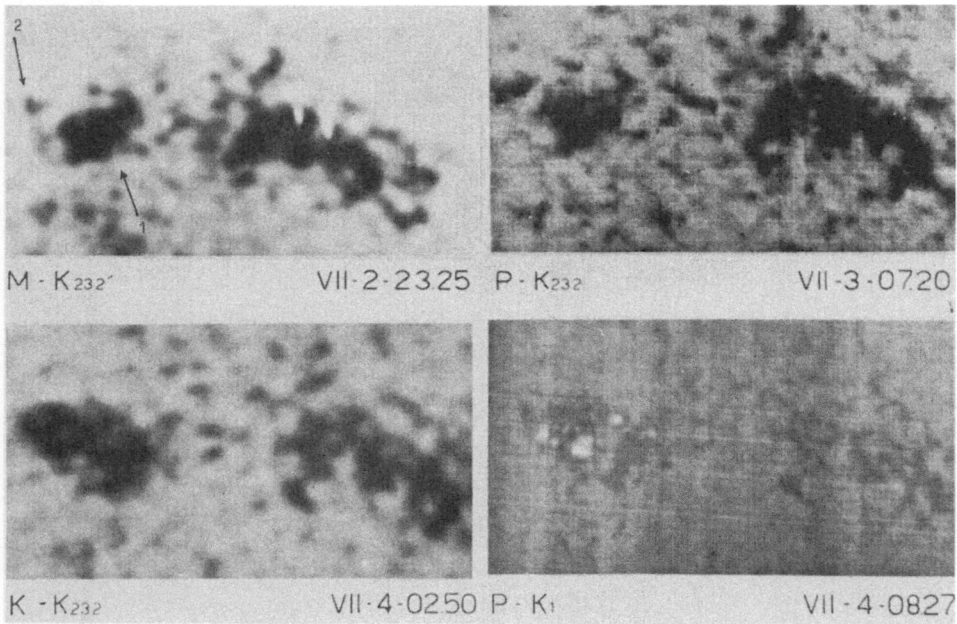

FIG. 1. *PF Region McMath 8362 (June–July 1966). 1 = First center born on the solar disk on June 28, 2 = Second center born on the solar disk on July 2, A = Arcetri; K = Kodaikanal; M = Manila; P = Paris.*

* Presented by T. Fortini.

*Kiepenheuer (ed.), Structure and Development of Solar Active Regions, 50–55. © I.A.U.*

of an old region on June 28; the second one, two degrees East in longitude, started in a less disturbed area on July 2. Due to the closeness of the two centers, the Eastern one, on developing from East toward West, merged with the first center. Therefore, after July 4, two plages looked like one entity (Figure 1). The whole sunspot group (Rome 4369) was initially composed by two different groups, born successively in the first and in the second center (Figure 2). It is probable in this case that the merg-

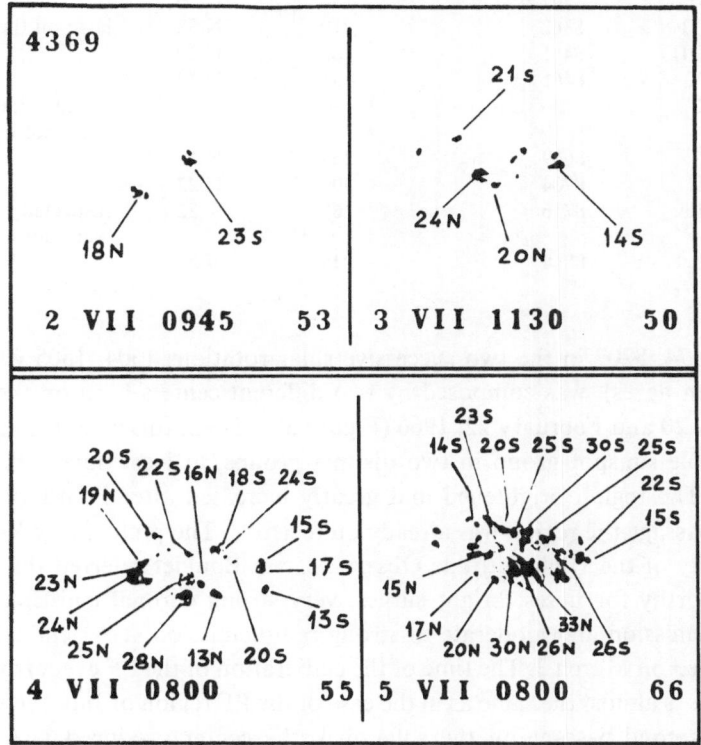

Fig. 2.  *PF Region McMath 8362. Sunspots magnetic configuration. Rome Observatory Magnetic Observations.*

ing of the two centers, maybe favoured by their particular inclination to the solar equator, has lead to the morphological and magnetic configuration of two rows of sunspots with opposite magnetic polarities characteristic of many proton centers (Avignon *et al.*, 1964). Therefore it was interesting to examine if the above circumstance, namely the merging of two or more active centers, is often or always verified during the formation of the PF regions.

With this purpose in mind, we have examined the PF regions occurred in the period January 1966 – May 1967, listed in Table 1. Only two regions were born on the solar disk; of these we have just now discussed the 1966 June–July region. The other region

**Table 1**

**List of polar cap absorption (PCA) events and associated proton flare (PF) regions
(1 January 1966 – 31 May 1967)**

| PCA | Region (McMath Number) | Coord. $\lambda$ | $\varphi$ | Remarks |
|---|---|---|---|---|
| 1966  3 III | 8174 | 130 | N 23 | born on the disk |
| 23 III | 8207 = 8174 | 138 | N 20 | |
| 7 VII | 8362 | 210 | N 32 | born on the disk |
| 28 VIII | 8461 | 185 | N 20 | |
| 2 IX | 8461 | 185 | N 20 | |
| 5 X | – | – | – | not identified (flare not observed) |
| 1967 28 I | 8659 | 269 | N 16 | |
| 2 III | 8704 | 300 | N 22 | |
| 11 III | 8716 | 165 | S 22 | suspected (flare not observed) |
| 28 V | 8818 | 228 | N 24 | |

(McMath 8174, 8207 in the two successive solar rotations 1504, 1505 where it produced proton flares), was composed by two different centers born on the solar disk on February 20 and February 22, 1966 (Figure 3). Also in this case it is easy to separate the whole sunspot group in two distinct groups, at least during the first days (Figure 4). The region brightened and greatly increased after February 26. At the West-limb passage the region was already quite active. The preliminary Report of the Solar Activity of the High Altitude Observatory at Boulder referred that the region was "noteworthy for flares, bright surges, very strong coronal emission at $\lambda$ 5303, yellow line emission and moderate to strong radio emission at 9·1 cm". The Proton Flare occurred on March 3. The time of the elaboration of the PF event from the birth of the centers is almost the same as in the case of the PF region of July 1966 (Figure 5). During the second passage on the solar disk the center produced again a proton flare. It is hard to guess what happened in the invisible hemisphere. We have not considered this late activity of the region.

All the other regions we have studied until now have not be seen at their birth. When observed after the East-limb passage, all of them presented a complicated magnetic configuration, which always could be interpreted as originated by the mixing of two or more components.

From the other point of view we have also examined if the merging of two or more different centers is determinant in the production of the PF's. For the time being we have examined all Calcium observations for the first six months of 1967. We have found only one case of two fairly close centers born on the visible solar hemisphere. They have not interacted. In other few cases we observed centers born in the proximity

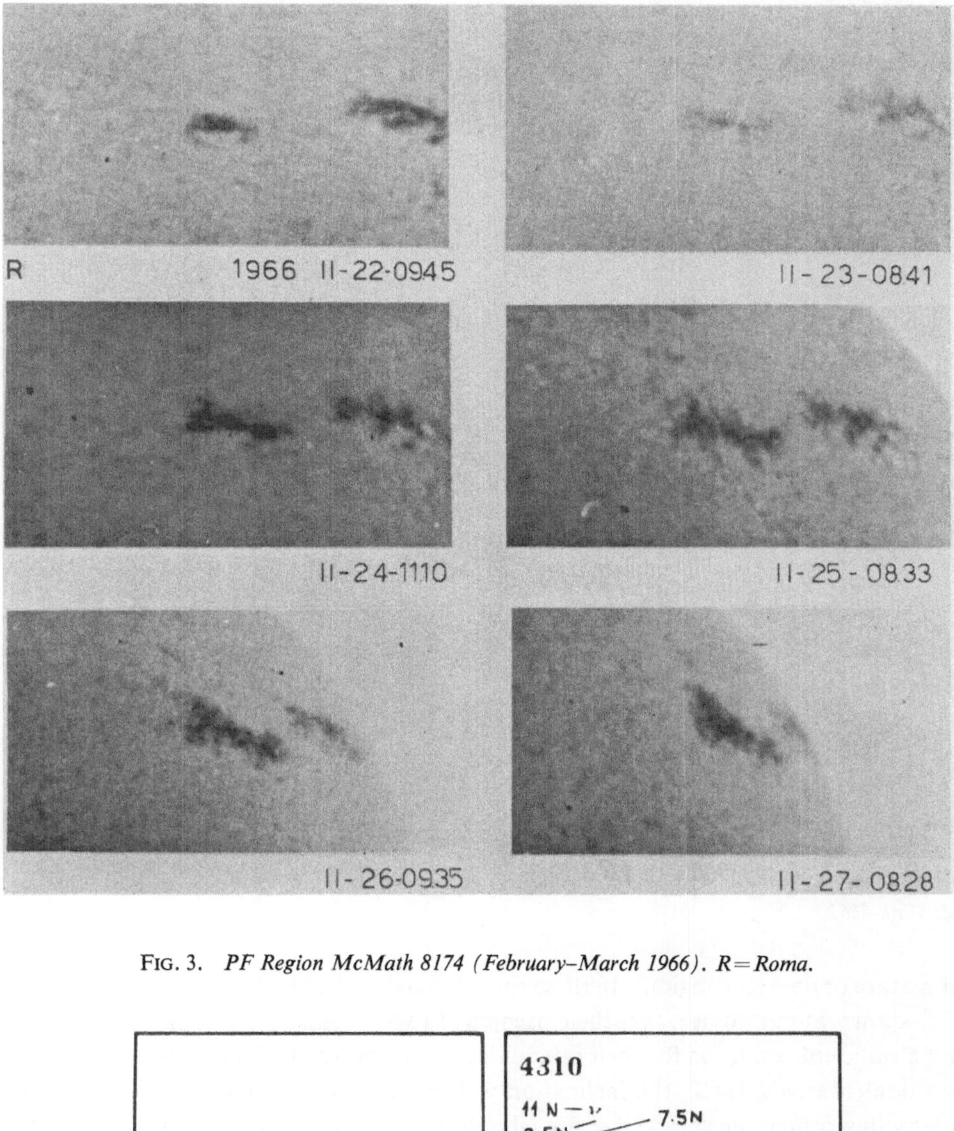

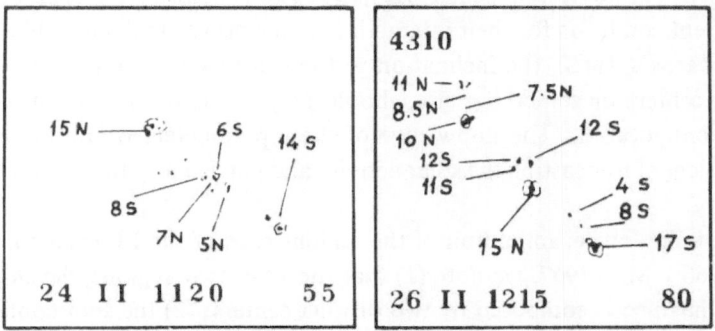

FIG. 3.   *PF Region McMath 8174 (February–March 1966). R=Roma.*

FIG. 4.   *PF Region McMath 8174. Sunspots magnetic configuration. Rome Observatory Magnetic Observations.*

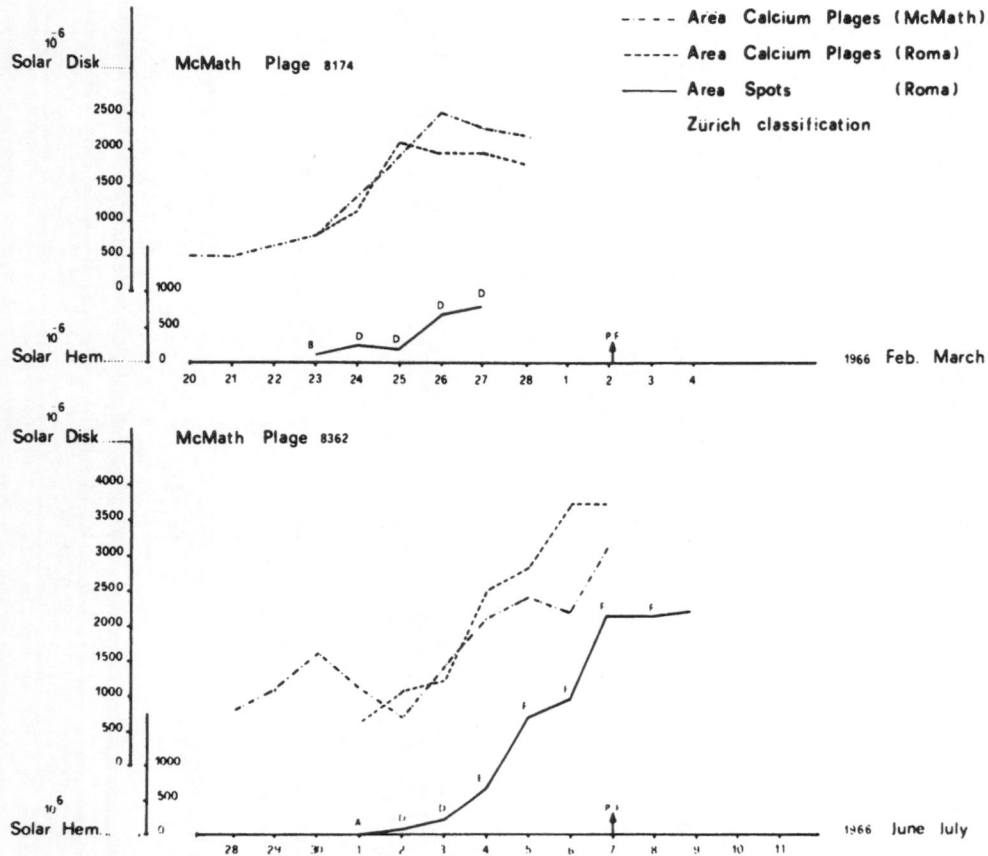

FIG. 5. *Development of the Calcium plages and sunspots for the PF Regions McMath 8174 and McMath 8362.*

of mature or decaying regions. In these cases too the centers have not interacted.

It shows, at our advice, that the closeness of two centers is, maybe, a necessary but not a sufficient condition for their interaction. Some other parameters, like the relative position (Martres, 1968), the inclination to the solar equator, the relative age of the interacting centers or something else, should play a role in the formation of the PF magnetic configuration. The knowledge of these parameters would be useful in the actual problem of forecasting the solar activity and particularly the proton flares.

On concluding our examination of the regions sites of the PF events in the period January 1966 – May 1967, we note (1) that the only two regions, developed on the solar disk, have been composed by two distinct centers; (2) the spot configuration of the regions not born on the visible hemisphere is not inconsistent with the fact that they have been originated by two or more separated centers; (3) the circumstance that the particular morphologic and magnetic configuration of the PF regions is due to the superposition of two or more active centers is also not inconsistent with the observation (Švestka, 1967) that the proton events occur in supercomplexes of regions. It is in fact more probable that the rather unusual fusion of two or more very close centers happen in the clustering of active regions, rather than in the less spotted areas of the solar surface.

## References

Avignon, Y., Martres, M.J., Pick, M. (1964)    *Ann. Astrophys.*, **27**, 23.
Fortini, T., Torelli, M. (1968)    Ann. IQSY, in press.
Martres, M.J. (1968)    in the present volume, p. 25.
Martres, M.J., Michard, R., Soru-Iscovici, I. (1966)    *Ann. d'Astrophys.*, **29**, 245.
Švestka, Z. (1967)    COSPAR, London.

# SOME PATTERNS IN THE DEVELOPMENT
# OF CENTERS OF SOLAR ACTIVITY, 1962–66*

HELEN W. DODSON and E. RUTH HEDEMAN

*(McMath-Hulbert Observatory, The University of Michigan, Pontiac, Mich., U.S.A.)*

ABSTRACT

A graphical representation of the 66 solar rotations (Carrington) between January 1, 1962 and December 31, 1966 has been prepared. It includes all centers of activity for which the calcium plage attained an area of at least 1000 millionths of the solar hemisphere and/or intensity 3 (McMath scale). In this study the antecedents, descendents, and neighbors of each region can easily be discerned. The work shows clearly that zones of activity, apparently closely related and much larger than single plages existed for long intervals of time. For example, the significant increases in solar activity in February, May, and October of 1965 occurred in a 'family' of calcium plages apparently related through similarities of position and strong radio emission.

The members of 'families' of centers of activity are found at systematically changing longitudes. For some 'families' the change of longitude appears to be primarily a consequence of differential rotation; for others, the pattern of formation of active centers dominates.

According to the data for 1962–66 a meaningful study of the development of a center of activity may require consideration not only of the past history of the zone of the Sun in which it occurs but also of the zone approximately 180° away on the opposite hemisphere.

A graphical representation of the 66 solar rotations (Carrington) between January 1, 1962 and December 31, 1966 has been prepared in an effort to gain insight into the course of solar activity in these years (see Figure 1). This graph includes all centers of activity for which the calcium plage attained an area of at least 1000 millionths of the solar hemisphere or intensity 3 (McMath scale). In this diagram the antecedents, descendents, and neighbors of each region can be discerned easily. Rotations are repeated both horizontally and vertically. It should be borne in mind that Carrington longitudes represent merely a convenient frame of reference corresponding to the mean rotation rate of 27·275 days. Other rotation rates or cadences can be recognized in the display by the lining up of phenomena in non-vertical lines.

The chart of Figure 1 shows that for the years studied zones of activity much larger than single plages existed for long intervals of time, in accord with similar reports by other observers (Becker, 1955; Bezrukava, 1963; Losh, 1939; Martres and Michard, 1965; Mori *et al.*, 1964; Saito, 1964; Vitinsky, 1960) for other time intervals. If this concept is correct, then the study of a single center of activity in 'isolation' represents consideration of only part of a larger and perhaps more fundamental phenomenon.

The chart of plages here presented suggests two types of order or pattern in the data.

---

* Presented by H. W. Dodson-Prince.

*Kiepenheuer (ed.), Structure and Development of Solar Active Regions*, 56–63. © *I.A.U.*

First there was an apparent concentration in 1962–66 of major activity in two zones, a primary zone with Carrington longitude $\sim 160°-230°$ and a secondary zone on the opposite side of the Sun at $320°-50°$. Secondly in these years, apparent families of plages can be traced through the successive returns of a region or by the formation of new regions in neighboring longitudes. Members of these families were often found at systematically changing Carrington longitudes. For some families, the systematic change of longitude was primarily a consequence of differential rotation; for others the pattern of formation of active centers appeared to dominate.

## 1. Families of Plages

The most conspicuous family of plages in the years 1962–66 is the one that appears to be rooted in major solar regions and phenomena in Carrington longitudes near 300° in January 1962. This family of plages drifted slowly with time to higher longitudes at rates corresponding to rotation appropriate to very low solar latitudes or to the solar equator itself – a rotation period of 27·0 days or slightly less. It should be noted that this 'family', extending over approximately 70° of longitude, included regions of both new and old cycle, high and low latitudes, Northern and Southern hemispheres. By the end of 1966 this family had drifted to $\sim 260°$ Carrington longitude. In 1962–64 the times of C.M.P. of this family of plages coincided with the times of the principal series of 27-day recurrent geomagnetic storms. In 1965 and early 1966 it included the major centers of activity of the new solar cycle.

The role of differential rotation appropriate to the latitude also can be seen in the recurrence of regions in this presentation. For example, the great March and April regions of 1966 broke out on opposite sides of the Sun in longitudes which can be traced back through differential rotation to the positions of the major regions of May and June 1965, likewise on opposite sides of the Sun. The relationship of the April 1966 region to that of June 1965 can be traced directly through a series of visible plages. For the March 1966 region, one must compute the track of differential rotation in order to establish a possible association with the May 1965 center of activity. This track is shown by a dashed line in Figure 1.

## 2. Adjacent Plages

There are many observations that suggest that the formation and development of a center of activity is not an independent or isolated phenomenon. Not only do plages tend to form in families, but certain active centers appear to beget other centers of activity in their close neighborhoods. This phenomenon is most conspicuous when the general level of solar activity is not too great.

Examples:

(a) On February 5, 1965 an electron-proton flare took place in the old cycle region

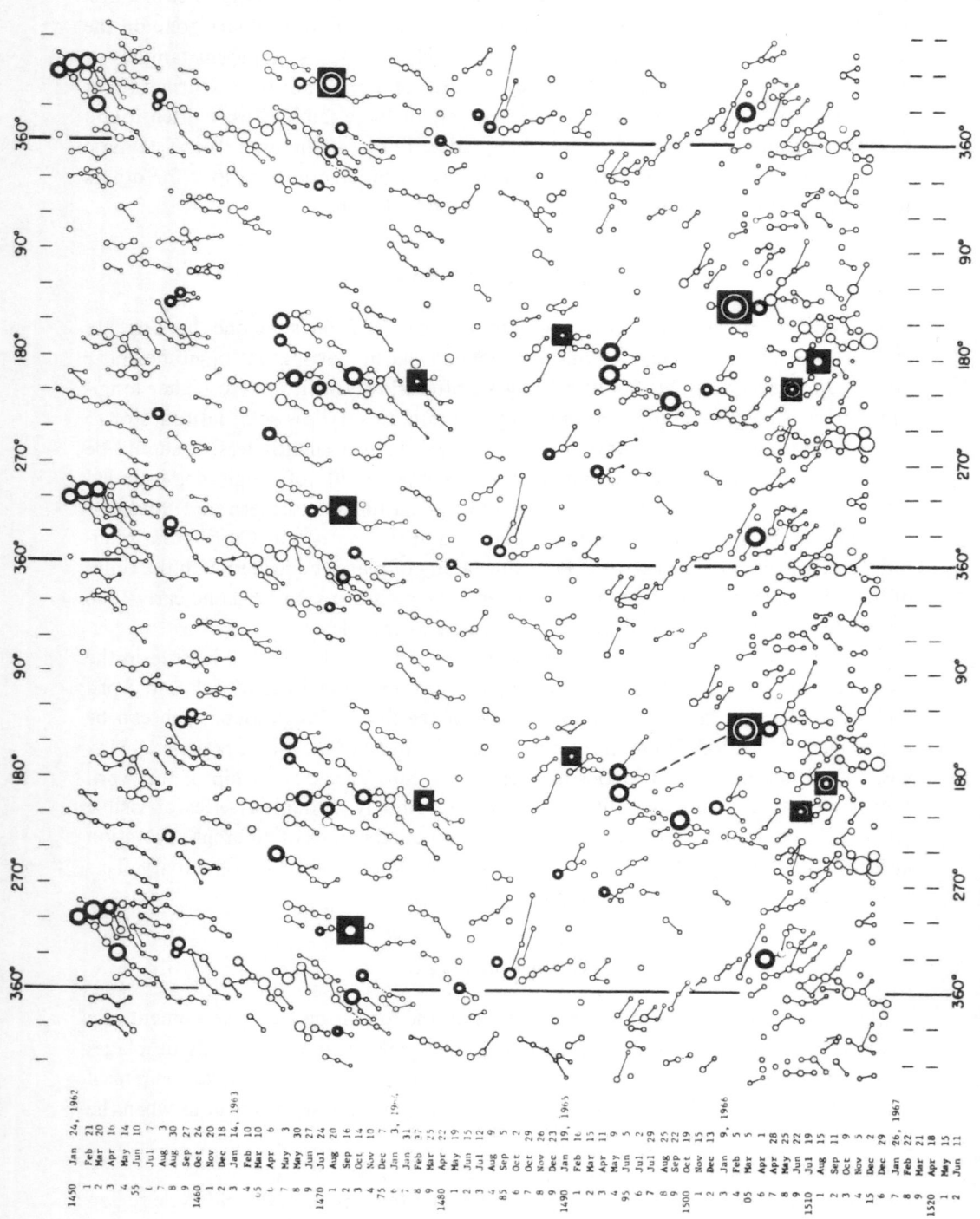

then located at N 08° W 25° ($L=160°$). In early March, the region returned, diminished in intensity and somewhat fragmented, but bounded on the North by a new cycle region, N 20° ($L=170°$), and on the South by an old cycle region S 04° ($L=175°$). (See Figures 2a and 2b.)

(b) A center of activity with a $\beta\gamma$ spot formed near the central meridian on December 25, 1965 at the low latitude of N 10° ($L=203°$). Polarity measurements confirmed its membership in the new cycle. It was the site of three electron-proton flares before West-limb passage. The plage returned in the next rotation diminished in activity, but surrounded by three newly formed regions. (See Figures 2c and 2d.)

(c) The formation of the two great flare-rich regions of March and April 1966 in Northern latitudes and on opposite sides of the Sun, was followed during the next two rotations by the development of a band of plages encircling the Northern hemisphere. Figures 3c and 3d show the March region, and the same solar longitudes two rotations later.

Either that which causes the formation of one center of activity, tends to cause numerous centers of activity in the near neighborhood, or the occurrence of a certain type of center of activity in some way sparks the formation of other active centers, perhaps different in kind.

It also can be reported that in 1963–65 *new-cycle* regions developed in close juxtaposition to major *old-cycle* regions and in the same longitudes that were and had been the favored zones for old-cycle activity (Dodson and Hedeman, 1967).

## 3. Centers of Activity on Opposite Sides of the Sun

Finally, according to the data for 1962–66 a meaningful study of the development of a center of activity may require consideration not only of the past history of the zone of the Sun in which it occurs, but also of the zone approximately 180° away on the opposite hemisphere. In 1964–66 there were numerous instances when the formation of a significant center of activity was followed within the course of a rotation by the development of another region on the opposite side of the Sun about 180° away (Dodson and Hedeman, 1967). Frequently in these years the two regions so placed formed the only significant regions on the sun (see Figure 3).

The question immediately arises as to the similarity or difference of regions that occur almost concomitantly on opposite sides of the Sun. According to our studies

---

Fig. 1.   *Plot by rotation number and Carrington longitude of all plages with area 1000 millionths of the solar hemisphere and/or intensity 3, 1962–66. The antecedents and descendents of each region are also included. – The size of the circle is an indication of the area of the plage. Heavy dark rings indicate especially flare-rich plages. A dark square outline identifies the regions associated with proton emission and PCA. Time runs from left to right and from top to bottom. Data are repeated for 2¼ rotations.*

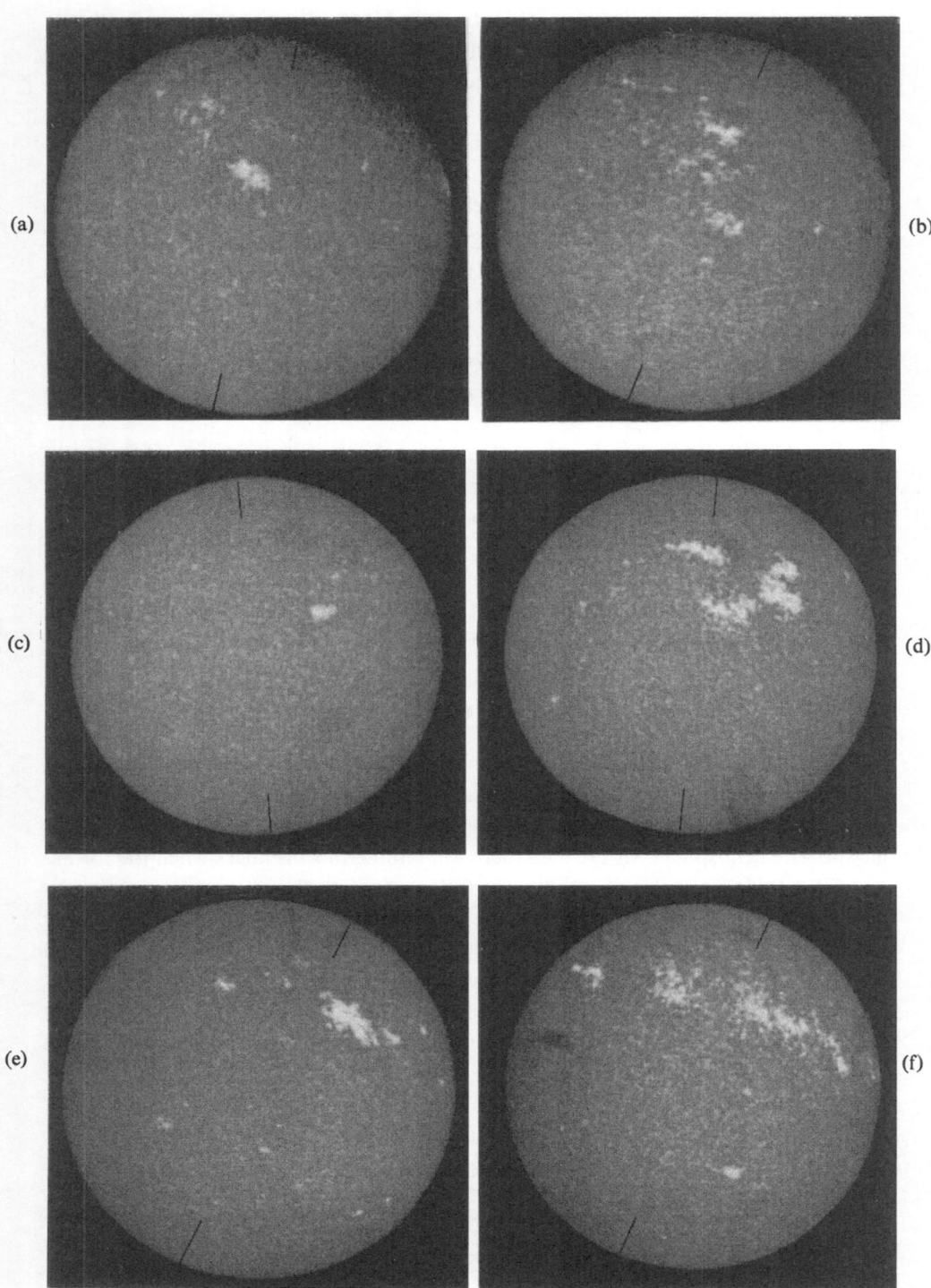

(a)

(b)

(c)

(d)

(e)

(f)

the answer to this question is far from clear, but in 1964–66 several of the major cases suggest a possible difference.

In Figure 3a, the region with C.M.P., May 19, 1964 was the most flare-rich region of 1964. The plage ~180° away was conspicuous but undistinguished.

The two regions that were near the center of the solar disk on May 21, 1965 (Figure 3c) marked a new high level of solar activity for solar cycle 20 according to both flare- and radio-frequency phenomena. The regions opposite them in early and late June were not sufficiently interesting to be included in the *Cooperative Study of Solar Active Regions, May–October* 1965.

The March 21, 1966 region (Figure 3e) was the site of numerous proton-electron flares, according to satellite data. Similar particle emission either did not occur with flares in the April 1966 region or such particles did not reach the neighborhood of the Earth.

Finally, the proton-emitting regions of July 7 and August 28 – September 2, 1966 were located near longitude 180° in late August and early September (Figure 3g). Opposite them, in longitude 354°, there developed in early October a large and conspicuous plage (with only minor spots). In comparison to the other major plages of the second half of 1966 the October region was markedly deficient in radio-frequency emission.

Studies of the gross aspects of centers of activity indicate that there may be recognizable patterns in the development of centers of activity and that there is great need for additional insight into the course of solar activity for the sun as a whole to supplement the fine observations and theoretical interpretation of chromospheric structure and magnetic fields.

## References

Becker, U. (1955)      *Z. Astrophys.*, **37**, 47.

Bezrukava, A.J.A. (1963)      *Publications of Pulkovo Obs.*, **23**, 57.

Dodson, Helen W., Hedeman, E.R. (1967)      'The History and Morphology of Solar Activity, 1964–65', in *IQSY-COSPAR Symposium, London 1967* (in press).

Losh, H.M. (1939)      *Publ. Obs. Univ. Michigan*, 7, 127.

Martres, M.J., Michard, R. (1965)      *C.R. Acad. Sc. Paris*, **261**, 4336.

Mori, S., *et al.* (1964)      *Rep. Ionosph. Space Res. Japan*, **18**, 275.

Saito, T. (1964)      *Rep. Ionosph. Space Res. Japan*, **18**, 260.

Vitinsky, J.U.I. (1960)      *Publications of Pulkovo Obs.*, **21**, 96.

FIG. 2.   *Calcium spectroheliograms showing the formation of a cluster, or band, of adjacent plages in the rotation following the formation of a center of activity.*

| | |
|---|---|
| a. 1965 Feb.  3 (L = 166°) | b. 1965 March  2 (L = 172°) |
| c. 1965 Dec. 26 (L = 182°) | d. 1966 Jan.  21 (L = 200°) |
| e. 1966 Apr.  4 (L = 319°) | f. 1966 May  2 (L = 309°) |

*East is at the left and North at the top of each picture.*

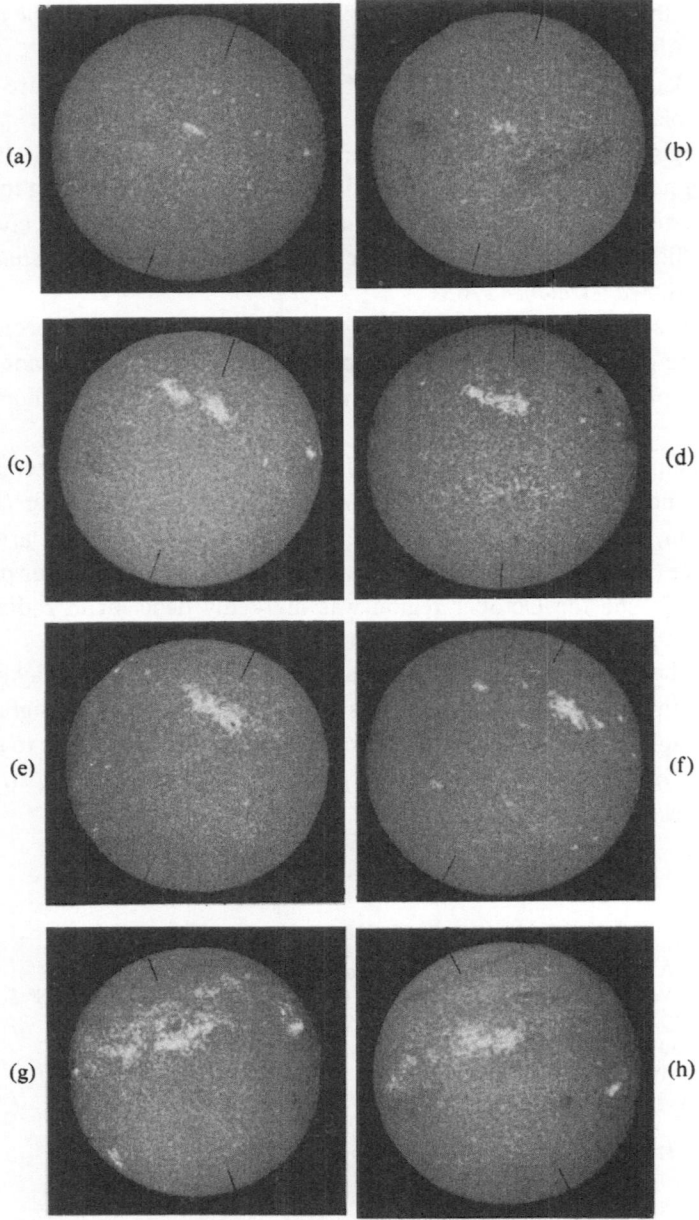

FIG. 3.    *Calcium spectroheliograms illustrating the formation of centers of activity on opposite sides of the Sun.*

| | | | |
|---|---|---|---|
| *a. 1964 May* | *19 (L = 359°)* | *b. 1964 June* | *1 (L = 185°)* |
| *c. 1965 May* | *21 (L = 195°)* | *d. 1965 June* | *30 (L = 26°)* |
| *e. 1966 March 21* | *(L = 144°)* | *f. 1966 Apr.* | *4 (L = 319°)* |
| *g. 1966 Aug.* | *28 (L = 188°)* | *h. 1966 Oct.* | *9 (L = 354°)* |

*East is at the left and North is at the top of each picture.*

# DISCUSSION

*Bumba:* I have some minor comments to your nice talk: Our two complexes of activity observed in 1962 are parts of your two main families of active centers at the beginning of your observational period. The same is true about the situation we showed today at our first slide demonstrating the simultaneous development of active centers of the new and old cycles of activity in the same heliographic longitude.

I saw a manuscript of a paper written by Dr. Berdicevskaja, which will be published in the Soviet *Astronomical Journal*. Dr. Berdicevskaja estimated the same behaviour of the solar activity distribution during all previous minima of activity, which is possible to study from the Greenwich Photoheliographic Results. Two maxima in the approximate distance of about 180° may be seen.

*Dodson-Prince:* I look forward to seeing the interesting manuscript by Dr. Berdicevskaja.

*Neupert:* We have studied the evolution of the major complex of active regions at longitude 180° observed in 1962 in conjunction with a study of the extreme ultraviolet emission of these regions as obtained by the satellite OSO-1 during that period. We find that the presence of major activity at this longitude can be traced back through 1960. The relative orientation of three recurring active regions (two in the Northern hemisphere, one in the Southern) is maintained over a period of three years (1960–62).

*Dodson-Prince:* It is interesting to have our observations in the optical range extended to the extreme ultraviolet, and to earlier years.

*Houtgast:* I have seen on the slides of Mrs. Dodson some of the $Ca^+$ plages surrounded by a dark area and others without such a phenomenon. May I ask for a possible explanation?

*Dodson-Prince:* The dark feature surrounding certain of the bright plages is the *circumfacule* identified by Deslandres and D'Azambuja decades ago. It is very conspicuous around some regions but is apparently absent from others. Since the circumfacule is primarily a $K_3$ phenomenon, its absence from spectroheliograms can mean merely an imperfection in instrumental adjustment. It is a solar feature worthy of more study.

*Beckers:* I observed the circumfacule (the dark region around K-line plages) simultaneously in $H\alpha$ and K. The circumfacule coincides with the $H\alpha$ vortex seen around active spot groups. Under good seeing such a vortex is also seen in the K line. The vortex filament may be the partial or sole cause for the circumfacule. Another cause may be the suppression of the calcium flocculi by the horizontal fields in the vortex.

*Sheeley:* Circumfaculae occur on $H\alpha$ core and $K_3$ spectroheliograms taken almost simultaneously (within about 60 sec). They also occur on $\lambda$ 8542 spectroheliograms (taken in the core of the $\lambda$ 8542 $Ca^+$ line), and when the seeing is good these circular dark regions appear striated. In the wings of $H\alpha$, say at $+0.7$ Å, these regions conspicuously lack the small absorbing features that are so characteristic of the $H\alpha$ wings. A good name for these regions visible on $H\alpha$ spectroheliograms at $\Delta\lambda = +0.7$ Å has been suggested by Beckers, and is 'the runways', where airplanes can land unobstructed by 'the spicule forest'.

*Rösch:* Trellis has found a correlation between the longitude of new centers and the longitude of Jupiter (based on Greenwich photoheliographic results for many cycles). The appearance of centers at 180° may be connected to such a type of tidal effect.

*Dodson-Prince:* It is something to think about.

# LAST PHASES OF DEVELOPMENT OF ACTIVE REGIONS*

V. Bumba, J. Kleczek,          and          J. Sýkora
J. Olmr, B. Růžičková-Topolová                (Astronomical Institute,
(Astronomical Institute, Ondřejov)        Skalnaté Pleso, Czechoslovakia)

ABSTRACT

A close relation of filament feet to the supergranular structure has been found. Green corona brightness, 1420 MHz flux, and the location of quiescent filaments depend upon the distribution of the photospheric magnetic fields. The development of polar maxima of the green corona and their relation to the following polarity have been studied.

This communication summarizes four different papers that are prepared for printing in the *Bulletin of the Astronomical Institutes of Czechoslovakia*. The results relate to the last phases of ARs as they are described for example, by Kiepenheuer (1953). The dissolution and migration of magnetic fields (facular regions) and the formation and growth of quiescent filaments are among typical features of the last phases. From the whole complex of related problems we concentrated only on the relations of photospheric magnetic fields, green corona, quiescent filaments and radio emission at 1420 MHz and the poleward shift of these features. Only the main results are given here, and for a description of the material, its treatment and a detailed discussion of the results, the reader is referred to the original papers in the BAC.

(a) A preliminary comparison of Hα and Ca$^+$ filtergrams and spectroheliograms by means of a blinkcomparator indicated that the feet of quiescent filaments are anchored in junctions of three or more supergranules.

(b) About one thousand measurements of distances of the feet of quiescent filaments have shown, that these distances between neighbouring feet (26000 km and 47500 km) correspond to the geometry of the supergranular network. Moreover there are indications that the distance is a function of the phase of the cycle.

(c) The relation of GCRs (=green coronal regions, emission maxima of green corona) in spot zones to the bipolar magnetic regions (BMR) and to the magnetic background field (MBF) is rather complicated. The maximum of a GCR is usually situated over an inclusion of the following polarity into the large area of the leading polarity.

(d) The emission maxima of 1420 MHz coincide in position with maxima of the

---

* Presented by J. Kleczek.

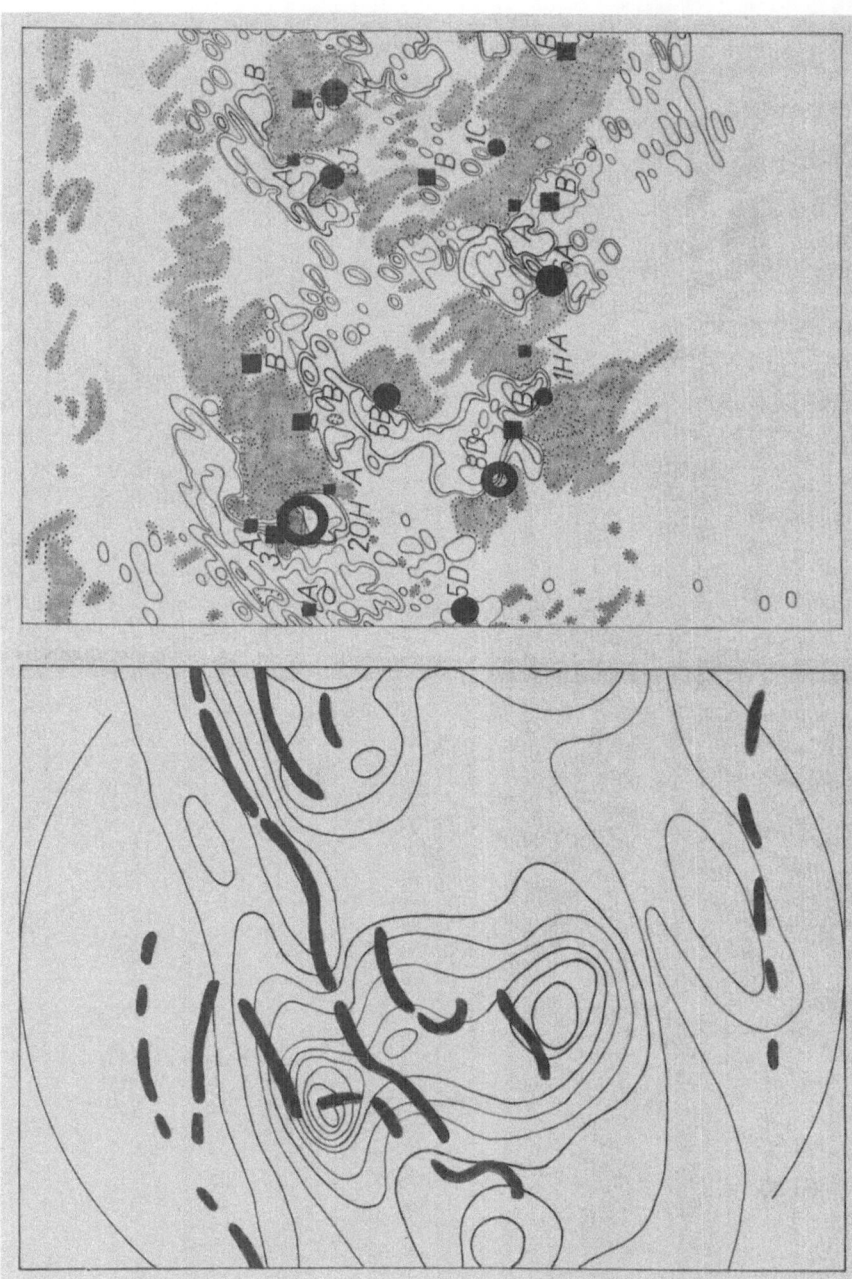

FIG. 1. *The left part is a daily (i.e. with foreshortening) 1420 MHz heliogram (Quarterly Bulletin) for September 11, 1960. Heavy lines represent filaments (Cartes synoptiques de Meudon). The right part is the corresponding section of the synoptic magnetic chart (Mt. Wilson). Dotted lines with shaded area are for minus polarity (i.e. following on the Northern hemisphere). Letters are Zürich types of ARs. Circles indicate ARs with flare activity (number of flares is assigned) while squares are ARs without flares.*

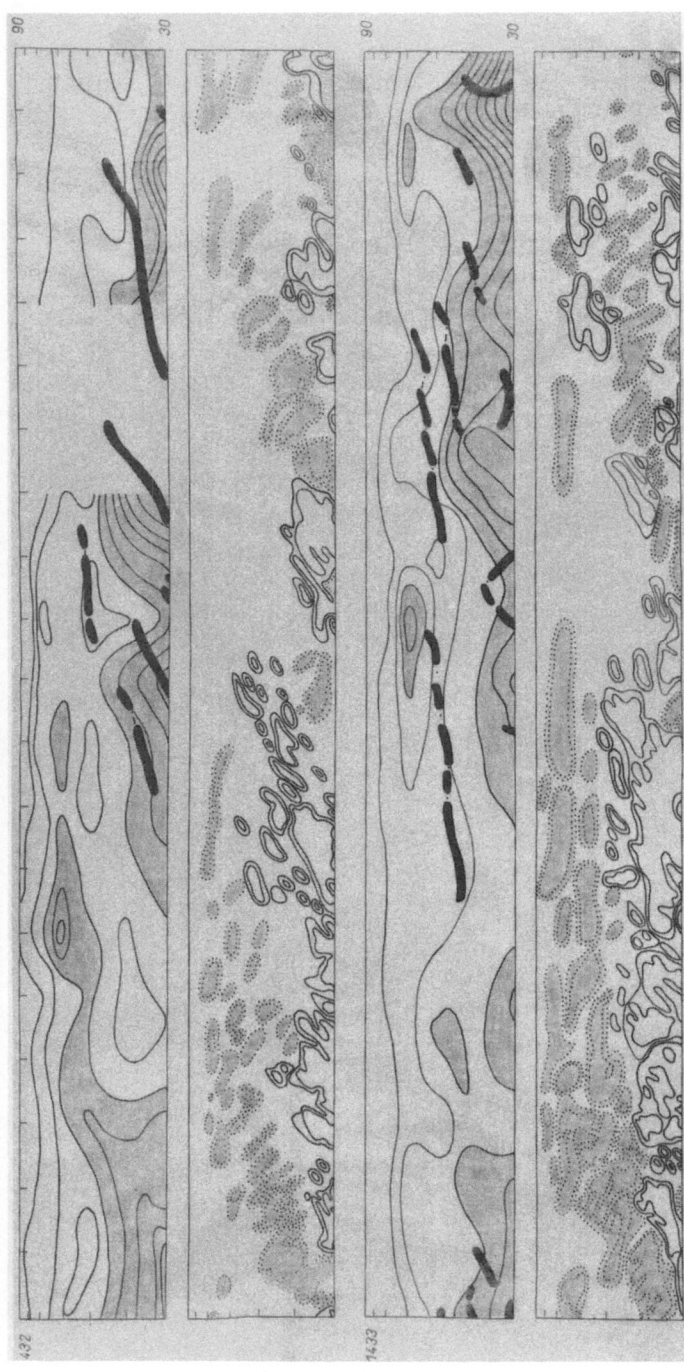

FIG. 2. *North polar zones of synoptic charts for green corona (Climax, Sacramento Peak) and photospheric magnetic field (Mt. Wilson) are compared for the Carrington rotations 1432 and 1433. The filaments are marked by heavy lines. Following polarity and maxima of coronal intensity are shaded.*

green corona. There are three different cases: (1) over simple BMRs (not connected with the main body of the background field, flux is practically balanced) the 1420 MHz is centered over the neutral line; (2) for the case of inclusion of the following polarity, the 1420 MHz radio emission is centered over the centre of gravity of the inclusion; (3) In rare cases, the leading polarity is included by the following polarity and the radio maximum is centered on the inclusion also (Figure 1).

(e) The polar maxima of green coronal emission are closely related to the tails of the following polarity pushed out from the spot-zone MBFs. Along the minimum dividing polar maximum from the spot-zone maximum there stretches a long quiescent filament (Figure 2). Although the poleward shift of the following polarity from the main zone of activity is observed, the polar maxima of the green corona appear only at about 60° of latitude. On our material, we could not trace any poleward shift for the polar maximum of the green corona.

(f) The polar maxima of the green corona are also followed by the secondary maxima of 1420 MHz emission. From the coincidence of 1420 MHz with the 5303 Å emission it follows, that also the 1420 MHz secondary maxima are located over the following polarity.

(g) Long quiescent filaments are marked by decrease of brightness at 5303 Å and 1420 MHz (Figure 1).

*Conclusion*: The changes of 5303 Å emission, of 1420 MHz brightness and of the quiescent filaments are closely related to the evolution of the magnetic-field pattern resulting from decay and interaction of several active regions. Some other dynamical factors (e.g. differential rotation and meridional circulation) may also participate in the interplay of solar plasma and magnetic fields called the 'last phases of active regions'.

## Reference

Kiepenheuer, K.O. (1953)    in *The Sun*, Ed. by G.P. Kuiper, The University of Chicago Press, p. 434.

# ON SOME PROPERTIES OF THE VELOCITY FIELD
# IN A DEVELOPED ACTIVE REGION

S. I. GOPASYUK

*(Crimean Astrophysical Observatory, Nauchny, Crimea, U.S.S.R.)*

Thirty-eight series of photoelectric recordings of radial velocities in a developed active region have been studied. The recordings were made in the Fe I line $\lambda$ 5250 Å from September 2–7, 1961, when the active region passed across the solar disk from E 30° to W 42°.

It is shown that the radial velocities averaged over many recordings for each day, i.e. 'quasi-stationary' velocities, differ from zero in the regions, where the magnetic-field strength is relatively high. Besides, the velocity can change the sign inside a region of a single polarity. There are found two types of 'quasi-stationary' motions for an active region as a whole: one type – with the predominantly horizontal motions, and the other type – with predominantly vertical motions (mainly downwards). Both types of motions are superposed by rapidly changing perturbations whose amplitude reaches (in some points) 1400 m/sec.

Predominantly horizontal motions are in the large spots and their vicinity. These are mainly the places where the stress of the longitudinal component of magnetic field is larger than that of the transversal component (it is also true for the surrounding regions). Predominantly downward motions occur in those parts of an active region where the transverse magnetic field predominates. These regions coincide with the places of the most rapid disintegration of an active region: disappearance of spots' penumbra and their magnetic fields.

In the regions with predominantly horizontal motions the descending flux of gas is practically totally compensated by ascending flux. On the contrary, the flux with predominantly downward motions is not compensated at all and may reach the value of $8 \times 10^{40}$ atom/sec.

## DISCUSSION

*Maltby:* How do you separate the horizontal and the vertical motion? Do you assume that the velocity pattern is constant as a function of time?

*Gopasyuk:* As the line of sight velocity changed sign during passage of a region from the East to the West limb, it was considered that the flow was predominantly horizontal. When the structure appears similar during the disk passage but the sign of the velocity changes, it is safe to assume that one is observing predominantly horizontal motion. When the velocity did not change sign, it was assumed that the motion was predominantly vertical.

*Schröter:* As far as I understood you get the result that motions in this spotgroup are often more or less perpendicular to the direction of the magnetic field. How did you get from your Doppler-shifts which give the line-of-sight component the spatial orientation of the motion vector?

*Gopasyuk:* The answer is the same as to the question by Maltby.

*Sturrock:* If gas is observed to be flowing downwards over a certain area, either gas must flow horizontally into this region, or gas must flow up in small regions of the area. In the latter case, the observed velocity profiles can be understood if the gas flowing downwards is hotter than the gas flowing upwards. Is this interpretation consistent with the observations?

*Gopasyuk:* In every case we observe that most of the gas moves downwards, so if the density of the upward- and downward-flowing gas streams are equal, there must be a net downward flow. This is one of the difficult problems in understanding the structure of active regions.

*Bumba:* If I understand correctly, large downward motions were found in places, where the disintegration of penumbra took place, this means in such a region where both polarities were fairly far apart each from the other. But in such place relatively weak transversal component of the field is normally observed in the photosphere. The transversal components are stronger only if the distance of both polarities is very small. Therefore I do not understand why you speak about motions going mainly perpendicular to the lines of force of the field.

*Gopasyuk:* At the location of the downward motion, the transverse component of the field was strong (1500–2000 gauss), and this was greater than the longitudinal field at this point. During the expansion of the region over several days, the magnetic fields weakened and the downward flow continued.

# THE MIGRATION OF SUNSPOT ACTIVITY
# ALONG SOLAR MERIDIANS AND PARALLELS

L. DEZSŐ, O. GERLEI, and ÁGNES KOVÁCS

*(Heliophysical Observatory of the*
*Hungarian Academy of Sciences, Debrecen, Hungary)*

## ABSTRACT

The places of local maxima of sunspot activity, i.e. the formation of solar active regions, seem to follow in a definite sequence of tracks over the Sun's surface. These sequences of activity apparently represent distinct continuous streams, whose duration may be of the order of magnitude of 1 year.

## 1. Introduction

It has been taken always as one of the crucial points in studying the nature and cause of the solar cycle to try to interpret the variation in the average heliographic latitude of sunspots, i.e., the butterfly diagram. But if the butterfly diagram has some fine structure in reality, then any theory concerning the solar cycle that ignores this fact becomes questionable.

We would like to call attention to the peculiarity that the butterfly diagram can be resolved into several tracks of bands with a time-scale of only 1 to 2 years, i.e., considerably less than 11 years. This peculiarity was indicated already by Lockyer (1904), using the data of observations from 1861–1902, and much more clearly by Kuleshova (1962), from Tashkent observations of solar cycles no. 17–19. One may find traces of this property even in one of Bell's papers (1960), who took into account some magnetic measurements too.

Since solar activity may be best characterized through the strengthening of local magnetic fields and the largest field strengths on the Sun are to be found in spot umbrae, it would therefore be most appropriate to investigate the problem on the basis of magnetic measurements relating to umbrae. But since we have no suitable data of this kind available, we used umbra areas, selecting those cases when the umbra of the spot group was growing. We did this because it was highly probable that during each such period the magnetic field was locally also growing.

## 2. Data used

Observations of the heliographic coordinates $(B, L)$ and the daily areal increases of umbrae $(\Delta U > 0)$, separately for each sunspot group, were taken from the Greenwich

*Kiepenheuer (ed.), Structure and Development of Solar Active Regions,* 70–76. © *I.A.U.*

Photo-Heliographic Results for the years 1889–1955, i.e., for 6 entire solar cycles (nos. 13–18). These Greenwich records contain in fact solar observations for every day, but daily only one. We set certain reasonable limits, neglecting certain observations, in order to avoid systematic measuring errors (Dezső, 1964). Thus we confined ourselves to certain values as regards the magnitude of $\Delta U$, the distance from the central meridian ($L_{CM}$) and the lifetimes ($u$) of the spot groups. When we speak of a spot group of $u$ days, it means that between its first and last (ultimate) observations there is an interval of $u-1$ days. Wherever we used an average value of the heliographic position, this is always an appropriate calculated weighted mean. (We used as weights the number of $\Delta U$ cases and the $\Sigma \Delta U$ values in Sections 3 and 4 respectively.)

While we studied data of observations over a 67-year period, the present communication gives only a few typical examples* from the solar cycle no. 18, to illustrate our preliminary results relating to certain migrations of sunspot activity.

## 3. Migrations shown by a Special Index of Spot Activity

Consequent to what was said above, we may introduce quite a practical and reasonable index of sunspot activity ($J$) by using the number of $\Delta U \geqslant 2$ cases of spot groups. To be quite clear: we counted for each day the number of those spot groups which by the next day has shown an umbra increase of at least 2 areal units ($10^{-6}$ of the solar hemisphere), and summed up these numbers for half-year periods, starting from the beginning of the calendar year for every subsequent 3 months. We determined this $J$ index of sunspot activity for areas of 8° latitude × 60° longitude of the solar surface, starting from the equator ($B=0°$) and Carrington's zero meridian ($L=0°$) for every 4th parallel and every 30th meridian. We used observations of a spot group for the calculations of our $J$ indices only when the central-meridian distance of the group was less than 50° ($|L_{CM}| < 50°$), and used a spot group only if its lifetime was at least 5 days ($u \geqslant 5$).

In the topmost part of Figure 1 a 2-year variation of sunspot activity ($J$) between Northern latitudes of 20° and 28° and longitudes of 60° and 120° (i.e., the center of the longitude zone is 90°) is shown. We note from the other similar curves of Figure 1, plotted for the longitude zones of 60°, 30°, 0° and 330°, that the $J$-maximum at 1945·5 shifted by 120° in about a year. Such behaviour, a shift in longitude of the maxima of local solar activity in any zone of latitude, is quite common. Sometimes a much smaller or greater shift occurs. But up to the present we have not been able to establish any definite regularities relating to this shift, anyhow it does not move only towards the decreasing longitudes but equally in the other direction as well.

If we determine the variation of spot activity in different zones of latitude, even by disregarding the distinction of longitudes, it is easy to recognize at once from several

---

* Several other examples were shown at the Symposium.

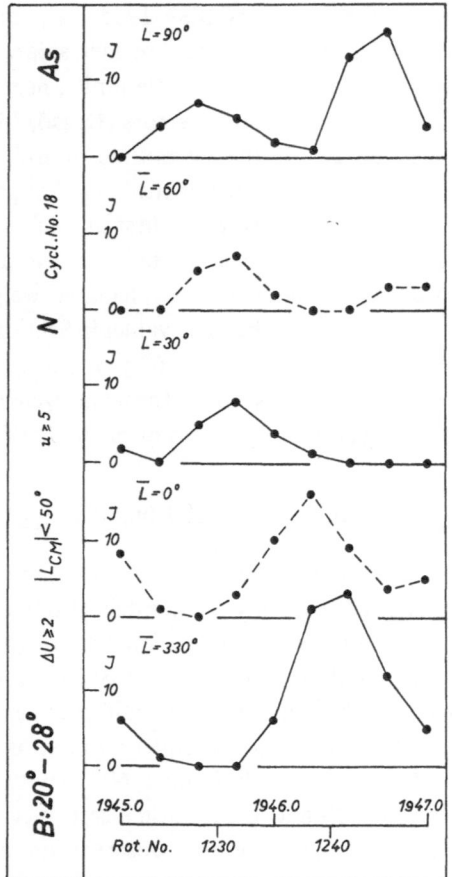

FIG. 1.   *Variations in sunspot activity on several small areas of the solar surface.*

well-defined maxima that the spot activity in general migrates with considerable speed towards lower latitudes. The velocity of these migrations is about one order of magnitude greater than the one which follows from the so-called Spörer's law. We would point out, however, that we noticed amongst the less important migrations of this kind a few which moved toward one of the poles. Nevertheless the dominant characteristics of the latitude migration of sunspot activity is a gradual descent to the lower latitudes. It is perhaps the most important feature of these migrations that before an activity track fades out, not far from the equator, a new one appears at higher latitudes. We may observe at least four main tracks of sunspot activity with fast latitude variation during every solar cycle in each hemisphere. Consequently, one should regard the butterfly diagram as also being built up by such tracks.

In Figure 2 we present examples to illustrate what we have said above. The three curves of sunspot activity relate to the 20°–28°, 16°–24° and 12°–20° latitude zones.

A part of the shift for two of the 'main' tracks are to be seen. The times of maxima of local activity and the related average latitudes, indicated according to the scale on the right-hand side of the figure, are shown by crosses.

We think that the curves of Figure 2, and other similar graphs, prove sufficiently well the rapid latitude variation of spot activity. But Figure 3, and some other similar

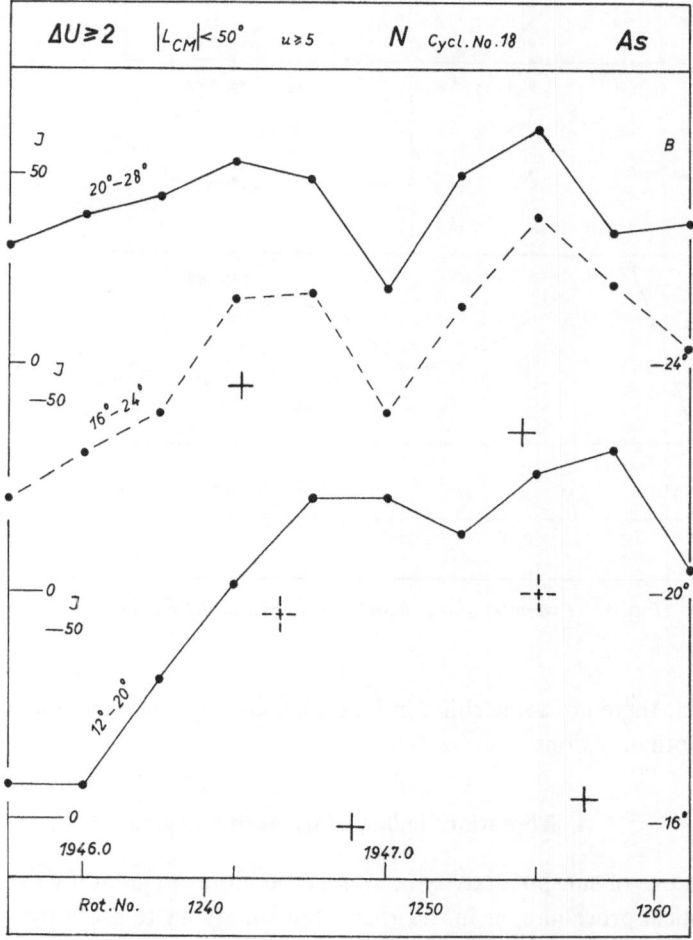

FIG. 2.   *Variations in sunspot activity in various belts of latitude.*

graphs, convince us of it even more. In Figure 3 we show how the spot activity is distributed along longitudes at the times of maxima for the earlier (1946) track of Figure 2. By comparing the three curves of Figure 3, we find that the local maximum has in fact descended to a lower zone of latitude, because during its shift (in 0.34 years) the distribution of activity in longitude on the whole remains the same; or to

L. DEZSÖ ET AL.

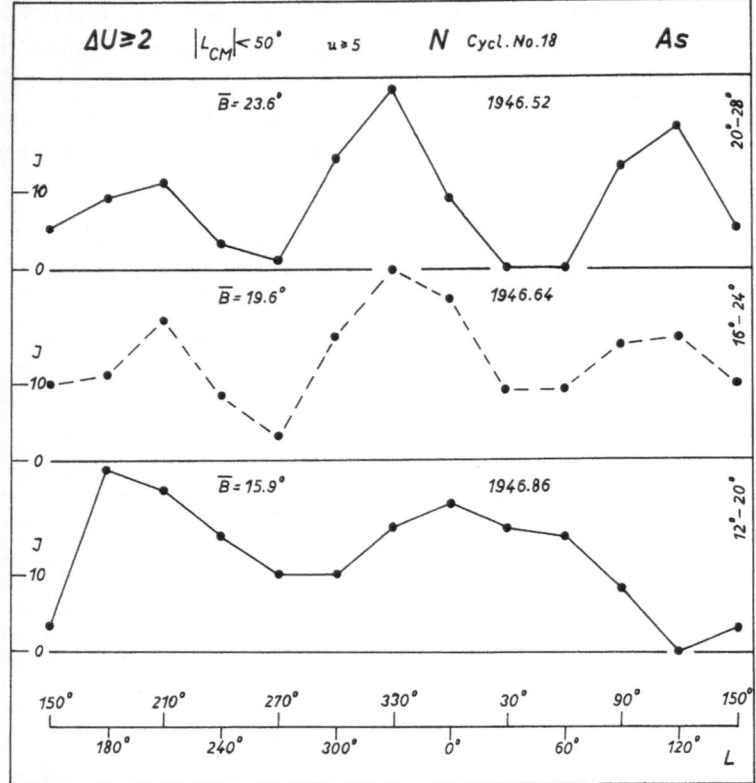

FIG. 3.   *Examples of distribution of sunspot activity in longitude.*

be more exact, there are some shifts in longitude also, but only comparatively small ones and in both directions.

## 4. Migrations indicated by Active Regions

The migrations of sunspot activity becomes even more apparent if we do not follow a strict statistical procedure, as in the above, but simply try to select the most important active regions. We chose $\Delta U \geqslant 5$ as a criterion of intense activity of a spot group. Thus we used only the spot groups that showed an umbral increase of at least 5 areal units from one day to the next within a distance of 60° longitude from the central meridian. Here, however, we used all spot groups whose lifetimes were at least 3 days ($u \geqslant 3$).

In the light of what we outlined in the foregoing section, it was not too difficult to pick out those active regions that are probably members of tracks which come into being through migrations of sunspot activity. In Figure 4 we show for two periods of approximately 1-year possible migrations of this kind, indicated by active regions of the Southern solar hemisphere.

The latitude and longitude variation of about a dozen tracks of solar active regions are plotted in Figure 4. The dots generally represent three and more spot groups (in some cases even seven) and in most cases $\Sigma\Delta U > 100$, the given positions of the regions being moreover mostly comparatively quite close to the actual measured position of the spot group. The related differences in latitudes and longitudes are nearly always less than 5° and 40° respectively. The times of observations that relate to the dots are all within three solar rotations approximately.

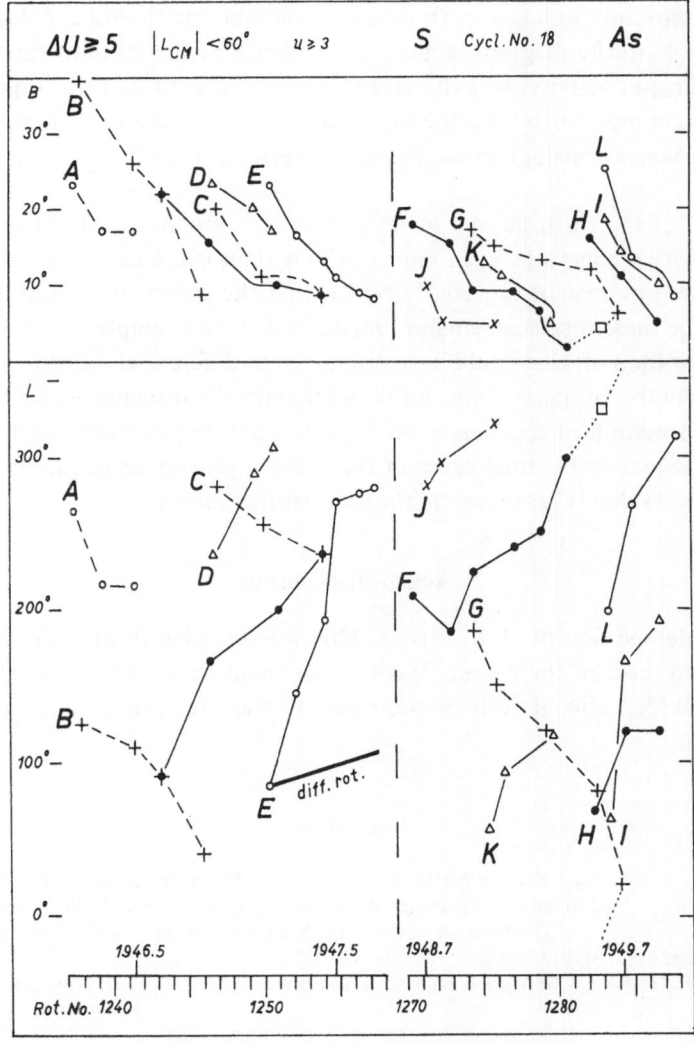

FIG. 4.  *Coordinates of solar active regions.*

The examples in Figure 4 represent 70 and 105 spot groups, i.e., 29% and 48% of those used, from the years 1946–47 and 1948–49 respectively. But we should point out that the corresponding percentages of the $\Sigma\Delta U$ values are 46 and 57, i.e., quite high. Accordingly it is justified to suspect that our samples in Figure 4, from the Southern hemisphere and the periods indicated, show in fact one of the main characteristics of sunspot activity.

## 5. Conclusions

Summing up what has been outlined, it may be taken as a fact that solar activity has a rapid variation in latitude. It is easy to see that one should conclude from this fact that the butterfly diagram is mere consequence of the latitude variation of migration in sunspot activity. It is therefore a prime necessity that this rapid migration should be accounted for by the theories. The velocity of the migrations in question has by no means a constant value, though nevertheless we find it very frequent for 1.5° per rot.

The speed of the shift, as well as its direction in longitude, of the migrations of sunspot activity seems to be even more variable than is the case with latitude variations. It is very important to note, however, that the differential rotation cannot be the reason for these displacements in longitude. (As an example, our calculations in Figure 4 show the shift that could have originated in differential rotation in the case of track E.) From this property alone, it follows that the disturbances which are responsible for the formation of sunspots must migrate quite deep-seatedly under the photosphere. Consequently the final cause of the sunspot phenomenon can hardly be considered a process that takes places on the solar surface alone.

## Acknowledgements

We are indebted first of all to Miss E. Horváth and also to Miss Gy. Gyertyános, who compiled most of the necessary data. Our thanks are also due to Dr. I. Guman and Mr. I. Ökrös, who actively participated in the first preliminary phase of this investigation.

## References

Bell, B. (1960)   'On the Structure of the Sunspot Zone', *Smithson. Contr. Astrophys.*, **5**.
Dezsö, L. (1964)   'Statistical Investigations of Sunspots by a New Method', *Publ. Debrecen Obs.*, **1**.
Kuleshova, K. F. (1962)   'The Fine Structure of the Spot-Formation Zones', *Astr. Zu.*, **39**, 273–277 (= *Soviet Astr. – AJ.*, **6**, 213–216).
Lockyer, W. J. S. (1904)   'Sunspot Variation in Latitude 1861–1902', *Proc. Roy. Soc.*, **73**, 142–152.

# CHROMOSPHERIC EXPLOSIONS AND
# SATELLITE SUNSPOTS*

DAVID M. RUST

*(Mount Wilson and Palomar Observatories, Pasadena, Calif., U.S.A.)*

## ABSTRACT

Observations of the longitudinal component of the photospheric magnetic fields near sunspots imply that surges and Ellerman bombs occur at neutral points in the magnetic fields (i.e. $|B|=0$) in the chromosphere. The neutral points appear above satellite sunspots, which are defined as polarity reversals (in $B_{\parallel}$) near the edges of large-spot penumbrae. A series of magnetograph observations shows a point of satellite polarity vanishing during a period of almost continuous surge activity. The lost magnetic energy is comparable to the energy release evidenced by the surge.

## 1. Introduction

Several researchers have used solar magnetographs to study the magnetic-field structure of large sunspots in active regions. They have found that very frequently these spots were surrounded by a number of 'satellite' magnetic features. The polarity of the satellites was opposite to that of the central spot (Bumba, 1960, 1962; Howard, 1959). In most cases these satellites were not visible on white-light photographs. However, Gopasyuk *et al.* (1963) obtained many observations of flare-producing regions which included large sunspots accompanied by small, visible spots imbedded in penumbrae. The smaller spots ('satellites') usually had polarity opposite to that of the main spot. I used the magnetograph at the Mount Wilson Observatory to map the longitudinal component (as indicated by Zeeman splitting in the photospheric FeI line at $\lambda 5250$) of the magnetic field of the active region of June 2–3, 1967. The region passed central meridian in the Northern hemisphere on June 3. The observations show many satellite sunspots (invisible in integrated light) near the penumbra of the largest spot in the region. The spacial resolution of the scans was either 5 or 10 sec of arc and the noise level was about 2 gauss. The observed size of the satellite magnetic features is at the resolution limit.

## 2. Observations

As the magnetic measurements were being made, personnel of the Lockheed Solar Observatory made large-scale Hα motion pictures of the region. Figure 1 is an Hα photograph of the region at 1947 UT on June 2. A surge is occurring at the penumbra

* Presented by R. Howard.

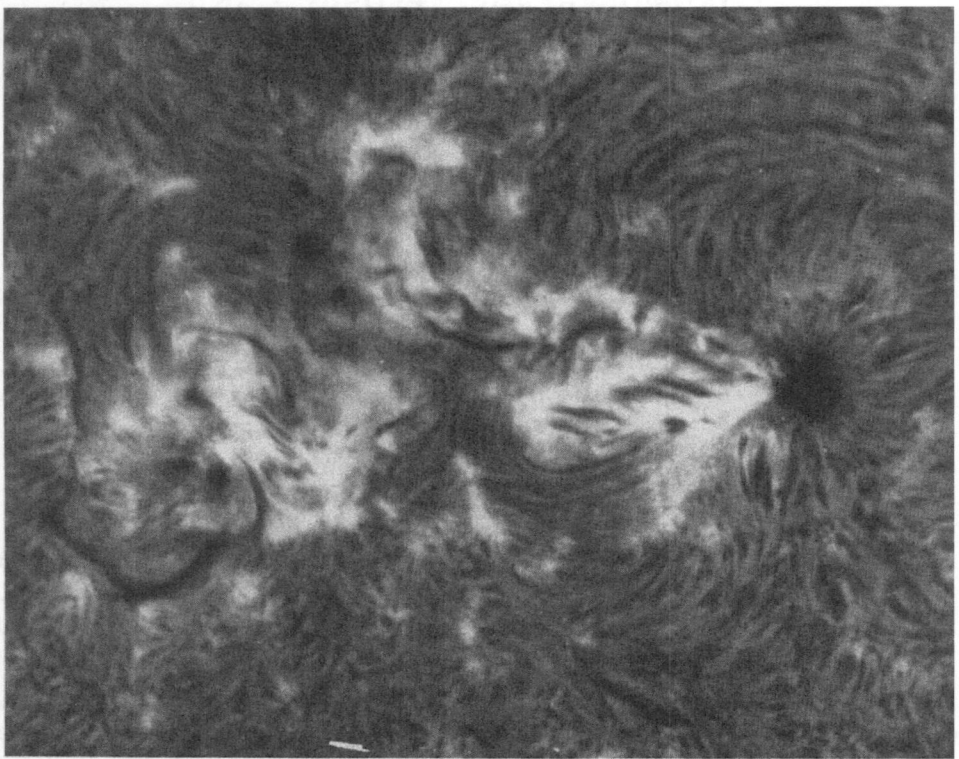

FIG. 1. *The active region of June 2, 1967. North is at the top and East is on the left. This photograph was made in the center of Hα at 1947 UT. During the period of observation (1635 UT, June 2, to 0116 UT on June 3) nearly continual surge activity took place on the Northwest side of the large leading spot. A small surge is shown in this picture. Many flare brightenings occurred in the center and left-of-center plages in the region.*

of the large, leading sunspot, and some traces of a flare brightening are evident in the middle of the picture. There were several flares in the region on June 2 and 3. The motion pictures show continual surge activity in the region. The positions of all the observed surges are shown in Figure 2, which is a magnetogram of the region as it appeared on June 3. Dashed lines inclose negative fields; solid lines inclose positive fields. Solid triangles show approximate surge size and direction. Solid dots are Ellerman bombs and the flare areas are hatched. *Every bomb, every surge, and many flare brightenings originate at the satellite sunspots.* Because of the relatively low resolution of the magnetic measurements, it is impossible to know the real size and the peak field intensity of the satellites. Even if their characteristic fields should be several hundred gauss and their cross-sections only 2 sec of arc, the magnetograph would transform them into broad 5 or 10 gauss features on the magnetic maps. All flaring regions of

FIG. 2.   Contour map of the longitudinal magnetic fields in the region at 1600 UT on June 3, 1967. Dashed lines inclose areas of negative field, and solid lines inclose areas of positive field. Contour levels are 5, 10, 20, 40, 80, and 160 gauss. Dark surges are indicated by solid triangles. Solid dots denote Ellerman bombs (moustaches) and areas of flare brightening are indicated by hatching. 'Valleys' in the field intensity are distinguished from 'hills' by small converging tick marks. A comparison of the visual sunspot polarities and some valleys on this map reveals that the valleys are really sites of polarity reversals not completely detected with the magnetograph. The resolution is 5 sec of arc.

June 2 have been included in the figure. Perhaps inclusion of only the points which showed a flash phase would give a higher correlation between satellite spot and flare-brightening positions. More research is needed on this point.

### 3. The Lines of Force

What do the fieldlines above satellite sunspots look like? To find this out, I used the method of computing the magnetic-field lines developed by Schmidt (1964). His method is to assume that the space above the photosphere is current-free. Potential theory may then be used to calculate the field from the observed flux distribution in the photosphere. Figure 3 shows the lines of force near the largest surges. *The surges follow the lines of force.** On this figure I have drawn the principal neutral lines

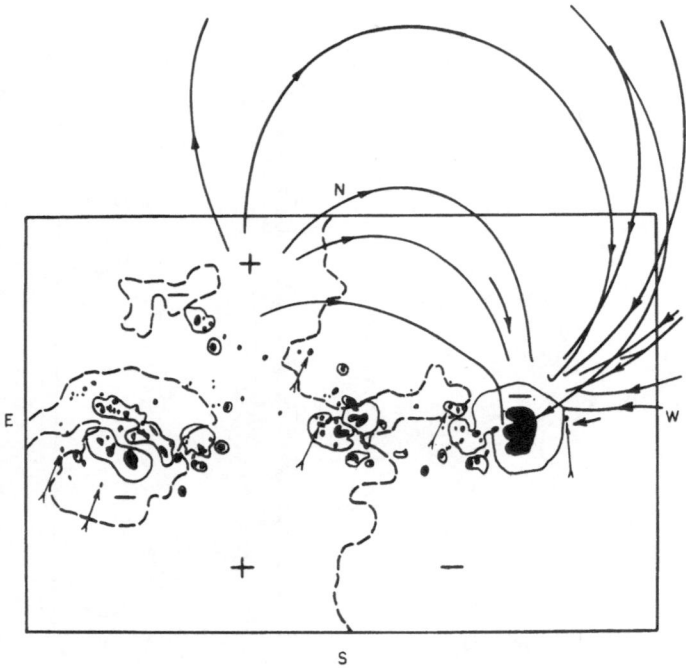

FIG. 3.   *Magnetic lines of force (heavy lines with arrows) near the region of most intense surge activity. A comparison with Figure 2 shows that the surges follow the lines of force. The fieldlines were constructed according to the potential theory by considering the chromosphere current-free. The sources of the field are fixed in the photosphere, which is assumed plane-parallel. Heavy dashed lines coincide with the major lines of zero longitudinal field in the region and the polarities on either side are indicated by symbols (+ and −). Thin arrows point out satellite sunspots detected by the routine Mount Wilson sunspot field measurements.*

* I do not intend to imply, however, that the magnetic field above the region was current-free. The surges alone are good evidence for currents. Nevertheless, it is apparent that the lines of force do not deviate radically from those calculated under the current-free approximation.

(heavy, dashed lines) in the measured longitudinal fields. The thin arrows point to the few visible satellite sunspots in the region. A flare occurred near the *visible* satellite in the center of the figure; however, the observed surges and Ellerman bombs all occurred near *invisible* satellites, i.e., near points of isolated polarity detected only with the magnetograph.

## 4. Neutral Points

The significance of the satellite sunspots for bombs and surges, at least, is revealed in Figure 4. There is most likely a neutral point (i.e. $|B|=0$) in the magnetic field in

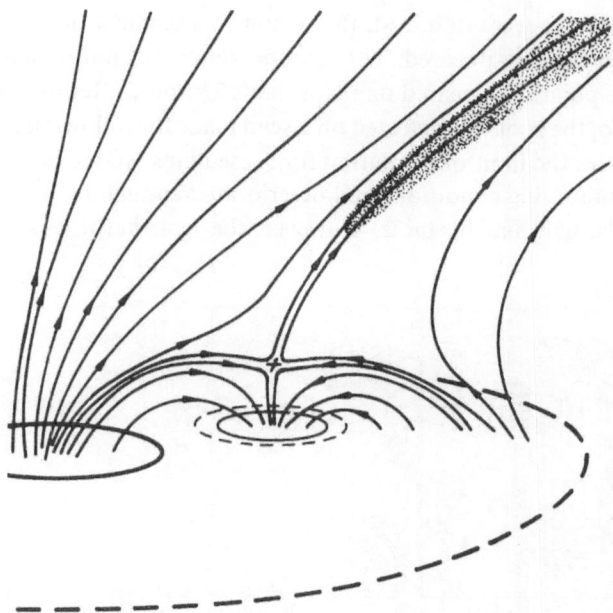

FIG. 4.    *Sketch of the fieldlines above a satellite sunspot. An X marks the neutral point where explosive events probably originate. The satellite spots are small features easily overlooked in the large scale (usually bipolar) pattern of the major regions, even though the satellites seem to be an integral feature of most large spots in growing regions (Bumba, 1960).*

the solar atmosphere above every satellite sunspot. It appears that the explosive events of June 2 and 3 originated at the neutral points that are implied by the geometry of the underlying fields. Severny (1965) and Koval (1965) at the Crimean Observatory have come to similar conclusions. It is clear from the figure that the larger the satellite spot relative to its parent, the higher the neutral point will be. For example, bombs are very small and they occur in the low chromosphere. Large flares occur much higher up. Apparently, bigger, higher events occur above larger satellite spots, i.e., those spots which have the more elevated neutral points, but more research is needed to establish whether this pattern is generally obeyed by explosive events.

Neutral points in the magnetic field have played an important role in almost all theoretical discussions of explosive events. At these points magnetic energy may be converted to observable radiative and kinetic forms.

## 5. The Field Changes

Figure 5 shows the appearance of a satellite spot under the surges before and after the most violent activity. The small spot of positive polarity is a stable feature of the first three maps, taken only 8 min apart. After a delay of 1 hour and 50 min, during which no observations were obtained, the region was scanned three more times. The satellite-spot field had disappeared. The disappearance was not permanent, however, since the satellite polarity appeared on maps made $2\frac{1}{4}$ hours after the final scan shown in the figure. Also, the spot was detected on a scan made the following day. The satellite spot was just above the limit of resolution for these maps (10 sec of arc), and it is not impossible that observing conditions had deteriorated enough by the second series of scans to make the field undetectable. However, the fact that it was detected several

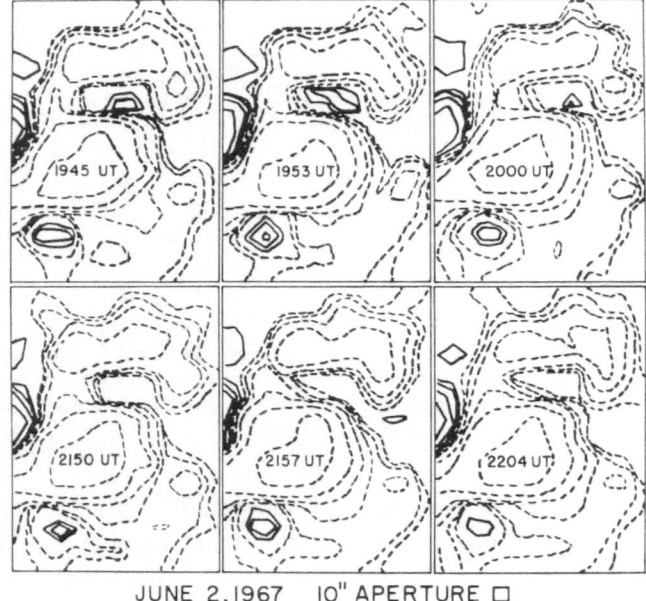

JUNE 2, 1967    10″ APERTURE □

FIG. 5.   *A sequence of magnetic maps of the most active surge region. Contour levels are 5, 10, 20, 40, 80, and 160 gauss. The top three maps were made before the largest surges occurred. They all show a small positive feature at the Northwest (upper right) edge of the penumbra of the large negative sunspot. The lower maps all lack the small feature which was at the location of the surges. Note that other features of comparable size are unchanged during these measurements.*

hours later under even worse seeing conditions, and the fact that similar features everywhere else on the magnetograms did not undergo a similar disappearance tend to support the conclusion: the field of the invisible satellite spot just under the largest surges decreased significantly during the most pronounced surge activity. The later recovery of the field is consistent with other observations of photospheric field changes during minor events in a growing active region (Michard *et al.*, 1961).

## 6. Energy for Surges

What was the magnetic energy of the vanished spot? The magnetic flux associated with the spot was about $5 \times 10^{18}$ maxwells. The lines of force symbolizing this flux must pass through the plane of observation twice (see Figure 4). If the field did vanish, we would expect both positive and negative regions to weaken or disappear. However, it is obviously much easier to detect a small change in the weak satellite spot than in the large parent spot which has extensive regions of strong fields. If I assume that there was a change in the negative fields to accompany the change in the small positive spot and that the annihilated flux extended upward to a height of 1000 km, then there was about $2 \times 10^{26}$ ergs available for the surges. If the radiative and kinetic energy released in a surge are comparable, we can find the total energy required to produce a surge from an estimate of the kinetic energy of the outward-moving material. The velocity of the surges is about 100 km/sec. If I assume a diameter of 1000 km, a length of 5000 km and a density of $10^{11}$ protons/cm$^3$, then I find that the surge had a kinetic energy of about $3 \times 10^{25}$ ergs. Thus, the kinetic and radiative energy of the surge was about $6 \times 10^{25}$ ergs. This is comparable to the vanishing magnetic energy. Perhaps then, we have a clear case of annihilation of magnetic energy near a neutral point with attendant brightening (at the base of the surge) and acceleration of chromospheric matter. Since virtually all events with pronounced flash phases occur in or near spot penumbrae, I think it is probable that we will find every such event taking place above the satellite sunspots found there.

## Acknowledgements

I am grateful to Mr. Thomas Cragg for securing the magnetic observations and to Dr. Robert Howard for critically reading the manuscript. I thank the personnel of the Lockheed Solar Observatory for the use of their films and for several helpful discussions.

## References

Bumba, V. (1960)     *Publ. Crim. Astrophys. Obs.*, **23**, 212.
Bumba, V. (1962)     *Bull. Astr. Inst. Czech.*, **13**, 48.
Gopasyuk, S.I., Ogir, M.B., Severny, A.B., Shaposhnikova, E.F. (1963)     *Publ. Crim. Astrophys. Obs.*, **30**, 15.

Howard, R. (1959)      *Astrophys. J.*, **130**, 193.
Koval, A. N. (1965)      *Publ. Crim. Astrophys. Obs.*, **34**, 278.
Michard, R., Mouradian, Z., Semel, M. (1961)      *Ann. Astrophys.*, **24**, 54.
Severny, A. B. (1965)      'Stellar and Solar Magnetic Fields', in *I.A.U. Symposium No. 22*, Ed. by
   R. Lüst, Interscience, New York, p. 255.
Schmidt, H. U. (1964)      AAS-NASA Symposium on the Physics of Solar Flares, *NASA Report
   SP-50*, p. 107.

# DISCUSSION

*Sturrock:* The fact that surges occur above circular neutral lines (zero vertical magnetic field) fits well with the fact that flares occur above neutral lines. However, if the surge energy is derived from magnetic energy, the pre-surge state cannot be current-free and therefore not of the form depicted by Dr. Rust.

*Howard:* This is true.

# STUDY OF AN ACTIVE REGION OF THE SUN DURING THREE ROTATION PERIODS*

CONSTANTIN J. MACRIS and T. J. PROKAKIS

*(Astronomical Institute, National Observatory of Athens, Greece)*

### ABSTRACT

The abnormal evolution of an active region during three solar rotations is studied. The high density of flares during the second and third rotation seems to be caused by the collision of new active centres with existing ones.

The increase of the activity is probably due to the disturbance of the magnetic field which became more complex because of the appearance of new centres near the original one.

## 1. General Description of the Centre of Activity under Study

### A. SPOTS, FACULAE, AND MAGNETIC FIELD OF THE CENTRE OF ACTIVITY

A centre of activity having as coordinates $\varphi = +21$, and $L = 144$ has been studied during three solar rotations including spots and magnetic fields and during four rotations for faculae. As this centre presented certain particularities, it was chosen to be studied.

On February 20th, 1966, a small facula was born in the N-E part of the visible solar hemisphere; it was the result of the creation of a bipolar magnetic field. Some hours later one of the poles of the bipolar centre appeared in the form of a class-A spot. The following day, February 21st, the centre quickly developed in a clear bipolic shape. The area and brightness of the plage increased swiftly. The intensity of the magnetic field of the spot group reached high values: 1000 gauss for pole S (leading spot) and 500 gauss for pole N (following spot).

After optical observations of the sunspots carried out at the National Observatory of Athens by D. P. Elias, and according to the Zürich Classification of Sunspots, the class of the spot group was normally evolved; and passing through the classes B, C, and D on February 25th, evolved into a class E, which it conserved steadily till the West-limb passage of the centre on March 1st.

The evolution of the faculae followed that of the spots. The plage remaining compact increased in area and brightness.

* An extract of this paper was presented at Budapest by G. Righini.

*Kiepenheuer (ed.), Structure and Development of Solar Active Regions,* 85–91. © *I.A.U.*

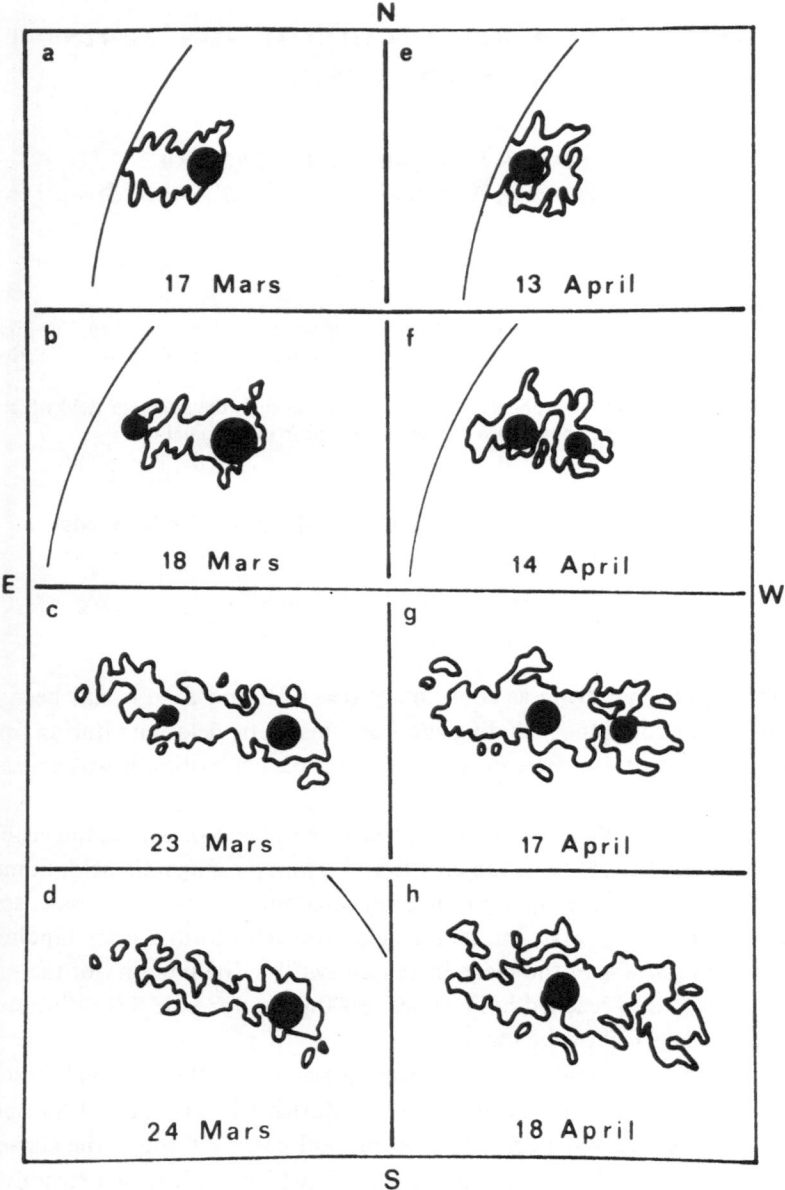

FIG. 1. *The appearance of new centres in the Active Region during the second and third rotations.*

The C.A. reappeared on March 14th, 1966, and its structure became more complicated because of the appearance of a new active centre in its Eastern part; that active
centre, incorporated with the former one, constituted a centre of multipolar shape
(Figure 1a, b, c, d). From a magnetic point of view, the type of the spots became

$\gamma$ from $\beta$, according to the Mount Wilson magnetic classification. The plage took a great area and its brightness continued to increase. The intensities of the magnetic field of the spot group reached high values: 3400 gauss for pole S and 2500 gauss for pole N.

The centre accomplished West-limb passage on March 28th, and appeared again on the East limb on April 11th. During the third rotation of the A.R. a new centre appeared in the Western part of the initial A.R., which suddenly reinforced the dwindling centre and made it more active than the usual (Figure 1e, f, g, h). The plage was extended to the West and the new spots, together with the previous ones, consti-tuted a unique Brunner type E group. The values of the intensity of the magnetic field of the former rotation remained quite high: 2400 gauss for pole S and 1700 gauss for pole N.

The C.A. passed over the West limb on April 25th, and reappeared on May 9th, while the spots were observed for the last day on April 23rd. The plage was conserved, appearing for a fourth time: it was, however, completely broken up, its brightness was weak and the pictures taken in the light of K showed that the region of faculae had begun to take the shape of a bright network. We had no observations of the magnetic field, which, however, most likely had to exist.

## B. FLARES

In the active centre under study, a total of 217 flares were observed. The centre lived in fact for 63 days. Observations were taken every day that it was upon the visible hemisphere of the Sun. So from its appearance to its West-limb passage (20 February – 1 March, 1966: i.e. 10 days) a total of 15 flares were observed. From the time of its reappearance on the Eastern limb of the Sun till its West-limb passage (14–28 March, 1966: i.e. 15 days) the flare activity was important. 162 flares had been observed – about one flare every 2 hours. During the third rotation of the centre (11–25 April, 1966: i.e. 15 days) the number of observed flares was 40.

Table 1 gives the distribution according to the importance of the observed flares for the three rotations.

### Table 1

**Number and importance of the observed flares in the C.A. under study**

| Rotation | S | Importance 1 | 2 | 3 | 4 | Total |
|----------|-----|-----|-----|-----|-----|-------|
| 1 | 7 | 7 | 1 | – | – | 15 |
| 2 | 94 | 56 | 8 | 4 | – | 162 |
| 3 | 31 | 8 | 1 | – | – | 40 |
| Total | 132 | 71 | 10 | 4 | – | 217 |

## C. RADIO EMISSION

The radio emissions followed the evolution of the centre. The flux density on 2980 MHz (Penteli Station, Athens), in 2800 MHz (Ottawa A.P.O. Station, Canada) and in 600 MHz (Humain Station, Belgium) that is to say in centimetric and decimetric frequencies, reached its highest value during the second rotation of the centre (14–28 March 1966). In its third appearance and in the three frequencies the value of flux density remained inferior to that of the second rotation but higher than that of the first one. Let us say that the values of the flux density are the average daily values, containing also the values during the bursts. As the centimetric bursts have their origin from thermal radiation connected with flares (De Jager, 1959) it is natural for the flux density to take higher values, as the number of flares during the second and third rotations is higher than that of the first rotation.

The contrary was observed in a frequency of 327 MHz (Bologna Station, Italy). The average values of the flux density during the first rotation are far higher than the corresponding values during the third rotation.

## 2. The Cause of the Abnormal Evolution of the C.A.

The behaviour of a C.A. concerning the appearance of flares is mostly the following: The number of flares and flare surges increased steadily from the 6th till the 13th day of its life. From the 14th till the 30th day the number of flares decreased, while from the 30th till the 60th day of its life, its activity was continuously decreasing.

The C.A. under study behaved in a way opposite to the expected one. So from the time of its appearance (20th February 1966) to its 10th day of life included (the C.A. set on March 1st, 1966) its activity was very restricted. Figure 2 shows that the greatest activity of the C.A. was observed between the 24th and 37th day of its life. Duirng the third rotation of the centre, its activity in flares remained inferior to that of the second rotation but much higher than the activity of the first one.

This increased activity of the centre from the 24th till the 37th day of its life, can be attributed to the appearance of a new C.A. in contact with the original C.A. and to the collision with it. Indeed, during the day when the A.R. under study rose performing its second rotation, a new C.A. emerged beside the A.R. and in its Eastern limb, and quickly joined together with the former one. (Figure 1b, c).

The collision of such active centres or magnetic fields, originated by moving or rapidly developing sunspots, may lead to a high concentration of electric-current density along the neutral lines of the field, with a great probability of flares occurring as Dungey (1958) and Sweet (1958) had shown.

Figure 2 shows that the intensity of the magnetic field during the second and the third rotations, as well as those of the C.A., reached high values. The flare production during the third rotation (between the 51st and 64th day of its life) can be explained in

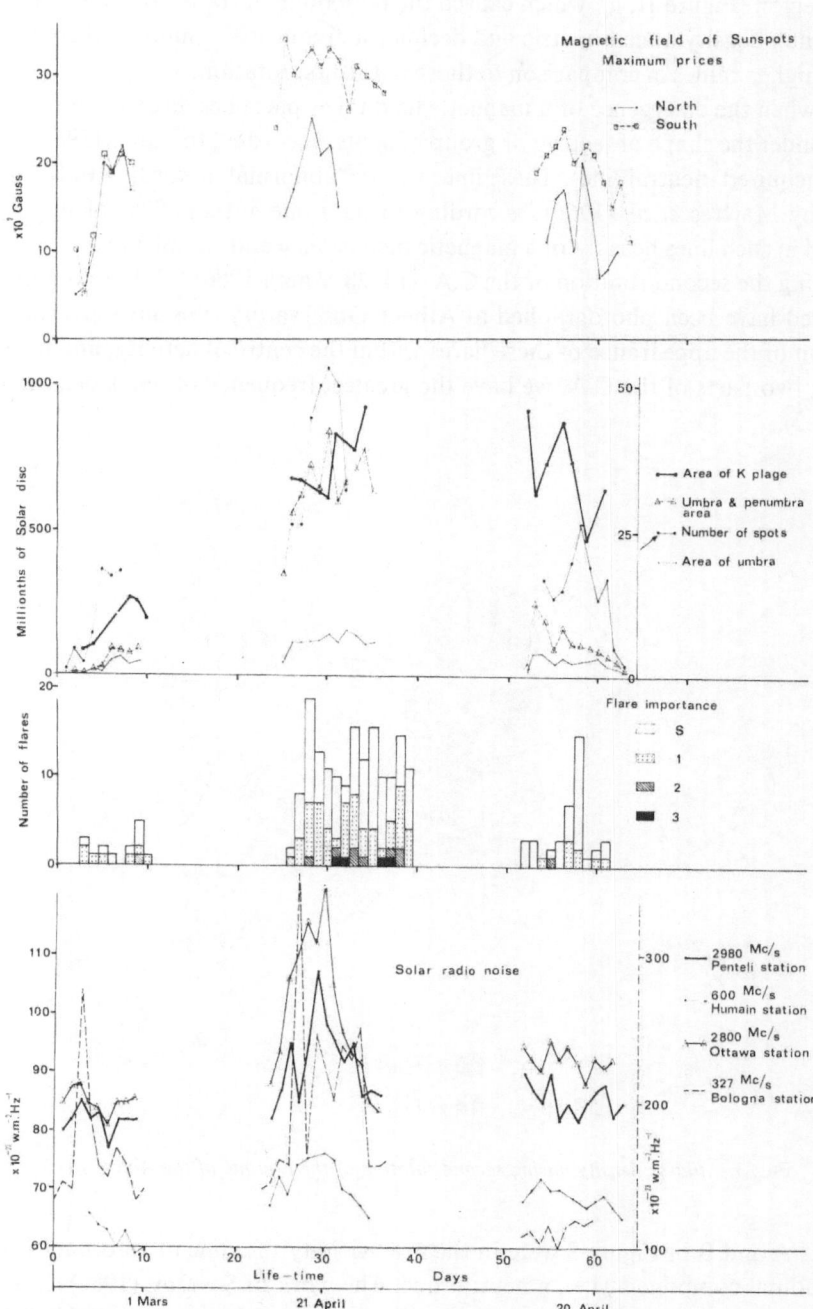

FIG. 2.   *Different phenomena – magnetic fields, flare importance, radio data, etc. – during the life-time of the Active Region.*

the same way. It appeared in the Western part of the centre and collided with a magnetic region (Figure 1f, g), which caused the rekindling of the centre. This caused spot areas, flux density in centimetric and decimetric frequencies, number of spots, etc., to reach higher values in comparison to those of the first rotation.

So, when the emergence of a magnetic field takes place beside or within an existing A.R. under the shape of faculae or group of spots, according to Banos (1967), we shall have 'acquired' neutral lines. These lines will be 'abnormal' according to the meaning given by Martres *et al.* (1966); according to the same authors 97% of the flares are created in such lines because of a magnetic inconstancy and complexity.

During the second rotation of the C.A. (14–28 March 1966), 45 flares out of the 162 observed have been photographed at Athens Observatory. We have investigated the position of the appearance of these flares within the centre of activity, and have found that in two parts of the C.A. we have the greatest frequency of flares occurrence (see Figure 3).

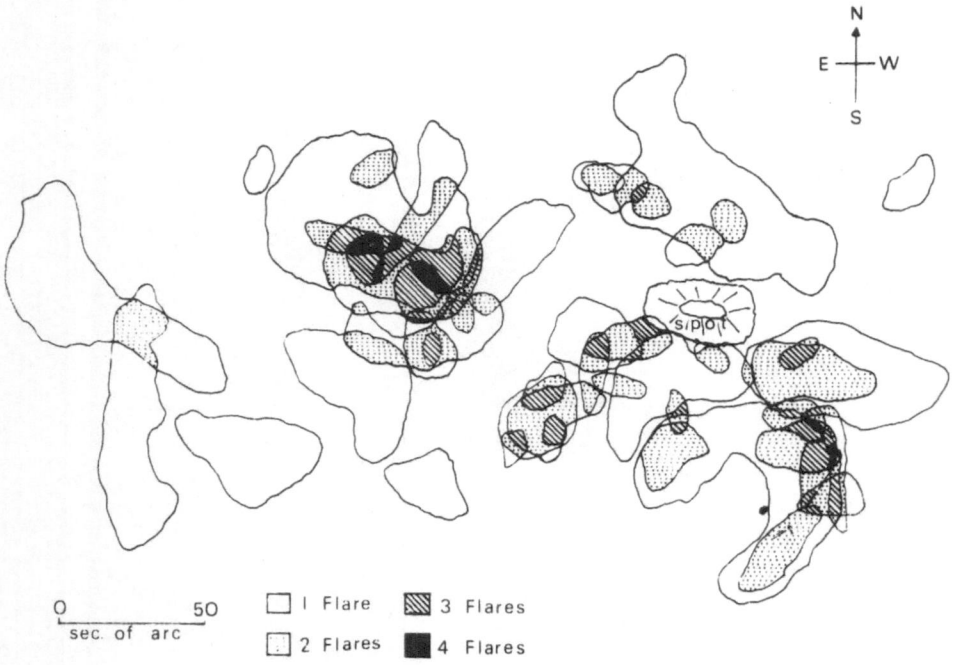

Fig. 3.    *Flares density during second rotation of the lifetime of the Active Region.*

Parts A and B in Figure 3 were in the near vicinity of a line of zero intensity of the longitudinal component, i.e., a neutral line. The work of Severny (1964) showed that the flares are connected with particular configurations of the magnetic field of the A.R.
Consequently the increase of activity in the A.R. during the second and third

rotations could be attributed to two separate magnetic arches, the interaction of which produce neutral lines characterized as 'abnormal', and it is well established that the majority of flares appear in the vicinity of such lines.

## References

Banos, G. (1967)     *Solar Phys.*, **1**, 397.
Dungey, J.W. (1958)     in *VI Symposium of I.A.U. (Cambridge)*, p. 135.
De Jager, C. (1959)     in *Handbuch der Physik* **52**, *Springer Verlag*, 283–322.
De Jager, C. (1965)     in *Introduction to Solar Terrestrial Relations* (ed. J. Ortner and H. Maseland) Reidel, Dordrecht, p. 55.
Kiepenheuer, K.O. (1959)     *Radioastronomia Solare, Varenna*, Corso XII, p. 39.
Martres, M.J., Michard, R., Soru-Iscovici, I. (1966)     *Ann. Astrophys.*, **29**, 245.
Severny, A.B. (1964)     *Ann. Rev. Astr. Astrophys.*, **2**, 384.
Sweet, P.A. (1958)     in *VI Symposium of I.A.U. (Cambridge)*, p. 123.

Part II

THEORETICAL ASPECTS

# MAGNETOHYDRODYNAMICS OF AN ACTIVE REGION

H. U. SCHMIDT

*(Max-Planck-Institut für Physik und Astrophysik, München, Germany)*

The dynamics of the magnetic fields which are imbedded into the non-stationary outer layers of the Sun show many facets of interest to observers and theoreticians alike. In a short review I can only deal with a small number of them and occasionally glance at some others. I hate to call these magnetic fields frozen into a matter which is rather in a boiling state, but the electrical conductivity in these layers is high enough to keep matter and magnetic flux together for rather long times, so that we can discuss the most important questions within the framework of magnetohydrodynamics with infinite conductivity. I will first talk mainly about the layers below the photosphere, where the matter controls the motion of the field, secondly about the intermediate state near the photosphere, where matter and field have comparable energy, and finally about the upper layers where the field controls the material motion.

## 1. Dynamotheory of the Solar Cycle

The cycle of solar activity ought to be explained as an oscillating dynamo with a period of 22 years. Some theories relevant to this problem have been put forward already many years ago by Cowling (1946), Elsasser (1950, 1956), Bullard (1955), and Parker (1955). The most detailed model for a theory of the solar cycle was derived by Babcock (1961) from observations with the use of the earlier theories. It was developed further by Leighton (1964), and also by Kopecký (1966), and Godoli (1966). Let me briefly discuss some aspects of the Babcock model and some of its theoretical problems and implications. I will also occasionally touch upon other theories.

Babcock calls the solar cycle a free-running oscillator, lacking stabilization and driven by the differential rotation. Leighton visualizes it as a relaxation oscillation. The fast rotation near the solar equator stretches any meridional flux within a few years into an almost toroidal topology, in which the fieldlines trail behind their equatorial vertex. This implies a strong amplification of the field, which ends when a hypothetical twist instability or the magnetic buoyancy brings substantial parts of the flux to the solar surface therewith producing the active regions of a new cycle. The follower spot in a bipolar group appears usually at a higher latitude than the corresponding preceding spot and this tilt of about 6° with respect to the equator is much larger than the average tilt of the almost toroidal field below the photosphere can be after years of differential rotation. The enhanced tilt may be due to Coriolis forces, as was supposed by Cowling (1966). Whatever its reason, this tilt is essential for the dynamo action in this model. If the stretches of fieldlines above the surface are tilted in one direction, the stretches below the surface must become tilted the other way by the same token. In the rapid formation of active regions at the peak of the cycle this process will piecewise and locally turn over the subphotospheric field into a tilt which is against the grain of the differential rotation. When this is accomplished the latter will rotate the field further by almost 180°. The rotating field at first relaxes into a roughly polar field and then it is again wound up into an almost toroidal field opposite to the original one and ready for the second half-cycle. It is by no means clear whether the intermediate new polar field has connected fieldlines all the way from pole to pole or not.

This model is so closely connected to some basic facts of the solar cycle which are well established by observation that it hardly can be completely wrong. But besides the problems posed by the spectacular large scale features discussed this morning by Dr. Howard and Dr. Dodson-Prince, this model encounters some theoretical problems, which I want to comment on.

(1) Obviously the flux is very buoyant only during the phase of amplification. Babcock (1961) argues that the differential rotation acts like a roller bearing on the flux and forms flux ropes. Excessive local twisting then causes an instability which brings a stitch of this rope up into the surface to form a new active center. This proposal has two difficulties. The twisted field in the Northern hemisphere has a right-hand thread which should show up in all Northern spots. But this is not observed. Furthermore the rising stitch in the overtwisted rope must relax the twist and therefore rotate anticlockwise in the Northern hemisphere. This is inconsistent with the observed clockwise tilt of the axis between newly formed bipolar spots. Contrary to the observed tilt a strong twist would tend to preserve the original poloidal component and thereby counteract the solar cycle quite severely.

But if the roller-bearing mechanism is ruled out what causes the formation of a fairly uniform buoyant flux unit of some $10^{21}$ maxwell responsible for the average active region? This question was answered by Weiss (1964a, b, 1966). He shows that throughout the convection zone the convection must expel the flux from the cells. The amplified toroidal flux beneath the sunspot zone is concentrated into toroidal tubes of some $10^{21}$ maxwell. The flux will be compressed by the convection until the field intensity suppresses the convection inside the tube almost entirely, which affords about 5000 gauss in the deep convection zone. The buoyancy of these tubes is proved for all parts which have a radius of curvature larger than 0·17 of the solar radius. Near the surface the buoyant tubes must diverge to intensities of a few hundred gauss, since neither convective nor thermal energies can balance more.

(2) The tilt of the axis of an active region should be caused by Coriolis forces. In the sunspot zone the horizontal Coriolis force needs a day of unimpeded uniform motion to rotate the resulting displacement by 6°. Such motions can be found in a super-

granulation cell which seems to last for 1 day (cf. Simon and Leighton, 1964). The same timescale holds for the appearance of a new active center. The circulation in a supergranule builds up the tilt of the axis between opposite base points of fieldlines in two steps. The upward motion in the center of the cell brings a stitch of the toroidal flux tube to the surface and subsequently the horizontal flow at the top of the cell moves the basepoints of the fieldlines apart toward the vertices of the cell. In both steps the rotation is clockwise in the Northern hemisphere and the total motion takes about 1 day.

(3) An oscillating dynamo can hardly operate without some kind of dissipation to dispose of the fields which are not in phase. But the usual dissipative processes which are due to particle interaction are probably not of primary importance in this context. The described dynamo should work for some periods even without these processes, i.e. for infinite conductivity because it can act on finite stretches of subphotospheric fieldlines in active latitudes just as well as on fieldlines extending all the way from pole to pole. It is rather the finite length of these stretches than the dissipation of flux which allows their permanent rotation. We may say that the rise of flux into the atmosphere provides the dissipative mechanism for the solar dynamo.

Of course the phases of activation of all the field stretches in the different latitudes would soon be put off their stroke if there occurs no reconnection between them. But solar cycles are certainly not completely in step for all latitudes, as Gnevyshev (1967) has just shown for the recent cycles. The many irregularities which solar activity exhibits, e.g. the so-called long-term cycles, or the very interesting fine-scale polarity changes in the polar caps as studied by Severny (1965), or even in moderate latitudes by Livingston (1967) are very encouraging for theoreticians because all these facts are evidence in favor of a long timescale for the final dissipation of solar magnetic flux crossing the photosphere. There is no reason why this timescale cannot be as long as some tens of years, just what it should be accounting for the diffusion and dissipation of flux by the surface convection. Diffusion timescales of this order were derived by Leighton (1964). If this concept is applicable, each solar latitude could act on its own for a timescale of years or so, and this seems to be in good agreement with observational findings.

(4) The flux crossing the surface is subject to the granular and supergranular convective motions. Leighton (1964) has shown that they act like a two-dimensional diffusivity on the footpoints of the fieldlines, at least if we consider scales large compared to a single cell. This random walk of the footpoints combined with their differential rotation causes the well-observed expansion, stretching and weakening of the aged bipolar magnetic regions, the cancellation of the polar cap field, and the equatorial leftovers called UM regions (cf. Bumba et al., 1966).

But the random walk also mixes fields of opposite polarity, so that they can be dissipated in small bits by local currents near to the photosphere where the conductivity has its minimum. In agreement with the local Cowling timescale this dissipation will reconnect the subphotospheric fields slowly but sufficiently to provide a weak coupling between the different latitudes and thereby a weak coherency of the phases in the solar cycle. The upper branches of the reconnected fieldlines will move into the corona and possibly into the solar wind.

(5) In Babcock's model solar activity takes its energy from the differential rotation, and this poses the problem where the latter gets its energy from. Kippenhahn (1963) has argued that this energy can be supplied from meridional circulations, which may be caused by anisotropic eddy viscosity in the convection zone. Plaskett (1959, 1962) and Weiss (1965) have discussed meridional differences of the superadiabatic gradient in the convection zone due to rotation as a possible driving mechanism for circulation and differential rotation. These authors argue that the solar magnetic field is too weak to interfere with the differential rotation, which otherwise should depend on the solar cycle.

(6) Another problem is posed by the amplitude of the oscillating magnetic activity which must be determined by a non-linear effect. The basic process which turns the tilt of the subphotospheric field over when part of the field rises above the surface would not contradict an ever-growing amplitude, since the newly formed reversed meridional flux can be produced from the strongly enhanced toroidal flux. The flux of the old cycle through the polar cap and the remnant UM regions are no match for this flux and they themselves must be proportional to the amplitude.

The most probable limiting mechanism is the following: The buoyancy of the flux-tubes must work against dragforces so that the upward drift becomes more rapid for the larger fluxtubes. Since the fluxtubes will grow by merger of neighboured tubes during the amplification, major activity at the surface will start, when the total intensity of subphotospheric flux has reached a certain critical level. Once this critical level is reached the further amplification of the meridional flux can be stopped by a reversal in the poloidal flux. Such a mechanism would be consistent with the fact that strong cycles have a short rise time.

(7) The most challenging problem for the Babcock model and for any dynamo theory of the solar cycle seems now to be posed by the recurrence tendency of active regions and by the simultaneous appearence of similar active centers at opposite longitudes which was demonstrated so perceptually by Dr. Dodson this morning. She pointed out that there is a systematic difference between the shifts in longitude for the individual active regions and for the location of recurrence itself. Obviously the individual active regions participate in the differential rotation, whereas the location of the recurrence does not. This difference points to very deep layers of the convection zone as for the cause of the recurrence. We may witness the structure of convection at the botton of the zone in the large dimensions, in the recurrence periods, and in the very long lifetimes of the system of recurrent centers, or 'giant granules' as Dr. Bumba calls them. But we also witness an ordered magnetic structure, probably a circular fluxtube, which seems to have a particularly unstable mode for two azimuthal nodes. This should have a simple explanation, but does not have it yet.

Before I leave the subject I should mention a few other dynamo models. The stochastic dynamo action of the Coriolis forces in the solar convection zone was studied in detail by Steenbeck and Krause (1966) and Krause (1965), who developed a very promising linear theory for a purely oscillatory dynamo driven by the differential rotation. In this theory the necessary dissipation is assumed to be homogeneous throughout the convection zone which may need modification, since the most important dissipation is provided by the rise of magnetic flux into the atmosphere and by reconnection of fieldlines in the photosphere. Iroshnikov (1965) has put forward an investigation on the instability of a pregiven stationary subphotospheric field aligned in a surface of constant angular velocity. He finds instabilities in modes somewhat similar to the observed behaviour during the first half of a cycle, but of course for small amplitudes, so that no field reversal can be described. Finally I should mention in this context the work of Lortz (1967), who has constructed an exact stationary solution of the dynamo equations in helical symmetry, which fulfills reasonable boundary conditions.

## 2. Formation of Plages and Spots

What happens to the magnetic flux when its own buoyancy and the convection bring it up into the photosphere to form a new active region? Naturally it must arrive at this interface as a horizontal field stretching over a limited region for which upward motion prevails. We expect that the flux of an active region, usually some $10^{21}$ maxwell, will appear distributed over an area of a very few supergranules within a few lifetimes of the latter, i.e. within a few days.

The first phenomenon of a new active region is a plage and it has roughly the described features. Stepanov and Grigorev (1966) with the new Sibizmir-Magnetograph found transversal and longitudinal fields of about 200 gauss in bright plages, and Dr. Bappu presented similar results in his paper. These observations are consistent with an equipartition of energy between the photospheric convective motions and the magnetic field as estimated by Weiss (1964b) and De Jager (1964). They also imply that the flux of a sizable spot can be found in one supergranulation cell. Though any new flux elements or fieldlines should arrive first in the rising middle part of the convection elements, their basepoints will immediately be driven to the border of these elements, where the convection currents sink into the Sun and leave the flux behind. The observational evidence for this process is overwhelming for the supergranulation and to a lesser degree for the granulation, as is well known from the work of Simon and Leighton (1964), Bumba and Howard (1965), Sheeley (1966), and many others. Theoretical work on the expulsion of magnetic flux from eddies was done by Parker (1963a) and in detailed and thorough computations by Weiss (1964a, 1966), who presented part of his spectacular results in a most beautiful movy. He as well as Simon and Leighton (1964) pointed out that in nearly polygonal convection cells the flux becomes even more concentrated in the polygon vertices than simply at the borderlines of the cells. The fieldlines which connect these basepoints through the chromosphere and corona will therefore cover the cells with a nearly horizontal field, which must be responsible for the obviously aligned fine-structure in Hα, which can be observed on high resolution filtergrams. It is also not surprising that Bumba and Howard (1965) found that the plages as well as the spots of an active region begin to appear just at the corners of supergranulation cells and probably in intergranular space. Bumba (1966) even found that active regions arise in the corners of still larger elements, which he calls 'giant granules'. It is interesting to note that the diameter of these structures of 400000 km corresponds nicely to the depth of the convection zone of about 100000 km if one applies the ratio for diameter and depth derived by Weiss (1966). But it is hard to imagine how so large an element can form an eddy.

The equipartition fields in the photosphere of about 200 gauss are not sufficient to reduce the convection in the deeper layers. But they are able to increase the exitation of sound and hydromagnetic waves near to the photosphere (cf. Kulsrud, 1955; Pikelner, 1960; Parker, 1964). In this way the enhancement of the non-radiative flux in an active region by an order of magnitude may be understood (cf. Livshitz, 1965).

The most obvious phenomenon in an active region is the sunspot. It is a local deficiency of temperature, radiative flux and pressure. The latter is balanced by the compressed magnetic field which may reach 5000 gauss. Various models by Deinzer (1965), Weiss (1964b), De Jager (1964), and Yun (1967) describe the observed spots quite well and explain the deficiency of radiative flow at the surface by a systematic reduction of the convective energy flow in the underlying convection zone. Modifying an early suggestion by Biermann (1941) they assume that such a reduction is caused by the magnetic field. According to these models the relative temperature or pressure deficiency becomes negligible at a depth of some $10^3$ km, so that the spot is a rather shallow entity. The only thing which must persist below this depth is the magnetic flux which, rather than being disordered by the convection, will be compressed by it into a flux tube containing all the parallel flux of the active region, as Weiss (1964b) has suggested.

The energy flux which is missing in the radiative flux escaping from the sunspot must get out somehow and somewhere. Danielson (1965) argues that this flux, which amounts to about $\frac{1}{3}$ of the total flux over the sunspot area, can hardly be non-radiative flux into the local chromosphere overlying the spot, because the missing flux of one big spot would already account for $\frac{1}{6}$ of the non-radiative flux supplied into the whole chromosphere. Therefore most of this missing flux will probably be distributed over such a large area that it can escape as radiation from the photosphere without being detectable as a surplus. Whether this energy can be transported away from the spot by magnetohydrodynamic waves is an interesting problem which is intrinsically connected with the problem of the modes of convection in a strong magnetic field, as became clear from the work of Danielson (1965, 1966). But convection itself may already divert the energy flow sufficiently.

The mechanism by which the missing flux is transported away from the magnetized upper layer of the convection zone underlying the umbra of a sunspot is rather efficient and selective, since the magnetic-field strength in an umbra falls into a narrow range between 1000 and 5000 gauss. As Zwaan (1965) has pointed out the cooling process seems to be sufficient for just this range but overly effective for intermediate field strengths, so that the field strength may flip flop between this range and the range below 200 gauss, which provides only a minor enhancement of the energy losses of the quiet photosphere (cf. Livshitz, 1965). Zwaan (1967) argues that the tiny magnetic regions with roughly $10^3$ gauss and diameters of about 1″ found by Sheeley (1966, 1967) and Beckers and Schröter (1966) should be considered 'invisible sunspots' (cf. Hale and Nicholson, 1938) which are stabilized by the same mechanism but do not have an umbra due to the inflow of radiation above the layer at which the cooling mechanism is effective.

To summarize, it seems to be conceivable that sunspots are formed by the supergranulation which concentrates the flux in its vertices, where the magnetized matter at the top of the convection zone is cooled by the radiative losses until the reduced convection can balance the losses. Obviously a sizable temperature difference is needed, which implies also a compression and a sizable increase in the magnetic intensity up to the typical sunspot values. In this sense it is possible to say that sunspots are not brought up from below but are formed at the top of the convection zone by the supergranular motion which can produce local detours in the flow of energy. Sunspots seem to be secondary phenomena which appear if the necessary unconcentrated flux through the solar surface is provided in an active region by the buoyancy of a larger flux tube. But they do not form until so much flux is concentrated at one point that the cooling mechanism can overcome the radiative flux from the neighboured subphotosphere.

Before I leave the subject of sunspots I should say a few words about their lifetime. For this lifetime we have three options:

First the Cowling timescale, which is very long and therefore should be ruled out. I dare to make this statement despite the results of Kopecký and Kuklin (1966) and Schröter (1966), who have computed comparatively low timescales from the local conductivity in the umbra. I do not think that such a local timescale is applicable for the decay of the whole sunspot field. Even if the local currents in the umbra decay, the result would simply be a small bulge in the field at that level, but certainly all the flux deeper down or high up in the corona cannot contract towards that tiny ring, it will rather induce some additional currents in the high conductivity of its surroundings and then stop worrying about the conductivity at the umbra. But I should emphasize that nevertheless these low conductivities in the umbra and also in the photosphere are of importance for the whole solar cycle, because in just this level occurs the dissipation and reconnection of opposite fluxtubes, which meet stochastically at the borderlines of supergranules and granules.

The next choice for a sunspot lifetime is the radiative relaxation time which has the right order of magnitude, as Weiss (1964b) has estimated. But an umbra does not seem to shrink smoothly, and the cooling mechanism which holds the spot stable will not stop unless sufficient flux has been transported away from the umbra.

The third possibility seems to be the timescale for destruction of the spot through the action of the supergranulation which may chip off the flux in small pieces, as Simon and Leighton (1964) have proposed. But this lifetime is hard to estimate, though the mechanism seems to be reasonably confirmed by observations.

## 3. Alignment, Energy Storage, and Release

We now have to deal with the active magnetic fields above the solar photosphere. Here the magnetic energy is predominant from the low chromosphere out into the solar wind for many solar radii. Since there is no serious competitor the magnetic field will be nearly 'force-free' in the sense that any sizable electrical currents which are able to change the magnetic field will flow along the field. But it would not be correct to say that the Lorentz force vanishes. This force is only small compared to what it could be if there would be comparable material energies present. The Lorentz force can be quite strong, however, in terms of the hydrostatic pressure gradient or gravity force. Therefore we will find many aligned structures in the solar atmosphere. The chromosphere is actually overcrowded with such aligned structures, as the excellent modern Hα filtergrams demonstrate.

In principle the force-free currents along the field can be deduced from the horizontal derivatives of measurements of the transverse field components, but such measurements do not often have sufficient accuracy. In many cases it seems to be more reliable to use the information available from fine-scale longitudinal magnetograms together with filtergrams. Especially for this purpose some years ago I described a method (Schmidt, 1964, 1965) to compute the full three-dimensional magnetic field from a longitudinal magnetogram under the fictitious assumption of vanishing currents. Such computed transverse fields can be directly compared with filtergrams. From the directional deviations one can infer electrical currents. The method has been used for aligned structures like prominences by Rust (1966), Hyder (1967), and Semel (1967), who on a large scale generally found good agreement between predicted field direction and aligned structure, which seems to imply that there are no large electrical currents of such scales. But it should be borne in mind that the deviations between predicted and observed directions are useful information about electrical currents. I should add one comment. We have not yet any observational proof of the nearly force-free nature of the field in the solar atmosphere. This is difficult to get because locally it can only be done if all three components are measured in a volume with sufficient accuracy that the data can be differentiated with respect to all three dimensions. Such a task seems to be almost hopeless and useless, too, because near the photospheric level of observation the field simply will not be force-free. A sensible test would therefore involve the integral structure over a large area, but that is difficult, too. Only if we are able to identify the two photospheric basepoints of an individual fieldline, we could test whether all the current leaving at one basepoint is arriving at the other end. Simple tests for the twisted nature of the field do not prove that the current is actually parallel to the field.

If the atmospheric fields are nearly force-free, what are the proper boundary conditions? Since on the fieldlines the ratio of current and field intensity stays constant, we have one real characteristic. The other two are imaginary as for current-free fields. We therefore have to prescribe the flux everywhere at the boundary and additionally the current either in that part of the boundary where the flux is positive or where it is negative. If we exclude closed flux tubes in the atmosphere, there will be only one solution. Unfortunately the real situation is much more complicated and the natural boundary conditions are quite different. Once a fieldline has crossed the surface it has two basepoints and these are independently transported by horizontal convection, but so that the relative topology of the fieldlines connecting the basepoints is preserved if we forget about the extremely slow dissipation processes in the high temperatures of chromosphere and corona. The somewhat faster dissipation in the photosphere merely causes a slow slip of the basepoints relative to the convective circulation and occasionally a reconnection by annihilation of two encountering basepoints of opposite polarity. If the field is to be force-free no further condition can be put on the direction of the fieldlines, at the boundary. These boundary conditions will leave enough freedom so that the solution is not determined unequivocally. There arises a stability problem which may be important for the flare problem, as Gold (1964) has pointed out. The convection will steadily put energy into the field by moving the basepoints around. Thereby it will induce currents along the field and work against Lorentz forces acting in the surface. This energy storage will be the source of the flare energy, as many would agree upon. Now there may be more than one solution to a given boundary condition and gradual change of the boundary conditions by convection may put the magnetic field into a metastable solution different from the one with the lowest energy. Then by a finite disturbance a sudden change into the latter solution may be triggered off, so that the energy difference between the two solutions is released without any dissipation of magnetic flux. The latter point is of importance, since the timescales for dissipation of flux are so long that Parker (1963b) argues against any flare mechanism involving such processes. In this context it is also important to note that a large part of the flare energy is released at the very beginning of the flare into non-thermal motions.

In Munich we have tried to find a simple model to prove the existence of such metastable force-free fields which after a finite disturbance can release energy within the timescale of an Alfvén wave, because there is a state of lower energy accessible. The type of solar convection which most readily transfers energy into force-free currents along the atmospheric fields, is probably found in the penumbral filaments, which might be convection rolls aligned by the field, as Danielson (1961) has argued. Flare activity starts with the formation of the penumbra, so it seems to be reasonable to investigate models which might simulate the flux tubes between penumbral convection rolls of opposite spots. Meyer (1965) has shown that for a force-free field of cylindrical symmetry bounded by a cylinder of finite length so that the basepoints of the field are frozen into the two plane disks at the ends of the cylinder, there exists a state of lower energy if the pitch angle of the original field is of order unity. This state of lower energy is of course asymmetric and has a counterhelix to the side. Since it is not force-free, the energy difference for a corresponding force-free solution is even larger. This looked very promising, but one has to prove also the stability of the original solution for infinitely small disturbances because otherwise the convection of the basepoints represented as a rotation of one of the limiting disks would never put the field into the symmetric solution to begin with. I guess everyone here would expect this solution to be stable in this sense for very small pitch angles. At least we did. But the opposite was proven recently by Anzer (1968). To his own surprise he was able to show that any force-free field of cylindrical symmetry is unstable for infinitely small disturbances of a helical symmetry where the helix runs with the pitch of the original field. This simply means that in real nature there are not any force-free fields of cylindrical symmetry. Now we are trying to find them for a rotational symmetric model which might be a more realistic approach to the flare problem anyhow. By the way, if we consider penumbral filaments to be convection rolls and if we consider a flare to be caused by the twist induced in the field by these rolls, then it might be worthwhile to look into the Evershed effect during a flare.

Let me add a few remarks on loop prominences and loop structures in general. These most beautiful structures pose very difficult and fundamental problems. With this remark I do not refer to the intricate radiative and mass balance which Jefferies and Orrall (1964) and Kleczek (1964a) have studied some years ago. There is another difficulty involved in these innocent-looking simple structures, which occur during the course of a flare event and which obviously must be aligned by the field. Bumba and Kleczek (1961) and Kleczek (1964b) have shown that these filaments do not show any significant change in diameter if one follows them from base to base. Bruzek (1964) has shown that they are an essential phenomenon in the course of many flare events, especially so because they are based in the flare filaments themselves which overlie a region of enhanced field intensity. How can these structures exist in an almost force-free magnetic configuration? E.g. a closed force-free loop cannot exist because it contradicts the virial theorem, as Cowling (1965) has noted. If one assumes as an initial configuration a toroidal loop imbedded in poloidal fields the loop must expand indefinitely and with Alfvén velocity because it cannot be balanced without Lorentz forces, arising e.g. from pressure gradients. Of course at the base of a solar loop structure there are Lorentz forces available which just hold the base in place or move it slightly with the horizontal convection. But I do not see how this can prevent the loop itself from rapid expansion into the corona which should take place with Alfvén velocity. Since this does not happen the loop may be compressed by some non-magnetic energy. If we assume only 100 gauss for the loop this energy must have a density of about 400 dyn/cm² or a total amount of about $10^{32}$ erg in a reasonable volume surrounding the loops. This seems to be a bit large even for flare conditions. If it were there it must expand also with high speed. So it would be extremely valuable to have reliable measurements of the toroidal magnetic field in a loop, because such measurements would inform us about the loop as well as about the post-flare conditions in the corona.

## References

Anzer, U. (1968)    *Solar Phys.*, **3**, 298.
Babcock, H.W. (1961)    *Astrophys. J.*, **133**, 572.
Biermann, L. (1941)    *Vierteljahresbericht der Astron. Ges.*, **76**, 194.
Beckers, J.M., Schröter, E.H. (1966)    *A.A.S. Meeting on Solar Physics, Boulder.*
Bruzek, A. (1964)    *Astrophys. J.*, **140**, 746.
Bullard, E.C. (1955)    in *Vistas Astron.*, Ed. by A. Beer, **1**, 685.
Bumba, V. (1966)    Varenna Summer School on Plasma Astrophysics.
Bumba, V., Howard, R. (1965)    *Astrophys. J.*, **141**, 1492.
Bumba, V., Kleczek, J. (1961)    *Observatory*, **81**, 141.
Bumba, V., Howard, R., Smith, S.F. (1966)    *Proc. of the Meeting on Solar Magnetic Fields, Rome*, p. 203.
Cowling, T.G. (1946)    *Mon. Not. R. astr. Soc.*, **106**, 218.
Cowling, T.G. (1965)    in *Stellar Structure*, Chicago, Vol. VIII, Chapter 8.
Cowling, T.G. (1966)    *Proc. IAU Symp. No. 22*, North-Holland Publ. Co., Amsterdam, p. 405.
Danielson, R.E. (1961)    *Astrophys. J.*, **134**, 289.
Danielson, R.E. (1965)    *Proc. IAU Symp. No. 22*, p. 314.
Danielson, R.E. (1966)    *Proc. of the Meeting on Sunspots, Florence*, p. 120.
De Jager, C. (1964)    *Bull. astr. Inst. Netherl.*, **17**, 253.
Deinzer, W. (1965)    *Astrophys. J.*, **141**, 548.
Elsasser, W.M. (1950)    *Rev. Mod. Phys.*, **22**, 1.
Elsasser, W.M. (1956)    *Rev. Mod. Phys.*, **28**, 135.
Gnevyshev, M.N. (1967)    *Solar Phys.*, **1**, 107.
Godoli, G. (1966)    *Proc. of the Meeting on Solar Magnetic Fields, Rome*, p. 289.
Gold, T. (1964)    *A.A.S.-NASA Symp. on the Physics of Solar Flares*, p. 389.
Hale, G.E., Nicholson, S.B. (1938)    *Magnetic Observations of Sunspots 1917–1924* (Publ. Carnegie Inst. No. 498).
Hyder, Ch.L. (1967)    *Solar Phys.*, **2**, 49.
Iroshnikov, R.S. (1965)    *Astr. Zu.*, **42**, 494.
Jefferies, J.T., Orrall, F.Q. (1964)    *A.A.S.-NASA Symp. on the Physics of Solar Flares*, p. 71.
Kippenhahn, R. (1963)    *Astrophys. J.*, **137**, 667.
Kleczek, J. (1964a)    *A.A.S.-NASA Symp. on the Physics of Solar Flares*, p. 77.
Kleczek, J. (1964b)    *Bull. astr. Inst. Csl.*, **15**, 123.
Kopecký, M. (1966)    *Proc. of the Meeting on Solar Magnetic Fields, Rome*, p. 285.
Kopecký, M., Kuklin, G.V. (1966)    *Bull. astr. Inst. Csl.*, **18** (in press).
Krause, F. (1965)    *Proc. IAU Symp. No. 22*, p. 426.
Kulsrud, R.M. (1955)    *Astrophys. J.*, **121**, 461.
Leighton, R.B. (1964)    *Astrophys. J.*, **140**, 1547.
Livingston, W.C. (1967)    *Science Journal*, **3**, 46.

Livshitz, M. A. (1965)     *Solnezn. Aktibn.*, **2**, 103.
Lortz, D. (1967)    *Physics of Fluids.*
Meyer, F. (1965)    private communication.
Parker, E. N. (1955)     *Astrophys. J.*, **122**, 293.
Parker, E. N. (1963a)     *Astrophys. J.*, **138**, 552.
Parker, E. N. (1963b)     *Astrophys. J.*, Suppl. **8**, 177.
Parker, E. N. (1964)     *Astrophys. J.*, **140**, 1170.
Pikelner, S. B. (1960)     *Astron. Zu.*, **37**, 616.
Plaskett, H. H. (1959)     *Mon. Not. R. astr. Soc.*, **119**, 197.
Plaskett, H. H. (1962)     *Mon. Not. R. astr. Soc.*, **123**, 541.
Rust, D. M. (1966)    Thesis Boulder 1966.
Schmidt, H. U. (1964)    *A.A.S.-NASA Symp. on the Physics of Solar Flares*, p. 107.
Schröter, E. H. (1966)     *Proc. of the Meeting on Sunspots, Florence*, p. 190.
Semel, M. (1967)    *Ann. Astrophys.*, **30**, 513.
Severny, A. B. (1965)     *Observatory*, **85**, 183.
Sheeley, N. R. (1966)     *Astrophys. J.*, **144**, 723.
Sheeley, N. R. (1967)     *Solar Phys.*, **1**, 171.
Simon, G., Leighton, R. B. (1964)     *Astrophys. J.*, **140**, 1120.
Steenbeck, M., Krause, F. (1966)     *Z. Naturf.*, **21a**, 1285.
Stepanov, V. E., Grigorev, V. M. (1966)     *Trudy Sib IZMIR.*
Weiss, N. O. (1964a)     *Phil. Trans.* **A256**, 99.
Weiss, N. O. (1964b)     *Mon. Not. R. astr. Soc.*, **128**, 225.
Weiss, N. O. (1965)     *Observatory*, **85**, 37.
Weiss, N. O. (1966)     *Proc. of the Meeting on Solar Magnetic Fields, Rome*, p. 299.
Yun, H. S. (1967)    *Astron. J.,* **72**, 838.
Zwaan, C. (1965)    *Rech. astr. Obs. Utrecht*, **7**, Part 4.
Zwaan, C. (1967)    *Solar Phys.*, **1**, 478.

# DISCUSSION

*De Jager:* Why do you rule out a sunspot life time in accordance to the low-conductivity values found by Kopecký and Schröter?

*Schmidt:* I do not think that such a local timescale is applicable for the decay of the whole sunspot field. Even if the local currents in the umbra decay, the result would simply be a small bulge in the field at that level, but certainly all the flux deeper down or high up in the corona cannot contract towards that tiny ring, it will rather induce some additional currents in the high conductivity of its surroundings and then stop worrying about the conductivity at the umbra. But I should emphasize that nevertheless these low conductivities in the umbra and also in the photosphere are of importance for the whole solar cycle, because in just this level occurs the dissipation and reconnection of opposite flux tubes which met stochastically at the borderlines of supergranules and granules.

*Davis:* It seems to me to be worthwhile to extend your brief remark on the possible mechanisms that might drive the solar differential rotation. Recent discussions show that the magnetic field in the solar wind exerts a substantial decelerating torque across the photosphere. If this decreases the angular momentum of the entire Sun, no spectacular effects are produced; but if it acts only on the convection zone, the rotation of the surface will be essentially completely stopped in a fraction of the life of the Sun. Thus the rotation of the surface must be driven by friction or magnetic coupling with a presumably slightly, or perhaps substantially, more rapidly rotating interior. The rotation rate of the surface layers is determined by a balance between the exterior torques exerted by the magnetic field and the interior torques. Since at present there seems no way to determine how either set of torques varies with latitude and with the surface-angular velocity, there is no way to predict what the surface-angular velocity should be. Consequently, no exotic mechanism is needed to produce differential rotation; and, if exotic mechanisms are present, there is no reason at present to expect that they can not be dominated by the torques mentioned above.

*Sturrock:* Since the solar wind must flow along magnetic-field lines, some field lines must be open. The boundary of two open flux tubes of opposite sense comprises a sheet pinch, at which the field will not be force-free.

*Schmidt:* I completely agree that there are exceptions to the general rule that the atmospheric fields are force-free. The most important examples are the current sheets in the quiescent prominences and these sheets are often underlying the sheets you were referring to, as one can see from the work of Dr. Newkirk's group.

# CONCENTRATION OF MAGNETIC FIELDS
# IN THE DEEP CONVECTION ZONE

G. W. SIMON        and        N. O. WEISS

*(Sacramento Peak Observatory,*        *(Dept. of Applied Mathematics*
*Sunspot, N.M., U.S.A.)*        *and Theoretical Physics,*
        *Silver Street, Cambridge, England)*

## ABSTRACT

The strong magnetic fields observed between supergranules indicate that there must be subphotospheric convection in cells with a preferred diameter of about 30000 km. Orthodox mixing length
theory assumes that the dimensions of cells are limited by the density scale-height. This is adequate
for explaining granules but cannot account for supergranulation. A model is therefore proposed in
which cellular motions extend over several scale-heights. In addition to granules and supergranules,
this model predicts a third characteristic scale of motion, with giant cells around 300000 km in
diameter. These cells may produce a pattern of magnetic fields like that suggested by Bumba and
Howard for complexes of activity.

## 1. Introduction

We shall first discuss the structure of convection in the Sun's hydrogen convection zone
and, in particular, the scale of the motions occurring in it. Then we shall briefly attempt
to relate this structure to magnetic-field patterns observed at the surface of the Sun.

Granules and supergranules show cellular motion on two scales in the solar atmosphere. The average diameter of a granule is around 2000 km, while the average diameter of supergranules is 30000 km. Moreover, the supergranulation is apparently
caused by subphotospheric convection. The photospheric magnetic fields appear to
have been swept aside and concentrated at the boundaries of supergranules; but the
kinetic energy of motion in supergranules corresponds to a field of about 20 gauss,
whereas Sheeley (1966) has measured strengths of several hundred gauss. Again, the
tendency of sunspots to form at junctions of supergranules (Bumba and Howard,
1965a), coupled with the size of sunspot fields and the scale of sunspot groups, indicates that supergranular motions must represent deep-seated phenomena. If, then,
we have a satisfactory model of correction in the Sun, it must explain not only
granules but also the supergranulation.

## 2. Convection Cells extending over Several Scale Heights

Let us consider first the orthodox theory of stellar convection. If we had only a

*Kiepenheuer (ed.), Structure and Development of Solar Active Regions,* 108–111. © *I.A.U.*

layer of incompressible fluid and could ignore the boundaries, we could describe convection in terms either of interacting bubbles or of eddies: in each case, the scale of motion would be comparable with the depth of the convective layer. However, in the Sun the gas is compressible and the density varies by a factor of order $10^5$ over the convective zone. Because of continuity the motion would become very complicated: a small motion at the bottom of the layer would require very rapid motion at the top. So Biermann and Schwarzschild suggested that the local density scale-height should be taken as the mixing length or, effectively, the characteristic scale for convective eddies. Thus we obtain a pattern of motion whose scale is determined by the local scale-height. When we study a model convective zone obtained in this way, we find that all its properties vary smoothly with depth. It is impossible to produce any single distance typical of the convective zone, let alone explain the predominance of eddies with a scale of 30000 km. The orthodox model accounts nicely for granulation but it cannot explain the existence of supergranules.

To cope with this difficulty we were therefore led to postulate that 30000 km is indeed in some sense a typical distance. We suppose that convection occurs in cells with a diameter of 30000 km over at any rate a substantial part of the convection zone.

We have discussed two crude physical models of convection in a compressible atmosphere, in order to have some quantitative results (Simon and Weiss, 1968). First of all, we considered convection cells extending over many scale-heights in an atmosphere whose scale-height was independent of depth. Owing to the stratification, the motion is dominated by rising blobs of gas. Very little of the fluid originating at the bottom of the cell will reach the top but since the fluid rises adiabatically it will carry a greater temperature excess than would be possible for blobs originating locally. The mass flux at the top of the cell is relatively small and the argument that continuity demands rapid motions does not apply. The superadiabatic gradient, $\beta_1$, required to transport a fixed convective flux can be crudely estimated by balancing non-linear dissipation against work done by the buoyancy forces. This can be compared with the superadiabatic gradient, $\beta_0$, obtained from the orthodox model. We found that

$$\beta_1 \approx \tfrac{1}{3}\beta_0$$

and cells extending over many scale-heights are therefore more efficient at transporting heat.

In the solar atmosphere the scale-height is not constant. Instead, it increases with depth, $z$, below the surface. As a result, horizontal velocities dominate the motion in the upper regions of a cell extending over many scale-heights. We considered a polytropic atmosphere with a relationship

$$P = K\rho^{\Gamma}$$

between the pressure, $P$, and the density, $\rho$, and a polytropic index

$$m = (\Gamma - 1)^{-1}.$$

Much of the solar convective zone is fairly well described by a polytrope with $m = \frac{3}{2}$. In such an atmosphere we would expect cells of diameter $2\,L$ to be generated where the local scale-height was comparable with $L$. But the level to which motions on this scale can penetrate is now limited. Using the same physical arguments, we would expect cells of diameter $2\,L$ to extend from $z \approx \frac{1}{10}L$ to $z \approx L$. Thus they would cover 3 to 4 scale-heights or a range of about 30 in the density $\rho$.

This crude physical treatment indicates that we might expect to find cellular motions extending over, say, four density scale-heights rather than just one as in the orthodox model. This result should be supported by detailed numerical experiments but we may still ask the question: What can we infer about the pattern of convection in the Sun?

## 3. Convection in the Sun

The solar convection zone is currently estimated to be about 150 000 km in depth; it is unlikely that its overall structure would be significantly affected by changes in the convection model used. The detailed motion will be dominated by rising blobs which spread out and interact to produce eddy-like patterns. If these typically cover four scale-heights, we might expect eddies generated at the base of the convective zone to penetrate to a depth of about 20 000 km and to have a characteristic diameter of around 300 000 km; motions originating at $z = 20\,000$ km might have a diameter of 30 000 km and penetrate to $z \approx 3000$ km; above that level radiative transfer becomes important, so all scales from 4000 km downwards might be visible at the surface.

We do not, of course, suggest that the convective zone is separated into three discrete regions. However, we predict that the distribution of cell sizes should show three distinct peaks: one at about 2000 km, corresponding to the granulation; another at 30 000 km, corresponding to supergranules; and a third around 300 000 km, corresponding to giant cells in the deep convective zone. That is to say, a convection model that explains both granules and the preferred dimensions of supergranules will also predict a third preferred scale, comparable with the depth of the convective zone itself.

For both supergranules and giant cells we can estimate typical values of the vertical component of the velocity, $w$, the horizontal component, $u$, and the timescale $\tau = L/w$. These values are shown in Table 1. In fact, the values of $w$ are effectively the same as those found using orthodox mixing-length theory near the lower boundaries of the eddies.

### Table 1

#### Cellular motions in the convective zone

|                | $L$ (km)          | $w$ (km/sec) | $u$ (km/sec) | $\tau$ (days) | $B$ (gauss) |
|----------------|-------------------|--------------|--------------|---------------|-------------|
| Supergranules  | $1.5 \times 10^4$ | 0·3          | 0·6          | 0·4           | 2000        |
| Giant Cells    | $1.5 \times 10^5$ | 0·1          | 0·2          | 20            | 6000        |

## 4. Interaction of Convection and Magnetic Fields

Over most of a cell the horizontal velocity acts in the same sense, outwards from the centre; in the upper part it will be greater than the vertical component. This pattern of motion will therefore be particularly effective in sweeping aside magnetic flux and concentrating it at the edges of the cell. This process will be halted when the energy of the concentrated field is comparable with the kinetic energy of the horizontal motion. This equipartition field varies slowly with depth for a given value of $L$. Typical values are given in Table 1.

The formation and development of individual active centres is indeed closely related to the supergranulation pattern. (Bumba and Howard, 1965a). If giant cells are present, they should also have observable effects on magnetic fields at the surface of the Sun. Some structure with a characteristic scale of about 300000 km should be present. As Professor Kiepenheuer has already pointed out, Bumba and Howard have indeed suggested that complexes of activity (Bumba and Howard, 1965b) do show such a structure and that it may persist over many solar rotations. If this pattern is present, the growth and migration of such complexes may be related to cellular motions at the base of the convection zone. Complexes of activity should form at junctions of giant cells, while individual active centres are controlled by supergranular motions. The complexes might be swept aside as the pattern of motions varies near the bottom of the convection zone. Close study of their development may thus enable us to investigate convection 100000 km below the surface of the Sun.

### References

Bumba, V., Howard, R. F. (1965a)    *Astrophys. J.*, **141**, 1492.
Bumba, V., Howard, R. F. (1965b)    *Astrophys. J.*, **141**, 1502.
Sheeley, N. R., Jr. (1966)    *Astrophys. J.*, **144**, 723.
Simon, G. W., Weiss, N. O. (1968)    to be published.

## DISCUSSION

*Kiepenheuer:* How does the lifetime of your model of a convective cell agree with the observed lifetime of a supergranulum (about one day)?

*Weiss:* The turnover time for our supergranular cells is about 10 hours; for the giant cells it is around 10 days. So a giant cell might have a lifetime of about 20 days, corresponding to the observed lifetime of supergranules, which is twice this turnover time.

*Acton:* How can a toroidal magnetic field as postulated in the Babcock theory be compatible with the radial convection field?

*Weiss:* The convection is a three-dimensional process and the horizontal components of the velocity may be greater than the vertical components. However, some concentration of the toroidal field below the bottom of the convection zone may occur, as was suggested by Spitzer.

*Beckers:* Have you estimated the temperature fluctuations in the photosphere resulting from your convection cells?

*Weiss:* Not precisely. One would expect fluctuations of, say, 200° at the top of supergranular cells, but the granulation would make any variation at the photospheric level much smaller.

# OSCILLATORY MODES OF ENERGY TRANSPORT IN
# SOLAR MAGNETIC REGIONS*

R. E. Danielson and B. D. Savage

*(Princeton University Observatory, Princeton, N.J., U.S.A.)*

ABSTRACT

It appears that it is necessary to look for oscillatory (i.e. overstable) instabilities as the cause of energy transport in sunspots. Making use of recent calculations on the hydromagnetic stability of thermally unstable layers with open boundary conditions, it is found that oscillatory modes can occur in the interior of sunspots. Some possible consequences of these oscillatory modes are discussed.

## 1. Introduction

A long-standing problem in solar physics is the energy balance in a center of activity. This problem is often referred to as the missing energy in a sunspot, i.e.: what has happened to the energy the Sun would have radiated if the sunspot had not been present? This problem is still unsolved and will probably remain so until the mechanism of energy transport in sunspots is understood. For this reason, this paper will largely be concerned with the method of energy transport in a sunspot.

It is often tacitly assumed that some form of convection takes place in sunspots and that the convection motions are the main mechanism of energy transport. Some arguments to support this idea are as follows. First, some fine structure such as the small umbral dots reported by Danielson (1964) and Beckers and Schröter (1967) and the larger umbral granules reported by Bray and Loughhead (1959) suggests the presence of convection. Second, the 'turbulent velocities' observed in sunspots are comparable with those observed in the undisturbed photosphere (Howard, 1958; Elste, 1963; Elsässer and Fricke, 1965; Brückner, 1965) and suggest that some type of convective or oscillatory motion takes place. And thirdly, it is very difficult (if not impossible) to produce a self-consistent model of a sunspot with only radiative transport (Chitre, 1963).

However, hydromagnetic stability calculations have failed to reveal any unstable convective modes in the sunspot if we assume the presence of a vertical magnetic field of about 2500 gauss. In general this is to be expected because $H_{eq}$, the equipartition magnetic field (i.e. the magnetic field for which the magnetic-energy density equals the convective-energy density in the normal convective zone), is generally smaller than the magnetic field in the sunspot.

---

* Presented by R. E. Danielson.

*Kiepenheuer (ed.), Structure and Development of Solar Active Regions, 112–125. © I.A.U.*

As shown in the appendix,

$$H_{eq} \leqslant 6600 \text{ gauss} \left( \rho \left[ \frac{gm}{cm^3} \right] \right)^{1/6} \left( \frac{l}{h} \right)^{1/3}. \tag{1}$$

It is evident from Equation (1) that $H_{eq}$ is quite insensitive to the value of the mixing length which one chooses. The value of $H_{eq}$ computed from Equation (1) is shown in Figure 1 as a function of depth in the normal convection zone. The values of $\rho$ which were used were taken from a table adapted by Böhm (1963) from a model (based on $l/h = 1$) computed by Böhm-Vitense (1958).

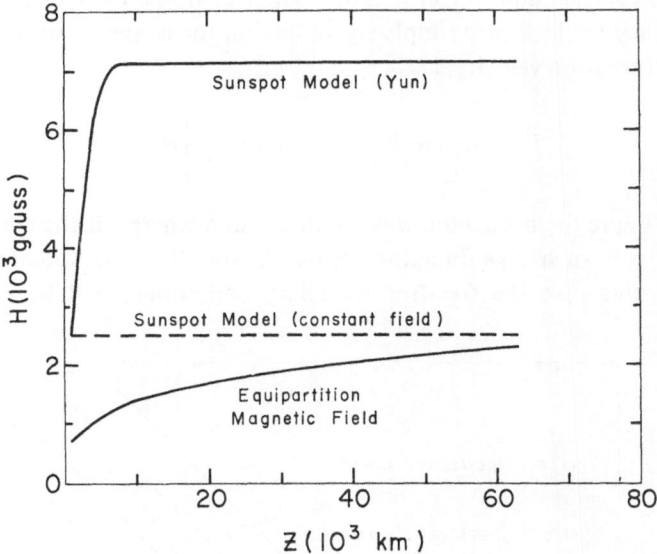

FIG. 1. *The equipartition magnetic field and the magnetic field in a sunspot model of Yun (1967) as a function of depth (z) into the sunspot. Also shown is a constant field (2500 gauss) sunspot model.*

It is clear from Figure 1 that the equipartition magnetic field is several times smaller than the field in a sunspot model computed by Yun (1967) following the methods of Deinzer (1965). Indeed for $l/h = 1$, $H_{eq}$ does not exceed 2500 gauss (the surface field in Yun's model) for any depth in the convection zone. However, if $l/h = 2$, $H_{eq}$ exceeds 2500 gauss at depths in excess of 30000 km, but never exceeds the magnetic fields in Yun's model. The values of $H_{eq}$ calculated by Marik (1966) based on an earlier model of the convection zone (Vitense, 1953) are somewhat larger than given in Figure 1, but they do not exceed the magnetic fields in Yun's model.

Thus, with the possible exception of some very special modes, one would in general not expect any convection modes in sunspots. No such special modes have been found and therefore it is necessary to look for oscillatory (i.e. overstable) instabilities as the cause of energy transport in sunspots. Section 2 of this paper summarizes our present understanding of these oscillatory instabilities, while some possible consequences of them in sunspots are discussed in Section 3.

## 2. Summary of Hydromagnetic Stability Calculations in the Presence of a Vertical Magnetic Field

Early studies of hydromagnetic stabilities in the presence of a vertical magnetic field (Danielson, 1961; Weiss, 1964; for a general summary see Chandrasekhar, 1961) assumed a superadiabatic plane-parallel layer of depth $d$ and employed the Boussinesq approximation (Spiegel and Veronis, 1960). The free-free boundary conditions were used because they led to a very simple eigenfunction for $w$, the vertical (or $z$) component of the perturbation velocity, i.e.

$$w = \left[ \cos k_x x \cos k_y y \sin \frac{\pi z}{d} \right] e^{nt} \qquad (2)$$

where $k_x$ and $k_y$ are the horizontal wave numbers and where $n$ is the complex growth rate. We will call such eigenfunctions 'closed', and the corresponding boundary conditions (in this case the free-free boundary conditions) will be referred to as

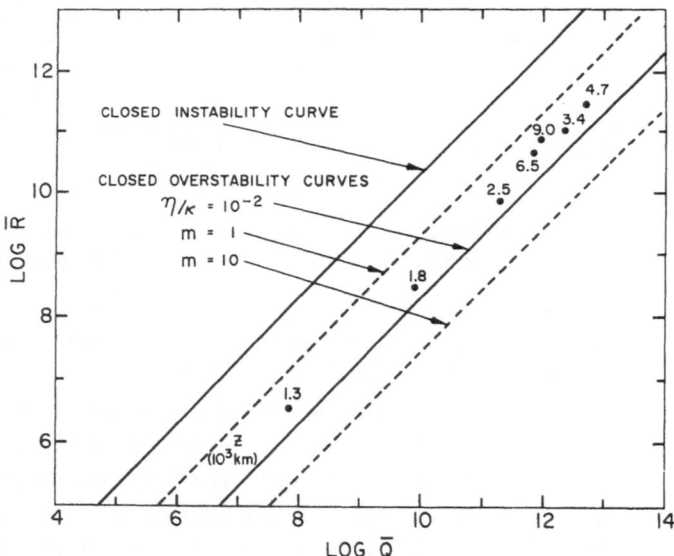

FIG. 2.    *An $\bar{R}$, $\bar{Q}$ stability diagram for a closed fluid layer (no outgoing hydromagnetic waves permitted). Shown are closed instability and overstability curves. The overstability curve marked $\eta/\kappa = 10^{-2}$ includes only resistive heating as a dissipation mechanism. The curves marked $m = 1$ and $10$ include the compressive heating mechanism of Kato (1966) as the means of dissipation. The points are the $\bar{R}$, $\bar{Q}$ coordinates of Yun's (1967) sunspot model at each pressure scale height into the model. The depth ($Z$) into the model at which these quantities were computed is indicated.*

'closed' boundary conditions since they prohibit hydromagnetic wave emission from the superadiabatic layer.

The results of the calculations with the free-free boundary conditions are summarized in Figure 2, where the curves separate regions in the $\bar{R}$, $\bar{Q}$ plane according to the character of $n$. The dimensionless numbers $\bar{R}$ and $\bar{Q}$ are defined by:

$$\bar{R} = \frac{g\alpha\beta d^4}{\kappa^2 \pi^4} \quad \text{and} \quad \bar{Q} = \frac{\mu H^2 d^2}{4\pi^3 \rho \kappa^2},$$

$$(3)$$

where $g$ is the acceleration of gravity, $\alpha \approx 1/T$ is the volume coefficient of expansion, $T$ is the temperature, $d$ is the depth of the superadiabatic layer, $\kappa$ is the radiative diffusivity, $\mu$ is the magnetic permeability, $H$ is the strength of the vertical magnetic field, $\rho$ is the mean density in the layer, and $\beta$ is the superadiabatic gradient defined by:

$$\beta = -\left(\frac{dT}{dz} + \frac{g}{c_P}\right).$$

$$(4)$$

In Equation (4), $(-g/c_P)$ is the adiabatic temperature gradient. The calculations assumed that the cell shape factor,

$$s = (k_x^2 + k_y^2)^{1/2} \frac{d}{\pi}$$

$$(5)$$

was equal to unity and that $\eta/\kappa = 10^{-2}$, where $\eta$ is the magnetic diffusivity. This value of $\eta/\kappa$ is approximately the mean value between $Z = 2000$ and $Z = 9000$ km in Yun's (1967) sunspot model (see Figure 3). The exchange of stabilities curve is not shown because of its minor physical significance (Danielson, 1961).

The $\bar{R}$, $\bar{Q}$ coordinates of Yun's sunspot model at each pressure scale height are also shown in Figure 2. That is, the pressure has increased by a factor of $e$ at the depth represented by each point and the characteristic depth $d$ has been set equal to the local pressure scale height at each point. It is evident that the points lie about a factor of 30 below the closed instability curve. Approximately the same situation occurs for a uniform 2500-gauss field superimposed on the normal convection zone supporting the general conclusion (based on the equipartition magnetic field) that there are no unstable *convective* modes in the sunspot umbra. However, the points do lie somewhat above the overstability curve for joule heating, based on $\eta/\kappa = 10^{-2}$, and therefore the possibility of overstable oscillations is not excluded. The overstability curve for compressive heating (Kato, 1966) is also shown for $m = V_S/V_A = 1$ and $m = 10$, where $V_S$ and $V_A$ are the sound and Alfvén velocities, respectively. The overstability curve for $m = 4$ coincides with the overstability curve for $\eta/\kappa = 10^{-2}$. However, Kato's simplified dispersion relation overestimates the dissipation for $m > 1$ and therefore the true curves lie below the indicated ones. Since $m$ is greater than 10 for $Z$ greater than 7000 km (see Figure 4), compressive heating is probably less important than joule heating as a source of dissipation in the bulk of the sunspot interior.

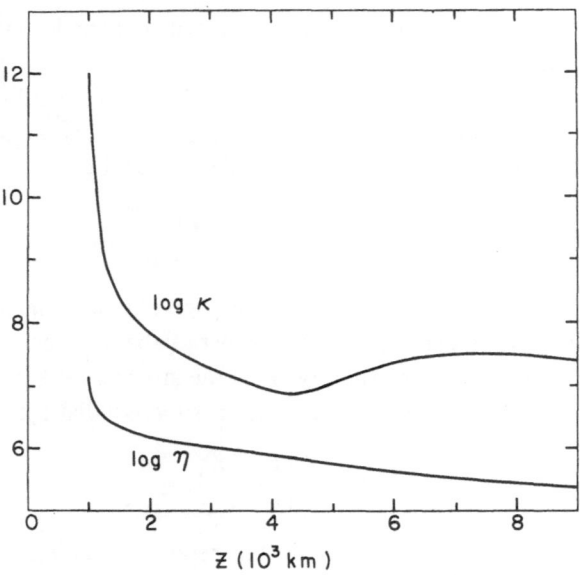

FIG. 3.   *Log $\kappa$ (the radiative diffusivity) and log $\eta$ (the magnetic diffusivity) as a function of depth in Yun's (1967) sunspot model.*

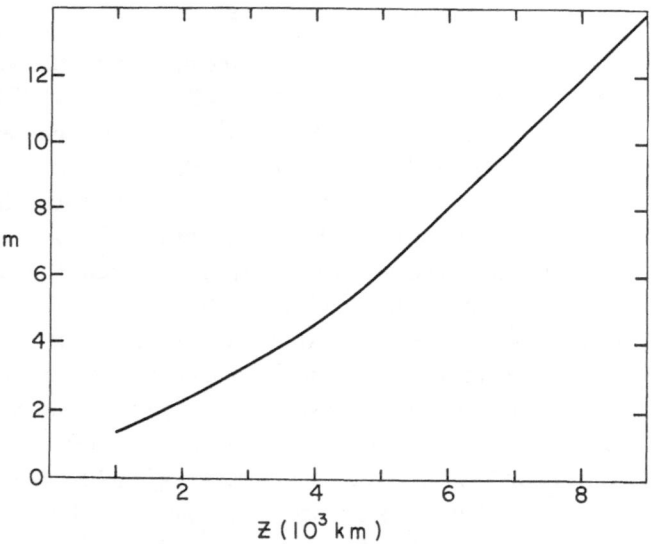

FIG. 4.   *$m = V_S/V_A$ as a function of depth in Yun's (1967) sunspot model.*

However, the closed boudary conditions upon which Figure 2 is based are probably rather unrealistic because they do not allow any energy to be emitted from the layer in the form of hydromagnetic waves. The first calculations with open boundary conditions (Danielson, 1966; Musman, 1967) are shown in Figure 5. Again in the

Boussinesq approximation, free-free boundary conditions were chosen for the bottom of the superadiabatic layer while the boundary conditions at the top of the layer were chosen in such a way as to allow hydromagnetic waves to propagate upwards into a semi-infinite adiabatic region.

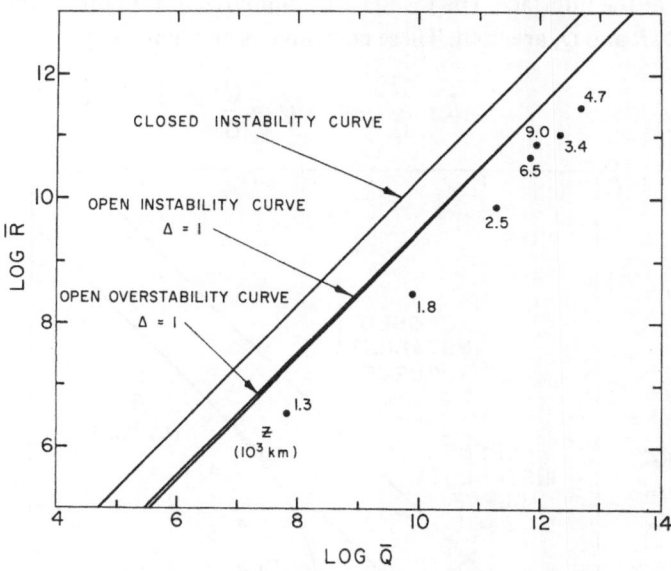

FIG. 5. $\bar{R}, \bar{Q}$ stability diagram for an open fluid layer (upward-propagating hydromagnetic waves are permitted). Shown are the open instability and overstability curves obtained by Musman (1967) for the case of no density discontinuity ($\Delta = \rho_{down}/\rho_{up} = 1$). The $\bar{R}, \bar{Q}$ coordinates of Yun's (1967) sunspot model are also shown.

The open instability curve in Figure 5 is lower than the closed instability curve due to the fact that the open eigenfunction in the superadiabatic layer is approximately sinusoidal with a wavelength equal to $4d$ while the closed eigenfunction has a wavelength of $2d$. The largest change takes place in the open overstability curve because the damping caused by hydromagnetic wave emission is so large that the open overstability curve lies nearly as high in the $\bar{R}, \bar{Q}$ plane as the open instability curve. Furthermore, attempts to reduce the wave emission from the superadiabatic layer by introducing a discontinuity in the hydromagnetic velocity (by means of a density discontinuity between the superadiabatic and adiabatic regions) did not significantly affect the results (Danielson, 1965). The reason for this is that very large density discontinuities are required before a large fraction of the hydromagnetic waves are reflected.

Figure 5 repeats the $\bar{R}, \bar{Q}$ coordinates of Yun's model and it is evident that they lie below the open overstability curve. Thus on the basis of these calculations, one would conclude that the sunspot umbra is stable not only to convective overturning but also to overstable oscillations.

However, a density discontinuity introduces the possibility of surface gravity waves. The effect of these gravity waves was not included in Musman's (1967) calculations, but Savage (1967) has extended the calculations to include the presence of standing gravity waves at the interface. His results are summarized in Figure 6 where a new set of coordinates, $\hat{R}$ and $\hat{Q}$, are used. These coordinates are defined by

$$\hat{R} = \frac{\bar{R}}{G} \quad \text{and} \quad \hat{Q} = \frac{\bar{Q}}{G} \tag{6}$$

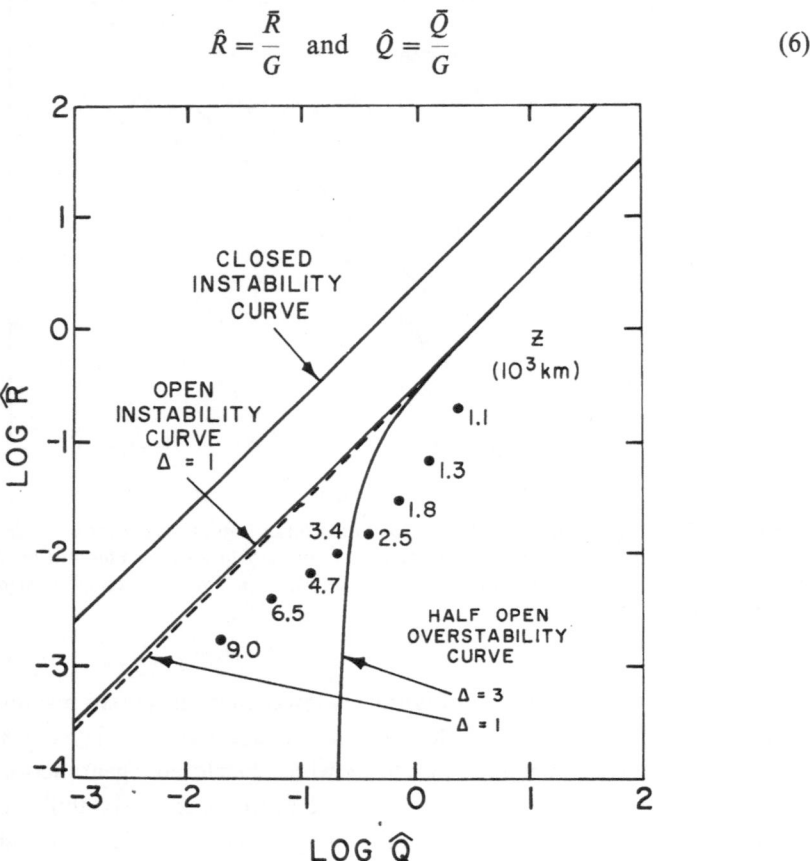

FIG. 6. $\hat{R}$, $\hat{Q}$ *stability diagram including standing gravity waves obtained by Savage (1967). The solid curves are open instability and overstability curves for* $\Delta = \rho_{\text{down}}/\rho_{\text{up}} = 1.1$, *3 and 10 and for* $G = 69.5$ *and* $s = 1$. *The upper short-dashed curve is the closed instability curve. The two long dashed curves are open* $(\Delta = 1)$ *instability and overstability curves. This stability plot is valid for the surface layer of a sunspot.*

where $G$, a dimensionless gravity, is given by

$$G = \frac{g d^3}{\kappa^2 \pi^3}. \tag{7}$$

The $\hat{R}, \hat{Q}$ coordinates eliminate the $\kappa^2$ dependence of both $\bar{R}$ and $\bar{Q}$ and emphasize the effect of gravity waves.

It is evident from Figure 6 (which is based on a typical value of $G$ for the surface layer of a sunspot) that the gravity waves produce a large change in the open over-stability curve for $\hat{Q} < 1$. The physical reason for this large change can be seen with the aid of Figure 7, which shows the fraction of the energy of an incident Alfvén wave

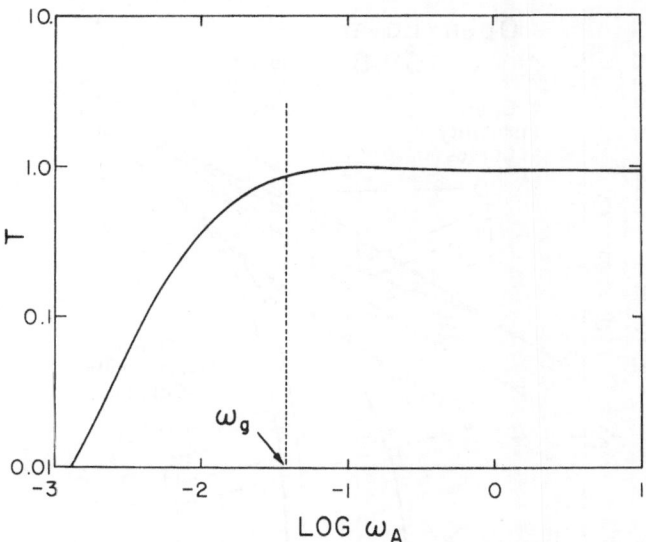

FIG. 7. *The energy flux transmission coefficient of Alfvén waves through a density interface as a function of* $\omega_A$, *the Alfvén wave frequency. We assume the horizontal wave number,* $k_H$ *is* $10^{-7}$ *cm*$^{-1}$, $V_A = 10^6$ *cm/sec (the incident Alfvén velocity) and* $\Delta = 3$. *The vertical dashed line indicates the gravity wave frequency.*

which is transmitted by the interface. For frequencies larger than $\omega_g$, the frequency of the gravity wave oscillations, the transmission is nearly unity. The slight difference from unity is a result of a small amount of wave reflection due to the change in the Alfvén velocity at the interface. For frequencies less than $\omega_g$, however, the amount of transmitted energy is very much smaller. For these frequencies, the gravity waves respond to perturbations at the interface caused by the incident Alfvén wave in a time which is short compared with the period of the Alfvén wave. This has the result that the reflection coefficient increases greatly and correspondingly produces a large decrease in the overstability curve. The reason that gravity waves affect the overstability curves for $\hat{Q} < 1$ can be understood in terms of the above discussion by noting that $\hat{Q}$ can be written as

$$\hat{Q} = \left(\frac{V_A}{V_g}\right)^2 \tag{8}$$

where $V_A$ is the Alfvén velocity and where $V_g$ is the velocity of gravity waves having a wavelength equal to twice the depth of the superadiabatic layer.

Thus for $\hat{Q} < 1$, the gravity waves reduce the energy loss by hydromagnetic wave emission and therefore tend to 'close' the layer. This 'closing' can also be seen in the

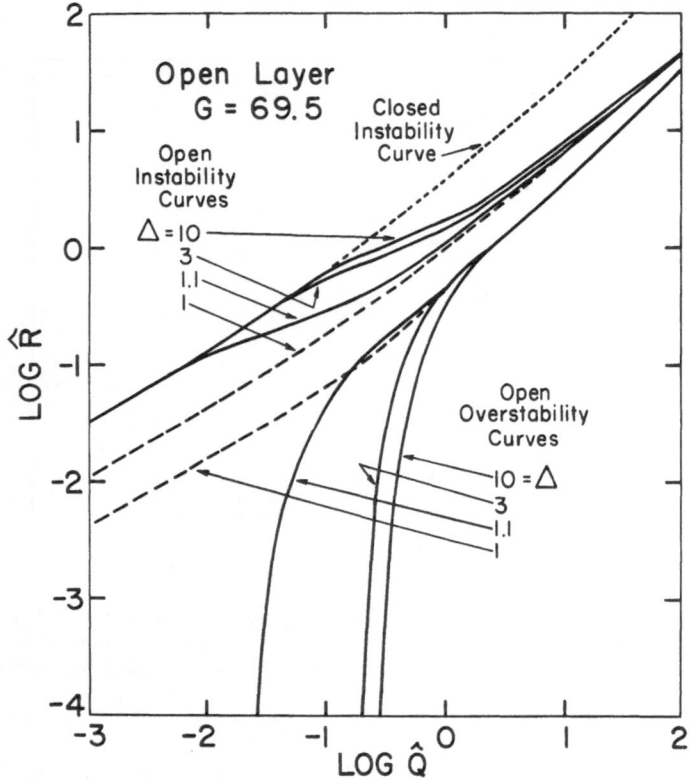

FIG. 8.    An estimated $\hat{R}$, $\hat{Q}$ stability diagram (including gravity waves) which is valid for deep umbral layers $(G \approx 10^{13})$. The $\hat{R}$, $\hat{Q}$ coordinates of Yun's (1967) sunspot model are included.

instability curves shown in Figure 6 where the transition from the open instability curve for $\hat{Q} > 1$ to the closed instability curve for $\hat{Q} < 1$ is evident.

Figure 8 shows an estimated $\hat{R}$, $\hat{Q}$ diagram valid for deep umbral layers $(G \approx 10^{13})$. The open overstability curve is for $\Delta = 3$, which may be the most appropriate curve for our purposes (Savage, 1967). The $\hat{R}$, $\hat{Q}$ coordinates for each pressure scale height in the model computed by Yun (1967) are also shown. One sees that for $Z \gtrsim 3000$ km (about 2000 km below the surface of the sunspot) one is in a regime of overstable oscillations. And if the characteristic depth $d$ for this stability theory is more than one pressure scale height (as may well be the case in a vertical magnetic field) the upper three scale heights may also lie in the overstable region.

It therefore seems that overstable oscillatory modes can occur in the interior of sunspots but that unstable convective modes do not occur. However, one should not forget that the existence of the overstable modes is based on a rather crude hydromagnetic model which, strictly speaking, should only be applied to the surface layer of the sunspot. The situation may alter a great deal when the stability of more realistic sunspot models are investigated. With these reservations, some possible consequences of overstable oscillations are discussed in the next section.

## 3. Some Possible Consequences of Oscillatory Modes in Sunspots

If oscillations occur in the interior of sunspots, one can compute some characteristic periods from the hydromagnetic velocity by choosing the vertical wavelength to be equal to 2 $d$, where $d$ is the characteristic depth in the stability calculations. This corresponds to nearly closed boundaries for the layer under consideration. And if we choose $d$ to be the pressure scale height, we find (as shown in Figure 9) that the predicted period of oscillation varies from about 1 min at the surface of the sunspot

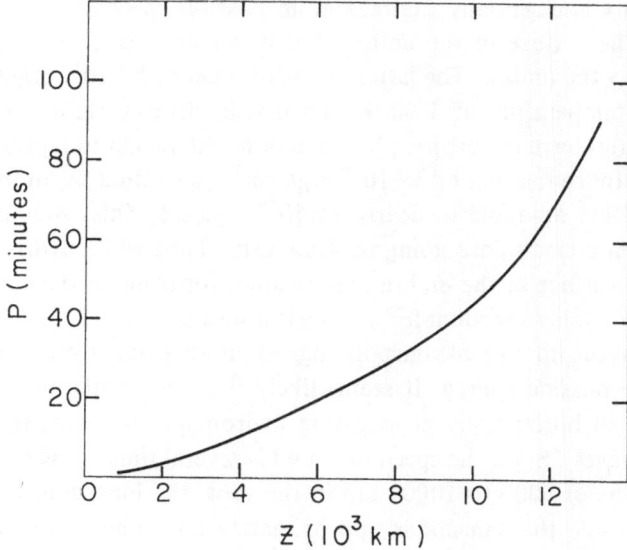

FIG. 9.    *Period of oscillation (in minutes) of umbral layers as a function of depth into the sunspot.*

($Z \approx 1 \times 10^3$ km) to nearly 1 hour at a depth of $10^4$ km ($Z \approx 11 \times 10^3$ km) below the surface. The latter depth is about equal to the umbral diameter for a sunspot having a total area of about $150 \times 10^{-6}$ of the solar hemisphere.

Thus at depths of $5 \times 10^3$ to $10^4$ km beneath the surface of the sunspot, one expects periods of the order of $\frac{1}{2}$ hour. It is interesting to note that the lifetime of the umbra dots is also about $\frac{1}{2}$ hour. This suggests the possibility that the umbral dots are caused by some sort of 'overshoot' from overstable oscillations deep in the interior of the sunspot.

Another possible consequence of oscillations in the interior of a sunspot is that they may play an important role in the missing energy of sunspots. Some of the missing energy is undoubtedly radiated upward at the surface of the sunspot in the form of hydromagnetic waves. Indeed if the observed 'turbulent velocities' in sunspots are interpreted as displacement velocities in a hydromagnetic traveling wave, the resulting non-thermal flux will be

$$F = \tfrac{1}{2} \rho V_{\max}^2 V_A \tag{9}$$

and if we choose the density ($\rho$) at which the maximum turbulent velocities ($V_{\max}$) are measured to be approximately $5 \times 10^{-7}$ gm/cm$^3$, we find that (for $H = 2500$ gauss) the Alfvén velocity ($V_A$) is approximately $10^6$ cm/sec. Then for $V_{\max} = 3 \times 10^5$ cm/sec (r.m.s. turbulent velocity $= 2 \times 10^5$ cm/sec), we find that $F = 2 \times 10^{10}$ ergs/cm$^2$ sec which is a significant fraction of the solar flux in the undisturbed photosphere (i.e. $6 \times 10^{10}$ ergs/cm$^2$ sec).

The above flux is essentially the maximum possible mechanical flux that can be radiated from the surface of the umbra, but it is still a factor of 2.5 less than the missing energy in the umbra. The latter quantity is about $5 \times 10^{10}$ ergs/cm$^2$ sec based on an effective temperature of 3700 °K and it is far from certain that the observed turbulent velocities have the proper phase relation with height to represent an upward traveling wave. Indeed, a flux of $2 \times 10^{10}$ ergs/cm$^2$ sec emitted by an umbra having a diameter of $10^4$ km amounts to nearly $2 \times 10^{28}$ ergs/sec. This amount of energy is sufficient to keep a small flare going continuously! Thus while hydromagnetic wave emission by the surface of the umbra may account for some of the missing energy, it does not seem possible to account for it all by this means.

Oscillations in the interior of sunspots suggest another possible way for accounting for some of the missing energy. It seems likely that the oscillations in the sunspot would give rise to horizontally propagating hydromagnetic waves (i.e. longitudinal magnetosonic waves). Since the speed of sound is several times larger than the Alfvén velocity at depths of 5000 to 10000 km in the spot, the longitudinal magnetosonic wave velocity inside the sunspot is approximately the same as the sound velocity outside the spot. At these depths, the speed of sound is of the order of 20 km/sec and the density is of the order of $5 \times 10^{-5}$ gm/cm$^2$. One may estimate the horizontal flux from an equation similar to Equation (9) and one finds the horizontal fluxes equal to $5 \times 10^{11}$ ergs/cm$^2$ sec (nearly 10 times the normal photospheric flux) for displacement velocities of 1 km/sec. And for this flux, the total energy emitted by a cylindrical surface having a diameter of $10^4$ km and a depth of $5 \times 10^3$ km is nearly $10^{30}$ ergs/sec.

This is approximately equal to the missing energy of an entire sunspot having a penumbral diameter of $2.5 \times 10^4$ km (and an umbral diameter of $10^4$ km) if one takes $\frac{1}{3}$ of the photospheric flux over the entire sunspot area to be the missing energy.

The numbers in the above paragraph are very speculative, but they may indicate one way to account for the missing energy in sunspots. And if this be the case, the lack of a substantial bright ring near sunspots could be understood if the horizontally propagating waves travel several times the penumbral radius of the sunspot before dissipating. Also some energy may be propagated away from the sunspot by means of surface gravity waves.

A third implication of oscillatory modes in the sunspot interior is that they may lend some support to the use of a reduced mixing length by Deinzer (1965) and Yun (1967) in computing models of the sunspot interior. In the absence of a magnetic field, the reason for using a mixing length (equal to the local-pressure scale height, for example) is that an upward-moving element goes roughly one scale height before stopping and giving up its excess thermal energy to its surroundings. In the case of oscillations in the presence of a vertical magnetic field, however, the upward-moving elements do not stop and mix with the surroundings, but reverse their motion instead. Thus the moving elements have a limited amount of time in which to exchange energy with respect to their surroundings and therefore the use of a mixing length which is only a fraction of a scale height seems like a plausible first approximation to estimating the non-radiative energy transport in a sunspot.

In the event that horizontally propagating hydromagnetic waves are emitted in the interior of the sunspot, a modification to the procedure used by Deinzer (1965) and by Yun (1967) is suggested. Namely, the total flux is not constant in the sunspot but decreases with height as energy is radiated horizontally. Thus, the flux should be varied with the depth in future models. One consequence of this is that the larger flux in the deeper portions of a sunspot would produce a larger superadiabatic gradient and, therefore, a greater instability to oscillations.

## Appendix

In mixing length theory, the convective flux $F_c$ is given by (Böhm-Vitense, 1958)

$$F_c = \tfrac{1}{2} C_p \rho T V \left(\frac{l}{h}\right) (\nabla - \nabla'),$$

where $C_p$ is the heat capacity per gram, $\rho$ is the density, $T$ is the temperature, $V$ is the mean velocity of a 'turbulent element', $l$ is Prandtl's mixing length, $h$ is the pressure scale height, $\nabla = (\mathrm{d}(\ln T))/(\mathrm{d}(\ln P))$, where $p$ is the gas pressure, and where $\nabla'$ is the same quantity in a 'turbulent element'. Similarly (Böhm-Vitense, 1958),

$$V^2 = \frac{RT}{8M} Q \left(\frac{l}{h}\right)^2 (\nabla - \nabla'),$$

where $R$ is the gas constant, where $M$ is the mean molecular weight, and where the parameter $Q = 1 - (\partial \ln M)/(\partial \ln T)$.

Eliminating $(\nabla - \nabla')$ from the above two equations and defining the equipartition magnetic field $H_{eq}$ by

$$\frac{\mu H_{eq}^2}{8\pi} = \tfrac{1}{2}\rho V^2$$

one obtains

$$H_{eq} = \left(\frac{4\pi}{\mu}\right)^{1/2} \rho^{1/6} Q^{1/3} \left(\frac{F_c}{4}\right)^{1/3} \left(\frac{R}{MC_p}\right)^{1/3} \left(\frac{l}{h}\right)^{1/3},$$

where $\mu$ is the magnetic permeability.

Since $F_c \leqslant F$, where $F$ is the total flux, and since $R/MC_p \leqslant 0.4$ (the value for a monatomic gas), one obtains

$$H_{eq} \leqslant 6600 \, \text{gauss} \left(\rho \left[\frac{\text{gm}}{\text{cm}^3}\right]\right)^{1/6} \left(\frac{l}{h}\right)^{1/3} Q^{1/3}.$$

The parameter $Q$ differs from unity only in the upper few thousand kilometers of the normal convection zone (where it takes on values of the order of 2). Therefore $Q$ may be set equal to unity for the purposes of this paper and

$$H_{eq} \leqslant 6600 \, \text{gauss} \left(\rho \left[\frac{\text{gm}}{\text{cm}^3}\right]\right)^{1/6} \left(\frac{l}{h}\right)^{1/3}.$$

## Acknowledgements

We wish to express our indebtedness to Mr. H. S. Yun of the University of Indiana for the use of his sunspot models prior to publication.

This paper is part of Project Stratoscope of Princeton University sponsored by NSF, ONR, and NASA.

## References

Beckers, J., Schröter, E. (1967)     private communication.
Böhm, K. H. (1963)     *Astrophys. J.*, **137**, 881.
Böhm-Vitense, E. (1958)     *Z. Astrophys.*, **46**, 108.
Bray, R., Loughhead, R. (1959)     *Austr. J. Phys.*, **12**, 320.
Brückner, G. (1965)     in *I.A.U. Symposium No. 22: Solar and Stellar Magnetic Fields*, Ed. by R. Lüst, North-Holland Publ. Co., Amsterdam, p. 293.
Chandrasekhar, S. (1961)     *Hydrodynamic and Hydromagnetic Stability*, Oxford University Press, London.
Chitre, S. (1963)     *Mon. Not. R. astr. Soc.*, **126**, 431.
Danielson, R. E. (1961)     *Astrophys. J.*, **134**, 289.
Danielson, R. E. (1964)     *Astrophys. J.*, **139**, 45.
Danielson, R. E. (1965)     in *I.A.U. Symposium No. 22: Solar and Stellar Magnetic Fields*, Ed. by R. Lüst, North-Holland Publ. Co., Amsterdam, p. 315.
Danielson, R. E. (1966)     in *Convegno Sulle Machie Solare*, Ed. by Barbera (Firenze).

Deinzer, W. (1965)        *Astrophys. J.*, **141**, 548.
Elsässer, H., Fricke, K. (1965)        in *I.A.U. Symposium No. 22: Solar and Stellar Magnetic Fields*, Ed. by R. Lüst, North-Holland Publ. Co., Amsterdam, p. 297.
Elste, G. (1963)        *J. Quant. Spect. Rad. Trans.*, **3**, 185.
Howard, R. (1958)        *Astrophys. J.*, **127**, 108.
Kato, S. (1966)        *P.A.S.J.*, **18**, 201.
Marik, M. (1966)        *Soviet Astr. – A.J.*, **10**, 315.
Musman, S. (1967)        *Astrophys. J.*, **149**, 201.
Savage, B.D. (1967)        Ph.D. Thesis, Princeton University.
Spiegel, E.A., Veronis, G. (1960)        *Astrophys. J.*, **131**, 442.
Vitense, E. (1953)        *Z. Astrophys.*, **32**, 135.
Weiss, N.O. (1964)        *Proc. Roy. Soc.*, **A256**, 99.
Yun, H.S. (1967)        private communication.

# DISCUSSION

*Severny:* Should there exist some phase shift in oscillations at different levels inside a sunspot?

*Danielson:* Yes, oscillations at different levels inside the sunspot will be coupled to some extent since they are traversed by the same magnetic field. However, the extent of this coupling will not be known until the actual modes in Deinzer's model are computed.

*Maltby:* Since you have mentioned the transport of energy by sound waves in the horizontal direction, I would like to draw attention to a paper by G. Eriksen and myself (to be published in *Solar Physics*) where the Evershed effect is explained as a wave phenomenon and the effect of energy transport is mentioned.

*Sturrock:* Sunspots normally begin as a small dark pore, giving the impression that, when the magnetic-field strength grows to a certain value, the normal convection pattern is suppressed and some other pattern is set up. Can your theory predict the field strength at which a pore would form?

*Danielson:* Our stability model predicts that the normal convection pattern begins to be suppressed at the instability curve. The magnetic field corresponding to the half open instability curve is of the order of 800 gauss, but this value should not be taken too seriously since our stability model is rather crude.

*Elske Smith:* Some very tentative results of my study of vertical velocities in the umbra of sunspots may have some bearing on Danielson's model. I have found downward velocities, relative to the photosphere, of the order of 0.3 km/sec in certain photospheric lines ($\lambda$ 5123, 5434, and 5576). There is an indication that these velocities show a time variation with a period of about 5 min. Hence the actual velocities range from almost zero to 0.6 km/sec or even higher, though almost always downward. Weaker lines, like $\lambda$ 5436 (Ni I), arising from greater depths in the umbra, have smaller velocities, on the average of the order of 0.1 km/sec.

# ON THE MAGNETIC STRUCTURE OF AN ACTIVE REGION*

F. A. ERMAKOV, E. I. MOGILEVSKY, and B. D. SHELTING

*(Institute of Terrestrial Magnetism, Ionosphere and Radio-Wave Propagation,
Academy of Sciences, Moscow, U.S.S.R.)*

## ABSTRACT

(1) A scheme of magnetic-field generation in an active region within the subphotospheric convective zone is discussed. Magnetic-field amplification is considered to result from the interaction of the local cyclonic motion and the turbulent motion of the subphotospheric plasma.

(2) A model of the magnetoplasma of an active region is considered consisting of a current-free magnetic field, whose lines of force stretch deeply under the photosphere in the form of subgranules (force-free fine-scale plasma elements). The explanation of some peculiarities in the development of an active region with the considered structure is discussed.

(3) Some peculiarities of the general structure of the magnetic field and the motion within photosphere and chromosphere obtained by the analysis of observations with the IZMIRAN two-channel magnetograph is discussed.

## DISCUSSION

*H. U. Schmidt:* What size do you have in mind for the subgranules? Why did you choose this size?

*Mogilevsky:* The calculated size of the subgranules is $\sim 10^7$ cm. This estimation of the high limit of the subgranule middle size was obtained from the equation of the integral magnetic flux over the whole active region by present scheme 'magnetisation' of the magnetoplasma.

* A detailed paper will be submitted to *Solar Physics*.

# OSCILLATORY CONVECTION IN STRONG MAGNETIC
# FIELDS AND ORIGIN OF ACTIVE REGIONS

S. I. SYROVATSKY          and          Y. D. ZHUGZHDA
*(Physical Institute of Lebedev,*          *(Institute of Earth Magnetism,*
*Moscow, U.S.S.R.)*          *Ionosphere and Propagation*
*of Radio-Waves,*
*Academy of Sciences, Moscow, U.S.S.R.)*

ABSTRACT

The convection in a compressible inhomogeneous conducting fluid in the presence of a vertical uniform magnetic field has been studied. It is shown that a new mode of oscillatory convection occurs, which exists in arbitrarily strong magnetic fields. The convective cells are stretched along the magnetic field, their horizontal dimensions are determined by radiative cooling. Criteria for convective instability in a polytropic atmosphere are obtained for various boundary conditions in the case when the Alfvén velocity is higher compared with the velocity of sound.

The role of oscillatory convection in the origin of sunspots and active regions is discussed.

A number of problems of solar physics are connected with the effect of the magnetic field on the convective transfer of heat. The main question is whether the magnetic field suppresses the convection in vertical field or the convective heat transport has place in the strong magnetic field too, because of an existence of some other type of convective motions. Both theoretical investigations of sunspots structure (Chitre, 1963; Deinzer, 1965), as well as observations of granulation in umbra (Danielson, 1964; Bray and Loughhead, 1964) show that the convection is not suppressed in spots.

The fact that the usually considered circulatory convection is suppressed in sunspots can be verified by using the criterion of stability, obtained in the most general form by Gough and Tayler (1966). It follows then that non-circulatory convection must occur in the strong magnetic field of spots.

Oscillatory convection in sunspots was investigated by Danielson (1965) and Danielson and Savage (1968) in the frame of the theory of convection in an incompressible fluid (Chandrasekhar, 1961). Results of Chandrasekhar's theory can be used for astrophysical objects only if the unstable layer is thin and can be approximated by the layer with constant density. But the solar convective zone is thick and we need therefore a theory of convection in non-uniform, compressible atmosphere.

The aim of the present investigation is to search for convection mode, which is not suppressed by any magnetic field. This problem was initiated in the paper of Syrovatskii and Zhugzhda (1967), and the properties of this convection are investigated by Zhugzhda in more detail (1968).

*Kiepenheuer (ed.), Structure and Development of Solar Active Regions,* 127–130. © *I.A.U.*

We consider convection in a compressible inhomogeneous atmosphere in the presence of a vertical magnetic field. The general linearized equations of magneto-hydrodynamics for a polytropic atmosphere can be simplified in the following way:

It is evident that a strong vertical magnetic field will not prevent oscillatory motion along lines of force. In a uniform medium the motion of this type corresponds to a slow magneto-acoustic wave, which propagates almost transversally to the field. We assume that oscillatory motions along the field in non-uniform medium also have the properties of a slow magneto-acoustic wave. Besides the equations can be simplified on the condition that the Alfvén velocity is large as compared to the acoustic one and convective cells are stretched along force lines. It appears that the obtained equation describes acoustic oscillations of gas along magnetic force lines in a non-uniform medium. The following equation was obtained:

$$z^2 \frac{d^2 w}{dz^2} + \left(n + 1 - \frac{q}{\gamma\sigma + q}\right) z \frac{dw}{dz} - \left(\frac{\sigma^2 (n+1)(\sigma+q)}{g(\gamma\sigma + q)} z + \frac{nq}{\gamma\sigma + q}\right) w = 0,$$

where $w$ = vertical component of velocity, $z$ = depth in atmosphere, $n = (\mu g/R\beta) - 1$ is polytropic index, $\mu$ = molecular weight, $g$ = acceleration due to gravity, $R$ = gas constant, $\beta$ = constant temperature gradient ($T_0 = \beta z$), $q$ = inverse time for radiative cooling (see Spiegel, 1957), $\sigma = i\omega + \varkappa$ = complex frequency, $\omega$ = frequency of convective oscillation, $\varkappa$ = increment of linear theory, and $\gamma$ = ratio of specific heats.

The equation solution is expressed by Bessel's functions.

This oscillatory motion in the strong magnetic field is unaffected by field strength. If heat transfer does not occur between adjacent cells ($q = 0$), the oscillatory motions are adiabatic and undamped. Oscillatory convective motions grow in time provided there is heat transfer between adjacent convective cells ($q \neq 0$). If $q$ does not depend

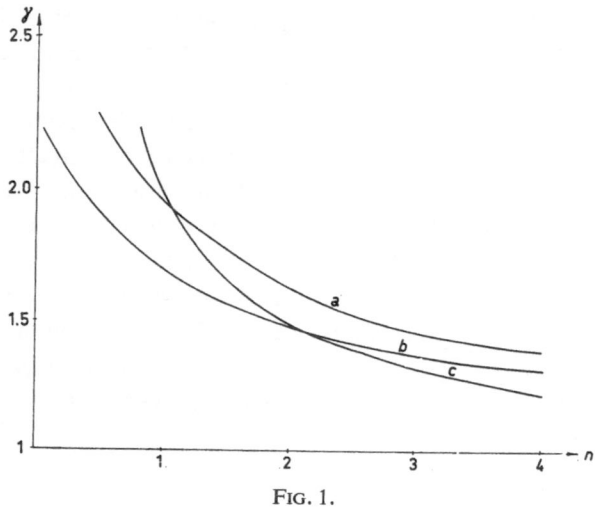

FIG. 1.

on the depth of cells ($q$=const) the position of instability-region boundary for fixed value of $q$ depends only on ratio of specific heats and polytropic index $n$.

The instability criterion ($\varkappa=0$) depends in general on the inverse time for radiative cooling to oscillatory frequency ratio. Consequently, in the case of optically thick cells the instability criterion depends on the cell size. There are two limiting cases. If convective cells are thick and the inverse time for radiative cooling is small compared with angular frequency ($q\ll\omega$) the convective oscillations are quasi-adiabatic. If the convective cells are thin and the inverse time for radiative cooling is large compared with angular frequency, the convective oscillations are quasi-isothermal ($q\gg\omega$).

The slide (Figure 1) shows criteria of instability (for rigid boundary conditions) for these limiting cases in the ($\gamma$, $n$) diagram. Let us consider an atmosphere which consists of gas with definite molecular weight and definite ratio of specific heats. If the temperature gradient increases, we move to the left along the line parallel to the abscissa axis. When we cross the curve labelled $a$, the oscillatory convection with very thin cells arises. When we cross the curve labelled $b$, the oscillatory convection with very thick cells arises. Instability boundaries for intermediate size of cells are placed between these curves. What size of cells must be realized? It can be shown that viscosity is not of importance here. The maximum increment corresponds to such size of cells at which the inverse time for radiative cooling appears to be of order of oscillation frequency. Simultaneously this size of cells corresponds to maximum heat flux. For this case the instability boundary is located approximately halfway between the curves labelled $a$ and $b$.

We have considered the case when the inverse time of cooling is unaffected by temperature and gas pressure. In this case the solution of the obtained equation is expressed by analytical functions. It is given also another method of obtaining criteria. This method consists of the energy-balance calculation. The criterion for atmosphere with an arbitrary dependence of inverse cooling time on temperature can be obtained by this method.

The criterion is strongly dependent on boundary conditions. With this in mind, the effect of the motion penetration into stable layers on instability criteria was investigated. This investigation permits to obtain criteria of instability for real astrophysical conditions. These criteria are affected by the structure of layers lying above and below the unstable layer. The main result is that the obtained criteria of instability are the same or more strict in comparison with the criteria demonstrated in Figure 1.

The heat transport is an important property of oscillatory convection. The flux of heat is equal to

$$I = \frac{\gamma\lambda K_\perp^2}{2\omega} T'v\xi,$$

where $\lambda$=thermal conductivity, $K_\perp^2$=square of horizontal wave vector, $T'$ and

$v$ = amplitudes of temperature and velocity oscillations, $\xi$ = factor which depends on phase differences of these oscillations.

There is also non-thermal flux of energy. This non-thermal flux transports the energy within unstable layers and out into adjacent stable layers, and thus changes an initial polytropic atmosphere.

The curve (Figure 1) labelled $c$ is Schwarzschild's criterion for non-conducting fluid. The origin of sunspots may be treated considering the fact that in atmosphere with ratio of specific heats around 5/3, the oscillatory convection arises at a larger temperature gradient compared with the adiabatic one. Consequently, the temperature gradient in sunspots is larger compared with the temperature gradient in convection zone. More careful consideration of spot origin was undertaken, which takes into account the penetration of motions into stable layers; some other factors were also taken into account. It is interesting that in the atmosphere with ratio of specific heats around unity, the oscillatory convection arises at a smaller temperature gradient compared with the adiabatic one. This may be a cause of a hotter region appearance. But the complete consideration of this problem demands a solution of the same problem for the more general case of intermediate values of magnetic-field strength, which has not yet been done.

## References

Bray, R., Loughhead, R. (1964)      *Sunspots*, Chapman and Hall Ltd., London.
Chandrasekhar, S. (1961)      *Hydrodynamic and Hydromagnetic Stability.*
Chitre, S. M. (1963)      *Mon. Not. R. astr. Soc.*, **126**, 431.
Danielson, R. E. (1964)      *Astrophys. J.*, **139**, 45.
Danielson, R. E. (1965)      in *I.A.U. Symposium No. 22*, Ed. by R. Lüst, North-Holland Publ. Co., Amsterdam, p. 314.
Danielson, R. E., Savage, B. D. (1968)      the present volume, p. 112.
Deinzer, W. (1965)      *Astrophys. J.*, **141**, 548.
Gough, D. O., Tayler, R. J. (1966)      *Mon. Not. R. astr. Soc.*, **133**, 85.
Spiegel, E. A. (1957)      *Astrophys. J.*, **126**, 202.
Syrovatskii, S. I., Zhugzhda, Y. D. (1967)      *Astr. Zu.*, **44**, 1180.
Zhugzhda, Y. D. (1968)      *Astr. Zu.*, in press.

# QUANTITATIVE ESTIMATIONS OF THE
# ANOMALOUS PLASMA DIFFUSION
# IN AN ACTIVE REGION*

M. KOPECKÝ** and G. V. KUKLIN

*(Siberian Institute of Terrestrial Magnetism, Ionosphere and Radio Propagation, Academy of Sciences, Irkutsk, U.S.S.R.)*

In some recent papers the interdependence of the gas and magnetic-field motions in the solar atmosphere was considered. Some results indicate the occurrence of gas motion along the magnetic-field lines combined with motion of the field line, but sometimes we have to assume an obvious gas motion across the magnetic-field lines. As one of the possible mechanisms explaining this fact the anomalous plasma diffusion may be proposed.

In this paper we attempt to estimate the Böhm-diffusion effect. We are inconsistent in a certain sense using the Böhm-diffusion theory for a fully ionized plasma (Galeev *et al.*, 1963) in the case of a partly ionized plasma. In an unbounded medium the Bohm-diffusion coefficient has a maximal value

$$D^B_{\perp M} = \frac{ckT}{2\pi eH} = \frac{6.91 \times 10^6}{\vartheta H} \tag{1}$$

and in a system with longitudinal and transversal sizes $L$ and $\rho L$ correspondingly it is

$$D^B_{\perp} = \frac{ckT}{2\pi eH} \frac{H_*}{H} \tag{1'}$$

where

$$H_* = \frac{c}{e}\left(\frac{v_{ei}m_i m_e kT_e}{\rho^4 L^2}\right)^{1/3} = 4.09 \times 10^5 \left(\frac{\mu_i r^2_{ei} P_e}{\rho^4 L^2 \sqrt{\vartheta}}\right)^{1/3}. \tag{2}$$

The Böhm diffusion is effective when it is not masked by the classical diffusion

$$D^B_{\perp M} > D^C_{\perp}. \tag{3}$$

Taking into account both the low ionization degree plasma and the almost fully ionized one we must write down in a rough approximation

$$D^C_{\perp} \approx D^A_{\perp} + D^L_{\perp}, \tag{4}$$

* Presented by G.V. Kuklin.
** On leave from the Astronomical Institute of the Czechoslovak Academy of Sciences, Ondřejov.

*Kiepenheuer (ed.), Structure and Development of Solar Active Regions*, 131–133. © *I.A.U.*

where $D_\perp^A$ is the ambipolar diffusion coefficient in the fully ionized plasma, and $D_\perp^L$ is the classical diffusion coefficient in the low ionized plasma. Then a critical value of the magnetic field strength $H_0$, when the relation (3) is correct, is

$$H_0 \geqslant 1.90 \times 10^{14} P_e r_{ei}^2 \sqrt{\vartheta} \left\{ 1 + 21.3 \frac{P_n r_{in}^2}{P_e r_{ei}^2} \sqrt{\frac{\mu_i \mu_n}{\mu_i + \mu_n}} \right\}. \qquad (5)$$

Using the data given in Appendix C of the paper by Zwaan (1965) we have computed $\log H_0$ in gausses (Table 1), $\log H_*(\rho^4 L^2)^{1/3}$ (Table 2), and $\log D_{\perp M}^B$ (Table 3) for some values of $\vartheta$ and $P_g$. We have taken $r_{in}^2 = 10^{-15}$ cm$^2$ and the values $D_\perp^B$ Max are computed for $H = H_0$ according to Equation (1). All designations are the same as in our previous papers (Kuklin, 1966; Kopecký and Kuklin, 1966, 1967).

### Table 1

$\log H_0$

| $\log P_g/\vartheta$ | 0.8 | 1.1 | 1.4 | 1.7 |
|---|---|---|---|---|
| 3.0 | 3.50 | 3.70 | 3.75 | 3.79 |
| 4.5 | 5.04 | 5.20 | 5.25 | 5.29 |
| 6.0 | 6.60 | 6.70 | 6.75 | 6.79 |

### Table 2

$\log H_*(\rho^4 L^2)^{1/3}$

| $\log P_g/\vartheta$ | 0.8 | 1.1 | 1.4 | 1.7 |
|---|---|---|---|---|
| 3.0 | 1.58 | 1.50 | 1.39 | 1.23 |
| 4.5 | 1.90 | 1.91 | 1.73 | 1.62 |
| 6.0 | 2.30 | 2.25 | 2.10 | 1.94 |

### Table 3

$\log D_\perp^B$

| $\log P_g/\vartheta$ | 0.8 | 1.1 | 1.4 | 1.7 |
|---|---|---|---|---|
| 3.0 | 3.44 | 3.10 | 2.94 | 2.82 |
| 4.5 | 1.90 | 1.60 | 1.44 | 1.32 |
| 6.0 | 0.34 | 0.10 | $-0.06$ | $-0.18$ |

Looking through the tables we can be convinced that within the physical parameter range of the sunspotss the photosphere, the low chromosphere, and the faculae, the Böhm-diffusion effect is negligible for the macroscopic regions. It is caused firstly by too high critical values of $H_0$, which seldom occur in reality; secondly by too low values of $D_\perp^B$, which can not explain usually observed discrepancies; and thirdly by too low values of $H_*(\rho^4 L^2)^{1/3}$. One can see that the filament in the sunspot penumbra with $\rho \approx 10^{-3}$ and $H_* \sim 1$ gauss cannot be longer than tens of kilometers. So the Böhm diffusion would have some meaning for systems of such sizes only, but it does not exclude the possibility for other anomalous diffusion types to be essential.

## References

Galeev, A. A., Moiseev, S. S., Sagdeev, R. Z. (1963)     Preprint, Novosibirsk.
Kopecký, M., Kuklin, G. V. (1966)     *Bull. astr. Inst. Csl.*, **17**, 45.
Kopecký, M., Kuklin, G. V. (1968)     *Solar Phys.*, **4**, in press.
Kuklin, G. V. (1966)     Resultaty nabl. i issled. v period. MGSS, **1**, 17.
Zwaan, C. (1965)     *Rech. astr. Obs. Utrecht*, **17**, 4.

# ÉTUDE CRITIQUE D'UN CHAMP 'CURRENT-FREE' DANS L'ATMOSPHÈRE SOLAIRE*

J. RAYROLE et M. SEMEL

*(Observatoire de Meudon (S. & O.), France)*

## ABSTRACT

To test the validity of the assumption of a 'current-free' magnetic field in the atmosphere above sunspots, we measured all three parameters of the magnetic field in an active region and compared it with the calculated current-free field according to the solution proposed by Schmidt. We found important differences between the calculated and the measured field.

We refer to an earlier paper where the objections to the 'current-free field' were that the calculated field has too simple a configuration to account for the fine structure observed in Hα and other solar observations.

Les observations ne fournissent qu'une information partielle sur la configuration du champ magnétique dans les centres actifs. Le paramètre le plus facile à mesurer est la composante du champ le long de la ligne de visée. De très grandes difficultés surviennent si l'on veut mesurer les trois composantes du champ au niveau de la photosphère et le problème devient presque insoluble au-dessus de la photosphère.

Plusieurs auteurs ont proposé des configurations du champ basées sur diverses hypothèses. Schatzman (1961) a calculé un modèle théorique de champ magnétique à force de Lorentz nulle, ayant une symétrie axiale. Oster (1964) a suggéré que le champ d'une tache unique peut être représenté par celui d'une boucle de courant située sous la surface photosphérique et centrée sur la tache visible. Godovnikov et Smirnova (1965) ont calculé le champ des taches solaires à partir d'un système de dipoles. Schmidt (1964) a montré comment on peut calculer le champ dans une atmosphère sans courant. Dans ce cas le champ dérive d'un potentiel scalaire qui peut être entièrement et univoquement déterminé à partir de la composante du champ perpendiculaire à une surface plane (photosphère).

Le but de ce travail est de discuter la configuration du champ 'current-free' obtenu avec le méthode proposée par Schmidt. Une première tentative a été faite pour un centre actif à son passage au centre du disque le 12 juin 1963 (Semel, 1967). Seule la composante longitudinale du champ était connue et il ne fut pas possible d'établir un test quantitatif de l'approximation 'current-free' puisque le champ calculé satisfait à cette hypothèse d'une part et aux observations de la composante longitudinale d'autre

---

* Présenté par J. Rayrole.

*Kiepenheuer (ed.), Structure and Development of Solar Active Regions*, 134–141. © I.A.U.

part. Une étude qualitative a pu cependant souligner quelques particularités non favorables à l'hypothèse faite:

(1) Contrairement à la composante longitudinale, le champ calculé présente des régions de champ transversal pur en dehors des facules photosphériques. Or l'observation d'un champ purement transversal n'est qu'une question de perspective et la corrélation champ-facules n'est pas affectée par une variation centre-bord.

(2) La configuration du champ calculé tend vers une simplification importante au fur et à mesure que l'on s'éloigne du plan où le champ a été mesuré. Déjà à l'altitude de 5000 km on ne saurait reconnaître la trace d'une structure fine et complexe généralement observée sur le champ magnétique au niveau de la photosphère, sur les spectrohéliogrammes Hα ou K et même sur les observations de la couronne. Le champ magnétique est certainement la cause des structures observées; leurs complexités suggèrent alors une structure complexe du champ et par suite la présence de courant électrique ayant une structure également complexe.

Cette première tentative nous a incités à pousser l'étude plus à fond et pour cela à comparer directement le champ 'current-free' calculé à la structure réelle du champ magnétique du centre actif observé. Cette comparaison a été faite pour le centre

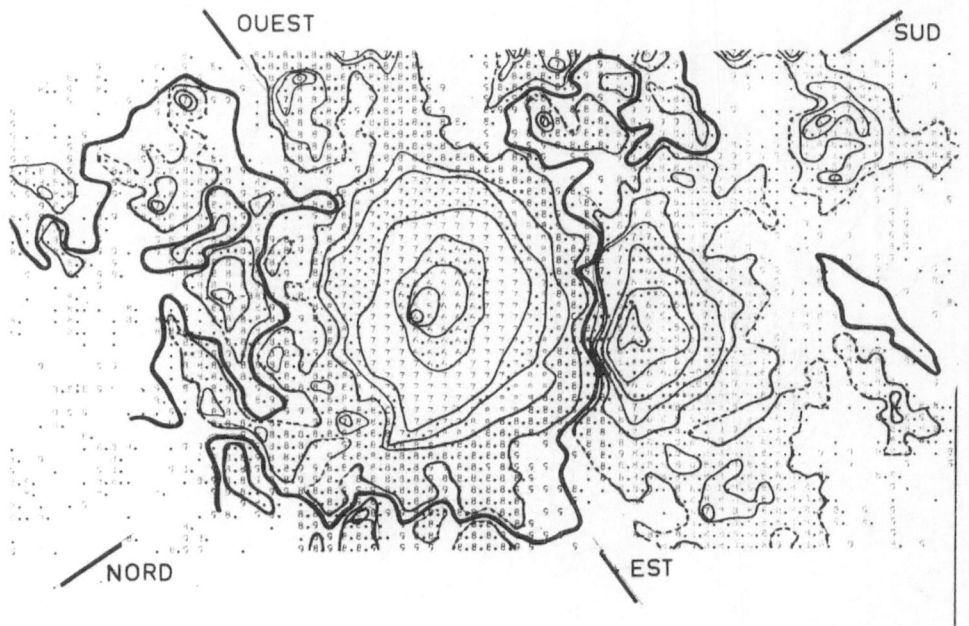

FIG. 1.   *Inclinaison mesurée du champ par rapport à la verticale. Les angles sont comptés de 0 à 90°*
*pour les deux polarités. Le gros trait noir indique les changements de polarité.*

| Symbole | 0 | 1 | 2 | 3 | 4 | 5 | 6 | 6. | 7 | 7. | 8 | 8. | 9 |
|---|---|---|---|---|---|---|---|---|---|---|---|---|---|
| $\psi \geqslant$ | 0 | 10 | 20 | 30 | 40 | 50 | 60 | 65 | 70 | 75 | 80 | 85 | =90 |

actif au centre du disque le 19 septembre 1966 et pour lequel nous avions les observations nécessaires pour mesurer tous les paramètres du champ. L'observation
nous fournit la composante longitudinale, l'intensité du champ total, l'angle $\psi$ du
champ avec la ligne de visée et l'orientation de la composante transversale (Rayrole,
1967). La composante longitudinale, ici normale à la surface solaire, est utilisée pour
obtenir le potentiel qui nous permet de calculer tous les paramètres du champ 'current-
free'. Pour effectuer la comparaison entre l'observation et le calcul, nous avons choisi
comme paramètres l'angle $\psi$ avec la ligne de visée et l'orientation de la composante
transversale. Ces deux paramètres sont en effet obtenus directement à partir des
observations, et leur détermination ne dépend que des erreurs de mesure du paramètre qui sert à les déterminer. D'autre part à l'intérieur des taches ils sont très peu
influencés par la présence de lumière photosphérique diffusée et par les variations
possibles des paramètres définissant le profil de la raie en absence de champ (Rayrole,
1967).

Les Figures 1 et 2 représentent respectivement les valeurs mesurées et calculées de
l'angle $\psi$ avec la ligne de visée. Nous ne considérerons que les résultats relatifs à
l'intérieur des taches, car plus le champ est fort, plus les valeurs mesurées et calculées
sont précises. Le champ calculé est en effet très sensible aux erreurs de mesure de la

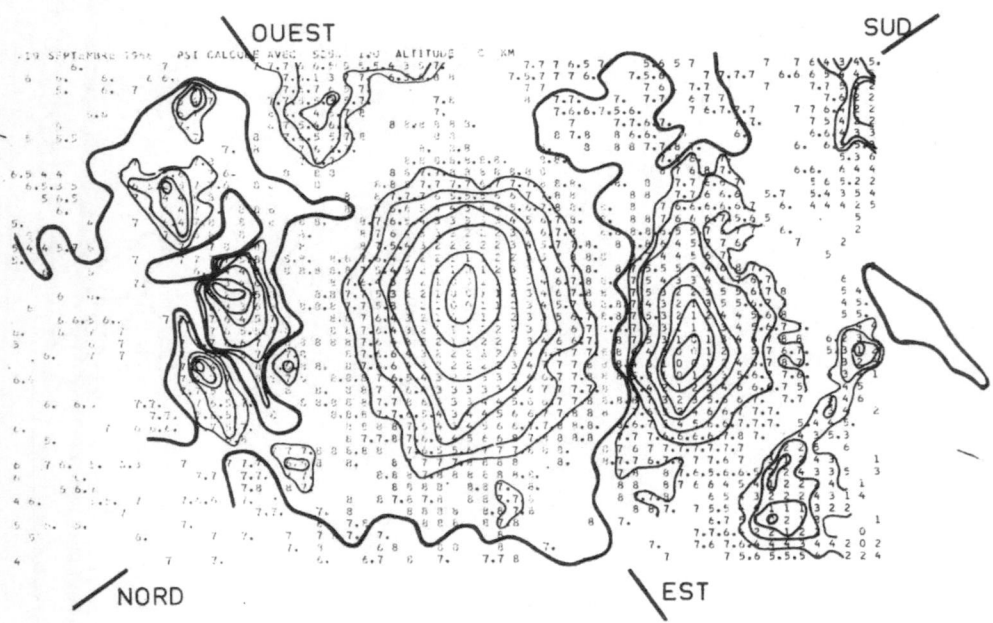

FIG. 2.  *Inclinaison calculée du champ par rapport à la verticale. Les angles sont comptés de 0 à 90°*
*pour les deux polarités. Le gros trait noir indique les changements de polarités.*

| Symbole | 0 | 1 | 2 | 3 | 4 | 5 | 6 | 6. | 7 | 7. | 8 | 8. |
|---------|---|---|---|---|---|---|---|----|---|----|---|----|
| $\psi \leqslant$ | 10 | 20 | 30 | 40 | 50 | 60 | 65 | 70 | 75 | 80 | 85 | 90 |

composante longitudinale lorsqu'elle est très faible. Il apparaît aussitôt que le champ observé est beaucoup plus incliné par rapport à la verticale solaire que le champ calculé; en particulier à la limite ombre-pénombre l'inclinaison des lignes de force par rapport à la normale à la photosphère est de 70° tandis que la valeur calculée est à peine 30°. Ce résultat est bien visible sur la Figure 3, qui représente les variations de

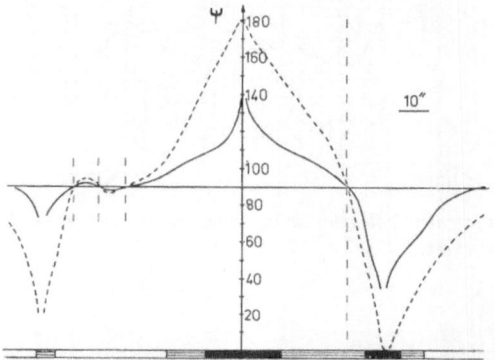

Fig. 3.    *Variation de l'inclinaison du champ par rapport à la verticale pour une section passant par le centre des deux taches.* ——— = *Inclinaison mesurée*, ----- = *Inclinaison calculée*.

l'angle $\psi$ observé et calculé pour une section passant par le centre des deux taches. Comme nous l'avons déjà mentionné la forte inclinaison des lignes de force observée ne peut être due à la présence de lumière diffusée dont l'influence est éliminée par la méthode utilisée.

De grandes divergences existent également sur l'orientation de la composante transversale dont les valeurs mesurées et calculées sont respectivement représentées par les Figures 4 et 5. Les différences de structure sont très nettes: si dans l'ensemble

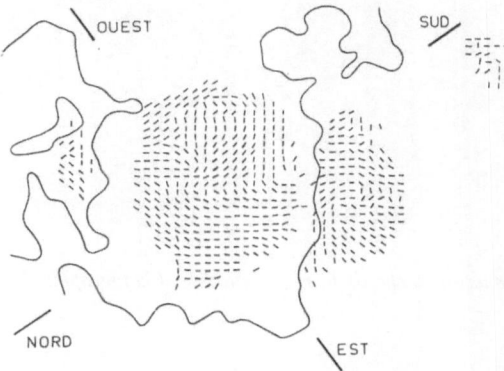

Fig. 4.    *Orientation mesurée de la composante transversale. Le gros trait noir indique les changements de polarité.*

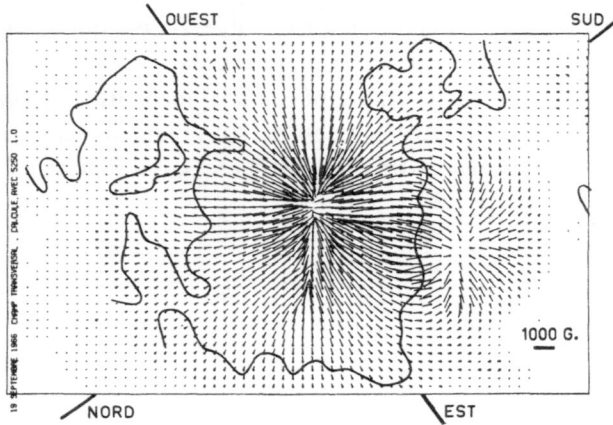

Fig. 5.  *Orientation et intensité calculées de la composante transversale. Les champs plus petits que 50 gauss ne sont pas représentés.*

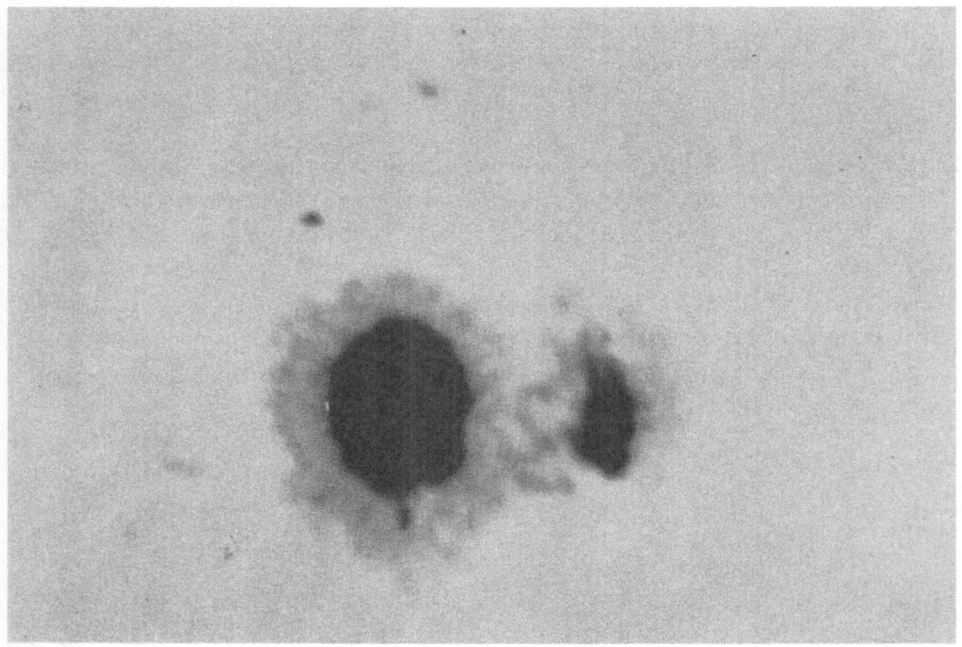

Fig. 6.  *Aspect en lumière blanche de la région étudiée. Photographie de l'Observatoire solaire de Debrecen, Hongrie.*

le champ calculé schématise assez bien la configuration d'un dipole, il n'en est pas de même pour le champ observé qui s'écarte nettement d'une structure radiale.

À ce stade il est indispensable de faire quelques remarques sur la morphologie de la région observée. En effet nous ne sommes pas en présence d'un groupe bipolaire classique. La plus grosse des deux taches est le retour de la tache de tête d'un groupe de la rotation précédente tandis que l'autre est la tache de queue d'un groupe jeune en période de développement dont la tache de tête, simple pore le 19 septembre, se développera par la suite. La Figure 6 donne l'aspect en lumière blanche de la région observée. Nous avons peut-être là une des causes de la différence entre le champ calculé et la structure réellement observée, mais il reste cependant certain que l'hypothèse d'une atmosphère sans courant conduit à une structure du champ magnétique incompatible avec les observations.

Comme pour le cas précédent (12 juin 1963), nous avons calculé la composante verticale du champ à l'altitude de 5000 km (Figure 7) et nous avons également obtenu une structure beaucoup trop simple par rapport aux détails chromosphériques observés avec la raie Hα. La composante verticale du champ observé au niveau de la photosphère montre par contre une structure très complexe en bon accord avec les structures observées dans la chromosphère. La Figure 8 représente, superposé à un héliogramme Hα, le champ vertical observé avec la raie λ 5250 Fe I. L'observation a

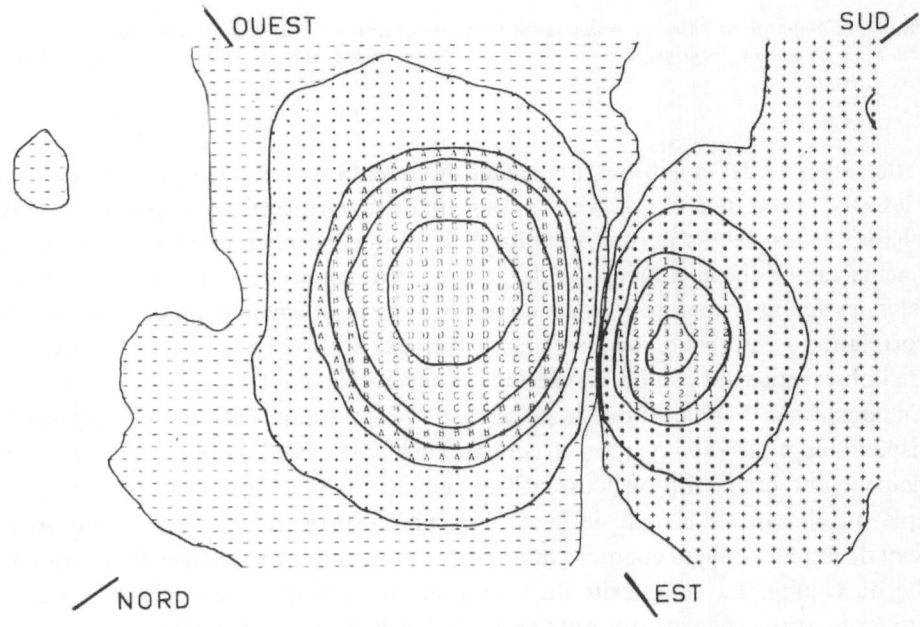

FIG. 7. *Composante verticale du champ calculée à une altitude de 5000 km.*

| Symboles | — | · | A | B | C | D | E | F | G | H | I | *polarité Sud* |
|---|---|---|---|---|---|---|---|---|---|---|---|---|
| | + | * | 1 | 2 | 3 | 4 | 5 | 6 | 7 | 8 | 9 | *polarité Nord* |
| $H \geqslant$ | | 5 | 50 | 200 | 300 | 500 | 1000 | 1500 | 2000 | 2500 | 3000 | 3500 | *gauss.* |

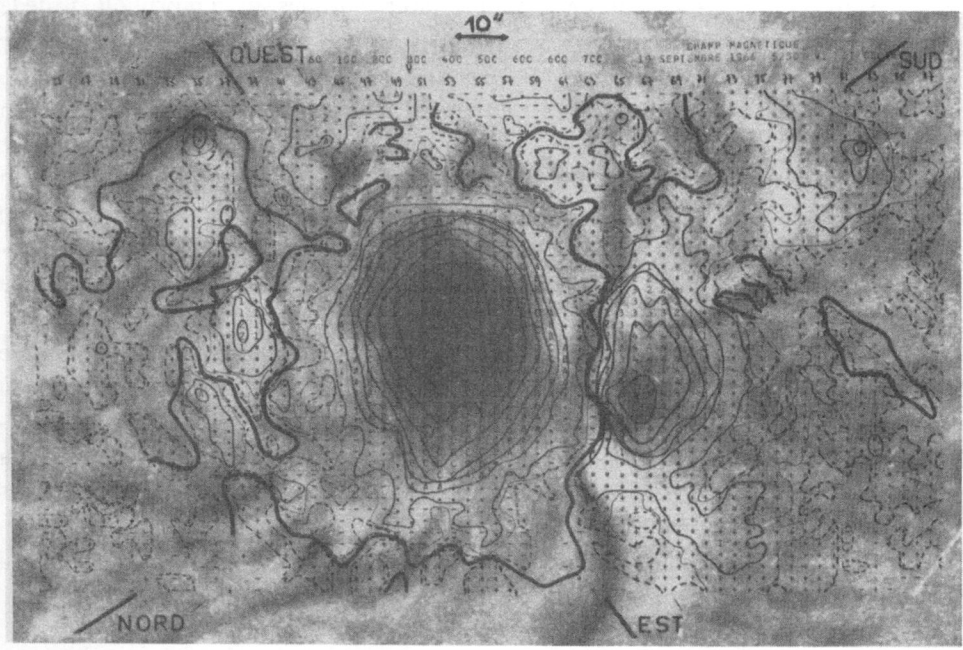

FIG. 8.   *Comparaison entre la composante verticale observée (9ʰ35 TU) et un héliogramme Hα (7ʰ50 TU). Symboles identiques à la Figure 7. Photographie Hα de l'Observatoire d'Ondřejov, Tchécoslovaquie.*

été effectuée avec des conditions atmosphériques très favorables; la résolution obtenue sur les spectres est de 2″ d'arc. Il est intéressant de remarquer la correspondance entre le champ et les détails chromosphériques et plus particulièrement la position des filaments sur les lignes d'inversion de polarité. En dehors de certains filaments très stables, les détails chromosphériques changent assez rapidement et on peut penser que la corrélation serait encore meilleur s'il n'y avait pas presque deux heures de différence entre l'observation du champ magnétique et l'héliogramme Hα.

En conclusion il apparaît que la présence de courant dans l'atmosphère solaire est indispensable pour expliquer les structures fines observées à tous les niveaux. Non seulement une configuration 'current-free' conduit à une structure beaucoup trop simple du champ au niveau de la chromosphère, mais encore son utilisation au niveau de la photosphère conduit à une image beaucoup trop éloignée de la structure réelle du champ. La complexité du champ et en particulier de l'orientation de la composante trasversale est souvent liée au déclenchement des éruptions. Au cours du passage du groupe, plusieurs éruptions ont commencé à la limite Ouest de la pénom- bre de la grosse tache où le champ transversal observé montre une grande complexité. Nous signalerons en particulier l'éruption observée le 20 septembre 1966 à 18ʰ00 TU par Tandberg-Hanssen (1967), qui se développe jusqu'au centre de la tache dans cette région particulièrement complexe alors que le champ calculé ne montre aucune particularité.

## Remerciments

Nous remercions les Docteurs L. Dezső et B. Valnicek pour les photographies en lumière blanche et avec la raie Hα qu'ils ont bien voulu nous communiquer.

## Références

Godovnikov, N. V., Smirnova, E. P. (1965)     *Izv. krym. astrofiz. Obs.*, **33**, 86.
Oster, L. (1964)     *Ann. Astrophys.*, **27**, 14.
Rayrole, J. (1967)     *Ann. Astrophys.*, **30**, 257.
Schatzman, E. (1961)     *Ann. Astrophys.*, **24**, 251.
Schmidt, H. U. (1964)     *NASA Solar Flare Symposium, Nasa SP-50*, 107.
Semel, M. (1967)     *Ann. Astrophys.*, **30**, 513.
Tandberg-Hanssen, E. (1967)     *Solar Phys.*, **2**, 98.

# DISCUSSION

*H. U. Schmidt:* This study gives reliable information about the electrical currents crossing the photosphere of an active region, and this is exactly what I had in mind when I proposed this method to evaluate longitudinal-field measurements some years ago.

# SUNSPOTS AND MAGNETOHYDRODYNAMIC FLOWS*

G. F. ANDERSON and D. H. MENZEL

*(Harvard College Observatory,
Cambridge, Mass., U.S.A.)*

ABSTRACT

Theoretical techniques are developed to study compressible, steady-state, magnetically aligned gas flows in sunspot regions. The flows are adiabatic and occur in a known streamline configuration. The non-linear parabolic partial differential equation describing the flow reduces to an ordinary linear differential equation. The solutions are briefly discussed.

## 1. Introduction

This paper develops some theoretical tools necessary to study compressible, steady-state, magnetically aligned gas flows in sunspots. The motions in the spot region take place along magnetic field lines, i.e.,

$$\mathbf{v}(\mathbf{r}) = \alpha(\mathbf{r})\,\mathbf{H}(\mathbf{r}), \tag{1.1}$$

where $\mathbf{v}$ is the velocity of a gaseous element, $\alpha$ a function of position, and $\mathbf{H}$ the magnetic field.

We specify, at the outset, the form of the magnetic field. We impose no restriction of irrotationality or incompressibility. Such constraints immediately define the streamline configuration. From the known streamline configuration we deduce the velocity $\mathbf{v}(\mathbf{r})$, mass density $\varrho(\mathbf{r})$, and pressure $p(\mathbf{r})$ distributions through the flow region by solutions of the MHD equations.

The magnetic field, which possesses cylindrical symmetry, must satisfy

$$\mathbf{V}\cdot\mathbf{H} = 0, \quad \frac{\partial \mathbf{H}}{\partial t} = -\frac{1}{\mu}(\mathbf{v} \times \mathbf{H}) = 0. \tag{1.2}$$

Specifically, we employ the dipole-like field of Menzel and Shore (1966) as given by

$$\mathbf{H}(r, z) = H_r(r, z)\,\mathbf{e}_r + H_z(r, z)\,\mathbf{e}_z, \tag{1.3}$$

where

$$H_r(r, z) = 3\,Mrza\,(a^2 + r^2 + z^2)^{-5/2}, \tag{1.4}$$

$$H_z(r, z) = Ma\,(2\,a^2 + 2\,z^2 - r^2)\,(a^2 + r^2 + z^2)^{-5/2}. \tag{1.5}$$

* Presented by G. F. Anderson.

The vector $\mathbf{e}_z$ is a vertical unit vector directed upward from the solar surface and $\mathbf{e}_r$ is the horizontal unit vector. In addition, we shall assume that the gas is perfectly conducting and obeys the adiabatic relation

$$p/p_0 = (\rho/\rho_0)^\gamma, \tag{1.6}$$

where $\gamma$ is the ratio of specific heats.

## 2. The Formulation of the Problem

We need two additional equations to determine the steady-state flow: the equations of momentum transport and of continuity. The first of these is

$$\nabla\left(\frac{v^2}{2} + \phi + \int \frac{\mathrm{d}p}{\rho}\right) = \mathbf{v} \times (\nabla \times \mathbf{v}) - \frac{1}{4\pi\rho} \mathbf{H} \times (\nabla \times \mathbf{H}), \tag{2.1}$$

where $\phi$ is the gravitational potential. Using the Equations (1.1)–(1.5) and taking the scalar product of (2.1) with $\mathbf{H}$, we get

$$\frac{\alpha^3}{2}(\mathbf{H}\cdot\nabla H^2) + \alpha(\mathbf{H}\cdot\nabla\phi) + (\mathbf{H}\cdot\nabla\alpha)(\alpha^2 H^2 - K\gamma\rho^{\gamma-1}) = 0, \tag{2.2}$$

where $K = p_0\varrho_0^{-\gamma}$. The continuity equation leads to the result

$$\nabla\cdot\rho\mathbf{v} = \nabla\cdot\rho\alpha\mathbf{H} = \mathbf{H}\cdot\nabla\rho\alpha = 0. \tag{2.3}$$

This equation tells us that the gradient of $\varrho\alpha$ is perpendicular to the magnetic field. In other words, $\varrho\alpha$ is constant along the stream lines.

We now seek solutions of (2.2) where the proportionality function $\alpha$ can be separated into a product of three functions

$$\alpha(\mathbf{r}) = E(\xi)G(\rho)S(H), \tag{2.4}$$

where $G(\varrho) = (\varrho/\varrho_0)^{(\gamma-1)/2}$, $S(H) = (H/H_0)^{-1}$ and $H_0$ is the value of the magnetic at the point $p_0$, $\varrho_0$. The form of $E(\xi)$ will be demonstrated shortly. Substitution of (2.4) into (2.2) leads to

$$
\begin{aligned}
&\mathbf{H}\cdot\nabla(\mathbf{H}\cdot\nabla\xi) + \frac{(\mathbf{H}\cdot\nabla\xi)^2}{\mu\xi} \\
&\quad \times \left\{\left(\frac{4}{\gamma+1}\right)\frac{\xi E'}{E} + \frac{K\gamma}{E^2}\left[\frac{1 + 2(\gamma-1)}{(\gamma+1)}\frac{\xi E'}{E}\right] - 1 + \frac{\mu[\mathbf{H}\cdot\nabla(\xi E'/E)]}{(\mathbf{H}\cdot\nabla\xi)(E'/E)}\right\} \\
&\quad + (\mathbf{H}\cdot\nabla\xi)\left[\frac{2h}{H^2}\frac{(\gamma-1)}{(\gamma+1)} - \frac{\mathbf{H}\cdot\nabla\mathscr{G}}{\mathscr{G}}\right] \\
&\quad + \frac{(\gamma+1-2\mu)}{4\mu H^2}\left(\frac{E}{E'}\right)\left[\mathbf{H}\cdot\nabla h - h\left(\frac{\mathbf{H}\cdot\nabla\mathscr{G}}{\mathscr{G}}\right) - \frac{2h^2}{(\gamma+1)H^2}\right] = 0,
\end{aligned} \tag{2.5}
$$

where

$$E' = dE/d\xi, \quad h = \mathbf{H} \cdot \nabla H^2, \quad \mu = 1 - K\gamma/E^2, \quad \text{and} \quad \mathcal{G} = \mathbf{H} \cdot \nabla \phi.$$

Equation (2.5) constitutes a second-order, non-linear, parabolic partial differential equation for $\xi(\mathbf{r})$. Convergent and stable numerical solutions to partial differential equations of this type depend upon the form of the variable coefficients as well as upon the nature of the integration scheme and the boundary conditions. Well-behaved numerical solutions for Equation (2.5) readily follow if we set the coefficient of $(\mathbf{H} \cdot \nabla \xi)^2$ equal to zero.

This assumption not only eliminates the second term from (2.5) but also leads to an ordinary differential equation for $E$ as a function of $\xi$,

$$\frac{E''}{E'} - \frac{E'}{E} \left[ \frac{(3\gamma - 1)}{\gamma + 1} - 2 \left( 1 - \frac{K\gamma}{E^2} \right)^{-1} \right] = 0. \tag{2.6}$$

Equation (2.6) integrates directly to yield

$$C_1 \xi + C_2 = E^{(4/(\gamma + 1))} \left[ \frac{(\gamma - 1)}{2} + \frac{K\gamma}{E^2} \right], \tag{2.7}$$

where $C_1$, $C_2$ are constants. Substituting (2.7) into (2.5), we find the equation for $\xi$

$$\mathbf{H} \cdot \nabla (\mathbf{H} \cdot \nabla \xi) + (\mathbf{H} \cdot \nabla \xi) \left[ \frac{2(\gamma - 1)}{(\gamma + 1)} \frac{h}{H^2} - \frac{\mathbf{H} \cdot \nabla \mathcal{G}}{\mathcal{G}} \right]$$
$$+ \frac{1}{H^2} \frac{(\gamma - 1)}{(\gamma + 1)} \left( \xi + \frac{C_2}{C_1} \right) \left[ \mathbf{H} \cdot \nabla h - h \frac{(\mathbf{H} \cdot \nabla \mathcal{G})}{\mathcal{G}} - \frac{2h^2}{H^2 (\gamma + 1)} \right] = 0. \tag{2.8}$$

Let us expand (2.8) in cylindrical coordinates into the canonical form (Koshlyakov *et al.*, 1964)

$$A(\eta, \lambda) \xi_{\eta\eta} + B(\eta, \lambda) \xi_\lambda + C(\eta, \lambda) \xi_\eta + D(\eta, \lambda) \xi = 0, \tag{2.9}$$

by means of the transformation equations

$$\lambda = \lambda(r, z), \quad \eta = \eta(r, z), \tag{2.10}$$

where

$$\xi_\eta = \partial\xi/\partial\eta, \quad \xi_{\eta\eta} = \partial^2\xi/\partial\eta^2, \quad \xi_\lambda = \partial\xi/\partial\lambda.$$

A parabolic differential equation has only one family of characteristic curves and is derived by integration of the equation of characteristics

$$H_r \, dz - H_z \, dr = 0. \tag{2.11}$$

The integration of (2.11) yields

$$\lambda(r, z) = r^2 (a^2 + z^2 + r^2)^{-3/2} = \text{constant}, \tag{2.12}$$

and provides us with the first of our set of transformation equations (2.10). The second relation $\eta(\mathbf{r}, z)$ is chosen independently. Let us select

$$\eta(r, z) = r. \tag{2.13}$$

The transformation is valid everywhere, provided its Jacobian,

$$\frac{\partial(\lambda, \eta)}{\partial(r, z)} = \frac{\partial\lambda}{\partial z} \times \frac{\partial\eta}{\partial r} - \frac{\partial\lambda}{\partial r} \times \frac{\partial\eta}{\partial z}, \tag{2.14}$$

is not zero.

The Jacobian is zero only for points on the $z$-axis and on the $z=0$ plane. When the differential Equation (2.9) is transformed by means of the relations (2.12), (2.13), we have the result

$$H_r^2 \xi_{\eta\eta} + \left\{ H_r \left[ \frac{2(\gamma - 1)}{(\gamma + 1)} \frac{h}{H^2} - \frac{\mathbf{H} \cdot \nabla \mathscr{G}}{\mathscr{G}} \right] + (\mathbf{H} \cdot \nabla) H_r \right\} \xi_{\eta}$$
$$+ \frac{h}{H^2} \frac{(\gamma - 1)}{(\gamma + 1)} \left[ \frac{\mathbf{H} \cdot \nabla h}{h} - h \frac{(\mathbf{H} \cdot \nabla \mathscr{G})}{\mathscr{G}} - \frac{2h}{H^2 (\gamma + 1)} \right] \xi = 0, \tag{2.15}$$

where $C_2 = 0$. The coefficients of $\xi_\eta$, $\xi_{\eta\eta}$ and $\xi$ are now functions of the new variables $\lambda, \eta$. The differential equation upon which the solution to our problem depends has been reduced to an ordinary, linear, second-order differential equation that can be solved by standard numerical methods.

## 3. The Solution on the $z$-axis and on the $z=0$ Plane

The integration over $\eta$ in Equation (2.15) is equivalent to an integration over $r$. We first deduce $\xi(\mathbf{r})$ on the $z$-axis. From these values we compute $\xi(\mathbf{r})$ away from the axis by the numerical integration of (2.15). Since Equation (2.15) is not valid on the axis, we must return to our original expression, (2.5). This equation reduces to

$$(a^2 + z^2)^2 \xi_{zz} - \frac{12(\gamma - 1)}{(\gamma + 1)} z(a^2 + z^2) \xi_z$$
$$+ \frac{6(\gamma - 1)}{(\gamma + 1)^2} [z^2(7\gamma - 5) - a^2(\gamma + 1)] \xi = 0, \tag{3.1}$$

and can be either integrated numerically or transformed into an equation whose solution is a Gaussian hypergeometric function. In the neighborhood of the origin $(0, 0)$, Equation (3.1) leads to

$$a^2 \xi_{zz} - \frac{6(\gamma - 1)}{\gamma + 1} \xi = 0, \tag{3.2}$$

whose general solution is

$$\xi(z) = \xi_+ e^{+mz} + \xi_- e^{-mz}, \tag{3.3}$$

where $m = \dfrac{1}{a}\sqrt{\left(6\dfrac{(\gamma-1)}{(\gamma+1)}\right)}$, and $\xi_{+,-}$ are constants.

Similarly, the form of the solution in the plane can be derived. In this instance we find it necessary to use a transformation of variables. The result is

$$\xi/\xi_0 = \left\{\frac{(r^2/a^2)^2\,[(r^2/a^2)-2]^4\,(|(r^2/a^2)-2|)}{[(r^2/a^2)+4]}\right\}^{(\gamma-1)/(\gamma+1)} \tag{3.4}$$

where $\xi_0$ is the value of $\xi$ at some reference point in the sunspot region. Near the origin, $\xi/\xi_0$ approaches zero. As a result, from (3.3),

$$\lim_{z\to 0}\xi = \xi_+ + \xi_- = 0, \tag{3.5}$$

and $\xi_- = -\xi_+$. Hence, the solution for $\xi(\mathbf{r})$ along the $z$-axis in the neighborhood of the origin is of the form $\xi(z) = \xi_+\,(e^{mz}-e^{-mz})$.

## 4. The Velocity Fields

The velocity is given by

$$v(\mathbf{r}) = \alpha H = E\,[\xi(\mathbf{r})]\,[\rho(\mathbf{r})]^{(\gamma-1)/2}. \tag{4.1}$$

In addition to $E[\xi(\mathbf{r})]$, we must specify the mass density, $\varrho(\mathbf{r})$, throughout the flow regime. The mass density is computed in the following manner. Along a streamline

$$\rho\alpha = [\rho(\mathbf{r})]^{(\gamma+1)/2}\,E\,[\xi(\mathbf{r})]/H(\mathbf{r}) = \text{constant}. \tag{4.2}$$

For our magnetic field all streamlines intercept the $z=0$ plane. Since we know the variation of $E$ and $H$ in the $z=0$ plane, we need only compute $\varrho(r, z=0)$ in order to evaluate the constant in (4.2) for any given streamline. After we evaluate the constant, we use (4.2) to compute $\varrho(\mathbf{r})$ along streamlines extending anywhere in the sunspot region. We determine $\varrho(r, z=0)$ from Equation (2.2). After some manipulation we find

$$\rho(r, z=0)/\rho_0 = -\left[\frac{(\gamma-1)\,\mathscr{G}}{C_1\xi_z}\,(E/E_0)^{-2/(\gamma+1)}\right]_{z=0}, \tag{4.3}$$

where $E_0 = E(\xi_0)$. The term $\xi_z(r, z=0)$ appearing in Equation (4.3) is calculated from $\xi(\mathbf{r})$.

Finally, once we have deduced the mass density, we can derive the pressure and temperature distributions from the polytropic condition (1.6).

## Acknowledgements

This work was supported in part by the U.S. Air Force contract AF 19(628)3322.

# References

Koshlyakov, N. S., Smirnov, M. M., Gliner, E. B. (1964)     *Differential Equations of Mathematical Physics*, North-Holland Publ. Co., Amsterdam.

Menzel, D. H., Shore, B. W. (1966)     'Sunspots, Magnetic Fields and the Structure of the Solar Atmosphere', in *Atti del Convegno sulle macchie solari, IV. Centenario della Nascità di Galileo Galilei*, Ed. by G. Barbera, Florence, p. 226.

Part III

# OPTICAL STRUCTURE OF AN ACTIVE REGION

# ON THE STATE OF THE PHOTOSPHERE BEFORE THE APPEARANCE OF SUNSPOTS

Z. B. KOROBOVA
(Tashkent Observatory, U.S.S.R.)

A. K. TCHANDAEV
(Gorki Institute, U.S.S.R.)

and

G. Y. VASSILYEVA
(Pulkovo Observatory, U.S.S.R.)

## 1. Observations

The object of our investigation was to clarify the behavior of the fluctuations in the brightness in photosphere on the scale of the supergranulation 1–2 days before the appearance of sunspots.

A detailed photometrical treatment and statistical analysis of areas with sizes $100'' \times 200''$ and $100'' \times 100''$ have been made.

Three groups near the center of the disk have carefully been studied at 8 moments of their development, using Tashkent daily white-light plates obtained by means of Macksutov's system photoheliograph ($D = 100$ mm, $F = 8200$ mm).

Printon plates with effective wavelength 4100 Å have been used.

## 2. Analysis

In order to reduce the influence of the measuring errors and errors of the observations, the control areas have also been measured on the same plates, on the same distance as the areas under study.

The fluctuations in the brightness have been recorded at about 3000 points on every area under study and at about 1500 points on every control area.

The next values were obtained at every area: $J =$ the average meaning of the brightness for the area on the whole in terms of the brightness at the center of the disk; $\Delta J / \bar{J} =$ the contrast at every point; two-dimension autocorrelation function

$$B(l, m) = \frac{1}{(L - l)(M - m)} \sum_{\lambda=1}^{L-l} \sum_{\varphi=1}^{M-m} \frac{\Delta J}{J}_{\lambda, \varphi} \frac{\Delta J}{J}_{\lambda + l, \varphi + m}$$

where $l = 0, 1, 2 \ldots 20$; $L = 80$. $m = 0, 1, 2 \ldots 20$; $M = 40$.

*Kiepenheuer (ed.), Structure and Development of Solar Active Regions, 151–160.* © *I.A.U.*

**Table 1**

| | | The areas under study (size ~ 200″ × 100″) | | | The control areas (size ~ 100″ × 100″) | | | | $J_s$ (%) | $J_c$ (%) | $\Delta J$ (%) | $\sqrt{\overline{\left(\frac{\Delta J}{J}\right)^2}}_s$ | $\sqrt{\overline{\left(\frac{\Delta J}{J}\right)^2}}_c$ |
|---|---|---|---|---|---|---|---|---|---|---|---|---|---|
| $N$ | Date U T | $\varphi$ | $\lambda$ | $\rho$ | $N$ | $\varphi$ | $\lambda$ | $\rho$ | | | | | |
| I | | | | | | | | | | | | | |
| 1 | 22/IX/61 $5^h11^m$ | +7 | −17 | 0·29 | 2 | −10 | 0 | 0·29 | 101 | 98 | 3 | 2·1 | 2·7 |
| 15 | 8 22 | +7 | −15 | 0·27 | 16 | −10 | 0 | 0·29 | 103 | 98 | 5 | 1·3 | 1·1 |
| 5 | 23/IX/61 5 08 | +7 | +1 | 0·07 | 6 | −10 | +13 | 0·39 | | Spots | | | |
| II | | | | | | | | | | | | | |
| 3 | 22/IV/59 $3^h35^m$ | +16 | +14 | 0·42 | 4 | −30 | 0 | 0·42 | 98 | 92 | 6 | 3·2 | 2·2 |
| 7 | 23/IV/59 6 19 | +16 | +27 | 0·53 | 8 | −30 | +13 | 0·47 | | Spots | | | |
| III | | | | | | | | | | | | | |
| 11 | 2/VI/65 12 04$^m$ | −10 | +7 | 0·30 | 12 | +10 | +7 | 0·30 | 97 | 98 | −1 | 7·3 | 3·7 |
| 9 | 3/VI/65 3 44 | −10 | +15 | 0·30 | 10 | +9 | +15 | 0·30 | 100 | 96 | 4 | 3·2 | 2·5 |
| 13 | 4/VI/65 4 58 | −11 | +28 | 0·50 | 14 | +11 | +28 | 0·50 | | Spots | | | |

$\Delta l = \Delta m = 1$ corresponds to $2''.6$ on the disk. Table 1 shows time of observations, the coordinates, the average meanings of the brightness and root-mean-square contrasts of the areas under study in comparison with the control areas.

## 3. Results and Conclusions

The appearance of sunspots is accompanied by a very strong photosphere disturbance, involving an excess of the brightness, a change of the structure, unusual time-dependent macroscopic fluctuations in the brightness.

Before the appearance of sunspots the brightness increases on the whole area at an average for some cases of about $3.5 \pm 2.4\%$. The additional radiation is observed on the place with size $200'' \times 100''$ for 1 day, the total extra energy being of the order of $10^{34}$ erg.

As to a structure, its changes are manifested in the enlargement of the macroscopic inhomogeneities and visible separation of the area under study on two parts (Figure 1a, b).

The new spots tend to be placed along the boundary, dividing lighter-than-average and darker-than-average regions (Figure 1c).

Such a location of new spots seems to be similar to the formation of the spots on the boundaries of the adjacent supergranules according to Bumba and Howard (1965). Besides the enlargement, changes of the structure are indicated by the forming of the dark lanes and the bright features, stretched along the equator. Spatial autocorrelation functions in Figure 2 convincingly show this phenomenon. The same photosphere phenomena with the appearance of the characteristic structure have been observed at the active regions before pores' birth by Loughhead and Bray (1961) and in photosphere between spots by Miller (1960).

The filamentary structure of the inhomogeneities in the longitude direction before the appearance of sunspots seems to be a distinctive trait in the development of the active regions. We found such a structure at nine more cases without special treatment. The typical character of the straightened elements in the photosphere is to be emphasized by the appearance of the elongated filaments in $H\alpha$ at the first stage of development of active regions.

The comparison of two pairs of areas under study at two different moments reveals definite time-dependent fluctuations in the brightness on a large scale. The lighter-than-average region turns into the darker-than-average one and on the contrary. For all that, the location of the dividing line is roughly kept.

Figure 3 shows this phenomenon at the area for 3 days of the development of active region. But another studied case reveals the same character of changes in the brightness for 3 hours, and we do not yet know the low limit of such fluctuations.

Possible rapid changes of the macroscopic fluctuations in the brightness ($\sim 20$ min) on the scale of the supergranules in the photosphere have been discussed by Beckers

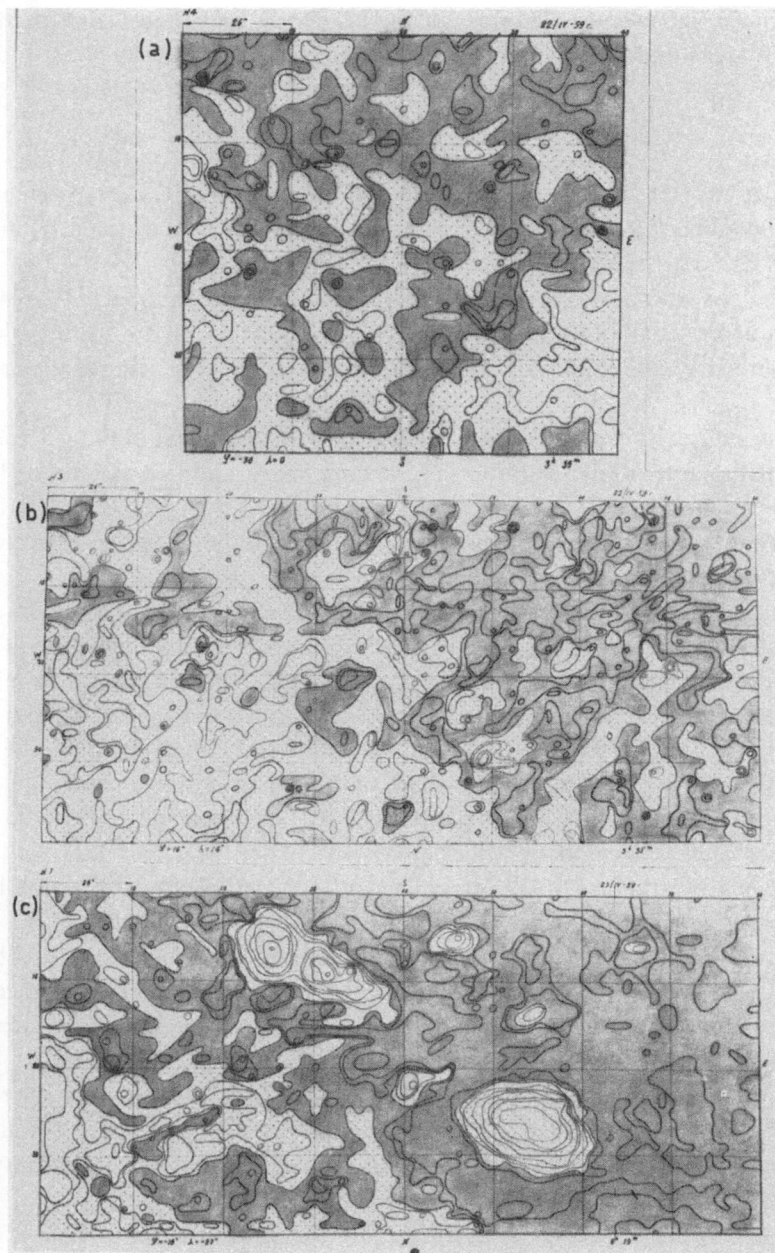

FIG. 1.   *The brightness maps of some areas in the photosphere with the meanings of the contrast*
$\Delta J/J$ *equal to 0·03; 0·05; 0·07; 0·1 in regions far from spots. (a) The undisturbed control area; (b) The
area under study before the appearance of sunspot group 22/IV/59; (c) The location of spots on the
same area next day, 23/IV/59.*

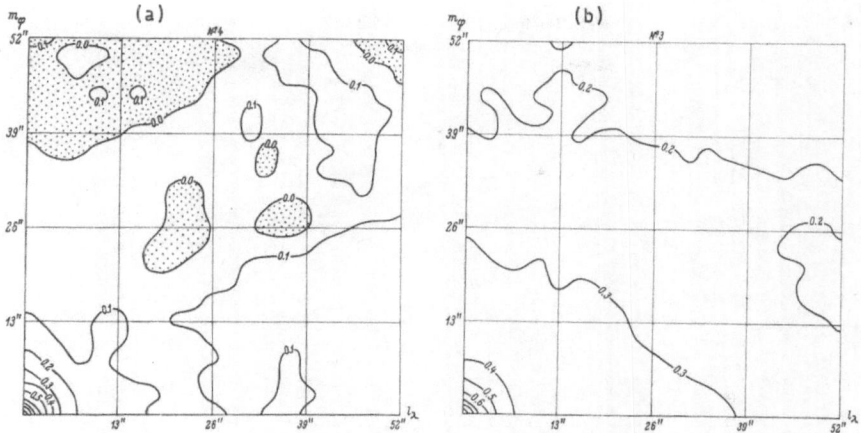

Fig. 2.   *Samples of the two-dimensional autocorrelation functions; (a) for the undisturbed control area (Figure 1a); (b) for the area under study 22/IV/59 (Figure 1b). l and m are longitude lag and latitude one accordingly.*

(1966) and Harvey (1965). As a result of our treatment, we obtained time-dependent changes of the areas under study which have no common features with those of the control undisturbed areas.

In conclusion we can say that the observed structure at disturbed photosphere seems to be caused by the rising of the magnetic loop with filamentary structure. This explanation as to pores was offered by Loughhead and Bray.

Time of the disturbance of the photosphere before the pores' and spots' birth is equal to $3^h$ and $20–30^h$ respectively.

If we accept the speed of the magnetic-field rising equal to $10^4$ cm/sec, according to Iksanov and Vitinsky (1966), we shall obtain the size of the cross-section magnetic loops of the order of $10^8$ cm and $10^9$ cm, which is in agreement with the size of pores and spots. Then it is to be noted that our results appear to be affected by the definite selection of the observations. We treated the areas with the brightest features, and the future groups turned out to be very unstable with the great flares.

So, our test sunspot groups make us conclude that the character of the inhomogeneities in the brightness before the appearance of spots reflects the conditions of sunspot genesis and reveals the possibilities of predicting their future history on the base of the investigation of the brightness in the photosphere.

## Acknowledgements

The authors wish to thank Dr. M. M. Kobrin and Dr. O. I. Judin, who offered them the opportunity to use the computer of the Gorki Institute, and Professor S. A. Kaplan, who helped us to accomplish this work.

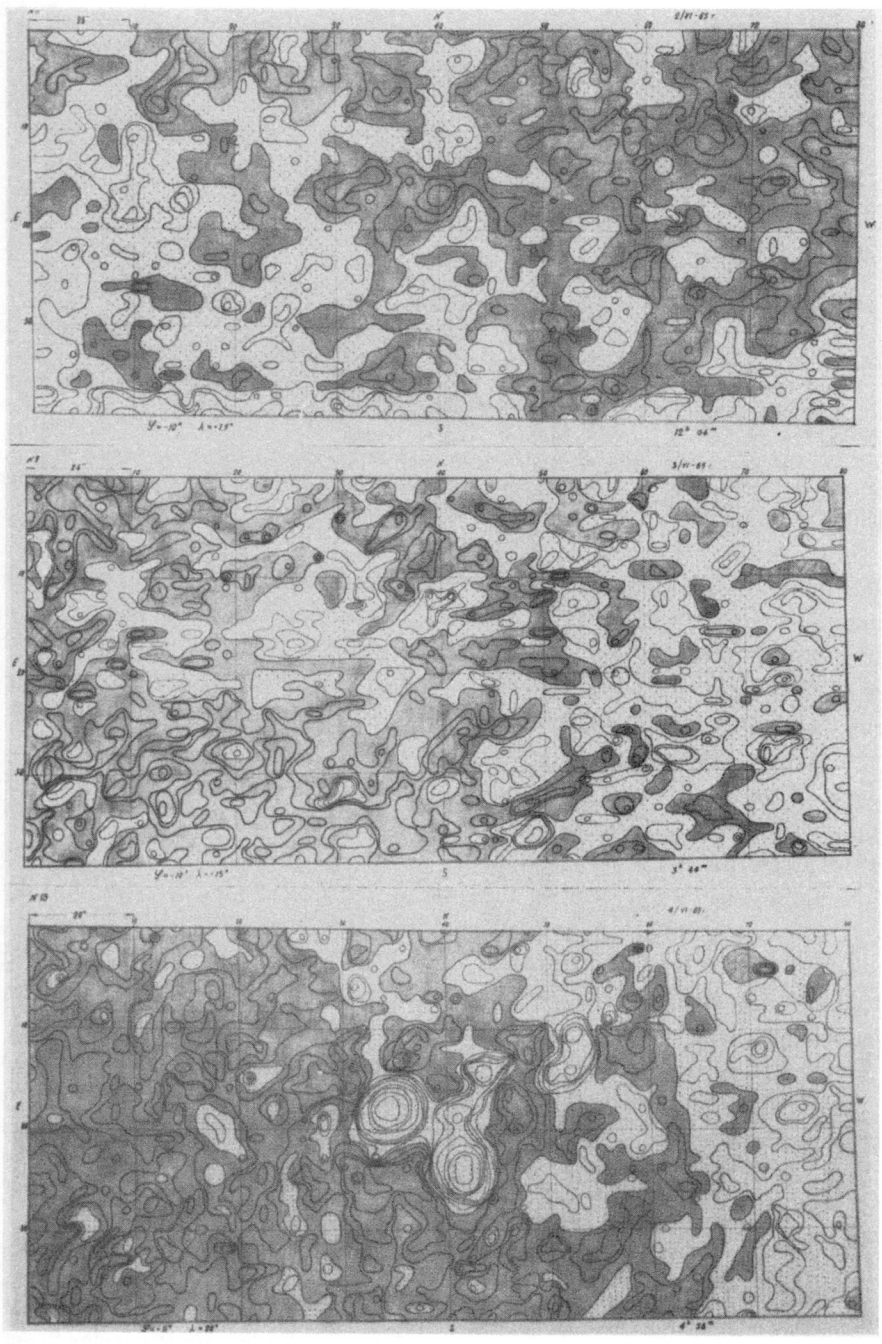

FIG. 3.    *The brightness maps of one area under study for 3 days: 2, 3, and 4/VI/65.*

## References

Beckers, J. M. (1966)    'The Fine Structure of the Solar Atmosphere', *Colloquium held at the German Solar Observatory, Capri, June 6–8*, p. 56.
Bumba, V., Howard, R. (1965)    *Astrophys. J.*, **141**, 1492.
Harvey, J. W. (1965)    *Publ. astr. Soc. Pacific*, **77**, 129.
Iksanov, R. N., Vitinsky, J. I. (1966)    *Izv. glav. astr. Obs.*, 180.
Loughhead, R. E., Bray, R. I. (1961)    *Austr. J. Phys.*, **14**, 347.
Miller, R. A. (1960)    *J. Brit. Astron. Assoc.*, **70**, 100.

# DISCUSSION

*De Jager:* Since you find that the brightness contrast in the granular field increases prior to the birth of sunspots, I wonder whether this effect is not simply due to the birth of the photospheric facular field. In its first stage this would signify an increased brightness of the supergranulation pattern.

*Vassilyeva:* We studied the properties of the brightness field in active regions before sunspot appearance and compared the results with what we obtained for the undisturbed photosphere. The cause of the difference in the brightness field was not investigated.

*McIntosh:* My white-light observations also show bright, distorted photospheric granules near positions where sunspots will soon form. I have not made quantitative studies of these but it appears that these brightenings are short-lived and not necessarily related to supergranules.

*G. W. Simon:* How do you determine the supergranular pattern from the photoelectric observations?

*Vassilyeva:* Autocorrelation functions of the photoelectrical records sometimes show not-accidental periodic component with a period coincident with sizes of the supergranules. Ordinary photometry of the photospheric region also shows existence of macroscopic inhomogeneities with noticeable fluctuations in brightness.

*G. W. Simon:* Do the autocorrelation functions show secondary maxima? If so, do these maxima occur at distances corresponding to supergranule sizes?

*Vassilyeva:* Two-dimension autocorrelation functions reflect non-isotropic character of the fields of the fluctuations in the brightness in disturbed and in undisturbed areas. The second maximum is observed with a lag of about 20″–30″.

*Rösch:* Je voudrais signaler, a l'occasion de cette communication, que nous avons récemment obtenu, a l'Observatoire du Pic-du-Midi (A. Carlier, F. Chauveau, M. Hugon, J. Rösch) une série d'images montrant la naissance d'un *pore*, dont certaines sont reproduites dans la Figure 4; l'intervalle entre les images 1 et 16 est de 26 min. Le Nord est en haut, l'Est à gauche. Objectif de 38 cm, coupole 'tourelle', film Kodak Microfile Ortho et filtre orange, pose 1/250 sec. Le cercle mesure 2″.

On remarquera, dans la région au Sud-Ouest du futur pore, des granules allongés, ressemblant plutôt a des éléments de filaments de pénombre, et suggérant l'existence d'un champ magnétique horizontal. Noter aussi le gros granule brillant, sur l'image 1, un peu au Süd-Ouest du futur pore, qui devient un fer-à-cheval sur l'image 4, puis un anneau de 5 à 8, et disparait par fragmentation à partir des images 9–10.

*Beckers:* Does Dr. Rösch believe that it is a common behaviour for a granule to explode or are the exploding granules exceptional cases?

*Rösch:* It may not be common to *all* granules, but we saw it quite often on the first movie-film we produced in 1959, and it appears more frequently on this new one, which is better in quality. Maybe one-third of the granules end in that way.

*Schröter:* I followed during the movie one granule. I saw that this granule dissolved by a very rapid change (decrease) of the central brightness, the boundaries remaining unchanged. Suddenly near the end of the film in the central part of this granule the brightness increased again. A new bright granule was the result. Is this observation an exceptional case or did you observe this behaviour on other granules too?

*Rösch:* I have noticed this case, indeed, but only the one you mention.

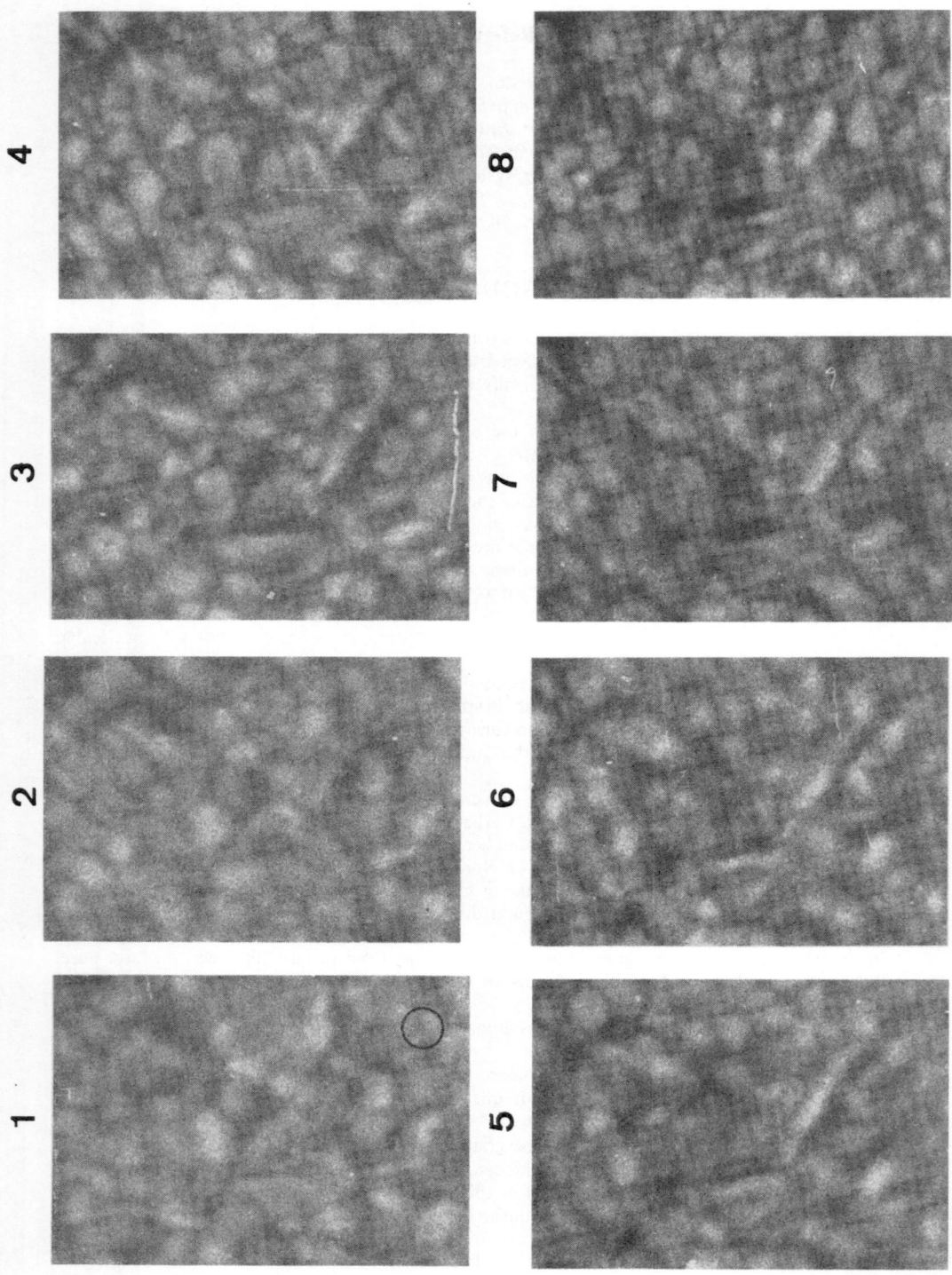

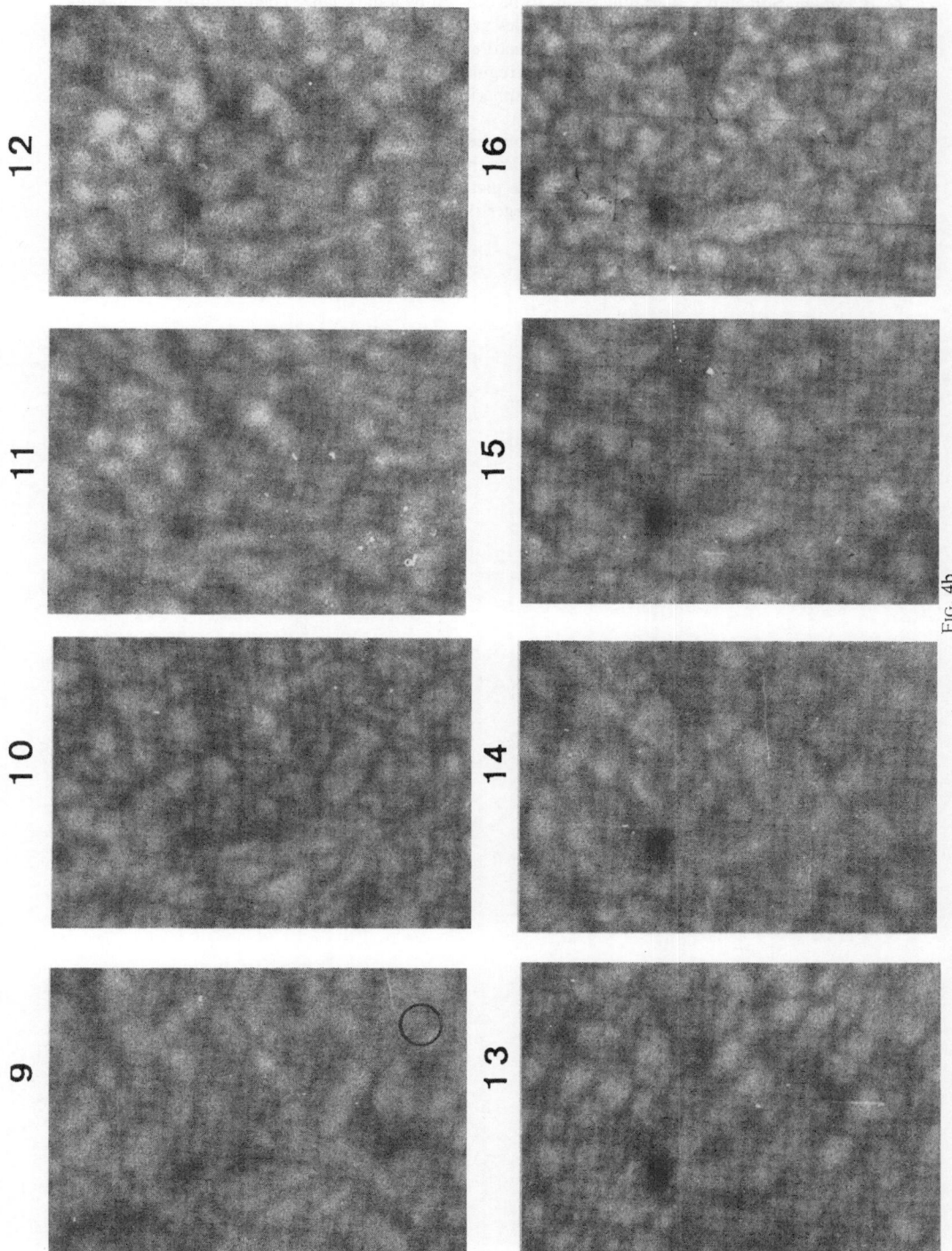

FIG. 4b

*G. W. Simon;* Sometimes a granule seems to be crossed by a dark lane. Then the granule suddenly divides along this lane and becomes two granules which then move apart from each other.

*Rösch:* This case is rather common; the structure is what I have called a 'coffee-grain'.

*Fokker:* There seemed to be some kind of a regular fluctuation or oscillation affecting the whole image. Is this effect of an instrumental or atmospheric nature, or can it be considered as of real solar origin?

*Rösch:* Because of its regularity, and rather long period ($\sim$ 300 sec perhaps!) I do not think it may be atmospheric. Of course, we should try to confirm that it has something to do with supergranulation.

*Righini:* What is the resolving power on these pictures?

*Rösch:* We have seen some small dots not larger than one-third of a second of arc.

# CORRELATIONS BETWEEN BRIGHTNESS FIELDS AND MAGNETIC FIELDS ON THE SUN*

G. A. CHAPMAN     and     N. R. SHEELEY, JR.

*(Steward Observatory,*     *(Kitt Peak National Observatory**,*

*University of Arizona,*     *Tucson, Ariz., U.S.A.)*

*Kitt Peak National Observatory,*

*Tucson, Ariz., U.S.A.)*

## ABSTRACT

In places on the solar surface where longitudinal magnetic fields are detectable using Leighton's photographic technique, spectroheliograms taken in the cores of many Fraunhofer lines show a *bright photospheric network* similar to, but with finer structure than, the familiar chromospheric network visible on $Ca^+$ $K_{232}$ spectroheliograms. This paper describes preliminary results of a study of the relation between the photospheric network and its associated magnetic fields.

## 1. Introduction

As Dr. Kiepenheuer mentioned in his introductory lecture yesterday, many optical phenomena seem to be *direct products* of magnetic fields. The $Ca^+$ $K_{232}$ emission and Hα brightenings are repeatedly referred to as indicators of magnetic fields, and spectroheliograms in these lines are commonly used to guess where these fields may be located on the solar surface. Figure 1 illustrates the close spatial relationship between the longitudinal magnetic fields and the $K_{232}$ emission.

We are not limited to chromospheric lines as indicators of magnetic fields, but can also use photospheric lines. We have known for some time (Beckers and Schröter, 1966; Sheeley, 1966b, 1967) that many photospheric lines are weakened in places on the solar surface where there are relatively strong magnetic fields. This is illustrated in Figure 2. As we shall demonstrate, spectroheliograms taken in the cores of these lines show a *bright photospheric network* similar to, but more delicate than, the familiar bright chromospheric network visible on spectroheliograms taken in the core of the $Ca^+$ K line and Hα. At the Kitt Peak National Observatory we are now using the newly installed spectroheliograph with the 82-cm solar image of the McMath Solar Telescope to study the relation between photospheric magnetic fields and the as-

---

* Presented by N. R. Sheeley, Jr.

** Operated by the Association of Universities for Research in Astronomy, Inc., under contract with the National Science Foundation.

*Kiepenheuer (ed.), Structure and Development of Solar Active Regions, 161–173. © I.A.U.*

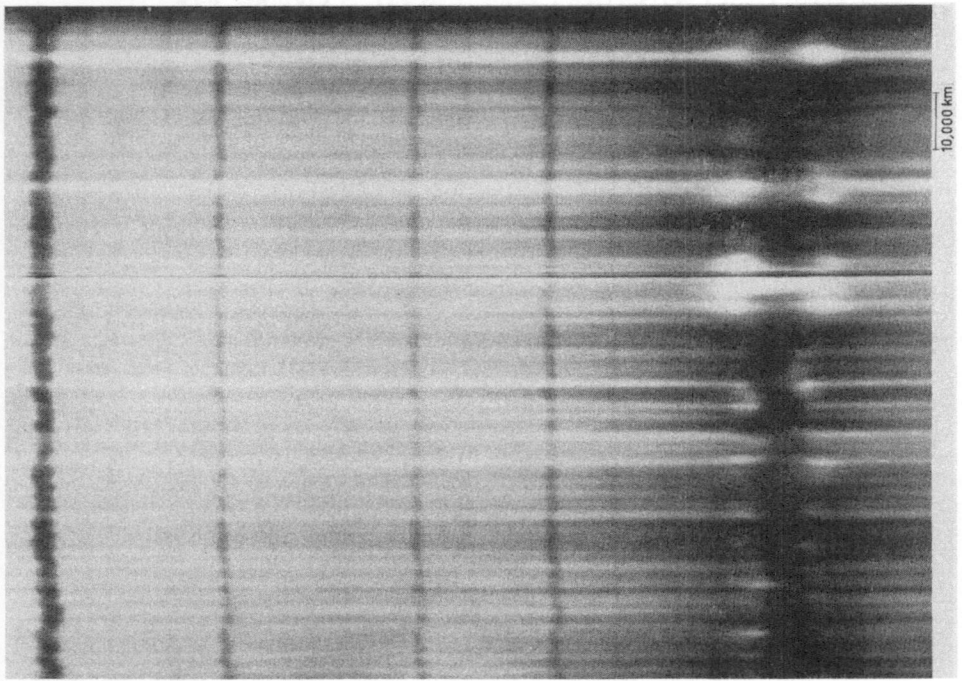

FIG. 1.    *Spectrogram of a magnetic feature taken on June 19, 1966. The magnetic feature is indicated by the pair of dark lines caused by hairs across the entrance slit. Note that the Ca$^+$ K line has bright $K_{232}$ emission at this point.*

sociated weakenings in Fraunhofer lines in hopes of finding out what causes the line weakenings.

## 2. Preliminary Results

Figure 3 shows a spectroheliogram taken in the core of the $\lambda$ 5131 line of Fe I together with a Zeeman photograph of the same region. The $\lambda$ 5131 line is a 'degenerate' Zeeman triplet arising from transitions between states having the same Landé g-factors. The magnetic sensitivity is $\Delta\lambda/B = 3.0 \times 10^{-5}$ Å/gauss. The slit widths were 25 $\mu$ corresponding to a band pass of 0·02 Å. The atmospheric seeing was excellent. The Zeeman photograph was made about $1\frac{1}{2}$ hours later during very poor seeing conditions, and is included here only for a rough comparison between the pronounced photospheric network visible in the $\lambda$ 5131 spectroheliogram and the longitudinal magnetic fields. In addition to the bright network visible in the $\lambda$ 5131 spectroheliogram there is a 'bright ring' visible in the penumbra of the large sunspot. Spectroheliograms in the core of the Zeeman-insensitive $\lambda$ 5124 line do not show such 'bright

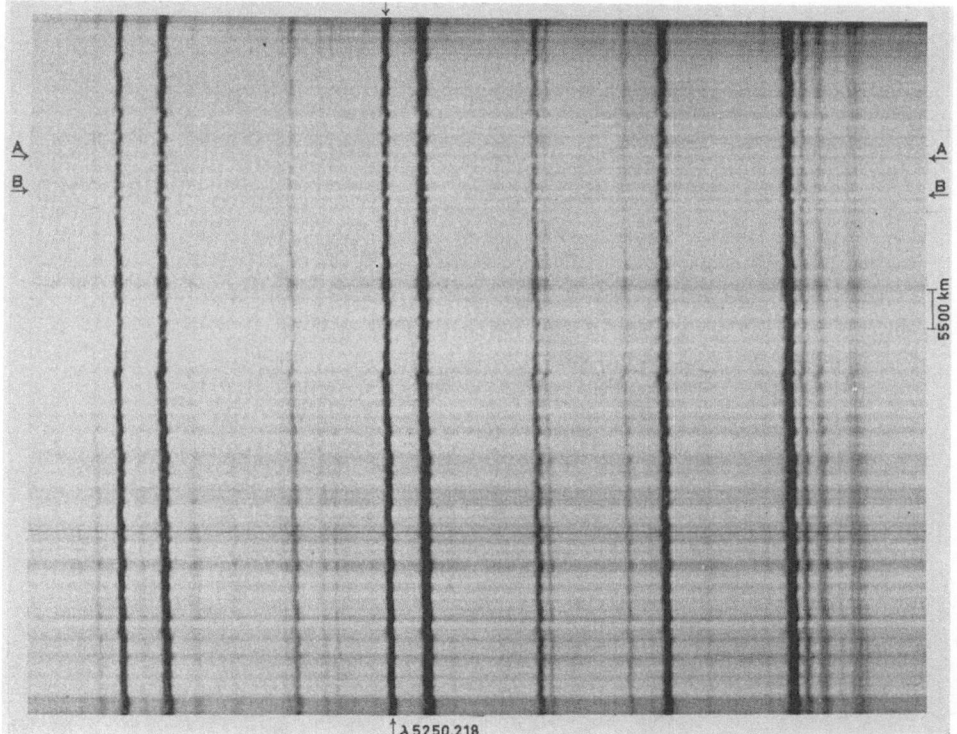

FIG. 2. *Spectrogram of the λ 5250 wavelength region taken on July 4, 1966. The arrow A indicates the position of a magnetic feature along the slit. The arrow B marks a position on the Sun both where the absorption lines have a characteristic arch-shaped appearance (the arch concave to the red) and where the continuum is locally bright, probably corresponding to a very bright solar granule. Notice that several Fraunhofer lines appear weakened at A, whereas they do not at B. Notice also that the continuum at A is locally dark, suggesting that the magnetic field occurred in a dark lane in the granulation field.*

penumbral rings', as one would expect if these rings are produced by separation of the magnetic components of a Zeeman-sensitive line by a magnetic field.

Figure 4 shows a Zeeman-step scan made using the Ca I λ 6103 line with an entrance slit of 75 μ and exit slits each of 50 μ. The left and right columns show uncancelled Zeeman spectroheliograms, made from left-circularly and right-circularly polarized light respectively for a given sign of the quarter-wave plate (λ/4). The bottom figures are doubly cancelled Zeeman photographs made from some of the above pairs. The longitudinal magnetic fields shown in the Zeeman photograph correspond to a bright network visible in the spectroheliograms made at $\Delta\lambda = -0.02$ Å, near the core of the line. Proceeding further into the wings of the line this bright network occurs alternately on spectroheliograms in the left or right column depending on the sign of the

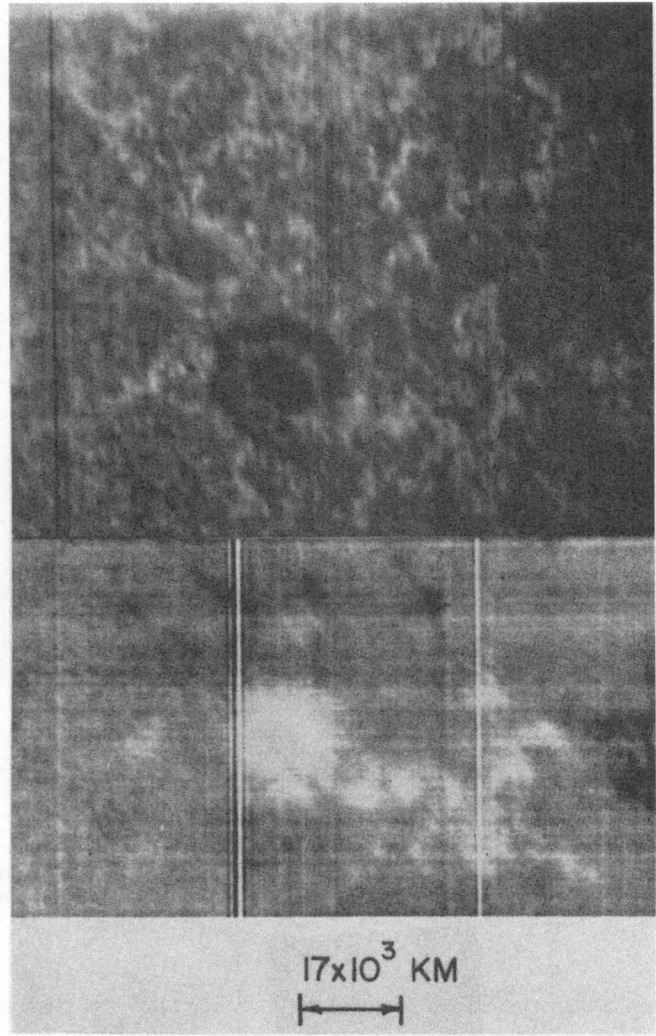

FIG. 3.   *Upper picture: SHG taken in the core of λ 5131 on August 2, 1967. Lower picture: Z-photo of low quality obtained on August 2, 1967 about 1 hour after the λ 5131 SHG.*

quarter-wave plate. We interpret this as an indication that wherever a line-of-sight component of magnetic field can be detected outside sunspots using Leighton's technique with the λ 6103 line, a locally bright feature can also be detected in the same solar position on a spectroheliogram taken in the core of this line.

Figure 5 shows a pair of spectroheliograms made simultaneously through two exit slits. The upper figure is a λ 5124 spectroheliogram and the middle figure is a λ 5131

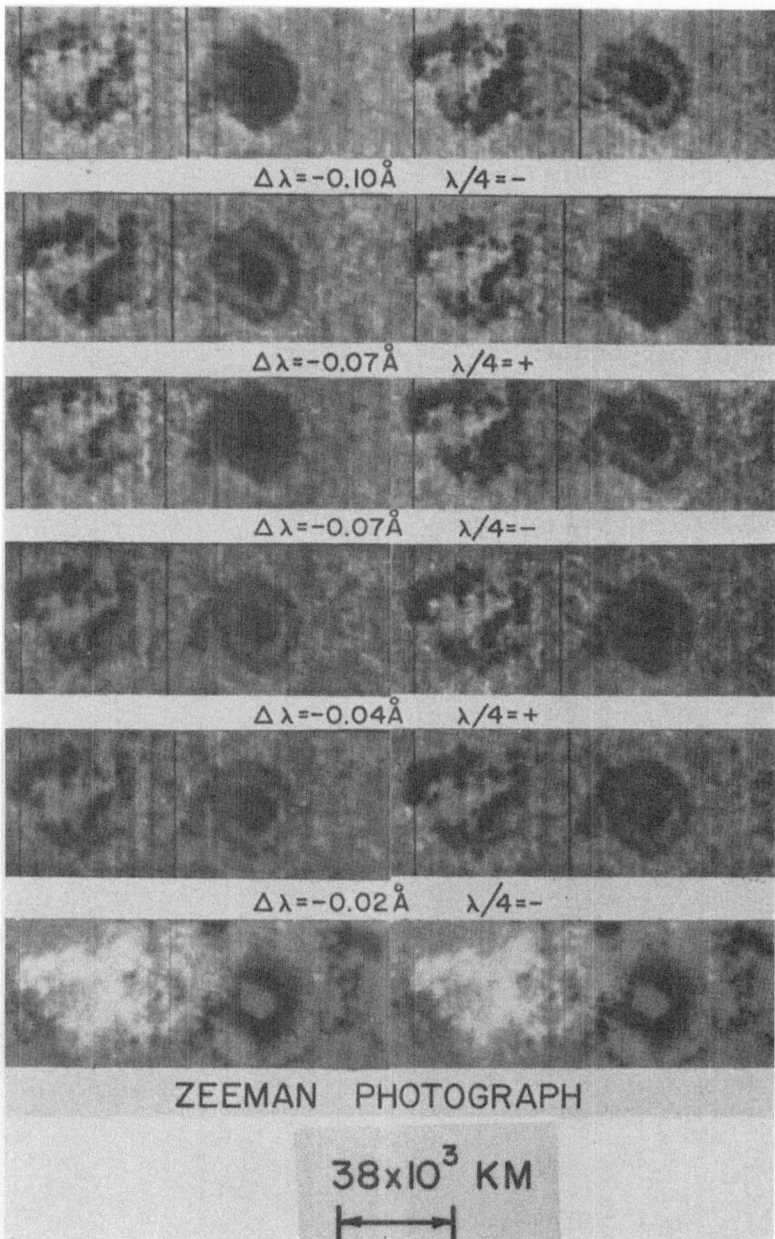

Fig. 4. *A sequence of uncancelled Zeeman spectroheliograms at different distances from the λ 6103 line core together with a doubly cancelled Z-photo for the same region (August 7, 1967).*

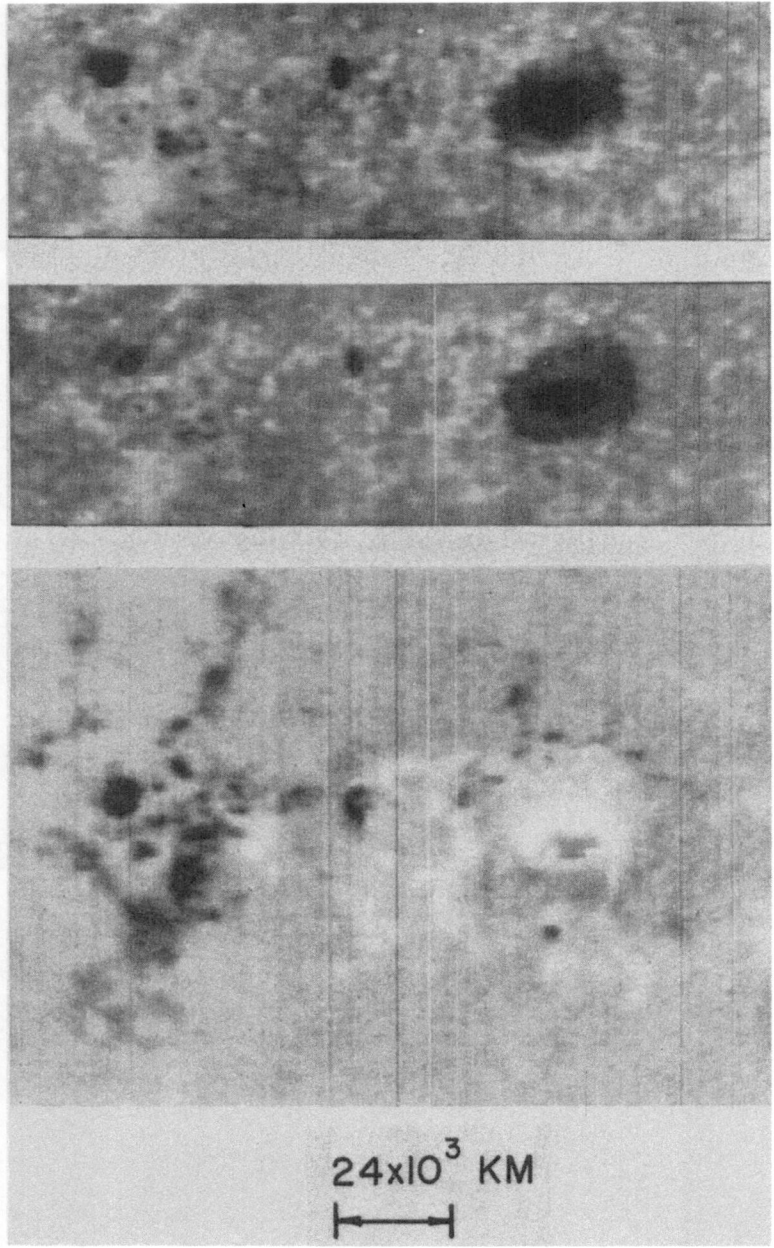

Fig. 5.    *Upper picture: SHG taken in the core of the Zeeman-insensitive line λ 5124. Middle picture: SHG taken simultaneously in the core of the strongly Zeeman-sensitive line λ 5131. Bottom picture: Doubly cancelled Z-photo taken about 1 hour earlier on June 27, 1967.*

spectroheliogram. Also included is a Zeeman photograph of the same region. A bright network corresponding roughly to the longitudinal magnetic fields is apparent in both the $\lambda$ 5124 and $\lambda$ 5131 spectroheliograms despite the fact that the $\lambda$ 5124 line is Zeeman-insensitive and the $\lambda$ 5131 is strongly Zeeman-sensitive. However, the bright network is definitely more pronounced in the $\lambda$ 5131 spectroheliogram than in the $\lambda$ 5124 one. Also the bright penumbral ring is evident in the $\lambda$ 5131 picture but not in the $\lambda$ 5124 picture. Outside sunspots, the field strengths vary from about 70 gauss for the fainter features to about 500 gauss for the more intense ones.

Figure 6 shows two spectroheliograms made simultaneously in the core of the $\lambda$ 5250·218 FeI line but through polaroids that were at right angles to each other. Note that the appearance of the bright penumbral ring is different in the two spectroheliograms, but other features in the two pictures appear the same, as shown in the

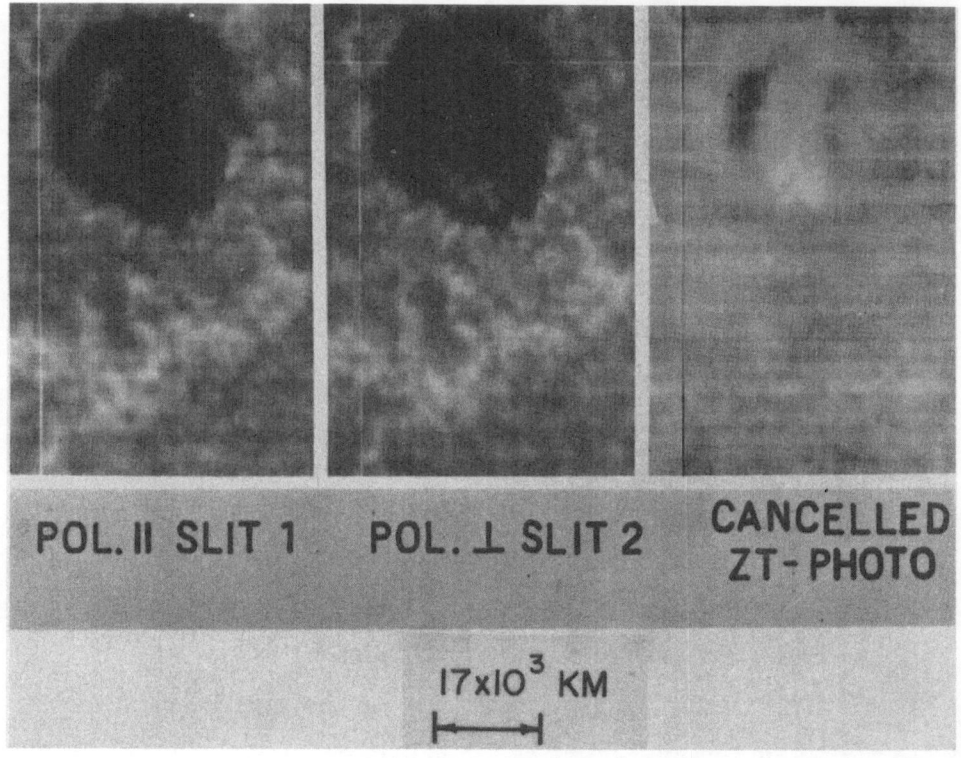

FIG. 6.   *Left picture: SHG taken in the core of $\lambda$ 5250 through a polaroid placed with its axis of polarization parallel to the exit slit. Middle picture; SHG taken in the core of $\lambda$ 5250 simultaneously with that in the left picture but with a polaroid placed with its axis of polarization perpendicular to the exit slit. Right picture: Cancelled transverse Zeeman photograph produced by placing a unity gamma contact print of the left picture in contact with the middle picture (June 29, 1967).*

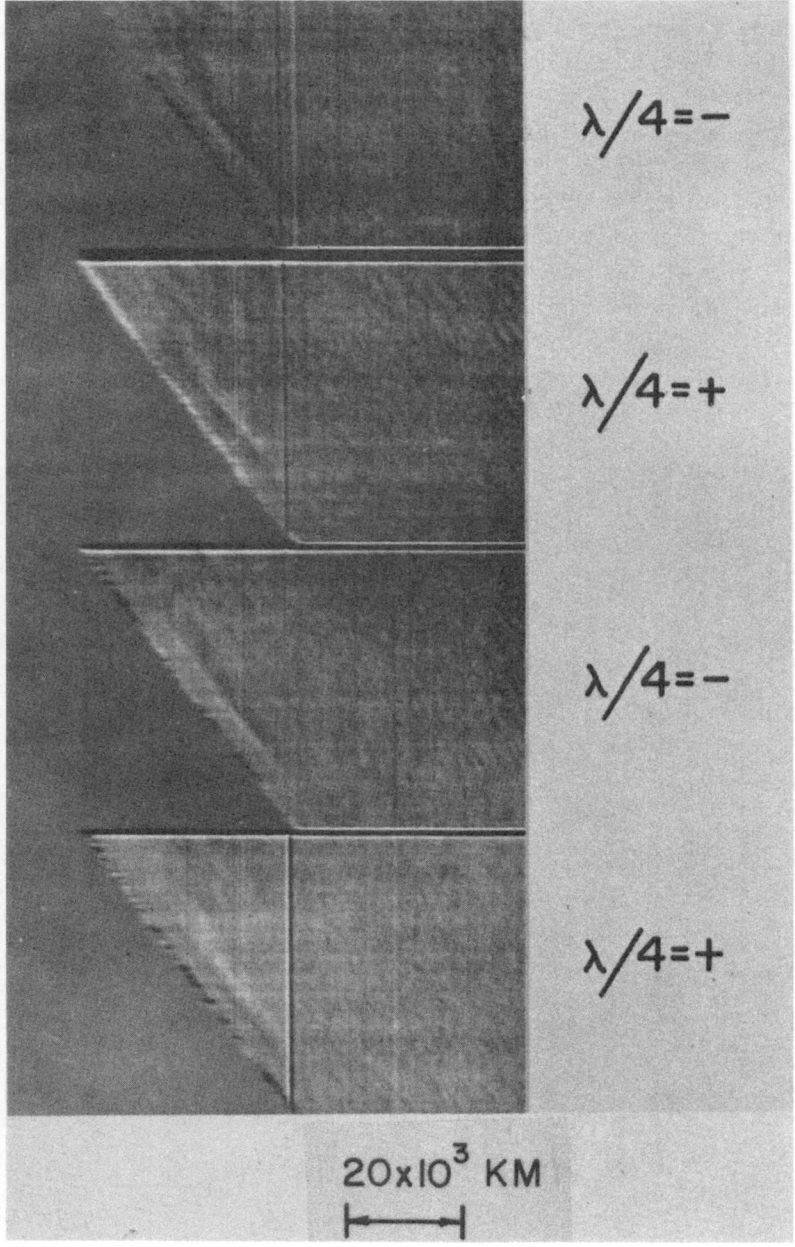

FIG. 7.    *A sequence of doubly cancelled Zeeman photographs showing the magnetic field in a facula near the East limb of the Sun on July 20, 1967.*

cancelled picture on the right. We interpret this result as an indication both that the component of magnetic-field transverse to the line of sight is directed approximately along the radius of the sunspot and that if such a transverse field exists in the nearby plage its field strength is less than our detection threshold of approximately 300 gauss.

Figure 7 shows a sequence of Zeeman photographs in which the quarter-wave plate was alternately $+45°$ and $-45°$ from the entrance slit. An elongated feature appears very close to the Southeast solar limb corresponding to the location of a large facula observed on the white-light image. This feature is bipolar in the sense that in the upper picture the limbward side is lighter-than-average whereas the centerward side is darker-than-average. This polarization reverses with each reversal of the quarter-wave plate, as shown in the figure. Calibration and measurement of this Zeeman photograph revealed that the line-of-sight component of the field was in the 70–100 gauss range. The 'neutral line' occurs at $85° 32'$ from disk center and the maximum field strengths occur at approximately $0° 26'$ (centerward) and $0° 21'$ (limbward) of this line. The range of angle over which the centerward side has detectable field is $1° 06'$, whereas the range of angle over which the limbward side has detectable field is $0° 36'$. These figures correspond to a feature which extends 13 000 km centerward and 7300 km limbward, measured along the solar surface. The polarity of the side of the facula nearest the disk *center* is the same (negative) as the polarity of the *following* spots of this same Southern hemisphere. This bipolar effect has often been seen in large sunspots near the solar limb and has been interpreted as a manifestation of an inclination of the sunspot field from the vertical (Hale and Nicholson, 1938; Leighton, 1960). Often the sunspots observed have been the long-lived leading spots that are typical of the remnants of sunspot groups, and the polarities of the centerward sides of these bipolar features have the polarities that the spot assumes as it comes into view on the solar disk. In the case described here, there were no spots visible when the Zeeman photograph was obtained and none came subsequently into view. We interpret this then as a measure of an inclination from the vertical of the lines of force associated with faculae by enough to produce a tangential field of roughly 70–100 gauss. Since the components normal to the surface are generally in the range 200–500 gauss, we estimate a divergence of the lines of force with inclinations in the range 10–20° from the normal to the solar surface.

Figure 8 shows two pairs of uncancelled Zeeman photographs obtained in the violet wing of the Na $D_1$ ($\lambda$ 5896) line. The first pair (upper row) has $\lambda/4 = +45°$ with respect to the entrance slit, and the second pair (lower row) has $\lambda/4 = -45°$ with respect to the slit. A doubly cancelled Zeeman photograph is included (twice) for comparison. These spectroheliograms were taken on May 3, 1967 during rather poor seeing, but in an active region where small flares and bombs were observed in a 1/4 Å bandpass H$\alpha$ filter. The appearance of the region in the H$\alpha$ filter looked considerably more like the doubly cancelled Zeeman picture than either of the uncancelled plates. There were filaments visible where the regions of opposite magnetic field are contigu-

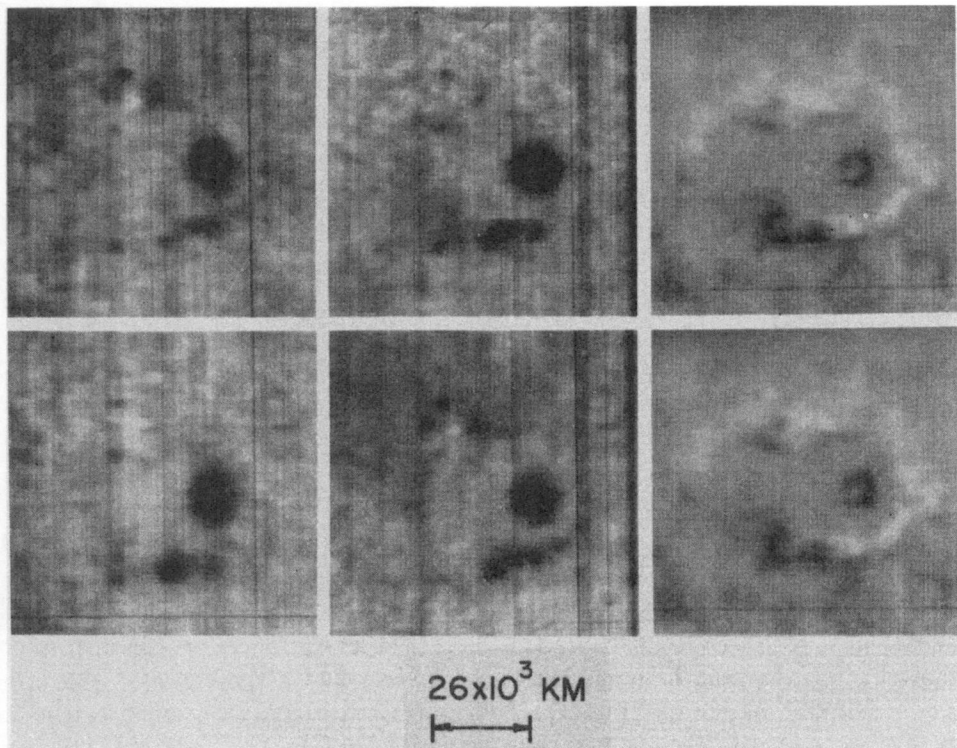

Fig. 8.    *Upper left: SHG taken in the violet wing of Na D$_1$ with a quarter-wave plate in front of the entrance slit and a polaroid parallel to the exit slit. Upper center: SHG taken simultaneously with that of the upper left in the violet wing of Na D$_1$ with a quarter-wave plate in front of the entrance slit and a polaroid perpendicular to the exit slit. Lower left and center: Same as upper left and center, except the axis of the quarter-wave plate was changed from +45° to the entrance slit to −45° to the entrance slit. Right column: Doubly cancelled Z-photos made from the four other pictures (poor seeing on May 3, 1967).*

ous. We see a bright feature in the left picture of the first pair (upper row) and the right picture of the second pair (lower row). This bright feature is also visible as a 'black' magnetic field in the doubly cancelled Zeeman photograph. We think that this Na D$_1$ brightening corresponds to a flare or bright bomb, and thus suppose that such features occur *in* regions of longitudinal magnetic field that are contiguous to regions having longitudinal magnetic field of opposite sign, but not *between* regions having fields of opposite sign.

Another example of such bright features in Na D$_1$ is given in Figure 9, which compares a spectroheliogram in the core of Na D$_1$ with a spectroheliogram taken in the Ca$^+$ K line on the same day.

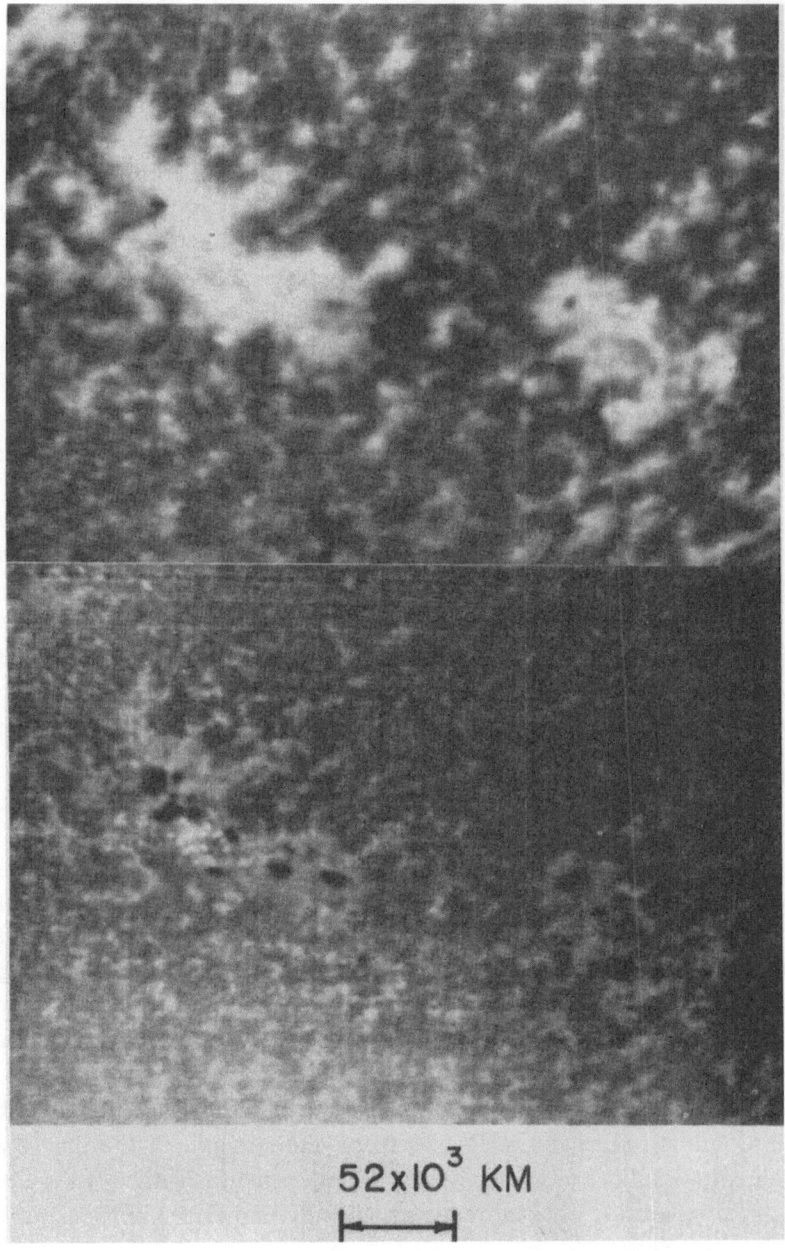

FIG. 9. *Upper: SHG taken in the core of the K line on January 30, 1967. Lower: SHG taken near the core of the Na $D_1$ ($\lambda$ 5896) line the same day.*

## 3. Summary

We summarize our preliminary results as follows:

(1) Although the photospheric network is visible on spectroheliograms in both Zeeman-sensitive and Zeeman-insensitive lines, it seems more pronounced on spectroheliograms taken in the cores of highly Zeeman-sensitive lines than on spectroheliograms taken in the cores of Zeeman-insensitive lines of the same strength. This suggests that the direct separation of magnetic components of Zeeman-sensitive lines by the magnetic field may play a significant role in producing the photospheric network. This is certainly true for the strong fields in sunspot penumbras where spectroheliograms in the cores of highly Zeeman-sensitive lines show 'bright penumbral rings' surrounding sunspot umbras while spectroheliograms in the cores of Zeeman-insensitive lines do not show such 'rings'.

(2) An attempt was made to see if the magnetic fields associated with the network deviate appreciably from the normal to the solar surface, first by photographically measuring the line-of-sight component near the solar limb, and second by photographically measuring the component transverse to the line of sight near the disk center. The measurements are consistent with a divergence of the lines of force from the network with inclinations of roughly 15° from the normal to the solar surface.

(3) Many bombs and flares that occur in Hα seem visible as corresponding brightenings on uncancelled Zeeman spectroheliograms made with the Na $D_1$ ($\lambda$ 5896) line, and produce detectable line-of-sight components of magnetic field on the cancelled Zeeman spectroheliograms.

## 4. Conclusion

We have begun to extend our study of the line weakenings or gaps associated with magnetic fields from the one-dimensional aspect provided by spectrograms to the two-dimensional view provided by spectroheliograms. We can easily find regions on the solar surface where the Fraunhofer lines are weakened, but we still do not know what causes these weakenings. Our preliminary observations suggest that they are produced by a combination of line formation under the physical conditions accompanying magnetic fields and the direct separation of Zeeman components by these fields, the dominant mechanism depending on the strength of the line. By taking spectroheliograms in lines of various strength and Zeeman sensitivity we are presently attempting to replace the gaps in our knowledge by knowledge of the gaps.

## References

Beckers, J. M., Schröter, E. H. (1966)    Communication at the Boulder A.A.S. Meeting, October.
Hale, G. E., Nicholson, S. B. (1938)    in *Magnetic Observations of Sunspots* (Carnegie Institution of Washington, Washington, D.C.), **1**, No. 498, p. 26 and Plate 7.

Leighton, R.B. (1959)     *Astrophys. J.*, **130**, 366.
Leighton, R.B. (1960)     unpublished observations.
Sheeley, Jr., N.R. (1966a)     *Astrophys. J.*, **144**, 723.
Sheeley, Jr., N.R. (1966b)     Communication at the Boulder A.A.S. Meeting, October.
Sheeley, Jr., N.R. (1967)     *Solar Phys.*, **1**, 171.
Simon, G.W., Leighton, R.B. (1964)     *Astrophys. J.*, **140**, 1120.

## DISCUSSION

*Bappu:* Could you elaborate on the techniques used, image size, dispersion, method of Zeeman shift detection, etc.?

*Sheeley:* The Kitt Peak spectroheliograph is similar to the one at Mount Wilson except that there are two exit slits with a prism-type beamsplitter just ahead of them at Kitt Peak, whereas at Mount Wilson there is a beamsplitter just ahead of the entrance slit and only one exit slit. The Kitt Peak solar image is 82 cm and the dispersion used was about 0.8 Å/mm. The techniques used were essentially those used by Leighton at Mount Wilson.

*Severny:* In my opinion the described effect of the increase of intensity in the core of metallic lines associated with strong magnetic field is essentially the same as that investigated by Dr. Tsap of Crimean Observatory. Tsap found well-expressed correlation between the strength of magnetic force ($H_\parallel$ as well as $H_\perp$) and the contrast $\Delta J/J$ in the core of the line in the sense that $\Delta J/J$ increases with increasing $H$. (See *Publ. Crim. astrophys. Obs.*, **35** (1966), 161.)

*Sheeley:* If I were asked how the fluctuations in line intensity might depend on magnetic field I would suggest that for sufficiently small fields the relation would be

$$\frac{\Delta J}{J} \sim 2B_\parallel{}^2 + B_\perp{}^2,$$

and since we might expect $B_\parallel$ to be roughly 3 times $B_\perp$, the longitudinal component, $B_\parallel$, would be expected to have the dominant effect. This relation comes from an expression of the form $(1 + \cos^2 \gamma)(\Delta\lambda)^2$, and does not include possible effects of line saturation.

*Krat:* I should like to remind that at first time after the beginning of regular magnetographic observations many 'invisible' spots with the field strength about of several hundred gauss were discovered. But an exact comparison of the results of magnetic-field scanning with direct photographs of the active regions with good resolution showed that always these 'invisible' spots coincide with very small sunspots which were not noticed at the process of scanning. The direct photographs of the Sun at high resolution obtained by Dr. Schwarzschild and his colleagues show many small spots in the vicinity of the greater ones. Such small spots cannot be seen on the photographs or spectrograms taken at lower resolution. As preliminary results I may mention that on the photographs taken by the first Soviet stratospheric station show spotlike objects of great contrast being not greater as 0″.4 in size. The resolution on Dr. Sheeley's photographs does not enable us to make definite statement that the magnetic fields observed by him in the intergranular space do not belong to small sunspots.

*Sheeley:* I do not agree.

*Kiepenheuer* (Question to Dr. Krat): Are your small structures of 0″.4 diameter really spots, or are they just dark structures?

*Krat:* (did not answer this question).

*G. W. Simon:* In one of the slides you showed an intensity-network pattern which you said originated in the photosphere, although the structure appeared to resemble closely the well-known chromospheric network structure. At what height in the atmosphere is the line formed which you used to obtain this spectroheliogram?

*Sheeley:* Although the $\lambda$ 5131 network occurs in roughly the same places on the solar surface where the $K_{232}$ network occurs, the $\lambda$ 5131 network has a more delicate or finer appearance than the $K_{232}$ network. Although we do not really know where the $\lambda$ 5131 line is formed, presumably it is well below the height at which the core of the K line is formed, and we interpret the finer structure of the $\lambda$ 5131 network as a further indication of this.

# THE SUPERGRANULAR PATTERN AND THE STABLE
# STAGES OF SUNSPOT GROUPS*

M. G. Dmitrieva, M. Kopecký**, and G. V. Kuklin

*(Siberian Institute of Terrestrial Magnetism, Ionosphere and
Radio Propagation, Academy of Sciences, Irkutsk, U.S.S.R.)*

Assuming the sunspot area to evolve smoothly and the sunspot-group lifetime distribution to be a monotonous decreasing function one can easily conclude that the function $N(S)$, determining the number of sunspot groups having the area S at this moment, must increase monotonously and smoothly with decreasing area $S$.

The real behaviour of the function $N(S)$ in the individual 11-year cycles no. 15–18 (according to the Zürich numbering), obtained by Kopecký (1964) using the Greenwich catalogue and analysed in more detail by Dmitrieva *et al.* (1967), is not in accordance with this theoretical conclusion. All principal peculiarities of the $N(S)$ behaviour can be illustrated by the $N(S)$-curve of cycle no. 17.

Firstly all $N(S)$-curves show a clear maximum at $S_0 = 2\text{--}5 \times 10^{-6}$ hemisphere. Secondly all $N(S)$-curves do not decrease smoothly and uniformly with the increase of $S$ at $S > S_0$. At several intervals of $S$ values the $N(S)$ decreasing rate is appreciably reduced or even becomes zero, i.e. $N(S)$ is then practically a constant within the range of these values (Figure 1). Therefore one may speak of two 'steps' in the $N(S)$-curve, the first one is within the area range from $8\text{--}13 \times 10^{-6}$ hemisphere and the second one within the area range from $60\text{--}200 \times 10^{-6}$ hemisphere. The reality of these steps may be confirmed by two other ways at least.

We can explain these peculiarities of the $N(S)$-curves in the following way. It is quite probable that a sunspot originates only when the magnetic field reaches some critical value. As the magnetic field occupies the solar surface for a *part* of definite size at this moment, so the originating sunspot cannot have infinitely small size and must have a definite minimal area. In that case the area $S_0$ corresponding to the maximum of $N(S)$ must be accepted to be the mean value of the minimal sunspot-group area in statu nascendi.

The presence of steps on the $N(S)$-curves indicates that a number of the sunspot groups with areas corresponding to these parts of the curves is more than it must be in accordance with the natural sunspot-group area distribution and consequently such sunspot groups occur more often. As at the $N(S)$ determination the sunspot groups of various evolution stages were used, so the mentioned steps on the $N(S)$-curves

---

* Presented by G. V. Kuklin.
** On leave from the Astronomical Institute of the Czechoslovak Academy of Sciences, Ondřejov.

*Kiepenheuer (ed.), Structure and Development of Solar Active Regions*, 174–177. © *I.A.U.*

correspond to the stable stages of the sunspot group evolution in the sense that in the evolution process with the continuous area change, the sunspot groups stay within the range of corresponding values of $S$ sufficiently longer than at other values of $S$.

It should be noted that the supergranule area is approximately equal to $160–230 \times 10^{-6}$ hemisphere, that is in quite good accordance with $S$ values of the second step on the $N(S)$-curve. Assuming the supergranules to fill the solar surface uniformly and closely one may easily calculate the area of the solar surface part between the adjoining supergranules to be of the order of $6–8 \times 10^{-6}$ hemisphere.

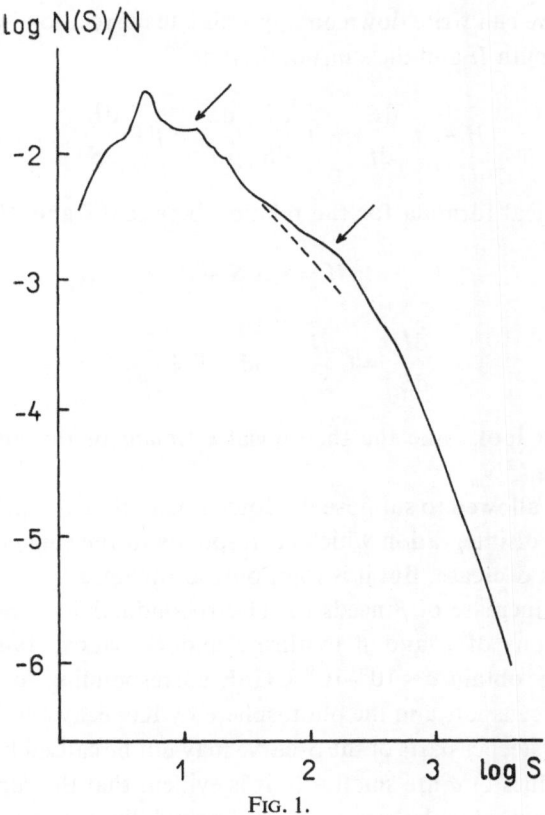

FIG. 1.

Using this fact and taking into consideration the results of papers by Bumba (1965), Bumba and Howard (1965), Kasinsky (1966), Bumba et al. (1966) we can suppose the following working hypothesis.

The sunspots as dark formations appear only at places where the magnetic field has reached a definite critical strength and a definite size. The originating sunspot has, on the average, a minimal area corresponding to the maximum of the $N(S)$-curve at $S = S_0$ close to the intersupergranular space area. The sunspots are born just in these spaces at the supergranular periphery. The leading and the following sunspots filling these spaces are more stable; that corresponds to the first step on the $N(S)$-curve. At the subsequent development the sunspot may fill the whole supergranule; that corresponds to another stage of higher stability of the sunspot groups, hence to the second step on the $N(S)$-curve.

Therefore we come to the conclusion that the supergranular pattern takes a determining part not only in the active-region structure but also in its evolution.

The linear and slow time variation of the sunspot area $S$ corresponds to stable stages of the sunspot evolution (Vitinsky and Ikhsanov, 1966; Ivushkina and Kuklin, 1967) $S = S^0 - \beta t$. Then we can write down an approximate correlation between the sunspot magnetic-field strength $H$ and the sunspot lifetime $T$

$$H \sim T \left| \frac{dH}{dt} \right| = T \left| \frac{dH}{dS} \right| \left| \frac{dS}{dt} \right| = \beta T \left| \frac{dH}{dS} \right|.$$

Using the empirical formula for the relation between $S$ and $H$ (Ikhsanov, 1966)

$$\lg H = c \cdot \lg S + d$$

we obtain

$$\frac{dH}{dS} = C \frac{H}{S} \quad \text{and} \quad T \approx \frac{S}{c\beta}.$$

The last formula looks like the theoretical estimate of the magnetic-field Joule dissipation time $T \approx L^2 / v_m$.

Therefore we are allowed to suppose the Joule dissipation as a principal mechanism of a magnetic-field disintegration which corresponds to the sunspot-evolution stages with the linear-area decrease. But it is right only at the linear parts of the $S$-variation curve, because the increase of $\beta$ needs the electroconductivity $\sigma$ decrease. Using the numerical estimations of $c$ and $\beta$ (Vitinsky and Ikhsanov, 1966; Bumba, 1963; Ikhsanov, 1966) we obtain $\sigma \sim 10^8$–$10^9$ CGSE corresponding to the lower limit of estimations for the sunspots and the photosphere by Kopecký and Kuklin (1966) and Kuklin (1966). The steeper parts of the $S$-curve may not be caused by Joule dissipation as the necessary values of $\sigma$ are small too. It is evident that the rapid decrease of the sunspot area is connected with some process of unstability which apparently occupies the whole active region and leads to an activity rise (non-stationary processes). But the sunspot is absolutely stable against macroscopic magnetohydrodynamic perturbations.

Hence the supergranular pattern in the active region is a stabilizing factor in the sense that the magnetic fields being inside the supergranular borders (inner and outer) have a more stable configuration, and their changes are caused only by Joule dissipation leading to a slow linear decrease of the sunspot-group area at this evolution stage. The magnetic fields not joined to this pattern are less stable and are being destroyed by the same pattern.

## References

Bumba, V. (1963)      *Bull. astr. Inst. Csl.*, **14**, 91.
Bumba, V. (1965)      in *Stellar and Solar Magnetic Fields*, Amsterdam, p. 192.
Bumba, V., Howard, R. (1965)      *Astrophys. J.*, **141**, 1492.
Bumba, V., Kopecký, M., Kuklin, G. V. (1966)      *Bull. astr. Inst. Csl.*, **17**, 57.
Dmitrieva, M. G., Kopecký, M., Kuklin, G. V. (1967)      *Izv. SibIZMIR*, **2**.
Ikhsanov, R. N. (1966)      *Izv. glav. astr. Obs. Pulkove*, **180**, 41.
Kasinsky, V. V. (1966)      *Issled. po geomagn. i aeronom*, Moskva, p. 196.
Kopecký, M. (1964)      *Bull. astr. Inst. Csl.*, **15**, 44.
Kopecký, M., Kuklin, G. V. (1966)      *Bull. astr. Inst. Csl.*, **17**, 45.
Kuklin, G. V. (1966)      *Rezultaty nabl. i issled. v. period MGSS*, vyp. I, Moskva, p. 17.
Vitinsky, Y. I., Ikhsanov, R. N. (1966)      *Izv. glav. astr. Obs. Pulkove*, **180**, 20.

## DISCUSSION

*H. U. Schmidt:* As I said yesterday, I find it hard to understand a decay of the whole sunspot field with the local timescale computed from the local low conductivity in the umbra because in such a decay the flux has to contract into the photospheric currents surrounding the umbra and that should take a time consistent with the high conductivity in the layers which the flux has to trespass. I think it might be possible to have the same decay behaviour which you describe caused by a flux diffusion due to the local structure of the convection surrounding the umbra.

*Kuklin:* We have not aimed to explain why the linear part of the sunspot-area variation curve is existing but only wanted to show the Joule dissipation may be one of possible mechanisms causing this fact.

# THE INTENSITY, VELOCITY, AND MAGNETIC
# STRUCTURE IN AND AROUND A SUNSPOT*

J. M. BECKERS          and          E. H. SCHRÖTER

(Sacramento Peak Observatory, AFCRL,     (Universitäts-Sternwarte, 34, Göttingen,
Sunspot, N.M., U.S.A.)                        Germany)

ABSTRACT

An observational study of the fine structure in a sunspot region led to the following results: (a) The photosphere around a sunspot is covered with the so-called Magnetic Knots. These features have a diameter of 800 km and a magnetic field of up to 1400 gauss. Although they coincide with dark, intergranular spaces, they are distinctly different from pores. We estimate some 300 to surround the sunspot. The magnetic field of the smallest pores were found to be 1500 gauss. (b) For the Umbral Dots we find a lifetime of 25 min. Their colour, as derived from simultaneous observations at 4700 Å and 6400 Å, was found to be identical to that of the photosphere. Assuming their brightness to be photospheric we derive a diameter of 160 km.

## 1. Observations

With the setup shown in Figure 1 we observed a number of large, fairly symmetrical sunspots during their passage across the solar disc in order to make a detailed study of the intensity, velocity, and magnetic field structure in and around a sunspot. We used simultaneously the 40-cm coronagraph and the 30-cm coelostat telescopes of the Sacramento Peak Observatory. The observations obtained at the West bench consist of high dispersion, simultaneous spectra (7 mm/Å) of the Fe lines $\lambda$ 5576 ($g=0$) and $\lambda$ 6173 ($g=2\cdot5$) exposed 10 sec at a rate of 15 sec. The Zeeman-line spectra were obtained through a $\lambda/4$ plate and a Wollaston prism. Slit-jaw photographs exposed simultaneously with the spectra allow a very precise determination of the slit position with respect to the sunspot fine structures. We placed the slit in many different positions across the spot covering also extensively the region around the spot.

The coelostat telescope fed the East bench camera which – through the use of a dichroic mirror – allows the simultaneous photography of two spot images in either red and blue continuum radiation ($\lambda$ 6400 Å and $\lambda$ 4700 Å resp.) or in H$\alpha$ and blue light. The East bench observations made at a 5-sec rate cover sometimes 1 to 1½ hours of intermittent good seeing. From these we are able to study both motions and changes of the spot fine structures. By comparison with the slit-jaw pictures we are moreover able to relate the fine structures visible in the spectra with details in the white-light images.

---

\* Presented by J. M. Beckers.

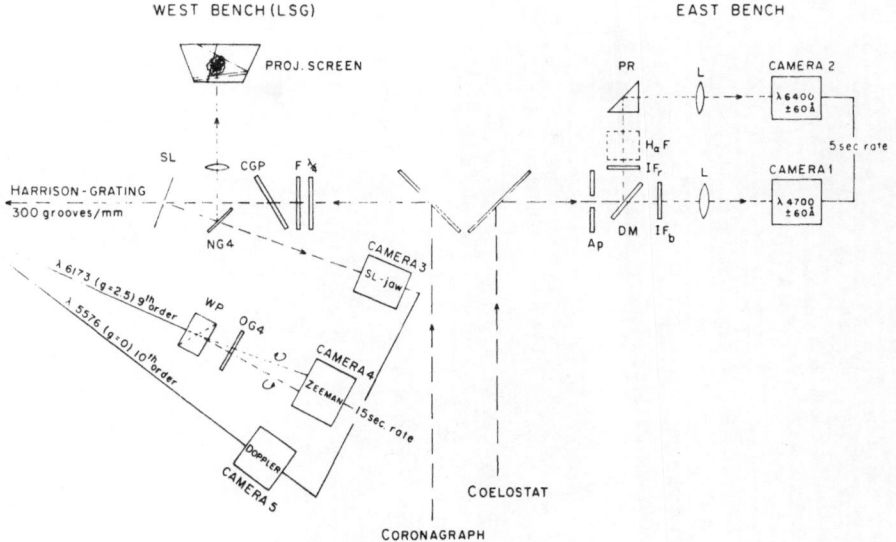

FIG. 1. *Diagram of the observational setup. – West bench: 40-cm coronograph with solar image 25 cm; LSG = Littrow spectrograph; F and OG 4 = colour filters; CGP = glass plate to correct for instrumental polarisation; SL = slit with reflecting jaws; NG 4 = neutral density filter; first surface is used to form an image of the solar region for visual monitoring; WP = Wollaston prism to separate the two opposite circular polarised spectra. – East bench: 30-cm coelostat with final solar image 25 cm; Ap = aperture defining the solar region to photograph; DM = dichroic mirror; IF_b and IF_r = blue and red interference filters; Hα F = Hα-Lyot filter with 0·5 Å bandwidth.*

A test of the performance of our photographic method for the magnetic field determination was made by investigation of the magnetic field in the quiescent photosphere. This showed the noise for our method to be 5 gauss r.m.s. for an effective scanning aperture of $(0.8 \text{ sec of arc})^2$. A recording time of 10 sec is needed for 600 of these resolution elements. Comparison with the noise characteristics of the Babcock magnetograph shows the noise of the photographic method to be less by about 1 order of magnitude. It takes about 30 times less time to compile a magnetic and velocity map for an active region with the photographic method.

The spectra of the undisturbed photosphere give after correction for noise an r.m.s. magnetic field of 7·5 gauss. This r.m.s. magnetic field is mainly due to a number of isolated regions a few seconds of arc in diameter in which the magnetic field is of order of 50 gauss. We find no significant correlations between the magnetic field and either continuum brightness or velocity.

Figure 2 shows a high-resolution photograph of the sunspot which we selected for our study.

## 2. Magnetic Knots

In the vicinity of spots we detected many isolated points (<1500 km) with strong

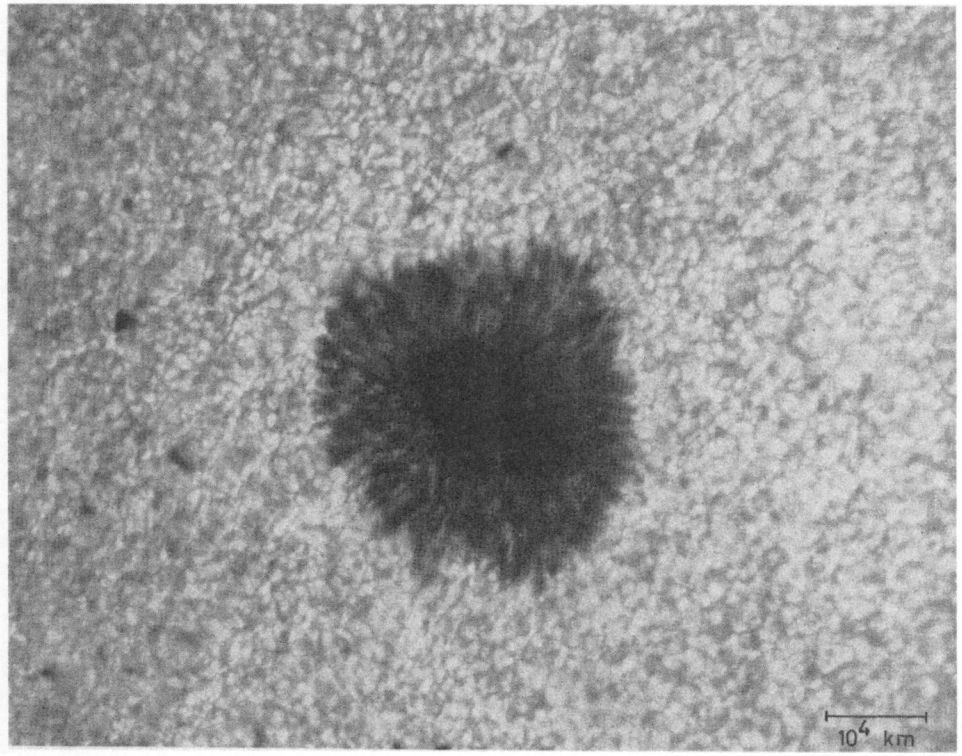

FIG. 2.    *White-light photograph of the sunspot selected for the study: July 25th 1966, exposure time*
*15 milliseconds on Kodak High Contrast Copy emulsion, λ = 4700 ± 60 Å.*

magnetic fields (see Figure 3). These 'magnetic knots' coincide with normal dark intergranular spaces and are clearly distinct from pores. They show a striking decrease in line depth; the equivalent width remaining unchanged.

An interpretation of the line profile measured in the magnetic knots in terms of the Stepanov-Unno theory (see Mattig, 1966) results in magnetic fields of about 1300 gauss. The inclination of the field with respect to the line of sight amounts to $\gamma \sim 75°$ when no correction is applied for light not originating in these magnetic knots. However, the inclination angle depends strongly on such a correction. Since the spatial resolution of our spectra is probably less than the actual size of the magnetic knots we expect the amount of this 'non-magnetic light' to be high. A vertical field ($\gamma = 0°$) in all magnetic knots is obtained for $\sim 60\%$ of the observed light to be blurred into the magnetic knots. This gives 800 km for the true diameter of the magnetic knots and 1400 gauss for the field.

The magnetic knots described here are probably identical to the so-called 'invisible sunspots' observed by Hale and Nicholson (1938), to the 'gaps' with apparent field

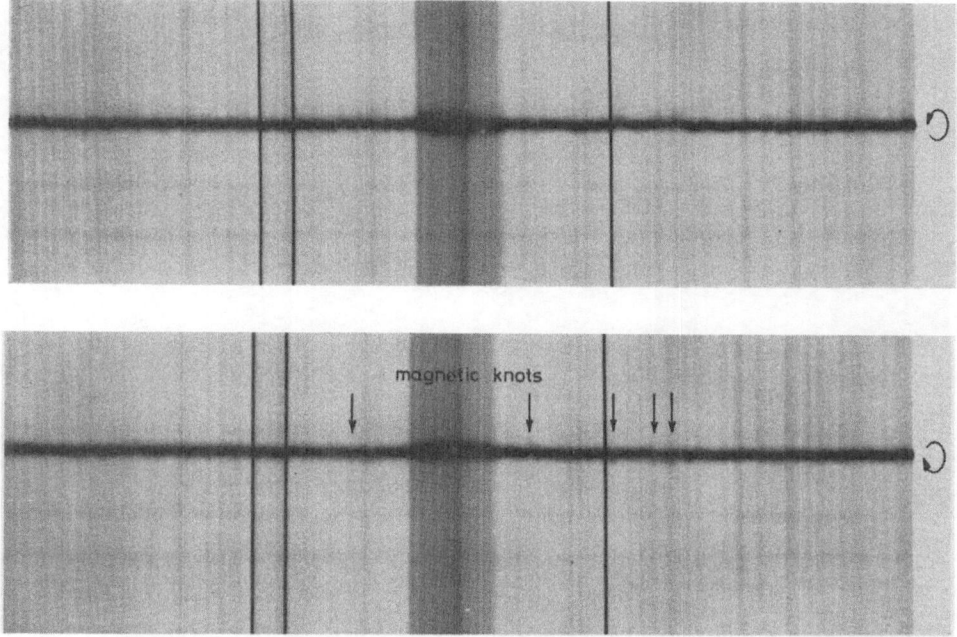

FIG. 3.  *Clockwise and anticlockwise circular polarised spectra around the Fe line λ6173 (Landé-factor 2.5 with the slit intersecting the penumbra. The arrows indicate a number of magnetic knots. Note: magnetic field produces a line shift of opposite direction, whereas Doppler effect gives a line shift of the same direction in both spectra.*

strength of 350 gauss observed by Sheeley (1967) and the 1000 gauss non-spot regions recently discussed by Steshenko (1967). We find the magnetic knots to have mainly a red shift in the non-magnetic line indicating a downward motion, and to be located in the Hα-plage regions. The non-magnetic line shows generally no weakening, confirming again that the change of the Zeeman line is of magnetic origin. The polarity of the knots varies at random, it can be both equal or opposite the spot polarity. We do not know yet the knot lifetime. We estimate our spot to be surrounded by 300 knots.

For some pores surrounding the spot we found a field strength of about 1500 gauss, so that about 1400 gauss appears to be the critical value for pore and spot formation.

### 3. Umbral Dots

We obtained some very good time sequences of umbral structures simultaneously at λ 6400 Å and λ 4700 Å covering 1½ hours in time at a rate of one photograph per 5 sec. On these photographs (see Figure 4) the umbra shows isolated bright points (<600 km). These umbral dots (called so by Danielson, 1964) form a pattern very

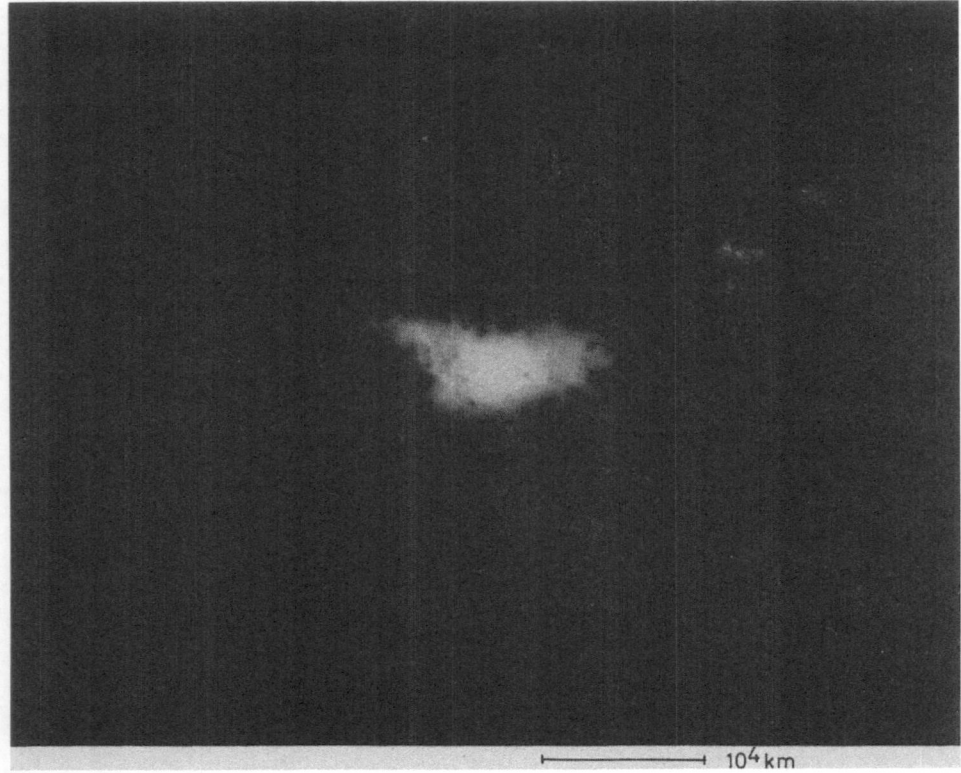

FIG. 4.   *Photograph of umbral dots at 4700 Å; July 20th 1966; exposure time 50 milliseconds on Kodak HCC-emulsion. Average observed diameter of the dots: 500 km.*

different from the photospheric granulation; we therefore disagree with the term 'umbral granulation' as used by Bray and Loughhead (1964).

In the particular spot under study the umbral dots were all located in the outer parts of the umbra. It is not sure yet if this is the case for all spots of this type. We did observe a number of sunspot umbrae in which the umbral dots are suppressed in parts of the umbra. This area of the umbra was generally the darkest.

We have determined a lifetime of 25 min for the structures from a good time sequence 1½ hours long. After 25 min no dots can be recognized except for statistical coincidences. This is a factor 2 smaller than that found by Bray and Loughhead (1964) for their umbral granulation. But it falls within the limits of 4 and 50 min given by Danielson (1964). We found the visibility curves of the individual dots to be asymmetrical in the sense that they show a faster increase to maximum brightness than the decrease thereafter.

For 10 umbral dots we measured an average diameter (halfwidth of the intensity peak) of 500 km, which is an upper limit for the true diameter of the features. The contrast with the umbral background is 44% at 4700 Å and 26% at 6400 Å, so that umbral dots are bluer than the umbra. The excess intensity expressed in terms of the photosphere is resp. 7·8% and 7·9%, meaning that the umbral dots have the same colour as the photosphere. In order to make the dot intensity equal to the photosphere intensity one would need a true diameter of 160 km. There is no way to separate the dot intensity of its diameter; we can give only limits. The true diameter lies between 160 km and 500 km, if one takes the photospheric brightness as the upper limit for the true dot intensity. No lower brightness limit can be given because we observed a number of dots with very low contrast.

We do not exclude the possibility that at least some of the umbral dots are of photospheric intensity and therefore 160 km in diameter since the blurring on our photographs could be as much as 500 km. In fact the white colour of the brighter dots is very suggestive for this possibility.

## 4. Magnetic- and Velocity-Field Measurements in a Sunspot

From each of the 9 days at which we observed the spot in Figure 2 we selected from ∼600 spectra the best 15–17 spectra, covering the entire spot area and the surrounding photosphere. After a lengthy investigation we found that there was no satisfactory short cut for the reduction of the magnetic spectra without losing a significant amount of information. We therefore undertook a detailed photometric reduction of the line profile of $\lambda$ 6173 in each circular polarisation at each point along the slit (every 600 km on the Sun). This reduction is not yet completed.

Figure 5 shows a flow diagram of our reduction procedure.

First the line center is determined by the elimination of the Doppler shift. The Doppler shift is obtained by shifting a mirror image of the one spectrum over the other spectrum, until the maximum correlation between the two spectra is obtained. The difference curve (the intensity difference between spectrum 1 and spectrum 2) is then fitted by least squares with the difference of two functions $f$. These functions have the same shape as the photospheric profile but are displaced by an amount $\Delta$ (separation of the $\sigma$-components). The quantity $\Delta$ is approximately equal to the Zeeman splitting of the line. The amplitude $A$ of the function is determined by the field angle $\gamma$ (the relative strength of both $\sigma$-components). We used the Stepanov-Unno theory for line formation in magnetic field to calibrate the $\Delta$ and $A$ in terms of $H$ and $\gamma$ by fitting the difference of theoretical line profiles of $\lambda$ 6173 in the same way.

With the $H$-value found from the difference curve the profiles of the individual spectra are analysed. By least squares we find the amplitudes of the two $\sigma$- and the $\pi$-components. These in turn give a more precise value for the field direction.

Figure 6 gives the result of such a reduction for one of our spectra.

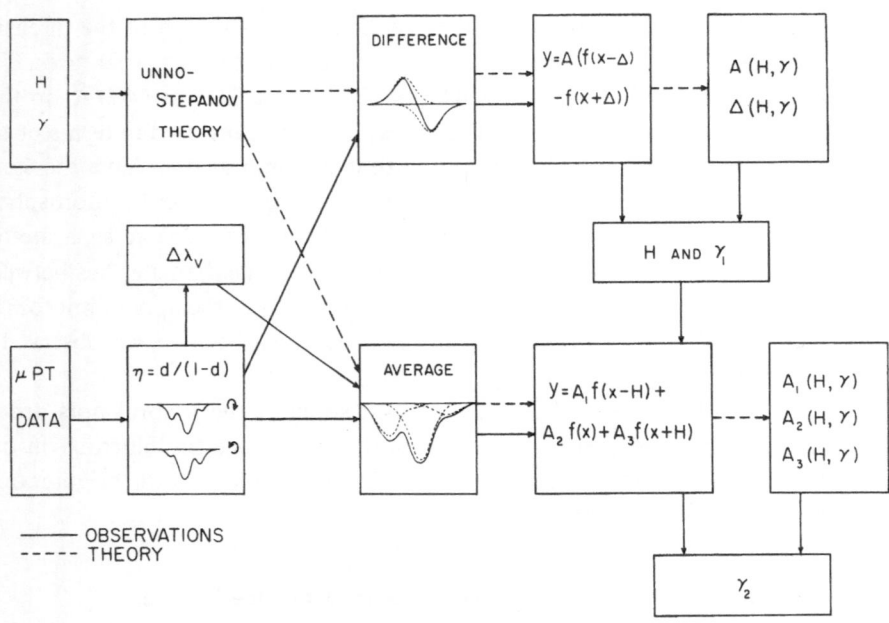

FIG. 5.  *Flow diagram of the interpretation of the Zeeman-line profiles in terms of the Stepanov-Unno theory.*

It is our intention to make reductions like the one given in Figure 6 for all magnetic and velocity spectra we selected from this spot and which we have traced already. Our observational data are of sufficient quality to relate the magnetic and velocity field and its fluctuation to the penumbral intensity structures.

The detailed publications will be submitted to *Solar Physics*.

## Acknowledgements

The work was done when one of us (E.H.S.) was guest investigator at the Sacramento Peak Observatory under contract AF 628-4078 with the National Center of Atmospheric Research, Boulder. We acknowledge the invaluable help of the Sacramento Peak big dome observing personnel.

## References

Bray, R. J., Loughhead, R. E. (1964)      *Sunspots*, Chapman and Hall Ltd., London, p. 85.
Danielson, R. (1964)      *Astrophys. J.*, **139**, 45.
Hale, G. E., Nicholson, S. B. (1938)      *Carnegie Inst. Washington*, **1**, No. 498, 37.
Mattig, W. (1966)      *Atti del Convegno sulle macchie solari*, p. 194, *Firenze.*
Sheeley Jr., N. R. (1967)      *Solar Phys.*, **1**, 171.
Steshenko, N. V. (1967)      *Izv. Krym. astrofiz. Obs.*, **37**, 21.

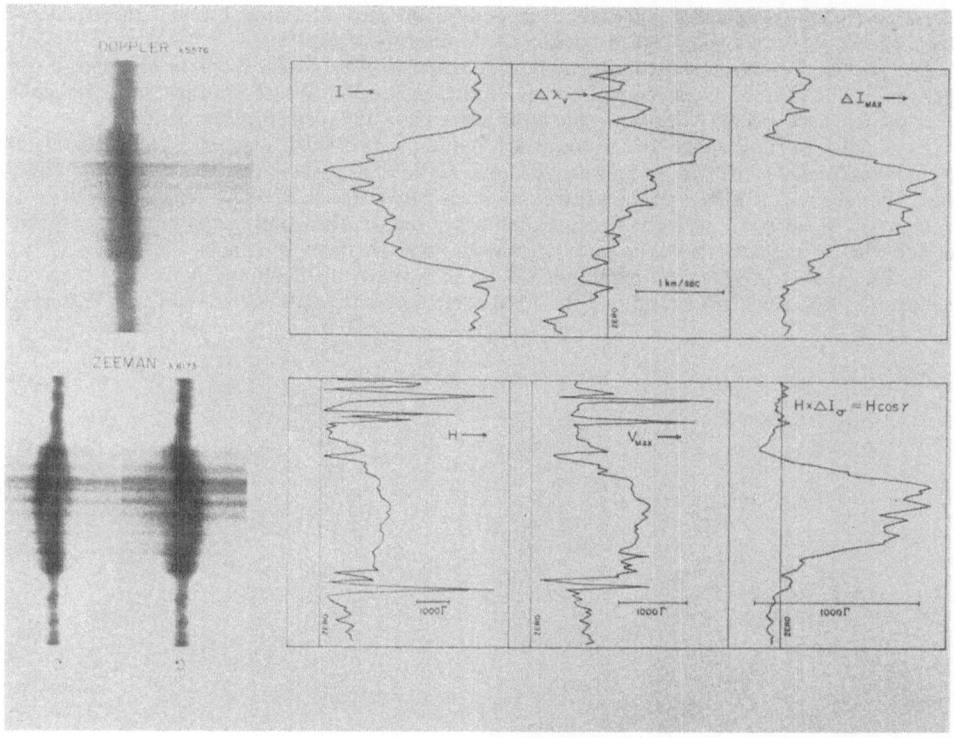

Fig. 6.  *Reduction of a Zeeman and Doppler spectrum obtained on 25th July 1966 (spot near disc center). I = continuum intensity across the slit; $\Delta\lambda_v$ = Doppler shifts across the slit (Evershed effect); $\Delta I_{max}$ = the maximum amplitude of the difference curve (see Figure 5) across the slit ($\Delta I_{max}$ is very close to $H\cos\gamma$): H = magnetic field across the slit: for small Zeeman splitting or for transversal field the position of the $\sigma$-components becomes very unreliable. This results in the large noise of the H-values outside the penumbra. Note the local fluctuations of I, $\Delta\lambda_v$ and $H\cos\gamma$. At points where the last curve intersects the zero line ($H\times\Delta I_\sigma = 0$) we have transversal magnetic field. Looking for these points in spectra exposed at different disc positions of the spot, we obtain the field configuration free from any theoretical assumptions.*

# DISCUSSION

*Rösch:* About umbral dots (French: 'points maculaires'): I agree that they should not be called 'granules' because they differ in nature from the photospheric ones. As for the distribution of these dots, there are at least cases where the whole umbra is filled with bright dots with different average level of brightness in different parts of the umbra. Before deciding that no dots appear, one should be sure that the exposure time puts them in the high-contrast part of the emulsion-characteristic curve. And of course, the stray-light plays an important role.

*Beckers:* We have exposures of various densities of the umbra of the spot in discussion, some of these give the high-contrast part of the emulsion in the darkest part of the umbra. These do not show umbral dots in this part of the umbra. Scattered light would indeed diminish the contrast of the umbral dots, we took this into account.

*Howard:* Can you from your counts of the number of 'gaps' in a plage estimate the fraction of the magnetic flux in the plage which exists in the form of these 'gaps'?

*Beckers:* We can estimate the magnetic flux contained in these knots. We have not made a comparison of this with the average magnetic flux of the plage region. We do not exclude the possibility that the magnetic knots (= gaps) are the basic elements of the plage magnetic fields.

*H. U. Schmidt:* You spoke about downward motion in the small magnetic knots outside the penumbra. Would it be consistent with your observations to assume that the downward motion surrounds the magnetic field or does it have to be exactly at the peak of the magnetic field?

*Beckers:* Because the magnetic knots are probably smaller than our resolution disk we cannot decide whether the downflow surrounds or coincides with the magnetic knots.

*Sheeley* (answer to Howard's question): One can see where the fields are by looking at spectroheliograms. The fields are not all in points, but sometimes lines and other complicated patterns.

# FINE STRUCTURE OF BRIGHTNESS, VELOCITY AND MAGNETIC FIELD IN THE PENUMBRA

W. MATTIG and J. P. MEHLTRETTER

*(Fraunhofer-Institut, Freiburg i.Br., Germany)*

As a preparation for the observational program of the first flight of the 'Spectro-Stratoscope' balloon-borne solar telescope, an attempt was made with the 35-cm aperture domeless Coudé refractor of the Capri Station of Fraunhofer Institute to obtain spectra of penumbral fine structure.

A polarizing image-splitting device (quarter-wave plate and Wollaston prism in front of the spectrograph slit) was used in order to obtain simultaneous spectra of the two states of circular polarization, containing the magnetically sensitive line $\lambda$ 6302·499 of FeI (Landé factor 5/2). This line has two sharp telluric lines ($\lambda$ 6302·000 and $\lambda$ 6302·764 of $O_2$) as immediate neighbors serving as wavelength reference.

The exposure time of the double spectra was 15 sec on Kodak 103a-E film. The image scale on the slit is 5″9 per mm, and the reciprocal dispersion 6·92 mm/Å.

A series of spectra was taken under very good seeing conditions, May 29, 1967 of the spot 17°S, 44°W, classification J1. The slit was set onto the penumbra at the limb's distant side, nearly tangential (deviation 20°) to the solar limb. The exposures were triggered by constantly watching the signal of the seeing monitor and thus selecting the periods of best seeing. Nevertheless, the individual spectra of this series have different resolution. So far, only the best spectrum, shown in Figure 1, has been analyzed.

Measurements of the spatial variation (along the slit height) were made of:

(a) the intensity $I(x)$ of the local continuum,

(b) the radial velocity $v(x)$,

(c) the component $H'(x) = H(x) \cos \gamma(x)$ of the magnetic field,

(d) the line asymmetry $a(x)$,

by the following two methods:

Firstly, from a total of 200 microdensitometer tracings covering the two spectra in steps 0″45 apart, the intensity of the local continuum, the wavelength of the line center, and the line asymmetry have been determined.

Secondly, an equidensitogram (Figure 2) has been made by photographic procedures (Sabattier effect), and the line position and line width have been measured in steps 0″35 apart. In this case, the apparent width of the telluric lines was taken as an equivalent to the intensity of the local continuum.

*Kiepenheuer (ed.), Structure and Development of Solar Active Regions, 187–192. © I.A.U.*

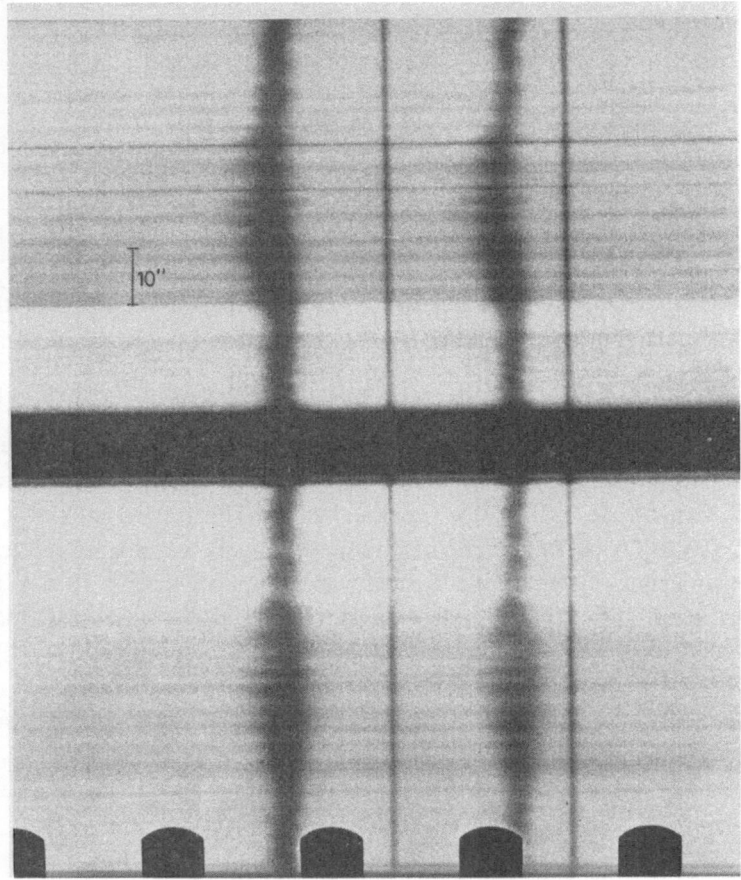

FIG. 1.   *Spectrogram showing penumbral fine structure, May 29, 1967, Spot 17°S, 44°W.*

The exact superposition of the two spectra and the measured values (wavelength) was accomplished by fitting the intensity curves (or line-width curves of the telluric lines) together.

Figure 3a summarizes the results obtained from the microdensitometer tracings, Figure 3b those from the equidensitogram.

Although in Figure 3a the velocity and intensity fluctuations are not well pronounced over the whole section of the penumbra, a good correlation appears between bright elements, smaller than average velocity, and stronger than average magnetic field.

In Figure 3b, the fluctuation of the magnetic field is small, but fluctuations of velocity and 'intensity' show the same correlation as in Figure 3a.

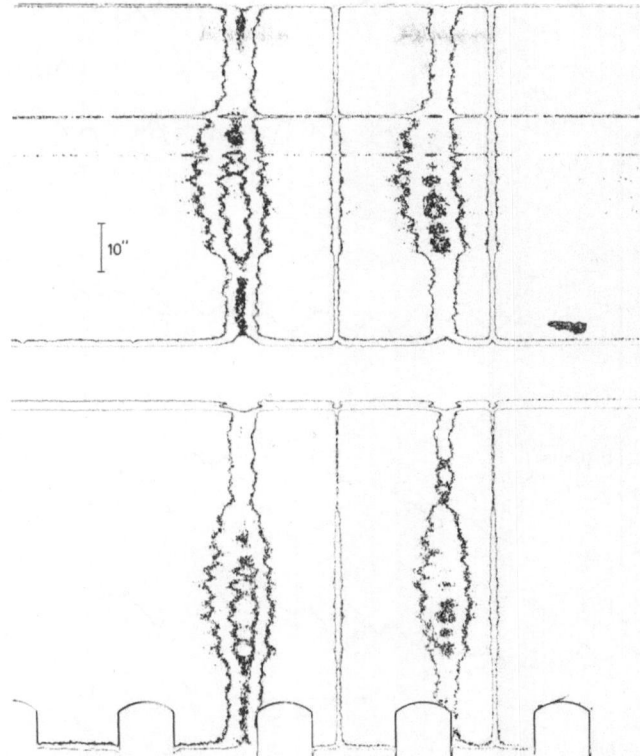

FIG. 2.   *Equidensitogram of the spectrum shown in Figure 1.*

We would conclude, therefore, that there is a real correlation in the penumbra of
(1) bright elements, strong magnetic-field component, small radial velocity,
(2) dark elements, weak magnetic-field component, high radial velocity.
Further results can be summarized as follows:

The mean penumbral intensity is 0·75 of the photospheric intensity. The measured
peak-to-peak fluctuation of intensity is about 10%.

The average radial velocity observed is 1 km/sec. The peak-to-peak fluctuation is
different depending on the height in the line contour. At the line center, $\Delta v$(Max.–
Min.) is 700–800 m/sec, somewhat higher in the contour (where the line width is about
200 mÅ), we find $\Delta v \approx 500$ m/sec, hence $\Delta v/v = 0.2$ to 0·4, depending on the line
contour.

The average magnetic-field component $H'$ is about 700 gauss, the fluctuation $\Delta H'$
about 200–300 gauss, hence $\Delta H'/H' = 0.3$.

Clearly, since the component $H(x) \cos \gamma(x)$ is observed, any fluctuation may be
due to either $H(x)$ itself or due to variation of the inclination angle $\gamma(x)$. Moreover,

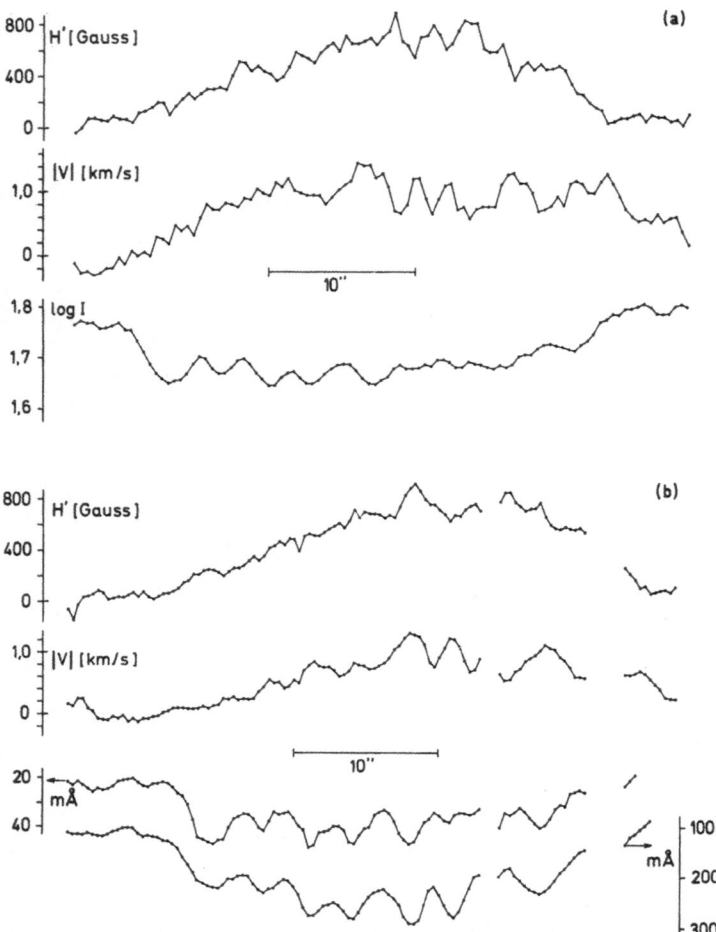

FIG. 3.    (a) *Results of analysis of microdensitometer tracings.* (b) *Results of analysis obtained from the equidensitograms.*

since the line is far from being completely split, the values given may be lower than the actual $H'$.

The asymmetry of the iron line was found to be essentially the same in both spectra, so we are allowed to conclude that the asymmetry is produced by unresolved velocity fluctuations rather than magnetic-field fluctuations. The same follows from the contour dependence of the measured $v$, and from the fact that the fluctuations in line width measured on the equidensitogram are approximately the same for both spectra.

In discussing the observation reported above, it should be stated that the spatial resolution (1″ to 2″) of the spectrum is far from being sufficient for detecting individual penumbral filaments (diameter <0.″4, distance 2″ or less). Most likely, the bright elements in the spectrum represent accumulations of penumbral filaments. This fact is supported by the behaviour of the center-line of the line contour, which occasionally shows distinct steps.

As far as the correlation between brightness and velocity is concerned, the results are in agreement with earlier findings of Beckers (1966), and in contradiction with the hypothesis put forward by Schröter (1965). Clearly, more and even better observations are needed in order to make a sound study of the phenomena possible.

## References

Beckers, J.M. (1966)    in *Convegno sulle macchie solari, Florence*, p. 186.
Schröter, E.H. (1965)    *Z. Astrophys.*, **62**, 228, 256.

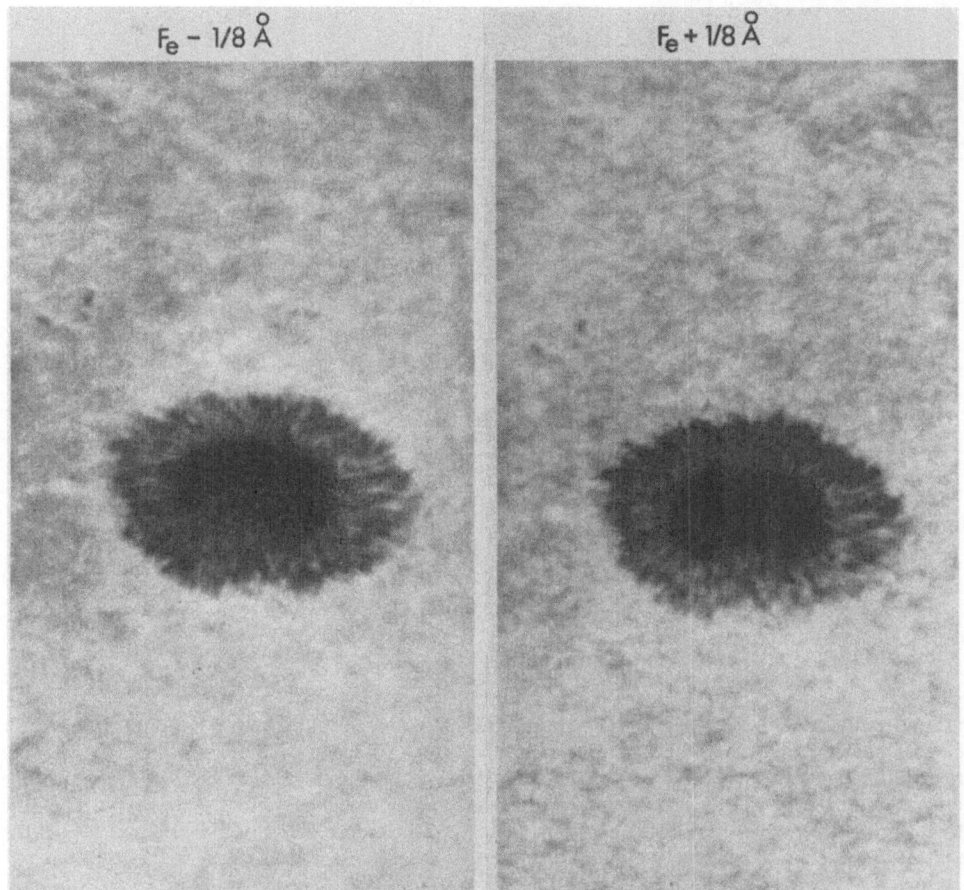

FIG. 4. *Sunspot filtergrams* $\lambda$ *6569·2 $\pm$ 1/8 Å at radius vector 0.7. Limb direction is vertically upward (Sacramento Peak Observatory).*

## DISCUSSION

*Beckers:* From the preliminary results reported in an earlier paper Schröter and myself found the dark penumbral regions to show a magnetic-field enhancement. This disagrees with your measurements. We want to investigate this further, however, before being definite. [See Figure 4.]

Observations which I made of the Evershed velocities with a Zeiss 1/4 Å filter shows the Evershed velocities to be located in the dark penumbral regions. This agrees with Dr. Mattig's measurements.

*Severny:* It is a pleasure to realize that the fine structures inside sunspots we found in 1959 (*Astr. Zu* **36**, 208) are now confirmed in the work of Dr. Beckers and Dr. Mattig. That time we found that the magnetic field even inside the umbra shows small peaks (or points, as calls them Dr. Beckers) about 30% of mean value and concentrated in regions comparable with the resolution of our magnetographic records (~2˝.5). [See Figure 5.] Magnetographic records have the advantage that they

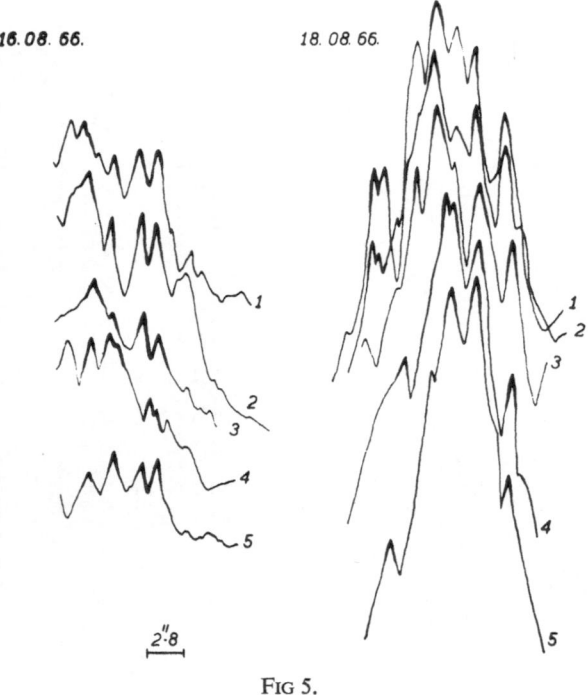

FIG 5.

are not influenced by stray light, while spectrographic data are influenced by stray light to higher extent. Since then we have made several times the records of the fine structure and the last ones were made in 1965–66 and they are shown at the slide (these last results are delivered also at the conference on magnetohydrodynamics in Sopot, Poland, Sept. 1966). The slide illustrates the repeated scans through the sunspot showing the same peaks of about 2˝ in extension. (It is important to repeat the scan to avoid the influence of trembling of images at the slit which may produce false peaks.)

# CONCERNING THE DEVELOPMENT OF THE EVERSHED MOTION IN SUNSPOTS*

Martin D. Altschuler, Yoshinari Nakagawa, and Carl G. Lilliequist

*(High Altitude Observatory,*
*National Center for Atmospheric Research,*
*Boulder, Colo., U.S.A.)*

## ABSTRACT

The non-linear, partial differential equations of magnetohydrodynamics are iterated simultaneously by computer to determine the time development of a single sunspot. Axisymmetry and incompressibility are assumed. The initial conditions are (1) zero velocity everywhere, and (2) the magnetic-field distribution of a ring current embedded in the photosphere. The initial magnetic field is then allowed to relax by magnetic diffusion and by the creation of a velocity field. It is shown that (1) Evershed motion outward from the sunspot will develop from reasonable initial parameters, and (2) the growth rate of the magnetic configuration depends on the strength of the initial magnetic field.

One of the most interesting problems concerned with the sunspot phenomenon is the origin of the Evershed motion. In an attempt to find some clues to this problem, we have examined the MHD equations in the non-linear, time-dependent form by numerical methods. The magnetic and velocity fields are strongly coupled with one another through the non-linear terms in the differential equations. Magnetic forces accelerate the fluid via the $\mathbf{J} \times \mathbf{B}$ term in the equation of motion and the fluid flow distorts the magnetic field via the $\nabla \times (\mathbf{v} \times \mathbf{B})$ term in the equation for magnetic field. We limit ourselves to an incompressible medium and to an axisymmetric sunspot. We neglect temperature gradients. Our primary concern is the acceleration of fluid by $\mathbf{J} \times \mathbf{B}$ forces and the distortion of the magnetic field by the fluid motions. Our initial conditions (i.e. at time $t = 0$) are (1) zero velocity everywhere, and (2) the magnetic-field distribution of a ring current embedded in the photosphere. The initial magnetic field is then allowed to dissipate by both magnetic diffusion and the creation of a velocity field.

Our abstraction from a real sunspot to an axisymmetric ring current in an incompressible fluid with no temperature gradients is somewhat severe. However, the advantage of a simple calculation is that we know exactly what physics has been included and what assumptions have been made. We can therefore determine in detail the interactions of a few physical parameters.

---

* Presented by M. D. Altschuler.

In the absence of temperature gradients and density gradients, there are no pressure gradients to restrain the $\mathbf{J} \times \mathbf{B}$ forces, and a magnetostatic model of a sunspot cannot evolve.

The equations that we wish to solve are (Gaussian units)

$$\frac{\partial \mathbf{B}}{\partial t} = \nabla \times (\mathbf{v} \times \mathbf{B} - \eta \nabla \times \mathbf{B}) \tag{1}$$

$$\frac{\partial (\nabla \times \mathbf{v})}{\partial t} = \nabla \times \left[ \frac{1}{4\pi\rho} (\nabla \times \mathbf{B}) \times \mathbf{B} - (\nabla \times \mathbf{v}) \times \mathbf{v} + \nu\nabla^2 \mathbf{v} \right] \tag{2}$$

$$\nabla \cdot \mathbf{v} = 0, \tag{3}$$

where $\eta$ is the magnetic diffusivity, $\nu$ is the viscosity (assumed constant), and $\varrho$ is the density. A constant force of gravity and a scalar pressure may be included in Equation (2), but they have absolutely no effect on the time development of the magnetic and velocity fields in a constant density (homogeneous) atmosphere, and vanish when the curl operator is applied. By the same token, only the tension due to the curvature of the magnetic field affects the velocity.

To simplify the geometry (and decrease the number of equations), we assume the sunspot to be axisymmetric, the solar plasma to be incompressible, and the azimuthal fields to be zero. Equations (1) and (2) can then be written in cylindrical coordinates in the scalar form (Chandrasekhar, 1956)

$$\frac{\partial P}{\partial t} = \eta \Delta_5 P - \frac{1}{r^3} \frac{\partial (r^2 P, r^2 U)}{\partial (z, r)} \tag{4}$$

$$r \frac{\partial \Delta_5 U}{\partial t} = \frac{1}{4\pi\rho} \frac{\partial (\Delta_5 P, r^2 P)}{\partial (z, r)} - \frac{\partial (\Delta_5 U, r^2 U)}{\partial (z, r)} + \nu F(U), \tag{5}$$

where

$$F(U) \equiv r \frac{\partial^2 (\Delta_5 U)}{\partial z^2} + \frac{1}{r} \frac{\partial^2}{\partial r^2} (r^2 \Delta_5 U) - \frac{1}{r^2} \frac{\partial}{\partial r} (r^2 \Delta_5 U) \tag{6}$$

$$\frac{\partial (f, g)}{\partial (z, r)} \equiv \frac{\partial f}{\partial z} \frac{\partial g}{\partial r} - \frac{\partial f}{\partial r} \frac{\partial g}{\partial z} \tag{7}$$

and

$$\Delta_5 \equiv \frac{\partial^2}{\partial r^2} + \frac{3}{r} \frac{\partial}{\partial r} + \frac{\partial^2}{\partial z^2}, \tag{8}$$

where $z$ corresponds to the sunspot axis. The magnetic and velocity vectors are related to the scalars $P$ and $U$ by the relations

$$\mathbf{B} = -r \left( \frac{\partial P}{\partial z} \right) \hat{\mathbf{r}} + \left[ \frac{1}{r} \frac{\partial}{\partial r} (r^2 P) \right] \hat{\mathbf{z}} \tag{9}$$

$$\mathbf{v} = - r\left(\frac{\partial U}{\partial z}\right)\hat{\mathbf{r}} + \left[\frac{1}{r}\frac{\partial}{\partial r}(r^2 U)\right]\hat{\mathbf{z}}. \tag{10}$$

Our task is to iterate Equations (4) and (5) simultaneously by numerical methods.

Our initial magnetic field is essentially that of a ring current with a radius of 1000 km, a current of $2 \times 10^{11}$ ampères, and a dipole moment of $6\cdot3 \times 10^{26}$ gauss cm$^3$ ($=6\cdot3 \times 10^{23}$ amp m$^2$). We have modified the field in the region immediately surrounding the ring to avoid the singularity. Figure 1 shows our assumed initial magnetic field. The highest field strength (at the point marked $H$) is 1800 gauss. The initial velocity field is chosen to be zero at every point. The density is taken to be $\varrho = 10^{-7}$ gm/cm$^3$, the viscosity to be $v = 10^{12}$ cm$^2$/sec, and the magnetic diffusivity to be $\eta = 10^{10}$ cm$^2$/sec. The time scale for the diffusion of magnetic field across 1000 km is $L^2/\eta = 10^6$ sec. Since we will be limited to only about 1 hour of development, magnetic diffusion does not affect the results of the calculation.

Figures 2a and 2b show the magnetic and velocity fields after 10 sec of development; Figures 3a and 3b show the fields after 100 sec; Figures 4a and 4b show the fields after

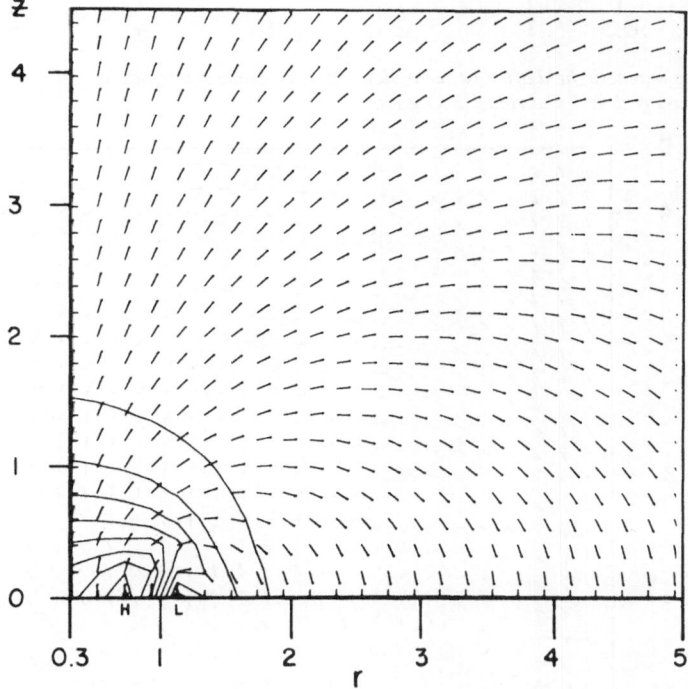

FIG. 1. *The initial magnetic field. There is symmetry around the z-axis ($r = 0$) and with respect to the z = 0 plane. The r- and z-axes are calibrated in units of $10^3$ km. The H (high) locates the maximum B of 1780 gauss; the L (low) locates the minimum B of 70 gauss. The contours extend to 1600 gauss in steps of 200 gauss.*

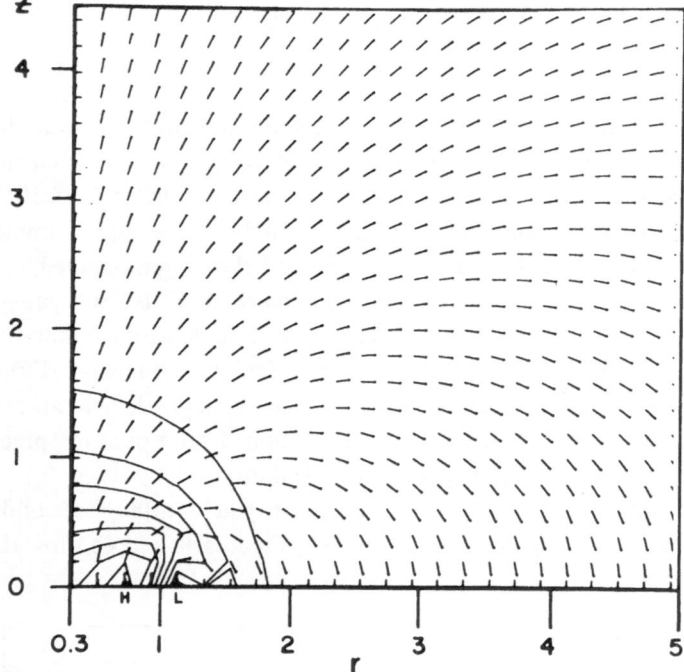

FIG. 2a. *The magnetic field after 10 sec of development. Same scaling as in Figure 1. At H the value of B is 1750 gauss and at L the value is 55 gauss.*

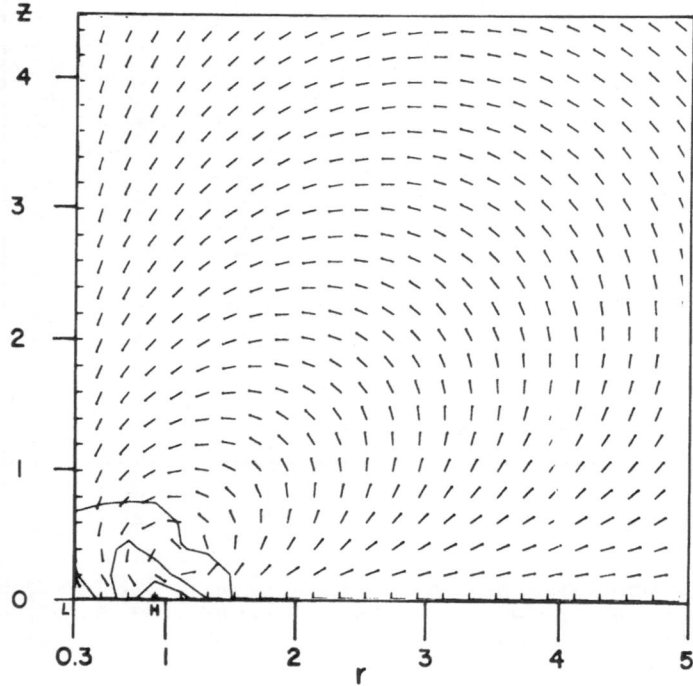

FIG. 2b. *The velocity field after 10 sec of development. The contours are in steps of 500 m/sec. At H the maximum velocity is 1970 m/sec; at L the minimum velocity is 250 m/sec.*

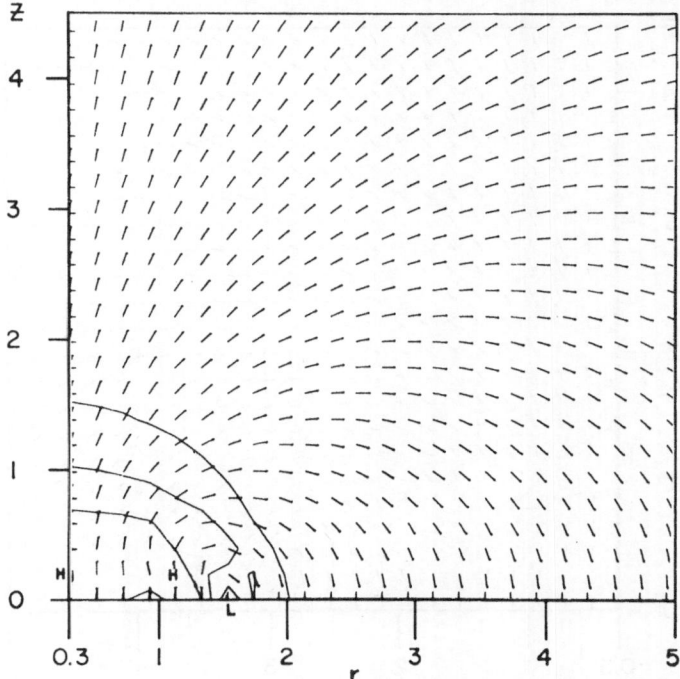

FIG. 3a.  *The magnetic field after 100 sec of development. Same scaling as in Figure 1. At H ( r = 0·3), B equals 760 gauss; at H ( r = 1·1), B equals 820 gauss; at L (z = 0), B equals 20 gauss.*

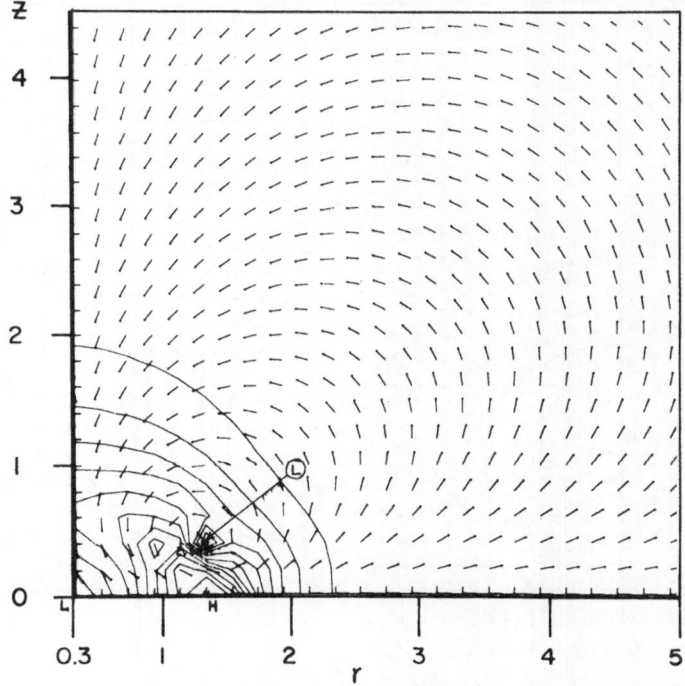

FIG. 3b.  *The velocity field after 100 sec of development. Same scaling as in Figure 2b. At H, v equals 7130 m/sec; at L ( r = 0·3) v equals 120 m/sec; at L ( r = 1·3), v = 210 m/sec.*

MARTIN D. ALTSCHULER ET AL.

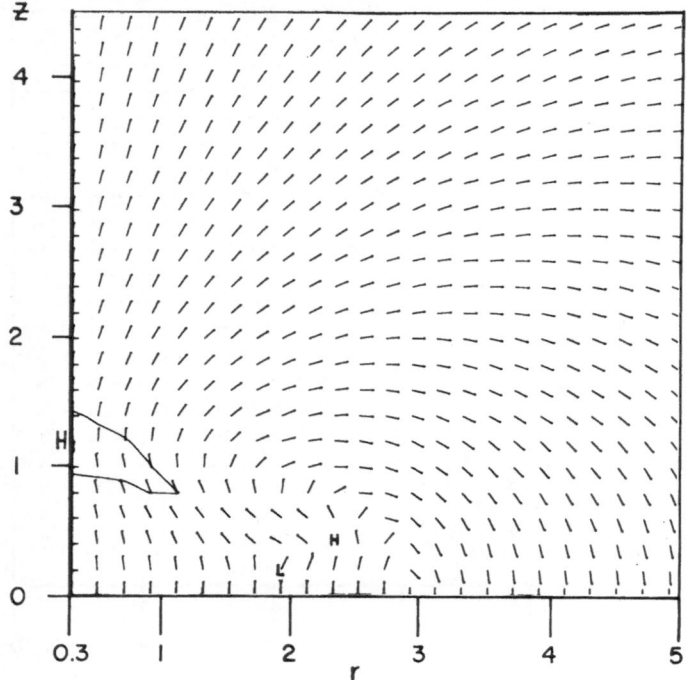

Fig. 4a. *The magnetic field after 1000 sec of development. Same scaling as in Figure 1. At* $H (r = 0\cdot3)$, *B equals 220 gauss; at* $H (r = 2\cdot3)$, *B equals 190 gauss; at L, B equals 40 gauss.*

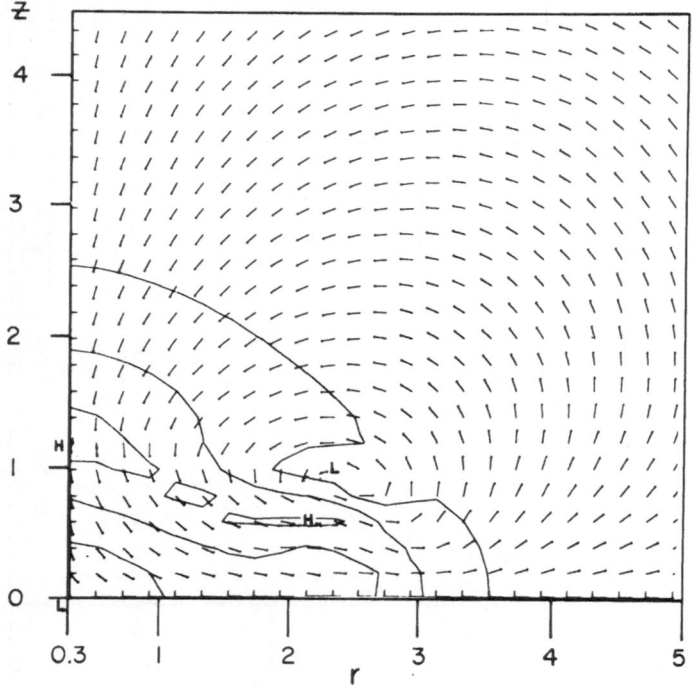

Fig. 4b. *The velocity field after 1000 sec of development. Same scaling as in Figure 2b. At* $H (r = 0\cdot3)$, *v equals 1630 m/sec; at* $H (r = 2\cdot1)$, *v equals 1590 m/sec; at L* $(r = 2\cdot3)$, *v equals 210 m/sec.*

1000 sec. Although we have carried the computation to $10^4$ sec, these early developments (in logarithmic time steps) show the actual development of the velocity pattern.

A careful examination of these diagrams indicates that the magnetic-field configuration is frozen to the moving fluid, and that the fluid velocity is approaching the Alfvén velocity in the ring-current region. The magnetic field, although perpendicular to the velocity field at small $|z|$, is swept along by the flow. Near the sunspot axis, the magnetic and velocity fields are parallel.

The flow is radially outward from the sunspot axis for $|z| < 1500$ km, and radially inward for $|z| > 1500$ km. This result is independent of the magnetic polarity of the sunspot, and is strongly suggestive of the Evershed motion. The flow pattern resembles a vortex ring. As the magnetic configuration expands, the vortex ring expands in step.

This calculation, which applies to a large magnetic field and negligible magnetic diffusion (i.e., to Lundquist number $Lu = v_A L/\eta \gg 1$), shows that a strong (non-linear) interaction between the magnetic and velocity fields achieves an equipartition between the kinetic and magnetic-energy densities. The time scale for the growth of the magnetic configuration is $L/v_A$, where $L$ is the scale length for magnetic variations ($\approx 1000$ km) and $v_A$ is the Alfvén velocity.

A similar calculation for weak magnetic fields and large magnetic diffusion ($Lu = v_A L/\eta \ll 1$) indicates that the drift of the magnetic field corresponds to a time scale $L^2/\eta$. Nevertheless, a small-magnitude Evershed flow pattern still results.

For different initial Lundquist numbers and field strengths, the magnetic field can develop toward dissimilar configurations.

Our calculations show that the existence of magnetic forces is a sufficient condition for the development of Evershed motion. Further work is needed to determine whether magnetic forces are also a necessary condition.

We conclude:

(1) The Evershed motion is probably the result of $\mathbf{J} \times \mathbf{B}$ forces which are radial to the axis of the sunspot in the photospheric region.

(2) The flow is perpendicular to the direction of magnetic field near the edge of the sunspot (where the current $\nabla \times \mathbf{B}$ is concentrated), and parallel to the magnetic field near the sunspot axis.

(3) The Evershed flow may be the most important factor in the decay of sunspots.

We have refrained from a detailed comparison of our results with observational data, because our assumption of incompressibility and the consequent omission of thermal pressure gradients are severe. Further development of our numerical model will eventually justify a thorough comparison of theory and observation.

## Reference

Chandrasekhar, S. (1956)     *Astrophys. J.*, **124**, 232.

# DISCUSSION

*Jäger:* In your calculations what did you assume for the magnetic-energy density compared with the energy density of matter?

*Altschuler:* The initial condition is a medium of constant density which is nowhere in motion. As the magnetic-field configuration relaxes, magnetic energy is converted into kinetic energy by the $J \times B$ forces.

*F. Meyer:* I would like to add a remark of caution, if I may, with respect to the direct comparison of your computations with real sunspots. In your model you basically compute the development of a ring current in an incompressible medium. At the outset this current is in unequilibrium with the surroundings and then expands to achieve a lower energy configuration. This all occurs in a few hours, comparable to the traveling times of Alfvén waves. The much longer life-times of real sunspots indicate that they are basically in equilibrium with their surroundings. The forces leading to motions and changes here seem to be of convective type as evidence by the granulation, supergranulation and the penumbral convection rolls of Danielson. Thus it seems important to include the energy transport in the treatment. That this is much more complicated I think we all know.

*Altschuler:* Your points are well taken. We agree completely.

*Sturrock:* Your theory assumes that the Evershed effect is a transient phenomenon peculiar to the decay phase of sunspots. I believe that observations do not support this assumption. If the Evershed effect can occur in a sunspot which is virtually in a steady state, one must look for a *power* supply, not an energy supply. I suggest that the 'missing power' mentioned by Dr. Danielson, associated with the reduced radiation from sunspots, may in fact be the power which drives the Evershed effect.

*Altschuler:* We did this calculation in order to understand as much as possible about the behavior of a few physical parameters, rather than to understand only a little about the behavior of a great many parameters. If we could understand in detail how the missing power from sunspots is converted into Evershed motion, we would have much of the problem solved.

# THE CONNECTION OF FINE-STRUCTURE PHOTOSPHERIC FEATURES IN ACTIVE REGIONS WITH MAGNETIC FIELDS

N. V. STESHENKO

*(Crimean Astrophysical Observatory,*
*Nauchny, Crimea, U.S.S.R.)*

1. The fine structure of the proton sunspot group of July 4–8, 1966 was studied on the basis of high-resolution heliograms. The comparison of the orientation between penumbral filaments and the transverse magnetic fields (observed by A. B. Severny and T. T. Tsap) shows that the direction of the filaments coincides in general with that of the magnetic field.

2. Measurements of the magnetic fields of smallest pores (1·5″–2″) showed that the pores are always connected with strong magnetic field (in average 1400 gauss), which is localized at the same small area as the pore.

3. Magnetic fields of faculae are concentrated in small elements with the dimension not exceeding 1·5″–3″. Magnetic-field strength $H_{\parallel}$ of about 45% of facular granules is within the limits of photographic measuring errors (approximately 25 gauss). For a quarter of all facular granules the strength $H_{\parallel}$ is from 25–50 gauss; about 30% of facular granules have $H_{\parallel} > 50$ gauss, and sometimes there appear faculae with field strength of about 200 gauss. The magnetic-field strength of facular granules, which are found directly above spots, is 10–20 times less than the field strength of spots. This field is 80–210 gauss only.

4. All observational data mentioned above show that the appearance of the fine-structure features in active regions is directly connected with the fine structure of magnetic field of different strength and different orientation. The study of high-resolution heliograms gives additional information about the fine structure of the magnetic field.

## DISCUSSION

*McIntosh:* Your slide of the sunspot group on July 6, 1966 showed penumbral structure between the strong spots aligned North to South and parallel to the magnetic-field lines. I have observations later that same day that show the penumbral filaments at the same location then oriented East to West. Did your magnetic observations show a corresponding change in the orientation of magnetic lines of force?

*Steshenko:* Observations on July 7 show the field orientation as East to West in that location.

# ON THE MAGNETIC-FIELD CONFIGURATION IN SUNSPOTS

O. KJELDSETH MOE*

*(Institute of Theoretical Astrophysics,
University of Oslo, Norway)*

ABSTRACT

During 1963–67 observations of the magnetic fields in sunspots have been obtained at the Oslo Solar Observatory. For the largest spots the detailed distribution of the magnetic-field strength is found. Based on calculations of line profiles made by the author (Kjeldseth Moe, 1967) also the direction of the magnetic field is derived. Observations of the magnetic field of the same spot at several positions on the solar disk give further information regarding the magnetic-field configuration. Our results are in fair agreement with those of Bumba (1962).

## 1. Introduction

During the years 1963–67 observations of the magnetic field in sunspots have been obtained at the Oslo Solar Observatory. For our observations we have used the Zeeman triplet $\lambda$ 5250 with a Landé factor $g = 3$. The entrance slit of the spectrograph is placed at the off axis focus, $f = 30$ m, of the solar telescope, giving a 27-cm image of the solar disk. The spectrograph has a focal length of 21 m and is equipped with a blazed Babcock grating with 600 rulings per mm. A more detailed description is given by Brahde (1967). The observations are made in the 5th order. This gives us a linear dispersion of 10·1 mm per Ångström and a resolving power of approximately 1/100 Ångström. In the direction perpendicular to the dispersion 1 mm in the focus of the spectrograph corresponds to 4·7 sec of arc on the solar disk. The exposure time was less than 15 sec.

A quarter-wave analyzer is placed in front of the spectrographic slit. The quarter-wave analyzer consists of a grid of mica strips followed by a Glan-Thompson prism. The principal axes of neighbouring mica strips are at right angles to each other, while the axis of the Glan-Thompson prism makes an angle of 45° to the axes of the mica strips. This arrangement enables us to determine the inclination, $\gamma$, of the magnetic-field vector to the line of sight. The direction of the azimuthal field component has, however, not been measured.

## 2. Sources of Error

Before we discuss the results of the observations we would like to mention some of

---

* Presented by P. Maltby.

*Kiepenheuer (ed.), Structure and Development of Solar Active Regions*, 202–210. © *I.A.U.*

the sources of error, particularly the instrumental polarization. The light will be polarized in the telescope because of reflections from metal surfaces, in our case from the coelostat- and off-axis mirrors. The amount of polarization in the telescope has been calculated for some typical settings of the coelostat, using data for the refractive index and the absorption coefficient for thin aluminium films.

The greatest contribution to the instrumental polarization comes from the spectrograph, especially from the reflection of light from the grating. The instrumental polarization in the spectrograph has been thoroughly measured by Dr. Maltby and the author for different spectral orders. The effect of instrumental polarization in the spectrograph may, however, be almost completely eliminated by placing a quarter-wave analyzer in front of the spectrographic slit. (The very small effect of instrumental polarization in the spectrograph in this case is seen from a comparison of the second and third row in Table 1.)

## Table 1

**Inaccuracy in the inclination of the magnetic field to the line of sight, $\Delta\gamma$, due to instrumental polarization and scattered light for inclinations in the neighbourhood of 30° and 60°, respectively**

|  | $\Delta\gamma$ | |
| --- | --- | --- |
|  | $\gamma \approx 30°$ | $\gamma \approx 60°$ |
| Instrumental polarization in the telescope | $\pm 5°.0$ | $\pm 13°.0$ |
| Imperfections in the quarter-wave analyzer | $\pm 2°.0$ | $\pm 4°.5$ |
| Imperfections in the quarter-wave analyzer together with instrumental polarization in the spectrograph | $\pm 2°.5$ | $\pm 5°.5$ |
| Scattered light (1 %) from the penumbra | $5°.0$ | $6°.0$ |

The line profile of a Zeeman triplet may be seriously distorted by instrumental polarization. It is, however, possible to find a correct value of the magnetic-field strength when the splitting of the components is complete, even without the use of a quarter-wave analyzer in front of the spectrograph. (For $\lambda$ 5250 the components are separated if the field is stronger than 2000 gauss.) The inclination, $\gamma$, of the magnetic-field vector to the line of sight may on the other hand only be measured from exposures made with a quarter-wave analyzer in front of the spectrographic slit. But even in this case we must consider the effect of instrumental polarization in the telescope.

Other sources of error, like imperfections in the polarization optics and scattered light, have also been considered. Possible imperfections in the retardation as well as the mounting of the quarter-wave plate may introduce inaccuracies in the observations. It is further possible that the Glan-Thompson prism is not a perfect polarizer. A small amount of light polarized perpendicular to the optical axis of the prism may be transmitted. The inaccuracies in the determination of the angle $\gamma$ are listed in Table 1.

The values are found from measurements of the quarter-wave analyzer used at the Oslo Solar Observatory.

To correct for scattered light is an extremely tedious task. We have to take into account the different states of polarization of the light scattered from different parts of the spot and from the photosphere. The values in Table 1 are the results of a rough estimate, where we have assumed 1% scattered light from the penumbra. The magnetic field strength in the umbra and the penumbra was set equal to 2400 and 800 gauss, respectively; while the inclination of the magnetic field to the line of sight was assumed to be the same in the umbra and the penumbra.

The influence of instrumental polarization on the inclination measurements will be greatest for near-transverse fields. This is shown in Table 1, where we have listed the maximum difference, $\Delta\gamma$, between the measured and the real value of the magnetic-field inclination to the line of sight for two values of the inclination: 30° and 60°. The values in Table 1 are maximum values, and in most cases the difference between the measured and the real inclination will be less than $\Delta\gamma$.

### 3. The Strength of the Magnetic Field

Regarding the results of the magnetic-field strength measurements we will stress in particular these points: We have found the magnetic-field strength to be constant over a greater part of the sunspot umbra. This area of maximum field strength is not necessarily situated at the center of the spot. Further the rate of decrease of the field strength is greatest well outside the umbra/penumbra boundary. It is therefore certainly a difference between the variation of the intensity and the variation of the magnetic-field strength across the spot. The general outline of the isogauss lines seem, however, to follow the umbra/penumbra boundary fairly well. These results are demonstrated in Figures 1 and 2. The only exceptions to these rules are small spots with an umbra diameter less than 5 sec of arc and spots observed under poor observing conditions. In these cases it is likely that the observations will be strongly influenced by scattered light.

In Figures 1 and 2 are shown two examples of isogauss maps prepared for the great spot of March 21 and 22, 1966. The outlines of the umbra and the penumbra are marked with a heavy line. The field strengths on the isogauss lines are given in hundreds of gauss.

In Figure 3 the measured value of $(H/H_{\text{Max}})^2$ is shown as a function of the distance from the spot center. This curve represents an average of the observations at March 21 and 22 for a number of different position angles in the spot. The magnetic-field strength is constant out to a distance of $0.4r_u$ from the spot center ($r_u$ is the umbra radius). Outside this distance the horizontal gradient of the magnetic pressure, $H^2/8\pi$, may be regarded as nearly constant. The magnetic pressure at the center of the spot is approximately $3.3 \times 10^5$ dyn/cm$^2$ compared with about $10^5$ dyn/cm$^2$ for the gas pressure (if the magnetic field is not force-free).

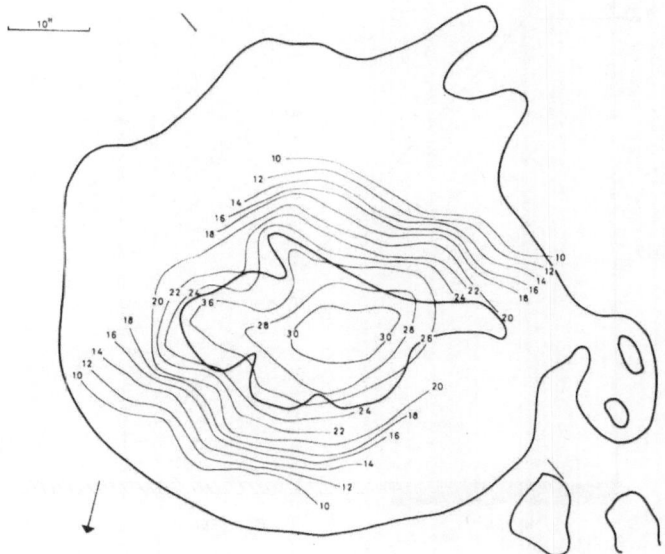

FIG. 1.  *Isogauss map for the spot of March 21, 1966. The outlines of the umbra and the penumbra are marked with a heavy line. The field strengths on the isogauss lines are given in hundreds of gauss. The arrow points in the direction to the center of the solar disk.*

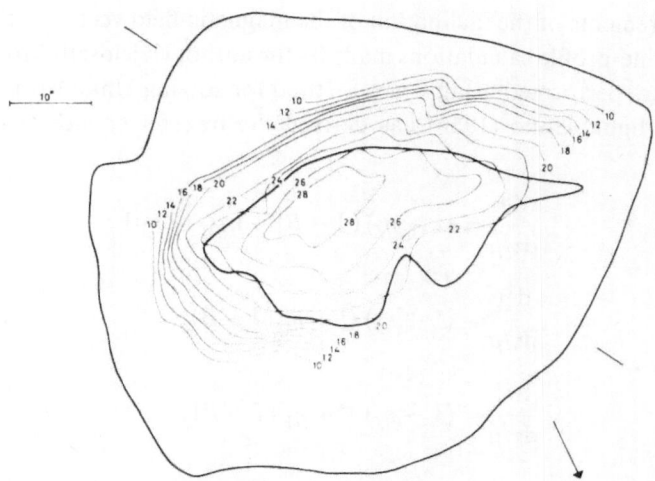

FIG. 2.  *Isogauss map for the spot of March 22, 1966 (see also text Figure 1).*

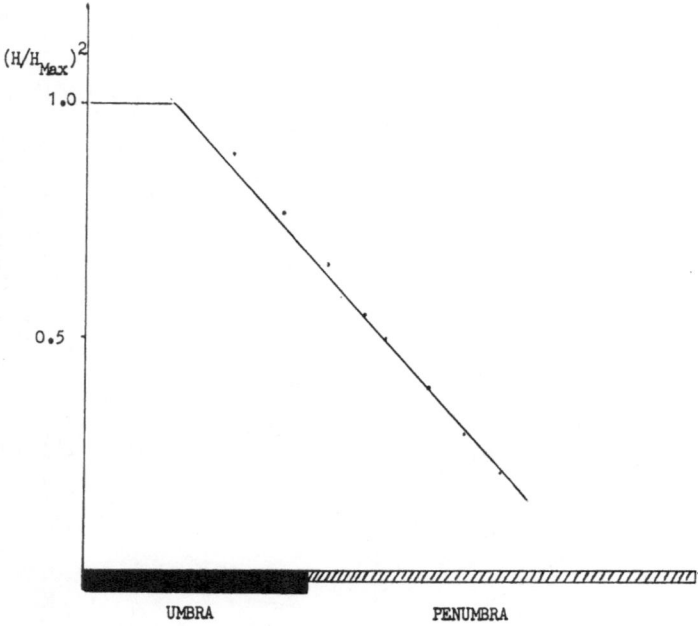

FIG. 3.   *The measured value of $(H/H_{Max})^2$ as function of the position in the spot. The values are the mean of the two days March 21 and 22, 1966, for a number of different position angles in the spot.*

## 4. The Direction of the Magnetic Field

The measurements of the inclination of the magnetic-field vector to the line of sight are based on line-profile calculations made by the author (Kjeldseth Moe, 1967). These calculations are performed using a new method for solving Unno's transfer equations for polarized light. Unno (1956) has derived the transfer equations for the Stokes parameters

$$\frac{dI}{d\tau/\mu} = (1 + \eta_I)(I - B) + \eta_Q Q + \eta_V V,$$

$$\frac{dQ}{d\tau/\mu} = (1 + \eta_I) Q + \eta_Q (I - B),   \qquad (1)$$

$$\frac{dV}{d\tau/\mu} = (1 + \eta_I) V + \eta_V (I - B).$$

Here $I$, $Q$ and $V$ are the Stokes parameters, $\tau$ is the optical depth in the continuum, $B$ the source function here set equal to the Planck function, and $\mu = \cos\theta$, where $\theta$ is the angle between the solar radius and the beam of light. The quantities $\eta_I$, $\eta_Q$ and

$\eta_V$ are defined as:

$$\eta_I = \frac{1}{\kappa_0}\left[\frac{\kappa_p}{2}\sin^2\gamma + \frac{\kappa_1 + \kappa_r}{4}(1 + \cos^2\gamma)\right],$$

$$\eta_Q = \frac{1}{\kappa_0}\left[\frac{\kappa_p}{2} - \frac{\kappa_1 + \kappa_r}{4}\right]\sin^2\gamma, \tag{2}$$

$$\eta_V = \frac{1}{\kappa_0}\frac{\kappa_r - \kappa_1}{2}\cos\gamma,$$

where $\kappa_p$, $\kappa_1$ and $\kappa_r$ are the selective absorption coefficients for linearly polarized light and for left-handed and right-handed circularly polarized light, respectively; $\kappa_0$ is the continuous absorption coefficient and $\gamma$ is the angle between the magnetic-field vector and the line of sight.

In our solution the variation of $\eta_0$ with depth has been taken into account. Here $\eta_0$ is the ratio of the line-absorption coefficient in the line center to the continuous absorption coefficient. We find

$$I = \frac{Z_1 + Z_2}{2},$$

$$Q, V = \frac{\eta_{Q,V}}{(\eta_Q^2 + \eta_V^2)^{1/2}} \times \frac{Z_1 - Z_2}{2}. \tag{3}$$

For $Z_1$ and $Z_2$ we have the expressions

$$Z_{1,2} = \int_0^\infty \lambda_{1,2}\, B \exp\left[-\int_0^\tau \lambda_{1,2}\, d\tau'/\mu\right] d\tau/\mu, \tag{4}$$

where $\lambda_{1,2} = 1 + \eta_I \pm (\eta_Q^2 + \eta_V^2)^{\frac{1}{2}}$. Equations (3) and (4) are special cases of a more general solution (Kjeldseth Moe, 1967). The magnetic field was supposed to be homogeneous and the contribution to the line formation from scattering was ignored.

Numerical calculations have been made for the Zeeman triplet $\lambda\,5250$ using a photospheric model atmosphere. The variation of $\eta_0$ with depth was found to be pronounced. Comparing different layers contributing to the line we find that the value of $\eta_0$ may change by a factor of the order of $10^2$.

From the calculations of the line profile the ratio $r_\pi/r_{\sigma_1}$ are found as a function of the inclination, $\gamma$, for $\lambda\,5250$ in the case of complete splitting ($H > 2000$ gauss). Here $r_\pi$ and $r_{\sigma_1}$ are the line contrasts in the $\pi$- and $\sigma_1$-components respectively. The $\sigma_1$-component is the strongest of the $\sigma$-components when the line is observed using a quarter-wave analyzer. The result is shown in Figure 4, where Seares' values $r_\pi/r_{\sigma_1} = \frac{1}{2}\sin^2\gamma/\frac{1}{4}$ $(1 + \cos\gamma)^2$ have been plotted for comparison. Comparing theoretical and observed values of $r_\pi/r_{\sigma_1}$ the inclination, $\gamma$, is found. It should, however, be emphasized that the values of $r_\pi/r_{\sigma_1}$ in Figure 4 are computed for a special model atmosphere (see Kjeldseth Moe, 1967).

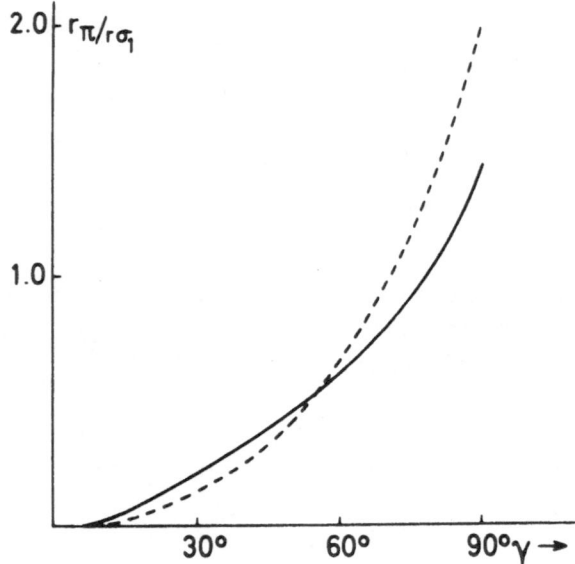

FIG. 4. *The ratio between the line depths in the π-component and the $\sigma_1$-component, $r_\pi/r_{\sigma_1}$, as a function of the inclination, γ, of the magnetic-field vector to the line of sight. The $\sigma_1$-component is the strongest of the two σ-components when the line is viewed through a quarter-wave analyzer. For comparison we have plotted the ratio $r_\pi/r_{\sigma_1}$ computed from Seares' formula (dashed curve).*

For high values of γ, $γ \geqslant 75°$, the weakest of the σ-components, the $\sigma_2$-component, will appear. In this case also the ratio $r_{\sigma_2}/r_{\sigma_1}$ may be used for the measurement of γ.

The results of the inclination measurements are shown in Table 2 and Figure 5. In Table 2 the inclination, γ, is listed as a function of the heliocentric angle, θ. The magnetic field is nearly transversal, $γ \approx 77°$, for all positions of the spot on the solar disk. The variation in γ with the heliocentric angle, θ, is very small. This result is in agreement with Bumba's (1962) observations. The values in Table 2 are based on inclination measurements for 10 spots in 30 different positions on the solar disk. In Figure 5 is shown a typical example of the variation in the inclination, γ, along a section through the spot center. The variations across the spot are within the limits of our accuracy.

### Table 2

**The inclination, γ, of the magnetic field vector to the line of sight as a function of the distance of the spot from the center of the solar disk, θ. Ten spots in thirty different positions on the solar disk have been inspected**

| Heliographic angle θ | 0°–35° | 35°–65° | 65°–90° |
|---|---|---|---|
| Inclination γ | 75° | 79° | 75° |

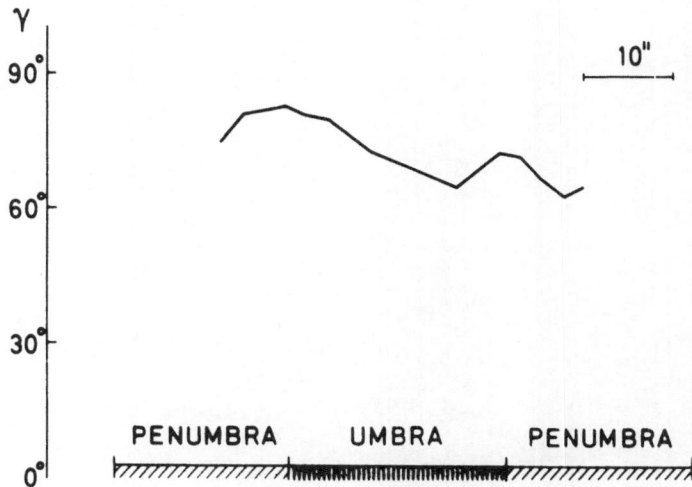

FIG. 5. *Variation of the inclination, γ, along a section through the spot center for the spot of March 22, 1966.*

## 5. Some Peculiar Spectra

In Section 4 the relative line contrasts in the π- and the σ-components have been interpreted in terms of the inclination, γ, of the field lines to the line of sight. This interpretation should be used with caution. Figure 6 is a reproduction of a spectrum taken June 9, 1967, using a quarter-wave analyzer in front of the spectrographic slit. The spot was situated close to the solar limb. During this exposure the spectrographic slit was placed across the central part of the umbra. In the same region there seemed to be a light bridge across the umbra.

In the umbra all the three Zeeman components are clearly visible and the two σ-components are of about equal strength, corresponding to a magnetic field perpendicular to the line of sight. Also the magnetic field in the umbra on each side of this section through the umbra center was of opposite polarity.

In the penumbra, however, the π-component is in some places absent while both the σ-components are present. This effect is observed only in this particular region of the spot and is conspicuous only on 2 of the about 50 spectra taken of the spot that day. This indicates that the effect is not instrumental. The same effect of a lacking π-component while both the σ-components are present have also been observed on other occasions, although not very frequently. (The effect is observed in less than 1 spot out of 10.) The reason for this may be that the effect in all cases seems to be limited to a rather small region in the spot. The other spots where the effect was observed were members of very complex groups with many small umbrae inside the same penumbra.

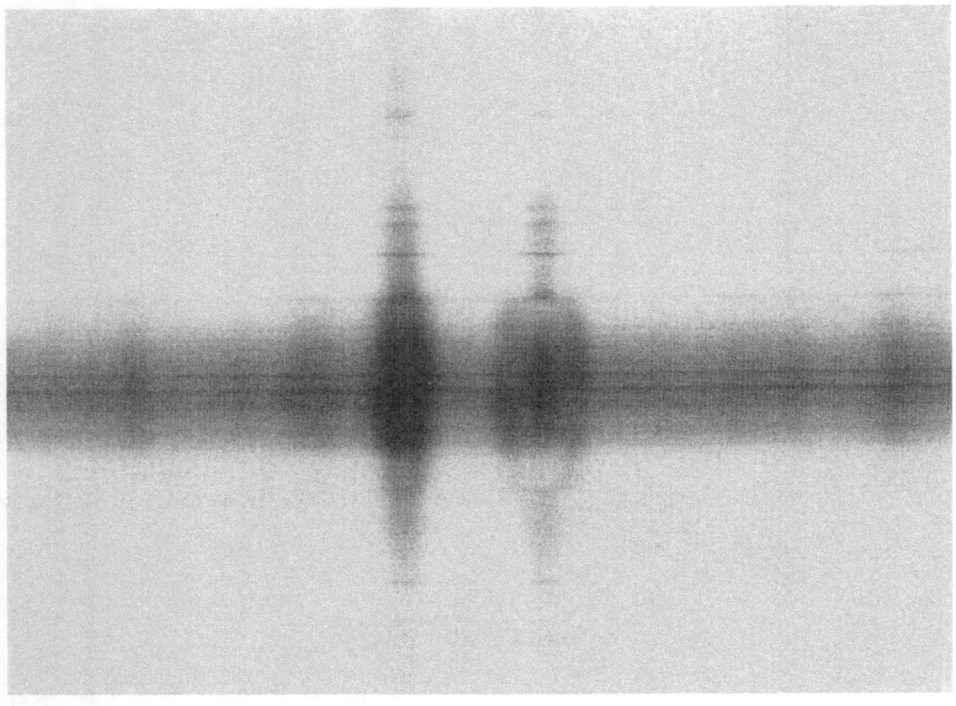

FIG. 6.    *A spectrum of λ 5250 taken June 9, 1967, using a quarter-wave analyzer in front of the spectro-*
*graphic slit. The spectrographic slit was placed across the central part of the umbra. The spot was situated*
*close to the solar limb.*

## Acknowledgements

The author wishes to thank Dr. P. Maltby and Professor E. Jensen for their helpful advice during the investigation.

This research has been sponsored in part by the Air Force Cambridge Research
Laboratories through the European Office of Aerospace Research, OAR, United
States Air Force, under Contract F61052-67-C-0070.

## References

Brahde, R. (1967)      *Institute of Theoretical Astrophysics, University of Oslo, Report No. 23.*
Bumba, V. (1962)      *B.A.C.*, **13**, 42.
Unno, W. (1956)       *Publ. astr. Soc. Japan*, **8**, 108.

# THE PROPER MOTIONS OF SUNSPOTS AND THE
# MAGNETIC FIELD OF ACTIVE REGIONS

G. V. KUKLIN

*(Siberian Institute of Terrestrial Magnetism,
Ionosphere and Radio Propagation, Academy of
Sciences, Irkutsk, U.S.S.R.)*

According to our program of sunspot proper motion investigations (Kuklin and Syklen, 1966) we study the interdependence of the sunspot proper motions inside the group and the magnetic field of the whole group or active region. This aspect of the dynamics of matter in disturbed regions of the Sun was not considered practically up to the last time.

Bumba (1964) came to the conclusion that the individual small umbrae move practically parallel to the magnetic field isolines (of the longitudinal component) and that the general character of the magnetic-field distribution in the active region changes very little during the motions. The magnetic field creeps in the active region according to many recent investigations of the magnetic-field configuration in active regions (for example Stepanov *et al.*, 1967). Therefore the small umbrae moving along the magnetic-field isolines must cross the magnetic lines of force quite perpendicularly, which is unlikely to be in accordance with the magnetohydrodynamics.

Gopasyuk and Moreton (1965) found that the transverse component of the magnetic field changes correspondingly to the main sunspot proper motions: the main sunspots pull out the magnetic lines of force during their motions and deform the magnetic field in such a way.

We have studied the proper motions in three sunspot groups during some days, taking into account the magnetic-field observations (three components) with the help of the solar magnetograph. We came to the following conclusions:

(1) The main sunspots in the active region move independently of the active region magnetic field but pulling the magnetic lines of force hardly fastened together and changing the magnetic field correspondingly.

(2) The satellites of the main sunspots (small umbrae and penumbra bits) accompany the main sunspots irrespectively of the magnetic field in general as if crossing the magnetic lines of force. It seems that their motions are more influenced by the changes inside the main sunspots (the magnetic field rebuilding, the sunspot rotation, the Evershed effect inhomogeneities, etc.).

(3) The small umbrae and penumbra bits placed far from the main sunspots move

*Kiepenheuer (ed.), Structure and Development of Solar Active Regions*, 211–213. © *I.A.U.*

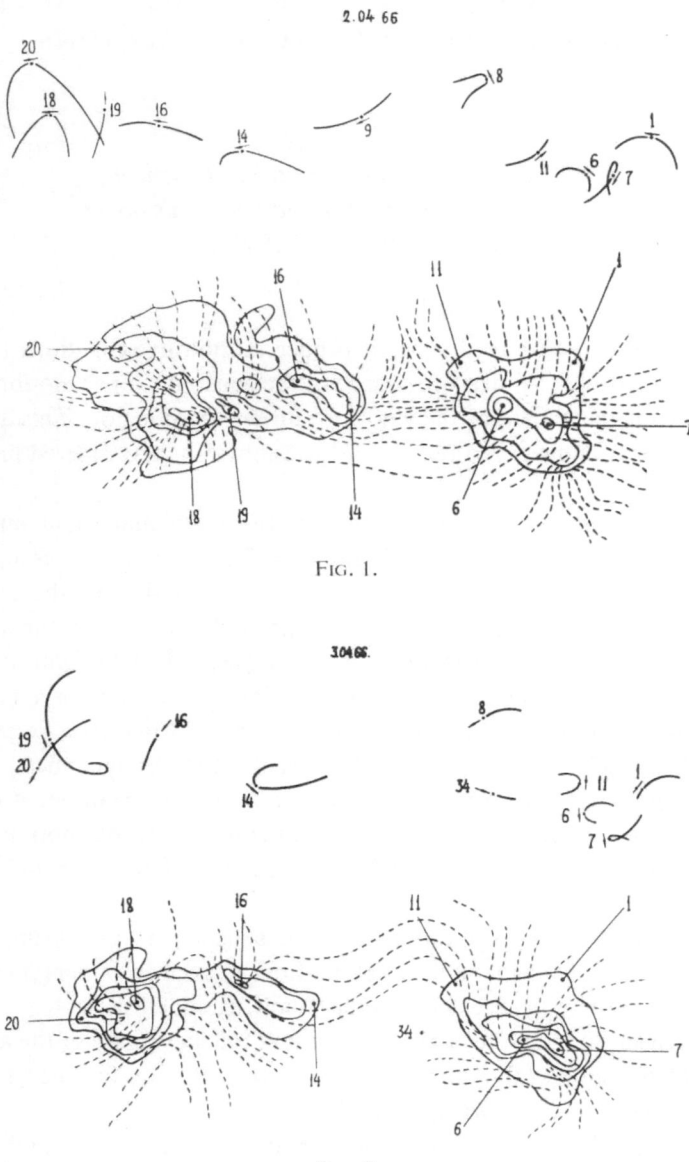

FIG. 1.

FIG. 2.

strongly along the magnetic lines of force, except for cases of close flares when it is impossible to determine that the cause of the line of force crossing by the sunspot is the magnetic-field rebuilding or the flare effect ('a shock wave').

Thus, if it is possible to make conclusions on the basis of such data we must suppose the following hierarchy of the dynamics (in the strict sense of the word) of matter in the active region:

(a) the main sunspot proper motions are caused by subphotospheric processes;

(b) the active region magnetic field is determined by the main sunspot magnetic fields and their proper motions;

(c) the small sunspot proper motions are corresponding to the active region magnetic-field structure.

## Acknowledgements

I am grateful to Miss A. E. Syklen, Miss S. A. Radnaeva and Miss L. E. Vilutis for help in measurements, and to Dr. V. E. Stepanov for data on the magnetic fields.

## References

Bumba, V. (1964)    *Prace Wrocławsk. towarz. nauk B,* **112**, 3–40.
Gopasyuk, S. I., Moreton, G. (1965)    Report at the Solar Study Commis. Plenum (KISO) in Kislovodsk.
Kuklin, G. V., Syklen, A. E. (1966)    *Rezultaty nabl. i issled. v period MGSS,* **1**, Moskva, p. 64.

## DISCUSSION

*H. U. Schmidt:* I do not see that the horizontal motions across the magnetic field which you deduce from changes in the flux distribution with time contradict the magnetohydrodynamic theory. Quite to the contrary, it seems to me that you successfully apply a fundamental result of this theory, i.e. the frozen-in field principle for a plasma of high conductivity. This principle can be used to deduce a detailed horizontal flow field even if one has no directly identifiable spots in subsequent magnetic flux charts. I comment on this in the written version of my paper.

*Kuklin:* Firstly I compared the sunspot motions with the transversal magnetic field without considering any changes of the flux distribution in time. Secondly I said the motions of small sunspots and penumbra bits across the magnetic field are in contradiction with the magnetohydrodynamic theory because of assuming the small sunspots and penumbra bits are not features with deep located bottoms and so their motions are controlled by the magnetic field of the high photospheric layers which is observable.

# FLUCTUATIONS OF THE MAGNETIC-FIELD STRENGTH OF SUNSPOTS WITHIN ONE DAY

H. KÜNZEL

*(Heinrich-Hertz-Institut für solar-terrestrische Physik, Sonnenobservatorium, Einsteinturm, Potsdam, Germany)*

ABSTRACT

In the period of May–June 1965, the magnetic-field strengths of twenty sunspots were measured in order to investigate fluctuations within one day. The results of spectrograms, which were taken in the interval of one hour, are given in graphs. The mean error of one value has the size of $\pm 169$ gauss. The graphs show the general tendency in the behaviour of field strength. Fluctuations of magnetic-field strength up to 800 gauss within a few hours were found. The short-time fluctuations shown in the graphs are mostly smaller than the mean error and therefore probably not real.

For more details see *Astron. Nachr.*, **289** (1967), 233.

# ON THE STRUCTURE OF THE MAGNETIC FIELD OF SUNSPOTS

E. I. MOGILEVSKY, L. B. DEMKINA, B. A. IOSHPA, and V. N. OBRIDKO

*(Institute of Terrestrial Magnetism, Ionosphere and Radio-Wave Propagation of the USSR Academy of Sciences, Moscow)*

## ABSTRACT

The model of the magnetic field of sunspots, taking account of fine structure of magnetic field in solar plasma, is considered. Small-scale subgranules with their own field form magnetic filaments in the external current-free field. The filaments are vertical in the umbra, while in the penumbra they run along the surface with sharp bends. In a number of spot umbra the relation between Doppler velocity and the field is established on polarized spectrograms. The $\pi$-component splitting in umbra is interpreted as a result of a weak background magnetic-field existence together with a large field of magnetic filaments. Spectrographic definition of the magnetic field in spot umbra is accomplished on the effect of magnetic-lines intensification and directly on spectrograms of low-excitation (Fe I, Ti I) and high-excitation (Fe II) lines. Magnetic field measured in low-excitation lines exceeds twice the field value obtained in high-excitation lines. This result has been considered in the light of the proposed model of sunspot field.

By studying the magnetic field of a sunspot one can obtain the most comprehensive information on the small-scale structure of the solar magnetoplasma. This is primarily due to the fact that the photographic method is the most efficient in measuring sunspot fields, since it yields a high resolution and good guiding. On the other hand, investigation of the magnetic-field distribution separately in the umbra and penumbra of a sunspot makes it possible to study the structural elements in two projections, as it were, i.e. from above in the umbra, where the field is predominantly vertical, and from the side in the penumbra, where the field is near horizontal. Finally, the sunspot field develops rather rapidly and evidently remains largely unaffected by the magnetic fields of neighbouring activity phenomena. A working model of a magnetic field of a sunspot, which takes into account the small-scale structure of the magnetoplasma and some spectral observations shedding light upon the fine structure of the magnetic field of a sunspot will be considered below.

The model of the magnetoplasma structure considered by Mogilevsky and Shelting (1968) implies the existence of small-scale ($\sim 10^7$ cm) elements (so-called sub-granules) with a closed magnetic field surrounded by a quasi-homogeneous current-free magnetic field $H = H_0 + \sum_i H_i$, where $H_0$ is the homogeneous magnetic field and $\sum H_i$ is the total field (at the given point) of the entire ensemble of sub-granules.

Due to the interaction between their magnetic moments, sub-granules tend to line up in long chains (magnetic filaments) whose space distribution determines the observed large-scale field of the spot. In the umbra we should observe a receding ensemble of vertical magnetic filaments which can group together into convection cells owing to convective sub-photospheric motions (which obviously take the form of oscillatory convection). The first appearance of a pore corresponds to the extending of a magnetic filament cell with a mean intensity of $\gtrsim 1200$–$1500$ oe. Neglecting the problem of the generation of sub-granules with a force-free magnetic field*, one may assume that the subsequent growth of the pore is a process of the successive extending of vertical chains of sub-granules grouped together into convection cells. It follows from observations (Vasilieva, 1963) that this process of the initial growth of the spot is preceded by the egress of the gas-dynamic shock-wave to the photosphere.** The wave, propagating partly upwards, carries the initial weak magnetic field of the future spot out into the chromosphere. Extending vertical magnetic tubes can extend into the chromosphere and corona through the initial field, because the diffusion of sub-granules over the gradient field is much more rapid (by a factor of the order of the Reynolds' magnetic number) than ordinary magnetic-field diffusion (Syrovatsky and Zhugzhda, 1967).

The umbra will continue growing until part of the outer magnetic tubes begin to deviate from the vertical, due to refraction when the field propagates into the chromosphere where the conductivity increases. This deviation will result in the bending of the outer filaments and later in their return into the photosphere (the outer boundary of the penumbra). These magnetic filaments, creeping through the photosphere, form a relatively thin ($\simeq 5 \times 10^7$ cm) translucent magnetic surface, the penumbra. The qualitative representation of this scheme is given in Figure 1. In this working model, the umbra of the spot is an inhomogeneous (due to convective perturbations) band consisting of sub-granule filaments whose vertical field may recede into the region of the chromosphere and corona and later close upon the umbra (or umbras) of other spots of the group, while the penumbra is a relatively thin layer of magnetic tubes detached from the umbra which yield a near horizontal field and close either under the photosphere or upon the dark intergranular areas of the photosphere. The essential point in this 'mushroom-like' schematic model of the spot is that the formation of the penumbra due to the bending of the magnetic tubes should be accompanied by substantial changes in the magnetic flux and by the generation of a ring current in the surface layer of the photosphere with a finite conductivity. The field of this toroidal current, located under the penumbra around the umbra, determines the stability of the sunspot umbra at the photosphere level. The ring-current energy, which is equal to or less than the energy of the penumbra field, may be estimated by calculating the

* That the field in sub-granules tends to become force-free follows from the condition of magnetic-energy extremum in the sub-granule.
** This is accompanied by the appearance of a bright-point floccule and then of a small ringlike one.

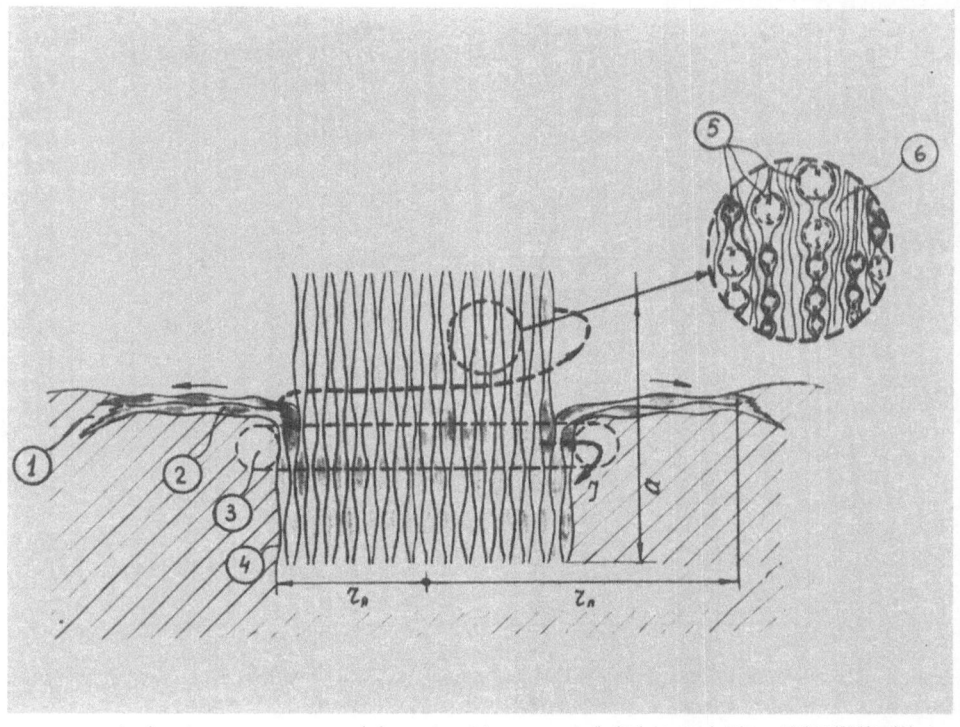

FIG. 1.   *Structural scheme of the magnetic field of a sunspot: 1 = photosphere; 2 = penumbra field; 3 = induced ring current restraining the umbra field; 4 = umbra field; 5 = sub-granules in various convection cells; 6 = current-free background magnetic field.*

field of the ring at the characteristic values given in Figure 2. It was assumed that the mean field value in the penumbra was $\lesssim 0.25\,H_{max}$, and therefore the ring was located at a depth of about 0·2 umbra radius (Kolpakov, 1966). The addition of the 'table-shaped' umbra field, the penumbra field and that of the 'holding' ring opposite in sign yields a sunspot field similar to the observed distribution (Figure 2). Here we only consider the magnetic-field intensity distribution. It is known, however (Severny, 1964) that the observed distribution of the sunspot-field vector is not approximated by the dipole field, nor is it a potential one, because the observations point to the existence of an azimuthal component. Restricting ourselves to the general structural scheme of the spot and without going into the nature of the field-vector distribution, we will note that the set of vertical magnetic tubes of the umbra should acquire the general azimuthal component due to the helical instability of the plasma with the field (Kadomtzev, 1962). In this case the general magnetic field of the umbra (especially in the chromosphere) will also be force-free (Mogilevsky and Shelting, 1966). With the growth of the umbra area ('surfacing' or 'sinking' of identical-type magnetic

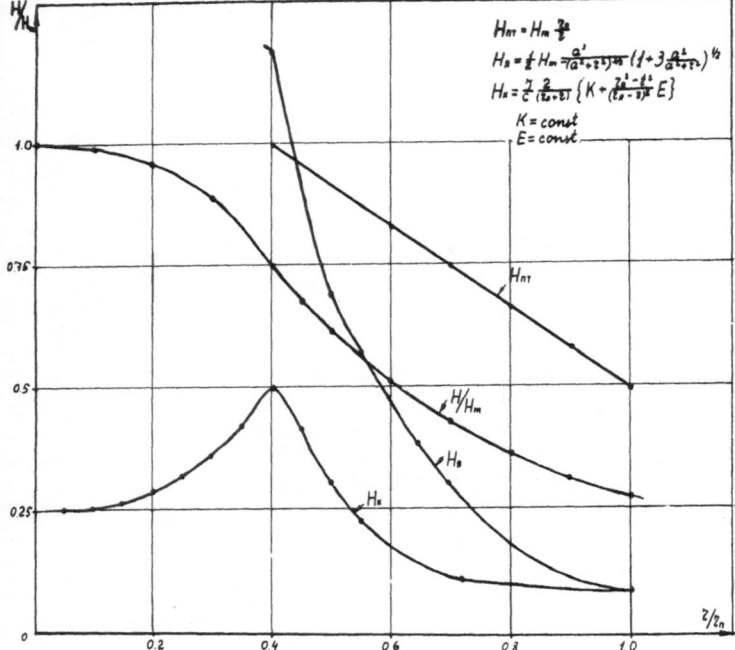

FIG. 2.    *Distribution of magnetic field H in sunspot:* $H_u$ = *umbra field,* $H_{nm}$ = *penumbra field,* $H_k$ = *ring-current field,* $H_{max}$ = *maximum intensity in sunspot umbra.*

filaments of the sub-granules) the specific tension (stress per unit length) of the ring is retained, which is in accord with its stability. The dissipation time of the ring current, primarily due to Joule losses (taking into account the conductivity according to Kuklin, 1966, in partly ionized sub-photospheric layers) will be of the order of $5 \times 10^6$–$10^7$ sec, which corresponds to the lifetime of the spot. The obvious implication of this sunspot model is the possibility of a considerable penetration of the umbra field into the chromosphere and the small height of the penumbra. The relatively small ($\lesssim 10^8$ cm) thickness of the penumbra naturally explains its sharply defined boundaries (photosphere–penumbra, penumbra–umbra), whose contrast does not vary with the shifting of the spot from the centre towards the limb. The filamentary structure of the penumbra aligned along the field and the near-round granules in the umbra, the nature of the gas motion within the umbra and penumbra, and finally the possibility of explaining the spot evolution, as well as some morphological features of complex (multi-nucleus) spots within the framework of this mode speak in favour of the working model under review.

To obtain additional information on the structure of the sunspot magnetic field, detailed photographic measurements of the magnetic-field distribution in a great number (about 50) of large (Sp $\gtrsim$ 250 units), predominantly quiet spots were carried out with the tower solar telescope IZMIRAN in 1966–67 (Zhulin *et al.*, 1966). For this purpose paired polarization spectrograms near line Fe I $\lambda$ 6302·508 Å were obtained by traversing the spot 15–25 times with the spectral slit. Simultaneously with taking the spectrograms, a series of pictures of the spot on the spectrograph slit were taken, and this enabled unambiguous 'guiding'. The observations were carried out in the Cassegrain focus of the telescope ($F=27$ m). On each pair of spectrograms of a given cross-section of the spot, measurements of the location of the $\pi$- and $\sigma$-components relative to the telluric line were made in 20–25 different points along the height of the slit. Thus, for each spot the magnetic fields and line-of-sight velocities were measured in several hundred points of the spot, which characterized the distribution of the $H$ and $V_D$ values comprehensively enough and with a high resolution. The results of the determination of line-of-sight velocities are as follows:

(a) For most of the umbras measured, with a field of 2000–3500 oe, the velocity of gas 'sinking' at the photosphere level was about 1 km/sec* (Figure 3). Such velocities were observed both for the $\sigma$ (0)- and $\pi$ ($\nabla$)-components.

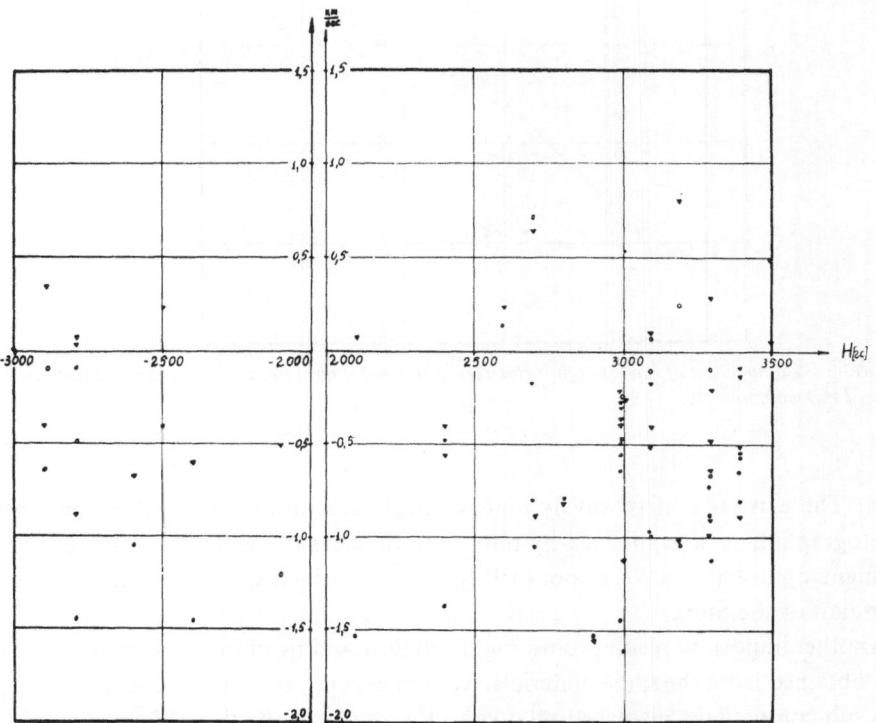

Fig. 3. *Dependence of line-of-sight velocities $V_D$ in sunspots defined from $\sigma$ (0)- and $\pi$ ($\nabla$)-components, on magnetic field H (taking account of polarity $+ S, - N$).*

* The sign of the velocity (descent or ascent) may depend on the phase of sunspot development. Since we used only spots of large area, we were able to observe mainly the descent of the gas which is characteristic of the period of the post-maximum development of the group.

(b) The feasibility of defining the line-of-sight velocities by the $\sigma$- and $\pi$-components is illustrated in Figure 4. A linear relationship of the type $V_\sigma = V_\pi - 0.26$ km/sec was found between these velocities ($V_\sigma$ and $V_\pi$). The ray shift of the lines indicates that, at least in the umbra, the role of the scattered light which would yield unshifted lines is not great.

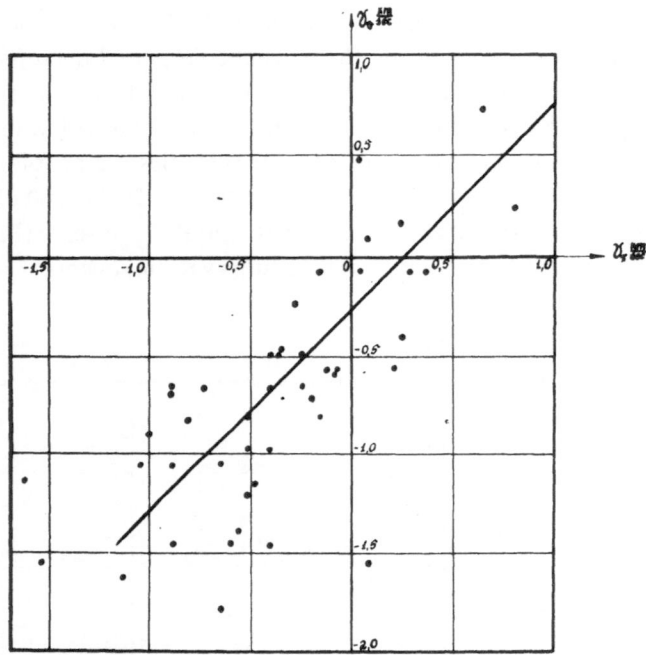

FIG. 4.   *Comparison of line-of-sight velocities in sunspot umbra obtained by measurements in $\pi$ ($V_\pi$) and $\sigma$ ($V_\sigma$)-components.*

(c) The existence of relatively high vertical velocities in the umbra, obtained by photographing, was confirmed by photoelectric measurements of the same spot with a magnetograph (for a large spot of July 19, 1966, which was located near the central meridian of the Sun).

Another important result promoting the understanding of the sunspot-field structure and obtained from the same materials, was the effect of the $\pi$-component splitting into two sub-components displaced relative to the centre of the line (Figure 5). A similar effect with a complete splitting of the line in the umbra was observed by a number of authors (for example, Severny, 1959; Bumba, 1962; Chistjakov, 1967; and others). It is also revealed by examining photocopies of spectrograms of large spots from the Monte Mario Observatory. This effect cannot be attributed to instrumental polarization, it remains at a considerable turn of the spectrograph grating, for instance

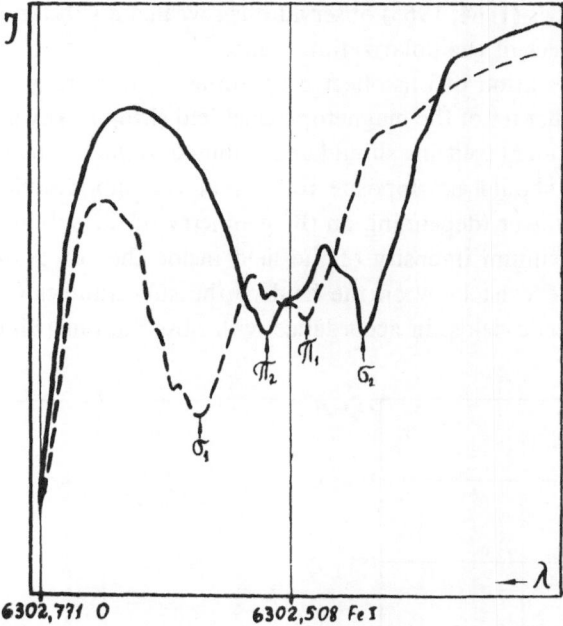

FIG. 5. *Example of registrogram of polarization spectra in the region of line Fe I λ 6302·5 Å:* *1 = polarization in spectrum* ↻, *2 = polarization* ↺.

upon transfer from the fourth right spectral order to the second left one (Chistjakov, 1967). Consideration of this effect of the π-component splitting in the umbras for 34 Southern-polarity and 10 Northern-polarity spots (observations of 1966–67) leads to the following conclusions:

(a) The value of splitting, obtained from a great number of points, is 0·008–0·020 Å, thus considerably exceeding the possible measurement error (r.m.s. error is 0·003 Å ± 10%).

(b) The displacement of the split π-component relative to the centre of the line in each spectrum is independent of the field sign and appears to be opposite to the displacement of the σ-component (Figure 6). The value of the π-component splitting into two sub-components was found to be about one order less than the distance between the σ-components.

As was shown in Rachkovsky's (1962) paper, such an effect of the π-component splitting could have been observed due to the rotation of the polarization plane in an inhomogeneous magnetic field. One must assume that the sense of the rotation of the polarization plane and, accordingly, the sign of the rotation of the field's tangential component, reverse with depth simultaneously with the change of the sunspot-field polarity. Besides, such an effect should be equal in value and sign for all umbra points, which often belong to a markedly inhomogeneous field of the umbra. At the same time, Severny's (1964, 1965) observations revealed a substantial inhomogeneity of the rotation effect of the polarization plane.

Another interpretation of this effect proceeding from more general possible conceptions of the structure of the magnetoplasma field (Mogilevsky and Shelting, 1968) is that the $\pi$-component splitting should be attributed to the real background magnetic field, whose sign should be opposite to that of the sub-granule field and whose intensity is much lower (depending on the geometry of the sub-granules, their sizes, etc.) than the maximum intensity of the field inside the sub-granules. Here, some relationship should exist between the field of the sub-granules ($H_\sigma$) and the background ($H_\pi$) magnetic fields, in accordance with observations (Figure 6), whereas in

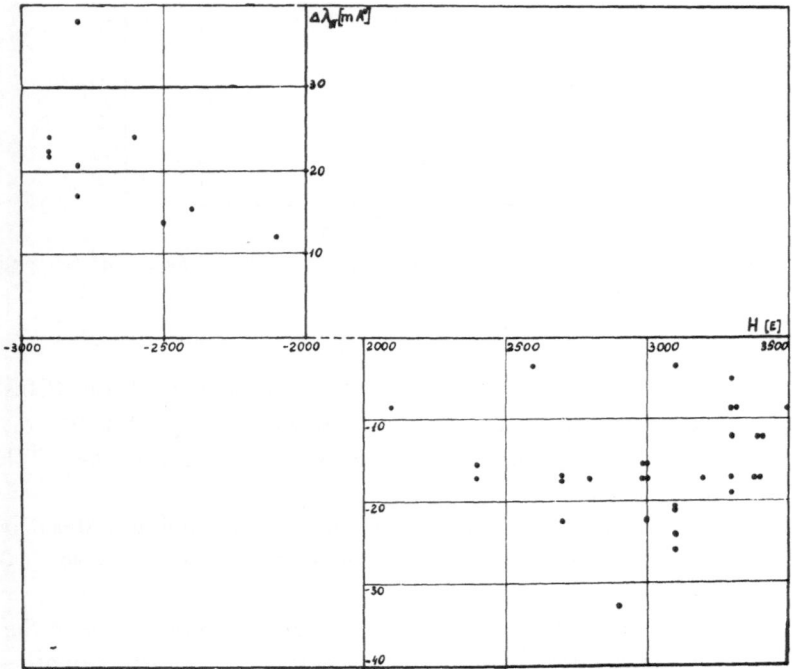

FIG. 6.    *Splitting of $\pi$-component of line Fe I 6302·5 Å as a function of sunspot umbra field H.*

the general case the rotation of the polarization plane is independent of the value and direction of the field.

One could try to clear up the question of the reality of the background field by considering the effect of the $\pi$-component splitting for the umbras of spots located at different distances from the central meridian of the Sun. Unlike all other possible effects, the value of the background field should depend only slightly on the longitude of the spot. Measurements in 44 spots (without distinction of polarities) were used.

The result of the statistical analysis is presented in Figure 7. The curve obtained by the least-squares method has the form $\Delta\lambda_\pi = 0.64\cos\theta + 0.03\sin\theta + 0.44$. In spite of the considerable scatter of the points one can see that, along with $\cos\theta$ terms, nearly half of the mean value of splitting is independent of the longitude.

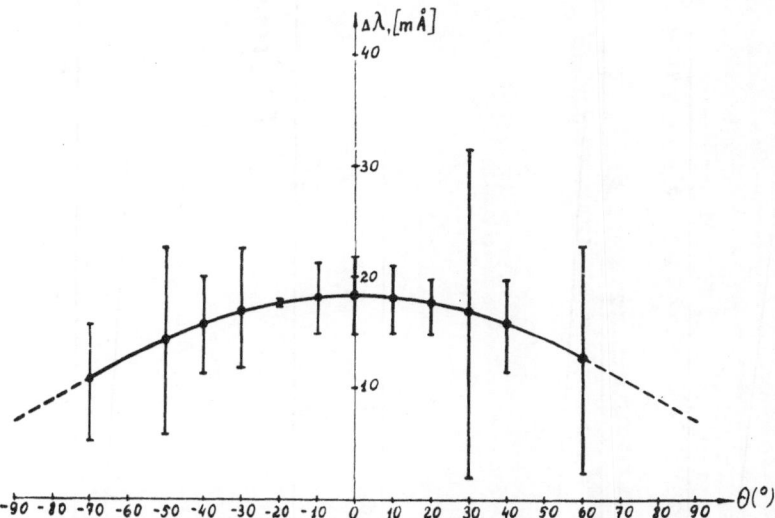

FIG. 7. *Dependence of $\pi$-component ($\Delta\lambda_\pi$) splitting in line Fe I $\lambda$ 6302·5 Å on location of spots in the disc.*

Note that the detection of such a relatively weak background field is possible only by the photographic method because of the averaging of these fields by the magnetograph.* Since the effect of the $\pi$-component splitting is of great importance for defining the sunspot-field structure, it is necessary to continue detailed investigation such as measurements within various lines and at various heights, the use of a photoelectric magnetograph for determining the variation of polarization along the contour of lines within the umbra of spots with a maximum intensity, etc.

Investigations of the magnetic field within the sunspot umbra using 'high-temperature' (Fe II, Sr II, Cr II, etc.) and 'low-temperature' (Ti I, Fe I, etc.)** lines may be another point in favour of the inhomogeneous-field model. To make sure that lines Fe II in the sunspot umbra are not associated with the scattered light from the photosphere as claimed by Zwaan (1965), a number of polarization spectrograms in a large

---

* The systematic understating of the field value on a magnetograph as compared with photographic determinations and which cannot be prevented by calibration may possibly be explained by the effect of averaging the fields $H_\pi$ and $H_\sigma$ in the magnetosphere.

** 'High-temperature' lines need practically photospheric conditions for excitation, whereas 'low-temperature' ones, 1500–2000 °K less.

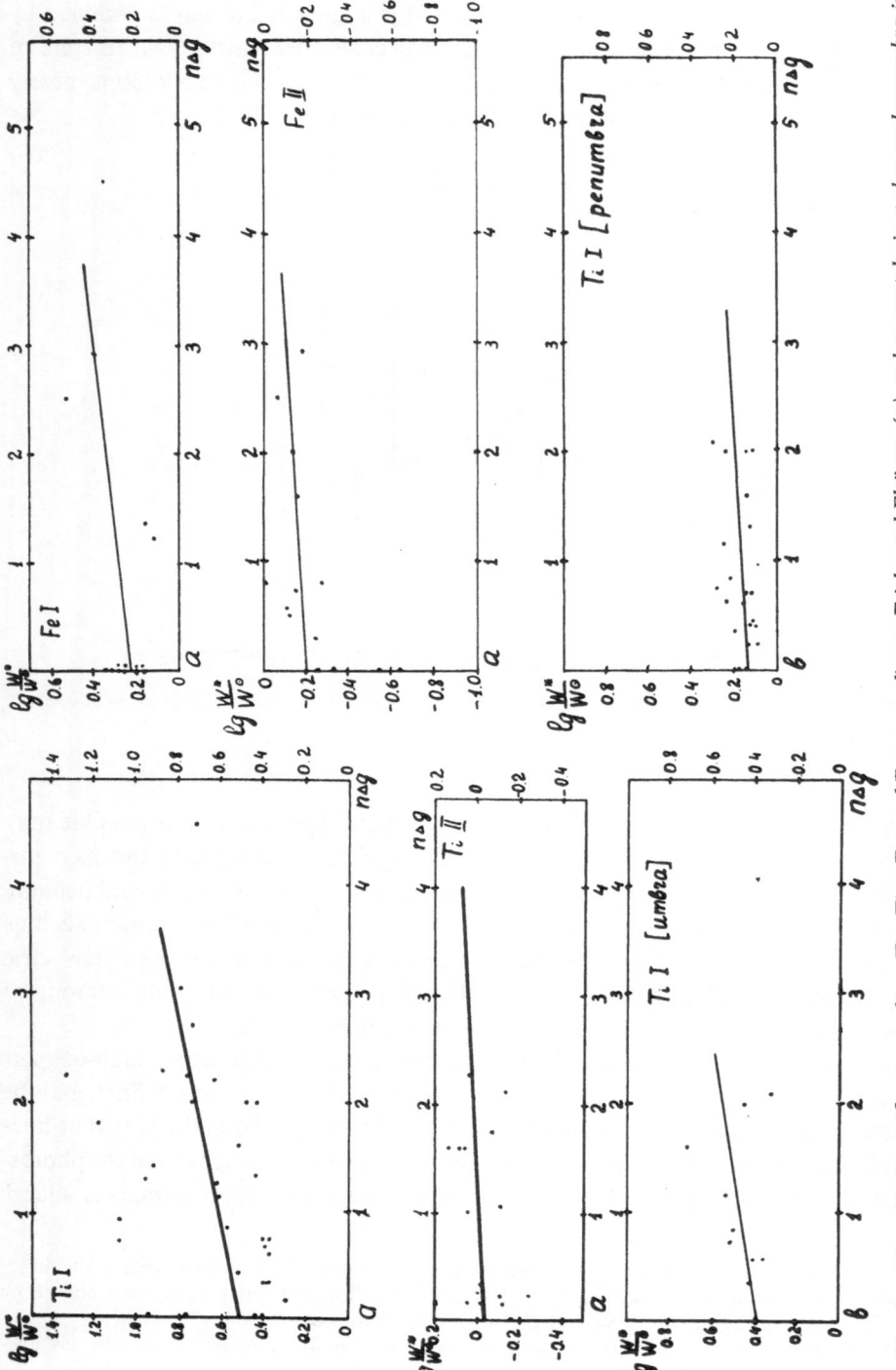

FIG. 8. *Magnetic intensification in lines Ti I, Ti II, Fe I and Fe II according to Fricke and Elsässer (a) and separately in umbra and penumbra in Ti I according to Makita (b).*

'quiet' spot were obtained with the IZMIRAN solar tower telescope (July–August 1967) in lines Fe II ($\lambda\lambda\,5018\cdot45$ and 4924 Å), Fe I ($\lambda\,5016\cdot86$ Å), and Ti I ($\lambda\,5020\cdot03$ Å) simultaneously. It was found that the magnetic field in lines Fe I and Ti I and Fe II are similar in the nature of splitting, showing maximum field values at the same points of the sunspot umbra. This indicates that, within the range of possible spatial resolution, lines with different excitation temperatures do coexist in the umbra, and the two-component model of the spot is quite promising (Makita, 1963). Therefore, one could expect that the magnetic fields determined by the lines Ti I and Fe I would be different as well. This question could be considered by investigating the magnetic intensification effect in the sunspot lines. According to Bojarchuk et al. (1960), the logarithm of magnetic intensification lg $W^*/W_\odot$ increases linearly with $n\Delta g$, the slope being proportional to the magnetic-field strength. Here $n=$ number of $\sigma$-sub-components, and $\Delta g=$ difference of the Lande factors of the upper and lower levels. Using the data of Fricke and Elsässer (1965) we plotted such a relationship for log $W^*/W_\odot$ (where $W$ are equivalent widths) for a number of lines Ti I, Ti II, Fe I, and Fe II.

For the purpose of control, a similar relationship was plotted for the umbra and penumbra for lines Ti I from the data of Makita (1963). The relatively large scattering of the points is obviously due not only to experimental errors or the difficulty of taking into account the scattered light, but also to the dependence of lg $W^*/W_\odot$ on the space orientation of the field, line strength, turbulent velocity, temperature and pressure relations in the umbra and the photosphere. A certain decrease in this scatter was achieved by selecting mean-intensity lines located in the middle part of the growth curve for spots.*

Figure 8 depicts the relationships thus obtained, and Table 1 gives the number of

### Table 1

| | Fricke and Elsässer 1965 | | | | Makita 1963 | |
|---|---|---|---|---|---|---|
| | Ti I | Ti II | Fe I | Fe II | Ti I, umbra | Ti I, penumbra |
| $N$ | 32 | 20 | 17 | 11 | 11 | 21 |
| $a$ | $0\cdot086\pm0\cdot045$ | $0\cdot031\pm0\cdot027$ | $0\cdot067\pm0\cdot012$ | $0\cdot041\pm0\cdot032$ | $0\cdot077\pm0\cdot035$ | $0\cdot024\pm0\cdot020$ |
| $b$ | $0\cdot553\pm0\cdot077$ | $-0\cdot050\pm0\cdot030$ | $0\cdot218\pm0\cdot019$ | $-0\cdot210\pm0\cdot043$ | $0\cdot395\pm0\cdot041$ | $0\cdot144\pm0\cdot024$ |

points used ($N$) and the coefficients $a$ and $b$ for expressing lg $W^*/W_\odot=an\Delta g+b$, which were obtained by the least-square method. The coefficient $a$ characterizes the magnetic intensification, and is proportional to the magnetic-field strength in the line-formation region, while $b$ defines the increase (or decrease) of the line strength neglecting the magnetic intensification. The intensification in lines TiI is close to that expected in a field of about 2500–3000 oe (Bojarchuk et al., 1960). In the penumbra

---

* For such lines magnetic intensification is determined with a high certainty and changes little with small changes in the line strength.

the intensification is approximately one-third that in the umbra, as could be expected. From the point of view of the present work, a comparison of the magnetic intensifications and the magnetic fields obtained from 'cold' (Ti I, Fe I) and 'hot' (Ti II, Fe II) lines, is of special interest. It can be seen from Table 1 that in 'hot' regions the magnetic field is approximately half that in 'cold' ones. Ignoring some weak photospheric lines Ti I (with an equivalent width of $\lesssim 10$ m Å), which cannot be measured with sufficient confidence, we obtain an even greater magnetic intensification for lines Ti I, and the field ratio will become about 3.* An analogous relationship (Figure 9) is observed in 9 lines of multiplet No. 38 Ti I ($a = 0.085$, $b = 0.35$) and, with less certainty, in 5 lines of multiplet No. 49 Fe II ($a = 0.020$, $b = -0.31$).

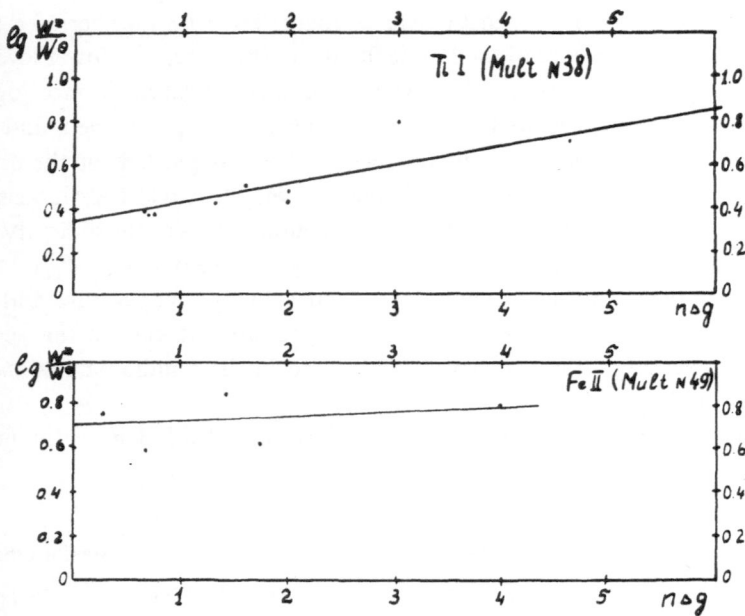

FIG. 9.   *Magnetic intensification from multiplets Ti* I *and Fe* II *according to Fricke and Elsässer.*

The same difference in the values of the magnetic fields in 'hot' and 'cold' lines was obtained by direct measurements of the magnetic field separately in lines Fe II $\lambda\lambda 5018\cdot45$ Å and $4923\cdot93$ Å, Fe I $\lambda\lambda 5016\cdot86$ and $4927\cdot78$ Å, and Ti I $\lambda\lambda 5020\cdot03$ and $5016\cdot17$ Å. For this purpose, polarization spectrograms were obtained with the IZMIRAN tower solar telescope for a number of sections in large spots (August 1, 2, and 7, 1967).

* An identical ratio is obtained (naturally, with a large scatter) by taking into account all the lines measured in the work (Fricke and Elsässer, 1965), as well as when using data uncorrected for scattered light.

All these lines were obtained in one spectrogram, thus permitting a comparison of magnetic fields measured from different lines at the same site. Measurements were carried out in 40 independent points of the sunspot umbra and penumbra. Fig. 10a displays the results of these measurements. It also shows the histograms for the occurrence of various values of the ratio $H_{Fe II}/H_{Ti I}$. Only in 4 of the 40 cases was $H_{Fe II} \geqslant H_{Ti I}$, whereas in 70% of the cases the result was $0.4 \leqslant H_{Fe II}/H_{Ti I} \leqslant 0.6$. The mean

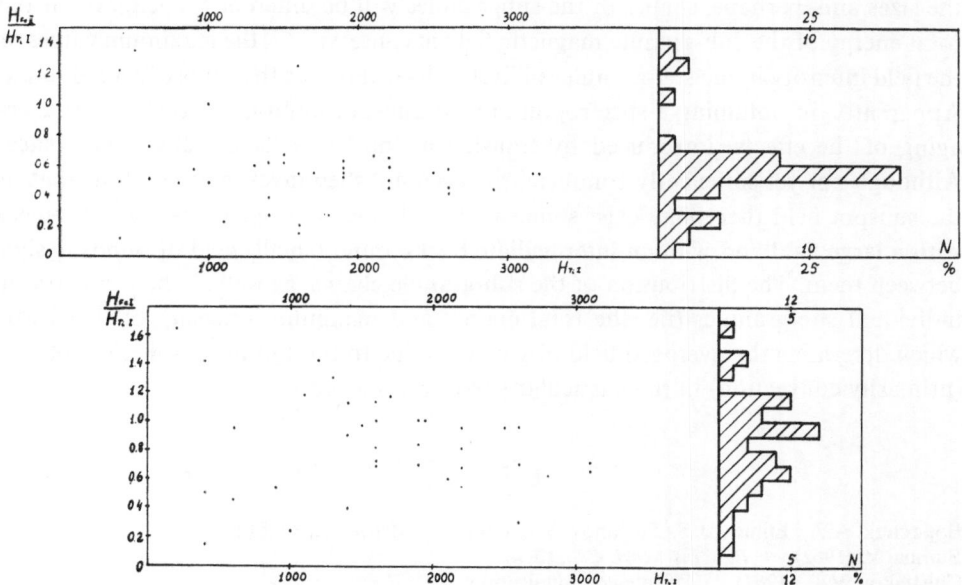

FIG. 10.   *Variation in magnetic-field ratio from lines Fe II, Ti I, Fe I in sunspots of August 1, 2, and 7, 1967.*

value of this ratio is 0.55. Thus, as well as in the case of determining the field from the difference in magnetic intensification, the magnetic field in 'hot' regions is, on the average, half that in 'cold' ones. We may also note that, despite the small number of points used, the ratio $H_{Fe II}/H_{Ti I}$ also exhibits a trend towards the grouping of points around a value of the order of 0.25, which means a four-fold excess of the field of 'cold' regions over that of 'hot' ones. This is close (with some scatter) to the above-mentioned values of the magnetic field measured by us from $\pi$- and $\sigma$-components in line Fe I $\lambda 6302.5$ Å. The ratios of the magnetic fields measured in the same spectrograms from lines Fe I and Ti I are presented in Figure 10b. Although in this case the scatter of points is considerable, in 70% of the cases $H_{Fe I}/H_{Ti I} \leqslant 1.0$, and the mean value is 0.83. Here, too, the histogram reveals that the predominant values are $H_{Fe I}/H_{Ti I} \approx 1$ and $H_{Fe I}/H_{Ti I} \approx 0.6$. From all these spectrophotometric data it follows that there is an appreciable inhomogeneity in the values of the magnetic fields, which have different values at the same points of the spot (mainly in the umbra), if the field is defined from 'cold' and 'hot' lines.

In the light of the model being considered, this result of spectrophotometric observations points to the existence of a space grouping of sub-granule chains within which temperature conditions are different. Such a difference in temperatures and densities (at equal gas pressure) may arise, for example, due to the oscillatory convection in sub-photospheric layers (Syrovatsky and Zhugzhda, 1967). In such convection cells the sizes and, perhaps, shapes of the subgranules will be different.* Then, even if the total energy of the sub-granule magnetic field is conserved**, the maximum values of the field intensity in the sub-granules of 'hot' cells will be less than those in 'cold' ones. Apparently, in obtaining a spectrogram in the sunspot umbra, a certain space averaging of the effects, introduced by translucent 'hot' and 'cold' cells, takes place. Although not yet sufficiently complete, the accomplished investigations show that in the sunspot field there may exist some small-scale magnetic elements (sub-granules) with a large field and with an intermediate background (small) field of opposite sign between them. The distribution of the sub-granule chains, as well as the geometry of individual sub-granules (i.e. the total energy and maximum intensity of the field), which determine the averaged field observed, is due to the sub-photospheric motions (primarily convection) in the particular site of the sunspot.

## References

Bojarchuk, A. A., Efimov, J. S., Stepanov, V. S. (1960)    *Astron. J.*, **37**, 812.
Bumba, V. (1962)    *Bull. astr. Inst. Csl.*, **13**, 42.
Chistjakov, V. F. (1967)    Private communication.
Fricke, K., Elsässer, H. (1965)    *Z. Astrophys.*, **63**, 35.
Kadomtzev, B. A. (1962)    *Nuclear Fusion*, **3**, 269.
Kolpakov, L. E. (1966)    *Solnechnie dannie*, **9**, 78.
Kuklin, G. V. (1966)    *The Results of the Observations and Investigations in the Period of the IQSY*, **1**, 16.
Makita, N. (1963)    *Publ. astron. Soc. Japan*, **15**, 145.
Mogilevsky, E. I., Shelting, B. D. (1966)    *Proc. of the Meeting on Solar Magnetic Fields and High Resolution Spectroscopy, Firenze*, Ed. by G. Barbera, p. 222.
Mogilevsky, E. I., Shelting, B. D. (1968)    *Solar Phys.*, **4**, in press.
Neuringer, J. L., Rosenzweig, E. R. (1967)    *Phys. of Fluids*, **7**, 1927.
Rachkovsky, D. N. (1962)    *Izv. Crimean Obs.*, **27**, 148.
Severny, A. B. (1959)    *Astr. Zu.*, **36**, 208.
Severny, A. B. (1964)    *Space Sci. Rev.*, **3**, 2.
Severny, A. B. (1965)    *Izv. Crimean Obs.*, **33**, 3.
Severny, A. B. (1967)    *Izv. Crimean Obs.*, **36**, 22.

* This follows from the generalized expression of the Bernoulli integral for the sub-granule chain in the outer field $H$. In our case, this relation will have the form $P + \rho v^2 + \rho g h + x H^2 = $ const, where $\rho = $ density, $g = $ gravity acceleration, $x = $ effective 'magnetization'.
** At relatively slow deformations the dissipation of the currents flowing inside the sub-granules may occur as well.

Syrovatsky, S. I., Zhugzhda, Yu. D. (1967)       'The Oscillatory Convection of the Conductive Gas in the Strong Magnetic Fields'. Preprint, No. 15, Phys. Inst. Ac. Sci., U.S.S.R., Moscow.
Vassilyeva, G. Y. (1963)       *Izvestija GAO* (Pulkovo), **23**, No. 171, 3.
Zhulin, I. A., Ioshpa, B. A., Mogilevsky, E. I., Obridko, V. N. (1966)       in *Proc. of the Meeting on Solar Magnetic Fields and High Resolution Spectroscopy, Firenze*, Ed. by G. Barbera, p. 155.
Zwaan, C. (1965)       *Recherches astron. Obs. Utrecht*, **17**.

# MAGNETIC AND DOPPLER OSCILLATIONS
# IN ACTIVE REGIONS

F. L. Deubner

*(Fraunhofer Institut, Freiburg i. Br., Germany)*

With the Capri magnetograph of the Fraunhofer Institute simultaneous measurements of sightline velocities and longitudinal magnetic fields have been carried out, in order to investigate the dependence of the 300-sec oscillation of sightline velocities on the magnetic-field strengths, and the oscillatory behavior of the latter, if there is any.

The principle of measurement was very similar to that used by Howard and Wilcox (1967) reported at the Prague meeting. The Sun has been scanned repeatedly along a fixed line parallel to the solar axis, with a scanning spot of 3″ diameter. The sensitivity was better than 20 m/sec and about 3 gauss (the time constant being 1 sec).

Since the complete work will be published elsewhere (Deubner, 1967), I shall simply report the results of this investigation without going into the details of the analysis.

*The influence of magnetic fields on the lifetime of single bursts of oscillations.* Whereas in plage regions as well as in undisturbed regions the most frequent lifetime seems to be about 15 min, a pronounced tendency for longer lifetimes has been found in plage regions.

It seems, that also the *mean linear distance of neighbouring oscillating elements* is influenced by the presence of magnetic fields. It is about 20% lower in plage regions than in undisturbed regions, where a value of 8″ has been found. One is reminded of the fact that the granular diameter seems to shrink in the presence of magnetic fields too (Schröter, 1962).

*The mean amplitude of 300-sec oscillations* as a function of the magnetic-field strength observed in the oscillating region. From 10 gauss to 100 gauss, the mean amplitudes decrease distinctly from about 600 m/sec to 400 m/sec. It cannot be decided whether the decrease continues towards still higher field strengths.

It seems safe to conclude from our observations, that the *period of oscillation* is independent of the field strength at least up to several 100 gauss.

It can be seen from velocity records of disturbed regions and simultaneous magnetic-field maps that regions, where the oscillations are superimposed by a strong (constant) downward-velocity component, often coincide with stronger magnetic fields. The mean-velocity level for different field strengths has been calculated. There is an unambiguous transition from upward velocities to downward velocities with increasing field strength. This correlation between Doppler shift and magnetic field cannot be

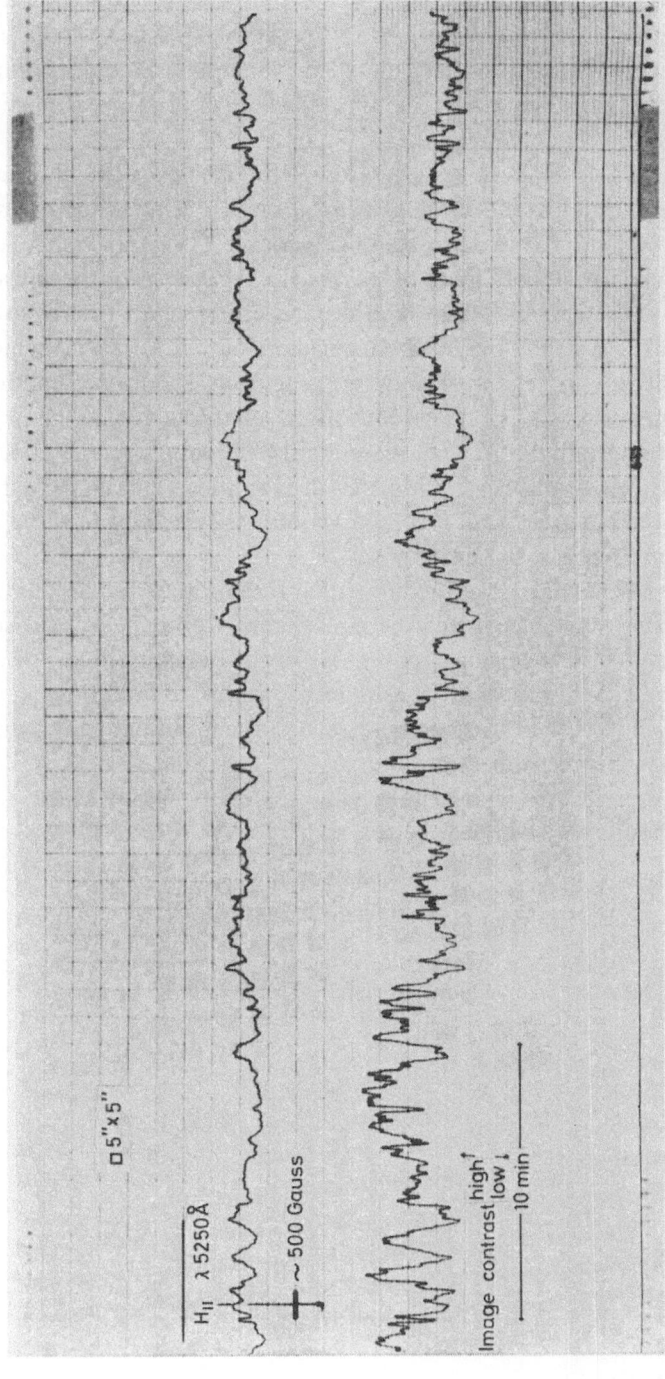

FIG. 1. *Simultaneous record of the longitudinal magnetic-field component and the image contrast. Note the strong correlation.*

due to the downward motion at the borders of the supergranular cells because of its low velocity. It therefore must be ascribed to the geometric effect which, by oblique projection, produces coincidence of regions of the maximum Doppler component of the horizontal motion directed away from the observer with regions of maxima of the longitudinal magnetic-field component.

This rough analysis is in good agreement with the idea that the magnetic flux is preferentially concentrated at the borders of supergranular cells by horizontal motion. This is especially true also for stronger field intensities $\geqslant$ 200 gauss in facular regions.

Finally let us turn to the possibility of *short-period variations of the magnetic field*. After eliminating instrumental effects which at times produced artificially 'oscillations' of the longitudinal field signal, only one type of short-period variations remained still apparent in the records. An example is given in Figure 1. The field strength has been recorded simultaneously with the image contrast, the latter being measured by means of a photocell with a small diaphragm, scanning the Sun in a circular pattern. There is a striking, nearly one-to-one correlation of both quantities. Low (high) field strengths are always measured in moments of low (high) contrast. The changes of the magnetic-field signal are sometimes of the order of 40%!

This effect may be explained if one assumes that the entrance diaphragm just covers a tiny magnetic structure, which may be considerably 'diluted' by bad seeing. To check this, I measured in the same way regions in the neighbourhood of old stable sunspots, where one could expect more homogeneous field distributions, not affected by seeing. There indeed the magnetic-field record was a straight line, only super-imposed by the known instrumental noise.

This example demonstrates once more, how carefully photoelectric records of magnetic fields should be examined, before turning them into results.

### References

Deubner, F. L. (1967)      *Solar Phys.*, **2**, 133.
Howard, R., Wilcox, J. M.      Transactions IAU (Prague Meeting, 1967).
Schröter, E. H. (1962)      *Z. Astrophys.*, **56**, 183.

# PRELIMINARY COMMUNICATION ON THE SHORT-PERIOD OSCILLATIONS OF SOLAR MAGNETIC FIELDS

A. B. Severny

*(Crimean Astrophysical Observatory, Nauchny, Crimea, U.S.S.R.)*

With the image of the Sun held fixed, time variations of solar magnetic field were recorded simultaneously in two lines ($\lambda 5250$ and $\lambda 6103$) with the aid of double magnetograph (for the description see Severny, 1966, examples of such records are on Figures 1 and 2). The resolution was $2''.5 \times 4''.5$. The records in both lines $\lambda 5250$ and $\lambda 6103$ covering a time interval of about 1 hour show oscillations with the period $\sim 7.4$ min and amplitudes 10–20 gauss for moderately active region (full lines on the top of Figure 2). The oscillations of the magnetic field at both levels (the heights of $\lambda 5250$ and $\lambda 6103$ differ by 260 km) are sometimes synchronous but now and then phase shift appears. For quiet regions near the centre of the disk the oscillations with the period $\sim 9.0$ min and amplitudes less than 5 gauss are recorded and presented in Figure 1 (full line). These 7–9 min oscillations are probably sometimes damped or interrupted by more short-period fluctuations (with the mean period 80s) of smaller amplitude, but these last oscillations could be due to the fluctuations of seeing.

Simultaneous records of radial velocities (Figure 2, bottom) for both levels at the same time show clearly Leighton's pulsations with the mean period 297s coinciding with that found in Leighton's (1963) paper. No appreciable changes in intensity (except usual fluctuations due to noises) of both lines recorded simultaneously were found.

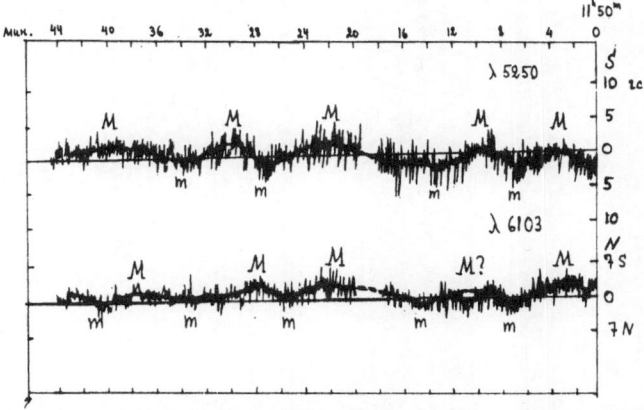

Fig. 1. *Fluctuations of magnetic field recorded in $\lambda$ 5250 (top) and $\lambda$ 6103 (below) for quiet regions near the centre of the disk.*

*Kiepenheuer (ed.), Structure and Development of Solar Active Regions, 233–235. © I.A.U.*

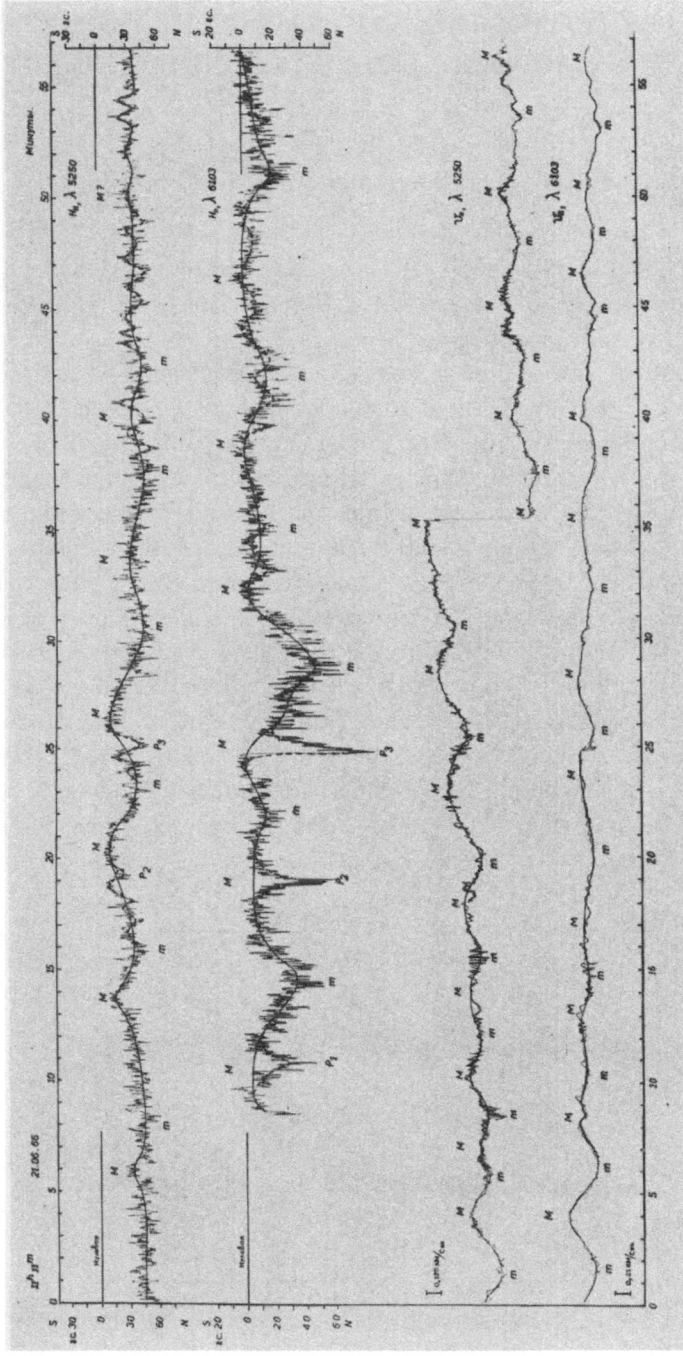

Fig. 2.　Oscillations of magnetic fields in active region recorded in λ 5250 (full line at the top of the figure) and in λ 6103 (the next one). At the bottom the fluctuations of radial velocities recorded in the same lines simultaneously are shown.

## References

Leighton, R. (1963)    *A. Rev. Astr. Astrophys.*, **1**, 19.
Severny, A. B. (1966)    *Astr. Zu.*, **43**, 465.

## DISCUSSION

*Howard:* Some of the oscillations you showed in the last slide indicate a variation of magnetic field which involved a change in sign of the field. Do you have an explanation for this? How can it happen?

*Severny:* We have not observed such oscillations in active regions, but in the case of general field (quiet regions) it is difficult to be sure that such changes of sign are reliable due to long-period noises. As regards to the explanation of such oscillations in general neither I nor, I think, Mr. Deubner are in possession of adequate explanation of the physical nature of phenomenon.

# SOME COMMENTS ABOUT CORRELATIONS BETWEEN MAGNETIC FIELD AND VELOCITY, MAGNETIC FIELD AND LINE INTENSITY IN THE UNDISTURBED PHOTOSPHERE

G. Y. Vassilyeva            and            A. K. Tchandaev

*(Pulkovo Observatory, U.S.S.R.)*            *(Gorki Institute, U.S.S.R.)*

## 1. Observations

Test cross-correlation functions between the magnetic field recordings and the sight-line velocity recordings with East and West relative lag for two regions near the centre of the disk have been computed. We have also found the deviations of the absolute value of the magnetic-field strength $|H|$ from the mean absolute value of magnetic-field strength $|\bar{H}|$ for the whole region. The same procedure for the velocity field has been made. The cross-correlation functions for these deviations have also been computed (Figure 1).

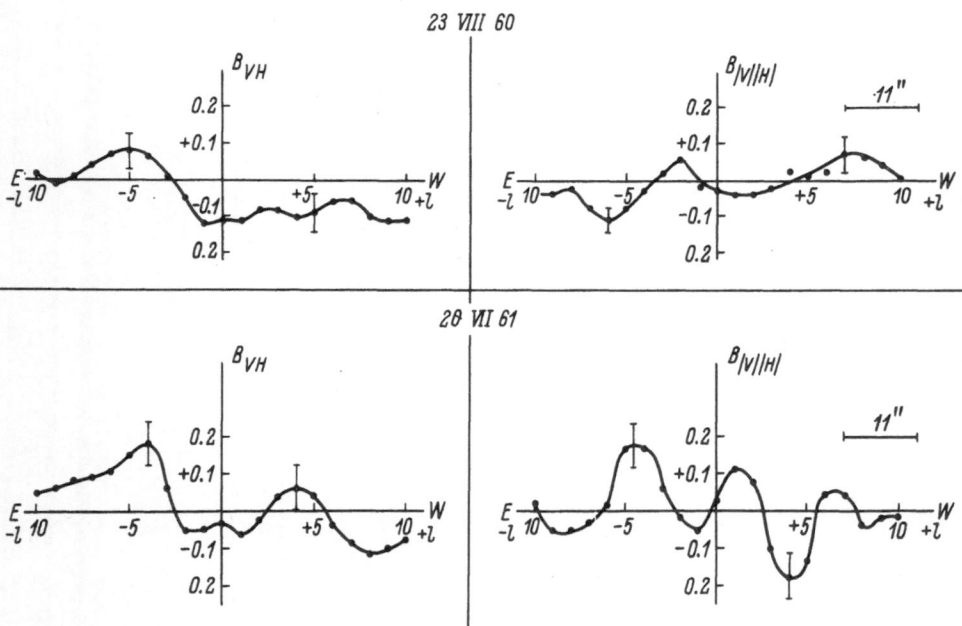

FIG. 1. *The cc-functions for the magnetic field recordings and velocity recordings, obtained in the undisturbed photosphere in Fe* I *5250.*

*Kiepenheuer (ed.), Structure and Development of Solar Active Regions, 236–239. © I.A.U.*

The material obtained during 1960 and 1961 by means of the Pulkovo magneto-graph in Fe I 5250 Å with a spatial resolution of $5''\!.5 \times 2''\!.2$ has been used for this purpose.

The cross-correlation functions between $H$ and the intensity in Fe I 5250 (Figure 2)

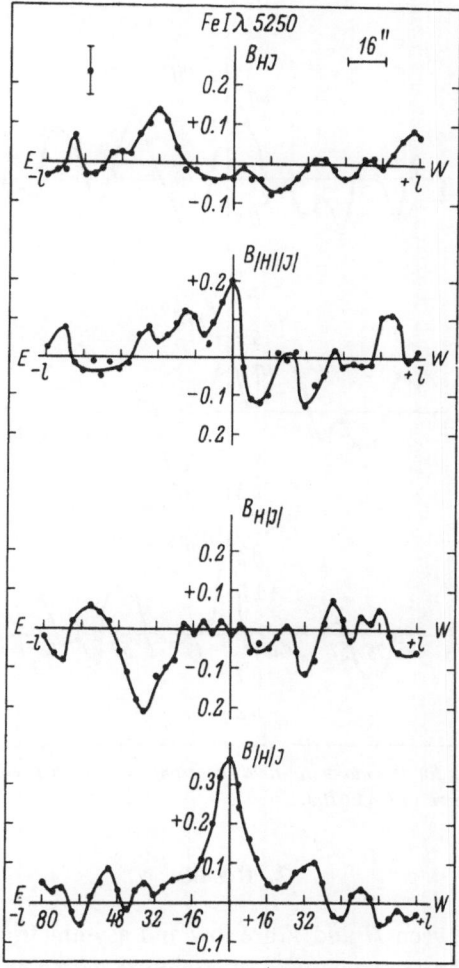

FIG. 2.   *The cc-functions for the magnetic-field recordings and intensity-in-line recordings, obtained in the undisturbed photosphere in Fe I 5250.*

and in Ca I 6103 (Figure 3) have been computed on the base of the treatment of Crimean magnetic and intensity records, kindly offered to us by Professor A. B. Severny.

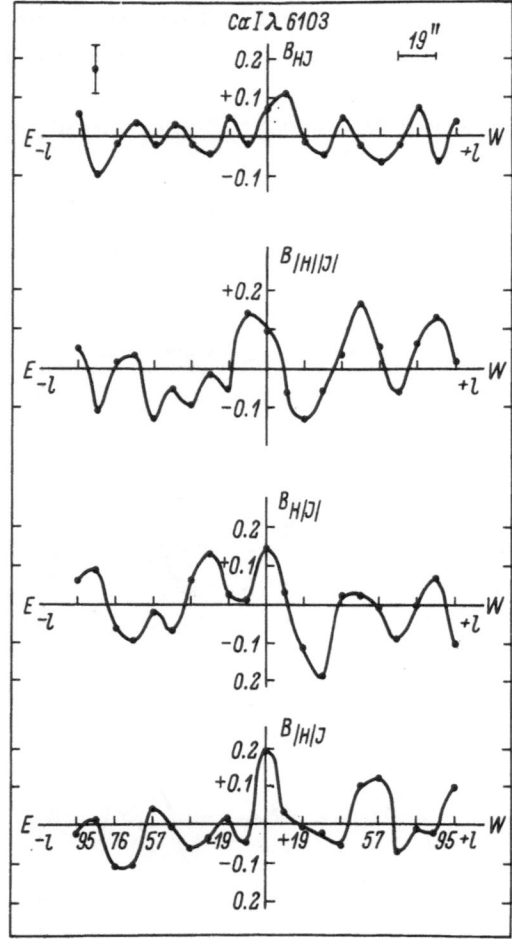

FIG. 3.  *The cc-functions for the magnetic-field recordings and intensity-in-line recordings, obtained in the undisturbed photosphere in Ca I 6103.*

## 2. Results

The cc-functions between $H$ and $V$ are low and asymmetric. For the $|V|\,|H|$-case cc-functions have a definite maximum and minimum with a change of the sign, when the lag is about equal to the radius of the supergranule.

This result means that the magnetic field at the centre of the supergranule seems to be connected with the velocity on the boundaries, and vice versa.

Besides, the change of the sign of the cc-functions indicates the different character of the velocity on the East and West sides of the magnetic hill.

The meaning of this asymmetry depends on the polarity and the magnetic-field strength. On the base of these results the mutual behavior of the magnetic elements of the background fields the same and the opposite, investigated by Bumba and Howard (1965) are proposed to be as a result of their dynamical interaction, but not a direct magnetic one.

The cc-functions between magnetic field and the brightness have nothing to do with those between magnetic field and velocity. They show maximum with lag equal to zero, which means the weakening of the lines at the points of the increasing magnetic field.

Thus, the locations of the places on the disk, where the brightness in lines and velocity are connected with the magnetic field, are not the same.

The lines FeI 5250 and CaI 6103 with different levels of formation at photosphere (the difference is about 160 km according to Severny, 1966) reveal the different correlation between brightness and magnetic field.

## References

Bumba, V., Howard, R. (1965)    *Astrophys. J.*, **141**, 1492.
Severny, A. B. (1966)    *Astr. Zu.*, **43**, 465.

# ON SOME SPECTROGRAPHIC OBSERVATIONS RELATED TO THE STRUCTURE WITH HEIGHT OF ACTIVE REGIONS AND PARTICULARLY SOLAR FLARES

YNGVE ÖHMAN

*(Stockholm Observatory, Saltsjøbaden, Sweden)*

ABSTRACT

A peculiar kind of double-peaked Hα lines in solar flare spectra is described. The most likely explanation to this phenomenon is an absorption in the centre of the line due to mottles and fibrils situated above the flare. Another explanation might be a rapid rotatory mass motion in the filamentary structure of the flare itself. In some other flare spectra a slight inclination of the Hα-emission streak has been found indicating as well that rotatory mass motions may be present in flares.

The writer has described before (1) a peculiar Hα spectrum of a flare of importance 1b observed in Anacapri on March 29, 1966 (UT 09·52) showing a double-peaked Hα line, similar to that observed in the low chromosphere at the limb. This interesting spectrum is presented again in Figure 1a. The emission streak is seen in the centre of the Hα line also but here the streak is not only weaker but also clearly narrower than in the two well-defined emission peaks. Therefore these peaks, though situated in the steep slope of the Hα photospheric line, can hardly be due to an apparent intensity increase produced by the increase in background intensity, but must be due to a real emission feature of the flare spectrum. The distance between the peaks is about 1·2 Å. The flare was situated very near a sunspot with the coordinates 26 N, 59 E.

In the same photographic recording of the solar spectrum a flare spectrum with moustaches (2) is seen. It is situated slightly above the same spot.

We have found several examples of the double-peaked Hα line-flare spectra. In Figure 1b a similar, though fainter object is seen, which was obtained at the Swedish Solar Observatory in Anacapri on July 8, 1966 at UT 12·56. The distance between the peaks is about the same as in Figure 1a, but the peaks are not quite so narrow resembling more or less a necktie.

Sometimes objects of this kind are connected with dark surges and may in fact be bright surges visible on the disk. Figure 2b obtained on April 5, 1967 at UT 10·07 would suggest a bright surge phenomenon combined with dark surges. But the emission features had a much longer duration than the dark ones, which is evident from Figure 2a obtained at UT 09·17 and Figure 2c at UT 10·23 and some later exposures. So at least the object in Figure 2 seems to have been more related to a flare than to a bright surge. In fact, it coincided with a small brilliant plage in 25° S, 40° E indicating

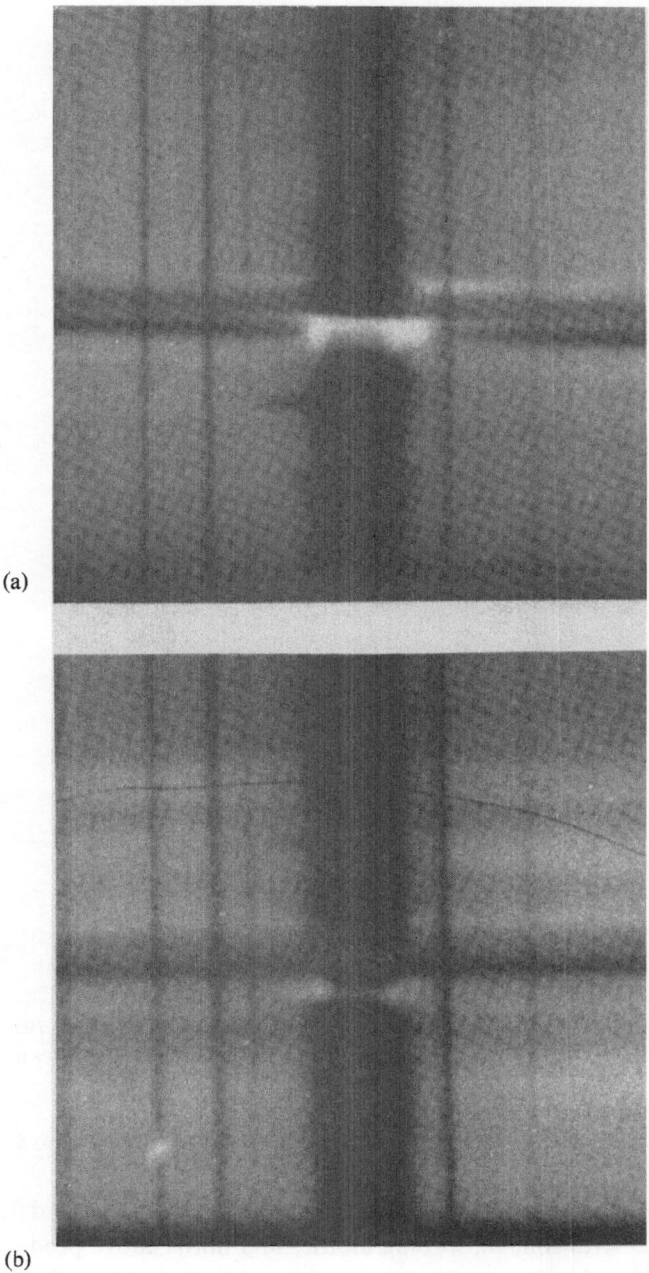

(a)

(b)

FIG. 1.   *Double-peaked Hα lines in solar flares observed on the disk. 1a was obtained with a Babcock grating spectrograph on March 29, 1966, whereas 1b was obtained on July 8, 1966. Note also the moustaches in (a) above the flare.*

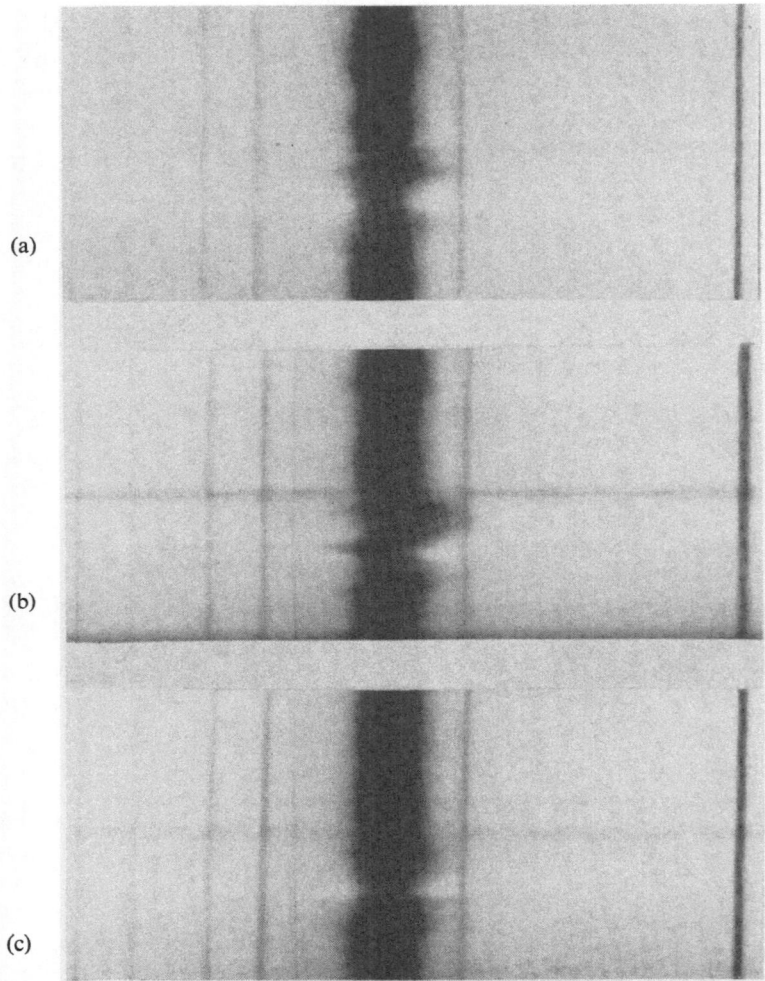

FIG. 2.    *Spectrum of a brilliant plage accompanying the appearance on April 5, 1967 of a new spot. 2a was obtained at UT 09·17, 2b at UT 10·07 and 2c at UT 10·23. Note the dark surge appearing in (b).*

the creation of a new spot. In a way it seems related to the moustaches too, but, if so, the moustaches are rather short.

In an earlier publication (**1**) I have suggested that the double-peaked flare in Figure 1a is a low object with some absorbing mottles and fibrils above producing a strong reversal of the emission line in the centre of the line. This kind of central reversal has been suggested by Giovanelli, Michard and Mouradian (**3**) to be present in the Hα spectrum of the low chromosphere. In fact dark features of even large size can sometimes be clearly traced in the chromospheric spectrum of the limb (**1**).

Though the spectra presented in Figure 1 and Figure 2 seem fairly well explained by assuming the objects to be low ones with absorbing mottles and fibrils above, we should not overlook the possibility of mass motions in different directions within the object imaged on the slit of the spectrograph. The possibility of such mass motions in solar flares has been considered before to explain the asymmetry of the Hα-emission line, and I want to draw attention particularly to a recent paper by Ballario (4). Mrs. Ballario has considered in this connection flares with loop structure.

One difficulty in explaining the double peak in the spectrum Figure 1a as a result of rapidly ascending and descending mass motions is the narrow-line profile of each constituent needed to obtain the well-defined peaks. The emission-line features of Figure 1b and Figure 2 are perhaps somewhat easier to explain as a result of mass motions only. Indeed, it is tempting to assume a rotatory mass motion with the axis of rotation more or less parallel to the slit of the spectrograph.

We have found recently some independent evidence of rotatory mass motions in flares, and, in fact, because of a slightly inclined Hα streak. Figure 3a and 3b show examples of this phenomenon appearing in an importance 1b flare observed by me in Anacapri on July 3, 1967 (20° S, 16° E). 3a was obtained at UT 07·48, 3b at UT 08·09, 3c, finally, shows an Hα-patrol image obtained by O. Gimse at UT 08·20 and with the approximate position of the slit. In addition to a faint inclination due to a remaining error of adjustment of the grating (changes had just been made) the emission streaks show a clear effect of intrinsic inclination. Similar effects have been observed and described by Severny (5). In my opinion it is difficult to explain the inclination of the streak without assuming a mass motion, and a rotatory motion is perhaps well possible. If so, the axis of rotation would be more or less perpendicular to the slit. It is interesting to note in Figure 2b that the inclination is stronger here than in Figure 2a. This may be a result of the reduced intrinsic line width, eventually combined with an increased separation of mass elements.

Inclined emission streaks are well known in the spectra of spicules, but have in general been attributed to effects of unresolved superimposed objects with different mass motions (6). But the possibility of a rotatory motion should perhaps be considered in the case of spicules too. Moreover, we have found some Hα spectra of prominences indicating rotatory mass motion, the spectra showing some features in common with Figure 1. That is to say, double emission peaks appear. Figure 4 shows such an object near the limb secured by us in Anacapri on July 2, 1967 at UT 13·31. In fact one of the circular Hα recordings shows a strong 'limb-brightening' not only to the left and right but also above. Self-reversal alone can hardly explain this structure. It is tempting therefore to consider mass motions as well and perhaps of a rotatory type. A study of a great quantity of material may give us a better clue to the understanding of these curious objects and their relation to flares and loopes. If possible, observations should be made simultaneously in different spectrum lines.

The possibility of rotatory mass motions should be considered also from a theoreti-

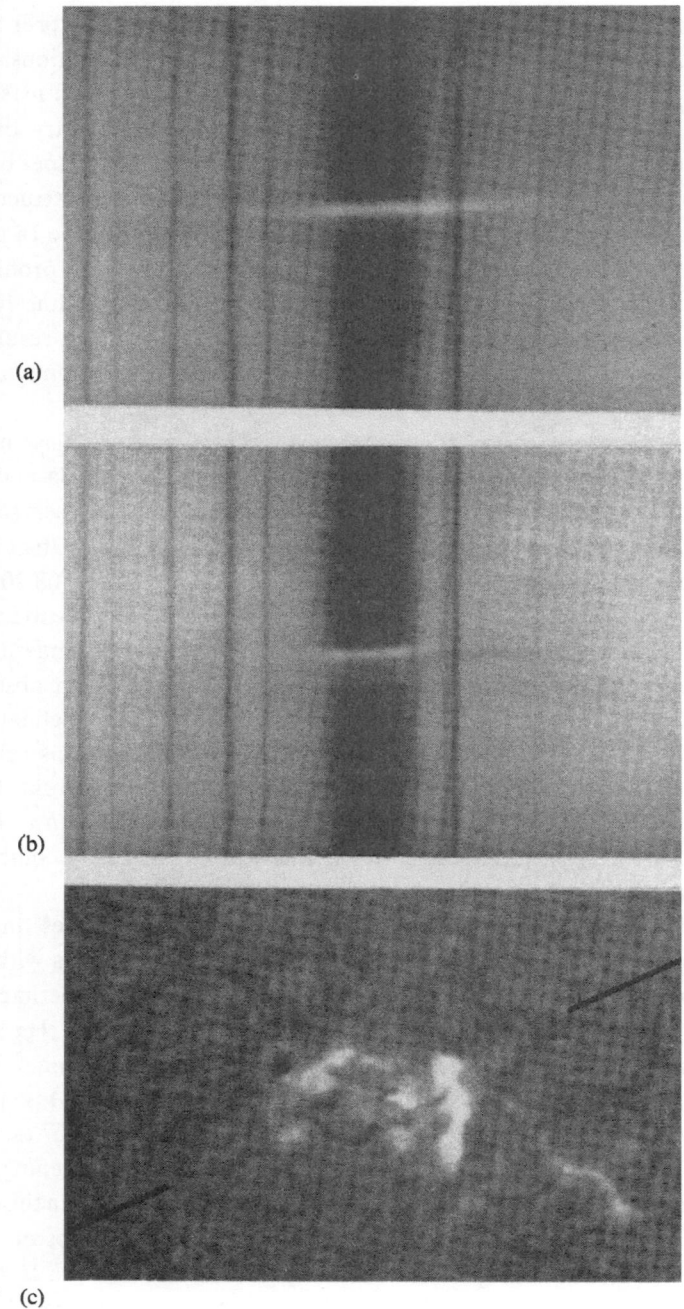

(a)

(b)

(c)

FIG. 3.    *Spectra of an importance 1b flare observed on July 3, 1967 at 20 °S, 16 °E. 3a was obtained at UT 07·48, 3b at 08·09. 3c is an Hα picture of the flare at UT 08·20. The inclination of the Hα streaks in (a) and (b) is partly real and indicates mass motion, perhaps by rotation.*

cal point of view. In fact Danielson (7) has considered already roll convection to be present in the fibrous structure of the penumbra in sunspots. Professor Alfvén has kindly informed me that rotatory motions may well be expected in flares according to his and Carlqvist's recent theory of flares (8).

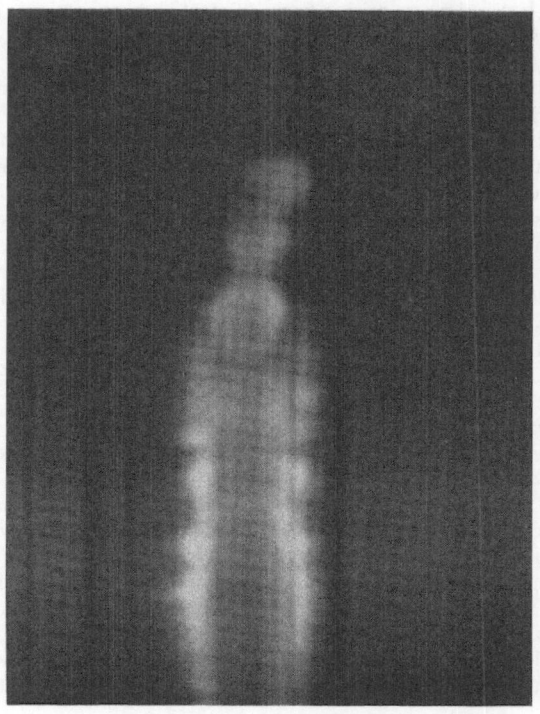

FIG. 4.  *Hα spectrum of prominences and the chromosphere at the limb obtained on July 2, 1967. Note the circular apparent shape of one of the prominences with a limb-brightening possibly due to mass motion of a rotatory type.*

### References

1. Öhman, Y. (1966)    *Stockholms Observatorium, Meddelande* No. 161.
2. Severny, A. B. (1957)    *Izv. krym. astrofiz, Obs.* **17**, 129.
3. Giovanelli, R.G., Michard, R., Mouradian, Z. (1965)    *Ann. Astrophys.*, **28**, 871.
4. Ballario, Maria Cristina (1963)    *Osservatorio Arcetri, Contr.* 78.
5. *Handbuch der Physik*, Springer, Band LII, Astrophysik, III, p. 206, Fig. 61*d*, 1959.
6. Beckers, J. M. (1964)    Dissertation, Utrecht, p. 72.
7. Danielson, R. E. (1962)    *Astrophys. J.*, **134**, 289.
8. Alfvén, H., Carlqvist, P. (1967)    *Solar Phys.*, **1**, 229.

## DISCUSSION

*Pasachoff:* I agree that many inclined spectral features are visible in limb observations. On spectra taken jointly with Drs. Beckers and Noyes at the Sacramento Peak Observatory and reduced at the

Harvard College Observatory, such inclined features are visible not only in Hα but also in the H and K lines of Ca II (J. M. Beckers, R. W. Noyes, and J. M. Pasachoff, *Astron. J.*, **71**, 1966, 255).

Our spectra were made with a dispersion of 0·15 Å/mm and a 25-cm solar image with the 16-inch coronagraph and 13-m Littrow spectrograph and sequences exist with exceptional seeing. We believe that we can satisfactorily resolve individual spicules and that many of the inclinations are real. Sometimes inclined features are visible at great heights, where the total number of features is small and the chances of accidental overlapping minimized.

We interpret these inclinations as arising from differential mass motions, perhaps from rotation around the spicular axes.

The inclinations at some features change with time. This can be interpreted as a real change in the velocity, a change in the orientation of the feature with respect to the line of sight or a change in the width of the line profile.

Further, our H and K spectra include a disturbed region in the chromosphere we used an image slicer to give us effectively two slits located approximately 4000 and 8000 km above the limb. This allows us to follow the relation of the simultaneous line-of-sight velocity at the two levels.

Exposures were 13 sec long for H and K and were taken every 30 sec. The disturbed region extended about 40 000 km along the slit.

For several minutes, the upper level is seen to contain a disturbed area, for there are many knots and short, highly inclined features indicating the presence of strong differential motions. The lower level has normal spicular features. Eventually it too begins to show abnormal structures.

Then, on the next frame, a feature becomes visible at the lower level with a line-of-sight velocity of about 50 km/sec. On this frame, the feature is very faint in the upper level and has a somewhat lower velocity, about 35 km/sec.

On the following frame, 30 sec later, the feature has become very bright in the upper level, with a large Doppler velocity of 60 km/sec. The velocity in the lower level has declined to about 15 km/sec, although the profile remains wide.

On the next frame, the velocity in the upper level increases slightly to about 70 km/sec, while the velocity in the lower level declines to about zero. This is, unfortunately, the end of the time sequence.

Note that the velocity of propagation implied is about 4000 km in 30 sec or 130 km/sec. This is far more than supersonic, but is comparable to the Alfvén speed for reasonable chromospheric densities and magnetic fields. Other events visible at both lower and upper levels do not usually show even a time difference of this magnitude, indicating still higher velocities.

*Maltby:* At the Oslo Solar Observatory we have observed inclined elements in Hα in the quiet chromosphere, close to sunspots as well as in flares. Several explanations, in addition to rotatory motion, are possible.

*Bappu:* I would just like to make a comment on the inclined structures seen in Hα. When one obtains Hα spectra of the centre of the solar disk with good image and spectrographic resolution, one can see many narrow inclined structures in absorption. Rotation of the mass element is an obvious interpretation, and one can postulate the mass to be either experiencing a helical motion or be in a form where one has ascending and descending columns side by side.

*Öhman:* I am glad to hear that several observers have found similar effects, and I am fully aware of the fact that different interpretations are sometimes possible.

# CHROMOSPHERIC HEIGHTS IN ACTIVE REGIONS*

M. K. V. BAPPU and K. R. SIVARAMAN

*(Kodaikanal Observatory, India)*

ABSTRACT

Spectra, with a radial slit in the K and Hα lines, of a sunspot close to the limb are used to determine chromospheric heights in an active region. Equidensity contours obtained with the Sabattier effect are used to measure the values of $h^*$ for different positions in the line core and wings. The spectra have also been utilized to determine similar heights in two very restricted regions of Ca$^+$ emission which show an arched emission structure in $K_{232}$ with bright continuum emission. At the K-line centre, unit tangential optical depth in the umbra, umbra-penumbra interface, and penumbra are at 1310, 1610, and 2330 km respectively. The corresponding value for the umbra in the Hα-line centre is 1080 km. The arches have a value of $h = 3000$ km at $\Delta\lambda = 0$ for the K line.

The extension of a sunspot into the chromosphere can be detected by two procedures normally available to the non-eclipse observer. One is by the detection of magnetic fields of sufficient intensity with the aid of lines that have a sizeable chromospheric emission contribution. Severny and Bumba (1958) showed that such magnetic fields have high values out to at least 2000 km. The variations of the high and low values with location indicate the retention of the characteristics of the group seen at lower levels. A second approach, first demonstrated effectively by Mattig (1958), is the shift that the core of a strong line experiences towards the limb when the spot is very close to the limb and is examined with a radial slit. This is an extremely simple and effective way of studying the properties of the spot region at normal chromospheric heights. It has been used by Mattig (1962) and by White and Wilson (1966) for the Balmer series, more especially the Hα line. We report in this study measures made in the CaII K line of a spot near the limb observed with the McMath solar telescope of the Kitt Peak National Observatory.

Our observations pertain to sunspot number Kodaikanal 12 399 which on May 25·7 had the heliographic co-ordinates 7°N, 81°24′W. The spot had an area of roughly 280 millionths of the visible hemisphere with a well-defined umbra of nearly 80 millionths. Both umbra and penumbra were of identical polarity. Figure 1 depicts the spatial variations of the longitudinal component of the magnetic field as derived at Kodaikanal from observations of May 23·2, just 60 hours before the K spectra were obtained at Kitt Peak.

---

* Presented by M.K.V. Bappu.

*Kiepenheuer (ed.), Structure and Development of Solar Active Regions, 247–254. © I.A.U.*

The K-line spectra were obtained in the 6th order with a dispersion of 0·134 Å/mm. These were taken with a radial slit and with three different settings across the spot. Since the image scale in the focal plane of the McMath telescope is 2·29 "/mm, 1 mm on the spectrogram in a direction perpendicular to the solar surface represents 1654 km on the solar surface. The single Hα spectrum taken of this spot had the radial slit pass directly through the umbra of the spot. The dispersion used is 0·174 Å/mm in the

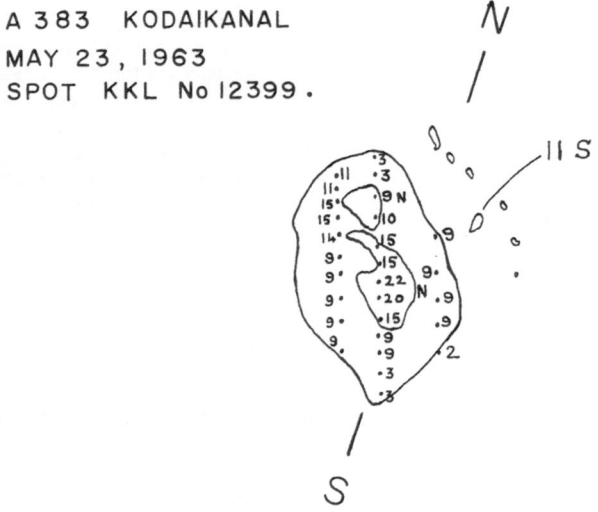

A 383  KODAIKANAL
MAY 23, 1963
SPOT  KKL No 12399 .

FIG. 1.   *The spatial longitudinal magnetic-field values of the spot of May 23, 1963.*

4th-order spectrum. The spectra were exposed during spells of good seeing. An estimate of the halfwidth of the seeing function from the shape of the solar limb profiles is 1000 km. A wire stretched across the slit provided a convenient reference mark on the spectra as well as the drawings made immediately after an exposure of the location of the slit on the spot.

On the spectrogram the spot spectrum is confined to a limited region in the form of an absorption band that traverses the entire spectrum parallel to the dispersion and which, while crossing the K or Hα line, experiences a deviation towards the limb that is a maximum in the line core. Other features of the spectrum display a similar tendency. On one of our spectrograms are two well-defined bright chromospheric arches that have a bright continuum emission. References in what follows to chromospheric arches pertain to these two features.

The measurements, in the line core and wings, of the deviation from a straight line parallel to the dispersion, indicate the different geometrical heights of unit tangential optical depth. Since the boundaries of the umbra and penumbra as seen on the spectrogram are seldom sharp enough to yield measures of precision, considerable care is necessary to derive these heights with some measure of accuracy. We have utilized the technique of equidensitometry exemplified by the use of the Sabattier effect. This method has been extensively utilized in astronomical research by Richter and his colleagues at Tautenburg. It was first employed in astronomy by Schröter (1958) when he successfully utilized it for studying the structure in the Balmer lines of chromospheric origin.

Our measuring technique has been to use twice Sabattiered prints of each spectrum (Figure 2) of the sunspot and the arches. The double contours of wire shadow and

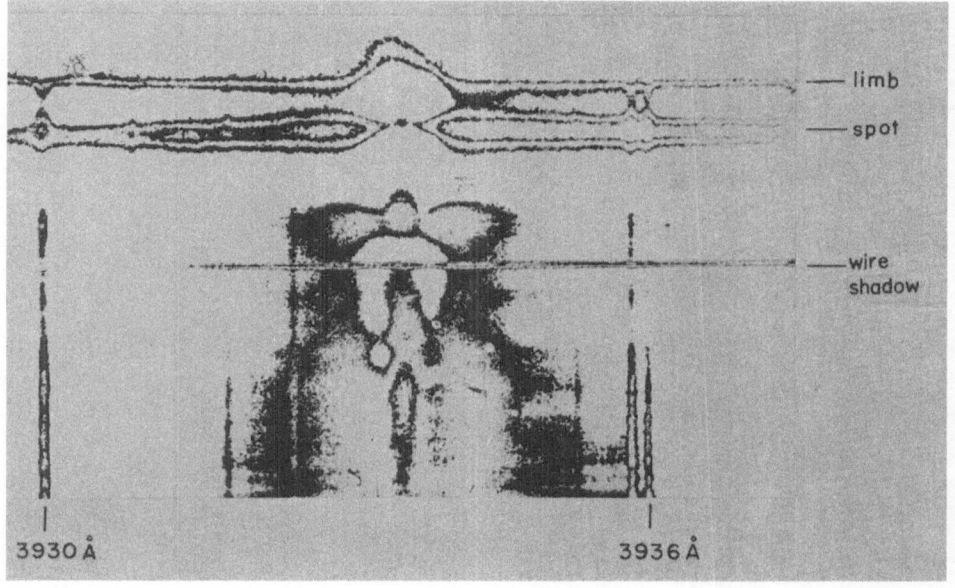

FIG. 2.  *A twice Sabattiered print of the K-line spectrogram with radial slit through the spot umbra (Position B).*

spectrum of the spot or inhomogeneity facilitate greatly the easy measurement with a two-co-ordinate measuring engine of the shifts in line wings and core towards the limb. The errors in measurement of the equidensity contours amount to about ± 25 km while the photographic reproducibility of the contours is within 75–100 km. Hence, we believe that our values of the shifts to limb are accurate to within 100 km on the solar surface.

The measures we make pertain to the displacement of the features caused by a

## Table 1

### Values of h*

$h^*$ (km)

| $\Delta\lambda$ (Å) | Position of radial slit on sunspot K line | | | Chromospheric arch 1 $\theta = 74°34'$ | Chromospheric arch 2 $\theta = 72°18'$ | H$\alpha$ line $\theta = 84°54'$ Radial slit over spot umbra |
|---|---|---|---|---|---|---|
| | A | B $\theta = 81°54'$ | C | | | |
| 0·0 | 1790 | 1490 | 2510 | 3410 | 3440 | 1230 |
| 0·1R | 1400 | 1250 | 1950 | 2510 | 3520 | 1200 |
| 0·1V | 1620 | 1390 | 2120 | 3380 | 3540 | 1010 |
| 0·2R | 1150 | 1070 | 1490 | 2400 | 2920 | 860 |
| 0·2V | 1100 | 940 | 1370 | 3500 | 2990 | 750 |
| 0·25R | 1050 | 940 | 1390 | 1390 | 1940 | 500 |
| 0·25V | 1040 | 840 | 1090 | 2730 | 2430 | 580 |
| 0·3R | 940 | 850 | 1320 | – | – | 440 |
| 0·3V | 970 | 750 | 680 | – | – | 510 |
| 0·4R | 750 | 670 | 1200 | – | – | 380 |
| 0·4V | 900 | 600 | 370 | 170 | – | 420 |
| 0·5R | 560 | 510 | 1000 | 240 | – | 330 |
| 0·5V | 830 | 530 | 280 | 90 | 240 | 340 |
| 1·0R | 300 | 530 | 230 | – | – | 150 |
| 1·0V | 600 | 420 | 320 | – | – | 120 |
| 1·5R | – | 250 | 100 | – | – | 70 |
| 1·5V | 120 | 170 | 90 | – | – | 50 |
| 2·0R | 0 | 170 | 50 | – | – | – |
| 2·0V | 0 | 0 | 80 | – | – | 150 |
| $\delta h$ (km) | 180 | 180 | 180 | 220 | 220 | 150 |

Radial slit positions A, B, C of the spot refer to locations of the slit over the penumbra-umbra interface, umbra and penumbra respectively.

projected geometrical difference $\Delta h$ in the heights between the positions of unit optical depth ($\tau_{tang}$) in the continuum and for any wavelength in the line. Hence, the actual chromospheric height difference is $h^* = \Delta h/\sin\theta$, where $\theta$ is the heliocentric angle of the inhomogeneity. This is a measure of the actual height difference between $\tau(\lambda) = \cos\theta$ in the line and $\tau_{5000} = \cos\theta$. We have measured these displacements for different values of $\Delta\lambda$ in the K line. We believe that such data would be of help in the future elaboration of inhomogeneous chromospheric models, especially those involving the absorption and source function in an active-region chromosphere. Table 1 gives the values of $h^*$ for different values of $\Delta\lambda$ for the three slit positions on the spot and the chromospheric arches.

Two corrections have to be incorporated to these values of $h^*$ in order to deduce the true heights of unit tangential optical depth in the chromosphere. One must first subtract the geometrical distance between $\tau_{5000} = \cos\theta$ and $\tau_{5000} = 0.004$. The values of this correction $\delta h$ obtained from Allen's (1963) table are also indicated in Table 1. A second correction factor needed for the sunspot chromosphere observations is the relationship between $\tau_{5000}$ in the spot and in the normal photosphere. This necessitates the use of data from a reliable model of a sunspot. Since our present knowledge of sunspots is inadequate to provide this information, we have refrained from making the required correction. This factor is, however, likely to be within 200 km, and needs to be subtracted from $h^*$ if we assume that the spot umbra is more transparent than the neighbouring photosphere.

We plot in Figure 3 the values of $h^*$ measured in the K line for the three positions

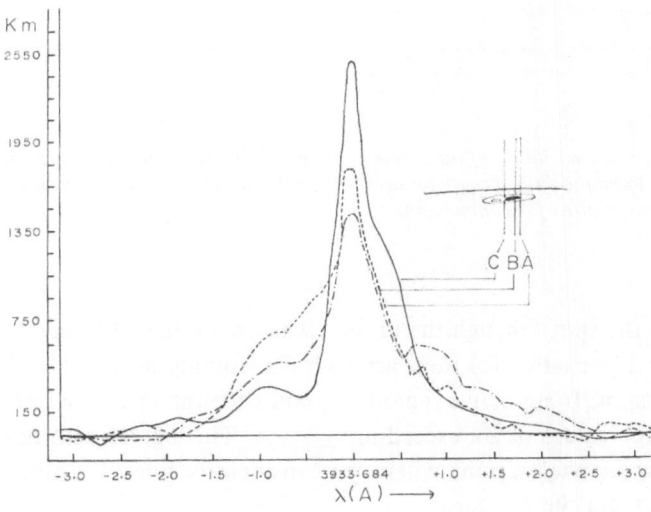

FIG. 3. *The variation with $\Delta\lambda$ of the values of $h^*$ of the sunspot in the K line. The dashed curve, dot and dashed curve and the continuous curve represent the variations for positions, A, B, C respectively.*

of the radial slit shown in the inset. A striking feature of this plot is the change in $h^*$ for different locations of the slit. We see farther down in the chromosphere in the centre of the K line in the sequence: penumbra, penumbra-umbra interface, and the umbra. This is indicative of a spatial opacity variation in a sunspot with the umbra being more transparent than the penumbra. Figure 4 is a plot of the $h^*$ values for the two chromospheric arches. Also plotted herein is the $\Delta\lambda$ variation of $h^*$ for the

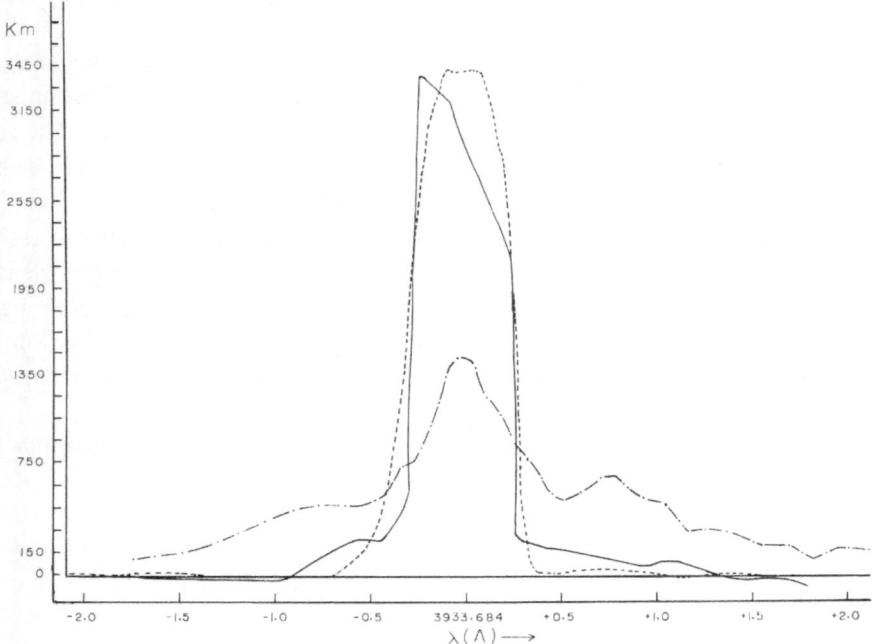

FIG. 4.    *The variation with $\Delta\lambda$ of the K-line values of $h^*$ for the chromospheric arches. The continuous curve is for arch 1 and the dashed curve for arch 2. The dot and dashed curve represents values of position B over the sunspot, plotted for comparison.*

slit setting on the spot through the umbra. The arches have $h^*$ values for $\Delta\lambda=0$ near 3000 km. The $\Delta\lambda$ variation for these arches is exceedingly steep unlike the slow changes seen over the spot. In an active region one still encounters $\tau_{tang}=1$ at chromospheric heights even for values of $\Delta\lambda$ exceeding $\pm 1.5$ Å. This agrees with our findings from Kodaikanal spectroheliograms where we can identify floccular outlines even as far away as 10 Å from the $K_3$ core.

In the case of H$\alpha$ the measures given may be compared with those of Mattig (1962) or White and Wilson (1966). The former finds a mean value of $h^*$ equal to 2160 km

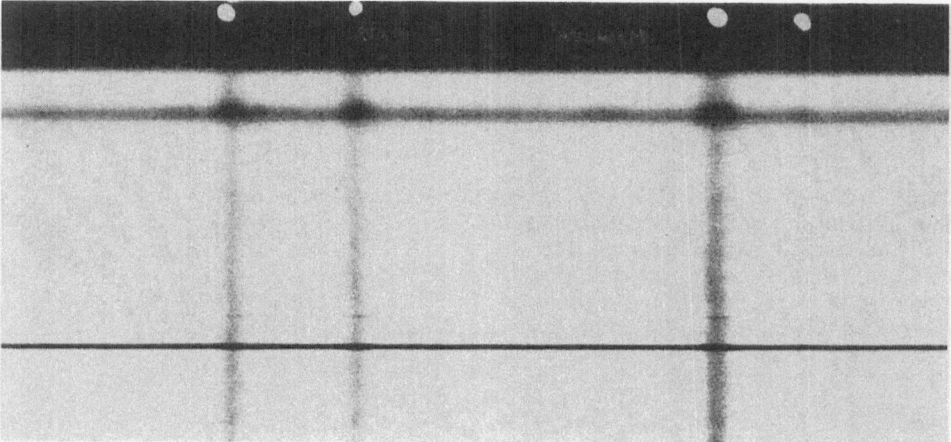

FIG. 5.   *This spectrum with the spot very close to the Sun's limb shows the displacement of the spot position in the three Ti lines near λ 5260 Å relative to the spot's continuum (Fraunhofer Institut, Anacapri station).*

which with the appropriate $\delta h$ correction reduces to 2000 km. The Sacramento Peak investigators obtain 1560 km for this corrected value while we obtain a figure of 1080 km. This is perhaps indicative of the variations from spot to spot and emphasizes the need for determining as many different parameters for a few well-studied spots than a mean value for a large number of spots. Our studies of the $\Delta\lambda$ variation of $h^*$ for K relate, therefore, to conditions in the spot which yield a value of $h^* - 150$ equal to 1080 km in the Hα-line core.

The chromospheric arches form the nearest approximation we have to the applicability of such a study to the quiet chromosphere. More specifically, in the current case we have studied the characteristics of a small-sized inhomogeneity and compared it with the corresponding case of an active-region chromosphere. However, we realize that the parameters from the arches form a lower limit to such values in the quiet chromosphere. The intense brightening in $Ca^+$ is indicative of a highly localized region that possesses a longitudinal magnetic field, which, while weak, may still be greater than in its immediate surroundings. Hence, while yet insignificant from the standards of an active region, it is by no means typically representative of the quiet chromosphere.

### Acknowledgments

The spectra studied herein were obtained by one of us (M.K.V.B.) during a tenure at the Kitt Peak National Observatory and the University of Arizona under the Foreign Visiting Professor's programme of the American Astronomical Society.

We are much indebted to Drs. Mayall and Pierce for the generous provision of telescope time, and to Messrs. Slaughter and Randall for considerable help at the telescope.

## References

Allen, C. W. (1963)      *Astrophysical Quantities*, Athlone Press, London, p. 164.
Mattig, W. (1958)      *Naturwissenschaften*, **45**, 104.
Mattig, W. (1962)      *Z. Astrophys.*, **56**, 161.
Schröter, E. H. (1958)      *Z. Astrophys.*, **45**, 68.
Severny, A. B., Bumba, V. (1958)      *Observatory*, **78**, 33.
White, O. R., Wilson, P. R. (1966)      *Astrophys. J.*, **146**, 250.

## DISCUSSION

*Mattig:* In the last year I have observed the same effect in 60 weak and middle-strong lines to determine the height of line formation in sunspots. The height differences between the formation of the line center and the continuum is always smaller than $1'' = 725$ km. There is a well-pronounced correlation between the height of line formation and the Rowland intensity; for stronger lines the heights are larger than for weaker lines.

Some further observations in the umbra and penumbra are in agreement with your observations. In the Na-D lines I found nearly the same heights in the umbra and penumbra. (See Figure 5.)

# THE STRUCTURE OF THE LOWER SOLAR CHROMOSPHERE IN UNDISTURBED AND ACTIVE REGIONS

E. DUBOV

*(Crimean Astrophysical Observatory, Nauchny, Crimea, U.S.S.R.)*

As observational material in this work we used spectroheliograms taken with the Crimean solar tower telescope in $K_{232}$ and $H\alpha$ filtergrams taken with the chromospheric telescope in Simeis. The $H\alpha$ birefringent filter was so adjusted, that by tuning the last polaroid we could take filtergrams in the centre of $H\alpha$ or combined filtergrams in the two wings at $H\alpha \pm 0.5$ Å. So the effect of Doppler shifts on image-brightness distribution was diminished. We compared the brightness distribution with that of spectroheliograms.

We also used the idea proposed by Thomas and Athay about the temperature jumps in the chromosphere. From our previous calculation it follows that such a thin boundary region in which temperature increases very rapidly would be connected

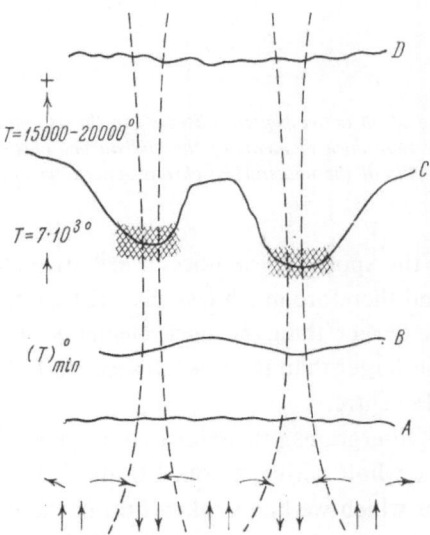

FIG. 1. *The structure of the undisturbed chromosphere. A – upper part of the convective zone, B – boundary between photosphere and chromosphere (level of temperature minimum), C – boundary between upper and lower chromosphere, D – transition to corona. The arrows indicate the matter movement. Dotted lines – magnetic lines of force. The regions bright in K Ca II line are dashed.*

*Kiepenheuer (ed.), Structure and Development of Solar Active Regions, 255–258. © I.A.U.*

with hydrogen ionization in the chromosphere. We regard this boundary as dividing the chromosphere into two parts – the lower chromosphere, where the temperature rises to some 7000° (its height is about 1500–2000 km) and the higher one, where the temperature rises to about 15000°.

The picture seen on spectroheliograms and filtergrams in Hα (the chromospheric network) might be accounted for as a result of different height of this boundary upon the middle or upon the border of supergranules. So at some heights hot and cold chromospheric elements coexist. The change of height is caused by different amounts of mechanical energy supply, depending on differences in the strength of magnetic fields. This is seen in Figure 1.

In active regions on filtergrams taken in both Hα wings (Hα ± 0·5 Å) we may note three types of areas (Figure 2):

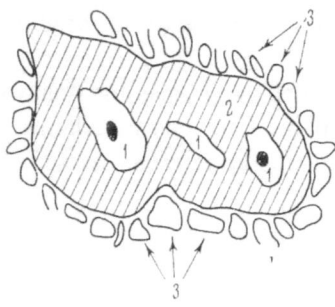

FIG. 2.    *Schematic picture of an active region, observed in the Hα wings (Hα ± 0·5 Å). 1 – bright areas, 2 – dark areas, darker than dark elements of the undisturbed chromospheric network, 3 – areas more bright than bright elements of the undisturbed chromospheric network.*

(1) Bright areas near the spots. These flocculi are less extended than Hα flocculi seen in the line centre and therefore much less than K Ca II flocculi.

(2) The dark regions, darker than the dark elements of the chromospheric net. Their areas are somewhat larger than the flocculi areas in K Ca II and therefore much larger than flocculi in Hα centre.

(3) Finally, on these filtergrams, there are sometimes chains of rather bright elements surrounding the whole active region. I think they are similar to those dark features in K Ca II about which we had spoken during the first session.

So the active region seems to be larger in area when we take the solar image in Hα ± 0·5 Å.

The observed picture may be interpreted if we accept that, in an active region, the amount of non-thermal energy supply increases as fast as the magnetic field increases toward the spot, and, that by this reason, at the same time, the boundary between the upper and lower chromosphere is lowered.

We may see from Figure 3 what happens when we approach the spot from outside. Curve 3 on the plot is the magnetic-field strength $H$. The lower boundary of the lower chromosphere (physically this boundary is the minimum-temperature level where the energy supply by mechanical waves exceeds the energy absorption from the radiation field) is depressed towards the spot (curve 1). The behaviour of the upper boundary of the lower chromosphere is quite similar (curve 2). The Hα brightness (curve 4) (we see corresponding depth levels only in the Hα wings) enhances when the boundary is

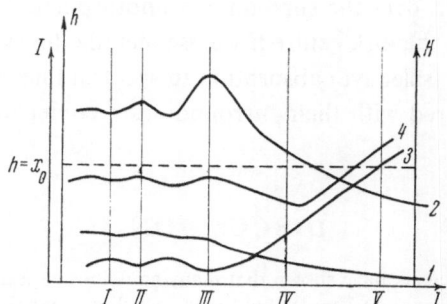

FIG. 3.    *Position of lower and upper boundary of the lower chromosphere in different regions of the solar disk. 1 – lower boundary of the lower chromosphere, 2 – its upper boundary, 3 – magnetic-field strength, 4 – brightness of the chromosphere in Hα wings.*

*Sections: I – dark element of the undisturbed chromospheric network, II – bright element of the undisturbed chromospheric network, III – border of the active region, IV and V – different places of the active region (still outside spots).*

higher – because the Hα-emitting slice is then thicker. The upper chromosphere emits less in Hα, in spite of its high temperature. This is connected with strong deviations from LTE conditions. But as the boundary is lowered the higher chromosphere begins to emit an increasing fraction of the total enhanced Hα emission arising in the chromospheric flocculi. The sections are denoted on the plot as follows: I – the dark part of the chromospheric net (the border of supergranules, dark in Hα), II – the bright elements of the chromospheric net, III, IV, and V – sections through several parts of the active region, corresponding to the areas 3, 2, and 1 on Figure 2.

An investigation of the brightness distribution on magnificent spectroheliograms taken in Meudon by Professor D'Azambuja and kindly made available to me by courtesy of Dr. Michard, does not contradict to such a picture of the structure of the lower chromosphere. These spectroheliograms were taken in several parts of 4202 and 4388 Fe I, 4227 Ca II, 4078 Sr II and 5890 Na I spectral line profiles. Most of the brightness fluctuations on these spectroheliograms are caused by Doppler shifts. In most cases the material is ascending near the supergranule border (except for line 5890 Na I, which behaves conversely); it is sometimes ascending, sometimes descending in other parts of the supergranules.

Furthermore, the network best seen in the wings of metallic line spectroheliograms does not correspond to chromospheric but rather to photospheric levels. The same might be said about the bright areas in active regions – that is, they are photospheric faculae but not chromospheric events. This conclusion is confirmed by the observational fact noted by D'Azambuja, that the wavelengths of best visibility of these faculae are on different distances from spectral line centres – the stronger the line the more distant is the wavelength of best visibility from the line centre. The corresponding levels have an optical depth of about $\tau_{5000} = 0{\cdot}3$ where the faculae have their maximum-temperature excess over the surrounding photosphere. So we may observe the photospheric faculae in the disk centre if we use metallic-line spectroheliograms. It is the result of additional (selective) absorption in spectral lines that we cannot see the deeper and – as compared with their surroundings – colder photospheric regions of the faculae.

## DISCUSSION

*Acton:* The preceding speakers have shown that many emissions appear to originate at physically higher levels over active regions. As I understand your last slide, your model predicts the opposite. Does this indicate a conflict between your model and observation or have I misunderstood your slide?

*Dubov:* I think that there is no contradiction. In active regions the upper boundary of the low chromosphere is at a lower level, and that only means that the temperature jump is at a lower level. Because in this case the upper chromosphere provides the major part of the total Hα emission of the chromosphere, and as the density gradient is small in active regions, we will observe that the whole Hα-chromosphere is optically thicker in active regions than in undisturbed ones.

# PROBLEMS IN THE INTERPRETATION OF
# POLARIZATION MEASUREMENTS IN ACTIVE REGIONS

## EBERHARD WIEHR
### (Göttingen, Universitäts-Sternwarte, Germany)

Difficulties and uncertainties are discussed concerning the calibration of the response of a solar vector magnetograph (linear and circular polarization degrees) in terms of field strength $H$ and angle of inclination to the line of sight $\psi$ using the theory of line formation in magnetic fields (Unno, 1956).

For such a calibration two main parameters have to be known quite precisely: $\eta_0 = \kappa_c/\kappa_0$ (the ratio of the absorption coefficient in the line center to the continuous absorption coefficient) and $\beta_0$ (the limb-darkening coefficient), both as used in the Milne-Eddington formula for the line profile:

$$r_\lambda = \frac{\beta_0 \cos \theta}{1 + \beta_0 \cos \theta} \times \frac{\eta_0 \, e^{-v^2}}{1 + \eta_0 \, e^{-v^2}}.$$

It is shown that even in the undisturbed photosphere these two parameters are quite uncertain; in a sunspot penumbra or umbra they are practically unknown.

The sensitivity of $H$ and $\psi$ to the choice of $\eta_0$ and $\beta_0$ has been investigated in detail. Table 1 shows values of $H$ and $\psi$ as obtained from measured polarization degrees $(P_{\text{lin}} = 1\%; P_{\text{circ}} = 5\%)$ using different parameters $\eta_0$ and $\beta_0$ for the Fe line $\lambda$ 5250 Å.

## Table 1

| | $\beta_0$ | $\eta_0$ | profile form | $H$ (gauss) | $\psi$ |
|---|---|---|---|---|---|
| 1 | $\gg 1$ | 1·5 | numerical | 970 | 61° |
| 2 | $\gg 1$ | 1·5 | $e^{-v^2}$ | 1140 | 61° |
| 3 | 2 | 10 | $e^{-v^2}$ | 680 | 45° |
| 4 | $\gg 1$ | 1·5 | numerical with 20% straylight | 1130 | 63° |

All profiles are normalized to an equivalent width $W_\lambda = 70$ mÅ and central depth $r_c = 0\cdot6$.

These computations show that in general it is not possible to determine the magnetic-field structure with a desirable accuracy even from very precise polarization measurements. Therefore a vector magnetograph has to be calibrated empirically.

*Kiepenheuer (ed.), Structure and Development of Solar Active Regions*, 259–260. © *I.A.U.*

This has been done recently at the Crimean Astrophysical Observatory (Severny, 1967).

Details will be published elsewhere in connection with the final results of measurements obtained in active regions on March 21–29 and May 20–29, 1967.

## References

Unno, W. (1956)      *Publ. astr. Soc. Japan*, **8**, 108.
Severny, A. B. (1967)      *Izv. Krym. astrofiz. Obs.*, **36**, 22.

## DISCUSSION

*Maltby:* I think that this paper brings out very clearly the need for a method of the type developed by Kjeldseth Moe, where the depth dependence of the line-absorption coefficient is taken into account.

*Wiehr:* Generally I agree with you, and I am looking forward for this paper. On the other side the difficulties originating from the depth dependence of the magnetic field and of the choice of the model of the various parts of an active region will remain.

*Mattig:* You have spoken about the strong depth dependence of the absorption coefficient for the magnetic sensitive line $\lambda$ 5250 Å, especially in sunspots. I want to say that in sunspots the absorption coefficient $\eta$ for the iron line $\lambda$ 6303 Å is nearly independent from optical depth. This is a result of new calculations with different sunspot models. So I think it is better to observe in the red than in the green line, if you want to interpret the magnetograms using the Unno theory.

*Wiehr:* The red line is difficult to use, because of the terrestrial blend on the decreased sensitivity of the multiplier. On the other hand, the depth dependence of the absorption coefficient can be calculated by the Moe method, as Dr Maltby pointed out. But here also one needs a model, which is unknown especially for the penumbra, where a vector magnetograph is most useful – in the photosphere the fields are too small, in the umbra one has saturation effects.

# ON THE MAGNETIC-FIELD STRUCTURE AROUND FILAMENTS

B. A. IOSHPA*

*(Institute of Terrestrial Magnetism, Ionosphere and Radio-Wave Propagation of the U.S.S.R. Academy of Sciences, Moscow)*

The results of an investigation of the magnetic-field structure in the chromosphere around filaments are briefly given in this paper. The hypothesis of the existence of a two-field system is suggested: an outer field which has the form of arches and supports the filament against gravity, and an inner one, directed along the filament.

The magnetic-field measurement was carried out with the magnetograph of the IZMIRAN Tower Solar Telescope (Zhulin *et al.*, 1965) in the photospheric line (FeI $\lambda$ 5250·2 Å) and the chromospheric line ($H\beta$). The study of the distribution of the longitudinal component in photosphere and chromosphere confirms Babcock's original conclusion that the filaments are preferably located near the boundary of the magnetic-field polarities. One can also see from a study of the magnetic maps that in a number of cases filaments near their ends cross the zero-line of the longitudinal field of the photosphere (it was also stated previously by Stepanov (1959) and by Howard and Harvey (1964)). They also pass through regions of strong chromospheric field of a single polarity (Ioshpa, 1967). Figure 1 shows the chromospheric (a) and the photospheric (b) field near the spot surrounded by the filament in the centre of the disk (August 4, 1963). For the same region all the photospheric field components were measured simultaneously (according to the method described by Ioshpa and Obridko, 1966). The distribution of the magnetic-field azimuths over the solar surface in that region for August 1 and 4 is presented in Figure 2 (a, b). One can see that the middle, the most 'massive' part of the filament, is preferably located across the magnetic field; but in those places, where the filament is pulled into the spot, the position of the filament appears to be along the direction of the lines of force. Comparison of the magnitudes of the longitudinal $H_{\parallel}$ and transversal $H_{\perp}$ components shows that in the region of the filament location the photospheric field is close to the horizontal one (the longitudinal field near the filament does not exceed several tens gauss, while the mean value of $H_{\perp}$ is about 200–300 gauss (Ioshpa, 1967). Thus in the region of a strong field the structure of which is mainly determined by the spot, the filament lies perpendicular to the nearly horizontal magnetic-field lines in the photosphere and at its ends – parallel to them.

* Presented by E. Mogilevsky.

*Kiepenheuer (ed.), Structure and Development of Solar Active Regions*, 261–266. © *I.A.U.*

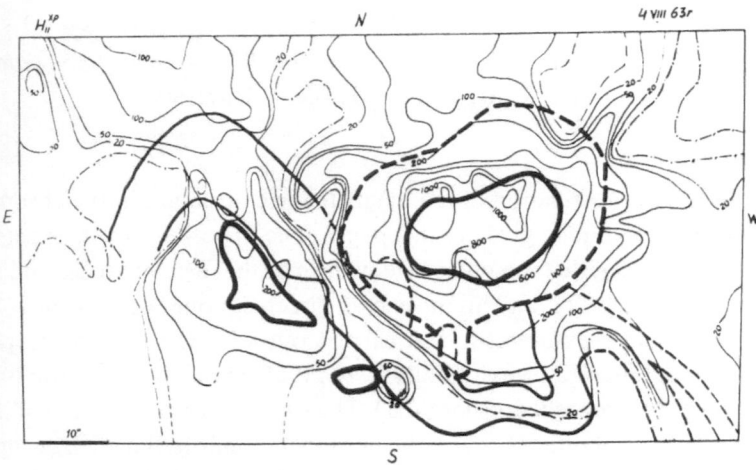

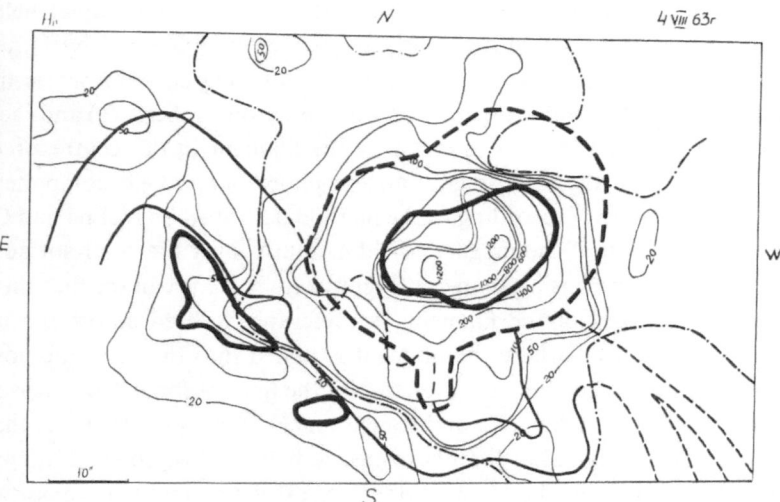

FIG. 1.  *The longitudinal magnetic-field maps near the centre of the disk in the region of the sunspot, surrounded by the filament; a – in the chromosphere (Hβ). b – in the photosphere (λ 5250·2 Å). Thick solid lines represent the sunspot umbrae, the hatched thick line – penumbra; the solid line of medium thickness – the filament – the thin solid lines, the lines of an equal field intensity; the thin hatched-dotted line represents the zero line of the longitudinal field (H∥ = 0). In the places where the measurements were not very certain, the isolines are hatched. In the S–W corner of the maps the chromospheric fibrils are shown by the hatched lines.*

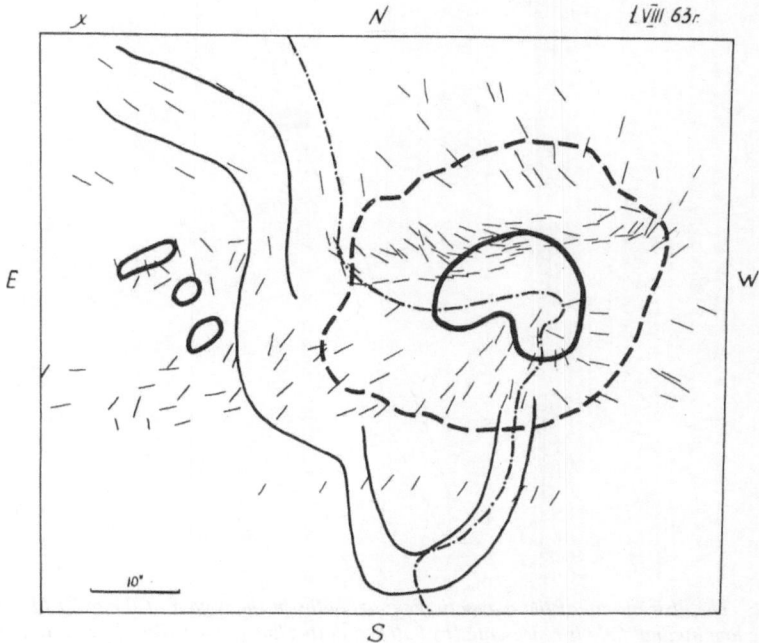

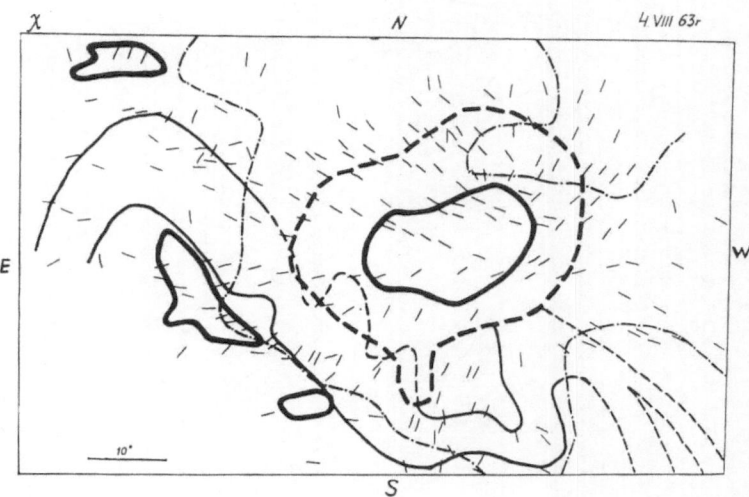

FIG. 2.    *The distribution of the magnetic-field azimuths in the same group as on the Figure 1. (a) for August 1; (b) for August 4, 1963. See caption to Figure 1.*

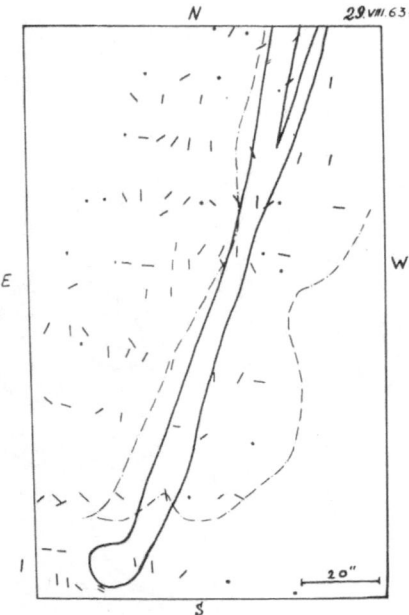

FIG. 3.   *The distribution of the magnetic-field azimuths in the region of the quiet filament. The zero line of the longitudinal field is represented by hatched-dotted line (–·–·). The places where the transversal magnetic field is less than the sensitivity of measurement (50–60 gauss) are shown by dots.*

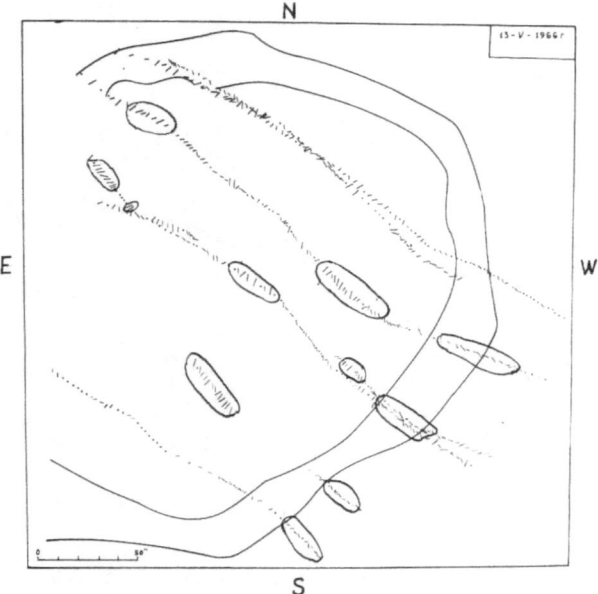

FIG. 4.   *The distributions of the magnetic-field azimuths in the region of the quiet filament near the centre of the disk. The most reliable values are encircled.*

Another structure of the magnetic field was found in the region of quiet filaments not associated with spots. Figure 3 shows the distribution of the magnetic-field azimuths in the region of the quiet chromospheric filament, obtained on August 29, 1963 (co-ordinates of the filament $\varphi \approx -10°$, $\lambda \approx 290°$). The filament was surrounded by flocculi. The field (from the comparison of $H_{\parallel}$ and $H_{\perp}$) is also nearly horizontal. The preferable direction of magnetic-field lines is along the filament. It should be noted that in this case the filament's position is also near the zero line of the longitudinal field. As some other examples, we present the maps of the magnetic-field azimuths obtained on May 18 and 19, 1966 near c.s.m. in the region of the two quiet filaments, which are arch-shaped (Figures 4 and 5). On Figure 4 the most reliable values are encircled. One can see, that the magnetic-field structure in the region covered by the filament is in a good correspondence with the shape of the filament. But near the filament and directly under it the direction of $H_{\perp}$ in some cases changes sharply and is perpendicular to the filament (it is seen most clearly on Figure 4).

ᵈᵈ.We have already suggested (Ioshpa, 1967) the existence of a two-field system near the filaments; an outer arched one on which the filaments lie, and an inner one directed

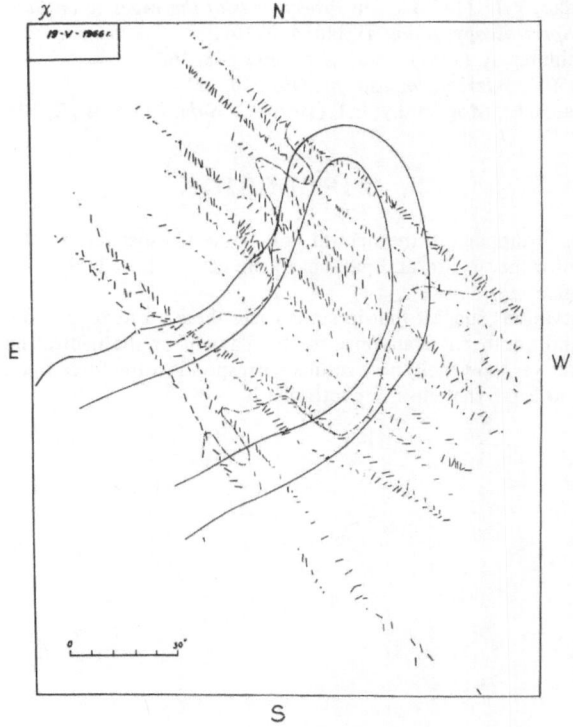

FIG. 5. *The distribution of the magnetic-field azimuths in the region of the quiet filament near the centre of the disk.*

along the filament. An outer arch field supports the filament against the gravity (Kippenhahn and Schlüter, 1957). The existence of such a field explains the filament's preferable location along the zero line of the longitudinal field. Our results also confirm the existence of such a field. At the same time a number of facts speak in favour of the existence of a field which is directed along the filament: in particular the discovery of a significant field ($\sim 60$ gauss) observed in some prominences not connected with an active group along their great axis; the field structure near quiet filaments; the field structure near the ends of sunspot filaments and some theoretical considerations on the filament stability (Ioshpa, 1967). The orientation of the filament, and consequently of the field in the filament, as it is shown in Figures 4 and 5, are closely associated with the field structure at the photosphere level. It seems that the presence of a two-field system near the filament corresponds to the presence of such systems of the field in the photosphere.

## References

Howard, R., Harvey, J. W. (1964)      *Astroph. J.*, **139**, 1335.
Ioshpa, B. A. (1967)      *Soln. Aktivnost*, **3**.
Ioshpa, B. A., Obridko, V. N. (1966)      in *Proceedings of the Meeting on Solar Magnetic Fields and High Resolution Spectroscopy, Roma*, II, No. 4, p. 161.
Kippenhahn, R., Schlüter, A. (1957)      *Z. Astrophys.*, **43**, 36.
Stepanov, W. E. (1959)      *Izv. Krym. astrofiz. Obs.*, **20**, 52.
Zhulin, I. A., Ioshpa, B. A., Mogilevsky, E. I. (1965)      *Soln. Aktivnost*, **2**, 108.

## DISCUSSION

*Sturrock:* If there is slippage at the neutral line at the photosphere the force-free magnetic field will have component transverse to and parallel to the neutral line. Is this interpretation compatible with the observations?

*Ioshpa:* The observations suggest that in some cases there are two components of magnetic field: one is parallel to and another is transverse to the filament or the neutral line in the photosphere. But I do not consider yet, how well these results correspond to the force-free model of the magnetic field. It seems that, at least, they do not contradict it.

# THE FORMATION, STRUCTURE AND CHANGES
# IN FILAMENTS IN ACTIVE REGIONS

SARA F. SMITH

*(Lockheed Solar Observatory, Calif., U.S.A.)*

During periods of good image quality the large-scale films of the Lockheed Solar Observatory obtained during 1966 and 1967 have proven to be especially useful for studying filaments and their relationship to the Hα fine structure observed in active regions. Structures with dimensions on the order of 1 sec of arc can frequently be resolved, as illustrated in Figure 1. The 'solar vortices' are clearly resolved into numerous individual fibrils. Bright Hα plage is resolved into a fine granular structure resembling the solar granulation. Filaments are usually seen as irregular dense

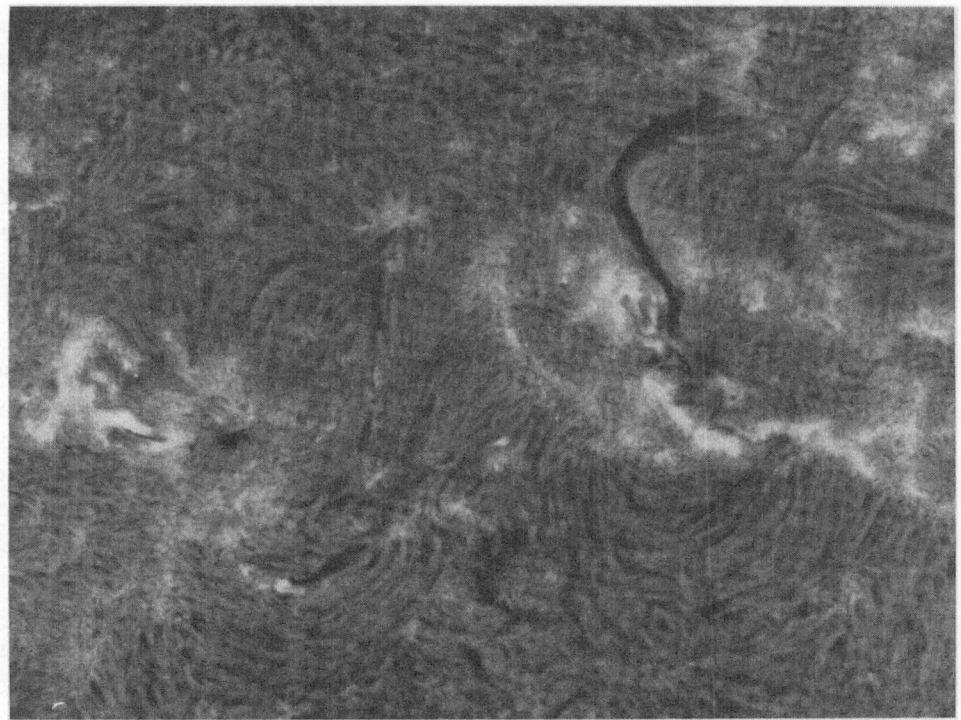

FIG. 1. *Hα fine structure in an active region, August 1, 1966. Bright plage is often resolved into fine mottles. The fine thread-like structures or 'fibrils' are shortest at the plage boundary.*

*Kiepenheuer (ed.), Structure and Development of Solar Active Regions, 267–279. © I.A.U.*

collections of material which are occasionally resolved into finer striations which run parallel to the path of the filaments.

Before discussing observed relationships between filaments and plage granules, fibrils and sunspots, it seems preferable to reiterate a few aspects of active regions that are already well known. It has been stated by Tsap (1963) that the structure and geometry of active regions are controlled by the configuration and strength of the magnetic fields of regions. Our interpretation of the observations is similar. We concur with other investigators in saying that fibrils appear to be a manifestation of the direction of the lines of force in the chromosphere.

From the work of Avignon *et al.* (1964), Howard (1959), Howard and Harvey (1964), Martres *et al.* (1966), and ourselves (1967), it is well established that filaments both inside and outside of active regions are invariably observed at the boundary between positive and negative polarities in magnetic fields. An example is presented in the right corner of Figure 2. In the lower half of Figure 2 is a fine scan magneto-gram of the longitudinal component of a region recorded with a 10 sec-of-arc reso-lution at the Mt. Wilson Observatory. Positive polarity is represented by solid lines and negative polarity by dashed lines. In the upper half of Figure 2 is a center Hα photograph taken at Lockheed on the same day, June 2, 1967. Although the difference in resolution between the magnetogram and filtergram is a factor of 10, the filament very nearly follows the border between positive and negative polarity. We often refer to this boundary between polarities as the line of $O$ longitudinal field or, sometimes, as the 'neutral' line. Lines of $O$ longitudinal field exist in all active regions observed near the center of the disk. They may also exist between two active regions or between remnants of old active regions. Correspondingly, filaments may be observed along any of these locations.

It has also been mentioned by Martres *et al.* (1966) that the boundary of polarity change may also be crossed by 'filamenteuses' or fibrils extending between adjacent plage of opposite polarity. These authors have also suggested that such fibrils may be related to the evolution of filaments. Examples of fibrils crossing the boundary between areas of opposite polarity are also shown in Figure 2. At other locations the boundary between opposing polarity is coincident with the path of a filament. In general, filaments and fibrils are not observed at the same location on a line of $O$ longitudinal field (Smith and Ramsey, 1966).

Since filaments very closely follow the boundary of a polarity change, we previously thought, along with other investigators, that the path of a filament ran perpendicular to a general pattern of lines of force in the chromosphere and low corona. In fact, this view of filaments has been so commonly held that a number of theoretical models of

Fig. 2.  *Hα filtergram and Mt. Wilson fine scan magnetogram, June 2, 1967. The fibrils in plage near the large spot join plage of opposite polarity.*

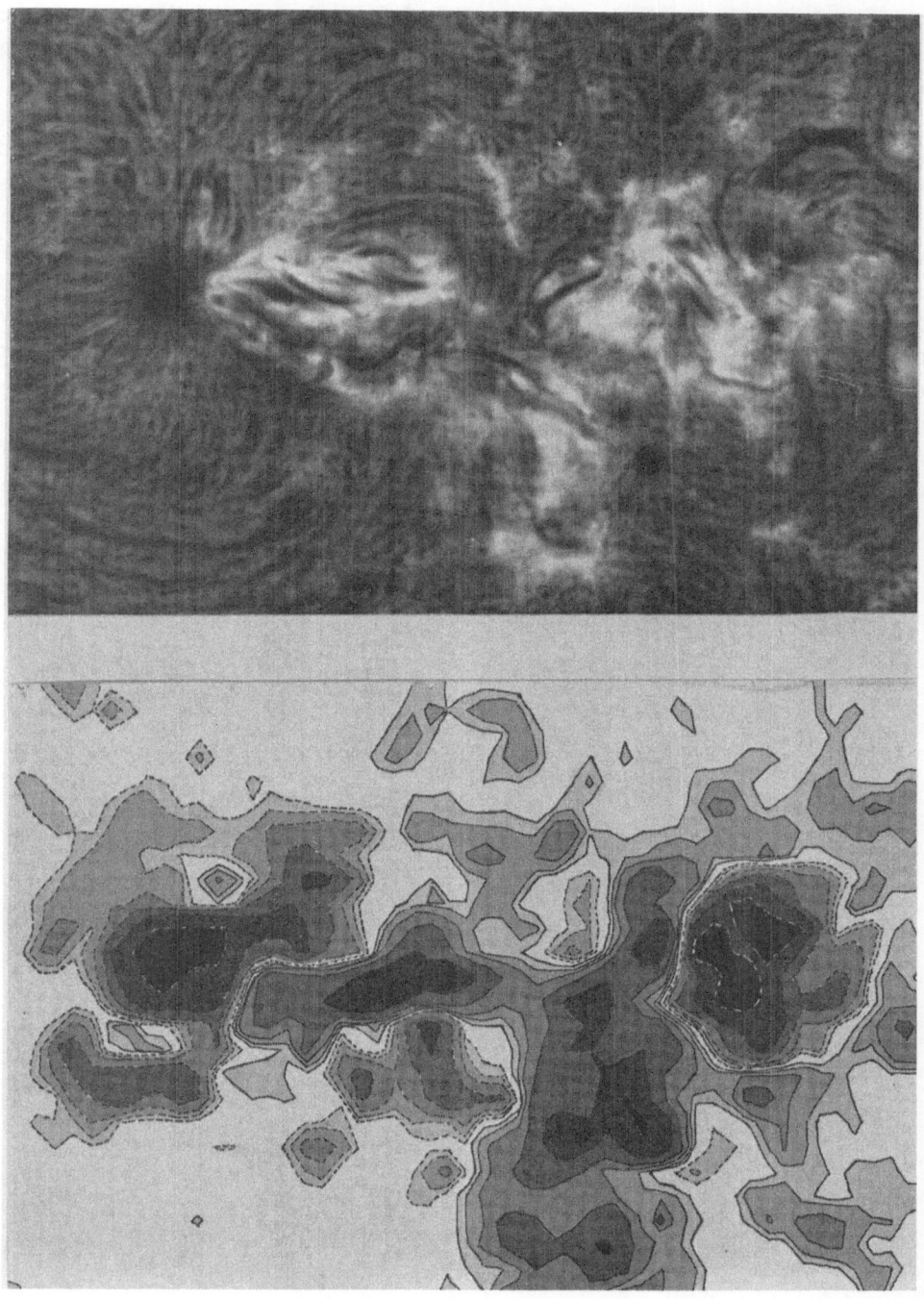

filaments have been put forth which require a deformation, sag, or distortion of the lines of force in order to make it theoretically possible for filaments to exist along lines of $O$ longitudinal field.

After studying the relationship of filaments to fibrils, plage granules, and sunspots, however, the Hα observations seem to be more consistent with the view that filaments follow a restricted set of lines of force in magnetic regions.

Figures 1, 3, and 4 illustrate the observed relationship of filaments and fibrils in active regions. When viewed near the center of the Sun, the fibrils immediately adjacent to a filament are aligned parallel or very nearly parallel to the path of the filament. A few seconds of arc away from the edge of the filament the fibrils may appear to bend away from the path of the filament, as seen in Figure 3. At the ends of filaments, two conditions have been observed. In some cases, the fibrils remain aligned parallel to the path of the filament. In other cases, a filament appears to terminate just before fibrils which appear to run nearly perpendicular to the end of the filament path. Very close inspection of these cases under conditions of good image

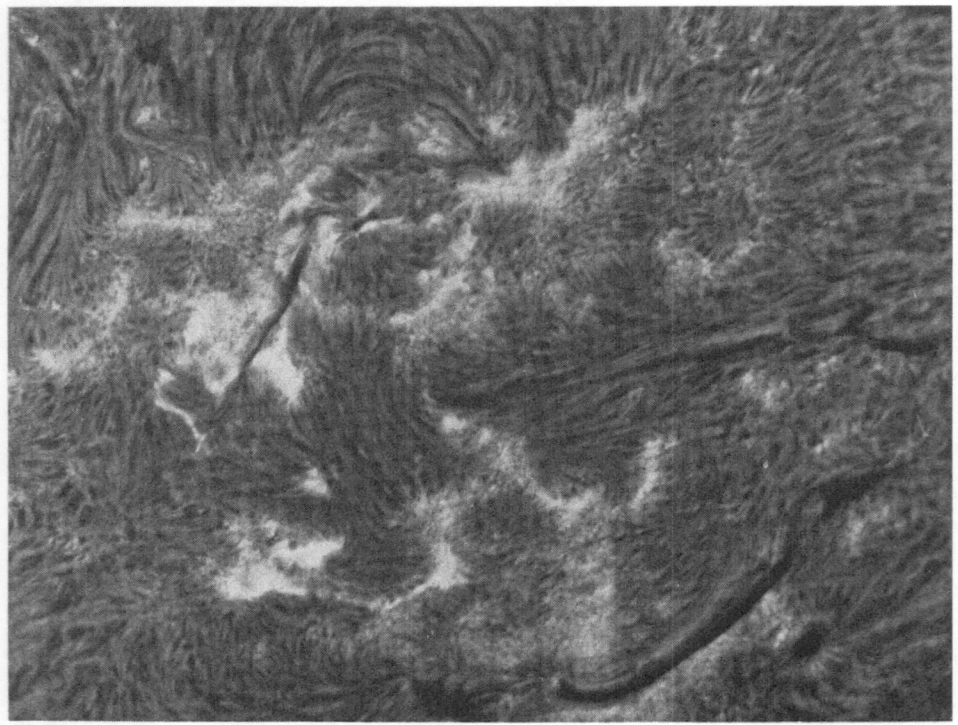

Fig. 3.    *Filaments and structures in an active region, August 3, 1966. The filament in the bright plage is bordered by plage granules rather than by fibrils.*

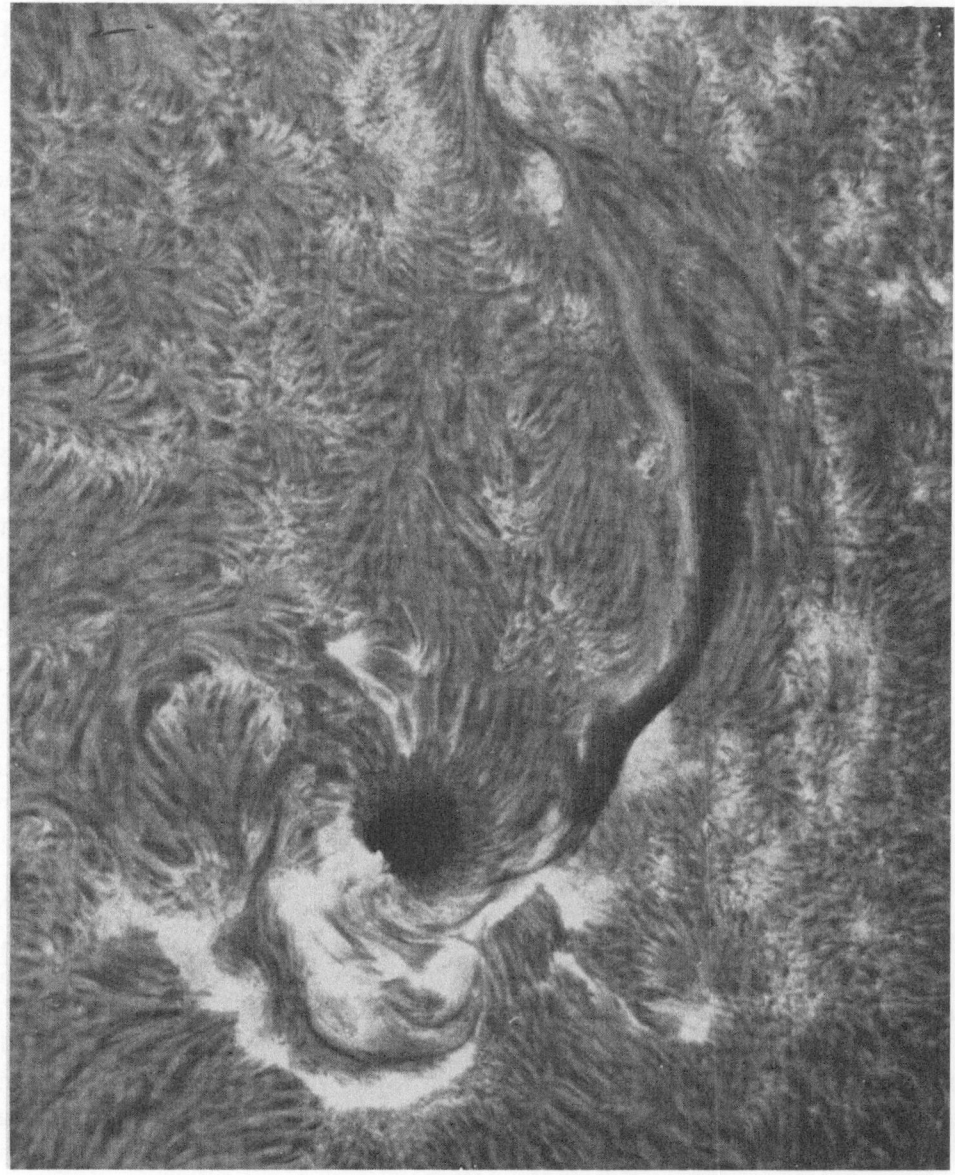

FIG. 4.    *Filament and adjacent fibrils in an active region, September 17, 1966. Internal structure in the filament appears to run parallel to the path of the filament.*

quality, shows that the end of the filament, in at least some cases, abruptly turns in the direction of the neutral-line fibrils. Note ends of filaments in Figure 1, 3, and 4. Generally, when atmospheric image quality is good all points along the path of filaments appear to be aligned in the same direction as adjacent fibrils. However, occasional exceptions have been observed attributable to differences in the height of filament and apparently adjacent fibrils or to effects of transient phenomena.

At some locations in active regions filaments are bordered by bright plage rather than by fibrils. The fibril structure also shows a definite patterned relationship to the plage granules, as well as to filaments. We have previously shown that the length of fibrils, except for those extending between areas of opposite polarity, show a general inverse relationship to the strength of the longitudinal component of the magnetic field (Smith and Ramsey, 1967). Except at locations where the longitudinal magnetic field changes polarity, the fibrils near plage are very short and those at the outer edge of a region tend to be longer (Figure 1). The geometry and sizes of the fibrils and plage granules are such as to suggest that the plage granules may be fibrils which extend approximately radially out from the chromosphere. Projection of the large-scale time-lapse films reveals continuous motion in the plage granules and fibrils, as well as in filaments. The observed motion is also consistent with the view that plage granules could be a cross-section view of bundles of fibrils which extend nearly radially out from the chromosphere although the time scale of the motion in plage granules and fibrils appears to differ. When viewed in the center of the disk, the motion in plage granules resembles the motion of the solar granulation. There is a relatively rapid appearance and disappearance of individual granules with corresponding brightness changes suggesting a strong component of motion in the line of sight. The motion in fibrils is invariably confined along the path of the fibril. In the non-activated or usual state, the motion in active-region filaments is also along the path of the filament. The motion in filaments is thus restricted to the same directions as the motion in fibrils immediately adjacent to filaments.

Figure 5 shows the same filament depicted in Figure 4 as a path which is favorable for the flow of material. During 7 days of observation, distinct flow of emission was observed 15 times along this filament. The flow along filaments occurs both in absorption and emission. The flow of material along filaments is occasionally seen in 10 or 15-sec interval time-lapse films projected at 16–24 frames per second. Experiments in projecting these films at rates in the range of 60–500 frames per second, however, reveal faint and subtle flow not readily detectable at slower projection rates. At projection rates around 100 frames per second and faster (1000 times real time), a continuous flow of material may be observed along the path of most filaments. Although not always readily visible, such flow, of course, must exist if filaments and prominences are still to be regarded as the same features.

From prominence observations, it has been demonstrated that prominence material may rain into sunspots. However, narrow-band Hα observations seldom reveal a

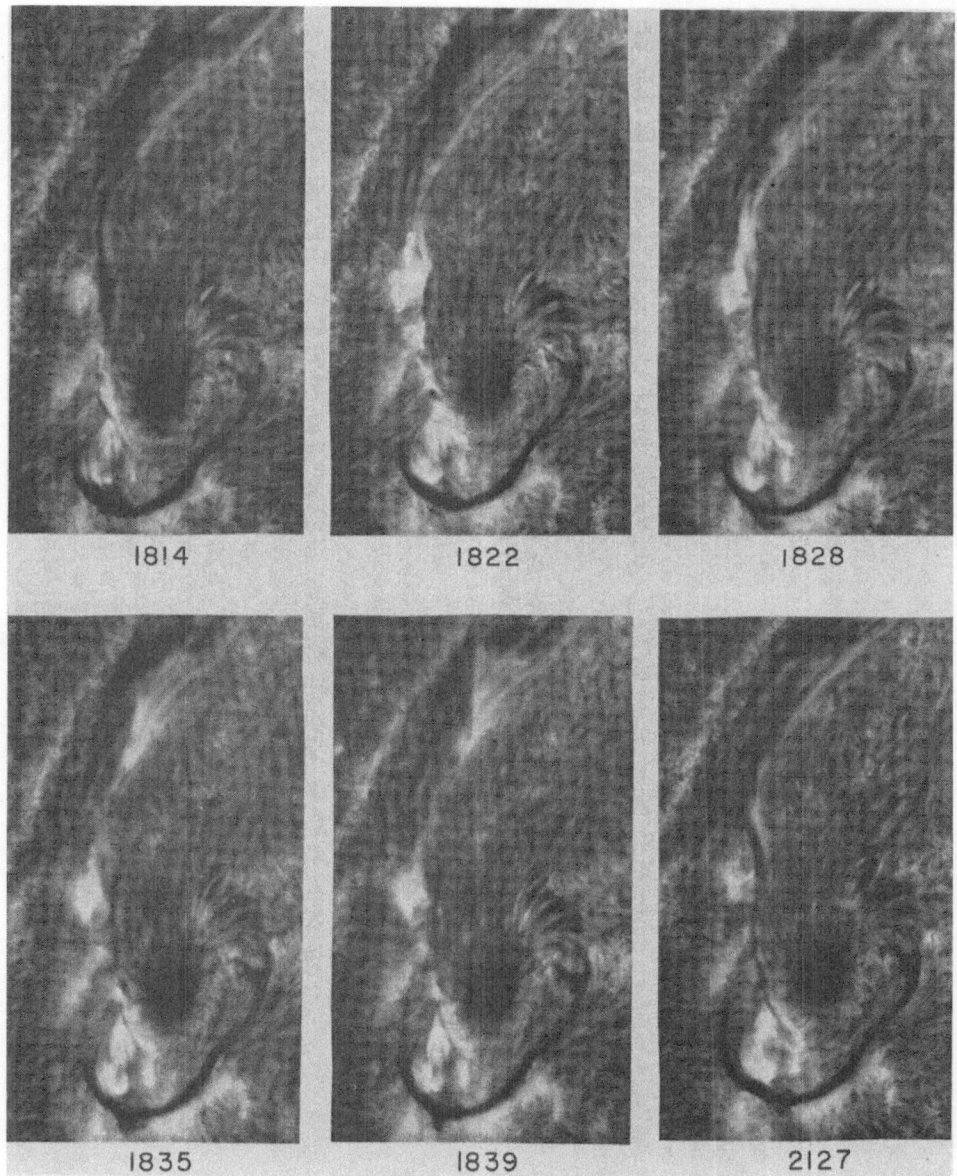

FIG. 5.  *Flow of emission along a filament, September 15, 1966. This flow of emission is initiated by a subflare.*

*21 September 1966*

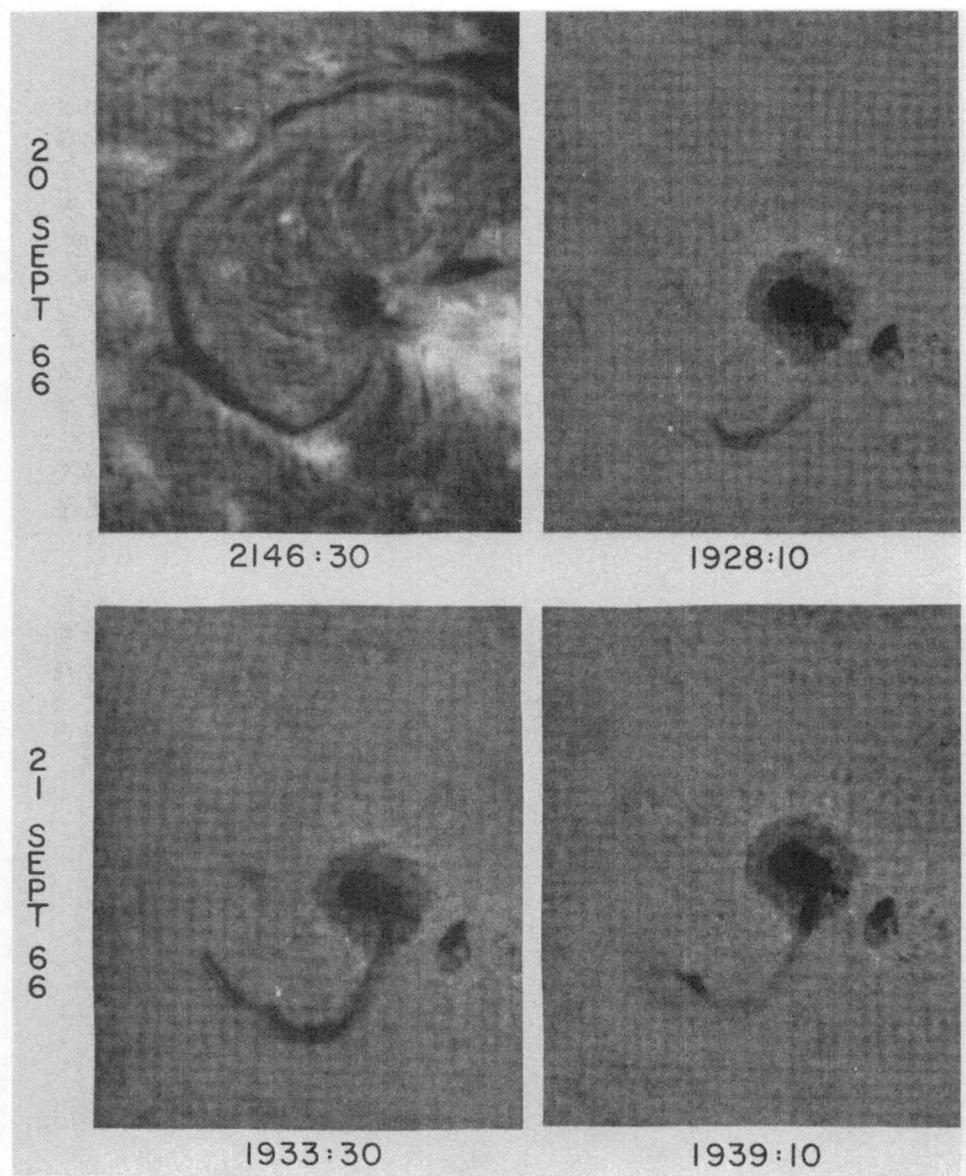

FIG. 6.    *Flow of filament material into a sunspot, September 20 and 21, 1966. The flow of material into the sunspot originates in the filament in the first frame.*

connection between flow of material in filaments and the flow of material into sun-spots. That this connection exists can be shown unambiguously by combining narrow-band and broad-band Hα observations, as presented in Figure 6. On September 21, we change from observing a ½ Å bandpass in the center of Hα to simultaneously observing two ½ Å wide transmission bands centered approximately 1 Å into each wing from line center. In the narrow-band (½ Å) observations on September 20, the filament appears to terminate an appreciable distance from the sunspot. In the broader band observations on September 21, an occasional flow of material continues from the location where the filament appears to terminate in centre Hα and flows directly into the sunspot umbra. This flow of material into the umbra of this spot was observed 18 times on 4 successive days of similar observations of this region. Additional obser-vations are needed to demonstrate the frequency of occurrence of visible flow of material from filaments to sunspot umbra. We conclude that flow from filaments into spots is probably the disk counterpart of prominence flow into sunspots which has previously been deduced by other investigators from observations at and near the limb.

A final demonstration of filaments as being preferred paths for material flow in active regions is shown in Figures 7 and 8, which illustrates an observed process of filament formation. The first stage of filament formation is the alignment of fibrils as in the left of Figure 7. This fibril alignment provides a path for material flow over long distances relative to the dimensions of individual fibrils. The flow is repre-sented in the center photograph by two dark strands which follow the general direction of the fibrils. The photo on the right in Figure 7 shows the filament after the strands of flowing material have increased and gradually merged together in the process of filament formation. The flow of material continues throughout the observed life of the filaments.

Figure 8 illustrates the formation of a section of a filament in plage where only very short fibrils or plage granules are observed. The observed process of formation, however, is similar to the situations in which long fibrils are observed. On April 30, 1967 only a few short-lived strands of absorption appeared at the location destined for the new filament section. On May 1, 1967, strands of absorption more frequently appeared, clearly travelled across a single path almost parallel to a nearby section of filament, and then disappeared. Initially most of the strands of absorption were short-lived, lasting only a few minutes. Some of the longer-lived and darker strands ap-peared to be ejected from small subflares originating near the sunspot in Figure 8. During the observing period on May 2, these moving strands of absorption became a new filament section with the flow of material continuing along the same path, as defined by the strands of absorption on the previous day.

Each of these observations presented, the alignment of fibrils and filaments, the direction of flow of material along fibrils and filaments, the flow of filament material into a sunspot, and the formation of filaments only along paths of aligned fibrils,

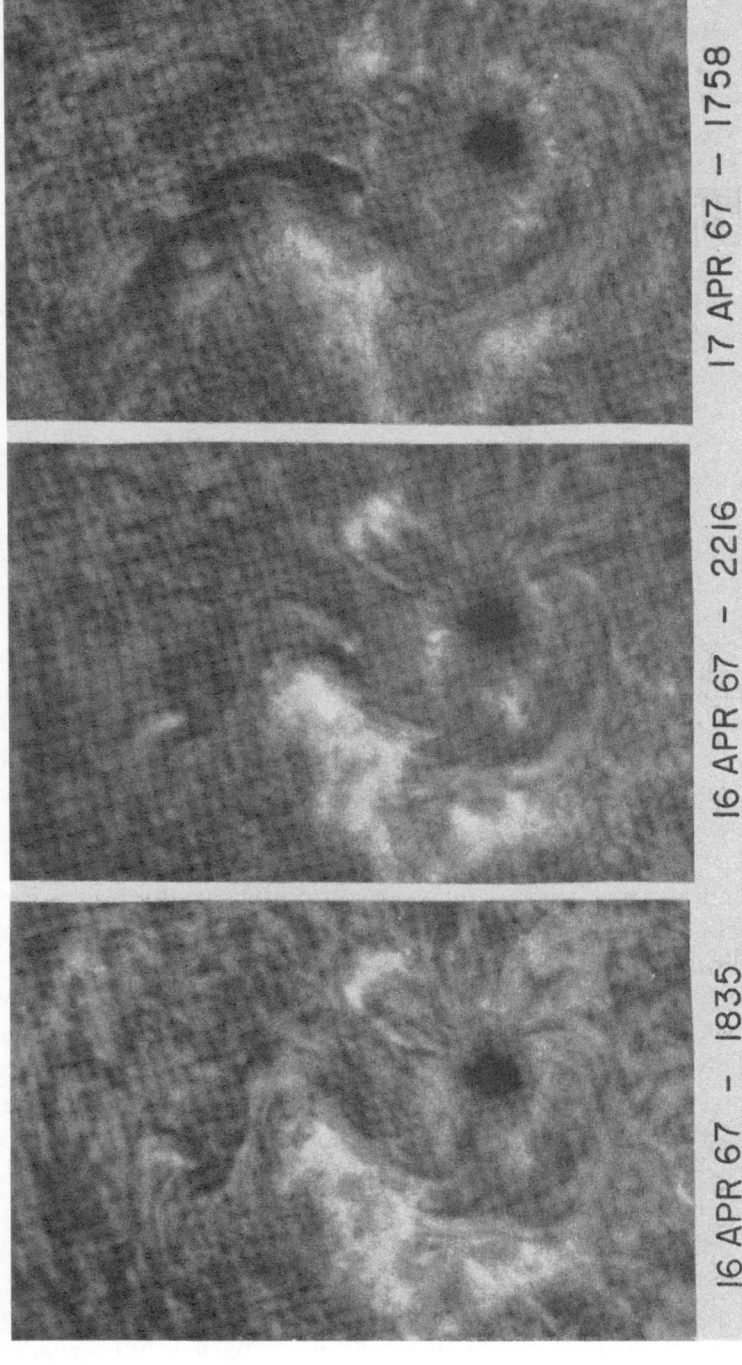

FIG. 7. *Formation of a filament among fibrils, April 16, 17, 1967. In the center picture the apparent darkening of fibrils represents the flow of material along the path which the filament later occupies.*

16 APR 67 — 1835   16 APR 67 — 2216   17 APR 67 — 1758

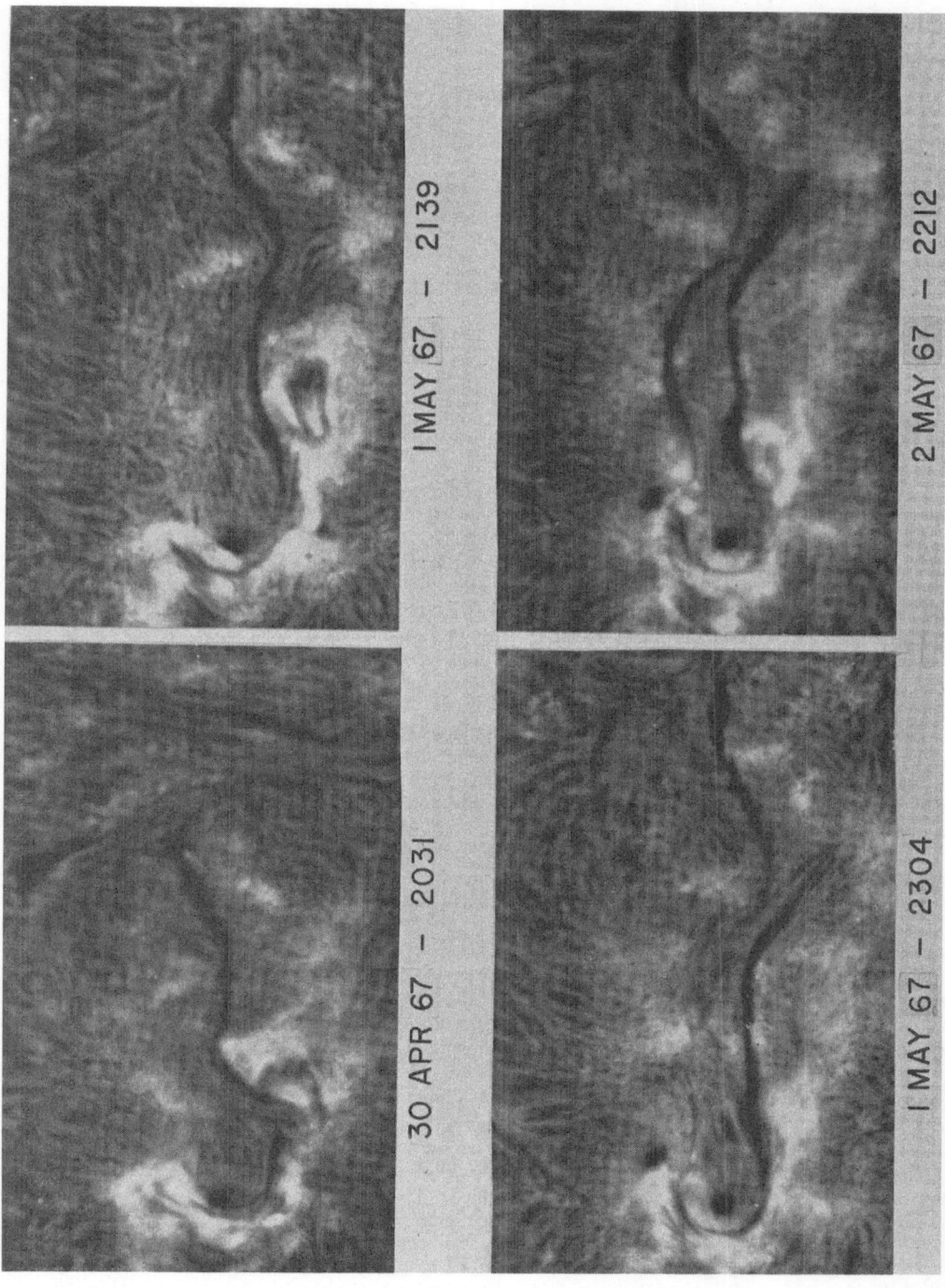

FIG. 8. Formation of a filament section in plage, April 30, May 1 and 2, 1967. The filament encircles the sunspot and repeatedly exhibits flow of material along a single path into the plage area. The material flow increases and becomes the filament seen in the 4th frame.

indicates that the lines of force of the magnetic field in filaments, are directed along the path of filaments.

## References

Avignon, Y., Martres, M.J., Pick, M. (1964)     *Ann. Astrophys.*, **27**, 23.
Howard, R.F., (1959)     *Astrophys. J.*, **130**, 193–201.
Howard, R.F., Harvey, J.W. (1964)     *Astrophys. J.*, **139**, 1328–35.
Martres, M.J., Michard, R., Soru-Iscovici, I. (1966)     *Ann. Astrophys.*, **29**, 245.
Smith, S.F., Ramsey, H.E. (1966)     *Lockheed Report 20444*.
Smith, S.F., Ramsey, H.E. (1967)     *Solar Phys.*, **2**, 158.
Tsap, T. (1963)     *Izvest. Krymsk. astrofiz. Obs.*, **31**, 200.

## DISCUSSION

*Falciani:* Have you measured the velocity of the mass motion in the filament and, in this case, does the measured velocity agree with the velocity measured by Doppler effect, for example, when the filament is near the limb?

*Sara Smith:* No, we have not yet made a study of the observable mass motion in filaments during the usual state of filaments, that is, when not activated by flares.

*Sturrock:* The apparent mass motion along filaments may in fact be the result of vertical up-down motion which propagates as an Alfvén wave along a filament. The speed of propagation seems more characteristic of the Alfvén velocity than of likely mass velocities.

*Kiepenheuer:* I have difficulties to understand, how mass motion in a filament can follow the so-called neutral line? This would imply, that the material is moving across the field.

Further I would like to mention that many if not all quiescent filaments are built on a fishbone-like structure, slightly inclined to the axis of the filament. Only for active filaments close to sunspots such transverse structures do not seem to occur.

*Sturrock:* The fine structure seems to be compatible with my suggestion that the magnetic-field pattern associated with a filament is a force-free field produced by slippage at the neutral line.

The fact that the magnetic-field patterns shown, which are associated with filaments, are bipolar provides evidence contradicting the Kippenhahn-Schlüter model, which requires a line-quadrupole magnetic-field pattern at the photosphere.

*De Jager:* Are filaments always oriented along the line of zero longitudinal field? Does your result agree with that of Ioshpa?

*Sara Smith:* Yes, within the accuracy of the Mount Wilson magnetograms that we have used for the determination of filament positions. In some cases, however, it may be that the end of a filament may deviate from the neutral line of a longitudinal field.

*H.U. Schmidt:* It seems to me that motions and aligned fine structure almost along a filament and along the neutral line must be due to a component of the magnetic field in the same direction. If such a component is present in the coronal matter condensing into the filament, it will be strongly enhanced in the process of condensation. Therefore it can be very weak in the beginning, and in many cases it will be due to differential rotation. This concept is consistent with the model of Kippenhahn and Schlüter, since there must still be a sufficient amount of flux crossing the filament overlying a neutral line. On the other hand, the absence of a quadrupole flux distribution below the filament indicates that this model needs some modification. It seems possible that the random walk of the base points of the magnetic flux near the neutral line can replace the stabilizing effect of a quadrupole flux distribution.

*Bumba:* I should like to mention two observational facts:

(1) The feet of filaments are going from junctions of several supergranules, this means from places where the concentration of magnetic field is observed, on both sides of the filament (different polarity on each side), and the feet joint the main body of the filament with the decreasing angle, going just before the junction with the filament practically parallel to the main body of the filament.

(2) Not only before the appearance of the filament but also after its disappearance it is possible to observe the dark fibrils elongated in the direction and on the place of the previous filament.

Both these observational facts seem to speak in favour of ideas mentioned in Mrs. Smith's talk.

*Newkirk:* Although it is not clear that the same type of prominence is discussed here, the work of Rust suggests that a major fraction of the magnetic fields in *quiescent* prominences are perpendicular to the axis of the filament, as would be required by the theory of Kippenhahn and Schlüter. A fairly large component of the field is, however, found perpendicular to the filament axis suggesting that a sheared magnetic configuration is present in quiescent prominences.

# PROMINENCES IN ACTIVE REGIONS

J. KLECZEK

*(Astronomical Institute Ondřejov, C.S.S.R.)*

As an interplay of plasma and magnetic field the solar prominences represent a nice example of cosmical hydromagnetics. Due to stronger magnetic fields in active regions the prominence processes are more violent there than in non-active regions. Observing with a coronagraph or watching prominence movies one recognizes some character-istic processes in prominences of active regions:

(a) *Condensation* of Hα material in coronal space without any direct observational evidence of the material supply. It seems that the material had been ejected in the form of corpuscular radiation, captured, thermalized and cooled down to $10^4$ °K, when it becomes observable in the Hα line.

(b) *Dissolution* of the Hα material. Knots, streamers of sprays, surges and puffs often undergo a reverse process: they dissolve and disappear when observed in Hα. Similar dissolution is typical for a great part of the material in eruptive prominences.

(c) *Spiraling* motion and *detwisting* of streamers are observed in surges, sprays, funnels and also in eruptive prominences. In the prominence movies of HAO Boulder, Sacramento Peak Observatory and Pic-du-Midi we have found many nice examples. In most cases (if not in all) the detwisting resulted in a marked simplification of the prominence structure.

(d) *Emerging magnetic fields* may be seen in form of small rising and expanding Hα loops (velocities about 50 km/sec). In this type of loops the material is dragged out of the chromosphere. A similar process with low velocities has been observed in 5303 Å movies of coronal condensations.

(e) *Capture* of a surge by transversal magnetic fields has been observed and photo-graphed on October 7, 1956. Surge material has been captured in a complex loopy structure and then streamed sunwards along the loops. This event shows that captur-ing of plasma streamers by transversal magnetic fields does occur on the Sun and that the material supply for *AS* prominences (see item (a)) would thus be possible.

(f) *Collisional processes* occur in different forms. One has been just described in item (e). However, the well-known prominence of Lyot from 1937 shows that a surge hits a prominence without being (completely) captured. The prominence which has been hit decays and the surge material seems to be reflected (at least partly) down-wards. Two colliding knots may bypass each other in complicated trajectories. Cases have been recorded of falling prominence knots being stopped and accelerated up-wards, with measured acceleration more than 10 solar g.

*Kiepenheuer (ed.), Structure and Development of Solar Active Regions*, 280–281. © *I.A.U.*

(g) *Oscillating prominences* at the limb have been photographed. The oscillations followed shortly after a nearby flare-spray event. This may be the counterpart of the disk phenomenon discovered by astronomers of Lockheed Observatory.

To explain the above-mentioned prominence processes, hydromagnetic laws are necessary. But before going into their theoretical explanation, more observational facts are needed. It has been the aim of my communication and of the movies demonstrating the processes described here, to call the attention of solar observers to the eventful prominence activity in solar active regions.

## DISCUSSION

*Sturrock:* I am intrigued by the statement that the plasma of a surge is captured in closed magnetic-field lines. It is difficult for a plasma to behave this way. Is there evidence that the field lines are closed before the surge, or only after the surge?

*Kleczek:* Yes, there is a clear observational evidence for that. The capturing of plasma in a transverse magnetic field has been proved experimentally (e.g. in laboratories of the U.S.S.R. and France) and the penetration of solar plasma into the Earth's magnetosphere is an example in cosmical scale.

*H.U. Schmidt:* I wonder whether it would not be possible to interpret these phenomena as a wave motion spread over a much larger solid angle than is covered by the surge material visible in $H\alpha$? I would expect that the downstreaming matter in the loops is rather compressed *in situ* by a wave than trapped from the surge.

*Kleczek:* But in this case we must explain where the surge material ($\sim 10^{14}$–$10^{15}$ grams) disappears. As far as I may estimate, the amount of the material ejected in the surge is equal to the bright ('captured') material streaming sunwards in the loops.

*Öhman:* I would like to give reference in this connection to some observations made by Dr. L. Liszka at our observatory on Capri. Liszka found some similar disturbances by surges, even when the surges did not reach the prominence.

*Kleczek:* Yes, also in the prominence film that I have shown, the spray had not touched the neighbouring quiescent prominences and they began to oscillate.

# THE 'DETWISTED' PROMINENCE OF SEPTEMBER 12, 1966

BORIS VALNÍČEK

*(Astronomical Institute, Czechoslovak Academy of Sciences, Ondřejov, C.S.S.R.)*

ABSTRACT

The eruptive prominence of September 12, 1966, is described. Detwisting of two filaments is demonstrated and compared with similar phenomena and with tornado prominences. A short discussion of mechanisms of the origin and subsequent development of the phenomenon is given.

## 1. Description of the Phenomenon

On September 12, 1966, 9·25 UT we saw a bright prominence on the East limb. During the next 2–3 min this prominence weakened, but in the lower part near the limb we observed a brightening which started to grow. After 9·30 UT it was clear that an interesting process had started. The outer part of the prominence was still growing, and at 9·45 a complicated inner structure began to develop.

After 9·50 UT we saw a very bright phenomenon on the coronagraph observations made in the Hα line with an 8 Å passband filter. It was composed of two interwoven spirals which straightened (detwisted) during the next 15 min. At 10·24 UT we found only one prominence filament remaining close to the limb. This filament did not undergo important and fast changes. The visibility of this prominence deteriorated during the day, but it remained visible.

The following day, September 13, we observed on the East limb a large spot group, with a filament crossing the limb. After this the group with the filament could be followed throughout the passage across the disk.

## 2. Analysis of the Observations

At first we studied motion effects in the prominence. Linear displacements of the mass are presented in the velocity curve, which shows three important phases:

(a) The sharp velocity maximum at 9·40 UT was in coincidence with strong radio emission at the wavelengths 3·2, 37, 56, and 115 cm.

(b) At 9·50 UT after a sudden decrease of the velocity there followed a new rise with maximum value, coinciding also with a radio-noise maximum at a wavelength of 115 cm.

(c) Before 10·00 UT a small decrease can be found followed by a fast rise to the maximum value.

*Kiepenheuer (ed.), Structure and Development of Solar Active Regions*, 282–286. © *I.A.U.*

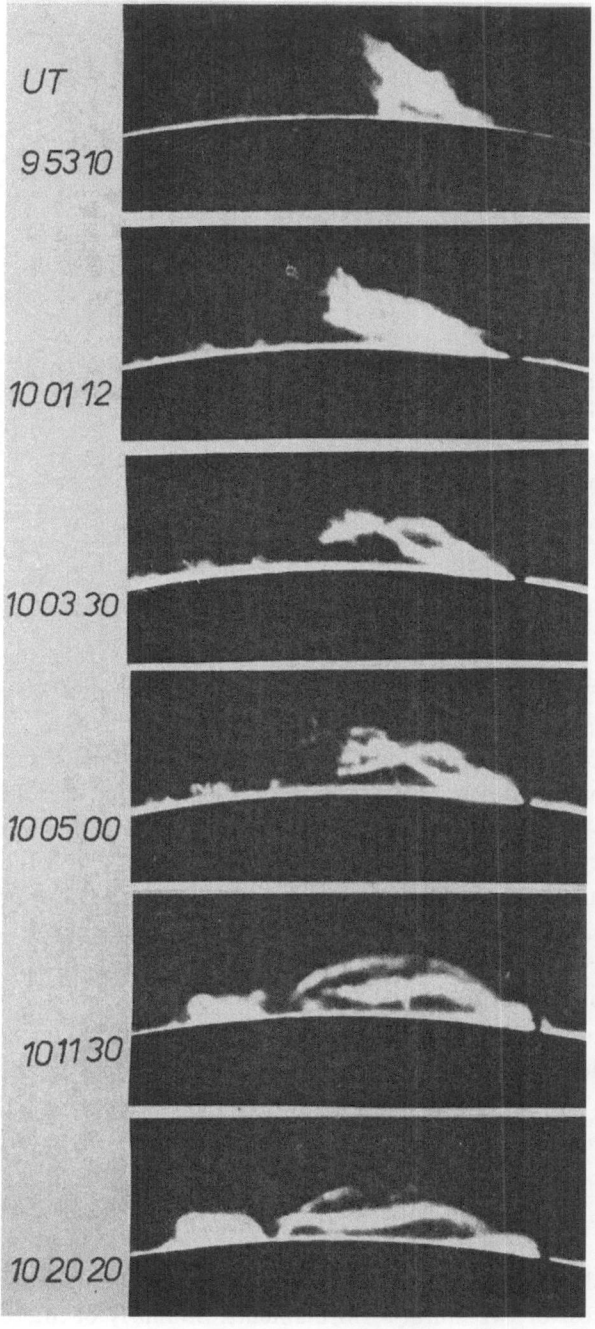

FIG. 1.    *Coronagraph observations of the detwisting phase of September 12, 1966. Ondřejov Observatory photograph.*

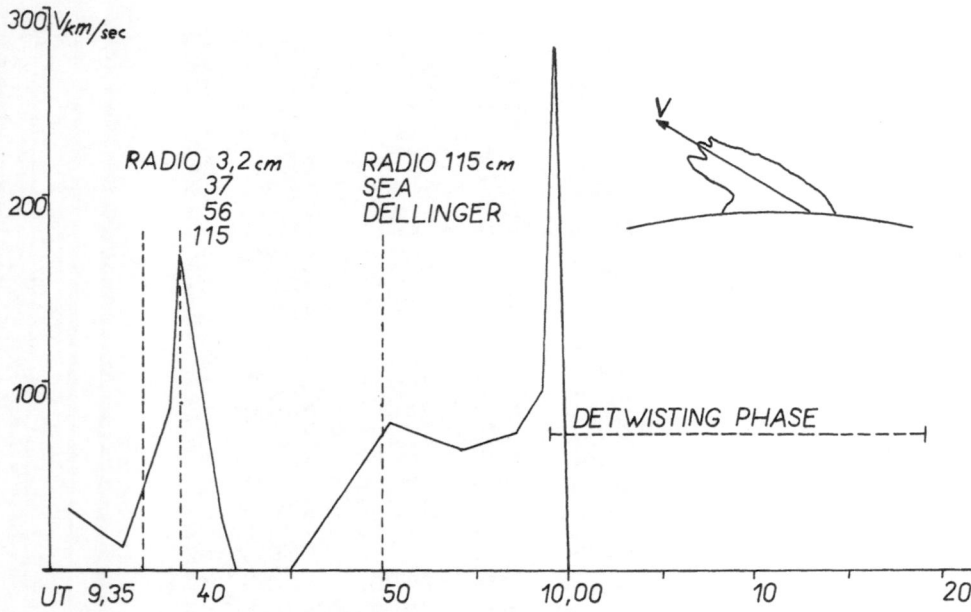

Fig. 2.  *Graphic representation of the velocity changes (full curve).*

After this last phase the detwisting of the interwoven spirals began. This finished at 10·20 UT.

It seems that the most energetic phase of the phenomenon was at 9·35–9·37 UT, when the radio emission was very important at all frequencies and a strong X-ray burst can be surmised. At this time the phenomenon was situated in deeper levels of the corona and in the chromosphere, as can be seen clearly on the photographs.

It is interesting that the second phase at 9·50 UT was the most important from the geophysical point of view. It was the time of maximum of the greatest effects of motion and very rapid brightening of the prominence. The division into two twisted filaments also started at this time.

For this period we also obtained with our flare-spectrograph a series of spectrograms clearly demonstrating for two times – 9·53 UT and 9·59 UT – Doppler shifts of both senses. This could be confirmation of the existence of a detwisting motion of both filaments. Doppler velocities measured here are of the order 200 km/sec.

## 3. Conclusions

The observed phenomenon of September 12 is most interesting because of the detwisting process of two spiral-form filaments. Similarly of interest is the fact that one of the detwisted spirals remained above the active region as a stationary filament during the entire passage of the region across the disk.

The development of this phenomenon resembles the mechanism tested by Bostick (1958) by firing one single 'plasmoid' across a magnetic-field. The result is of the same shape as the second stage of the prominence which was observed. From this point of view we can suppose that a very energetic process in the low chromosphere at the moment of maximum radio emission, 9·40 UT, caused the emission of matter which provoked the second stage – the double-spiral system of the prominence after 9·50 UT.

This thought is in agreement with ideas contained in the work of Carmichael (1963) and our observation is further evidence for this.

The detwisting phenomenon can be explained as the product of an unstable state in twisted magnetic fields following conditions for flux-tube instability, described by Alfvén (1955) and Gold (1963).

Such detwisting motions seem to be very general phenomena, especially in cases closely connected with very energetic eruptive processes. We have observed a very similar process, e.g., in the great prominence phenomenon of July 11, 1966. This effect occurred in the region of the proton-flare, included in the PFP, and we have given a detailed analysis of this phenomenon in the PFP-papers (Valníček *et al.*, 1968).

A certain analogy can probably be found in the phenomenon of Petit's (1950)

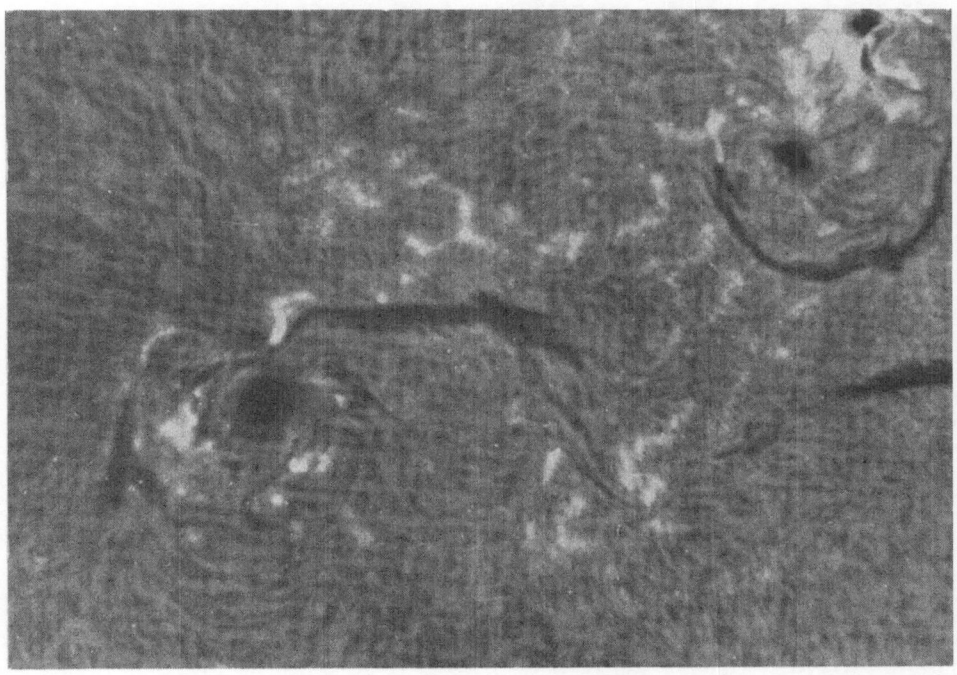

FIG. 3.    *Detail of the chromospheric structure on September 19, 7·50 UT. The orientation is the same as in Figure 1. Note the whirling structure of the filament. Ondřejov Observatory photograph.*

tornado prominences. Only the duration of both effects seems to be important in judging the difference between them: the lifetime of tornado prominences is of the order of days or weeks, while for the eruptive type described here it is of the order of 1 hour. But it seems that the filament, observed during its passage on the disk after September 12, resembles very closely the one observed by Richardson and Hickox and described in Petit's paper.

So it can be concluded that the tornado prominence is the later stage of a detwisted flare prominence of the type observed by us on September 12, 1966. This prominence – in its filament form – is also studied in the present volume by Sara Smith and by Rayrole (see pp. 267 and 134 respectively).

The future study of this active center could be very interesting in connection with other materials. The magnetic-field chart, presented by Rayrole, gives only the center of the active region, the position of the Northern part of the filament which could be seen on September 19. Nothing is known about the flare stage of the active region. But because the spots have not been observed here, the probability of obtaining a detailed magnetic-field chart is very small. Only the chaotic chromosphere structure indicates the present turbulent motions and the complicated magnetic structure.

## References

Alfvén, H. (1955)     in Cosmical Electrodynamics, p. 117.
Bostick, W.H. (1958)     Rev. mod. Phys., 30, 1090.
Carmichael, H. (1963)     AAS-NASA Symposium on the Physics of Solar Flares, p. 451.
Gold, T. (1963)     AAS-NASA Symposium on the Physics of Solar Flares, p. 389.
Petit, E. (1950)     Publ. astr. Soc. Pacific, 62, 144.
Valníček, B., Godoli, G., Mazzucconi, F. (1968)     IQSY-Annals, PFP-Project.

# LOOP-PROMINENCE SYSTEMS AND
# PROTON-FLARE ACTIVE REGIONS

## Z. ŠVESTKA

*(Astronomical Institute of the Czechoslovak Academy of Sciences, Ondřejov, C.S.S.R.)*

### ABSTRACT

It is shown that proton flares and loop-prominence systems form in the same type of active regions. Quite often both these phenomena appear simultaneously, but there are also many cases, when only the proton flare or only the loop-prominence system fully develops. The proton-flare active regions are not randomly distributed on the solar disk, but they tend to occur in complexes of activity which stay on the solar surface for many months and even years. Attention is called to the peculiar clustering of proton-flare regions on the Southern hemisphere, where two sources of activity, at a longitudinal distance of about 180°, seemed to move on the solar disk between 1956 and 1962 opposite to the solar rotation, shifting in the longitude at about 70 heliographic degrees per 10 solar rotations.

In my contribution, I would like to discuss very briefly two problems – the occurrence of loop-prominence systems in active regions which produce proton flares, and the occurrence of such proton-flare active regions in complexes of activity.

In 1964, Bruzek called attention to the fact that all loop-prominence systems observed on the disk were associated with proton flares, and he concluded that this association of loop-prominence systems with proton flares was a general characteristic of these two active phenomena (Bruzek, 1964).

We have tried to verify this conclusion of Bruzek using the catalogue of flares associated with type-IV radio bursts prepared by Olmr and myself (1966). This catalogue, containing 174 events, can also be considered for a list of proton flares which appeared on the Sun from 1956 to 1963, and we compared it with 65 loop-prominence system occurrences, taken from lists prepared by Bruzek (1964) and Kleczek (1967). We have verified that all 24 loop-prominence systems observed on the disk were preceded by proton flares listed in the catalogue, in full agreement with Bruzek's results. Of course, one must not forget that Bruzek tried to find loop prominences in this type of flares and, therefore, he might perhaps have missed some other events. For the limb-prominence systems the situation is quite different. Only 9 events of the 40 observed limb systems were clearly preceded by proton flares. It is true that, due to the directional sensitivity at long wavelengths, the classification of type-IV bursts for flares close to the limb is difficult, but one can hardly believe that we could have missed 31 type-IV bursts out of the total number of 40. And this is also supported by the fact that 14 events of these loop-prominence systems were observed on the Western solar limb without any PCA effect, which accompanies the proton flares close to the Western solar limb in about 80% of the cases (Fritzová and Švestka, 1966).

Finally, Figure 1 presents a proof that loop-prominence systems need not be necessarily associated with proton flares. It shows loop prominences accompanying the flare of July 9, 1966, which was no proton flare – no particles were observed in the space – but it formed in the same active region as the proton flare of July 7, studied in detail in the Proton Flare Project.

Fɪɢ. 1.    *Loop prominences following the flare of August 9, 1966 (picture taken by B. Valníček at Ondřejov at 5ʰ51ᵐ UT).*

Therefore, we have to conclude that, in fact, loop-prominence systems also occur without any simultaneous proton-flare appearance. But we also have found – and this, I think, is important – that the vast majority of limb-prominence systems, even if not directly connected with proton flares, appear in active regions, which produce proton flares on the Sun.

When preparing another report, for the COSPAR meeting in London in July 1967, we have studied the development of these proton-flare regions, according to the McMath classification as published in the Compilation of Solar-Geophysical Data (Švestka, 1968). That is, we have followed each active region from its first appearance on the solar disk, through the subsequent solar rotations, up to its final decay and disappearance; and we have plotted the development in a graph, which is shown in Figure 2. The left part refers to the Northern, the right part to the Southern solar

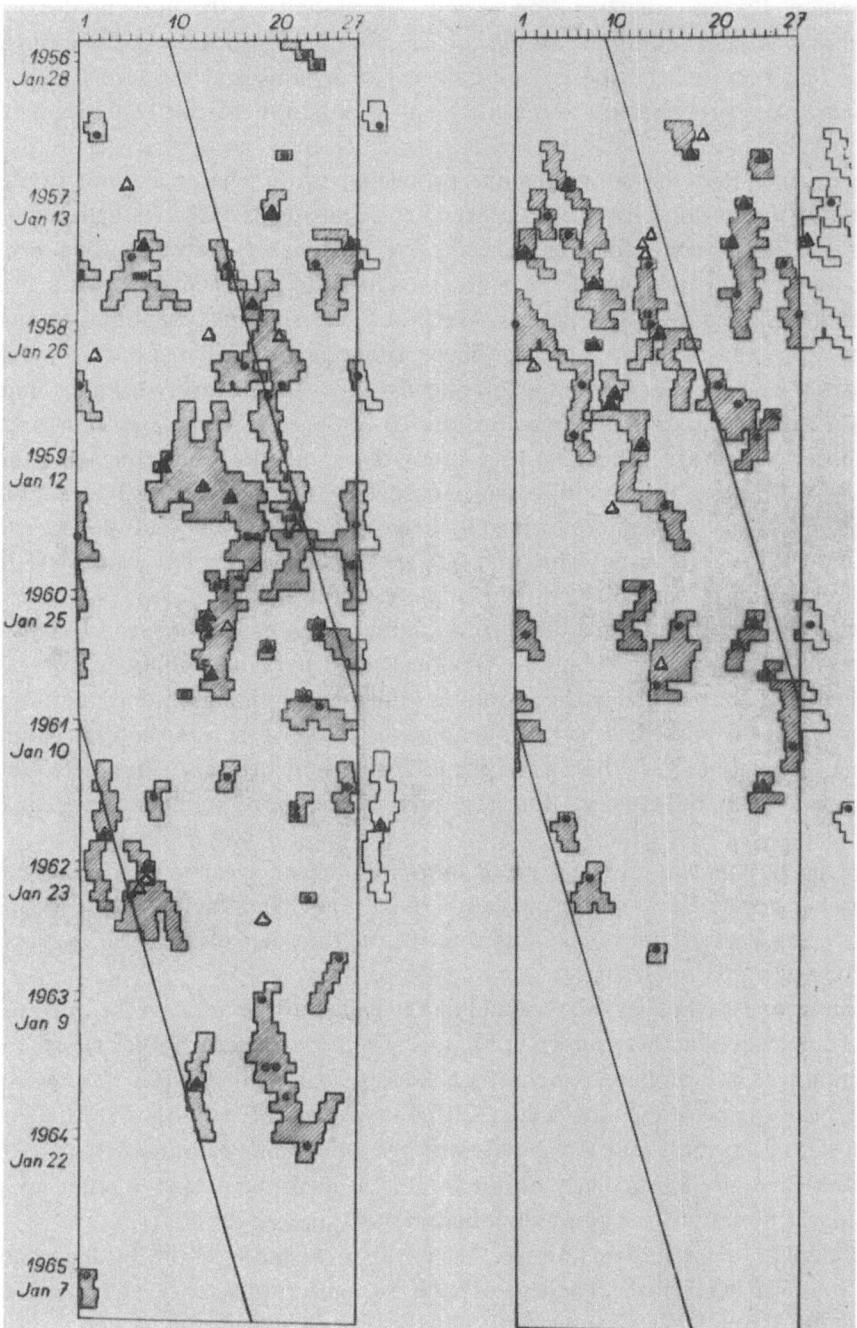

FIG. 2. *Longitudes on the Sun occupied by proton-flare active regions. For details see text.*

hemisphere. Bartels' solar rotation periods are plotted on the horizontal axis and the subsequent solar rotations on the vertical one. The dashed areas show the areas on the Sun occupied by the proton-flare active regions and the circles show the rotations, when proton flares were actually observed on the visible hemisphere of the solar disk.

We observe that the active regions producing proton flares are not randomly distributed on the solar disk, but they tend to occur in complexes of activity, which stay on the solar surface for many months, and in some cases even for several years.

The loop-prominence systems were observed in 41 different active regions and these are marked in Figure 2 by triangles. We find that about 70% of loop-prominence systems appeared in active regions, which produced proton flares during the same transit of the region over the visible solar hemisphere. And the overwhelming majority of the other events occurred in centres located in the same complexes of activity or very close to them. It is clear that we surely miss some proton flares and some of them also appear on the invisible solar hemisphere so that one can believe that these loop-prominence systems, too, formed in active regions capable of producing proton flares. There are only very few events outside the complexes, but even these lie in heliographic longitudes in which the complexes form.

Therefore, we can conclude that proton flares and loop-prominence systems form in the same type of active regions. Obviously, the particular configuration of the magnetic field characterizing the proton-flare regions, is the necessary condition for the formation of both these active phenomena, proton flares and loop-prominence systems. Many times both these phenomena appear simultaneously, but there are also many cases when only the proton flare or only the loop-prominence system fully develops.

This means that any observation of loop-prominence systems on the solar limb must be considered a very strong indication that the associated active region is capable of producing proton flares during its transit over the solar disk, which can well be used for proton-flare forecasts.

In the second part of my talk I would like to call your attention to the distribution of the complexes of activity shown in Figure 2. Some time ago, Warwick (1965) found a grouping of proton-flare regions in heliographic longitude. The line drawn in Figure 2 shows one of the longitudes (330°), for which Warwick found the maximum proton-flare occurrence. It corresponds with the great complex of activity of 1957 to 1960, but generally, particularly on the Southern hemisphere, such a tendency of a grouping in prescribed heliographic longitudes is quite small.

I would like to emphasize, however, the peculiar clustering of proton-flare regions in the Southern hemisphere. There seem to be two sources of activity, at a longitudinal distance of about 180°, in agreement with Warwick's conclusion, which, however, move on the solar disk in the direction opposite to the solar rotation. If we unfold the diagram in several subsequent rotations, we can follow these two active areas in the Southern hemisphere from 1956 to 1962 (Figure 3). If this were a real effect, it would show that there were two sources of activity in the Southern hemisphere, which rotated more slowly than phenomena visible on the solar surface, shifting in the longitude at about 70 heliographic degrees per 10 solar rotations. When looking

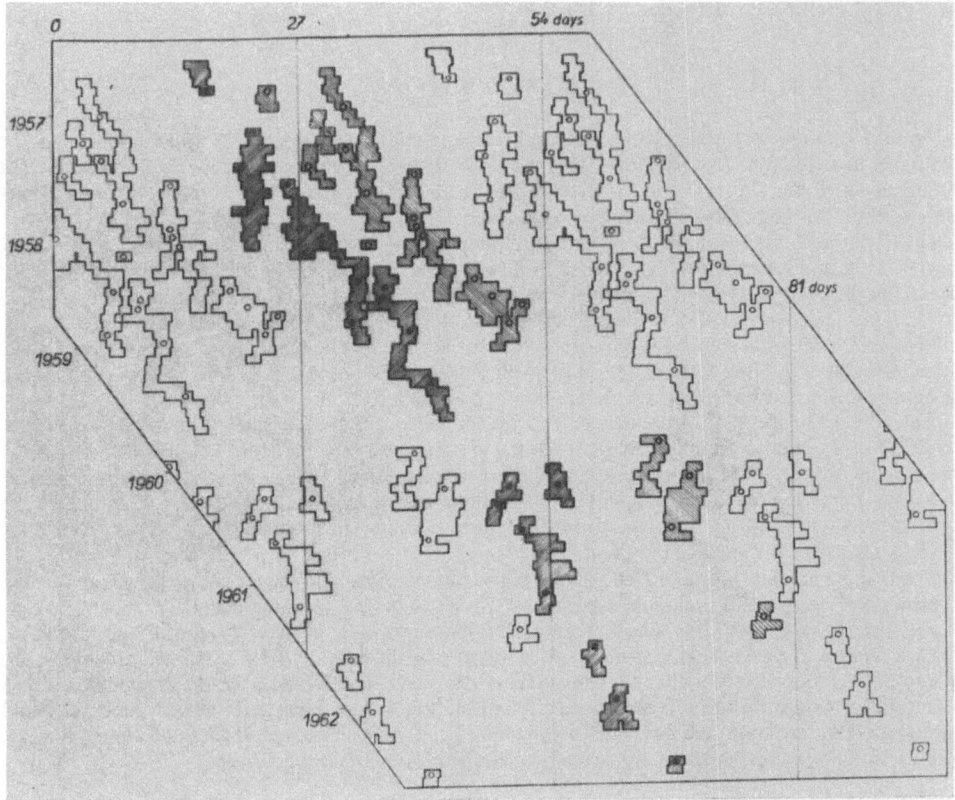

FIG. 3. *The right part of Figure 2 (the Southern hemisphere) unfolded in several subsequent rotations. It shows a shift of two active longitudes in the Southern hemisphere from 1956 to 1962.*

at the diagram, it is clear that as soon as an active region forms in the solar atmospheric layers, it rotates with the normal velocity of the solar rotation (one can see the vertical columns); but entirely new active regions or new complexes of activity always form in shifted positions, in accordance with the position of the source of activity which rotates slower.

It seems that this shift, if proved real, might give us some information on the variation of the rate of solar rotation with the depth.

## References

Bruzek, A. (1964)        *Astrophys. J.*, **140**, 746.
Fritzová, L., Švestka, Z. (1966)        *Bull. astr. Inst. Csl.*, **17**, 249.
Kleczek, J. (1967)        private communication.
Švestka, Z. (1968)        *Solar Phys.*, **4**, in press.
Švestka, Z., Olmr, J. (1966)        *Bull. astr. Inst. Csl.*, **17**, 4.
Warwick, C. (1965)        *Astrophys. J.*, **141**, 500.

## DISCUSSION

*Bruzek:* I completely agree with your results on the relation between LPS and proton flares. In my paper in *J. geophys. Res.* (**69**, 1964, 2386)* I showed only that almost all PCA effects were preceded by flares with LPS. From the tables given in *Astrophys. J.* (**140**, 1964, 747)** you notice that a part only of the LPS-flares was followed by proton events.

On the other hand, we have to be cautious when stating that a certain solar event did not emit particles because it can happen that – due to unfavorable propagation conditions – emitted particles could not arrive at the Earth. Even when using type-IV bursts as indicators for proton flares we have to keep in mind that type-IV bursts are less observable from flares occurring near the Sun's limb.

*Švestka:* I agree that some type-IV bursts close to the limb can be missed. But 14 loop-prominence systems on the West limb without any subsequent PCA seem to prove that there exist loop-prominence systems which are not directly associated with proton flares.

*Kundu:* (1) Do the loop prominences you mentioned occur before, during, or after the proton flares? (2) If they occur simultaneously, then the situation seems to be a little confused, because if I remember correctly, Drs. Jefferies and Orrall presented a paper in 1963 saying that loop prominences usually occur after proton flares. Indeed it has been suggested from time to time that these loop prominences act as storage regions of protons which continue to produce polar-cap absorption for several days after the start of the event.

*Švestka:* According to Dr Bruzek, the loop prominences probably form at the same time as the proton flare but one can usually observe them only several tens of minutes later.

*Jefferies:* In reply to a comment by Kundu, Dr. Kundu is quite correct. Dr. Orrall and I, in 1963, drew attention to the close association between loop prominences, type-IV bursts and proton flares. We found the loops to form after the proton flares (typically 30 min later). We interpreted our results as indicating that loop prominences were a manifestation of fast particles stored in the corona and placed there as a result of the flare. Our results were based on a small sample (about 8 cases). Bruzek, from a study of about 50 events, subsequently confirmed our basic conclusions.

---

\* 'Optical Characteristics of Cosmic Ray and Proton Flares'.
\*\* 'On the Association between Loop Prominences and Flares'.

# BRIGHT POINTS (MOUSTACHES) AND ARCH FILAMENTS IN YOUNG ACTIVE REGIONS

A. BRUZEK

*(Fraunhofer Institut, Freiburg i. Br., Germany)*

## ABSTRACT

A brief survey is given on the main characteristics of 'bright points' and of arch filaments in young bipolar spotgroups. Observations show that both features are closely associated, thus indicating that they have a common cause. It is suggested that the propagation of expanding and increasing active-region magnetic fields through the chromosphere might be that cause.

## 1. Introduction

The present paper is concerned with two apparently quite different phenomena of solar activity which, however, as will be shown, are closely associated in young active regions. These two phenomena are the 'bright points' (which are also called 'moustaches' or 'bombs') and a particular type of small dark filaments which will be called 'arch filaments'. The results presented here were derived from observations carried out with a Halle 0·5 Å Hα filter at the domeless Coudé refractor at the Anacapri station of the Fraunhofer Institut in summer 1966. Photographs of about 20 active regions of different sizes and types were taken during their disk passage in the Hα center and at Hα±0·5 and Hα±1·0 Å. Exposure times on Eastman Kodak 4E were 1/15–1/8 sec. The diameter of the Sun's image on the 35-mm negative is 150 mm.

## 2. Bright Points

Bright points have been investigated before by a number of authors (Beckers, 1964; Howard and Harvey, 1964; Koval, 1962, 1964, 1965, 1966; Lyot, 1941; McMath *et al.*, 1960; Severny, 1957, 1959; Severny and Koval, 1961). Their main characteristic is a Balmer-line profile with a minimum at the center of the line, a maximum at about +1·0 and −1·0 Å distance from the line centre and wings 5–10 Å wide. Due to that profile the bright points are invisible in the line centre and best visible at ±1·0 Å. If the Halle filter is tuned to the 1·0 Å position two 0·5 Å bands centered on Hα+1·0 and Hα−1·0 Å respectively are transmitted concurrently. This most favorable position was used in general in photographing the bright points at Anacapri.

The results of the analysis of the Anacapri filtergrams partly agree, partly disagree with those of previous investigations. A summary only is given here.

It was found that bright points occur in all active stages of active regions and are absent around old, strictly unipolar spots and in decaying plage regions. They are very conspicuous and numerous in large, complex spotgroups, where they are found at the outer edge of the penumbra of large spots, often at the ends of surgelike dark filaments, and also close to smaller spots (Figure 1). Bright points are a prominent feature also in the interspot region of many small and medium-sized bipolar spotgroups (Figure 2). They even occur in the very first stage of development of an active region when it still consists of a small bright Hα plage without a spot.

FIG. 1. *Bright points in a large spotgroup near the West limb; (Hα ± 1 Å)-filtergram, August 29, 1966, 1040 UT covers 11 × 16 (10⁴km)² of the sun.*

The diameters of bright points as measured on the negative or on a hard print range from 1–5 sec of arc (a typical value is 2″.5). These values are determined apparently rather by the resolving power of the image-forming system (including the Earth's atmosphere) than by the true dimensions of the source of emission which probably are less than 1″. During periods of excellent seeing a very large number of tiny bright points become visible in the eyepiece, as was also observed by Beckers (1964). Their number may exceed 100 by far in larger spotgroups and their size certainly is ≪1″. In one large spotgroup up to 40 bright points of medium size and up to 10 of large apparent size and brightness may be present simultaneously.

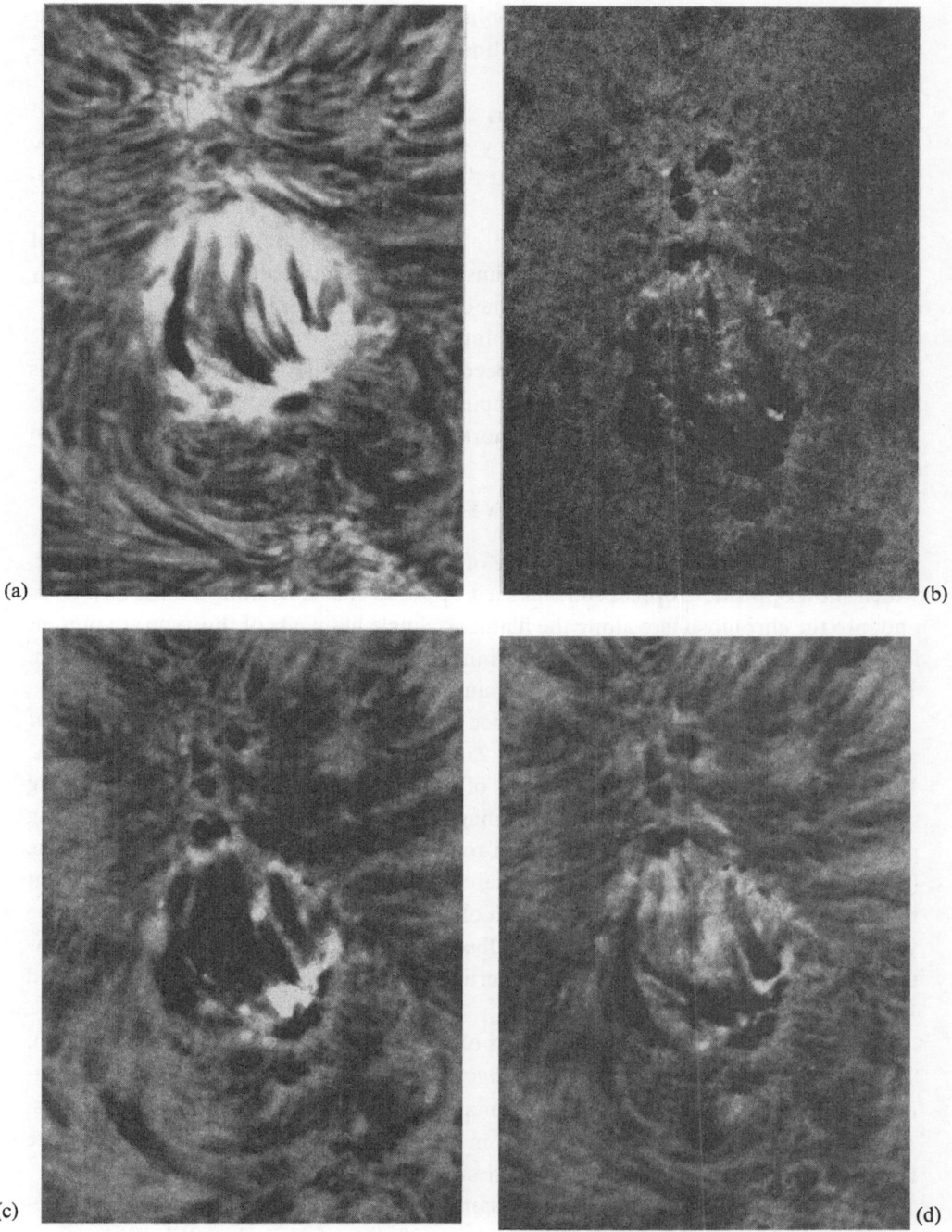

(a)

(b)

(c)

(d)

FIG. 2.   *Bright points and arch-filament system in a small bipolar spotgroup (type B) August 12, 1966, 1035 UT; photographs taken (a) in Hα, (b) at Hα ± 1 Å, (c) at Hα − 0·5 Å, (d) at Hα + 0·5 Å. The spots are visible in the 'wing filtergrams' (b)–(d); bright points are seen best in the ± 1 Å filtergram (b). Mass flow inside the dark filaments directed from the top to the bottom of the frame and an expansion of the arches towards the observer are indicated by the differences between the − 0·5 and + 0·5 Å pictures (frames (c) and (d)). The frames cover 75 × 110 (10³ km)² of the Sun's disk.*

The brightness of well-defined bright points as measured on calibrated photographs was 115–150% of the brightness of the undisturbed surrounding in the same wavelength (typical 125%).

The typical lifetime of bright points is 25 min; extreme values found were 4 and >120 min. No strict recurrence tendency was found. New bright points frequently appeared quite close to but not exactly at the place of a vanished bright point; sometimes close pairs were formed.

My observations do not confirm the preference of bright points for the vicinity of the neutral line which Koval (1966) claims to have found. I did not find either any evidence for a direct relation between the occurrence of flares and bright points.

Furthermore, occurrence of bright points in young spotgroups does not indicate whether or not the active region will become a large one: many and conspicuous bright points were observed in fast developing regions, which became large spotgroups within a few days as well as in very short-lived regions which disappeared after a lifetime of a few days.

## 3. Arch Filaments

In young active regions a peculiar type of active, arch-shaped filaments is observed, which show opposite Doppler shifts at their opposite ends indicating a mass flow from and into the chromosphere along the filament. Single filaments of this type are present in the very youngest active regions containing a small plage and a few spots just born. They were studied by Waldmeier (1937) and Ellison (1944).

Groups or systems of many arch-shaped filaments occur in the interspot region of small and medium bipolar spotgroups of Zurich type B, C, and sometimes D (Bruzek, 1967). They connect the innermost spots of opposite magnetic polarity, thus crossing the neutral line and obviously following magnetic-field lines. Sometimes the ascending part of the individual filaments is found to be much larger than the descending one (Figure 2). That obviously means that there is not only a mass flow along the filament but also an expansion or material growth of the individual arches. The corresponding velocities are about 50 and 20 km/sec. Observations at and near the solar limb show that the arches are rather flat. Their vertices are 2000–7000 km high; they exceed 10000 km in exceptional cases only.

The lifetime of the individual filaments of a system may be estimated to 20–30 min. They undergo, however, noticeable changes within 10 min, whereas the general configuration of the system as a whole remains almost unchanged for several hours. Arch-filament systems may be present in a spotgroup for several days. They are modified as soon as the spot configuration changes.

It must be pointed out that the arch-filament systems are quite different from loop-prominence systems investigated by Dodson (1961) and Bruzek (1962, 1964). They differ in origin, development, inside motions and final height, they are similar in their connection with the magnetic field.

## 4. The Relation between Bright Points and Arch Filaments

In Figure 2 a small bipolar spotgroup of type B is photographed in four wavelength bands. It shows that bright points occur almost only in the interspot region covered or bridged by the arch filaments. The bright points are situated frequently near the ends of the arches but are also visible between the filaments; it is likely that a number of bright points is even hidden below the arches. There is hardly a bright point found outside the spotgroup.

From a number of observations it became evident that bright points do not occur in young spotgroups unless there are arch filaments. In one case two bipolar groups of approximately the same age and size were present on the solar disk. In one of them both features, bright points as well as arch filaments existed, in the other one none of them was visible at the same time. In other cases a spotgroup contained neither arch filaments nor bright points on one day, but both of them on the following day.

The close connection between bright points and arch filaments is also true for the very first stages of an active region. A few bright points were found near a small arch filament or at its ends in two tiny active regions which had just appeared several hours ago. The active region consisted of no more than a very small bright plage (rapidly growing) and a short arch filament; no spot was yet visible.

## 5. Conclusions

The close relation between bright points and arch filaments suggests a common cause of both features. It is very likely that the arch filaments are related to the combined transport of matter and increasing magnetic field into higher levels during the growth of the spotgroup (Bruzek, 1967). We may therefore conclude that occurrence of bright points depends on either the propagation of the field through the low chromosphere or the ejection of matter. The latter suggestion is supported by the association of bright points with surgelike features in large spotgroups. The possible coincidence of bright points and surges with 'magnetic dots' as suggested by observations of Beckers and Schröter (1968) may be an important point for the understanding of the phenomena. A thorough discussion on the nature of bright points and their relation to the arch filaments has to be postponed, however, until more data are available.

## References

Beckers, J. M. (1964)    Thesis Utrecht, p. 33.
Beckers, J. M., Schröter, E. (1968)    the present volume, p. 178.
Bruzek, A. (1962)    Z. Astrophys., **54**, 225.
Bruzek, A. (1964)    Astrophys. J., **140**, 746.
Bruzek, A. (1967)    Solar Phys. **2**, 451.
Dodson, H. W. (1961)    Proc. nat. Acad. Sci. Am., **47**, 901.
Ellison, M. A. (1944)    Mon. Not. R. astr. Soc., **104**, 22.

Howard, R., Harvey, J.W. (1964)        *Astrophys. J.*, **139**, 1328.
Koval, A.N. (1962)       *Izv. Krim. astrofiz. Obs.*, **28**, 241.
Koval, A.N. (1964)       *Izv. Krim. astrofiz. Obs.*, **32**, 32.
Koval, A.N. (1965)       *Izv. Krim. astrofiz. Obs.*, **33**, 138.
Koval, A.N. (1966)       *Izv. Krim. astrofiz. Obs.*, **34**, 278.
Lyot, B. (1941)       *Ann. Astrophys.*, **7**, 48.
McMath, R.R., Mohler, O.C., Dodson, H.W. (1960)       *Proc. nat. Acad. Sci. Am.*, **46**, 168.
Severny, A.B. (1957)       *Izv. Krim. astrofiz. Obs.*, **17**, 129.
Severny, A.B. (1959)       *Izv. Krim. astrofiz. Obs.*, **21**, 131.
Severny, A.B., Koval, A.N. (1961)       *Izv. Krim. astrofiz. Obs.*, **26**, 3.
Waldmeier, M. (1937)       *Z. Astrophys.*, **14**, 91.

# DISCUSSION

*J. Evans:* You have shown that points and arches tend to occur in the same areas within new active centres. I want to be quite clear, however, about whether or not the locations of points are related to the specific geometry of individual arches. Do arches usually or always begin or terminate in a point? Or is there any other such relation?

*Bruzek:* Many of the arches begin and terminate, or either alone, in 'bright points', but far from all. On the other hand there are many 'bright points' between arches and at no foot of a filament at all.

*Krüger:* I should like to ask you if somewhat is known about the frequency of occurrence at bright points during the 11-year solar cycle; especially, is this phenomenon more characteristic for the beginning of a new solar cycle or not?

*Bruzek:* There is no known dependence of occurrence of 'bright points' on the 11-year cycle.

# SOLAR ACTIVE REGIONS IN MG II LIGHT

KERSTIN FREDGA*

*(Royal Institute of Technology, Stockholm, Sweden)*

ABSTRACT

During 1965 two rockets were launched in order to obtain monochromatic pictures of the Sun in the Mg II line at 2802·7 Å. The Mg II filterheliograms have been compared with simultaneous Hα and Ca II K spectroheliograms. An important observation from both flights concerns the relative intensities of different active regions. In the Hα pictures there are old fairly faint active regions and newly formed more intense regions. In the Mg II pictures, however, the older regions appear more intense than the younger regions.

Different explanations for this intensity-reversal effect have been considered. The effect has also been looked for in broadband Ca II H filterheliograms, with negative results, however.

## 1. Introduction

NASA Aerobee rockets were launched on April 12, 1965 and December 2, 1965, which carried instruments designed to obtain monochromatic pictures of the Sun in the singly ionized Magnesium line at 2802·7 Å. The instrument package was prepared at Goddard Space Flight Center and consisted of a Cassegrain-Maksutov telescope, behind which was placed a Šolc-type birefringent filter and an automatic camera. The birefringent filter had a spectral bandpass of 4·0 Å in the first flight and 3·5 Å in the second flight. The instrument was flown with a biaxial solar pointing control permitting a resolution of 1·0 arc-minute in the pictures.

## 2. Results

Figures 1 and 2 show a comparison between one Mg II picture from each flight and spectroheliograms in Hα and Ca II K obtained from Sacramento Peak Observatory at the same time as the flights. All active regions in the Mg II pictures correspond very well to the plages visible in Hα and Ca II K spectroheliograms. Even the detailed structure shows very good correspondence. A coarse mottling network is visible in the Mg II pictures.

An interesting observation from both flights, pointed out earlier by Fredga (1966), concerns the relative intensities of the different active regions. In the Hα picture taken on April 12, the small plage just coming around the East limb is considerably brighter than the more extended area in the Northern hemisphere. In the Mg II picture of the

---

* From January 1, 1964 to June 30, 1966, NASA–NAS, NRC Postdoctoral Research Associate at Goddard Space Flight Center, Greenbelt, Md., U.S.A.

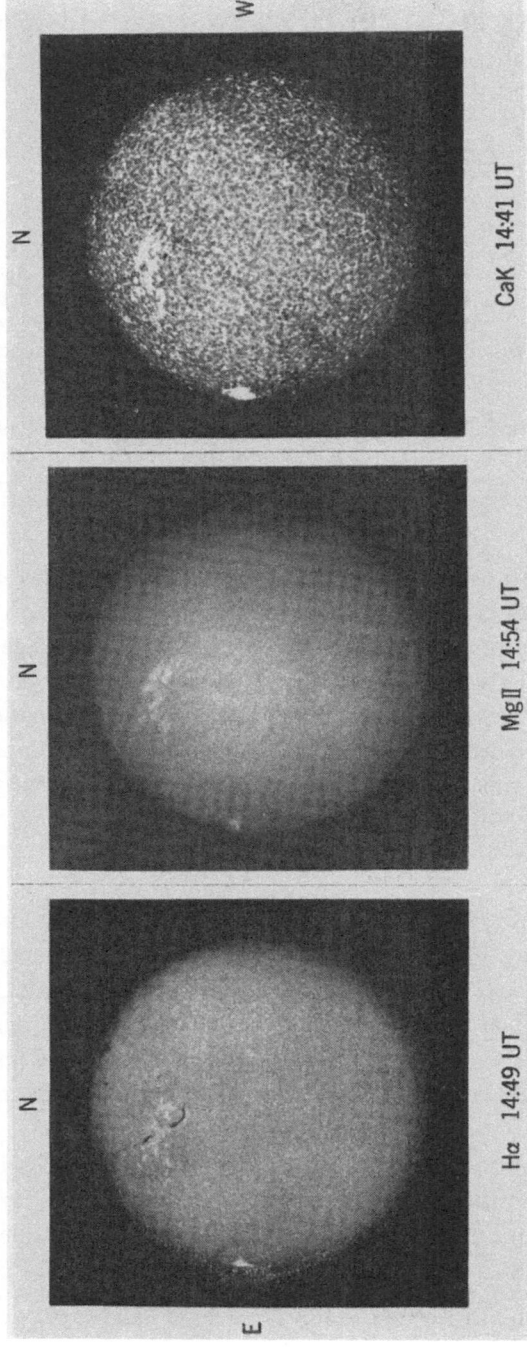

FIG. 1. *Three monochromatic pictures of the Sun obtained on April 12, 1965. (a) Hα filtergram at 14·49 UT, bandwidth 0·5 Å; (b) Mg II filtergram at 14·54 UT, bandwidth 4·0 Å; and (c) Ca II K spectroheliogram at 14·41 UT, slit width 0·5 Å. (Hα and Ca II K pictures by courtesy of Sacramento Peak Observatory, Air Force Cambridge Research Laboratory, New Mexico.)*

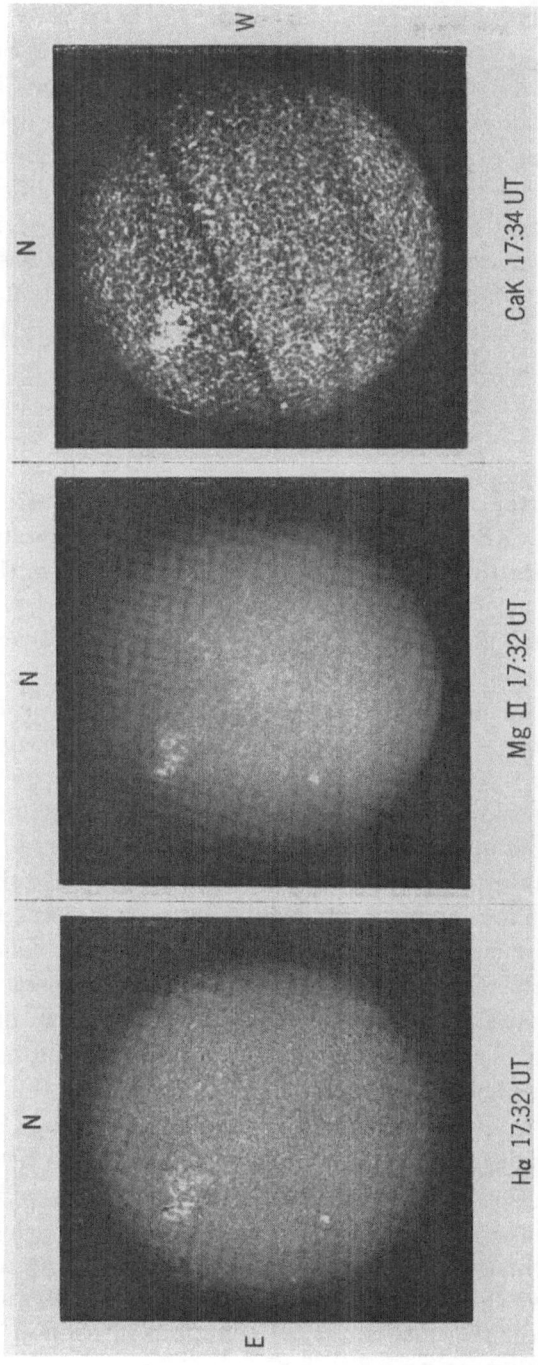

FIG. 2.  *Three monochromatic pictures of the Sun obtained on December 2, 1965. (a)* Hα *filtergram at 17·32 UT, bandwidth 3·5 Å; and (c)* Ca II K *spectroheliogram at 17·34 UT, slitwidth 0·5 Å. (b)* Mg II *filtergram at 17·32 UT, bandwidth 0·5 Å; (b)* Mg II *filtergram at 17·32 UT, bandwidth 0·5 Å. (Hα and* Ca II K *pictures by courtesy of Sacramento Peak Observatory, Air Force Cambridge Research Laboratory, New Mexico.)*

same day the intensities of the two areas are approximately equal. In the December 2 picture, there is also one large active region fairly faint in Hα (at 27 N, 27 E) and a small region of higher intensity (at 18 S, 33 E). In the Mg II picture, however, the larger area is the more intense of the two. The relative intensities of the Ca II K plages fall approximately between those of Hα and Mg II but are more similar to those of Hα.

This effect may be connected with the life history of the active regions. Both the large plage areas are old active regions visible one solar rotation earlier, while the small active regions are newly formed. The larger area shown in the April 12 picture was formed on or before March 11, 1965. The smaller area, at the East limb, was probably formed only a few days before it became visible at the East limb. The larger area shown in December 2 picture was formed on or before October 31, 1965. The smaller area in the same picture was only 3 days old.

## 3. Discussion

Different explanations for this intensity-reversal effect have been considered. Instrumental effects (i.e. vignetting) have been checked, and calibrations of the instrument before and after the flights show that this cannot explain the observed effect.

There is a considerable smearing in the Mg II pictures due to pointing errors during the exposure time. In general, a smearing will influence small and large areas in a different way; the small areas lose more in contrast than the big areas. In this case, however, the so-called large areas consist of individual parts of approximately the same size as the small areas. The individual parts of the large areas do not overlap each other during the exposure time. The smearing in the Mg II pictures probably cannot explain the observed intensity reversal.

The contrast between an active region and the undisturbed solar disk is affected by the bandwidth of the filter. However, the relative intensity between two regions should not be affected by the spectral bandpass. This conclusion is valid only with the assumption that the emission-line profiles of old and young active regions are approximately equal. For example, if the emission-line profile of an old region is much broader than the profile of a young region, the difference in bandwidth between the Mg II filter and the Hα filter could give the observed intensity-reversal effect.

In order to find what influence the broad bandwidth may have, Ca II H filter-heliograms have been obtained during the summer of 1967 at the Swedish Solar Observatory at Anacapri. The Ca II H birefringent filter which was used has a bandwidth (full width at half intensity) of approximately 2 Å. The filterheliograms have been compared with simultaneous Hα pictures to search for intensity reversals between old and young active regions of the type mentioned above. So far the results of this study have been negative; that is, the active regions in Hα and broad band Ca II H show the same relative intensity distribution.

The intensity reversal has so far only been found in the Mg II pictures. More material is needed for confirmation and interpretation of this effect. A future rocket experiment (hopefully with improved pointing) will take place during the time of maximum solar activity (1968). At this time one can expect to have several active regions on the Sun of different ages and in different stages of development.

## Acknowledgements

The author is very much obliged to Goddard Space Flight Center, Greenbelt, Md., and all the individuals there who took part in the rocket experiments. Thanks are also due to the Swedish Solar Observatory at Anacapri.

## Reference

Fredga, K. (1966)       *Astrophys. J.*, **144**, 854.

## DISCUSSION

*Sheeley:* What about time changes? If the K spectroheliogram and Mg II spectroheliogram were taken at different times perhaps the relative intensities of the two plages had changed in the intervening time? We know of such cases in Hα: Flares, of course, are extreme cases, and we know that the bright network varies in time also.

*Fredga:* It depends on what time scale you have in mind. For the pictures taken on April 12, 1965 the difference in time between the Hα picture and the Mg II picture was 5 min. The Hα and Mg II pictures from December 2, 1965 were secured within the same minute. No flares were reported.

*Kiepenheuer:* What was your exposure time?

*Fredga:* The exposure time was 1/8 sec for the picture taken on April 12, 1965, and 1/9 sec for the picture taken on December 2, 1965.

Part IV

# COOPERATIVE STUDY
# OF SOLAR ACTIVE REGIONS (CSSAR)

# COOPERATIVE STUDY OF SOLAR ACTIVE REGIONS

*Introductory Report*

R. MICHARD

*(Observatoire de Paris-Meudon, France)*

## 1. Introduction

The 'Cooperative Study of Solar Active Regions' (or CSSAR) originated from a suggestion of E. R. Mustel at the IQSY General Assembly in Rome, 1963. Later on it was also endorsed by IAU Commission 10, and a special Working Group of this Commission, under chairmanship of the writer, had the duty to organize this project, an experiment in both solar physics and international cooperation.

Our general purpose was to study in detail the evolution of a small number of selected Active Regions (AR) using the material obtained throughout the world, by all techniques of optical and radio-astronomy (and possibly also indirect information from interplanetary and geophysical effects). It was thought that 1965 would be a convenient period because the small number of ARs would limit interferences between their respective evolutions, particularly in their declining phase. Also the International Years of the Quiet Sun would form a natural frame for this cooperative project.

A total of 72 observing stations from 27 different countries announced their readiness to take part in CSSAR. Collection of the material was undertaken by the various World Data Centers in the field of solar activity, each of them taking care of a particular type of data. A final collection was achieved in Meudon in February 1967.

In such a project, one of the most difficult tasks is to make the material available to interested scientists in such a form and at such a time that it is effectively used. It was, at first, thought that some kind of publication of the data, for instance by circulation of microfilms, would be the best way. Later on, we realized that this would not be feasible due to enormous amount of work involved, and to the great loss of information in repeated copying of the photographs which form the bulk of CSSAR data. Therefore we tried to encourage direct study of the original and/or first copies. The 'collectors' of each sort of material were asked to analyse it, if they thought valuable. Further a large number of astronomers were invited to come to Meudon during spring 1967 for examining the whole collected information.* A total of 7 colleagues from 5 institutions accepted this invitation; a few local scientists also took part in the study of the CSSAR data.

* We thank the Centre National de la Recherche Scientifique, Paris, and the Director of the Observatoire de Paris for their generous support in this programme.

**Table 1**

| CSSAR n° | QBSA n° | McMath n° | KAO n° | Mt. Wilson n° | Coordinates | CMP 1st transit | Birth | Remarks |
|---|---|---|---|---|---|---|---|---|
| 1 | 1492-1 | 7718 | 76 | 15903 | 348, 30 N | March 16 | March 11 on disk | Cycle 20, well isolated, 12 flares in 1st transit, traceable at 3rd transit |
| 2 | 1493-4 | 7771 | 79 | 15909-12 | 277, 4 N | April 18 | ? | Cycle 19, well isolated at 1st transit. 15 flares at 1st transit. Very doubtful return of plage QBSA 1492-3 (McMath 7736) |
| 3 | 1494-6 | 7809 | 87 | 15918 | 200, 23 N | May 21 | ? | Cycle 20, large AR, close to n° 4, 15 flares. Small new region born west of it at 3rd transit |
| 4 | 1494-9 | 7812 | 88 | 15919 | 173, 25 N | May 23 | ? | Cycle 20, rather large, close to n° 3, 30 flares. Traceable at 3rd transit |
| 5 | 1494-17 | 7840 | 91? | 15923 | 15, 10 S | June 3 | June 1 on disk | Cycle 19. Nice case for growing phase of isolated AR. 7 flares |
| 6 | 1495-4 | 7863 | 95 | 15926 | 210, 28 S | – | June 16 on disk | Cycle 20, 10 flares, 3rd transit perturbated by new AR West of n° 6 |
| 7 | 1496-4 | 7886 | 102 | 15935-6 | 270, 20 N | July 8–9 | July 5–6 on disk | Cycle 20, 10 flares. Born on disk in existing old plage. Faint, well-isolated plage at 3rd transit |
| 8 | 1498-5 | 7971 | 112-3 | 15953 | 185, 26 N | September 8 | September 4 on disk | Cycle 20, 14 flares at 1st transit. Well isolated at 2nd transit |
| 9 | 1499-4 | 8005 | 118 | 15957 | 218, 20 N | October 2 | ? | Cycle 20, 27 flares at 1st transit. Large AR young at east limb. |
| 10 | 1499-1 | 8012 | 120 | 15961 | 230, 18 S | – | October 2 on disk | Cycle 20, 3 flares. Large flare on October 4 |

It would be clear that this study could start only in February 1967 and is still continuing in some cases. Therefore the various scientific reports presented below represent only preliminary results of the CSSAR project.

## 2. Review of CSSAR data

On the basis of their intrinsic interest and of the world-wide availability of good observations, 10 ARs were selected for data collection. These are listed in Table 1.

Sunspots photographs were also collected for a few other regions of interest at the initiative of WDC-C for sunspots (Eidgenössische Sternwarte in Zürich).

The available material for the ARs of Table 1 may be summarized as follows (with name of data 'collector').

(a) Sunspot photographs with average time interval of 1 hour (Waldmeier, Zürich).

(b) $K_2$ spectroheliograms with average time interval of 3–4 hours (Godoli, Florence).

(c) Daily Hα spectroheliograms, plus nearly continuous cinematographic Hα patrol for days of important flare activity in the selected regions (Michard, Meudon).

(d) Miscellaneous prominences photographs (Kleczek, Ondřejov).

(e) Mount Wilson whole-disk magnetograms, detailed maps of AR magnetic fields, sunspot magnetic data (Severny, Crimean Astrophysical Observatory).

(f) K-coronameter records and white-light photographs of corona (Newkirk, High Altitude Observatory).

(g) Coronal lines records (Leroy, Pic-du-Midi).

(h) Radioastronomical data (Fokker, Utrecht).

(i) X-rays flux and miscellaneous geophysical data (Miss Lincoln, ESSA, Boulder).

Part of the data, specially for topics (h) and (i), is published in the *Quarterly Bulletin on Solar Activity* and in the *Solar Geophysical Data* of ESSA. A large part will be conserved in Meudon, and requests for communication should be addressed to the writer. However, original photographs, such as part of Hα-patrol films, are being sent back to their owners.

## 3. Remarks about the CSSAR data and the 1965 solar patrol

CSSAR may have been the first opportunity to review the completeness and average quality of the worldwide survey of solar phenomena. The writer was interested in the following points:

(a) The many white-light pictures of the photosphere obtained at various stations allow the study of the evolution of sunspot groups with an average resolving power superior to that of other types of observations, such as magnetographic or spectroheliographic. More studies of the detailed evolution of sunspots would be feasible and useful.

(b) While the usual Hα cinematographic patrol is satisfactory for the detection and evaluation of solar flares – that is for its main purpose – its average resolving power is often insufficient for the detailed study of ARs and of flare events. As often advocated by Kiepenheuer and Giovanelli, efforts are needed to obtain better Hα surveys of selected solar regions.

(c) The detailed study of longitudinal magnetic fields in ARs is still very difficult. Only two stations, K.A.O. and Meudon, produced a significant number of isogauss maps for the CSSAR project (although a few others were available from Mount Wilson 'fine scans' and from IZMIRAN). An unwanted number of errors was found in the Meudon maps, while the saturation of the photoelectric magnetograph is a significant limitation in the study of ARs.

Progresses are still needed in order to produce detailed magnetic maps in much larger number, of better quality, and easier to compare between themselves or with other solar information. In the future it seems likely that two different approaches should be undertaken:

(1) on the one hand the mass production of rather qualitative magnetic 'pictures' for patrol purposes (and flare prediction) with techniques more or less based on image substraction like Leighton's one.

(2) on the other hand the obtention of precise quantitative maps for a lesser number of cases, with techniques allowing the measurement of a large *range* of fields and preferably also of the three field parameters.

## 4. Conclusion

Although the above remarks mainly deal with some limitations of the CSSAR data, it will be seen from forthcoming papers in this volume (and other publications) that significant scientific results could be obtained from it! This was due to the generous cooperation of many observatories providing their photographs and records of solar phenomena, to the 9 data 'collectors' who, in many cases, spent much work to arrange the raw material in an homogeneous way, to the excellent advice received from the CSSAR Working Group. I express to them the gratitude of the scientists who used the data, and also my own.

# FLARE ACTIVITY AND SPOTGROUP DEVELOPMENT

V. Bumba, L. Křivský          and   M. J. Martres, I. Soru-Iscovici
*(Astronomical Institute of the Czechoslovak   (Observatoire de Paris-Meudon, (92)*
*Academy of Sciences, Ondřejov, C.S.S.R.)*          *Meudon, France)*

ABSTRACT

The flare activity of all active regions having ten or more flares from August 1959 until December 1961 was investigated. We constructed flare-activity curves drawn with the aid of Kleczek's $q$-index. The characteristic magnetic situation on the Mt. Wilson synoptic charts of photospheric magnetic fields in which the flare-rich active regions developed is described. A flare activity rate 'quantization' was found. From the CSSAR magnetic observational material we studied the reorganization of the magnetic fields of active regions which correlated with sudden changes in the rate of flare activity.

## 1. Introduction

During recent years more and more effort has been spent in the investigation of solar flares. But usually the extremely large flares in extremely complicated magnetic situation of complex active regions are studied. We think that the systematic study of small, common flares and their relation to the normal common sunspot groups with relatively simple magnetic fields may have even greater importance.

Therefore we started to investigate flare activity and its changes in dependence on the development of the magnetic situation in all active regions covering a time interval of about two and half years. Our communication summarizes the first preliminary results obtained.

## 2. Method of Study and Observational Material

To visualize the frequency of flares of various importance we constructed curves of flare activity for each single active region from August 1959 till December 1961, which has the number of flares $n \geqslant 10$. For this period of time the Mt. Wilson synoptic charts of photospheric magnetic fields are available. For each of these active regions the curves of the development of the area of the sunspots were also drawn.

For the construction of flare activity curves we use the method of summation of $q$-indices defined by Kleczek (1952). This $q$ $(q = i \cdot \bar{t})$-index characterizes each flare by its importance $i$ and the mean duration $\bar{t}$ estimated for that importance, which means that the $q$-index may be in a certain degree proportional to the energy emitted by the flare. The flare-activity curve therefore can show the rate of flare-energy production given by the slope of the curve (Figure 1). The value proportional to the total flare-

*Kiepenheuer (ed.), Structure and Development of Solar Active Regions,* 311–317. © *I.A.U.*

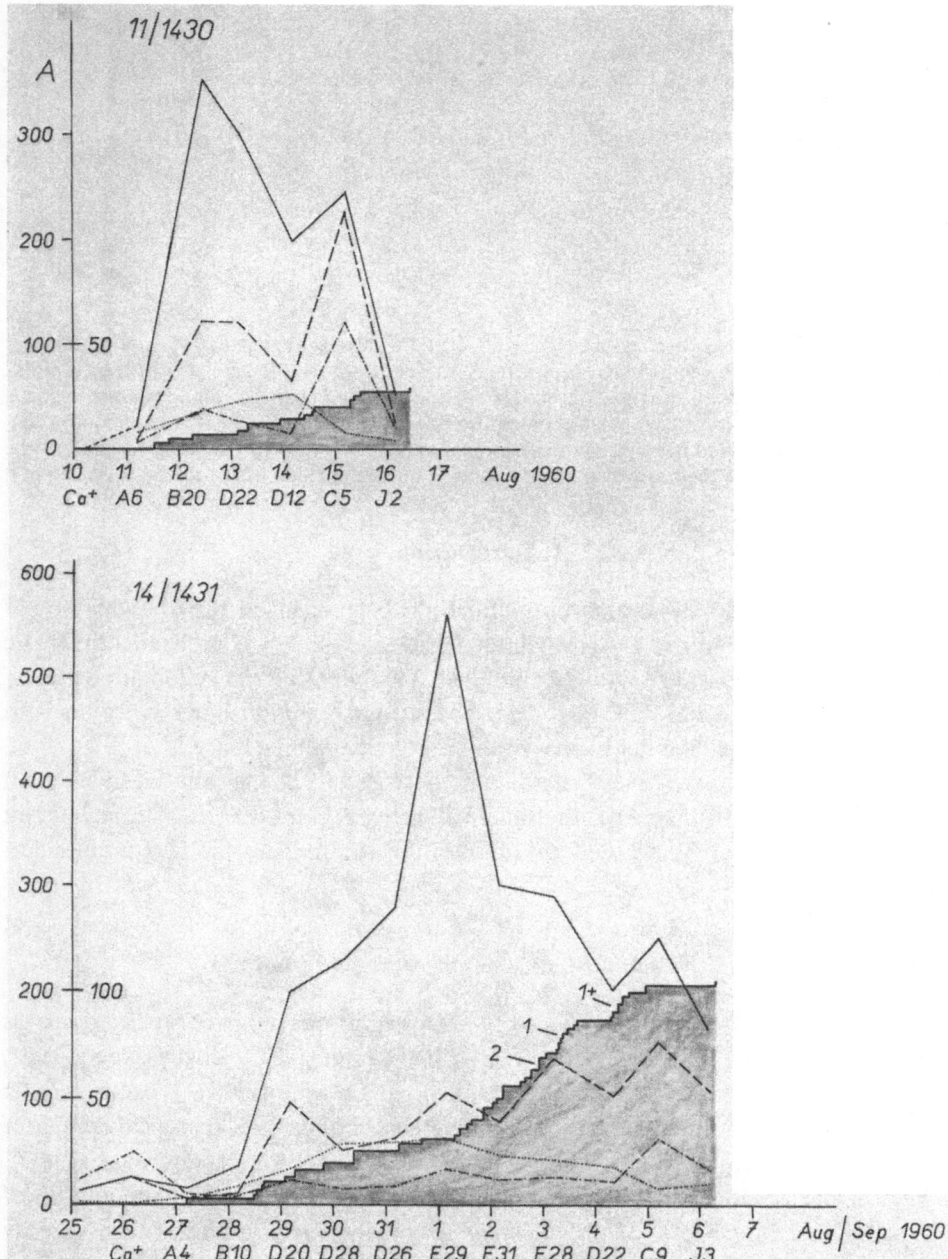

FIG. 1.  *Two examples of curves of flare activity: On the top for AR No. 11/1430. On the bottom for AR No. 14/1431 (Quarterly Bulletin). – For each AR the daily values of the total spotgroup corrected area A (solid lines) and umbral area (dashed lines) daily number of spots in the group n (dotted lines) and the 'mean area of one spot' n/A (dash-dot lines) are given. The pattern of development from Fraunhofer Institute daily data is presented below each graph. For better visualisation of the flare-activity curves, the areas between them and the horizontal axis are shaded. The values of the q-index for importance 1, 1+, and 2 are indicated.*

energy output of the active region is represented by the final coordinate of the curve on the $q$-axis.

## 3. Results

From the investigation of our observational material which includes about 115 active regions the following preliminary results were made:

(1) All active regions with the number of flares $n \geqslant 10$ develop in a very characteristic magnetic situation: they are usually bound to the places where the new developed magnetic field of the following polarity is strongly compressed between the large-scale patterns of the magnetic field of the leading polarity (Figure 2). Because the new formations as a rule are a part of a complex of activity (Bumba and Howard, 1965) and because only rarely is one active region apart from the main body of the background field observed, usually two or more flare-active regions are combined with one of such inclusions of following magnetic field mostly surrounded by the large body of the leading field. This magnetic situation is as a rule followed by a large maximum of green-line coronal emission and radio emission (1420 Mc/s) sitting above the inclusion (Bumba *et al.*, 1968*b*).

(2) Looking at the curves of flare activity we may see that their parts may be approximated by straight lines with certain value of inclination. If we plot the frequency curve of the slopes of these straight lines, we obtain several maxima in their distribution. This fact may be visualized by the help of the overlapping of individual flare-activity curves in the same way the curves of growth or the photographic characteristic curves are constructed (Figure 3). By this method we are able to divide all our active regions in 6 (7) groups with different values of flare-production rate. The first group of active regions has the slope of the main part of its flare-activity curves corresponding closely to the production of one flare of importance 1 per 24 hours, the second one to the rate of flare production of one flare of importance $1^+$ per 24 hours, and so on. (The maximum for $1^-/24$ hours is visible from the other observational material CSSAR.)

Very often one curve of flare activity for a single active region may be approximated by several straight lines, the inclination of which usually agrees with one of the main directions.

The duration of flare activity with high values of the rate of flare production is usually much shorter (1–3 days) (with the exception of the largest regions) than that of the small rate (5–10 days), so that the sum of $q$-indexes for each of the active regions does not differ very substantially from the mean. In the other hand the time of flare activity may differ within a much larger interval.

(3) As was found earlier from the observational material concerning the CSSAR-Period, the quality and methods of magnetic field measurement do not yet allow us to search for magnetic-field configuration changes connected with the occurrence of individual small flares. But the sudden changes of the rate of flare activity, as shown

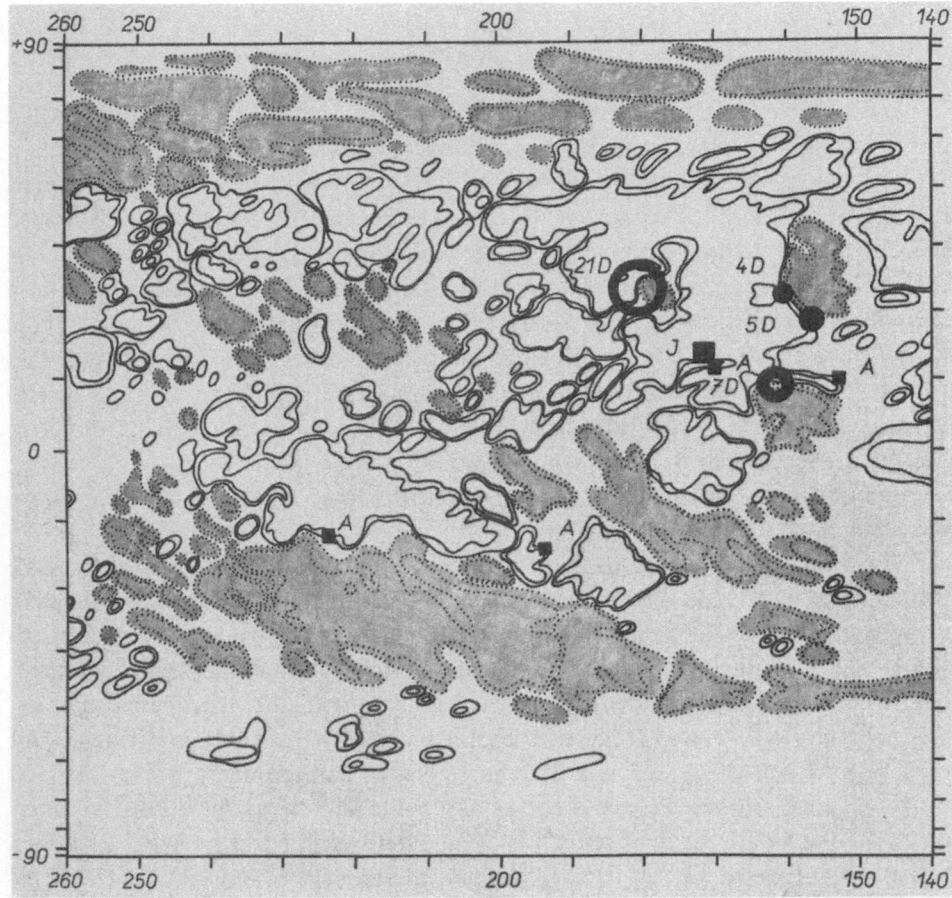

FIG. 2.   *Example of the magnetic situation in which flare-active centra develop. A part of the Mt. Wilson synoptic chart representing photospheric magnetic fields from November 1960 is given. Solid lines represent isogauss for the positive (leading in the Northern hemisphere) polarity, dotted lines for the negative polarity. Active regions with flare activity are represented by circles and the Waldmeier type of the maximum phase of evolution together with the total number of flares. The regions represented by squares did not have flares observed. The situation in the left part of the figure represents the old magnetic-field areas, the appearance of which is in strong contrast to the inclusions of following polarity of new formed ARs in the right upper part of the figure.*

by the changes of flare-activity curve slopes at the beginning of the main flare-activity phase, may be correlated very well with the reorganization of the magnetic field of the active region: as soon as the magnetic situation (as a rule in the centre of the group at the boundary of polarities) becomes more complicated the rate of flare activity becomes greater, and vice versa. The greater magnetic activity is a character-istic process taking part during the phase of growth or renewal of activity in the

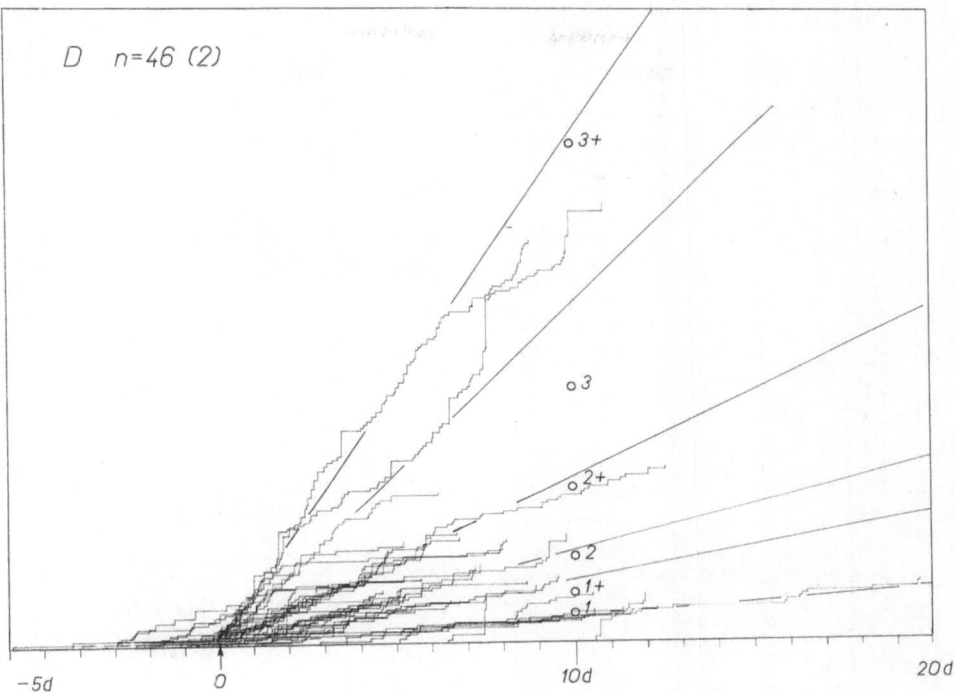

FIG. 3.  *Overlapped flare-activity curves for ARs developed during the studied time interval on the visible disk. The concentration of curves around certain values of slopes is seen. The theoretical values of slopes for frequencies one flare of importance 1 per 24 hours and so on are given.*

centre of the active region, where one observes the formation of a gulf on the boundary between both polarity regions or islands of opposite polarity connected with the appearance of new spots (Figure 4) (Bumba *et al.*, 1968*a*). The flare activity reaches its maximum rate in most groups 2–3 days before the maxima of their sunspot areas. The decay of the magnetic activity in the centre of the group is followed by a decrease of the flare activity after a certain interval of time.

It seems that in some active regions there may exist a secondary increase of flare activity connected, for example, with magnetic changes during the dissipation of sunspots in the group, but this process needs more observational material.

## 4. Discussion of Results

The division of active regions into different classes depending upon their flare-activity rate certainly depends upon the method of classification of flares, and it may be influenced by the irregularity of geographic longitudinal distribution of observational stations. But a detailed study of flare activities shows that the most important

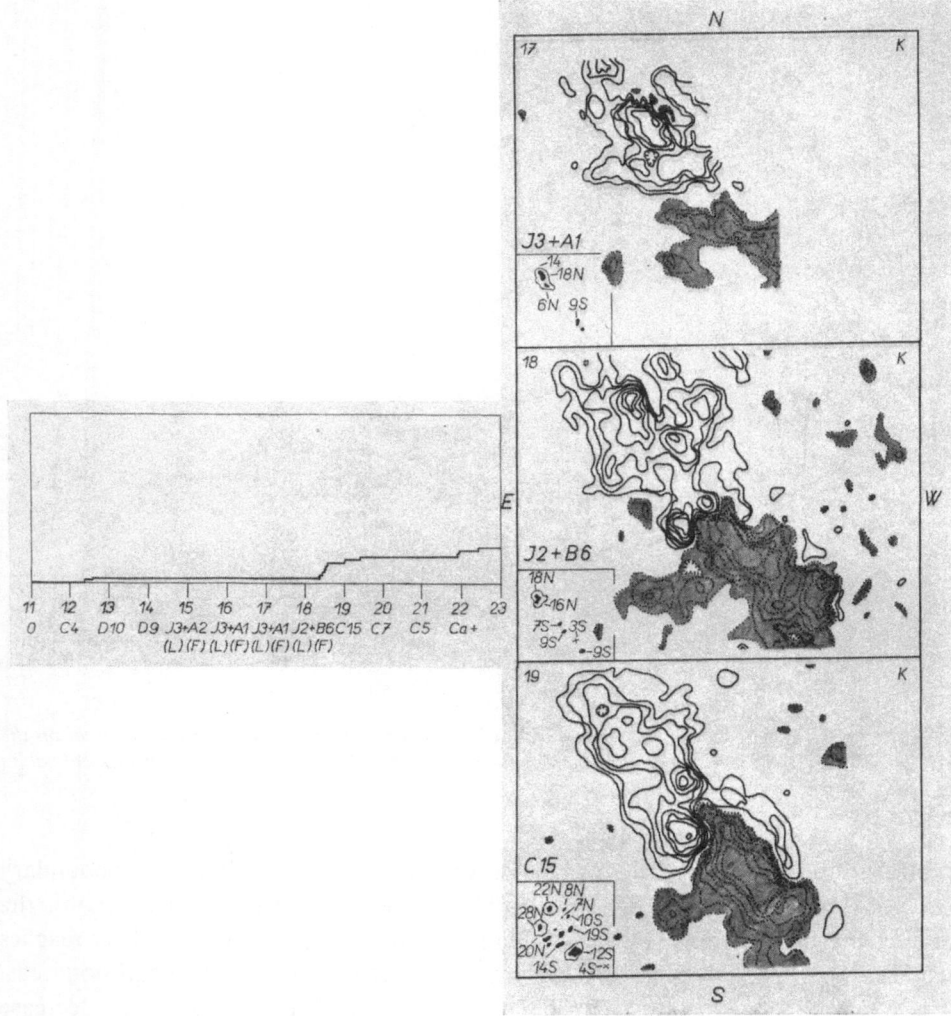

FIG. 4.  *A series of Crimean magnetic maps for AR No 1/1492 during the days of larger flare activity is shown. On the left part the flare-activity curve for the group is given. The simple magnetic situation on March 17 with no flare activity, and the complication of the magnetic situation at the boundary of the polarities accompanied by increased flare activity on March 18 and 19 is evident.*

factor influencing the slope of the curves is for a given part of the curve the relatively regular frequency of small flares of importance 1 and $1^+$.

If we try to give some explanation to the fact of the flare-activity rate 'quantization', we may point to the existing hierarchy in the sizes of the elements of the magnetic-field distribution in the solar atmosphere depending on the hierarchy of the dynamical elements of the atmosphere. This means that there exist tubes of magnetic lines of force with certain relatively constant values of their diameters given probably by the conditions of stability in individual atmospheric dynamical elements. Certainly the consumption of such magnetic fields by flare activity will be accompanied by the dissipation of whole tubes of lines of force involved.

The future study of magnetic changes in the centre of spotgroups which seem to be connected with the reorganisation of the supergranular pattern may give us more information about the physics of such processes.

## References

Bumba, V., Howard, R. (1965)     *Astrophys. J.*, **141**, 1492.

Bumba, V., Howard, R., Martres, M.J., Soru-Iscovici, I. (1968a)     in the present volume, p. 13.

Bumba, V., Kleczek, J., Olmr, J., Růžičková-Topolová, B., Sýkora, J. (1968b)     in the present volume, p. 64.

Kleczek, J. (1952)     *Publ. de l'Institut Central d'Astronomie*, Praha, **22**, 12.

# A STUDY OF THE LOCALIZATION OF FLARES
# IN SELECTED ACTIVE REGIONS

M. J. MARTRES, R. MICHARD, I. SORU-ISCOVICI, T. TSAP

*(Observatoire de Paris-Meudon, France and Crimean Astrophysical Observatory Nauchny, Crimea, U.S.S.R.)*

ABSTRACT

From the material gathered during the 'Cooperative Study of Solar Active Regions', we studied the flare locations in AR magnetic structure, and flare relations to changes in the magnetic fields and spot configurations. Besides a confirmation of previous results, we find that flares are often associated with two features of the spot configuration evolving in opposite senses, one growing, the other declining.

Previous work has shown that flares tend to occur at privileged locations in the magnetic configuration of solar AR. Severny (1958, 1963) pointed out that most flares start near the line of separation of polarities of the longitudinal field, specially if the horizontal 'gradient' of $H_\parallel$ is strong. It was soon recognized that this also means a strong transversal field (Michard *et al.*, 1961). Later on Severny emphasized the significance of points of 'bifurcation' of lines of force as a seat of flares (Severny, 1964). Finally Martres *et al.* (1966) pointed out that in case a flare is made up of two (or more) bright features these seem to appear at places of opposite polarities.

As flares show preferred locations in *space*, they also show preferred locations in *time*, being obviously associated with AR which are changing rapidly from the point of view of spot area (Giovanelli, 1939), spot configuration (Gopasyuk *et al.*, 1963) and therefore also magnetic structure. It seems that the pre-flare magnetohydro-dynamic configuration in the chromosphere and corona is not an unstable one, but has to be destroyed by a finite change in the photospheric field: in other words, it is a metastable configuration.

Accordingly the 'location' of flares should be studied simultaneously from both an historical and topographical point of view; a rather difficult task, which has not been attempted so far.

Some of the AR selected for the CSSAR project had a fair production of flares. Thanks to the generous cooperation of the participating observatories, it was possible to collect in Meudon:

(1) a continuous sunspot history with about one picture per hour,

(2) a nearly continuous Hα cinematographic record for days of significant flare activity,

(3) fairly complete series of calcium spectroheliograms,

(4) a number of detailed magnetic maps of AR from K.A.O. and Meudon, which, although significantly larger than hitherto achieved, proved still insufficient in many cases.

The study of this material from the point of view of flare location in space and time was started in Meudon rather recently, and only preliminary results can be given here. Various approaches to the problem were tried in succession.

*1st approach:*

This consisted in comparing the Hα pictures of flares with the magnetic maps obtained on the same day (and as close as possible in time), in order to find the location of the starting-points of flares. In this comparison, a necessary step is to locate correctly *sunspots* on both the magnetic and the Hα pictures: errors of 5 to 10″ are not unusual, specially when the time interval of both observations exceeds 2 hours.

Such comparisons were performed for 36 flares of regions CSSAR 4, 8, 9 and 10, and a statistic was made of the distances to the inversion lines of $H_{\parallel}$ of 154 bright points associated with the flares: in 40% of the cases these points are practically on the inversion line, while in 67% their distance is $\leqslant 10''$. We found two exceptional cases, two flares which contained only a single bright point far away from any inversion line. However, the magnetic map had been obtained 6 to 7 hours before the flares: the comparison with the magnetic data of the following day showed that a new inversion line had developed at some time between the two observations, possibly before the flares themselves.

The tendencies for flares to occur at places of strong horizontal gradient of $H_{\parallel}$ was confirmed. Also it was confirmed that if more than one bright point occurs (31 cases out of 36) they are distributed on both sides of the inversion line: exceptions to this rule were few and occurred when the time interval between the magnetic observation and the flare was large.

Examples of flare locations are given in Figure 1.

*2nd approach:*

An attempt was made to trace quantitative changes of the fields and to relate these to flare occurrence. For this purpose, the integrals $F_N$ and $F_S$ were calculated.

We have

$$F_N = \int_A H_{mN} \, dS,$$

where $H_{mN}$ is the *measured* longitudinal field of N polarity, $S$ is the element of area of the solar photosphere, and $A$ is the explored part of the Sun. Further we calculated the indices

$$\Sigma = |F_N| + |F_S| \quad \text{and} \quad k = \frac{|F_N| - |F_S|}{\Sigma}.$$

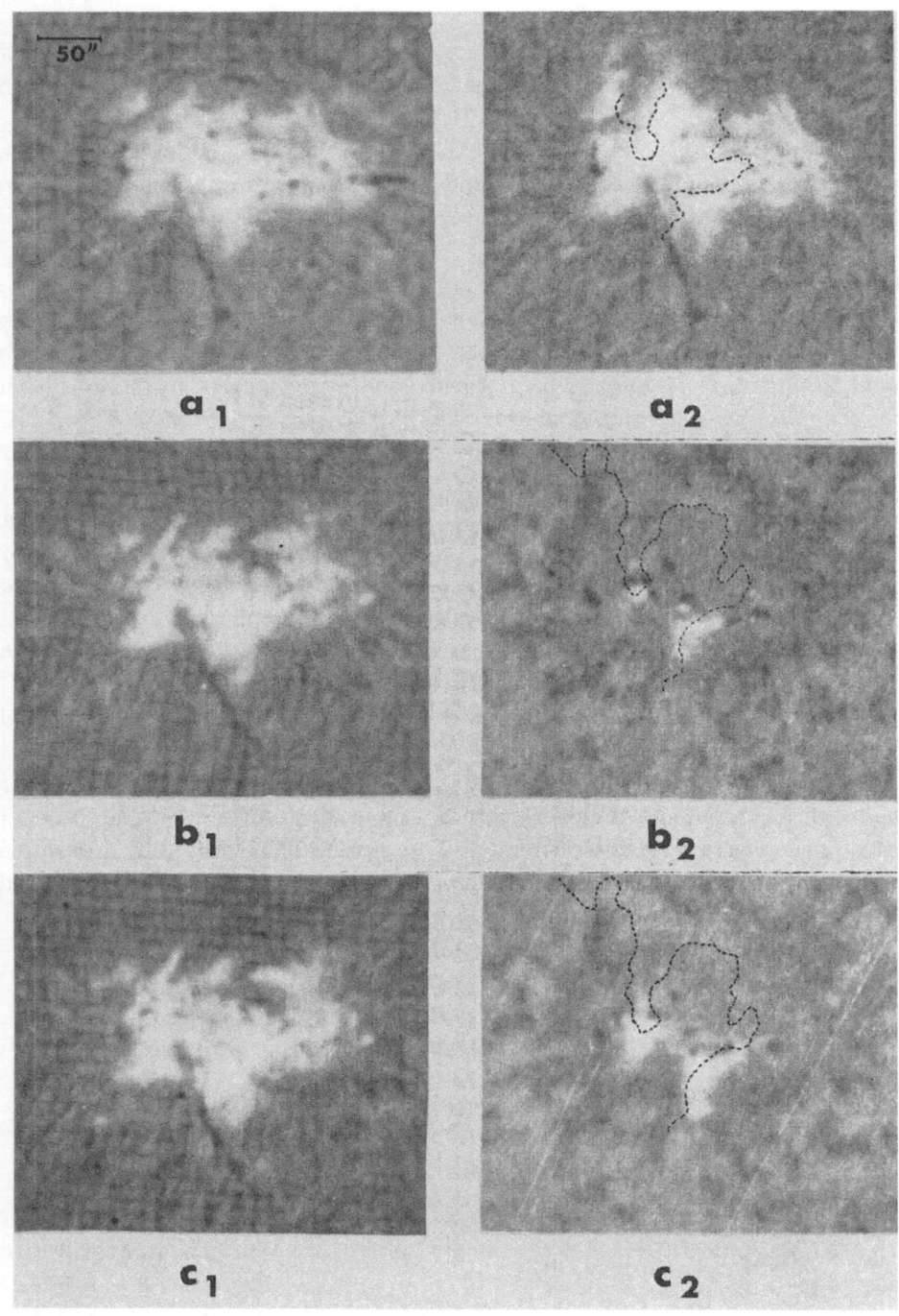

In cases where the AR is at the centre of the disk, where the measured field equals the longitudinal field and *where A is sufficiently large*, we could more or less identify $F_N - F_S$ with the flux of **H** through a closed surface and expect it to be zero. These three conditions are never fulfilled: the calculated quantities $\Sigma$ and $k$ are nothing more than possibly useful indices of total field strength and distribution.

Figure 2 gives the variation of $\Sigma$ and $k$ for the region CSSAR 9 between September 28 and October 5, 1965. Homogeneous magnetic data from K.A.O. was available and could be supplemented by Meudon data using an empirical relation between $H_m$ (K.A.O.) and $H_m$ (Meudon). It seems that flare occurrence was more frequent when the index $\Sigma$ was large and the index of 'unbalance' $k$ rapidly changing (see other comments in figure caption).

On the other hand the same indices where evaluated for these cases where two magnetic maps were available, one before, the other after a flare (interval 2–5 hours). Six such cases were found in the CSSAR material and two more were added from publications by Severny (1960) and Gopasyuk *et al.* (1963).

In this material there were indications that $\Sigma$ decreased after flares: 6 cases out of 8, and that $k$ changed, indifferently decreasing (3 cases) or increasing (5 cases).

The significance of these results will have to be tested by the study of more and better observations, and by objective evaluations of errors in the determination of the used indices. We believe that in about half the 8 cases under study, the changes of $\Sigma$ and $k$ following a flare did not exceed possible errors in the evaluation of these parameters, so that we cannot conclude.

*3rd approach:*

In this case it was tried to relate flare occurrence to the geometry and evolution of sunspot groups. With the CSSAR material it is possible to follow sunspot evolution in a fairly continuous way. Sunspots represent the distribution of peaks of the total field strength (not the longitudinal component). Their magnetic polarities are often known from patrol visual observations, even when no detailed field maps are available. Therefore good sunspots observations are a very useful substitute to the present incomplete magnetic observations.

The study of an AR like CSSAR 7 showed that new sunspots are continuously born

---

FIG. 1. *Example of location of flares relative to the inversion line of longitudinal field: region CSSAR 9, October 2, 1965. In each case a pre-flare Hα picture (1) and a picture during the flare (2) are given. Case (a): Magnetic map from K.A.O. obtained 0735–1000 UT. Hα images from Culgoora at 0350 and 0413. Case (b): Magnetic map from K.A.O. obtained 1230–1410 UT. Hα images from Sacramento Peak at 1338 and 1414 (off-band picture). Case (c): Same magnetic data as in (b). Hα images from Sac. Peak at 1555 and 1620 (off-band picture). – Note that the fit of flare knots and filaments to the inversion line is the best when the time interval between magnetic observations and flare is the shortest.*

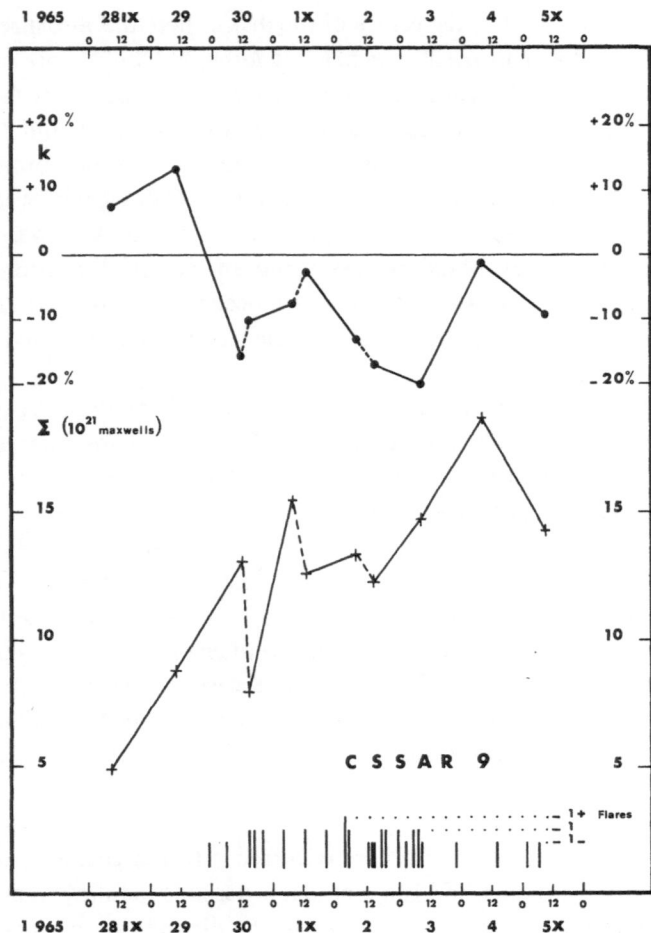

FIG. 2. *The run of magnetic flux indices Σ and k (see text) for AR CSSAR 9 (September 28, October 5, 1965) as derived from 9 magnetic maps from K.A.O. and 2 from Meudon. These indices are corrected for foreshortening on the assumption that they are proportional to cos θ. We observe a general increase of Σ in agreement with the increasing size and spots number of the region, and irregular variations of k: the most prominent is a change of sign between September 29 and 30, the N polarity showing the larger flux before and the smaller after. This change may be related to the birth of new important spots in the region.*

and dying during the history of the group. Therefore the concept of 'evolutionary features' was introduced: an 'evolutionary feature' (EF) is a sunspot or aggregate of spots inside the group, with well-defined birth and disappearance. Except for small relative motions it keeps its location in the overall structure. Figure 3 shows a summary of the history of sunspot group CSSAR 7.

The study of flare location relative to spotted EF in CSSAR 7 during its rise to maximum gave the following indications (see Figure 4).

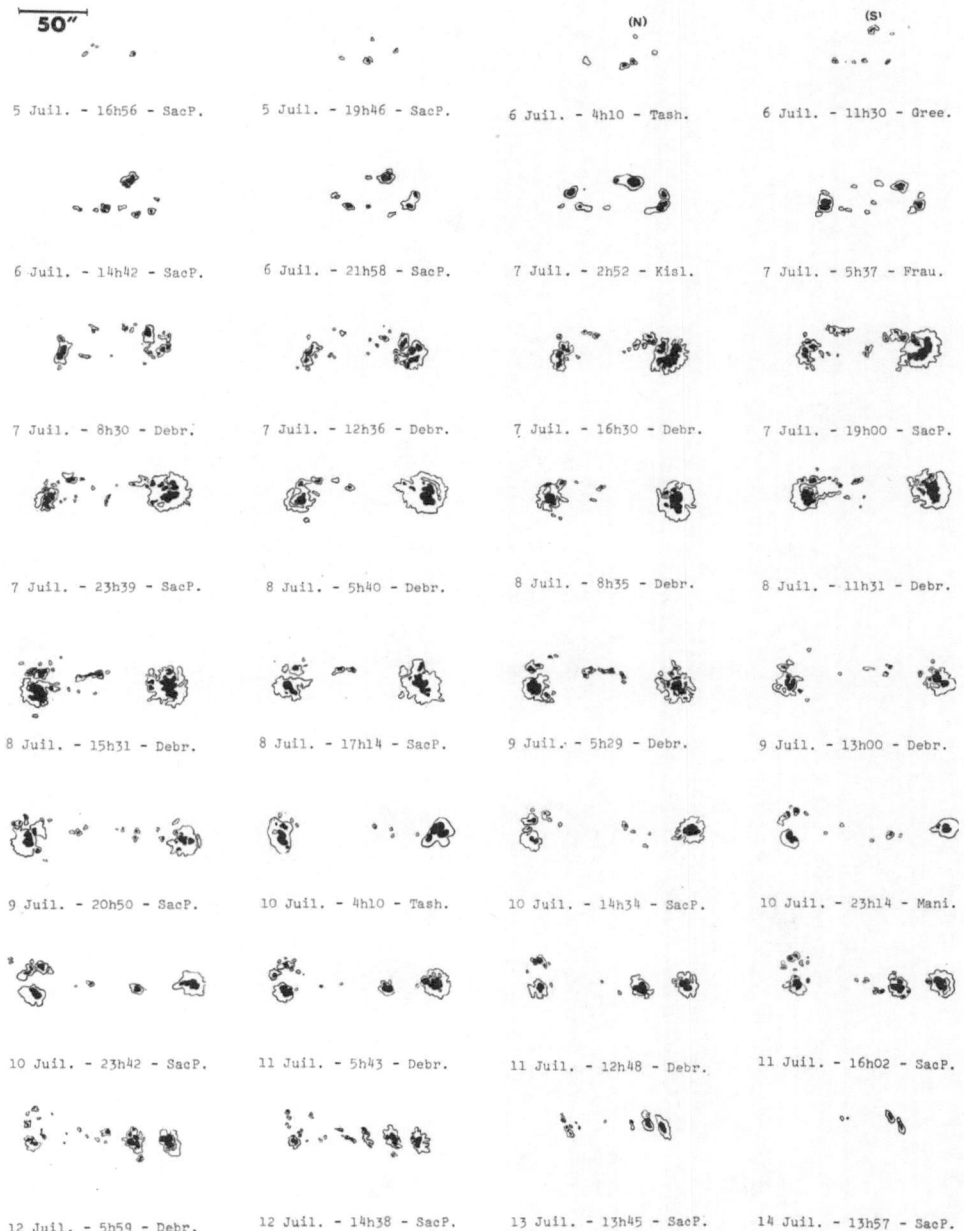

FIG. 3. *History of sunspot group associated with CSSAR 7 summarized by 32 sketches (or 1/6 of the total number of drawings used in the present study!). The names of observatories contributing the original photographs are abbreviated by their first four letters. East is at the left, North at top. – Note the frequent formation, evolution and disappearance of minor spots, or spots aggregates referred to in text as 'evolutionary features'.*

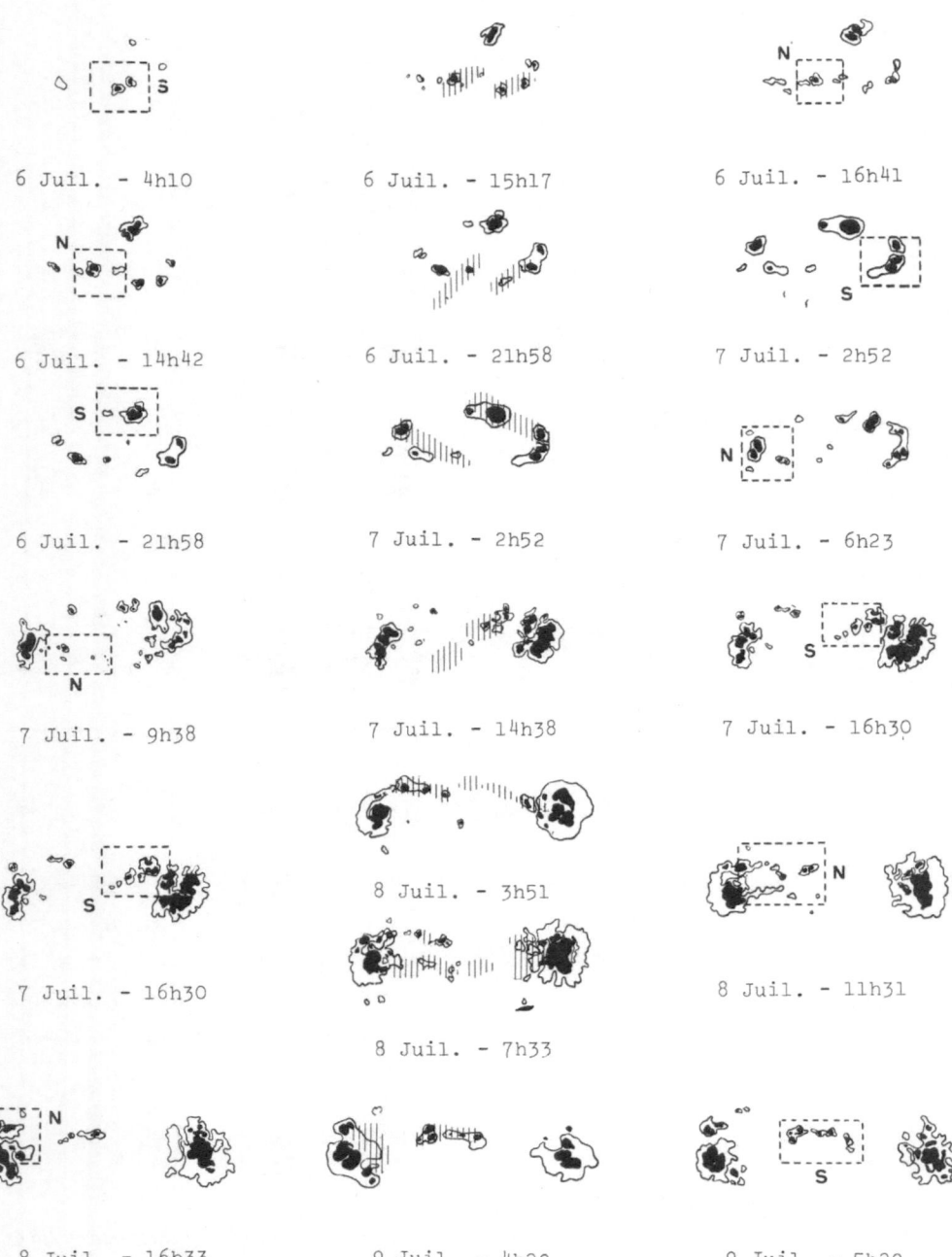

FIG. 4. *Sketches of 7 flares occurring in region CSSAR 7 (central column) in association with sunspot structure some time before (left) and after the flare (right). Each flare knot seems associated with a distinct spotted EF, each of opposite polarity and opposite sense of evolution. The involved features in each case are enclosed in a dashed square, the decreasing EF on the left pre-flare drawing, and the increasing EF on the right post-flare drawing.*

(1) Flares were made up of two brightenings, each covering a different EF.

(2) The two spotted EF involved in flares were of opposite magnetic polarities, a result in agreement with our previous finding on the distribution of flare bright knots on both sides of an inversion line.

(3) *Of two spotted EF involved in a flare one was increasing in area and the other decreasing.* This interesting new result may be related to our finding of variations in the index $k$ of unbalance of polarities. It emphasizes the probable relation of flares with *finite changes* of the magnetic structure. One might visualize the pre-flare situation as a system of coronal particles trapped in the tube of force connecting two poles: if the strengths of these two poles varies in opposite directions, the trap might eventually collapse and large precipitation phenomena, i.e. flares, will occur.

When the region CSSAR 7 had reached its maximum, flares began to occur outside spots; the brightenings were located on opposite polarities close to the inversion line of $H_{\parallel}$, but the available magnetic maps did not permit to evaluate variations of these polarities in order to check the above result (3).

It should be noted that the location of the bright details of the Hα plage (even outside flares) is closely related to sunspot evolution, at least in the pre-maximum phase of the AR. Clear changes of the Hα structure are *preceded* by the appearance of a new EF in the spotgroup. Since the brightest features in Hα have the same relationships to magnetic structure as flares proper, and also are the seat of flares, it is tempting to suggest that they are due to a permanent precipitation process of which the flare is a spectacular enhancement.

## References

Giovanelli, R. (1939)  *Astrophys. J.*, **89**, 555.
Gopasyuk, S., Ogir, M., Severny, A., Shaposhnikova, E. (1963)  *Izv. krymsk. Astrofiz. Obs.*, **29**, 15.
Martres, M.-J., Michard, R., Soru-Iscovici, J. (1966)  *Ann. Astrophys.*, **29**, 245.
Michard, R., Mouradian, Z., Semel, M. (1961)  *Ann. Astrophys.*, **24**, 54.
Severny, A. (1958)  *Izv. krymsk. Astrofiz. Obs.*, **20**, 22.
Severny, A. (1960)  *Izv. krymsk. Astrofiz. Obs.*, **22**, 12.
Severny, A. (1963)  *Izv. krymsk. Astrofiz. Obs.*, **30**, 161.
Severny, A. (1964)  *Izv. krymsk. Astrofiz. Obs.*, **31**, 159.

# EVOLUTION OF CA PLAGES OF THE CSSAR ACTIVE REGIONS

G. GODOLI          and          B. C. MONSIGNORI FOSSI

*(Astrophysical Observatory,*          *(Astrophysical Observatory,*
*Catania, Italy)*                       *Arcetri, Italy)*

ABSTRACT

For each plage of the CSSAR period the daily projected areas have been determined and plotted versus time. Also the correction for foreshortening has been taken into account.

Daily fluctuations, of the plage corrected areas, certainly greater than the errors, have been found. Also area fluctuations of shorter periods have been observed.

## 1. Material Available

During the IGY, IGC, CIG, IQSY campaigns, the World Data Center (WDC) C at the Arcetri Astrophysical Observatory collected and reduced material concerning Ca plages.

On the basis of the material collected for the IGY and IQSY campaigns a basic set was built as homogeneous as possible of at least one negative $K_{2,3,2}$ Ca heliogram per day with the diameter of the solar image of 63 mm (the size of the Arcetri spectroheliograms).

From this set the daily maps already published were deduced (Godoli, 1961a, b, 1962, 1965, 1966). The maps concerning the IGY period were also included in the IGY solar activity maps $D_1$ (Ellison, 1961).

Therefore for the CSSAR period the Arcetri WDC already had at its disposal at least one $K_{2,3,2}$ heliogram per day. This material was sufficient for the study of the first CSSAR problem concerning the general evolution of Active Regions (AR).

A standardized file of at least one $K_{2,3,2}$ heliogram per day is now available at the Meudon Observatory for the period March 11 – December 1, 1965.

For the second problem concerning the birth and fast changes of AR during the first transit we have asked the observatories of Abastumani, Catania, Crimea, Fraunhofer, Ikomasan, Kodaikanal, Manila, McMath Hulbert, Meudon, Mount Wilson, Rome, Tokyo, Wendelstein, Zürich to send all the $K_{2,3,2}$ material available for the periods:

| | |
|---|---|
| 11 March – 22 March | 1 July – 4 July |
| 14 May – 26 May | 5 July – 14 July |
| 16 May – 27 May | 2 September – 13 September |

*Kiepenheuer (ed.), Structure and Development of Solar Active Regions, 326–337. © I.A.U.*

| 1 June | – 9 June | 27 September – | 8 October |
| 16 June | – 22 June | 1 October | – 6 October |

Every collaborating observatory has been asked to send to Arcetri negative prints or film (diameter of the solar image 63 mm) or heliograms in the size and photographic presentation available.

## 2. General Evolution

For the study of the general evolution of each CSSAR Ca plage recognized in its successive transits in the visible hemisphere, we plotted versus time the daily projected areas $Ap$ and corrected areas $Ac$, taking into account the maximum deviations of the measurements.

The $Ap$ areas determined on the maps drawn by contact on transparent paper directly from the heliograms and expressed in $10^{-4}E$ ($E$ indicates the area of the solar hemisphere).

The drawing of the plage maps was the most difficult phase of the work: as is well known, the plages outlines are often disjointed and mixed with the chromospheric granulation; moreover, it is necessary to keep constant the examination of the brightness limit under which the phenomena are neglected. The estimation becomes complicated, moreover, by the non-linearity of the curve expressing the calibration of the photographic plate.

As far as the reproducibility of the plage maps is concerned, it was found that, if independent drawings were made by two different capable draftsmen using homogeneous heliograms, the differences of the areas within the plage outlines did not exceed 10%. If the drawings were made by the same draftsman using inhomogeneous heliograms (made with different apparatus) the differences might be much higher. Therefore we choose for the basic set, as far as possible, heliograms of the same quality. We also tried to eliminate by interpolation some obvious differences among heliograms of consecutive days.

The projected areas $Ap$ were transformed into corrected areas $Ac$ according to the well-known formula

$$Ac = Ap \sec h, \tag{1}$$

in which $h$ is the heliocentric angle of the plage.

We remember that the correction for foreshortening according to Equation (1) is based on the assumption that the effective plage layer is bidimensional and normal to the Sun's radius and that the values of corrected areas $Ac$, for the same plage, do not depend on the distance from the centre of the disk. Since the plage area generally increases when the height on the photosphere of the effective layer increases, and therefore when $h$ increases (Anichini and Godoli, 1967), the second assumption is not fulfilled. This may be the reason why, as has been shown in a research recently carried out at Arcetri Observatory (Godoli and Monsignori Fossi, 1967a, b), the correction for foreshortening according to (1), although it is in fairly good agreement with the observations for 'great' plages with

$$Ac \geqslant 20 \times 10^{-4}E,$$

it is no longer acceptable for 'small' plages with

$$Ac < 20 \times 10^{-4}E.$$

The difference between these two classes of plages can be ascribed to a probable different slope of the $Ap(x)$ curve (with $x$ indicating the height on the photosphere).

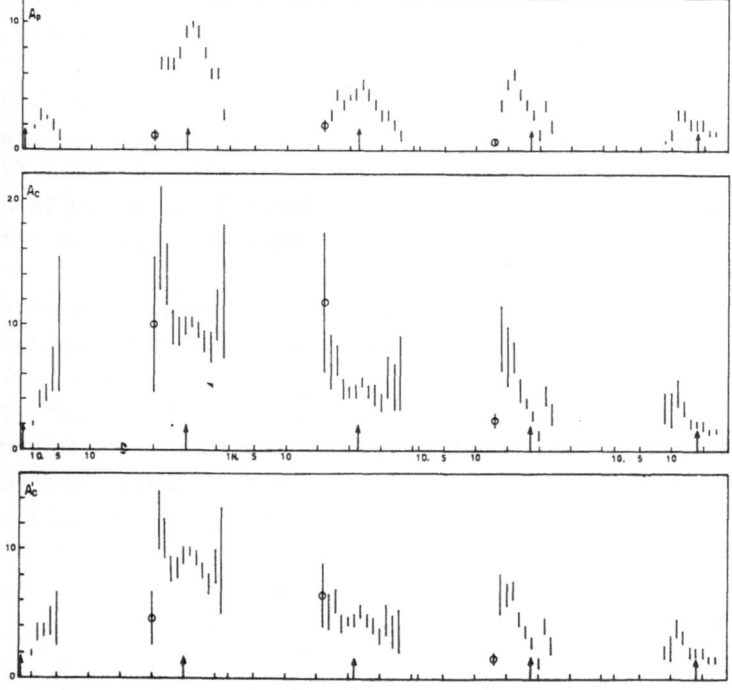

FIG. 1.   *Evolution of the small plage: 1964 Arcetri 18 (Godoli and Monsignori Fossi, 1967a).*

In Figure 1 the evolution of a plage of the second class is shown. It is evident that correction (1) is not satisfactory. The $A'c$ values, obtained by a statistical method, are more satisfactory (Godoli and Monsignori Fossi, 1967a, b). We notice that CSSAR plages are generally of the first class and therefore correction for foreshortening according to (1) was accepted.

### 3. Birth and Fast Changes of AR

The $K_{2,3,2}$ material especially requested for the second problem and received from cooperating observatories was as specified in Table 1. Only variations with time scale of $1^h$ have been considered during the period listed above, in Section 1.A.

### Table 1

### Material received

| Observatory | Material | | Diameter (mm) |
|---|---|---|---|
| Abastumani | pos. copy | film | 20 |
| Arcetri | neg. copy | plate | 63 |
| Freiburg | orig. | plate | 45 |
| Kodaikanal | pos. copy | film | 19 |
| Locarno | orig. | plate | |
| Manila | orig. | film | 68 |
| McMath Hulbert | pos. copy | paper | 83 |
| Meudon | pos. copy | film | 64 |
| Mount Wilson | pos. copy | film | 84 |
| Rome | pos. copy | film | 14 |
| Tokyo | pos. copy | paper | 63 |

When possible, at least one heliogram per 2 hours was reported to negative plates with one or two photographic inversions. The diameter of the solar image was reported at 63 mm. This standardized file of heliograms is now available at the Meudon Observatory.

Actually, variations of the time scale of $1^h$ were pointed out from this file.

### 4. The Evolution Curves

In Figures 2–12 daily *Ap* and *Ac* values are plotted versus time for the $K_{2,3,2}$ Ca plages of the eleven AR selected for the CSSAR.

The maximum deviations of measurements are indicated. The circles indicate plages only partially visible on the disk. Arrows pointing from above indicate variations of the time scale of the hour. As already pointed out, these variations were studied only for the periods concerning the second problem. Arrows pointing from below indicate Central Meridian Passage.

The most interesting feature of these evolutive plottings are the daily variations of the areas clearly out of the limits of error.

*Plage 1 (30 N):*
Born on disk near the East limb.
Daily fluctuations are present during the second transit.

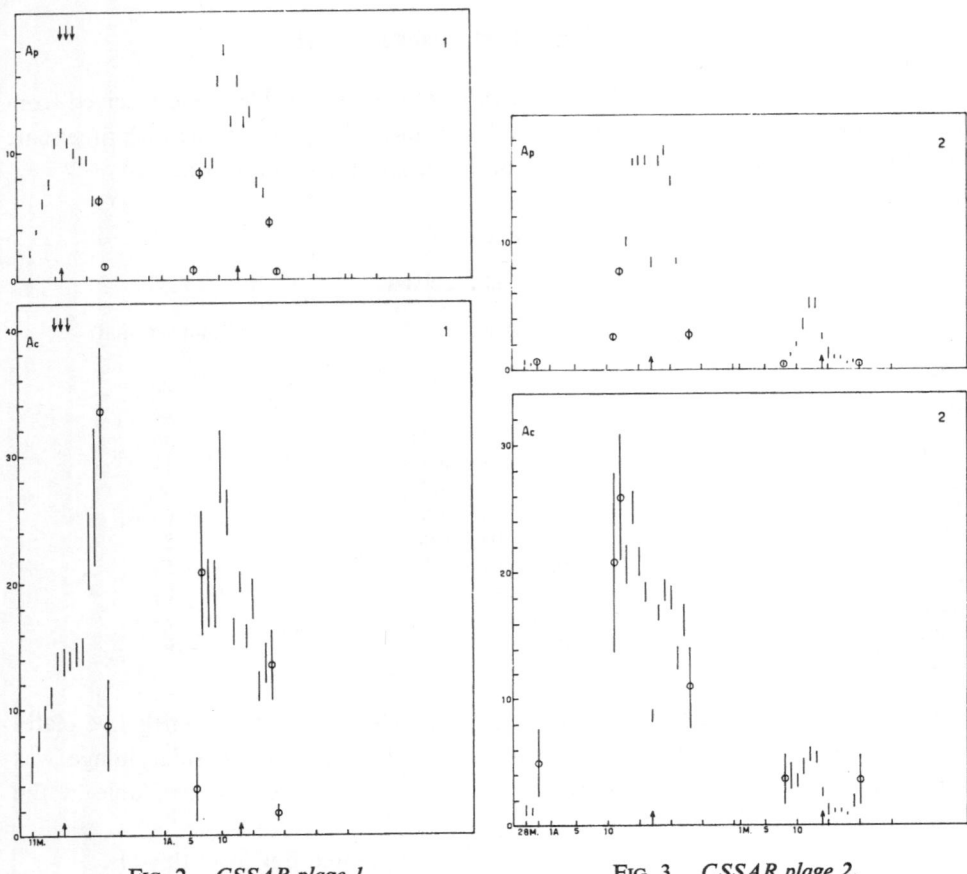

FIG. 2.   *CSSAR plage 1.*          FIG. 3.   *CSSAR plage 2.*

During the third transit the plage is still visible but under the compactness and brightness limit generally accepted.

Fast changes are present between

$$\text{March } 16^d \, 09^h \, 04^m - 16^d \, 15^h \, 05^m$$
$$16 \quad 15 \quad 05 \; - 16 \quad 23 \quad 33$$
$$17 \quad 10 \quad 08 \; - 17 \quad 23 \quad 36$$
$$17 \quad 23 \quad 36 \; - 18 \quad 00 \quad 54$$
$$18 \quad 08 \quad 24 \; - 18 \quad 15 \quad 10$$
$$18 \quad 15 \quad 10 \; - 18 \quad 23 \quad 32 \;.$$

*Plage 2 (4 N):*
   Born on disk.
   Daily fluctuations are present during the second transit.

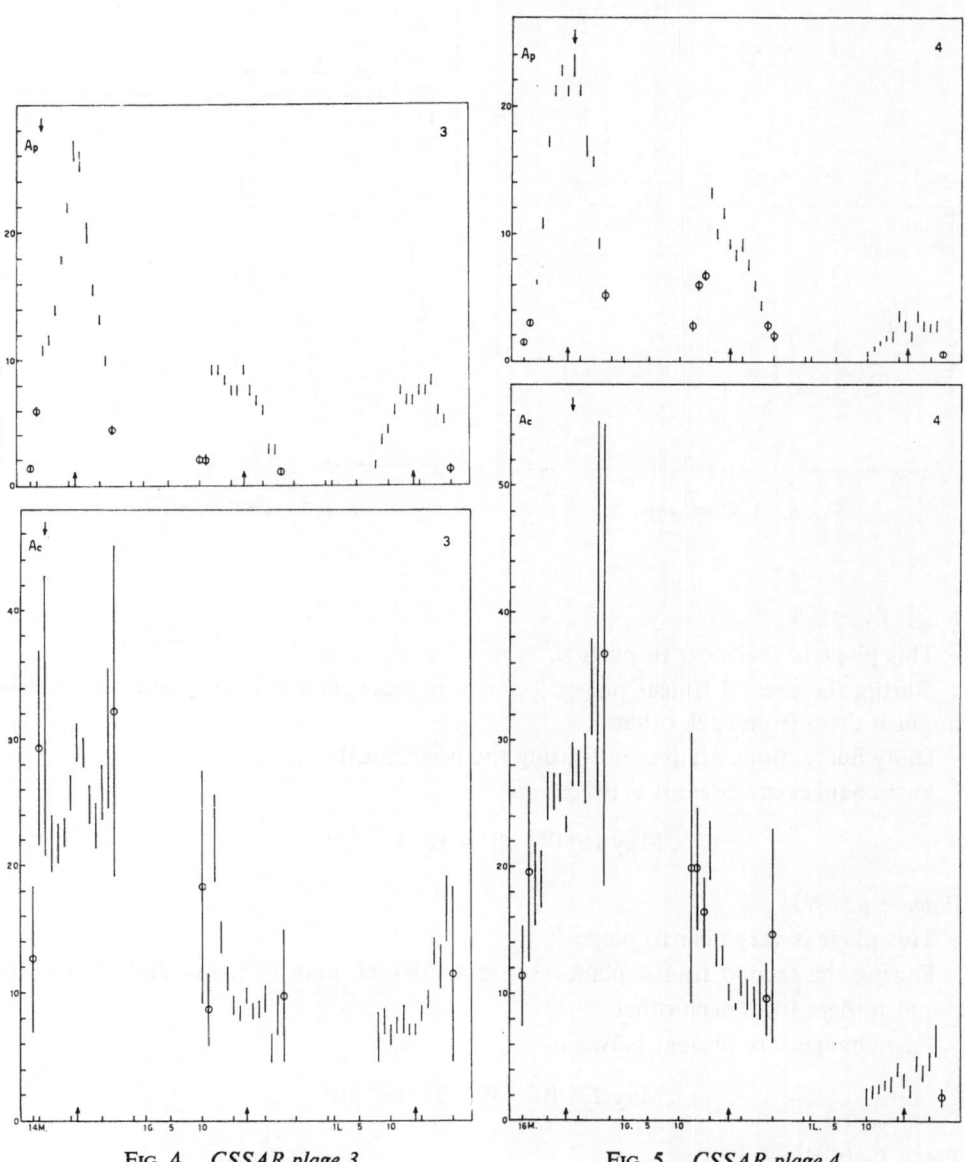

FIG. 4.    *CSSAR plage 3.*                    FIG. 5.    *CSSAR plage 4.*

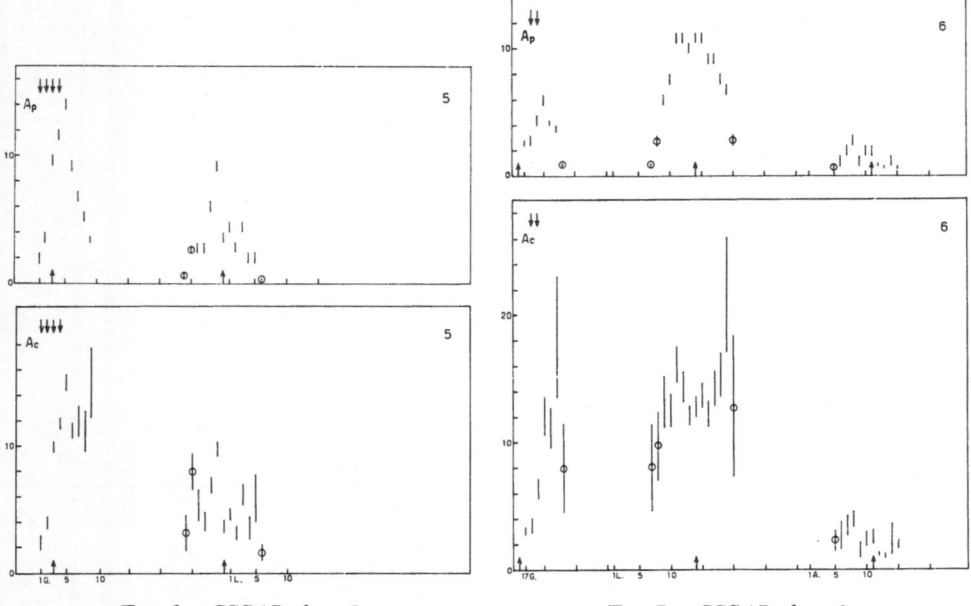

FIG. 6.   *CSSAR plage 5*.          FIG. 7.   *CSSAR plage 6*.

*Plage 3 (23 N):*

This plage is very near to plage 4.

During the second transit plages 3 and 4 interact, and it is very difficult to distinguish them from each other.

Daily fluctuations are present during the first transit.

Fast changes are present between

$$\text{May } 16^d\ 05^h\ 28^m - 16^d\ 13^h\ 53^m.$$

*Plage 4 (25 N):*

This plage is very near to plage 3.

During the second transit plages 3 and 4 interact, and it is very difficult to distinguish them from each other.

Fast changes are present between

$$\text{May } 23^d\ 10^h\ 45^m - 24^d\ 03^h\ 36^m.$$

*Plage 5 (10 S):*

Born on disk.

During the second transit the plage is not compact.

Fluctuations are present during the two transits.

Fast changes are present between

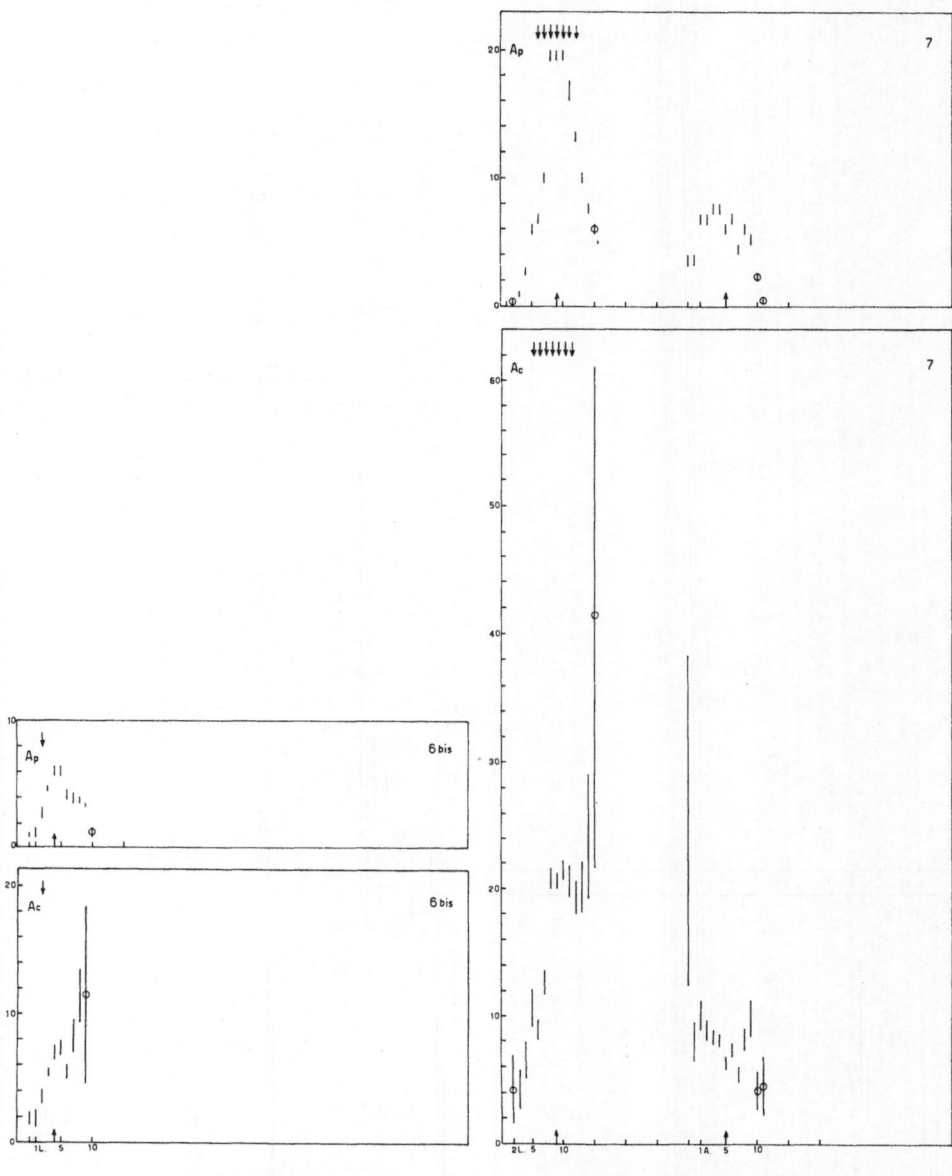

FIG. 8.   *CSSAR plage 6bis.*          FIG. 9.   *CSSAR plage 7.*

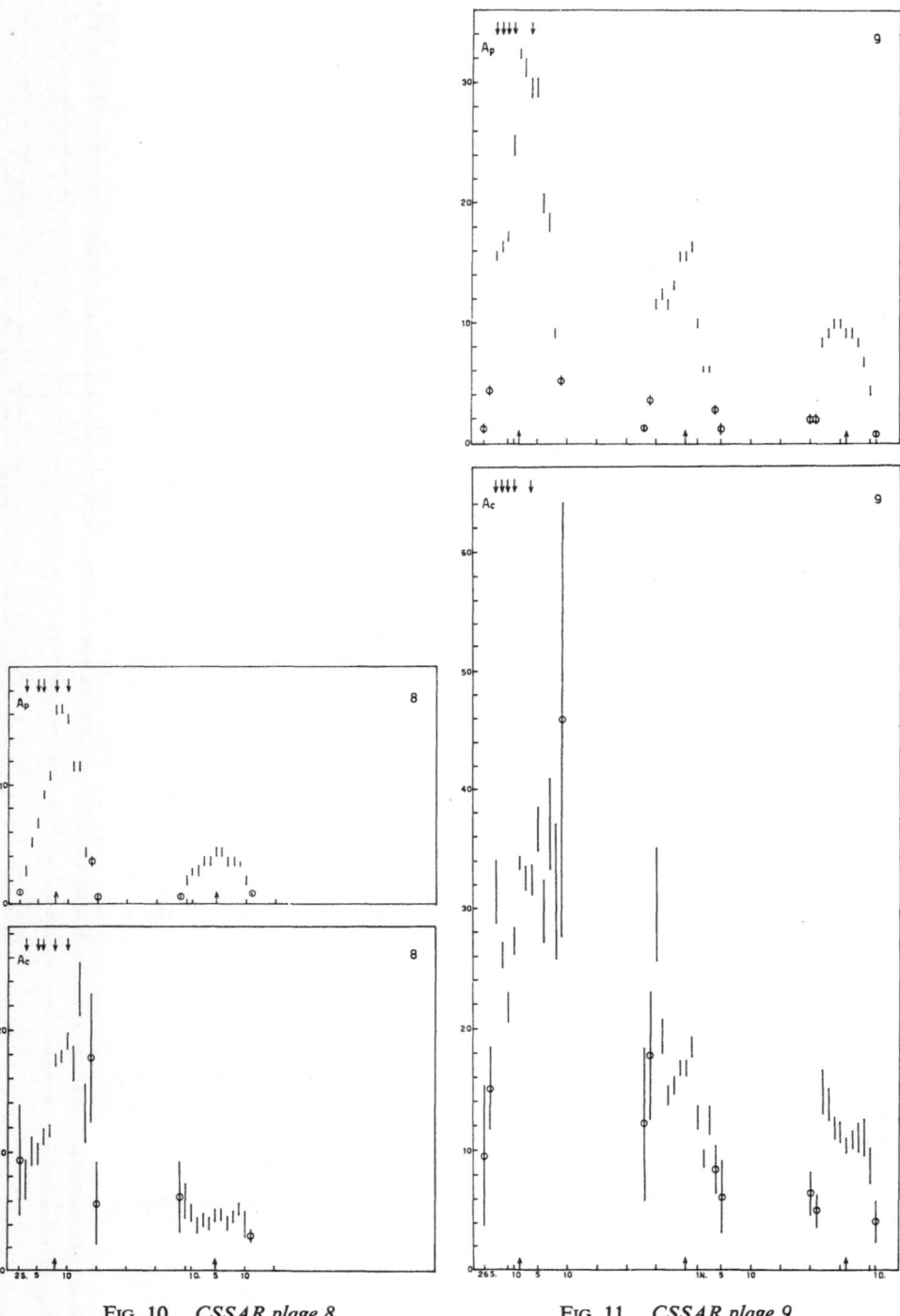

FIG. 10.    *CSSAR plage 8.*          FIG. 11.    *CSSAR plage 9.*

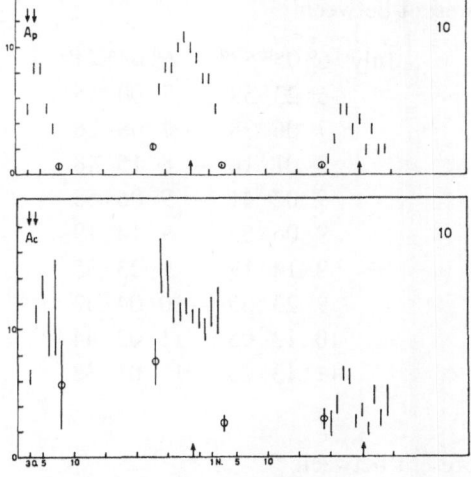

FIG. 12.   *CSSAR plage 10.*

$$
\begin{aligned}
&\text{June } 1^d\ 13^h\ 20^m - 1^d\ 16^h\ 15^m\\
&\quad\ 1\ \ 16\ \ 15\ -2\ \ 00\ \ 05\\
&\quad\ 2\ \ 09\ \ 02\ -2\ \ 14\ \ 41\\
&\quad\ 2\ \ 14\ \ 41\ -2\ \ 23\ \ 46\\
&\quad\ 2\ \ 23\ \ 46\ -3\ \ 02\ \ 22\\
&\quad\ 3\ \ 12\ \ 15\ -4\ \ 01\ \ 58\ \ .
\end{aligned}
$$

*Plage 6 (28 S):*

   Born on disk.

   During the third transit the region is disturbed by a new AR in the West side.

   Fluctuations are present during the second transit.

   Fast changes are present between

$$
\begin{aligned}
&\text{June } 17^d\ 12^h\ 30^m - 18^d\ 02^h\ 45^m\\
&\quad\ 18\ \ 05\ \ 43\ -18\ \ 09\ \ 17\\
&\quad\ 19\ \ 00\ \ 18\ -19\ \ 10\ \ 12.
\end{aligned}
$$

*Plage 6bis (32 N):*

   Born on disk.

   Fast changes are present between

$$
\begin{aligned}
&\text{July } 1^d\ 12^h\ 35^m - 2^d\ 00^h\ 29^m\\
&\quad\ 2\ \ 03\ \ 09\ -2\ \ 08\ \ 27\\
&\quad\ 2\ \ 08\ \ 27\ -2\ \ 23\ \ 31.
\end{aligned}
$$

*Plage 7 (20 N):*

   During the first four days the plage is not compact.

Fast changes are present between

$$
\begin{array}{llll}
\text{July} & 6^d\ 05^h\ 52^m\ - & 6^d\ 08^h\ 44^m \\
& 6\ \ 23\ \ 59\ - & 7\ \ 00\ \ 45 \\
& 7\ \ 06\ \ 38\ - & 7\ \ 08\ \ 26 \\
& 8\ \ 01\ \ 16\ - & 8\ \ 15\ \ 48 \\
& 9\ \ 02\ \ 41\ - & 9\ \ 06\ \ 55 \\
& 9\ \ 06\ \ 55\ - & 9\ \ 14\ \ 19 \\
& 9\ \ 14\ \ 19\ - & 9\ \ 23\ \ 35 \\
& 9\ \ 23\ \ 35\ - & 10\ \ 04\ \ 52 \\
& 10\ \ 13\ \ 05\ - & 11\ \ 02\ \ 44 \\
& 11\ \ 13\ \ 25\ - & 12\ \ 01\ \ 38\ \ .
\end{array}
$$

*Plage 8 (26 N):*

Fast changes are present between

$$
\begin{array}{llll}
\text{Sept.} & 3^d\ 02^h\ 01^m\ - & 3^d\ 14^h\ 00^m \\
& 5\ \ 00\ \ 40\ - & 5\ \ 02\ \ 53 \\
& 5\ \ 02\ \ 53\ - & 5\ \ 16\ \ 35 \\
& 5\ \ 16\ \ 35\ - & 6\ \ 00\ \ 00 \\
& 7\ \ 08\ \ 45\ - & 7\ \ 23\ \ 30 \\
& 10\ \ 03\ \ 19\ - & 10\ \ 10\ \ 54\ \ .
\end{array}
$$

*Plage 9 (20 N):*

Fluctuations are present during the first two transits.
Fast changes are present between

$$
\begin{array}{llll}
\text{Sept.} & 28^d\ 01^h\ 25^m\ - & 28^d\ 16^h\ 50^m \\
& 28\ \ 16\ \ 50\ - & 29\ \ 00\ \ 48 \\
& 29\ \ 06\ \ 13\ - & 29\ \ 23\ \ 31 \\
& 29\ \ 23\ \ 31\ - & 30\ \ 00\ \ 26 \\
& 30\ \ 09\ \ 39\ - & 30\ \ 23\ \ 28 \\
\text{Oct.} & 3\ \ 02\ \ 19\ - & 3\ \ 23\ \ 33\ \ .
\end{array}
$$

*Plage 10 (18 S):*

Born and died on disk.
Fast changes are present between

$$
\begin{array}{llll}
\text{Oct.} & 3^d\ 02^h\ 19^m\ - & 3^d\ 07^h\ 59^m \\
& 3\ \ 07\ \ 59\ - & 3\ \ 14\ \ 35 \\
& 3\ \ 14\ \ 35\ - & 3\ \ 23\ \ 33 \\
& 3\ \ 23\ \ 33\ - & 4\ \ 00\ \ 46 \\
& 4\ \ 00\ \ 46\ - & 4\ \ 08\ \ 44 \\
& 4\ \ 08\ \ 44\ - & 4\ \ 09\ \ 10\ \ .
\end{array}
$$

## References

Anichini, M., Godoli, G. (1967)    *Mem. Soc. astr. ital.*, **38**, 259.

Ellison, M. A. (ed.) (1961)    IGY Solar Activity maps $D_1$, *Annals of the International Geophysical Year*, **21**, Pergamon Press, New York.

Godoli, G. (1961*a*)    *Osserv. Mem. Oss. astrofis. Arcetri*, **73**.

Godoli, G. (1961*b*)    *ibid.*, **75**.

Godoli, G. (1962)    *ibid.*, **76**.

Godoli, G. (1965)    *ibid.*, **82**.

Godoli, G. (1966)    *ibid.*, **87**.

Godoli, G., Monsignori Fossi, B.C. (1967*a*)    *Lincei, Mem.*, Serie VIII, **8**, 85.

Godoli, G., Monsignori Fossi, B.C. (1967*b*)    *Solar Phys.*, **1**, 148.

# DISCUSSION

*Rigutti:* As Falciani, Righini Jr. and myself showed in Prague, if one estimates flare areas from contours judged by eye, one overestimates the areas of the limb flares with respect to the flare at the centre of the disk also for a factor of 1 or 2 orders of magnitude. If this were true also for plages, I wonder if your results could be affected by this uncertainty in fixing the areas.

*Godoli:* I do not think that the results found for flares near to the limb can be applied to plages: the flare contours must be generally estimated in a perturbed region, while the plage contours refer to the perturbed region as a whole. In *Mem. Soc. astr. ital.*, **38**, 75, 1967, you found that for plages near to the limb the area into contours estimated by eye is greater than the area into the isodense contours. I think that this fact can be explained, taking into account the instrumental limb darkening that is not taken into account with the isodense method you apply. Could this explanation also apply to your new flare measurements?

*Rigutti:* No, the measurements on flares to which I refer have been made in terms of the flare near the undisturbed chromosphere. So, any apparatus effect was the same for the flare region and for the background.

# CORRELATION BETWEEN CA PLAGES AND LONGITUDINAL MAGNETIC FIELDS OF THE CSSAR ACTIVE REGIONS

V. BUMBA and G. GODOLI

*(The Astronomical Institute of the*
*Czechoslovak Academy of Sciences,*
*Ondřejov, C.S.S.R.)*

*(Catania Astrophysical Observatory,*
*Italy)*

ABSTRACT

The shapes of $K_{2,3,2}$ Ca plages and longitudinal magnetic fields are compared for CSSAR active regions during their first transit on the solar disk.

The bipolar magnetic regions follow the Hale polarity law.

Often the region of inversion of the magnetic field corresponds to a gap in the Ca plage structure.

Bright patches of plages may coincide with magnetic inclusions, magnetic hills, and occasionally also with regions of inversion of the magnetic field. The outline of Ca plages follow well the isogauss of 20–40 oersted.

Histograms of the distances of individual magnetic field intensity peaks do not only correspond to the geometry of the supergranular network but also seem to indicate a difference in the organization of these peaks between the leading and following polarities.

## 1. Introduction

The existence of a high correlation between Ca plages and longitudinal magnetic fields is well established (Howard, 1959; Tsap, 1967).

Notwithstanding that, we have found it interesting to perform a comparison for the CSSAR Active Regions (AR) between Ca plages and longitudinal magnetic fields using the observational material collected at the Meudon Solar Service.

(a) For the Ca plages we had at our disposal a standardized file of at least one negative $K_{2,3,2}$ heliogram per day, with a solar image diameter of 63 mm (the size of the Arcetri spectroheliograms), available at the Meudon Observatory for the CSSAR period 1965, March 11 – December 1 (Godoli and Monsignori Fossi, 1968).

(b) For the magnetic fields we had at our disposal a standardized file of Crimea or Meudon magnetic maps with the scale 1 mm = 1″ also available at the Meudon Observatory.

The agreement between Crimea and Meudon magnetic maps is fairly good. The time of observation of the magnetic maps are reported in Table 1.

Unfortunately, all the Crimea and Meudon magnetic material refers only to the first transit of the AR.

We also had at our disposal Mount Wilson small-scale magnetic maps.

*Kiepenheuer (ed.), Structure and Development of Solar Active Regions, 338–345.* © *I.A.U.*

## Table 1

### Magnetic material available from Crimea and Meudon

| AR | Date 1965 | Time of observation | Observatory | AR | Date 1965 | Time of observation | Observatory |
|---|---|---|---|---|---|---|---|
| 1 | March 17 | 07 30 – 10 25 | C | 7 | July 12 | 10 25 | M |
|  | 18 | 12 05 – 13 10 | C |  | 12 | 12 00 | M |
|  | 18 | 13 15 – 14 21 | C |  |  |  |  |
|  | 19 | 06 40 – 07 50 | C | 8 | Sept. 04 | 11 55 – 13 45 | C |
|  | 19 | 10 00 – 11 15 | C |  | 05 | 05 37 – 07 10 | C |
|  |  |  |  |  | 05 | 09 45 – 11 00 | C |
| 3 | May 16 | 08 10 – 09 53 | C |  | 06 | 05 35 – 07 55 | C |
|  | 19 | 10 40 – 11 22 | C |  | 06 | 08 20 – 11 00 | C |
|  | 20 | 08 05 | M |  | 06 | 08 20 – 11 00 |  |
|  | 21 | 08 55 – 10 00 | C |  | 07 | 10 40 – 14 30 | C |
|  | 22 | 07 50 – 09 30 | C |  | 10 | 09 00 – 10 50 | C |
|  | 23 | 17 45 | M |  | 10 | 07 10 | M |
|  | 24 | 13 57 | M |  | 11 | 07 50 – 10 45 | C |
| 4 | May 20 | 08 19 | M | 9 | Sept. 28 | 07 45 – 09 10 | C |
|  | 21 | 10 10 – 11 22 | C |  | 29 | 09 30 | M |
|  | 23 | 17 55 | M |  | 30 | 11 00 – 12 50 | C |
|  |  |  |  |  | 30 | 14 00 – 14 55 | C |
| 5 | June 02 | 15 05 | M |  |  |  |  |
|  | 04 | 13 00 – 14 20 | C |  | Oct. 01 | 07 00 – 08 55 | C |
|  | 04 | 13 30 | M |  | 01 | 12 00 – 13 50 | C |
|  | 05 | 12 00 – 12 45 | C |  | 02 | 07 35 – 10 00 | C |
|  | 05 | 08 07 | M |  | 02 | 12 30 – 14 10 | C |
|  | 06 | 12 55 – 13 40 | C |  | 03 | 07 30 | M |
|  | 08 | 14 53 | M |  | 04 | 06 25 – 09 00 | C |
|  |  |  |  |  | 04 | 09 35 | M |
| 6 | June 19 | 10 15 | M |  | 05 | 09 00 – 10 18 | C |
|  | 20 | 09 20 | M |  | 05 | 08 44 | M |
|  |  |  |  |  | 05 | 14 20 | M |
| 6bis | July 02 | 06 30 – 06 57 | C |  | 07 | 07 50 | M |
|  | 04 | 08 55 – 10 10 | C |  |  |  |  |
|  | 05 | 08 00 – 09 00 | C | 10 | Oct. 04 | 09 20 – 11 20 | C |
|  |  |  |  |  | 04 | 12 00 – 13 27 | C |
| 7 | July 09 | 05 00 – 15 30 | C |  | 04 | 09 27 | M |
|  | 10 | 08 10 – 09 10 | C |  | 05 | 10 18 – 10 50 | C |
|  | 11 | 12 00 | M |  | 05 | 08 24 | M |

## 2. Method of Comparison and General Results

With a suitable camera we projected the Ca heliograms on the Crimea and Meudon magnetic maps and performed a comparison between plages and magnetic-field structures.

Unfortunately this comparison could be made only for the first transit on the disk of the CSSAR AR.

A comparison between the area within the isogauss lines and the plage areas was made using the Crimea magnetic observations only. For all the eleven CSSAR AR the results of these comparisons are examined in Section 3.

Here we notice that the bipolar magnetic regions follow the Hale polarity law. For ARs 2 and 5 belonging to the old 19th cycle the preceding polarity in the Northern hemisphere is North, while for the ARs 1, 3, 4, 6, 6bis, 7, 8, 9, 10, belonging to the even 20th cycle, the preceding polarity in the Northern hemisphere is South.

Often the region of inversion of the magnetic field corresponds to a gap in the plage structure.

Bright patches of plages may coincide with magnetic inclusions, magnetic hills, and occasionally also with regions of inversion of the magnetic field.

The outlines of plage follow well the isogauss of 20–40 oersted.

## 3. Detailed Results of the Comparison

*AR no. 1:*

This is a Northern region of the 20th cycle, visible for two or three transits on the solar disk. It was born on the disk.

It is an extended bipolar region with small peripheral inclusions. The polarity of the preceding part is South.

On March 17 the region of inversion of the magnetic field corresponds to a gap in the plage structure.

On March 18 bright patches of the plage coincide with a region of inversion of the magnetic field and with a magnetic hill.

The isogauss line of 30 oersted follows the outlines of the plage well.

*AR no. 3:*

This is a Northern region of the 20th cycle, visible for three transits on the solar disk.

It is a bipolar region with peripheral inclusions. The polarity of the preceding part is South.

The region of inversion of the magnetic field corresponds to a gap on the plage structure.

On May 20 a bright patch of the plage coincides with an inclusion; bright patches of the plage on May 22 and 24 coincide with regions of inversion of the magnetic field.

The isogauss line of 30 oersted follows the outline of the plage well.

*AR no. 4:*

This is a Northern region of the 20th cycle, visible for three transits on the solar disk.

It is a bipolar region with very small inclusions. The polarity of the preceding part is South.

The region of inversion of the magnetic field corresponds to a gap on the plage structure.

On May 21 bright patches of the plage coincide with magnetic hill and an inclusion. The isogauss line of 30 oersted follows the outline of the plage well.

*AR no. 5:*

This is a Southern region of the 19th cycle, visible for two transits on the solar disk. It was born on the disk.

It is essentially a bipolar region with peripheral inclusions. The polarity of the preceding part is predominantly South, but in front of the South region there is a small North region.

The region of inversion of the magnetic field corresponds to a gap in the plage structure.

On June 5 bright patches of the plage coincide with a magnetic hill and an inclusion. The isogauss line of 40 oersted follows the outline of the plage well.

*AR no. 6:*

This is a Southern region of the 20th cycle, visible for three transits on the solar disk. It was born on the disk.

It is a small bipolar region with the two magnetic polarities well separated. The polarity of the preceding part is North. An inclusion of North polarities in the South part is visible.

A bright point of the plage coincides with a magnetic hill; another with an inclusion.

*AR no. 6bis:*

This is a Northern region of the 20th cycle, visible for one transit. It was born on the disk.

It is a small bipolar region with the two magnetic polarities well separated. The polarity of the preceding part is South.

The region of inversion of the magnetic field corresponds to a gap in the plage structure.

Bright patches of the plage coincide with magnetic hills. The isogauss line of 40 oersted follows the outline of the plage well.

*AR no. 7:*

This is a Northern region of the 20th cycle, visible for two transits on the solar disk.

It is a bipolar region. Excluding some very small inclusions, the two magnetic polarities are well separated. The polarity of the preceding part is South.

The region of inversion of the magnetic field corresponds to a gap in the plage structure.

The isogauss line of 20 oersted follows the outline of the plage well.

*AR no. 8:*

This is a Northern region of the 20th cycle, visible for two transits on the solar disk.

It is a bipolar region with the two magnetic polarities well separated. The polarity of the preceding part is South.

The region of inversion of one magnetic field corresponds to a gap in the plage structure.

The isogauss lines of 30–40 oersted follow the outline of the plage well. The increase of area of September 8–11 is not followed by a corresponding enlargement of the isogauss lines.

*AR no. 9:*

This is a Northern region of the 20th cycle, visible for three transits on the solar disk. It was born on the disk.

It is substantially a bipolar region, which shows a remarkable inclusion of South polarity during the period of September 28 – October 2; on October 5 a small inclusion of South polarity and a North one are present. The polarity of the preceding part is South.

Some parts of the region of inversion of the magnetic field correspond to gaps in the plage structure.

On September 29 and October 2 and 5 bright patches of the plage coincide with inclusions. It is remarkable that on October 3 and 4 bright patches are present in the place where on October 5 the inclusion will appear.

On September 30 and October 4 bright patches of the plage coincide with regions of inversion of the magnetic field.

The isogauss line of 20 oersted follows the outline of the plage well. The sharp increase of the areas within the isogauss between September 28 and 30 is not followed by a corresponding increase of the area of the plage.

*AR no. 10:*

This is a Southern region of the 20th cycle, visible for three transits on the solar disk. It was born on the disk on the NW side of an existing region.

It is a complex magnetic region. The polarity of the preceding part is North with remarkable inclusions of South polarities.

For October 4 the plage shows gaps coinciding with regions of inversions of the magnetic field. On October 5 these gaps are no longer visible, perhaps because of the nearness of the plage to the Sun's West limb.

Brightenings of the plage coincides with inclusions.

The isogauss line of 20 oersted follows the outline of the plage well.

## 4. The Histograms of Distances of Magnetic-Field Intensity Peaks

Looking at the magnetic maps of the CSSAR active regions we may see in the distribution of magnetic fields the very well pronounced characteristic structure of the supergranular network (see e.g. Figures 1 and 2). Therefore one may expect that from the measurements of distances of individual peaks of magnetic-field intensity in each active region, one will obtain the same characteristic values as were found from measurements of Ca plages (Rogerson, 1955; Simon and Leighton, 1964).

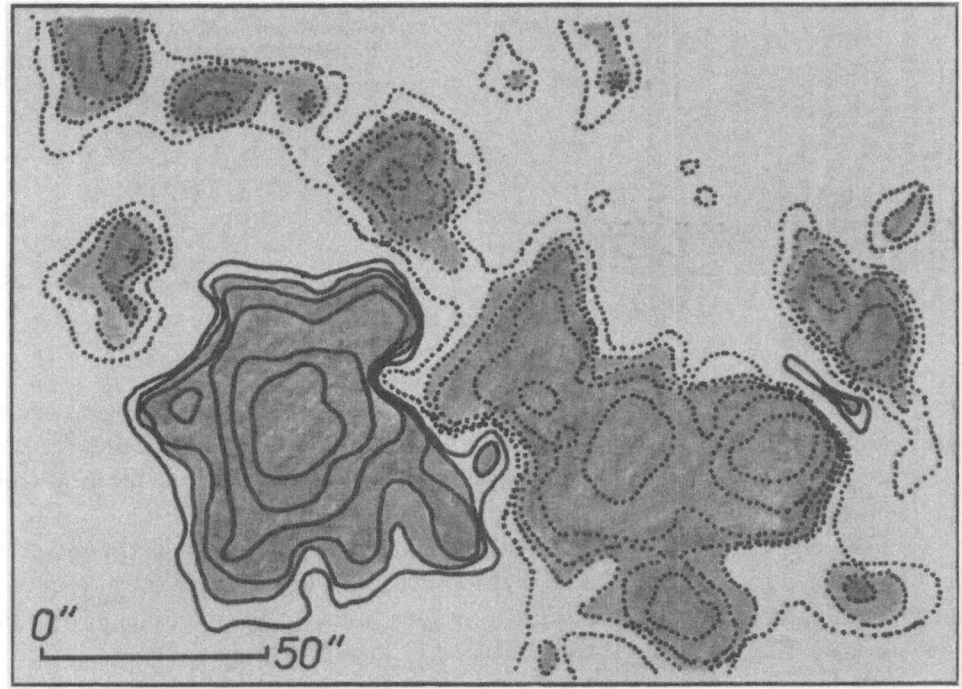

Fig. 1.   *Crimean magnetic map of the longitudinal component of the photospheric magnetic field for AR no. 7 (July 7, 1965). The positive polarity is indicated by solid lines, the negative polarity by dotted lines. Areas with intensity H ⩾ 20 gauss are shaded.*

For these measurements of distances between magnetic-field intensity peaks, the 38 Crimean magnetic maps were used. The distances were estimated between all peaks with intensity greater than 20 gauss – separately between plus-plus peaks, minus-minus peaks and plus-minus peaks. We took a value of 100″ as an upper limit for the distance. The total number of measurements exceeded 3000 individual values.

The main results in the form of histograms of measured values are presented in Figure 3. Because of the small number of individual peaks in some active regions, measured values obtained for smaller active regions are combined in one histogram (Figures 3a, d, g). The histograms of the largest active region, No. 9, are presented

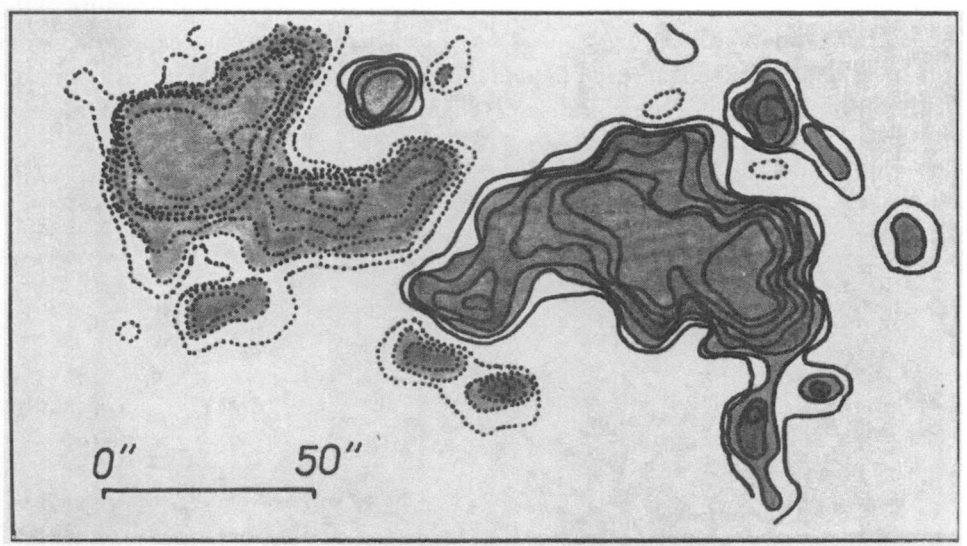

FIG. 2. *The magnetic map for AR no. 5 (June 4, 1965).*

separately as for region No. 10. Figures 3a, b, c show the distribution of distances between the peaks of the leading polarity, Figures 3d, e, f show the same for the following polarity, and Figures 3g, h, i, show the distances between the peaks of different polarities.

From Figure 3, it seems obvious that the character of the distribution of the distances of the leading polarity is different from that of the following polarity. The histograms of the leading polarity show two distinct maxima (even for the opposite polarity of AR no. 10 in the Southern hemisphere). The histograms for the following polarity are much broader with not such well-pronounced maxima.

The values of the characteristic distances of the leading polarity magnetic-field intensity peaks (30″–40″ and 60″–70″) coincide well with the values found for the supergranular network (Rogerson, 1955; Simon and Leighton, 1964), as well as with the values estimated for the distances of the feet of quiescent filaments (Sýkora, 1967). These facts may have some importance for the future study of differences between the roles of the leading and following polarities in solar activity development.

The smallest value of the distances between the magnetic peaks of the leading and following polarities which corresponds to about 30″–35″, also seems to be of interest.

## Acknowledgements

We are indebted to Dr. Michard, to Mrs. Martres and to the other colleagues of the Meudon Solar Service for their kindness and help during our stay at Meudon.

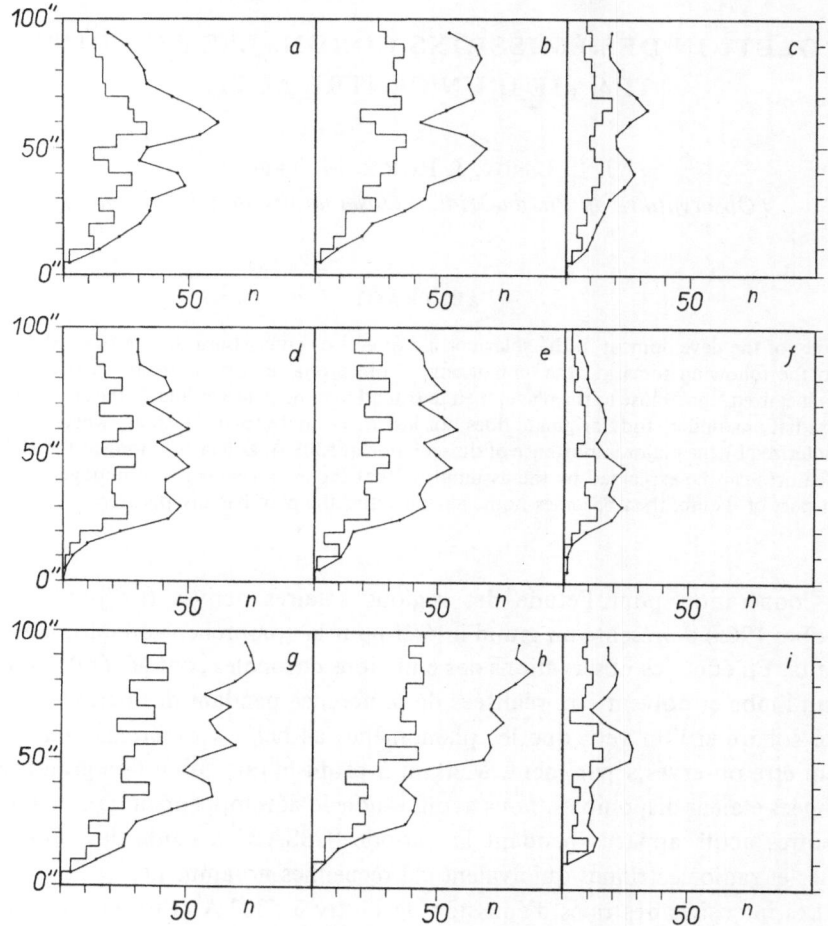

FIG. 3.   *Histograms of the distances between the peaks of the photospheric magnetic-field longitudinal component (H ⩾ 20 gauss) for CSSAR Crimean magnetic maps. On the vertical axis the distances are in seconds of arc; on the horizontal axis the frequency of individual values is given. The curves for the summation of values of distance in two intervals: 5″ and 10″ are drawn. The curves a, b, c represent the measurements of distances of peaks of the leading polarity (a) for ARs no. 1, 3, 4, 6bis, 7, 8, (b) for AR no. 9 all from the Northern hemisphere, (c) for AR no. 10 from the Southern hemisphere. The curves d, e, f show the same measurements for the following polarities respectively. The curves g, h, i are, in the same order, histograms of distances of peaks of different polarities.*

## References

Godoli, G., Monsignori Fossi, B.C. (1968)      in the present volume, p. 326.
Howard, R. (1959)      *Astrophys. J.*, **130**, 193.
Rogerson, Jr., J. (1955)      *Astrophys. J.*, **121**, 204.
Simon, G.W., Leighton, R.B. (1964)      *Astrophys. J.*, **140**, 1120.
Sýkora, J. (1967)      *Bull. astr. Obs. Csl.* (in print).
Tsap, T.T. (1967)      *Izv. krym. astrofiz. Obs.* (in print).

# ÉVOLUTION DES ÉMISSIONS CORONALES AU COURS DE LA VIE D'UN CENTRE ACTIF

J. L. LEROY, J. RÖSCH, M. TRELLIS

*(Observatoire du Pic-du-Midi et Observatoire de Nice, France)*

ABSTRACT

A study of the development in the solar corona of active centers born during the CSSAR period leads to the following remarks: the 'enhancement' of coronal emissions seems to take place first within a localized 'core' close to the plage, then to extend to a much larger 'halo'; the core from which all the radiations under study originate does not last much more than the spots, whereas the 'halo' is characterized by the major importance of the emission at 5303 Å, and lasts as long as the K 3 plages. These features can be explained by the assumption that the enhancement is inhomogeneous during the first part of its life, then becomes homogeneous after the core has disappeared.

La Coopération pour l'étude des régions solaires actives (CSSAR, 10 mars – 30 octobre 1965) a présenté un grand intérêt pour la compréhension des phénomènes coronaux. En effet, les observations des émissions coronales sont généralement effectuées au limbe et doivent être réalisées de préférence pendant des périodes de faible activité solaire si l'on veut que les phénomènes associés à différents centres actifs puissent être observés séparément. Mettant à profit le fait qu'un très grand nombre de données étaient disponibles, nous avons étudié le développement dans la couronne des centres actifs apparus pendant la période 'CSSAR' à l'aide des observations optiques et radioélectriques qui avaient été recueillies notamment:

(1) Les intensités des raies d'émission de Fe XIV à 5303 Å et de Fe X à 6374 Å.

(2) Les mesures de la brillance de la couronne blanche déduites d'observations polarimétriques qui fournissent la grandeur du produit $pB$ ($p$ = proportion de lumière polarisée, $B$ = brillance de la couronne) moyennant certaines corrections.

Ces deux catégories d'observations sont disponibles de façon presque quotidienne pour la basse couronne (altitude de 40000 km environ). On peut admettre qu'elles ont une résolution dans l'espace de l'ordre de 1′.5 et une résolution dans le temps de 24 heures.

(3) Les radiohéliogrammes obtenus tous les jours à Stanford sur 9·1 cm et à Fleurs sur 21 cm avec une résolution de 3′ environ.

La réduction à une même échelle d'intensité des différentes mesures des raies 5303 et 6374 Å a été effectuée à l'Observatoire du Pic-du-Midi: ce travail a été mené de façon empirique en comparant entre elles les couronnes semestrielles moyennes obtenues par chacun des observatoires participant à la Coopération. Dans la suite

*Kiepenheuer (ed.), Structure and Development of Solar Active Regions, 346–355.* © *I.A.U.*

de cet exposé, les intensités seront donc exprimées dans une seule échelle qui est celle du Pic-du-Midi.

Pour la couronne blanche, l'échelle adoptée est celle de Haleakala dont les observations étaient de beaucoup les plus nombreuses. Les cartes ont bénéficié de quelques observations de Norikura et du Pic-du-Midi qui ont été ajustées de façon très approximative à l'échelle de Haleakala.

Les radiohéliogrammes ont été relevés dans le bulletin des *Solar Geophysical Data* (CRPL.F., Part B). Aucun problème d'homogénéité des mesures ne se posait puisque toutes les observations proviennent d'un seul observatoire pour une longueur d'onde donnée.

Enfin, les données relatives aux centres actifs photosphériques et chromosphériques ont été déduites des spectrohéliogrammes K1 et K3 de l'Observatoire de Meudon et des cartes chromosphériques du même Observatoire.

Dans la suite de ce travail, nous suivrons la terminologie proposée par Billings (1966) et nous désignerons du nom de *renforcement coronal* ('enhancement' de Billings) l'ensemble des régions de la couronne qui sont perturbées par un centre actif. On réservera le terme de *condensation* aux régions les plus concentrées du renforcement coronal qui apparaissent parfois de façon éphémère quand le centre actif est au maximum de son développement.

## 1. Présentation des cartes synoptiques

Les observations disponibles n'ont pas été suffisamment continues pour que l'on puisse suivre sans interruption l'évolution de tous les centres actifs de leur naissance jusqu'à leur disparition. Nous illustrerons donc l'évolution de ce que nous considérons comme un centre actif moyen en présentant des exemples relatifs à des centres actifs *différents*, en l'occurence ceux dont le stade d'existence considéré a été le mieux observé.

Les illustrations qui constituent ce paragraphe sont des cartes synoptiques dressées à partir des *observations au bord*; elles tiennent donc "autant du film que de la carte" (Trellis, 1957). Ce mode de présentation est imposé dans le cas de la couronne blanche et des radiations 5303 et 6374 Å; nous l'avons également adopté pour les émissions radioélectriques, bien que dans ce cas de véritables cartes soient également disponibles, pensant qu'il valait mieux comparer entre elles des cartes synoptiques qui présentent – sinon les mêmes – du moins des altérations analogues dûes à la perspective. Il résulte de ce choix que les cartes synoptiques des émissions radioélectriques sont sans doute imprécises à cause des effets de réfraction notamment.

Outre les données relatives à la couronne blanche et aux émissions à 5303 et 6374 Å, nous présentons sur les Figures 1 à 4 des cartes synoptiques des émissions à 9·1 cm; nous les avons préférées aux observations à 21 cm, bien que ce soient ces dernières qui soient observées en principe à la même altitude que les radiations visibles étudiées,

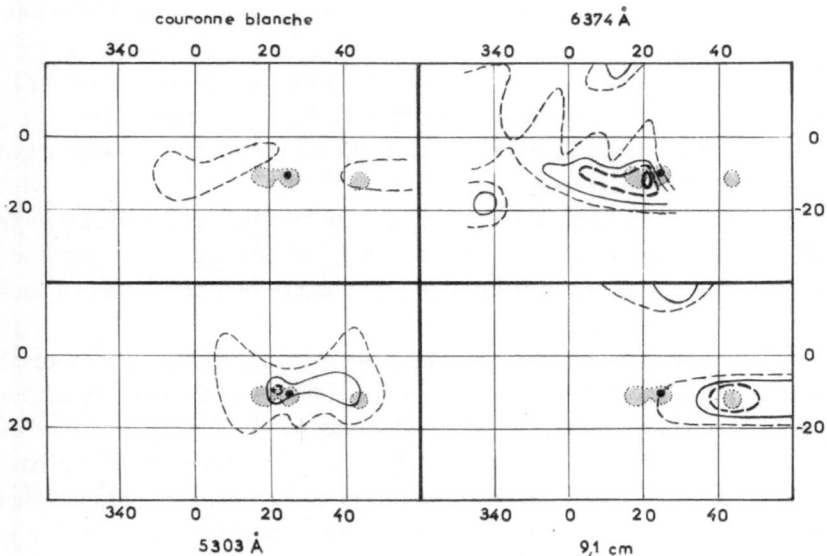

FIG. 1.   *Cartes synoptiques établies à partir des observations effectuées au bord ouest du 8 au 14 juin 1965, 10 jours après la naissance de CSSAR no. 5. La signification des isophotes est précisée dans l'annexe aux figures.*

parce que les cartes que nous avons établies ont montré que l'apparition d'un centre actif au bord solaire est beaucoup plus sensible à 9 cm qu'à 21 cm.

## 2. Evolution du renforcement coronal

Nous avons dressé des graphiques qui donnent la variation au cours du temps de l'intensité maximale observée, pour chaque type de rayonnement étudié, à proximité de la plage faculaire (de façon plus précise, l'intensité maximale obtenue, après lissage des mesures de diverses provenances, dans un rayon de 10° héliographiques autour du centre de gravité des plages faculaires K3). Nous avons porté en ordonnée l'intensité maximale plutôt que l'étendue du renforcement coronal parce que les altérations dûes aux effets de perspective doivent être beaucoup plus réduites pour cette variable.

La Figure 5 donne ces graphiques sur lesquels on remarque tout de suite une très grande dispersion des points dûe en partie aux deux raisons suivantes: (1) L'âge des centres actifs est mal déterminé dans le cas des centres nés sur l'hémisphère invisible du Soleil. (2) Il n'est pas absolument légitime de mélanger les observations provenant de différents centres actifs puisque l'on sait par ailleurs dans le cas des émissions à 5303 Å que la durée de vie n'est pas indépendante de l'intensité maximale atteinte (Trellis, 1959). Pour limiter la dispersion dûe à cette dernière cause nous avons exclu

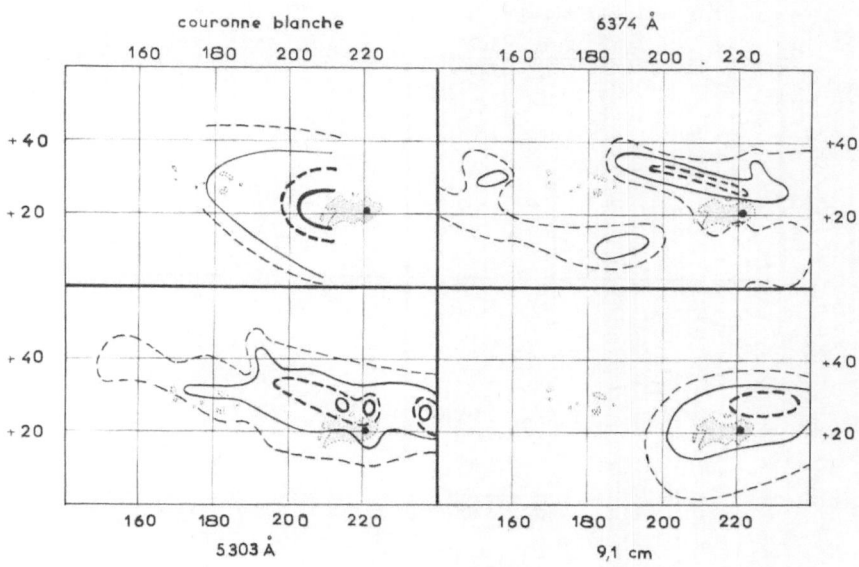

FIG. 2. *Cartes synoptiques établies à partir des observations effectuées au bord ouest du 9 au 16 octobre 1965, 20 jours environ après la naissance de CSSAR no. 9 (à droite). Au centre gauche des cartes on voit les restes de CSSAR no. 8 âgé de 40 jours. La signification des isophotes est précisée dans l'annexe aux figures.*

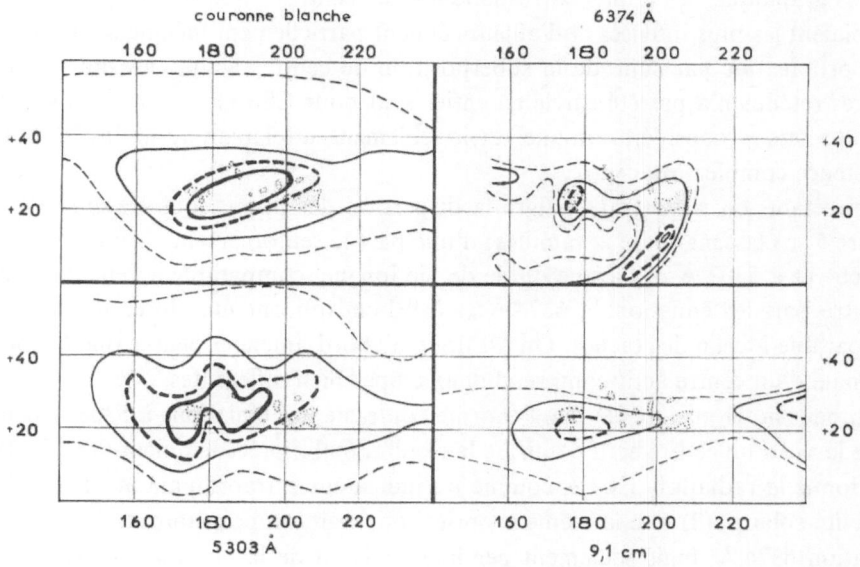

FIG. 3. *Cartes synoptiques établies à partir des observations effectuées au bord est du 8 au 15 juin 1965, 35 jours environ après la naissance de CSSAR no. 3 (à droite) et no. 4 (à gauche). La significa-tion des isophotes est précisée dans l'annexe aux figures.*

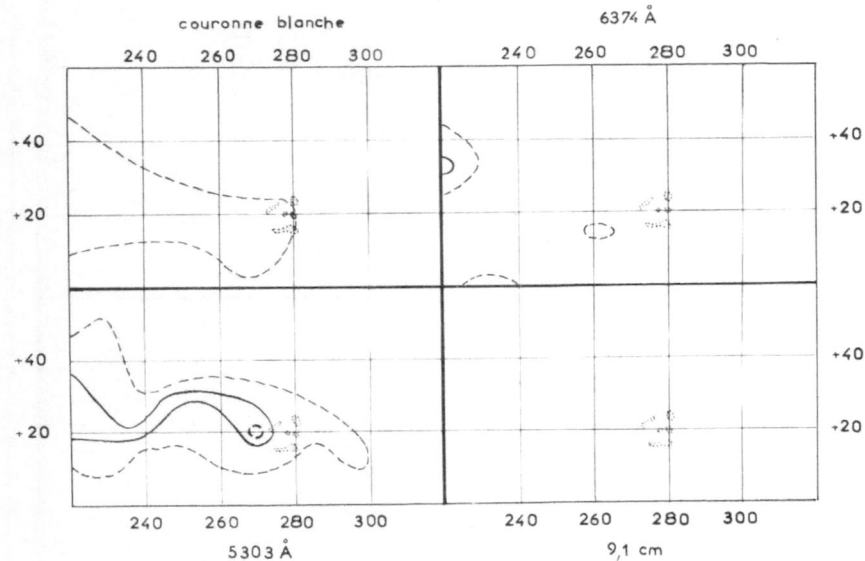

FIG. 4.   *Cartes synoptiques établies à partir des observations effectuées au bord est du 25 juillet au 2 août 1965, 25 jours après la naissance de CSSAR no. 7. La signification des isophotes est précisée dans l'annexe aux figures.*

de nos graphiques les points correspondant aux centres CSSAR no. 3, 4, 8, et 9, qui semblaient les plus intenses et d'ailleurs étaient partiellement mélangés. Il faut noter à ce propos que par suite de la superposition de centres actifs parasites la vie des centres retenus n'a pas été suivie en entier sauf pour CSSAR no. 5; dans six autres cas on a pris en considération une partie seulement, atteignant au moins 30 jours, de l'existence complète du centre.

Ceci étant, on s'aperçoit, malgré la dispersion des points, que les courbes de la Figure 5 se classent en deux familles: d'une part le renforcement coronal en lumière blanche et à 5303 Å qui a une durée de vie longue, comparable à celle des facules. D'autre part les émissions à 6374 Å et à 9·1 cm qui ont une durée de vie courte, comparable à celle des taches. On est donc d'abord amené à penser que l'évolution coronale d'un centre actif comprend deux étapes bien différentes.

On peut noter ensuite (1) que l'énorme contraste des émissions à 5303 Å pendant toute la vie d'un centre actif confirme le résultat établi précédemment (Trellis, 1959), qui donne la radiation 5303 Å comme un indicateur extraordinairement sensible de l'activité solaire, (2) que la même propriété apparaît, un peu moins accusée, pour la radiation 6374 Å, mais seulement pendant le début de la vie d'un centre actif; on pourrait dire que la radiation 6374 Å est un bon indicateur des centres actifs jeunes, (3) que bien que le renforcement coronal observé en lumière blanche ait à peu près la même durée de vie que les émissions à 5303 Å, son faible contraste en fait un indice

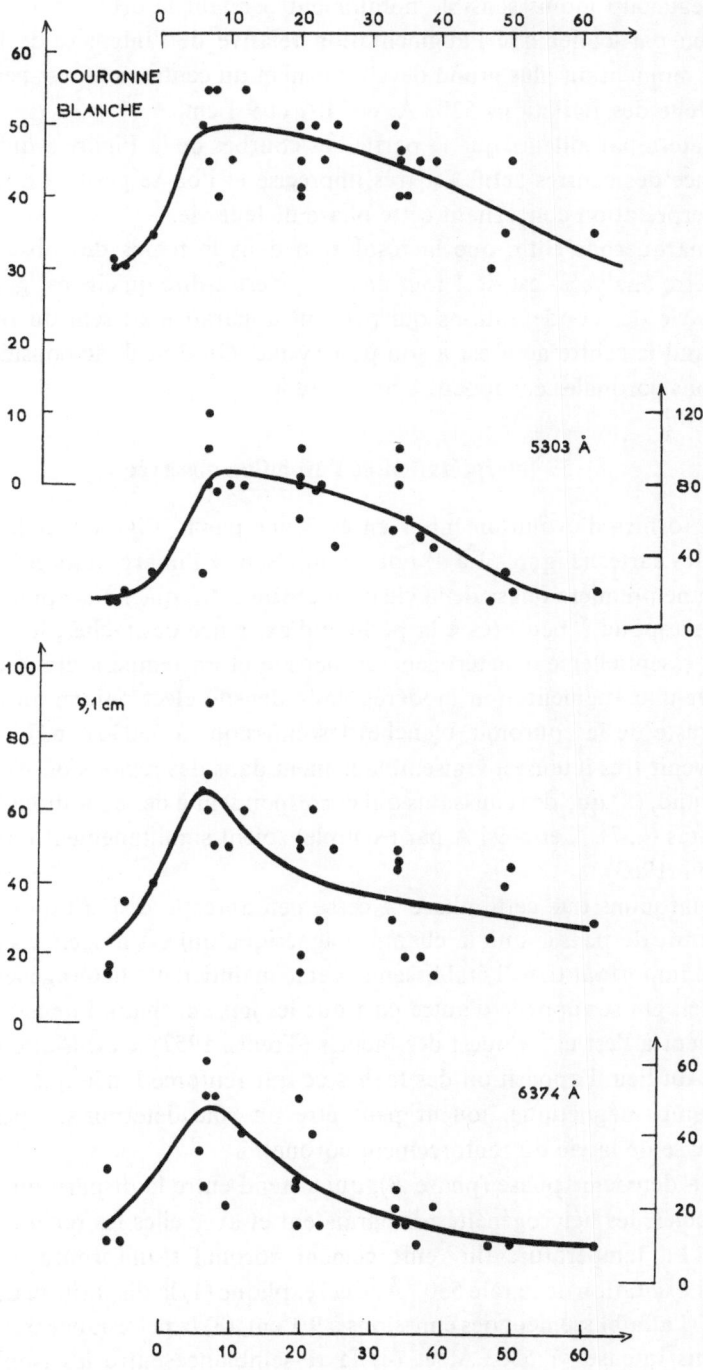

FIG. 5. *Evolution de l'intensité des émissions coronales associées à un centre actif en fonction du temps. On a porté en abscisse l'âge des centres actifs compté en jours et en ordonnée l'intensité des diverses émissions considérées exprimées dans les unités qui sont explicitées dans l'annexe aux figures.*

d'activité beaucoup moins sensible notamment pendant le déclin d'un centre actif; on notera en particulier que l'augmentation relative de l'intensité de la couronne blanche, au moment du plus grand développement du centre actif, est beaucoup plus petite que celle des radiations 5303 Å, 6374 Å et 9·1 cm.

On constatera par ailleurs que la partie des courbes de la Figure 5 qui est relative à la naissance des centres actifs est très imprécise et l'on se gardera donc d'en tirer quelque interprétation concernant cette phase de leur vie.

Nous remarquerons enfin que la résolution dans le temps des observations qui viennent d'être analysées est de 1 jour environ, c'est-à-dire qu'elle est grande devant la durée de vie des condensations qui peuvent apparaître au sein du renforcement coronal quand le centre actif est à son paroxysme. On doit donc considérer que les condensations coronales échappent à notre étude.

## 3. Interprétation de l'évolution observée

Les deux formes d'évolution mises en évidence par la Figure 5 et les différentes propriétés des cartes (Figures 1 à 4) nous conduisent à l'interprétation suivante:

Pendant une première phase de la vie d'un centre actif, que nous appellerons phase $A$ et qui correspond à peu près à la période d'existence des taches, le renforcement coronal est essentiellement hétérogène en densité et en température. Cela explique (1) que pour une augmentation modérée de la densité électronique moyenne (cf. le faible contraste de la couronne blanche) les émissions à 5303 Å, 6374 Å et 9·1 cm puissent devenir très intenses, vraisemblablement dans des régions où $Ne^2$ est localement très grand, (2) que des émissions qui correspondent à des conditions d'excitation très différentes (6374 Å et 5694 Å par exemple) soient simultanément intenses (Nishi et Nakagomi, 1963).

Nous remarquons que cette phase $A$ cesse peu après la disparition des taches ce qui permettrait de penser que le champ magnétique qui est associé à ces dernières joue un rôle important dans l'établissement et le maintien des hétérogénéités qui sont envisagées ici. On se rappelle d'autre part que les jets coronaux intenses s'observent le plus souvent à l'est et à l'ouest des facules (Trellis, 1957), c'est-à-dire précisément à proximité du lieu d'apparition des taches ce qui renforce l'idée que ces dernières, ou leur champ magnétique, jouent peut être un rôle déterminant pendant cette première phase de la vie du renforcement coronal.

Pendant la deuxième phase (phase $B$), qui s'étend entre la disparition de taches et celle des facules, les hétérogénéités disparaissent et avec elles les régions où $Ne^2$ est très grand. La température du renforcement coronal s'uniformise à une valeur favorable à l'excitation de la raie 5303 Å. Ceci explique (1) la disparition des émissions à 6374 Å, (2) l'affaiblissement des émissions à 9·1 cm, (3) la persistance très importante des émissions intenses à 5303 Å, et (4) la ressemblance entre les isophotes de la couronne blanche et celles de la radiation 5303 Å.

Ces deux grandes phases de la vie d'un renforcement coronal peuvent encore être divisées en plusieurs périodes, dans lesquelles on retrouve les propriétés aperçues sur les Figures 1 à 4 et également les caractéristiques de plusieurs modèles classiques de renforcement coronaux.

Au début A1 de la phase A, toutes les émissions optiques et radioélectriques apparaissent ensemble dans un *noyau* bien concentré à proximité immédiate de la facule (cet exemple est à peu près illustré par la Figure 1). On peut penser que ce noyau marque la limite des régions où existent de fortes densités électroniques et des températures non-uniformes.

La fin A2 de la phase A correspond au maximum d'activité du centre actif. (Figure 2). On peut alors distinguer: (1) Un noyau central hétérogène qui fait suite au noyau de la phase A1 où apparaissent simultanément les émissions à 5303 Å, 6374 Å et 9·1 cm. (2) Un *halo* périphérique (Waldmeier, 1963), qui peut être homogène et à une température relativement élevée et où l'on observe principalement une grande extension des émissions à 5303 Å. Cette structure doit entraîner que les maximums d'émission à 5694 Å, 6374 Å et 9·1 cm sont situés dans le noyau central à proximité immédiate de la facule alors que les maximums à 5303 Å peuvent se situer plus loin en bordure de cette dernière. Telle est bien la situation moyenne (non réalisée sur la Figure 2 mais visible dans le cas de CSSAR no. 4 sur la partie gauche de la Figure 3) décrite par de nombreux auteurs (Billings, 1966) pour la radiation jaune et par Trellis (1957) pour les maximums verts et pour certains maximums rouges.

Le début B1 de la phase B est marqué par la disparition progressive du noyau hétérogène qui entraîne l'affaiblissement des émissions à 6374 Å et à 9·1 cm auprès des facules. Par contre le halo se développe et avec lui les émissions à 5303 Å qu'il favorise. De plus, sa bordure extérieure est parfois marquée par des émissions à 6374 Å, dont l'existence avait déjà été signalée par Trellis (1957), qui doivent jalonner les régions où l'accroissement de densité électronique a précédé l'élévation de température. Le centre CSSAR no. 3, dans la partie droite de la Figure 3, illustre assez bien cette phase.

A la fin B2 de la période B les hétérogénéités ont complètement disparu et la température s'est uniformisée à une valeur favorable à l'émission à 5303 Å, qui restera donc seule intense notamment là où la densité est un peu plus élevée (Figure 4).

## 4. Conclusion

Il convient, pour terminer, de rappeler que l'interprétation précédente est fondée sur l'étude d'une douzaine de centres actifs seulement; c'est dire qu'il faut la considérer comme un modèle de travail susceptible d'être largement amélioré ou modifié dans l'avenir.

Nous avons signalé à plusieurs reprises au cours de cette étude les limitations qui étaient introduites par le manque de pouvoir séparateur dans l'espace et dans le temps des observations de routine qui ont servi de base à notre travail. Une telle limitation est particulièrement regrettable quand on est amené à introduire la conception d'un noyau hétérogène pour expliquer les observations. Sans préjuger des progrès instrumentaux qui pourront marquer, d'ici le prochain minimum solaire, les observations de la couronne blanche, celles de l'ultra violet coronal et celle des émissions radioélectriques, on peut déjà remarquer que l'usage extensif de filtres monochromatiques, qui permettent d'obtenir des images détaillées des structures coronales dans la lumière de diverses radiations visibles, serait sans doute d'un grand secours pour une meilleure compréhension des phénomènes coronaux associés aux centres actifs.

## Annexe aux figures

Les unités utilisées sont:

Pour les émissions à 5303 Å et à 6374 Å le millionième de la brillance de 1 Å du spectre du centre du Soleil au voisinage de la longueur d'onde étudiée.

Pour la couronne blanche le cent millionième de la brillance du centre du Soleil dans l'intervalle spectral considéré.

Pour le rayonnement à 9·1 cm, une température de brillance de 1000 °K.

Sur les cartes des Figures 1 à 4 les isophotes sont dessinées successivement en tirets fins, en trait continu fin, en tirets épais et en trait continu épais quand on va des petites aux grandes intensités.

La représentation utilise quatre isophotes, qui correspondent aux intensités suivantes:
Couronne blanche: $pB = 40$–$50$–$60$–$70$; unité $10^{-8}$ $B_\odot$ dans l'échelle d'Haleakala.
Radiation 5303 Å: $B = 25$–$50$–$75$–$100$; unité $10^{-6}$ $B_\odot$ dans l'échelle du Pic-du-Midi.
Radiation 6374 Å: $B = 20$–$30$–$40$–$50$; unité $10^{-6}$ $B_\odot$ dans l'échelle du Pic-du-Midi.
Rayonnement à 9·1 cm: $T = 40$–$60$–$80$–$100$; unité $10^3$ °K dans l'échelle de Stanford.

Les plages faculaires chromosphériques sont indiquées par une zone grise délimitée par une ligne en pointillés et les taches sont figurées par un point noir.

## Bibliographie

Billings, D. E. (1966)      *A Guide to the Solar Corona*, Academic Press, New York.
Nishi, K., Nakagomi, Y. (1963)      *Publ. astr. Soc. Japan*, **15**, 56.
Trellis, M. (1957)      *Suppl. Ann. Astrophys.*, 5.
Trellis, M. (1959)      *Ann. Astrophys.*, **22**, 845.
Waldmeier, M. (1963)      *Z. Astrophys.*, **58**, 57.

## DISCUSSION

*Nussbaumer:* You spoke of a strong enhancement in both the green and the red lines with only a slight increase in the mean electron density. Could you tell us to which increase in inhomogeneity one would have to conclude from these observations?

*Rösch:* I am sorry but I have no figures to answer that question.

*Fokker:* A possibility to check the hypothesis, that coronal structure becomes more homogeneous as the active centre ages, may be the measurement of angular diameters of radio type-I stormbursts. The angular diameter is largely due to scattering on coronal irregularities. If coronal structure, late in the centre of activity's lifetime, were more smooth, angular diameters of stormbursts may be expected to decrease.

*Rösch:* This is an interesting possibility.

*Elske Smith:* I would like to point out the similarities of these observations with the extreme ultraviolet data which will be reported on by Dr. Neupert this afternoon. The enhancements of the XUV permitted lines are similar to that shown by the visible and radio data, and therefore fit fairly well with the model of inhomogeneous structures in the corona as suggested by Leroy and Trellis. I am, however, surprised at the rapid decay of the red line in comparison with the green line. This is divergent from the changes in the lines of different ionization in the ultraviolet. Perhaps, as you say, this may be attributed to a more rapid change in density than in temperature.

Part V

# CORONAL AND INTERPLANETARY STRUCTURE OF AN ACTIVE REGION

# OBSERVATION DES JETS ET CONCENTRATIONS DE LA COURONNE AU-DESSUS DES RÉGIONS ACTIVES

AUDOUIN DOLLFUS

*(Observatoire de Paris-Meudon, 92, France)*

### ABSTRACT

The 'K-Coronameters' in use since 1956 at Pic-du-Midi and Meudon Observatories for the study of the white-light corona were replaced in 1963 by new improved instruments.

The coronal activity was continuously recorded from 1964 to 1967. The electron density in the low corona and its variation with height are summarized by curves and maps.

A study is given of the corona streamers above the proton flares area in July and September 1966.

## 1. Premières études de la couronne solaire par polarimétrie photométrique

Un polarimètre photoélectrique fut installé en 1956 sur le coronographe du Pic-du-Midi, pour déceler et mesurer l'intensité de la lumière diffusée par les électrons dans la couronne solaire. Les premières observations furent commencées en septembre 1956 (Dollfus, 1958*a*). Les grands jets coronaux pouvaient être suivis jusqu'à 35′ du bord solaire.

Un instrument plus développé fut ensuite préparé à l'Observatoire de Meudon en 1957. Le polarimètre photoélectrique fut adapté à une lunette conçue pour n'introduire aucune lumière polarisée nuisible (Dollfus, 1958*b*). La sensibilité permit de déceler les jets de la couronne sans coronagraphe et sans le recours aux stations d'altitude. L'instrument peut fonctionner sous le ciel ordinaire d'un observatoire de plaine.

La lunette fut montée à l'Observatoire de Meudon (altitude 170 m) en 1958; les premières observations de la couronne y furent obtenues en avril. Les séries d'observations de 1959, 1960 et 1961 montrèrent les jets jusqu'à 12′ du bord solaire (Dollfus, 1959*a*). Des études furent combinées entre les deux instruments de Meudon et du Pic-du-Midi lors de l'éclipse du 15 février 1961 (Dollfus *et al.*, 1961; Dollfus, 1963).

Ces observations permirent d'établir les propriétés morphologiques suivantes (Dollfus, 1959*b*, 1963; Leroy, 1960):

(1) Les jets et les concentrations de la couronne ne sont pas des phénomènes éphémères, mais persistent plusieurs semaines et quelquefois plusieurs mois. Leur durée de vie est au moins comparable à celle des centres actifs.

(2) La plupart des régions de forte densité électronique dans la couronne sur-

montent les centres actifs présentant des taches, des facules et des émissions radioélectriques centimétriques. D'autres concentrations sont associées à des protubérances
quiescentes isolées et des filaments. Certains jets se manifestent jusqu'à des latitudes
plus élevées que les centres actifs et ne correspondent à aucun détail chromosphérique
(Dollfus, 1959b); ainsi, un grand jet polaire a été étudié en 1959 par J. L. Leroy (1960)
à la latitude de +60° pendant plus de 3 mois consécutifs.

(3) Les concentrations coronales sont solidaires de la rotation différentielle de la
surface solaire et conservent une position fixe par rapport à la photosphère. La
Figure 1 donne par exemple la vitesse de rotation de la couronne à l'altitude de

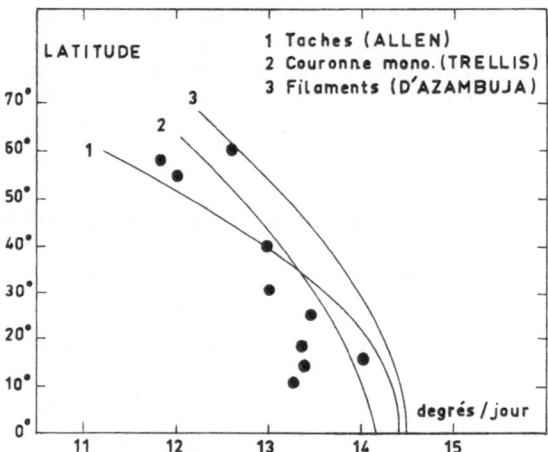

FIG. 1. *Rotation des jets de la couronne, identifiés alternativement aux bords est et ouest.*

60000 km, exprimée en degrés par jours, pour les formations coronales persistantes
les mieux identifiées alternativement aux bords est et ouest. Une légère tendance à
un mouvement plus lente que la photosphère près de l'équateur n'est probablement
pas significative.

(4) Les grands jets coronaux que l'on peut déceler jusqu'à 1 million km au-delà
du bord solaire (soit trois rayons solaires du centre) surmontent souvent les mêmes
régions photosphériques pendant plusieurs semaines consécutives sans déformations
appréciables. Cette permanence implique un renouvellement fréquent de la matière,
ou l'existence de forces plus importantes que celles de la gravitation, telles par exemple que des actions d'origine magnétique.

## 2. Nouvelle lunette polarimétrique pour l'étude de la couronne

Après ces premières études, nous avons réalisé à l'Observatoire de Meudon, en
1962, l'instrument final permettant de mesurer la quantité de lumière polarisée

calibrée dans une échelle absolue et constante, liée à la densité électronique en différents points de la couronne et dans les jets.

L'objectif coronographique mesure 15 cm de diamètre et 400 cm de distance focale. L'observation s'effectue toujours sur l'axe optique. L'orifice d'entrée du polarimètre est placé au centre du champ, il sous-tend au choix 60″ et 120″ sur le Soleil. La lunette décrit par rapport au berceau fixe un mouvement conique permettant de balayer tout le tour du disque solaire à une distance du bord choisie. Simultanément, la lunette tourne sur elle-même autour d'un axe passant par les centres de l'objectif et de l'orifice du polarimètre, afin de mesurer toujours la composante de lumière polarisée dans la direction parallèle aux bords solaires (Figure 2). Selon cette disposition, la faible polarisation résiduelle que peut introduire l'instrument reste constante et indépendante de l'azimuth visé; on l'annule complètement par compensation pour le centre du disque solaire.

Les mesures se font par compensation, selon une méthode de zéro, en annulant la quantité de lumière polarisée observée à l'aide d'un flux lumineux auxiliaire réglable, polarisé en sens inverse. Cette lumière auxiliaire provient du centre de l'image du disque solaire. Les mesures sont ainsi calibrées dans une échelle absolue de luminance. Les mesures sont indépendantes de la stabilité de l'électronique, exemptes de dérive, et elles se rapportent à une source invariable, le centre du disque solaire.

Afin de déceler les variations de la polarisation que peut donner la lumière du ciel, les mesures sont généralement effectuées tout autour du Soleil, une première fois de 10° en 10°, puis une deuxième fois de 5° en 15°, etc...; l'accord des deux séries, garantit la fidélité des mesures.

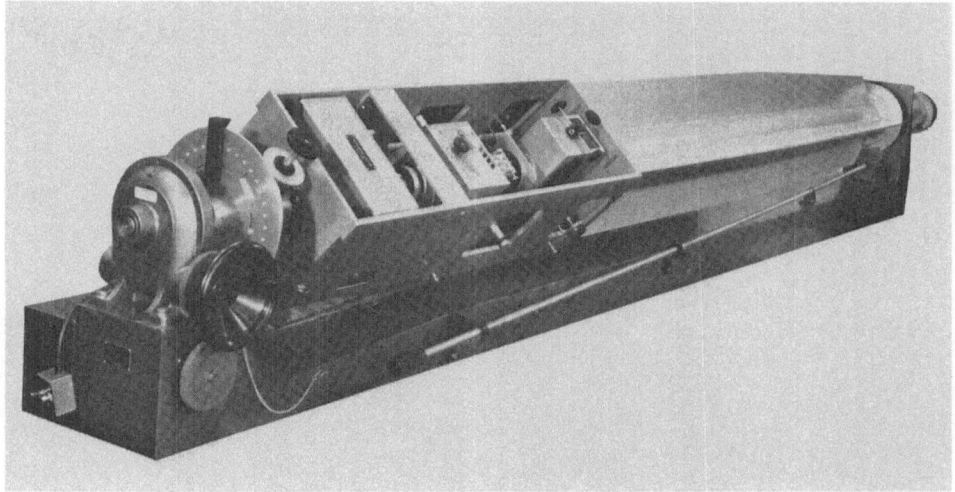

FIG. 2.  *La lunette polarimétrique solaire des observatoires de Meudon et du Pic-du-Midi (réalisation Ets. R. Danger, Paris).*

Le prototype, achevé et essayé à l'Observatoire de Meudon en 1963, a été réalisé en plusieurs exemplaires par les établissements R. Danger en France. Deux de ces nouvelles lunettes ont remplacé à la fin de 1963 les anciens instruments en service au Pic-du-Midi et à Meudon; d'autres lunettes ont été construites pour différents observatoires (Figure 2).

### 3. Etude de l'activité coronale pendant la période de l'AISC (IQSY)

Les observations régulières ont été commencées avec ces nouveaux instruments en décembre 1963 dans le cadre de la contribution française au programme des 'Années Internationales du Soleil Calme' (IQSY).

Les observations au Pic-du-Midi ont été assurées grâce à des séjours successifs de collaborateurs de l'Observatoire de Meudon.

De décembre 1963 à septembre 1967, environ 500 balayages de la couronne ont été recueillis. Les Figures 3 et 4 reproduisent deux feuilles d'observations quotidiennes dans leur présentation habituelle. Les balayages sont effectués à 1·5′ du bord solaire (60 000 km) avec l'orifice explorateur de 60″ de diamètre; les intensités de la lumière polarisée sont données en unités de $10^{-8}$ fois la luminance du centre du disque solaire et reportées en coordonnées radiales à partir d'un cercle figurant le contour du Soleil. Le graphique du 12 avril 1964 (Figure 3) correspond à l'activité coronale minimum. Le 10 juillet 1965 (Figure 4) donne l'exemple d'une activité moyenne. Les Figures 15 et 16 montrent des activités exceptionnelles.

En raison de la lente rotation du Soleil, les mesures relevées de la sorte permettent

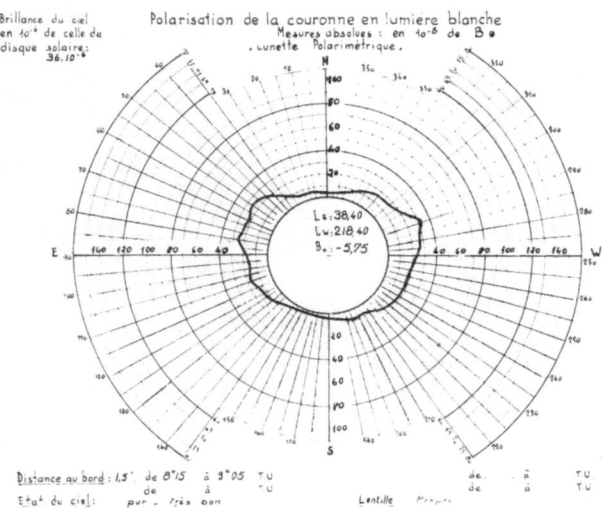

Fig. 3.   *Exemple d'observation à 1′.5 du bord, période du minimum d'activité (12 avril 1964).*

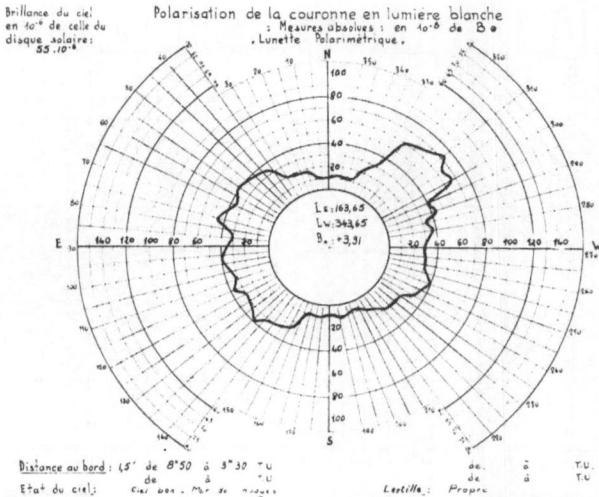

FIG 4.   *Exemple d'observation à 1'.5 du bord, activité coronale moyenne (10 juillet 1965).*

de dresser des cartes synoptiques en coordonnées héliographiques pour chacun des deux bords est et ouest. Ces cartes donnent la position des condensations coronales stables sur le disque solaire.

La carte de la Figure 5 (janvier 1964) montre la répartition des formations coronales à 60 000 km, peu avant le minimum de l'activité solaire; les dates des observations sont reportées au-dessus de la carte, aux emplacements correspondant à leurs longitudes. La Figure 6 (août 1964) correspond au minimum de l'activité. La Figure 7 (juin 1965) donne le début de la reprise de l'activité. La Figure 8 illustre le grand déploiement de l'activité coronale dans l'hémisphère nord en juillet et août 1966, tandis que cette activité reste modérée dans l'hémisphère sud.

Les intensités moyennes de la lumière polarisée coronale ont été calculées pour chaque mois. La Figure 9 donne les valeurs mensuelles de décembre 1963 à septembre 1967, séparément pour les deux hémisphères nord et sud et pour les deux bords est et ouest. Le minimum de l'activité s'est étalé d'avril 1964 à juin 1965. Puis, quelques concentrations coronales sont apparues, principalement dans l'hémisphère nord. L'activité devint ensuite très forte dans l'hémisphère nord, dans la seconde partie de l'année 1966.

La Figure 10 donne les répartitions moyennes des luminances coronales autour du disque au cours des mois successifs, pour les années 1964, 1965 et 1966. Les couronnes du minimum solaire, en 1964, sont régulières et symétriques.

## 4.  Etude de jets coronaux

Au-dessus des régions de fortes densités électroniques, on peut déceler des jets de la couronne. Des balayages supplémentaires à des distances croissantes du bord solaire

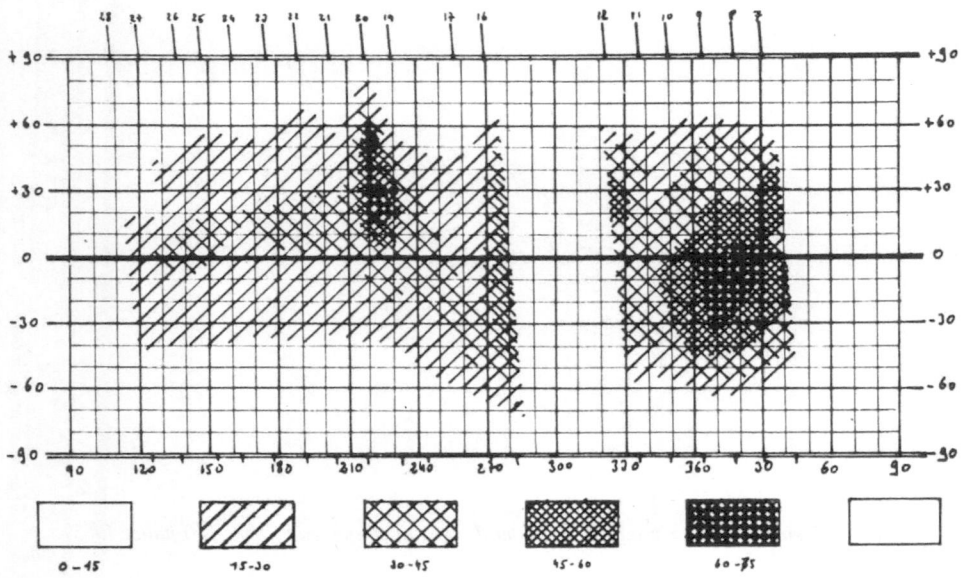

FIG. 5.  *Exemple de carte synoptique de la couronne en lumière blanche à 60000 km au-dessus de la photosphère: janvier 1964 au bord ouest: avant le minimum d'activité.*

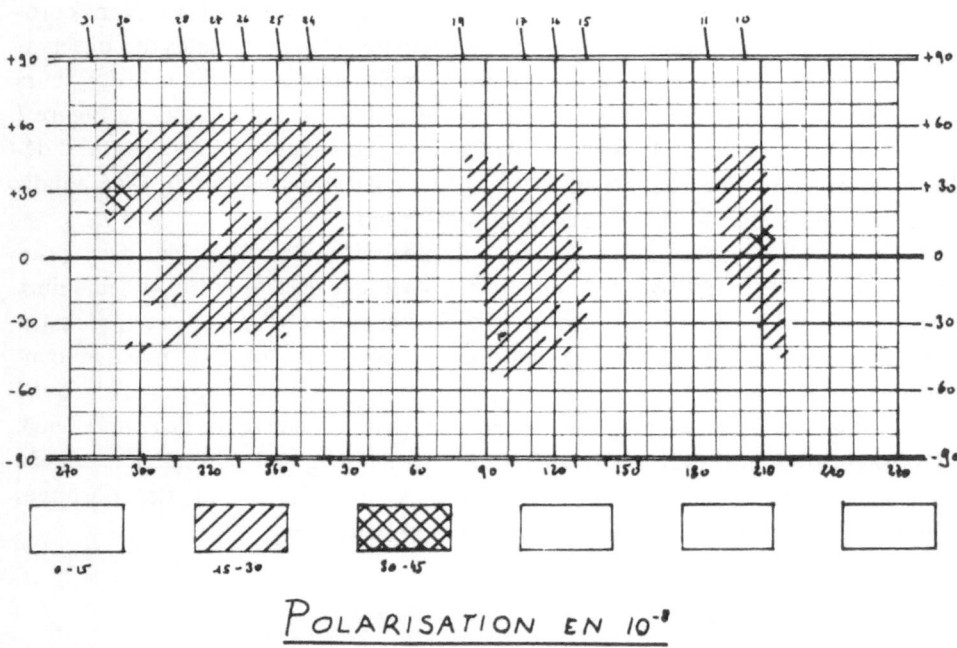

FIG. 6.  *Idem, août 1964 au bord est: au minimum d'activité.*

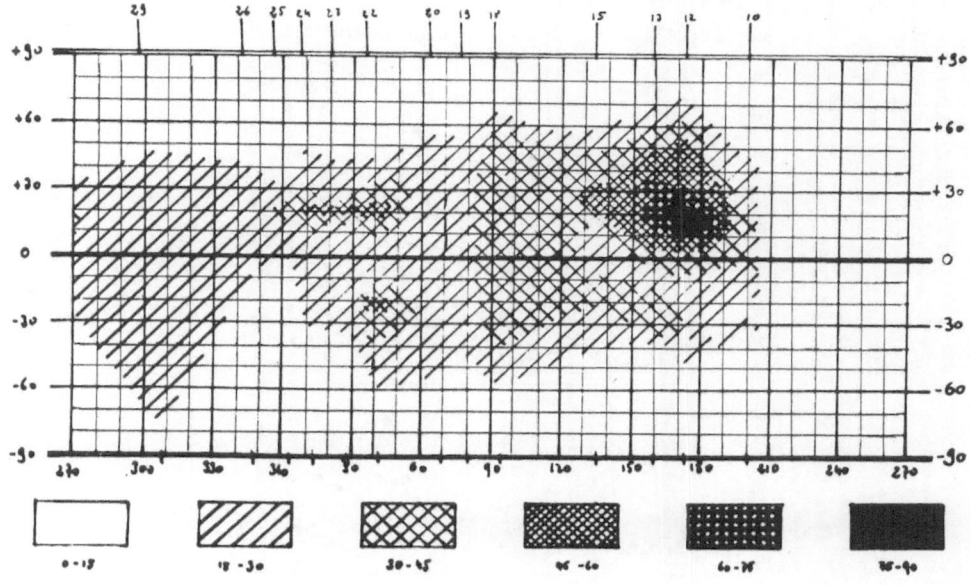

FIG. 7.   *Idem, juin 1965 au bord est: après la reprise de l'activité.*

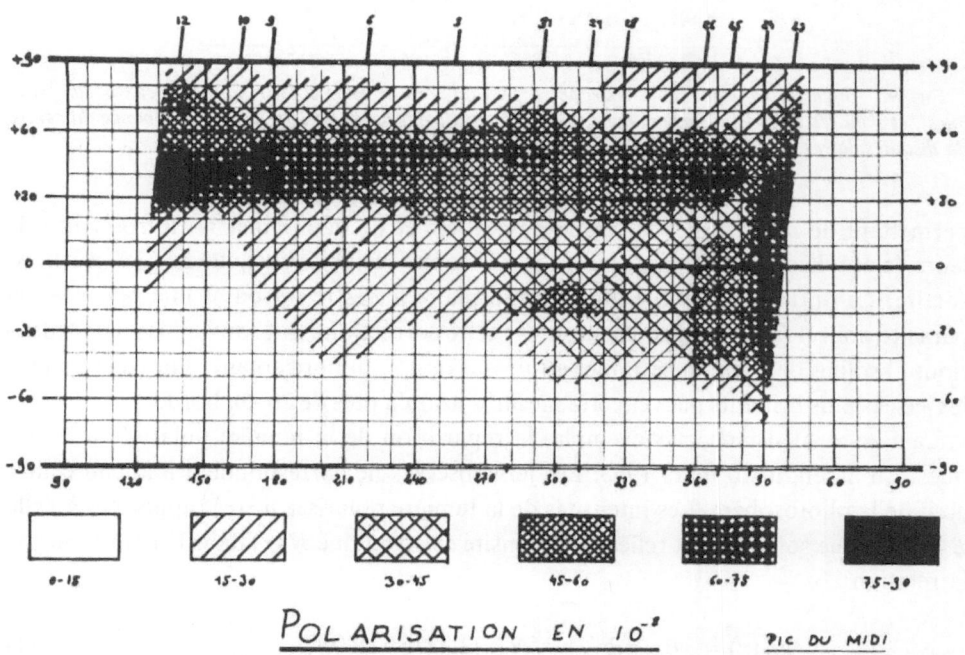

FIG. 8.   *Idem, juillet et août 1966 au bord ouest: forte activité dans l'hémisphère nord.*

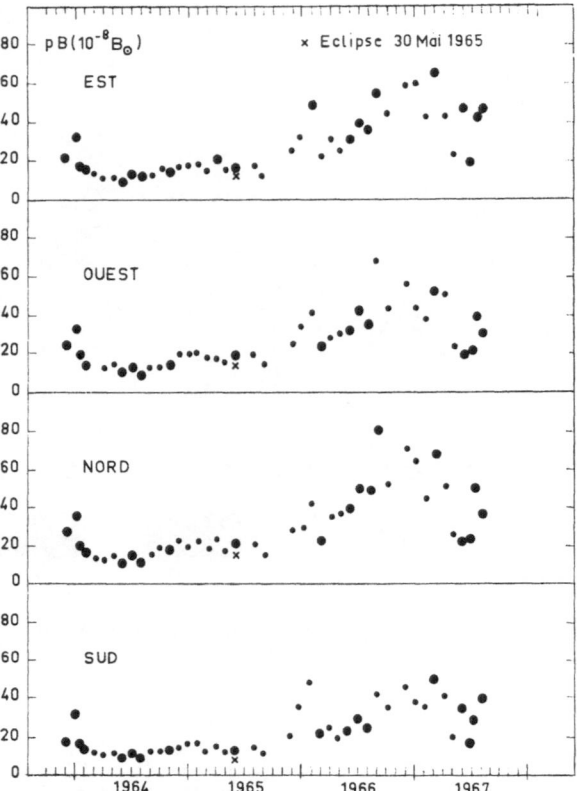

FIG. 9. *Intensité moyenne de la couronne à 1'.5 du bord de 1964 à 1967 pour les deux bords est et ouest, et pour les deux hémisphères nord et sud (intensité exprimée en $10^{-8}$ fois la luminance du centre du disque solaire).*

permettent de reconstituer la forme de ces jets et de déterminer la décroissance de leurs éclats. Les jets brillants et concentrés sont décelables jusque vers 35' du bord solaire. La diffusion multiple dans l'atmosphère terrestre introduit une polarisation parasite dont il faut corriger les mesures; cette correction rend souvent les détermina-tions absolues de l'éclat incertaines au-dessus de 15' du bord solaire. Par ciel très pur, les jets les plus brillants peuvent être mesurés jusqu'à plus de 20' du bord.

La Figure 11 donne, par exemple, la répartition de la lumière polarisée au bord ouest du Soleil, le 10 mars 1965. Les jets observés ne correspondent à aucun centre actif de la photosphère. Les intensités de la lumière polarisée $pB(r)$ rapportées à celle $B_\odot$ du disque solaire, sont reliées à la densité électronique $N(r)$ dans la couronne par la relation:

$$pB(r) = \tfrac{3}{4}B_\odot \; \sigma R_\odot \int\limits_{r_0}^{\infty} r_0^2 N(r) \left[A(r) - B(r)\right] \frac{\mathrm{d}r}{r\sqrt{r^2 - r_0^2}} \qquad (1)$$

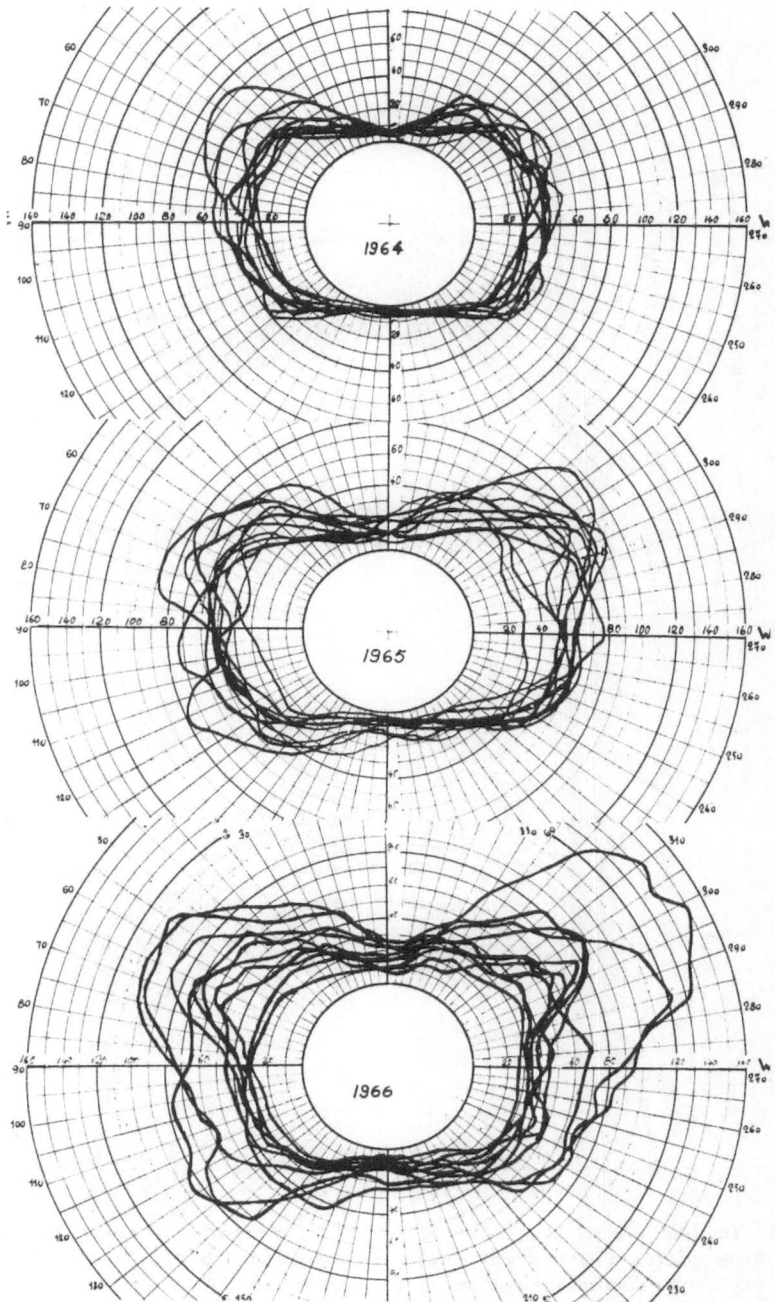

FIG. 10.   *Contours moyens de la couronne à 1′. 5 du bord du disque. Chaque courbe est la moyenne des observations pour 1 mois. En haut: courbes mensuelles pour 1964: minimum de l'activité solaire. Au centre: courbes mensuelles pour 1965: début du nouveau cycle. En bas: courbes mensuelles pour 1966: forte activité dans l'hémisphère nord.*

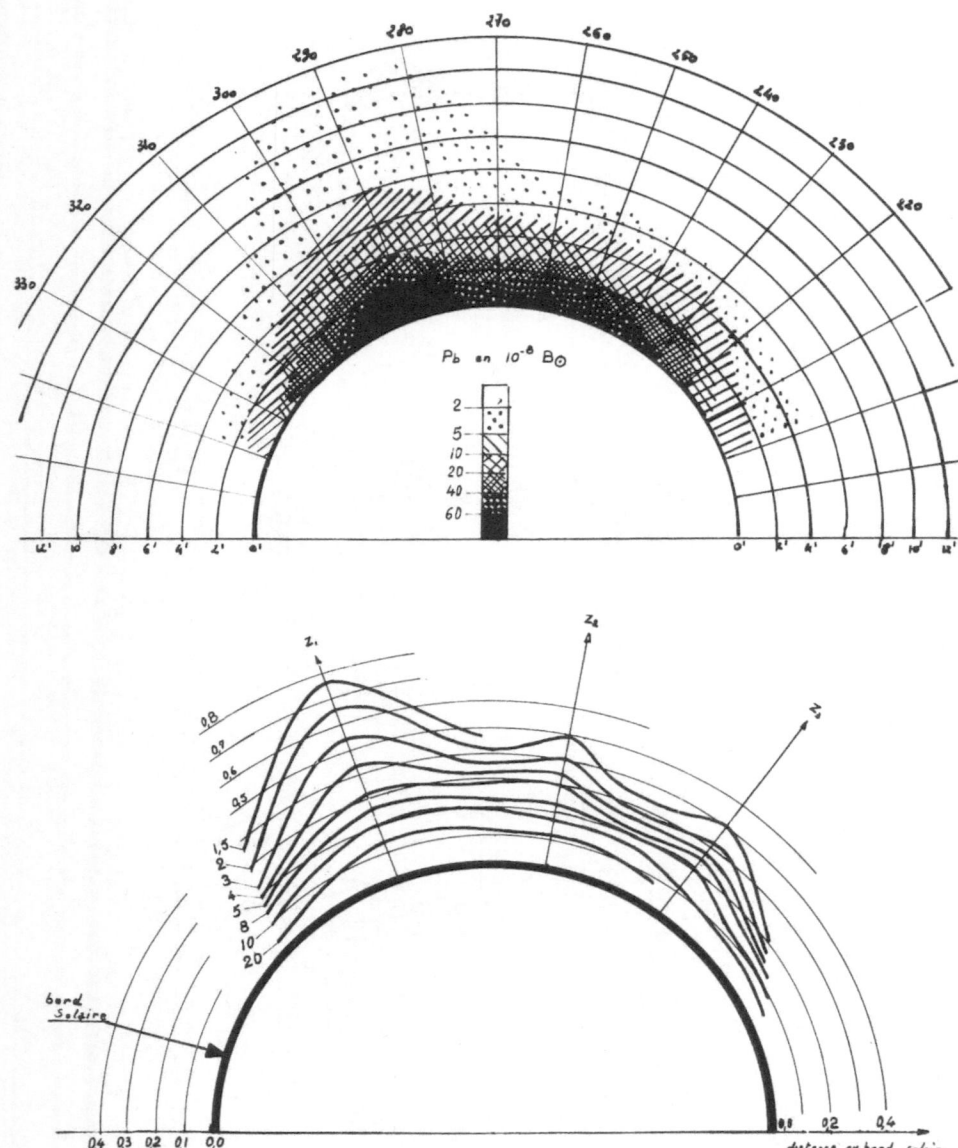

Fig. 11. *Isophotes d'un jet au bord ouest le 10 mars 1965 (intensité exprimée en $10^{-8}$ fois la luminance du disque solaire). Courbe d'égale densité électronique dans le plan du limbe, pour le jet du 10 mars 1965 (densité électronique exprimée en $10^7$ électrons par $cm^3$).*

dans laquelle $\sigma$ est la section de choc Thomson caractérisant la diffusion par les électrons, soit $0.6655 \times 10^{-24}$, $A(r)$ et $B(r)$ sont les fonctions introduites par Van de Hulst, $r_0$ est la distance de la ligne de visée au centre solaire exprimée en unités du rayon solaire, et $r$ est la distance réelle d'un point courant le long de la ligne de visée.

Les déterminations de la densité électronique en chaque point des jets observés sont effectués à l'Observatoire de Meudon, soit par traitement numérique de l'intégrale, soit, lorsque le jet est exactement au bord, en supposant des répartitions du type:

$$N(\rho, \xi, \eta) = N_0(\rho) \times e^{-\left(\frac{x}{\xi}\right)^2} \cdot e^{-\left(\frac{y}{\rho}\right)^2} \tag{2}$$

$\rho$ est la distance au centre du Soleil le long de l'axe du jet, et $\xi$ et $\eta$ sont les extensions du jet dans les directions $x$ et $y$ parallèle et perpendiculaire à la direction de la ligne de visée.

A titre d'exemple, la Figure 11 donne, le 10 mars 1965, la répartition de la densité électronique dans le plan de section contenant le contour apparent du Soleil. La densité vaut: $20 \times 10^7$ électrons par cm$^3$ à la base du jet; elle devient $1.5 \times 10^7$ électrons par cm$^3$ à $12'$ du bord dans le jet principal. Nous étudions à l'Observatoire de Meudon une quarantaine d'observations de cette nature.

Lors du minimum de l'activité solaire, le gradient de décroissance de l'éclat dans la couronne fut très régulier; la Figure 12 donne la moyenne des mesures relevées de mars à juin 1964 au-dessus des pôles et de l'équateur.

Des mesures obtenues à la même période au Japon par K. Nishi, à l'aide d'un instrument du type étudié par G. Newkirk (K Coronameter), s'accordent avec les nôtres. Toutes ces mesures s'accordent à peu près aussi avec les déterminations moyennes déduites des éclipses par H. C. Van de Hulst.

Les jets individuels manifestent une dispersion plus grande, dont la Figure 13 fournit quelques exemples. Les courbes donnent les intensités de la lumière polarisée le long des axes apparents des jets observés dans le plan du limbe. Une étude plus élaborée permettra de corriger les effets de la perspective et de déterminer la décroissance radiale réelle de l'éclat et de la densité électronique dans l'axe des jets. Il apparaît déjà que les densités et leurs gradients s'écartent peu, dans l'ensemble, des modèles moyens déterminés antérieurement par H. C. Van de Hulst et par G. Newkirk.

## 5. Activités exceptionelles de juillet à septembre 1968

Trois événements solaires consécutifs importants, accompagnés d'émissions de protons, ont été enregistrés en 1966, respectivement les 7 juillet, 28 août et 2 septembre. L'activité du 28 août n'a pu être observée. Les centres actifs liés aux deux autres évènements apparurent l'un et l'autre au voisinage de la longitude 180° et de la latitude +25°. Nos observations, étudiées à Meudon par B. Fort, précisent le comportement de la couronne dans ces deux cas.

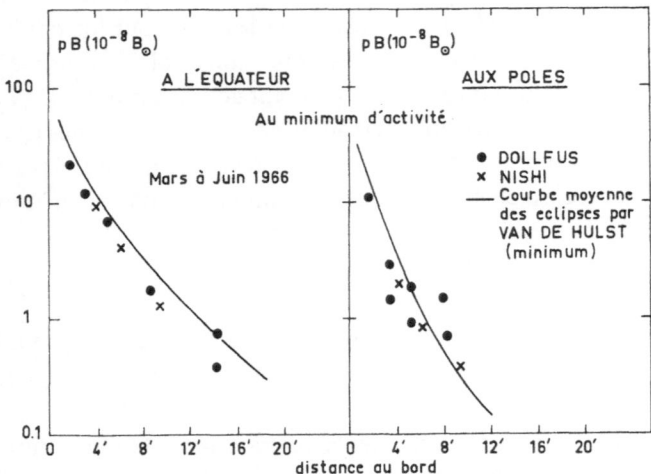

FIG. 12. *Intensité de la couronne solaire en fonction de la distance au bord du disque, à l'équateur et aux pôles, lors du minimum de l'activité solaire. Comparaison avec les observations japonaises et avec la courbe moyenne des éclipses.*

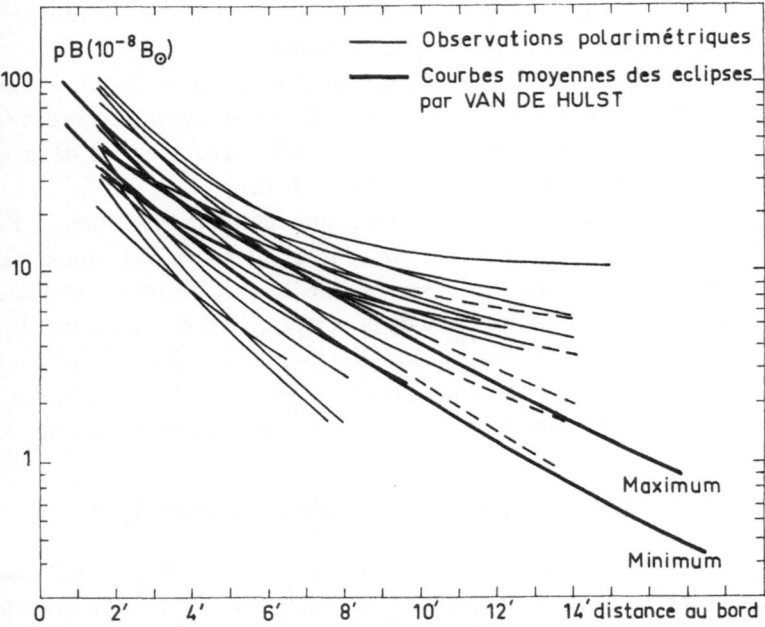

FIG. 13. *Intensité de la couronne solaire en fonction de la distance au bord du disque, le long de l'axe apparent des jets en projection dans le plan du limbe.*

La Figure 14 donne les cartes synoptiques successives de cette région de la surface solaire observée alternativement aux bords est et ouest, d'après les mesures relevées à 1ʹ5 du limbe.

Les Figures 15 et 16 reproduisent quelques-unes des observations à 1ʹ5 du bord.

## A. ÉVÉNEMENT DU 7 JUILLET 1966

Le centre actif avait pour longitude 200° et pour latitude +35°; il est passé au bord ouest le 10 juillet. Avant l'éruption, la couronne ne présentait aucune concentration particulière (Figure 14, cartes no. 1 à 4 de mai et juin 1966), et l'intensité de la lumière polarisée ne dépassait pas $70 \times 10^{-8}$ fois la luminance $B_\odot$ du centre du disque solaire.

Le 9 juillet (Figure 15, diagramme no. 1) un jet tout à fait exceptionnel était visible au bord ouest; sa lumière polarisée donnait, à 1ʹ5 du bord, une intensité de $160 \times 10^{-8} \times B_\odot$ comprenant environ $40 \times 10^{-8} B_0$ pour la couronne non-perturbée en avant et en arrière du jet, et $pB_K = 120 \times 10^{-8} \times B_\odot$ pour le jet lui-même. En supposant le jet dans le plan du limbe, et de révolution autour de son axe, la densité électronique peut s'écrire sous la forme de l'Equation (2).

L'observation du 9 juillet à 1ʹ5 du bord donne l'extension du jet en latitude $\xi = 0\cdot2\,R_0$.

L'intégration de l'Equation (1) donne:

$$pB_K = \tfrac{3}{8}\sigma\,10^8 R_\odot N\,[A - B]\,\sqrt{\pi}\,\xi\,e^{-\left(\frac{\rho}{\xi}\right)^2} \qquad \begin{array}{l} pB_K \text{ exprimé en } 10^{-8} \times B_\odot \\ \rho \text{ en rayon solaire } R_\odot \end{array}$$

on a la valeur numérique $[A - B] = 0\cdot210$ à 1ʹ5 du bord, d'où finalement la densité électronique $N_0 = 9 \times 10^8$ électrons par cm³. Cette valeur, qui correspond à l'altitude de 80 000 km, est la plus forte que nous ayons enregistrée à ce jour.

Ce jet resta intense jusqu'au 13 juillet (diagramme no. 2) puis la rotation solaire l'entraîna derrière le disque. Les 25 et 26 juillet il réapparaissait au bord est avec la même intensité (diagrammes no. 3 et 4). Les 9 et 10 août, il passait au bord ouest avec une intensité réduite de moitié (diagrammes no. 5 et 6). Les 23 et 24 août il était attendu au bord est, les observations ne furent possibles que les 25 et 26 août (diagrammes no. 7 et 8) et ne montrent pas de regain de l'activité.

## B. ÉVÉNEMENT DU 2 SEPTEMBRE 1966

Six observations à 1ʹ5 du bord ont été recueillies entre le 1er et le 8 septembre; elles sont reproduites dans la Figure 16 et permettent de construire la carte no. 9 de la Figure 14.

L'éruption se produisit le 2 septembre à 05h 28m TU. Deux observations furent recueillies à 09h 30m et 09h 50m TU respectivement à 1ʹ5 et 14ʹ du bord solaire. Le centre actif était alors à 55° à l'ouest du méridien central et à +23° de latitude nord.

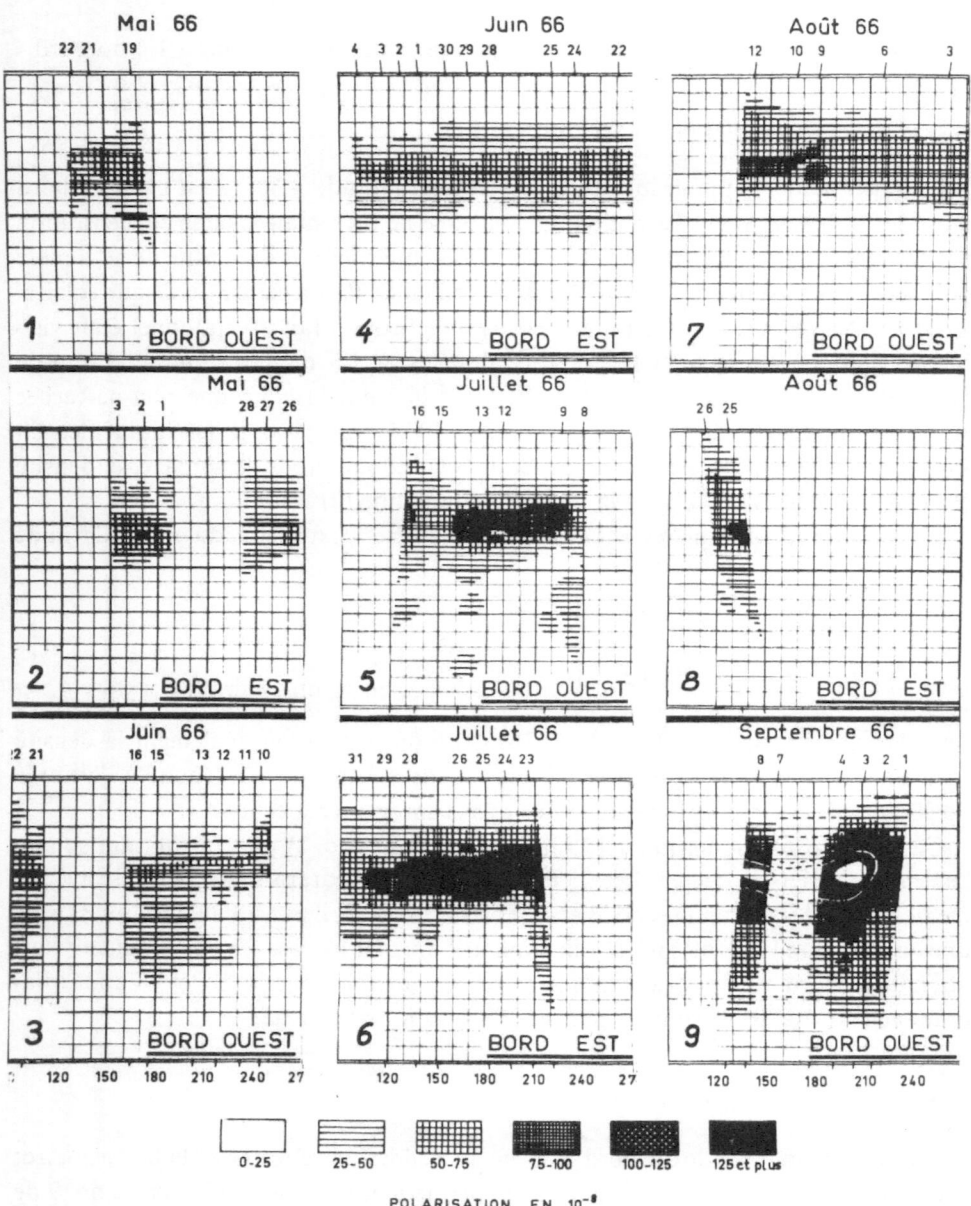

FIG. 14. *Extraits des cartes synoptiques de la couronne à 60 000 km au-dessus de la photosphère, de mai à septembre 1966: Evolution des configurations coronales au-dessus d'un centre actif à protons.*

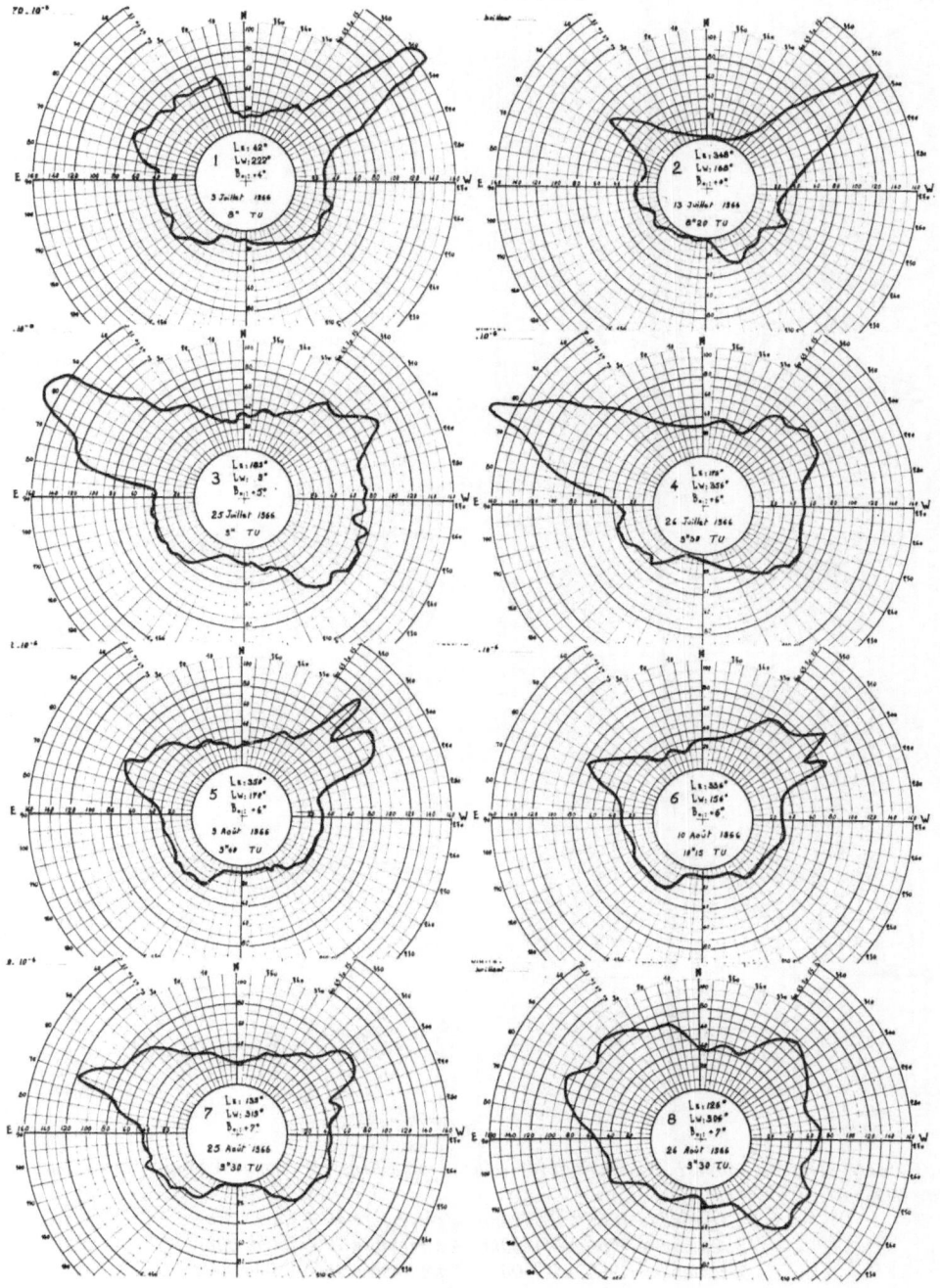

Fig. 15. *Quelques observations à 1'.5 du bord du jet de la couronne liées à l'événement à protons du 7 juillet 1966.*

*Au bord ouest:  9 juillet 1966  08h 00m  et  13 juillet 1966  08h 20m*
*Au bord est  : 25 juillet 1966  09h 00m  et  26 juillet 1966  09h 30m*
*Au bord ouest:  9 août   1966  09h 40m  et  10 août   1966  10h 15m*
*Au bord est  : 25 août   1966  09h 30m  et  26 août   1966  09h 30m*

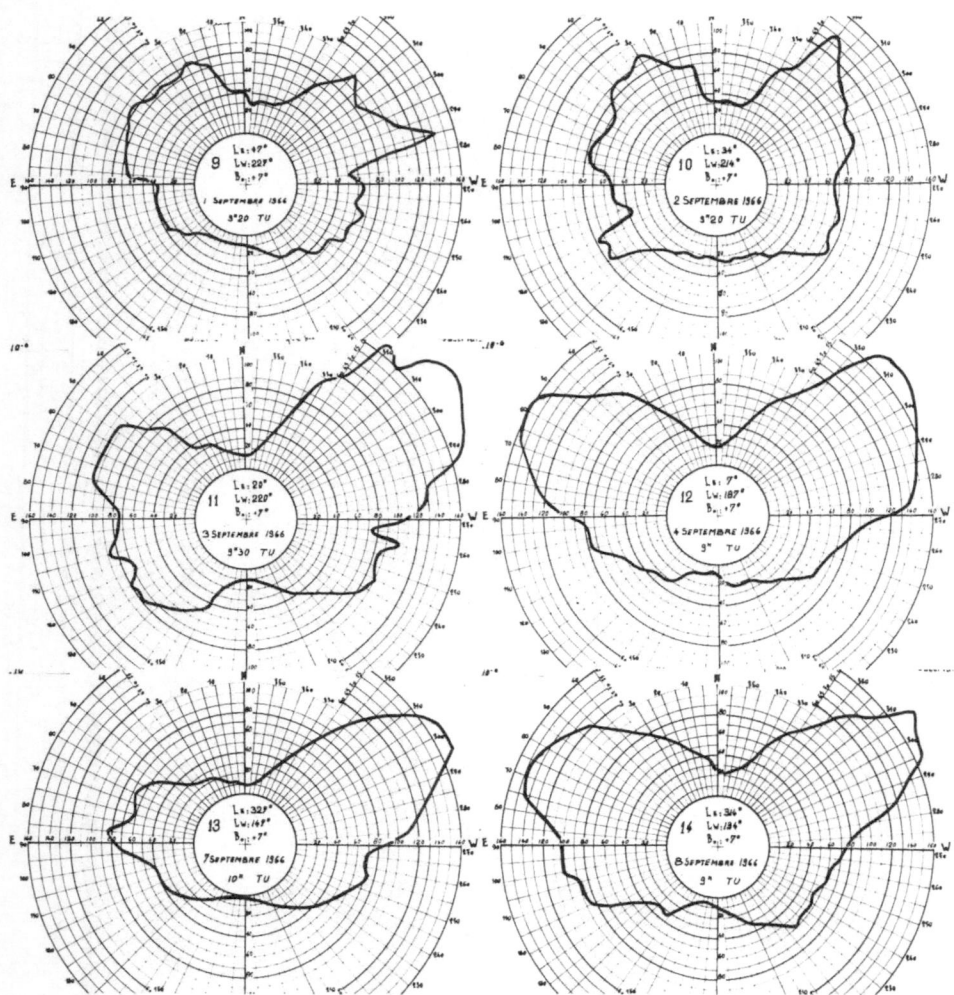

FIG. 16.    *Observations à 1'.5 du bord du jet de la couronne liées à l'événement à protons du 2 septembre 1966:*

1 septembre 1966 à 09h 20m    2 septembre 1966 à 09h 20m
3 septembre 1966 à 09h 30m    4 septembre 1966 à 09h 00m
7 septembre 1966 à 10h 00m    8 septembre 1966 à 09h 00m

La configuration est représentée dans la Figure 17; le jet dépassait le contour du Soleil et se trouvait balayé par les deux observations respectivement à 1·22 et 2·00 rayons du centre du Soleil. Les balayages recueillis les jours suivants à 1ʹ5 du bord coupent le jet à différentes hauteurs.

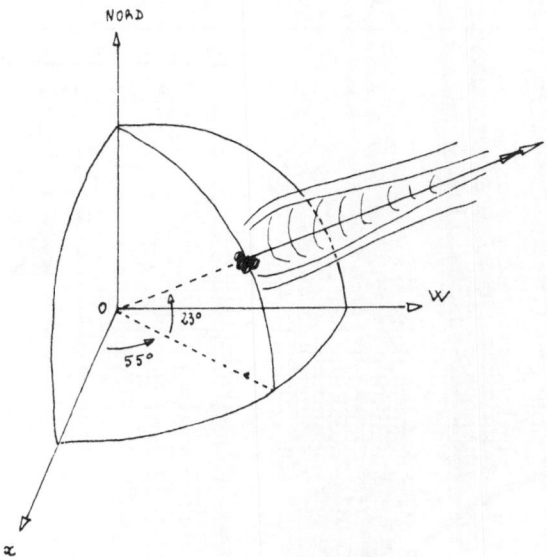

FIG. 17.  *Présentation du jet coronal surmontant l'émission de protons le 2 septembre 1966.*

L'examen des isophotes de la carte héliographique no. 9 (Figure 14) permet de montrer que:

(a) le jet s'élevait au-dessus du centre actif à peu près perpendiculairement à la surface solaire;

(b) l'intensité s'est brusquement accrue entre les deux observations du 1er septembre à 09h 20m et du 2 septembre à 09h 20m;

(c) l'intensité est ensuite restée à peu près constante jusqu'à la dernière observation du 8 septembre.

Les différents balayages donnent les densités électroniques suivantes:

| Date | 4 sept. | 3 sept. | 7 sept. | 8 sept. | 2 sept. |
|---|---|---|---|---|---|
| distance au bord: | 1ʹ5 | 1ʹ5 | 1ʹ5 | 1ʹ5 | 14ʹ |
| hauteur dans le jet, en $R_\odot$: | 1·10 | 1·20 | 1·25 | 1·25 | 2·0 |
| Densité électronique en $10^7$ électrons/cm³: | 50 | 25 | 20 | 9 | 0·8 |

La décroissance de la densité électronique avec la hauteur paraît suivre la loi habituelle.

Toutefois, l'observation du 8 septembre révéla une anomalie singulière. L'observateur eut l'attention attirée par l'intensité extrême de la lumière polarisée à grande

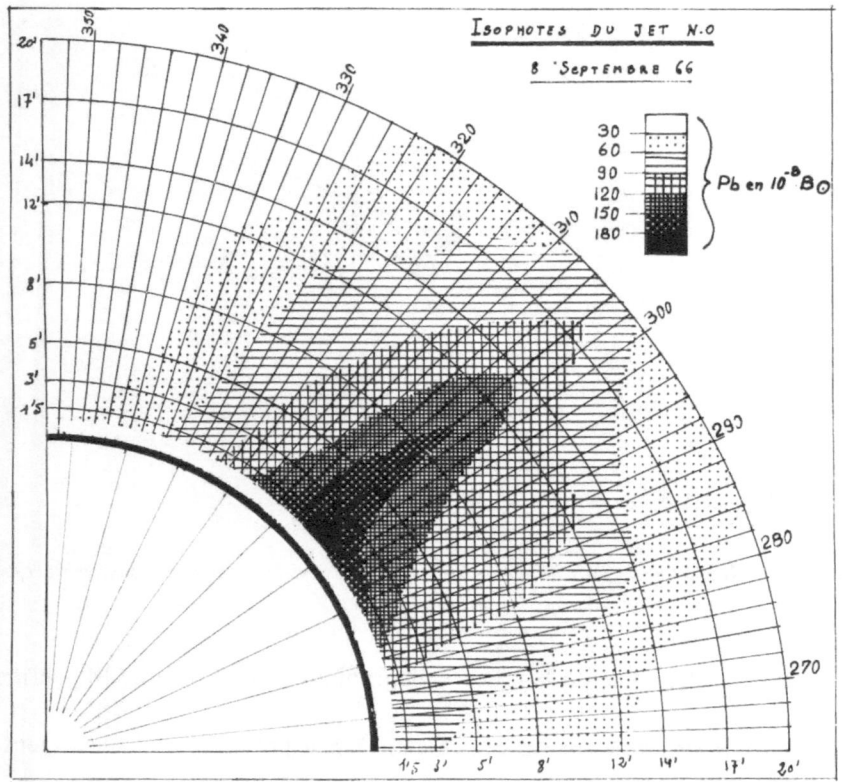

FIG. 18. *Isophotes du jet au bord ouest le 8 septembre 1966 (intensités exprimées en $10^{-8}$ fois la luminance du disque solaire).*

distance au-dessus de ce jet. Des balayages successifs ont été entrepris à $1'.5$, $3'$, $5'$, $8'$, $12'$, $14'$ et $20'$ du bord et ils donnèrent la répartition de l'éclat représentée dans la Figure 18. La décroissance radiale de l'intensité y est exceptionnellement faible; à $20'$ du bord l'éclat reste supérieur à $60 \times 10^{-8} B_{\odot}$ et y est au moins 10 fois plus élevé que les valeurs normales.

Le gradient de l'éclat est incompatible avec les modèles habituels, des hypothèses complexes sur la géométrie du jet ne permettent pas de résorber l'éclat. L'examen attentif des observations et de l'instrument ne révèlent pas d'anomalies. Si les observations ne sont pas viciées par une cause cachée, il faut alors envisager un phénomène solaire inhabituel.

## 6. Collaborations

Les études et une soixantaine d'observations ont été effectuées à l'Observatoire de Meudon, placé sous les directions successives de A. Danjon et de J. F. Denisse. Environ 500 observations furent réalisées à l'Observatoire du Pic-du-Midi que dirige J. Rösch. Le Comité Français des AISC (IQSY) a apporté un appui déterminant dans ce programme. Les collaborateurs suivants ont participé aux mesures: MM. Baudroit, Brel, Champion, Leroy, Lucas, Poulain, Maurice, Verdet, Zéau. Les propriétés de la couronne ont été dégagées grâce à la collaboration de J. L. Leroy, E. Maurice, et B. Fort.

## Références

### A. PUBLICATIONS ANTÉRIEURES PAR LES AUTEURS

Dollfus, A. (1958*a*)    'Procédé permettant d'observer la couronne solaire et ses jets jusqu'à une grande hauteur', *C.R. Acad. Sci.*, **247**, 42.

Dollfus, A. (1958*b*)    'Premières observations avec le polarimètre solaire', *C.R. Acad. Sci.*, **246**, 3590.

Dollfus, A. (1959*a*)    'Observations de la couronne solaire à l'Observatoire de Meudon en 1959', *C.R. Acad. Sci.*, **249**, 2273.

Dollfus, A. (1959*b*)    'Quelques propriétés des jets de la couronne solaire, observeés en lumière blanche', *C.R. Acad. Sci.*, **249**, 2722.

Dollfus, A. (1963)    'Etude des grands jets de la couronne solaire observés en lumière blanche sans coronographe', dans *The Solar Corona*, éd. par J. Evans, Academic Press, New York, p. 243.

Dollfus, A., Leroy, J.L., Marin, M. (1961)    'Observation de la couronne solaire avant, pendant et après l'éclipse totale du 15 Février 1961', *C.R. Acad. Sci.*, **252**, 3402.

Leroy, J.L. (1960)    'Observation d'un jet coronal persistant', *Ann. Astrophys.*, **23**, 567.

### B. TRAVAUX ANALOGUES AUX U.S.A.

Newkirk, G. (1959)    'A Model of the Electron Corona with Reference to Radio Observations', dans *Paris Symposium on Radio Astronomy*, éd. par R. N. Bracewell, p. 149.

Newkirk, G. (1961)    'The Solar Corona in Active Regions and the Thermal Origin of the Slowly Varying Component of the Solar Radio Radiation', *Astrophys. J.*, **133**, 983.

Newkirk, G., Axtel, J., Wlerick, G. (1957)    'Studies of the Electron Corona', *Astron. J.*, **62**, 95.

Newkirk, G., Curtis, G.W., Watson, K. (1958, 1962)    'Observation of the Solar Electron Corona', *I.G.Y. Solar Activities Report Series*, **4** et **16**.

Wlerick, G., Axtel, J. (1957)    'A New Instrument for Observing the Electron Corona', *Astrophys. J.*, **126**, 253.

### C. TRAVAUX ANALOGUES AU JAPON

Nagasawa, S. (1964, 1965, 1966)    'Observation of the Solar Electron Corona at the Norikura Corona Station', *Tokyo Astronomical Observatory. Bull. of Solar Phenomena*, **16**, 75; **17**, 83; **18**, 87.

Nishi, K., Nagasawa, S. (1964)    'On the Observation of the Electron Corona at the Norikura Corona Station', *Pub. astron. Soc. Japan*, **16**, 285.

### D. TRAVAUX ANALOGUES EN U.R.S.S.

Karimov, M.G. (1961)    'Polarization Measurements of the Solar Corona outside Eclipse', *Izv. Astrofiz. Inst. Akad. Nauk Kaz. S.S.R.*, **11**, 64.

## DISCUSSION

*Houtgast:* For the determination of the electron density you need to know the dimension of the streamer in the line of sight. What did you assume for this?

*Dollfus:* Généralement les jets sont allongés dans le sens est–ouest. Nous avons calculé la densité électronique en supposant que le jet avait une section elliptique, et que le rapport des axes était de 2·0.

*Rigutti:* How can you estimate the sky brightness during your balloon-borne experiments?

*Dollfus:* Nous avons un dispositif d'étalonnage qui permet de comparer la brillance du ciel et de la couronne à celle du centre du disque solaire. La brillance du fond du ciel, sur les clichés obtenus à l'altitude de 30000 m, valait $2 \times 10^{-9}$ fois celle du centre du disque solaire; elle était deux ou trois fois plus faible que celle prévue, ce qui est très favorable.

*Kundu:* How fast can you construct the coronal density isophotes? Can you construct them instantaneously, say 1 per minute, so that they can be compared with fast radio maps on meter wavelengths?

*Dollfus:* La mesure de la couronne tout autour du disque à 1ʹ.5 du bord solaire demande environ 30 min. Des balayages supplémentaires sont ensuite effectués au dessus des centres actifs pour déterminer la forme et la décroissance de l'éclat des jets. Une analyse complète de la couronne jusqu'à 30ʹ du bord demande un peu plus de 2 heures d'observation.

# INFLUENCE OF MAGNETIC FIELDS ON THE STRUCTURE
# OF THE SOLAR CORONA*

G. Newkirk, M. D. Altschuler, and J. Harvey

*(High Altitude Observatory, National Center for Atmospheric Research,
Boulder, Colo., U.S.A.)*

### ABSTRACT

A current-free approximation to the coronal magnetic field is calculated from measured photospheric magnetic fields (Mt. Wilson) and traced by computer. The calculated field structure is then compared to a white-light photograph of the November 12, 1966 eclipse.

It is generally agreed that solar magnetic fields play a dominant role in determining the gross structure of the corona. However, the nature of the spatial correlation between the magnetic fields which penetrate the corona and such observed coronal features as streamers and rays is uncertain. The purpose of our still incomplete investigation is to determine the correspondence between the magnetic-field configuration and the density structure of the solar corona.

The basic data for our calculation were the daily Mt. Wilson maps of the line-of-sight magnetic field at the solar surface for the period October 29 – November 26, 1966. We divided the solar surface into 24 latitude zones ($\Delta \sin \lambda = 0 \cdot 0833$ in latitude $\lambda$) and 27 longitude sectors ($\Delta \phi = 13 \cdot 3°$ in longitude $\phi$), thereby creating 648 surface elements of equal area. From the Mt. Wilson data we estimated the average (line-of-sight) magnetic field in each of the surface elements. When magnetic measurements for a given element were available over several days, the assigned field was taken to be the mean of the individual daily observations weighted according to the following scheme:

*Distance of Sector from*

| *Central Meridian (days)* | *Assigned Weight* |
|:---:|:---:|
| 0, $\pm 1$ | 1 |
| $\pm 2$ | 0·9 |
| $\pm 3$ | 0·7 |
| $> |3|$ | 0 |

The net magnetic flux through the solar surface, as determined by the data, is not necessarily zero because (1) the measurements were taken over a solar rotation rather

---

\* Presented by G. Newkirk.

*Kiepenheuer (ed.), Structure and Development of Solar Active Regions, 379–384. © I.A.U.*

than instantaneously, (2) only the line-of-sight component was measured, and (3) the accuracy was limited to $\sim 0\cdot 5$ gauss. This difficulty may be overcome by a simple calibration. Let $D_{ij}$ represent the assigned line-of-sight magnetic field obtained from the Mt. Wilson data for region $(i, j)$, where $i = 1, \ldots, 24$ specifies the latitude zone and $j = 1, \ldots, 27$ specifies the longitude sector. Since we require that the net magnetic flux $\Phi$ through the solar surface be zero, we let

$$\Phi = \Delta A \sum_{i=1}^{24} \sum_{j=1}^{27} (D_{ij} - \delta) \sec \lambda_i = 0, \tag{1}$$

where $\Delta A$ is the area of the surface element, $\lambda_i$ is the latitude of the midpoint of the element, and $\delta$ is the correction term. Thus,

$$\delta = \frac{\sum_{i=1}^{24} (\sum_{j=1}^{27} D_{ij}) \sec \lambda_i}{27 \times \sum_{i=1}^{24} \sec \lambda_i}, \tag{2}$$

which we find to be $\sim 5 \times 10^{-2}$ gauss for the data used. The corrected line-of-sight magnetic field in each region $(i, j)$ is then

$$D'_{ij} = D_{ij} - \delta. \tag{3}$$

Under the assumption that the corona is current-free, the magnetic field above the solar surface is completely specified by the distribution of the normal component of the photospheric field and the requirement that the field vanish at infinity. Since only the line-of-sight fields are known, we assume that the total field $B_{ij}$ is *normal* to the photosphere and thus is related to the corrected line-of-sight field $D'_{ij}$ by

$$B_{ij} = D'_{ij} \sec \lambda_i. \tag{4}$$

Moreover, since $\nabla \times \mathbf{B} = 0$ above the photosphere, we can represent $\mathbf{B}$ as the gradient of a scalar potential $\psi$, so that $\mathbf{B} = -\nabla \psi$. Also $\nabla \cdot \mathbf{B} = 0$, so that $\nabla^2 \psi = 0$. The solution of this Laplacian equation in the region $r \geqslant R_\odot$ is (Chapman and Bartels, 1940)

$$\psi(r, \theta, \phi) = R_\odot \sum_{n=1}^{\infty} \sum_{m=0}^{n} \left[ \left( \frac{R_\odot}{r} \right)^{n+1} (g_n^m \cos m\phi + h_n^m \sin m\phi) P_n^m(\theta) \right], \tag{5}$$

where the $P_n^m$ are the associated Legendre Polynomials, $\theta$ is the colatitude, and $g_n^m$ and $h_n^m$ are constants to be determined from the data.

At the solar surface, the radial component of magnetic field

$$B_r(r = R_\odot, \theta, \phi) = -\left. \frac{\partial \psi}{\partial r} \right|_{r = R_\odot} = \sum_{n=1}^{\infty} \sum_{m=0}^{n} (n+1)(g_n^m \cos m\phi + h_n^m \sin m\phi) P_n^m(\theta) \tag{6}$$

corresponds to $B_{ij}$ of Equation (4) and is therefore known. The coefficients $g_n^m$ and $h_n^m$ can then be calculated from

$$\begin{Bmatrix} g_n^m \\ h_n^m \end{Bmatrix} = \frac{2n+1}{4\pi(n+1)} \int\limits_0^\pi \int\limits_0^{2\pi} B_r(R_\odot, \theta, \phi) \, P_n^m(\theta) \begin{Bmatrix} \cos m\phi \\ \sin m\phi \end{Bmatrix} \sin\theta \, d\theta \, d\phi, \qquad (7)$$

in which we have used the Schmidt normalization (Chapman and Bartels, 1940). In practice, the double integration of Equation (7) is replaced by a summation over the 648 elementary regions, and $B_r(R_\odot, \theta, \phi) \sin\theta$ in the integrand is replaced by $D'_{ij}$ of Equation (3). The coefficients $g_n^m$, $h_n^m$ together with Equation (5), then determine the field at any point $(r, \theta, \phi)$ where $r \geqslant R_\odot$ by means of

$$B_r = -\frac{\partial\psi}{\partial r}; \quad B_\theta = -\frac{1}{r}\frac{\partial\psi}{\partial\theta}; \quad B_\phi = -\frac{1}{r\sin\theta}\frac{\partial\psi}{\partial\phi}. \qquad (8)$$

Since Equation (5) is a rapidly converging polynomial expansion, the series may be restricted to order $n = 5$ with little compromise in accuracy.

Using this technique, we have traced those magnetic-field lines which originate at the centres of the elementary surface areas. The magnetic field out to a distance of $3R_\odot$ from the centre of the Sun was calculated in this manner. For comparison with the observed corona, the calculated field lines were then projected against the plane of the sky (Figure 1a) for a given longitude $\phi_0$ of the centre of the solar disk.

Several comments regarding this presentation of the magnetic-field lines in the corona are in order. First, the density of field lines is in no way related to the strength of the magnetic field (as in the usual representation), since the foot points have been chosen *geometrically*. Second, the three-dimensional structure of the coronal magnetic field cannot be determined from a simple projection of the field lines, say for some single central longitude $\phi_0$. To achieve tri-dimensionality, we may examine the projections for two different central longitudes in a stereo-viewer or may produce a motion picture in which the central longitude changes with time. As displayed at this meeting, the resultant computer-drawn motion picture provides the three-dimensional structure of the field and reveals the presence of hitherto unnoticed arcades of magnetic loops elongated over 60° to 100° of the solar surface. Last, the presence of a solar wind of velocity $v$ and density $\rho$ invalidates our assumption of zero current in the outer corona, so that the magnetic field there cannot be calculated from potential theory. In our simple model we assume that wherever $B \geqslant B_c$, where

$$\frac{B_c^2}{8\pi} = \tfrac{1}{2}\rho v^2, \qquad (9)$$

the magnetic lines are accurately approximated by potential fields, and we draw the lines (Figure 1a) as solid. Wherever $B < B_c$ we draw the field lines as dashed to indicate that they are probably completely distorted by the solar wind. The value of $B_c$ was

FIG. 1b.  *The solar corona of November 12, 1966 photographed with a radially symmetric, neutral density filter in the focal plane of the camera to compensate for the steep decline of coronal radiance with increasing distance. The latitude of the print is further extended by use of the Fluor-o-dodge process. The overposed image of Venus appears in the NE quadrant.*

chosen empirically by comparing the calculated magnetic-loop configurations with similar loops actually visible in the corona at the same location (at the bases of the SE and SW streamers). To produce the transition from closed (solid) to distorted (dashed) structures at the observed height of about one solar radius above the limb requires

$$B_c \sim 4 \cdot 7 \times 10^{-2} \text{ gauss} .$$

With a streamer density $N_e \sim 10^7 \text{ cm}^{-3}$ at $2R_\odot$ (Malitson and Erickson, 1966) Equation (9) gives for the velocity of the solar wind

$$v \sim 28 \text{ km/sec} .$$

Such a velocity roughly agrees with the few empirical estimates available (Newkirk, 1967), but is considerably higher than that predicted by the theoretical models (e.g. Meyer and Schmidt, 1966; Whang *et al.*, 1966).

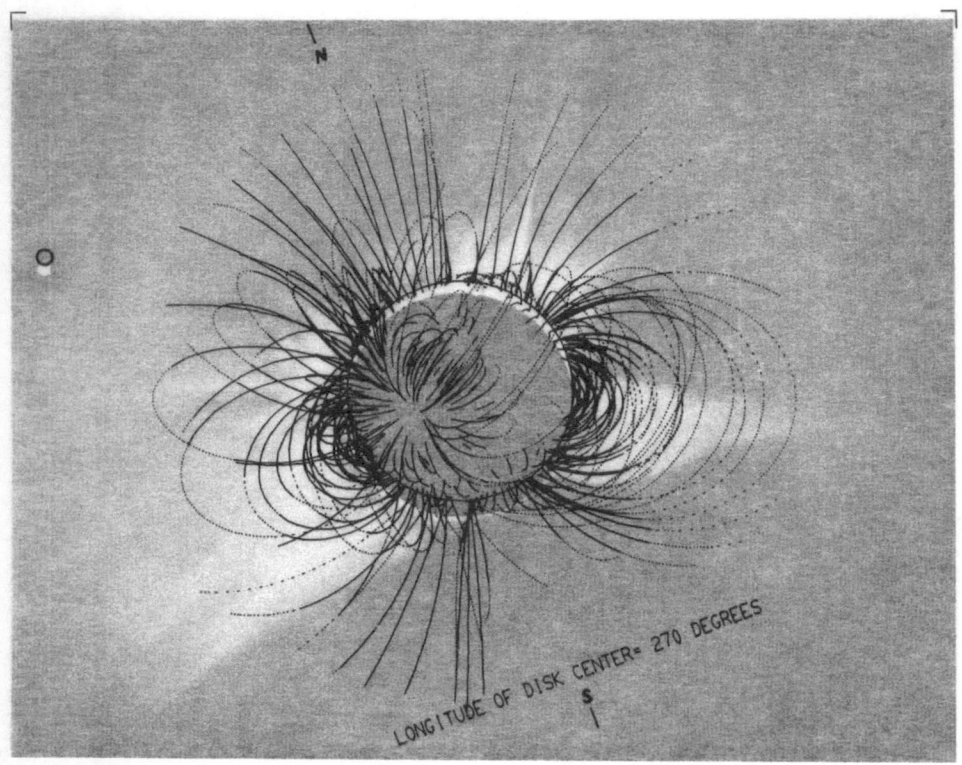

FIG. 1a. *Magnetic field lines in the solar corona projected against the plane of the sky for November 12, 1966. The foot points of the field lines are chosen geometrically, and lines originating at locations where the field is less than 0·5 gauss are not displayed. Dashes indicate field lines for which the kinetic-energy density of the solar wind exceeds the energy density of the field.*

Since our determination of the three-dimensional locations of the various coronal structures is incomplete, we can, for the present, compare only the appearance of the projected magnetic-field lines and the corona. Such a comparison is, however, suggestive. Of particular note are the following correspondences:

(1) The SW and SE streamers appear to be located over arcades of magnetic loops.

(2) The bush-like structure at the NE limb appears to be associated with magnetic loops passing over the limb. Similar loops appear within the corona.

(3) The outer corona above the Northern hemisphere displays a complex of gently curved rays whose shapes match those of the magnetic-field lines.

(4) The coronal condensation in the NW quadrant appears at a location where many field lines converge, although other similar convergences show nothing remarkable in the corona.

Our preliminary analysis suggests that the potential magnetic fields below $2R_{\odot}$ yield a reasonable approximation to the true fields present in the corona and that such fields are, in fact, a most important agent for determining the structure of the lower corona.

## Acknowledgments

The photograph of the corona used in this study was produced with the close cooperation of H. Hull and L. Lacey of High Altitude Observatory. We are indebted to R. Howard of Mt. Wilson Observatory for the use of his measurements of the surface fields and to F. Meyer, H. U. Schmidt, and J. D. Bohlin for stimulating discussions.

**Note added in proof:** In the method presented here, the Legendre coefficients are calculated in Equation (7) on the assumption of a completely radial surface field. Since this talk was presented, the Legendre coefficients have been recalculated by a least-mean-square fit with respect to the line-of-sight magnetic field. This new method calculates the best possible potential magnetic field for the data available. The qualitative results for $n=5$, however, are essentially the same as presented here in Figure 1a. When the Legendre Polynomial is expanded to $n=9$ or higher order, even more striking similarities arise between the calculated magnetic field and the observed coronal structure. These new results will be the basis of a forthcoming paper.

## References

Chapman, S., Bartels, J. (1940)      *Geomagnetism*, Oxford University Press, London.
Malitson, H. H., Erickson, W. C. (1966)      *Astrophys. J.*, **144**, 337.
Meyer, F., Schmidt, H. (1966)      *Mitt. Astron. Geos.*, **21**.
Newkirk, G. (1967)      *Annual Rev. Ast. and Ap.*, **5**, 213.
Whang, Y. C., Liu, C. K., Chang, C. C. (1966)      *Astrophys. J.*, **145**, 255.

## DISCUSSION

*H. U. Schmidt:* Why do you choose the ratio between undisturbed magnetic and total material energy density as a parameter to describe the break-up point of fieldlines, i.e. the lower end of current sheets separating zones of opposite polarity in the solar wind and overlying such zones in the solar photosphere? I think one should rather choose the ratio between magnetic and solar wind energy density only. This parameter has quite a different dependence on the height than your parameter, and I think it can be determined from the Parker model with a sufficient accuracy.

*Newkirk:* I agree that the comparison with the energy density of the solar wind would have been a better criterion. However, we wished to use a condition which could also tell us whether the field could support low-elevation density structures as well. (**Note added in proof:** The final manuscript makes just this comparison.)

*Davis:* Your very beautifully presented calculations omit one factor which I feel to be important. It is a reasonable approximation to use potential fields, but not to apply boundary conditions on the flux at the solar surface only. Over the equatorial regions at least, and presumably over the entire Sun, the solar wind sweeps a substantial fraction of the total flux out to infinity. This will substantially modify the calculated field and, in particular, is likely to substantially reduce the number of large arches.

*Newkirk:* Our criterion of comparing the energy density of the material and in the magnetic field is admittedly only a crude way of indicating what portions of the field lines are valid. Our pictures of the dashed lines indicate field lines which are probably strung out by the solar wind. At present we cannot solve the complete problem and calculate the form of the lower portion of the lines as modified by the flow above.

# PHOTOGRAPHS OF CORONAL STREAMERS
## FROM A ROCKET ON MAY 9, 1967*

R. TOUSEY, G. D. SANDLIN, and M. J. KOOMEN **

*( E. O. Hulburt Center for Space Research, U.S. Naval Research Laboratory, Washington, D.C., U.S.A.)*

ABSTRACT

A photograph of the white-light corona from a rocket on May 9, 1967 showed many streamers, all straight and nearly radial, extending across the field of view, 3 to $9R_\odot$. At the North there were three spectacular streamers, making angles with the polar axis of 21° E, 20° W and 27° N. Projected inward, they were radial and passed through the probable positions of the principal plages on the back side.

The most recent flight of the U.S. Naval Research Laboratory's white-light corona-graphs took place on May 9, 1967. Two externally occulted, small Lyot coronagraphs were built into the instrument compartment of the biaxial pointing-control section of an Aerobee-150 rocket. The instrument and some earlier results have been de-scribed by Tousey (1965), Purcell *et al.* (1967), and Koomen *et al.* (1967). In the present note attention is directed to the straight character of the streamers and to the possible origins of the three which appear to come from regions not far from the North Pole.

Simultaneous photographs from the two instruments are reproduced in Figure 1, as unretouched, positive prints. The dark central region is the shadow cast by the external occulter; the dark sectors, in the NW of one and in the SE of the other, are the shadows of the arms supporting each occulter. At the very center of one a white disk has been introduced to show the location and size of the Sun. The other instru-ment became slightly misaligned during launch, which accounts for the bright crescent at the edge of the occulter's shadow. The decrease in brightness of the corona and streamers with increasing *R* was compensated by the vignetting action of the external occulter. The useful field covered the range 3 to $9R_\odot$ in the plane perpendicular to the line of sight.

A few artifacts are present, caused by specks on a lens; and the short, bright streak in the NW of the one photograph is the trail of a sunlit particle just beyond the occulter. Mercury is seen almost out of the field in the West, and drawn into a short arc by the precession-caused rotation of the instrument around the solar vector.

* Supported by the National Aeronautics and Space Administration.
** Presented by R. Tousey.

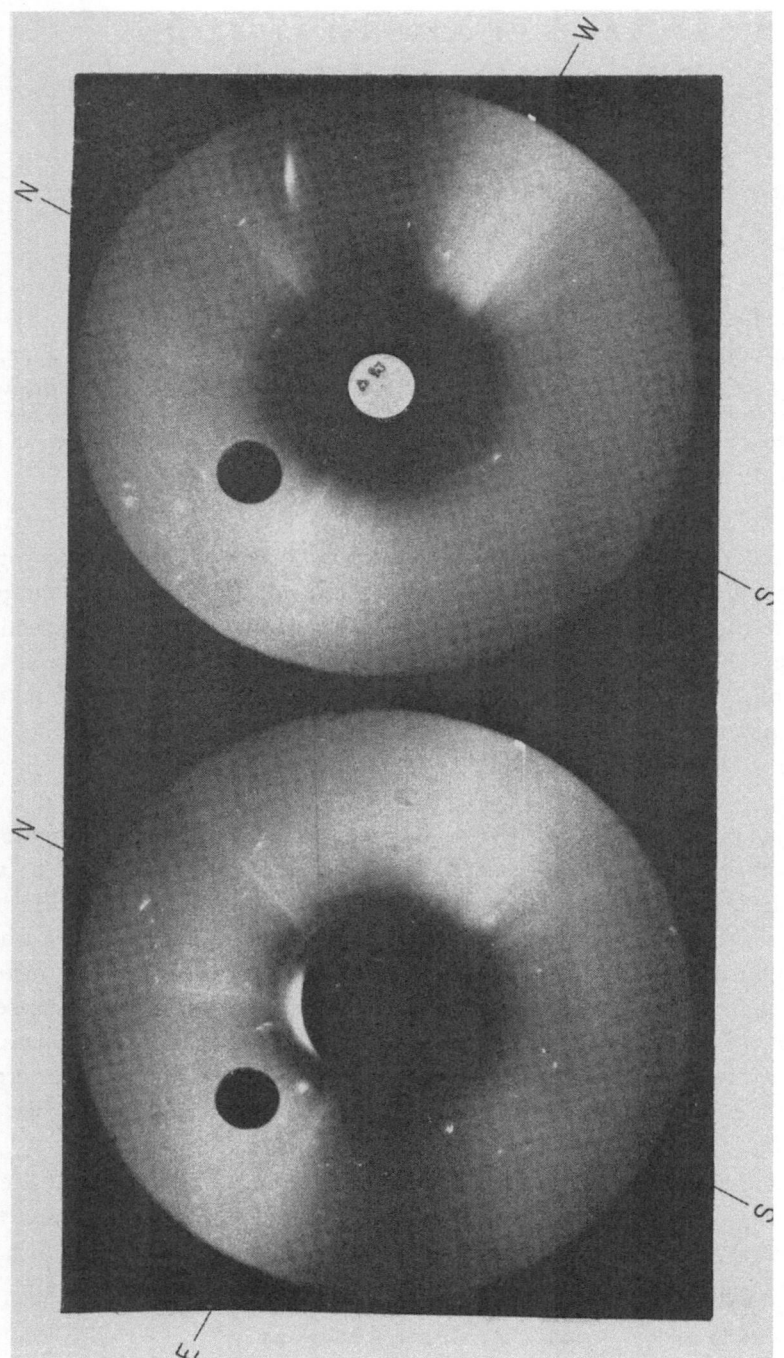

FIG. 1. *Simultaneous photographs of the white-light corona, made with a pair of externally occulted Lyot coronagraphs, flown from the White Sands Missile Range on May 9, 1967 at 1618 UT. Introduced into the central shadow cast by the occulter of one instrument is a disk showing the size and location of the Sun, and the approximate locations and shapes of the plages, believed to be on the back side of the Sun. The sharp black disk is the Moon, and the bright, short arc at the edge of the field is Mercury. The bright arc in the left image is evidence of slight misalignment; the streak in the right image is a particle trail, out of focus because it was not far beyond the occulter.*

The small, black disk East of the Sun is the Moon. The rocket was launched intentionally on May 9 at 1618 UT in order to have the Moon in the field of view. This aspect of the program has been described by Koomen *et al.* (1967). The black character of the shadow of the occulting disk and of the Moon shows that the instrumental stray-light level was negligible. The Moon's features are clearly visible in earthshine on the original film.

There are many conspicuous coronal streamers, especially at low solar latitudes. All of the streamers are straight; therefore, the range beyond $3R_\odot$ is largely outside the region where the corona is controlled by complex magnetic fields. It will be difficult to locate the origins of the low-latitude streamers because of the great amount of activity and because of the overlapping and merging of the streamers.

The streamers at high North latitudes can be studied in more detail. The NE streamer is single and appears to form a 21° angle with the solar axis; the NW streamer is double with angles of 20° and 27° to the axis. It appears unlikely that they really originated at high latitudes, because there were no large prominences in suitable positions and because sunspots and intense active centers were not found at those latitudes. They could very well have come from low-latitude centres of activity on the front or back side and thus be seen in projection. On the front side, however, there were no features from which the streamers conceivably could have arisen.

An investigation was made to determine whether or not intense active regions on the back side might have served as origins for these streamers. Examination of the Ca − K spectroheliograms for April 25 and May 23, 14 days earlier and later than May 9, showed plages in favorable locations, at latitudes where the solar rotation period is about 28 days. These regions were very large and intense, containing large spots at coordinates 31 °E, 26 °N under plage # 8785, at 19 °E, 35 °N under plage # 8793 which first appeared on April 28, and at 22 °W, 23 °N under plage # 8778. On May 24 the same regions persisted on the Sun, remaining very intense and at roughly the same positions but having grown markedly in area; the Eastern region, designated # 8818 in this rotation, covered about 8500 millionths of the hemisphere, and the Western region covered about 2500 millionths of the hemisphere. The Eastern region at 247° longitude was the most active region on the Sun in May, and was the site of the proton-flare events in the last week of May. Average positions for these regions were taken and transferred to the back of the Sun. Their locations and approximate forms are indicated on the white disk in Figure 1 as a view seen through the Sun.

A calculation was made of the apparent angle between each streamer and the solar axis, using the assumed positions of the active centers, making the assumption that the streamers were radial, and taking into account the tilt angle of the Sun and the perspective. Table 1 gives the results.

These values are in very good agreement, in consideration of the errors of measurement and of the uncertainties involved in estimating the centers of the plages.

The conclusion is that these streamers came from active centers that were 50° to 53°

## Table 1

### Apparent angle between streamer and solar North

| Plage #/Streamer | 8778/NE | 8785/NW | 8793/NW |
|---|---|---|---|
| Observed | 21° | 27° | 20° |
| Calculated | 20° | 29° | 17° |

heliocentric angle behind the limb, and that they were radial. Because of perspective, they were really recorded to about 15 $R_\odot$. Although they were favorably located to show any twisting, there is no indication of a 'garden hose' effect because they are straight, and project inward along radii passing through the large centers of activity present on the rear side.

## Acknowledgements

We are greatly indebted to Dr. Helen Dodson-Prince, and Miss E. Ruth Hedeman of the McMath-Hulbert Observatory for making available their solar data and for their discussions of the histories of the active regions.

## References

Koomen, M. J., Tousey, R., Seal, R. T. Jr. (1967)    in *Space Research*, VIII.
Purcell, J. D., Tousey, R., Koomen, M. J. (1967)    in *Space Research*, VIII.
Tousey, R. (1965)    *Ann. Astrophys.*, **28**, 600.

# NEW ASPECTS OF THE ROLE OF DEVELOPMENT AND STRUCTURE OF SOLAR ACTIVE REGIONS IN THE ARRANGEMENTS OF THE CORONA BASED ON ITS GEOMAGNETIC DISPLAYS

B. BEDNÁŘOVÁ-NOVÁKOVÁ and J. HALENKA*

*(Geophysical Institute of the Czechoslovak Academy of Sciences, Prague, C.S.S.R.)*

## ABSTRACT

The coronal plasma, which is the cause of geomagnetic storms, can impinge on the Earth only when some coronal formation is pointing towards the Earth. If no such formation is directed towards the Earth, a period of geomagnetic calm follows.

It was found that certain coronal formations, governed clearly by the appropriate local magnetic fields, correspond to the individual stages of development of active centres. Above the sunspot groups, there occur either conical rarifications or cylindrical, condensed fluxes dependent on a relatively non-variable, or unstable magnetic field. The filaments which outlive the sunspot period, provided they are located in floccular fields, are appropriate to streamers of helmet shape. The final stage of the active region, characterized by the presence of filaments only (called 'free filaments' by the authors), does not change the arrangement of the corona appreciably, i.e. the normal shape, or so-called minimum type. On the basis of the mentioned relations, it is possible to determine the shape and direction of so far currently unobservable coronal structures from chromospheric situations.

The mentioned facts enable a unified interpretation of the geomagnetic activity during the whole of the solar cycle to be made. The differences given for some types of storms (sporadic and recurrent, sudden and gradual commencement), may be explained by the various arrangements, occurrences or absences of local magnetic fields. The paper presents examples of chromospheric situations which were used, in some cases, at the Geophysical Institute in Prague for forecasting geomagnetic activity.

## DISCUSSION

*Fokker:* I should like to inquire whether the author's results apply to magnetic storms *with sudden commencement*. There is general agreement between solar scientists that sudden commencements can, for the greatest part, be related to certain individual flares, notably such flares as produce type-IV radio outbursts. I wonder therefore why it should be necessary to invoke filaments as being directly connected with sudden commencements of geomagnetic storms.

*Halenka:* Our results do apply to magnetic storms with sudden commencements, as is explicitly stated in the text (see 'Unstable Filaments' and Appendix). We forecast SSCs successfully on the basis of spectrohelioscopic observations by Bednářová-Nováková (*Travaux géophys.* 1964, NČSAV Praha 1965, 277).

In our opinion, the solar process – apparently changes in magnetic fields –, the consequence of which is the emission of the corpuscular stream responsible for a geomagnetic storm, may also be accompanied by a flare if the conditions for its formation are favourable. The flare in itself is not the necessary condition for the emission, which may be seen from the fact that a great number of even SC storms are not preceded by any flare (Bednářová-Nováková, B., *Travaux géophys.* 1966, Academia Praha 1967, 477). A comparison of the occurrence of radiobursts and disappearing filaments shows that the emission of particles producing geomagnetic storms need not be apparent in the form of radiobursts (Halenka, J., *Travaux géophys.* 1965, Academia Praha 1966, 395).

* Presented by J. Halenka.

*Kiepenheuer (ed.), Structure and Development of Solar Active Regions,* 389. © *I.A.U.*

# ACTIVE REGIONS AND THE INTERPLANETARY MAGNETIC FIELD*

JOHN M. WILCOX,
*(Space Sciences Laboratory, University
of California, Berkeley, Calif., U.S.A.)*

NORMAN F. NESS
*(Goddard Space Flight Center,
Greenbelt, Md., U.S.A.)*

and

KENNETH H. SCHATTEN
*(Space Sciences Laboratory, University of California, Berkeley, Calif., U.S.A.)*

## ABSTRACT

The relation of solar active regions to the large-scale sector structure of the interplanetary field is discussed. In the winter of 1963–64 (observed by the satellite IMP-1) the plage density was greatest in the leading portion of the sectors and lesser in the trailing portion of the sectors. The boundaries of the sectors (places at which the direction of the interplanetary magnetic field changed from toward the Sun to away from the Sun, or vice versa) were remarkably free of plages. The very fact that since the first observations in 1962 the average interplanetary field has almost always had the property of being either toward the Sun or away from the Sun (along the Archimedean spiral angle) continuously for several days must be considered in the discussion of large-scale evolution of active regions. Using the observed interplanetary magnetic field at 1 AU and a set of reasonable assumptions the magnetic configuration in the ecliptic from 0·4 AU to 1·2 AU has been reconstructed. In at least one case a pattern emerges which appears to be related to the evolution of an active region from an early stage in which the magnetic lines closely couple the preceding and following halves of the region to a later stage in which the two halves of the region are more widely separated.

Spacecraft observations from the time of the first extended observation of the interplanetary medium by Mariner 2 in 1962 to the present time have indicated that the sector structure is almost always a prominent feature of the interplanetary magnetic field. The sector property of the interplanetary field is defined to be its tendency to remain for several consecutive days predominantly directed either away from the Sun or toward the Sun. The average direction of the field is usually close to the theoretically predicted Archimedes spiral. Large dynamic deviations from the Archimedes spiral are frequently observed; nevertheless the field can usually still be clearly characterized as being either directed predominantly toward the Sun or away from the Sun.

---

* Presented by J. M. Wilcox.

*Kiepenheuer (ed.), Structure and Development of Solar Active Regions,* 390–394. © *I.A.U.*

The interplanetary sector structure observed by IMP-1 in the winter of 1963–64 during the decline of the last 11-year sunspot cycle was a quasi-stationary pattern during several solar rotations. This is presumably related to the low level of solar activity and the resultant small amount of perturbing influences on the sector pattern. The sectors observed at this time had on the average an internal structure such that the efflux of solar wind plasma and the strength of interplanetary magnetic field were largest in the preceding portion of the sectors. The position of active regions on the Sun with regard to the interplanetary sector pattern has been investigated in the following way. A cross-correlation analysis by Ness and Wilcox (1966) showed that the average time lag from the appearance of a photospheric magnetic feature at central meridian to the observation of this feature by the spacecraft at 1 AU was about $4\frac{1}{2}$ days. Thus, if a sector boundary is observed by a spacecraft at 1 AU on a certain date, the date at which this boundary on the Sun was near central meridian can be obtained by subtracting this $4\frac{1}{2}$-day lag. When the date at which a sector boundary is near central meridian on the Sun has been determined by this method, the Fraunhofer Institute daily map of the Sun for this day provides the location of plages on the disk. A photographic superposition is obtained of the plage locations for all days

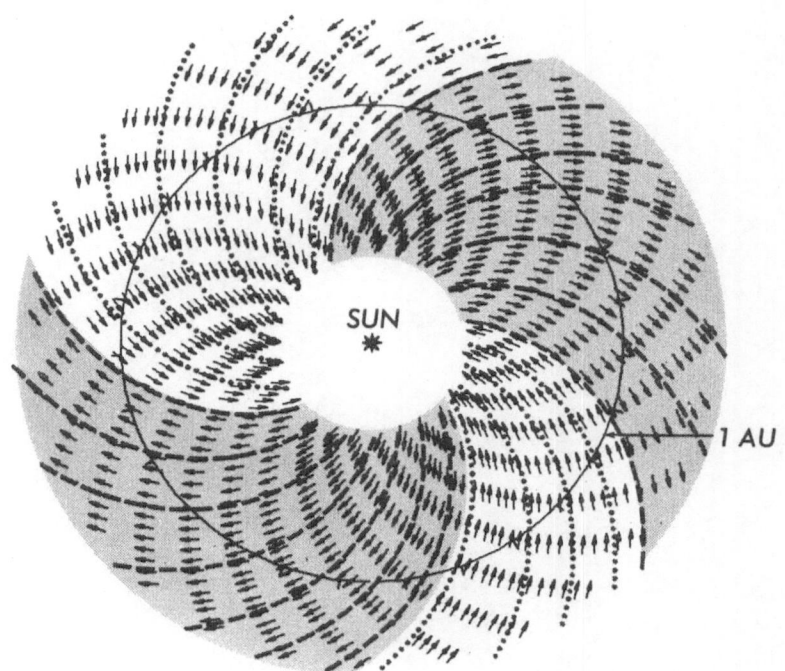

FIG. 1.   *Map in ecliptic plane of average interplanetary magnetic-field sector structure observed by IMP-1. The extrapolation technique is described in the text. Each arrow represents an equivalent magnetic flux of 5 gamma for 6 hours.*

on which a sector boundary was near central meridian. This superposition indicates
that the probability of occurrence of a plage is greatest in the preceding portion of a
sector, and declines in the trailing portion of a sector, with the sector boundaries
being relatively free of plages. Details of the method have been given by Wilcox and
Ness (1967); see in particular their Figure 4.

The birth of an away sector within a large toward sector has been observed with
the following method. A map of the configuration of the interplanetary field projected
onto the ecliptic in the radial interval from 0·4 to 1·2 AU has been constructed by
extrapolating satellite measurements at 1 AU. The radial component of the field is
assumed to scale as $1/R^2$ and the azimuthal component of the field is assumed to scale
as $1/R$. An Archimedes spiral is a special case of this assumption, as shown in Figure 1,
which is a map of the average sector structure observed by IMP-1. Figure 1 is included
to assist in explaining the method of analysis. It can be seen in Figure 1 that if the
observations at 1 AU are assumed to consist of field directed at 45° to the Earth–Sun
line the Archimedes spiral is recovered, and the sector boundaries remain as sharp
discontinuities in the direction of the field. The extrapolation is meaningful for those
features of the interplanetary field which do not change significantly in the 2- or 3-day

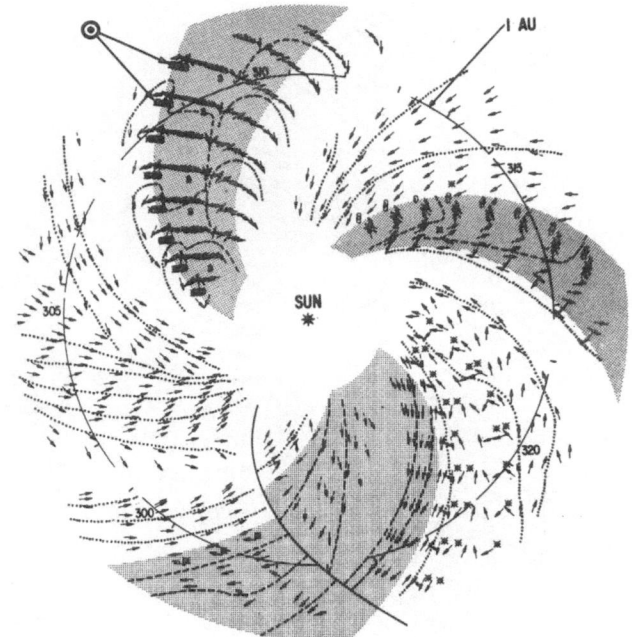

Fig. 2. *Map in ecliptic plane of interplanetary magnetic field observed by IMP-3 during Carrington
solar rotation 1500. Intervals in which the field is directed more than 45° from the ecliptic are indicated
with small circles. Intervals in which the field fluctuates to a large extent within the averaging interval
are indicated with small stars.*

interval required for the solar wind to flow from 0·4 AU to 1 AU. Since the sector pattern has often been quasi-stable over several solar rotations it can be expected that there will be significant features of the interplanetary field which will be invariant over a 2- or 3-day interval.

The method described above has been used to construct a map of the interplanetary field actually observed by IMP-3 during solar rotation 1500 in 1965, and the results are shown in Figure 2. The arrows are produced by the scaling process described previously, and the dashed lines are drawn by hand to represent typical field lines. The absolute magnitude of the field vector is represented by the spacing of the arrows, not by their length. Sector areas with field directed predominantly away from the Sun are shaded. The solid curved line near the bottom of the figure represents the cut between the beginning of observations on October 24 and the end of observations on November 20.

The away-from-the-Sun (shaded) sector near the top of Figure 2 is of interest for the present discussion. This represents a new away sector, since in the previous solar rotation this region was part of a large toward sector. The field configuration within this new sector is in the form of magnetic loops which are being convected away from

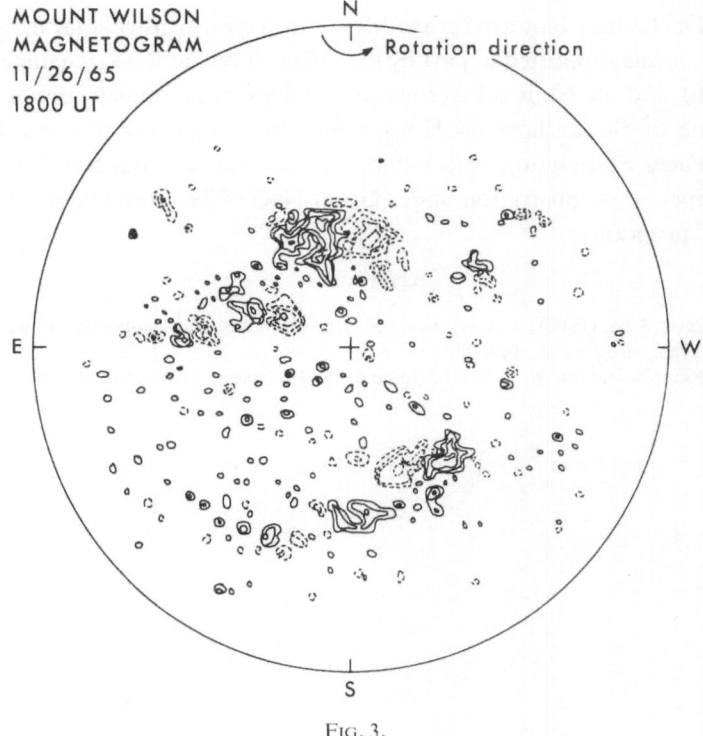

FIG. 3.

the Sun by the solar wind plasma. In the subsequent solar rotations this sector develops into a substantial away sector having on the average an Archimedes spiral configuration. The loop structure in this sector is observed only in the rotation shown in Figure 2. The central portion of the loop structure, which has magnetic flux directed away from the Sun, is centered at 1 AU on day 310, 0 hours.

Seven days earlier on day 302, 2300 UT, Mt. Wilson Observatory obtained the daily solar magnetogram shown in Figure 3, which displays the line-of-sight component of the photospheric magnetic field. The contours levels are 6, 12, 20 and 30 gauss with solid lines representing fields directed out of the Sun. The large bipolar region near central meridian extending from 15°N – 30°N appears to be the source of the magnetic loops shown in Figure 2. This bipolar region first appeared on the Sun two rotations before the time of Figure 3, and was observed in the next solar rotation to be somewhat dispersed and expanded. The amount of the magnetic flux convected away from the Sun in the loop structure of Figure 2 appears to be in reasonable agreement with the magnetic flux contained in and near the bipolar region shown in Figure 3. A more detailed discussion will be given in a paper presently under preparation.

## Acknowledgements

We thank Dr. Robert Howard for the Mt. Wilson Observatory solar magnetogram.

This research was supported in part by the Office of Naval Research under Contract Nonr 3656(26), and the National Aeronautics and Space Administration under Grant NsG 243. One of the authors (K.H.S.) would like to express his gratitude for a National Sciences Foundation Fellowship and for support from the National Aeronautics and Space Administration under Grant NsG 695 for the University of Maryland summer program.

## References

Ness, N. F., Wilcox, J. M. (1966)    Extension of the Photospheric Magnetic Field into Inter-planetary Space, *Astrophys. J.*, **143**, 23.
Wilcox, J. M., Ness, N. F. (1967)    Solar Source of the Interplanetary Sector Structure, *Solar Phys.*, **1**, 437.

# EXTREME ULTRAVIOLET OBSERVATIONS OF
# ACTIVE REGIONS IN THE SOLAR CORONA

W. M. BURTON

*(Culham Laboratory, Abingdon, Berks., England)*

ABSTRACT

The coronal features associated with solar active regions can be observed by recording images of the Sun at extreme ultraviolet (XUV) wavelengths. Pinhole cameras have been flown on stabilized sun-pointing 'Skylark' rockets to obtain broad-waveband XUV solar images. These images show localised emission from high-temperature regions located in the corona above calcium-plage areas. An improved design of pinhole camera, which uses a plane-diffraction grating to give increased spectral resolution, has recorded spectroheliograms in several intense solar lines including He II (304 Å), Fe IX–XI (180 Å), and Si X–XII (50 Å). Estimates are made of the size and brightness of the coronal emission region associated with a developing calcium-plage area.

## 1. Introduction

The structure of a solar active region can be studied by recording spectroheliograms in various spectral lines which are formed at different temperatures in the Sun. Each line then defines a different level in the solar atmosphere, so that a detailed picture can be obtained showing the three-dimensional structure of the active region.

Spectroheliograms recorded in the hydrogen Hα and Ca K lines are valuable for studying the chromospheric structure of active regions, but they provide no information about the high-temperature coronal features which may be associated with the calcium plages.

In order to study the coronal regions which extend above a solar centre of activity (CA), the observations should be made using a wavelength at which the brightness of the active region is greater than that of the undisturbed solar disk. This criterion is satisfied in two distinct wavebands, firstly at 10-cm radio wavelengths and secondly at much shorter wavelengths in the X-ray and extreme ultraviolet (XUV) region of the spectrum.

This paper describes some observations of the coronal features associated with active regions, obtained by recording solar images in the 20–400 Å XUV wavelength region.

## 2. Broad-band XUV Solar Images

A simple pinhole camera can be used for recording photographic images of the Sun at XUV wavelengths (Blake *et al.*, 1963). The camera used for the present observations

*Kiepenheuer (ed.), Structure and Development of Solar Active Regions, 395–402. © I.A.U.*

had an optical length of 25 cm and so formed a solar image 2·3 mm in diameter. The angular resolution of the camera was about 2 min of arc as determined by the pinhole aperture of 0·15 mm. Special XUV sensitive photographic film (Kodak-Pathé SC5 and SC7) was used to record the solar images.

The cameras have been carried on stabilized 'Skylark' rockets launched from Woomera, South Australia. A peak height of about 200 km can be reached, which provides an exposure period of 200 sec during the part of the flight when atmospheric absorption is reduced sufficiently for XUV studies of the Sun. An attitude-control system stabilized the rocket payload relative to the Sun, giving spatial resolution of about 1 min of arc on the solar disk, together with a few degrees of disk rotation during the exposure period (Cope, 1964).

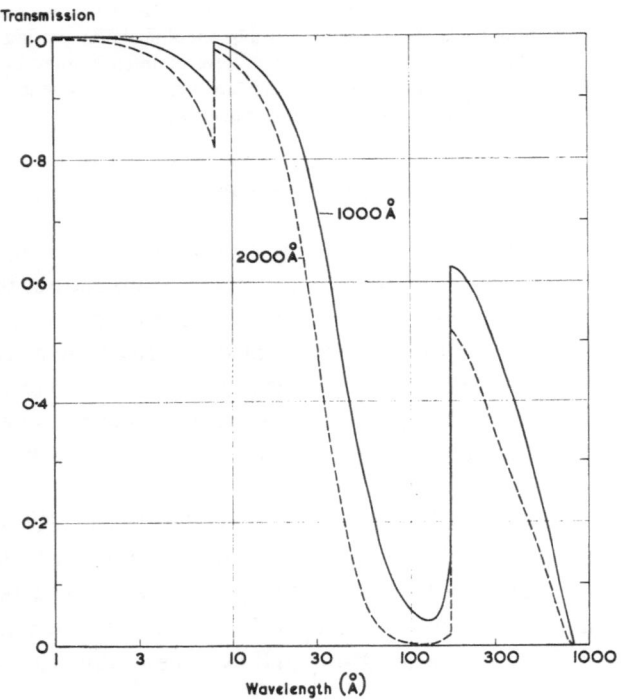

Fig. 1.   *Calculated transmission curves for aluminium filters of 1000 Å and 2000 Å thickness.*

The spectral response of the camera is set by a thin film of aluminium which is placed over the pinhole aperture to exclude visible light while transmitting XUV radiation. Figure 1 shows the variation of transmission with wavelength for typical aluminium filters. A 2000-Å thickness filter is opaque at wavelengths longer than 800 Å, but has two useful transmission bands, one from 800 Å to the L-edge at 170 Å and the other extending from 80 Å down to the X-ray region. In practice, the observed wavebands are limited further by residual atmospheric absorption and by the form of the solar emission spectrum. The effective wavebands producing the solar image for this type of camera are consequently 20–70 Å and 170–400 Å.

A camera with this broad-band response was flown on December 17, 1964 and again on April 9, 1965. Solar images were formed by a blend of XUV emission lines, of which the most important were spectral lines of Six–xii (40–60 Å) and Feix–xii (170–220 Å). These ions have their maximum equilibrium population in a thermal plasma at a temperature of about $2 \times 10^6$ °K and the emission therefore defines regions of enhanced density at this typical coronal temperature.

The XUV image recorded on December 17, 1964 is shown in Figure 2 together with the Fraunhofer Institute solar map for the same day. The image is displayed

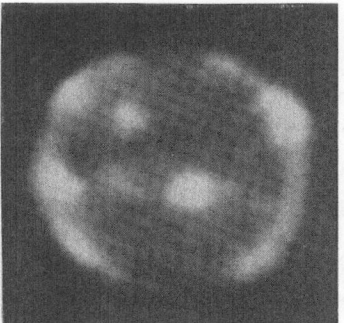 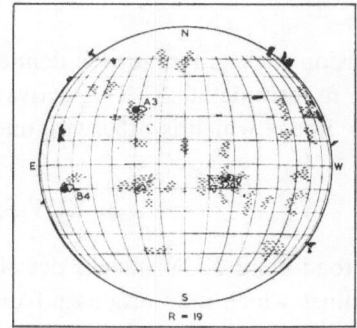

FIG. 2. *Solar XUV image recorded on December 17, 1964, together with Fraunhofer Institute solar map for the same day.*

with North at the top and the East limb to the left. Although near the minimum in the solar cycle, the Sun was relatively active at the time of observation. Strong limb brightening is present at middle latitudes but the polar regions are very dark. Several distinct active regions can be seen in the XUV image, with some localised darker areas between them. All of the XUV emission regions correspond with calcium-plage areas. The age range of these plages is very large, suggesting that the coronal emission region is a persistent feature with a lifetime similar to that of the calcium plage.

The XUV emission region on the NE limb is associated with a calcium-plage area which was not visible on the solar disk until the day following this observation. At the time of the rocket flight the plage area was located about 12° behind the NE solar limb. This implies that the coronal emission source was located at least 10000 km above the solar photosphere. Similarly, the XUV source on the NW limb persists strongly although part of the underlying calcium plage has moved behind the limb.

On April 9, 1965 a similar camera was used to record the XUV image which is shown in Figure 3 with the solar map for that day. The Sun was rather inactive on

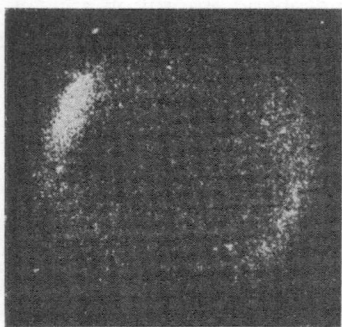 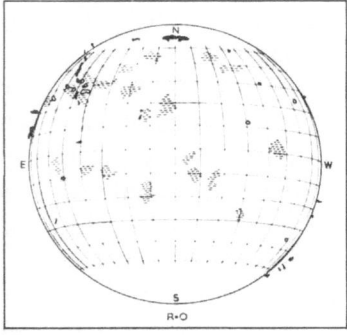

FIG. 3.   *Solar XUV image recorded on April 9, 1965, together with Fraunhofer Institute solar map for the same day.*

this occasion with only one well-defined plage area, which was located close to the NE limb at latitude 30 °N. This active region coincides with an emission feature in the XUV image which is otherwise underexposed.

### 3. XUV Spectroheliograms

The broad-band XUV images described above are formed by a blend of several spectral lines which may originate from different levels in the solar atmosphere, so complicating the analysis of the images. In an attempt to increase the spectral resolution, an improved pinhole camera has been developed which uses a plane diffraction grating at grazing incidence to provide wavelength dispersion (Burton, 1965). An outline diagram of this grating-pinhole camera is shown in Figure 4. The camera

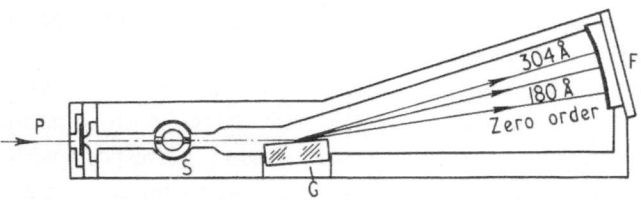

FIG. 4.   *Grating-pinhole camera forming dispersed XUV solar images. P = pinhole aperture covered by aluminium filter. S = shutter to control exposure duration. G = 600 line/mm plane diffraction grating. F = photographic film to record images.*

length is again 25 cm, so that the solar image diameter is about 2·3 mm. Since the reciprocal dispersion is only 20 Å/mm, the camera forms overlapping images in spectral lines, which are spaced by less than 50 Å. The grating is blazed for optimum efficiency at 180 Å, the wavelength of a group of intense spectral lines of highly ionized iron which are prominent in the solar XUV spectrum.

The grating-pinhole camera forms a broad-band XUV image at the zero-order position and also a first-order spectrum between 40 Å and 400 Å in which solar images are formed by each strong emission line. Figure 5 shows composite spectro-heliograms recorded by this instrument on two flights: April 9, and October 20, 1965. Wavelength dispersion is in the direction of the solar North-South polar axis and the West limb is at the top of each dispersed spectrum. Spectral resolution is limited by overlapping images but several separate features can be seen. The He II resonance line at 304 Å which is formed mainly in the chromosphere, shows a uniform disk image

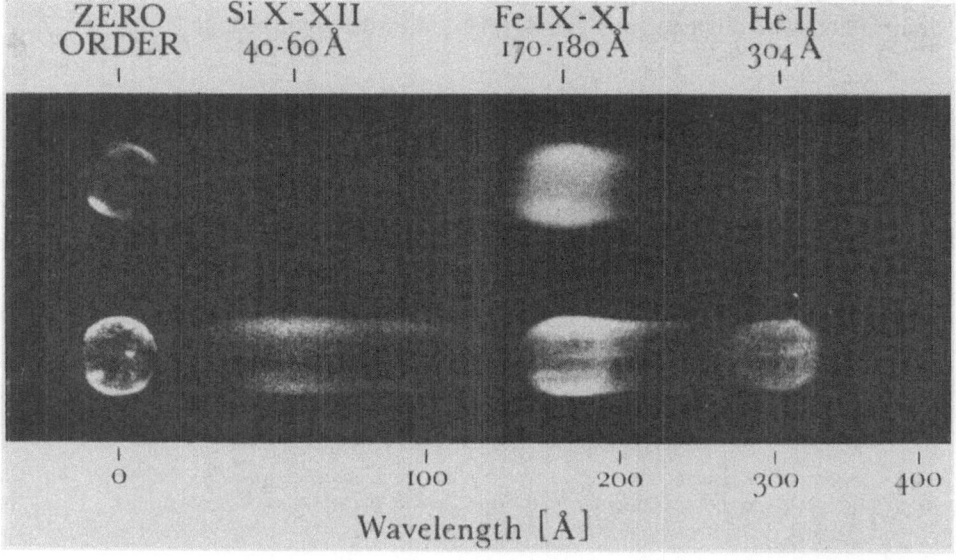

FIG. 5.    *Composite spectroheliograms (40-400 Å) from grating-pinhole camera. (a) April 9, 1965. (b) October 20, 1965.*

with slight limb brightening at this resolution. The most intense of the dispersed images is that formed by the group of intense ionized iron lines Fe IX–XI at 180 Å. Between 40 Å and 60 Å emission is seen from overlapping images in spectral lines of Si X–XII.

Several active regions can be seen in the October 20, 1965 images, all of which coincide with calcium-plage areas. Between the bright regions, the background-disk intensity is not uniform but shows localised dark regions which are comparable in area to the active regions. The bright source close to the disk centre (McMath plage no. 8032) is a developing CA, which was the site of a small solar flare recorded on the day before these XUV observations were made. In the He II 304 Å image, this is the only active region which can be detected with certainty, but in the other lines, formed at higher levels, this CA is similar in intensity to the other active regions.

The zero-order image is formed primarily by the group of ionized iron lines at 180 Å and is shown in more detail in Figure 6. Equatorial limb brightening is very strong since the emission comes mainly from the corona which is optically thin in these wavelengths. Emission can be seen from most of the solar disk but is particularly prominent across the middle latitude region in the Northern hemisphere. Several localised emission regions can be seen in this image and their size can be estimated by correcting the observed size for effects of camera resolution and pointing errors during the exposure period. The measurement is difficult for active regions close to the limb where the effects of limb brightening are important, but for the CA close to the disk centre (plage 8032) the size of the XUV source can be estimated as less than 1 min of arc, which is equivalent to $4 \times 10^4$ km in the solar corona.

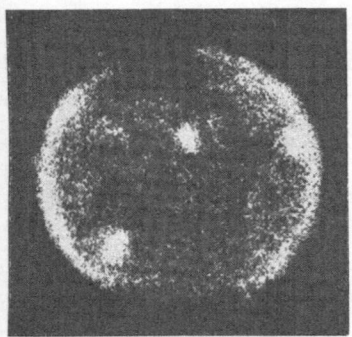 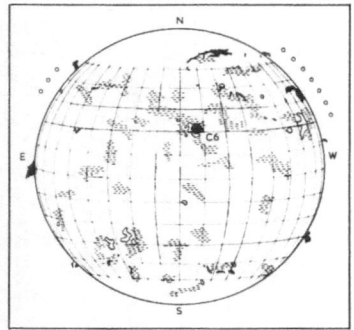

FIG. 6. *Solar XUV image recorded on October 20, 1965, together with the Fraunhofer Institute solar map for the same day. (Zero-order image from grating-pinhole camera.)*

By densitometer measurements of the photographic image, the XUV brightness of this active region is estimated to be about five times greater than the mean quiet solar disk background. This figure is an order of magnitude lower than the values obtained in similar observations using X-rays in the wavelength region below 60 Å (Blake *et al.*, 1963; Pounds and Russell, 1966). The observed brightness is related to the coronal electron density ($N_e$), and for each spectral line is proportional to $(N_e)^2 h$, where $h$ is a characteristic height for the emission region. The quantity $(N_e)^2 h$ is apparently greater in the active region than in the quiet undisturbed corona by a factor 5 in the XUV lines characteristic of a temperature of about $2 \times 10^6$ °K, and by a factor 50 in X-ray emission, which is formed at a rather higher temperature.

This result could be interpreted in terms of the density variation within the coronal active region if a reliable model of the temperature structure of a CA were available, but at the present time there is insufficient information to interpret the intensity data in this way.

## 4. Standardised Measurements on XUV Images

Solar images at XUV and X-ray wavelengths are now obtained regularly by several laboratories. The analysis of these images is not simple, and the data are seldom made available in a convenient form for use by those interested in solar physics. A standard procedure for measurement and definition of the XUV brightness of solar active regions has recently been proposed (Allen, 1967). The measurement of CA flux for particular XUV or X-ray wavebands is made in terms of 'solar disk units' to provide dimensionless numbers which can be used to describe the XUV importance of active regions in a similar manner to the use of the McMath calcium-plage index or the Zurich sunspot number.

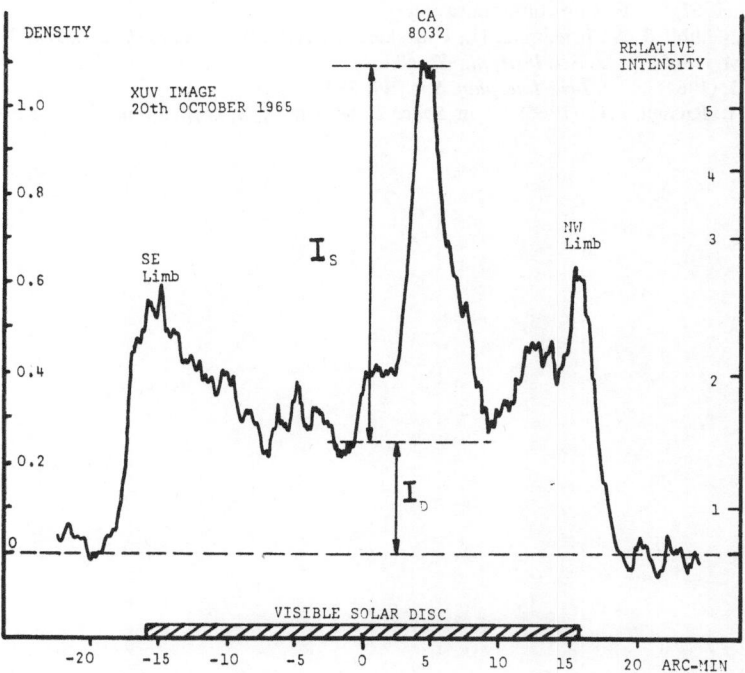

FIG. 7.   *Densitometer scan of XUV solar image, showing method of measurement and definition of CA flux in solar disk units.*

The measurement procedure is shown in Figure 7 with reference to a densitometer scan across the solar image for October 20, 1967 shown in Figure 6. Calibration of the photographic film is required to determine a characteristic curve which is used to convert the density measurements to relative intensities. The flux from each CA source then is estimated from the expression $F_s = \int I_s \, dA$, where $I_s$ represents the intensity of the source in excess of the quiet disk background intensity. The quantity

$F_s$ is standardised by measuring $I_s$ in units of $I_D$, the intensity at the centre of the quiet solar disk, and by measuring the apparent source area $dA$ in units of the projected solar disk radius $R_\odot$. Units based on $I_D$ and $R_\odot$ are then called 'disk units'.

For the images described in this paper the quantity $F_s$ for the XUV emission from active regions individual CA sources varies between 0·01 and 1·0. For X-ray measurements the value of $F_s$ in disk units is greater due to the increased contrast between the CA intensity and quiet-disk intensity. At shorter X-ray wavelengths the value of $I_D$ may be too low for reliable measurements, in which case the total solar flux would provide a better reference intensity for the standardisation of $F_s$ measurements.

## References

Allen, C. W. (1967)      Private communication.
Blake, R. L., Chubb, T. A., Friedman, H., Unzicker, A. E. (1963)      *Astrophys. J.*, **137**, 3.
Burton, W. M. (1965)      *J. Sci. Instrum.*, **42**, 477.
Cope, P. E. G. (1964)      *J. Brit. Interplan. Soc.*, **19**, 285.
Pounds, K. A., Russell, P. C. (1966)      in *Space Research*, VI, Spartan Books, New York, p. 34.

# PROTONS ASSOCIATED WITH CENTRES OF SOLAR ACTIVITY AND THEIR PROPAGATION IN INTERPLANETARY MAGNETIC-FIELD REGIONS CO-ROTATING WITH THE SUN*

C.Y. Fan, M. Pick, R. Ryle, J.A. Simpson, and D.R. Smith

## ABSTRACT

The Pioneer-6 and Pioneer-7 space probes carried charged-particle telescopes which measure, for the first time, both the direction of arrival and differential energy spectra of protons and alpha particles. The intensity changes, directional distributions and energy spectra of proton fluxes associated with solar activity are investigated. The data were obtained in the beginning of the new solar cycle (no. 20), when it is possible to unambiguously associate proton-flux increases with specific solar active regions. The origin, possibly long-term storage, and propagation of these proton fluxes are investigated. It was observed that enhanced 0·6–13 MeV proton fluxes associated with specific active regions were present over heliographic longitude ranges as great as ~ 180°. These enhanced fluxes exhibit definite onsets and cut-offs which appear to be associated with the magnetic-sector boundaries observed by Ness on Pioneer-6. Discrete flare-produced intensity increases extending in energy to more than 50 MeV are observed, superposed on the enhanced flux. These increases displayed short transit times and short rise times. Both the enhanced and flare-produced fluxes propagate along the spiral interplanetary magnetic field from the Western hemisphere of the Sun. From these observations we are led to a model in which the magnetic fields from the active region are spread out over a longitude range of 100–180° in the solar corona. The existence of strong unidirectional anisotropies in the initial phases of flare-proton events implies that little scattering occurs between the Sun and spacecraft. However, the gradual approach to an isotropic flux at late times indicates that the decay phase is controlled by the interplanetary magnetic field.

## DISCUSSION

*Acton:* Have you been able to decide where these low-energy protons are accelerated? Is it in the chromosphere, corona, or in interplanetary space?

*Pick:* The association of solar active regions with enhanced proton fluxes at low energies is a strong evidence for the solar origin (chromosphere or corona) of these protons. By another way, their energy spectrum is similar to solar flare-proton spectra.

*Noyes:* It was my understanding that there was no obvious solar activity associated with the proton-emitting region in December 1963. Was, in fact, some related activity detected?

*Pick:* This problem is not resolved. It is very difficult to know near the solar minimum cycle if there is a correlation between active regions and proton-flux enhancements. In December 1963, the situation was confused but we cannot say that there was no related solar activity.

* Presented by M. Pick, Observatoire de Meudon, France. To be published in full in *J. Geophys. Res.*, **73**, 1968.

# THE SOLAR CORONA ABOVE ACTIVE REGIONS:
# A COMPARISON OF EXTREME ULTRAVIOLET LINE EMISSION
# WITH RADIO EMISSION

WERNER M. NEUPERT

*(Goddard Space Flight Center, Greenbelt, Md., U.S.A.)*

ABSTRACT

The observations of extreme ultraviolet (EUV) emission lines of Feıx through Fexvı made by OSO-I have been applied to a study of the solar corona above active regions. Ultraviolet and radio emission are determined for several levels of activity classified according to the type of sunspot group associated with the active region. Both radio emission and line radiation from Fexvı, the highest stage of ionization of Fe observed, are observed to increase rapidly with the onset of activity and are most intense over an E spot group early in the lifetime of the active region. As activity diminishes, radiation from Fexv and Fexıv becomes relatively more prominent. Preliminary X-ray data from OSO-III obtained during a flare are introduced. These indicate that radiation from the highest stage of iron thus far observed, Fexxv, reaches a maximum first in an X-ray burst and that maxima in lower stages of ionization follow, with delays from 2 to 15 min.

## 1. Introduction

With the advent of rockets and satellites, it has become possible to make observations of permitted emission lines originating in the solar corona above active regions as these regions pass across the visible solar disk. The extreme ultraviolet (EUV) observations made by the first Orbiting Solar Observatory (OSO-I), launched on March 7, 1962, presented the first opportunity for such a study over an extended period of time. Contained in the scientific payload of this satellite was a spectrometer (Behring *et al.*, 1962) for recording the solar spectrum in the wavelength range from $\lambda$ 150 to $\lambda$ 400. The coronal emission lines of many elements were recorded for a time interval greater than 1 year and during a wide range of solar activity (Behring *et al.*, 1963). The observed 27 day variations in line intensities attributable to this activity have been discussed previously (Neupert, 1965).

EUV spectra of the entire Sun have recently been supplemented by slitless spectra (Tousey, 1965), which demonstrate directly the tendency for emission lines from the most highly ionized ions to be most intense over active regions. Because such studies have been made from rockets, no sequential study over the lifetime of any one region has been possible. The present paper attempts to describe the EUV and radio emissions of several specific active regions as they develop with time.

*Kiepenheuer (ed.), Structure and Development of Solar Active Regions*, 404–410. © *I.A.U.*

## 2. Active Regions observed by OSO-I

Our ability to discuss the development of active regions using the OSO-I data is limited by the fact that these observations provided no spatial resolution. The OSO-I EUV spectrometer recorded radiation from the entire visible solar hemisphere at all times, so that the observed counting rate in any spectral line represents the combined emission from all active centers as well as any radiation emitted by the undisturbed solar atmosphere. We have been able to discriminate between these two components by comparing data taken when the Sun was devoid of any active regions to data taken when one or more active regions were present on the visible solar disk. Minimum EUV line intensities coincided with a Zürich Relative Sunspot number near zero and a minimum in the Ca II plage area which occurred on March 9–11. All other EUV data are normalized to the observations of March 9–11. We refer to any increases above a value of one as 'relative enhancements' and assume that they are due to the presence of active regions on the solar disk.

A comparison of these EUV data with concurrent radio observations suggests that the variations of the highest stages of ionization of Fe which we have observed – Fe xvi and Fe xv – resemble most strongly the radio observations at frequencies around 2000 MHz. Emission lines from lower stages of ionization co-relate less well with radio fluxes and show no distinct association with either the higher or the lower range of frequencies although it appears that some fluctuations occurring over a period of only a few days can be traced through successively lower stages of ionization and also through the entire range of radio frequencies.

From these data it is possible to extract two sequences of observations showing changes from month to month in the EUV and radio emission from the corona above active centers. In each of these sequences the active regions were sufficiently similar so that the increases in EUV and radio fluxes could be associated with specific regions of a group of similar regions. The method used will be published in detail in *Solar Physics*. Results are given in Figure 1. The EUV increase attributed to active regions are again given in units of the flux from the Sun as observed on March 9–11. It should be reiterated that we have used several similar solar regions in arriving at these results. The time scale used in Figure 1 therefore does not apply to any one particular region observed throughout its entire lifetime, although it is correct for the development of one region which was observed in the Southern hemisphere.

The time dependence of the radio emission, as shown in Figure 1, generally confirms earlier observations made by Vauquois (1959). The highest reported frequency (9400 MHz) is emitted almost exclusively during the period of highest activity, while the lowest frequency persists even after sunspots have disappeared. Only weak, or, in most cases, no polarization of the radio emission was observed during the interval of March–May 1962.

Of all the EUV radiations observed by OSO-I, it appears that the fluxes from Fe xvi

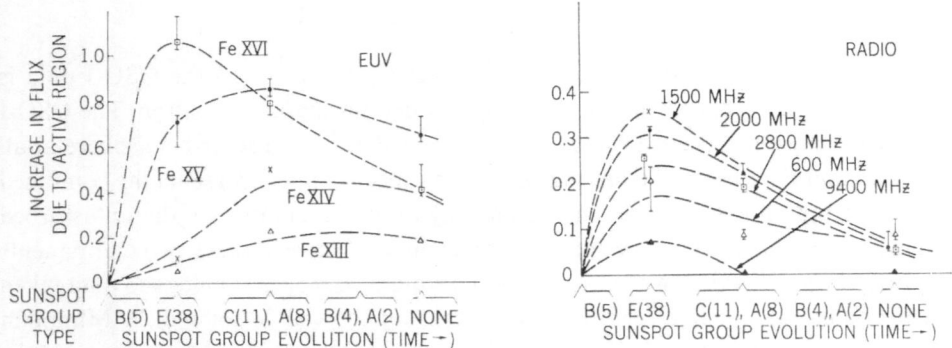

FIG. 1.   *The EUV and radio fluxes attributed to an active solar region at various stages in its develop-ment, as derived from OSO-I observations. The increases in emission are expressed in terms of the total solar emission observed on March 9–11, 1962, in the absence of appreciable solar activity. Note that maximum emission in the highest observed stage of ionization coincides with maximum complexity and size of the sunspot group. The time scale shown is typical for the type of region being discussed. Other such regions may have different rates of development.*

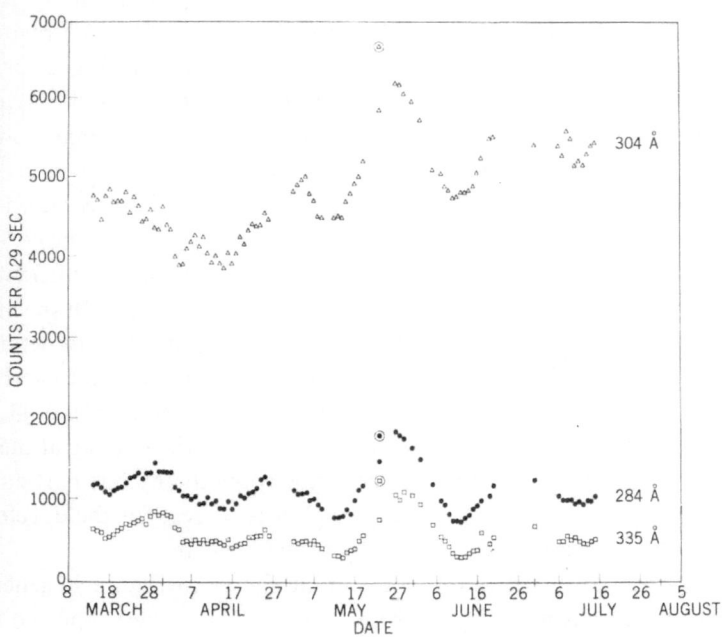

FIG. 2.   *Preliminary OSO-III data on the counting rates observed for emission lines of Fe*xv, *Fe*xvi *and He*II *for nearly 5 months of observation starting in March 1967. During this interval, the Zürich Provisional Sunspot Number reached a minimum of 17 on May 10 and on June 10 and 11. A maximum of 197 was attained on May 28. The circled data were obtained during a flare of importance 3 on May 23, 1967.*

and Fexv show the greatest increases over active regions. More precisely, it appears that the greatest increase is associated with the highest stage of ionization, at the time of maximum development of the sunspot group. The similarity of the Fexvi and decimetric radio results extends also to solar flares accompanied by ionospheric effects (Neupert, 1965). In that case the time of maximum emission and total duration of Fexvi emission coincides with the 'gradual rise and fall' observed at decimetric wavelengths. The emission lines from Fexiv and Fexiii associated with an active region are observed to reach a maximum after the peak of flare activity. Below Fexiii, the ultraviolet increases are small and less well-correlated with the appearance of active centers.

The time dependence observed for Fexiv agrees well with data presented by Le Roy and Trellis at this conference. We ourselves have studied the coronal limb observations at λ 5303 for March–May 1962, and observe that the forbidden line intensity confirms the dependence on stage of activity as deduced from the EUV data. The results obtained from this work indicate a two-fold advantage in using the EUV observation: (1) Higher stages of ionization, which are more sensitive to solar activity can be

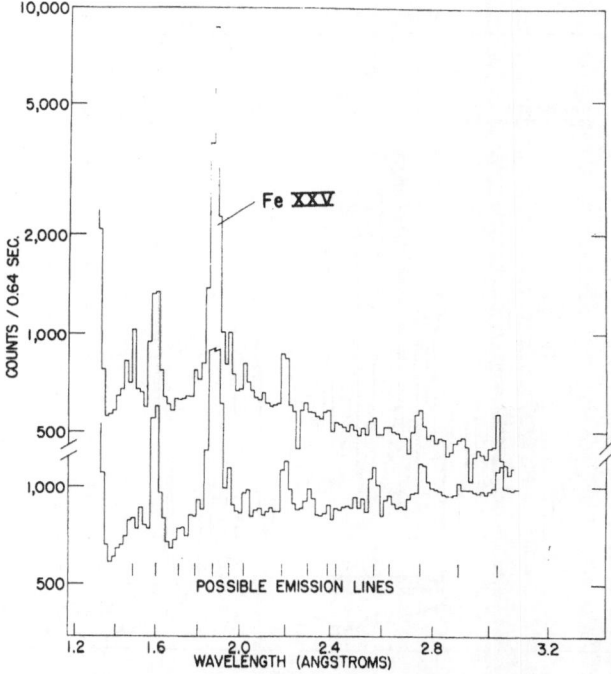

FIG. 3. *Two spectral scans in the region 1·3 Å to 3·1 Å obtained during the increasing phase of a solar X-ray burst on March 22, 1967. Apparent differences in spectral distribution are due to the increase in intensity of the X-ray burst in the time (5 min) required to make the two scans. The onset of increased count rate at 1·34 Å coincides with the detector position at which it begins to be illuminated directly by the Sun.*

observed in the EUV than in the visible region, and (2) The EUV emission can be observed for regions on the disk of the Sun so that the relationship between coronal emission and sunspot activity can be clearly observed for the first time.

### 3. Preliminary XUV Observations made by OSO-III

The Orbiting Solar Observatory III, launched in March, 1967 carries spectrometers, supplied by the Goddard Space Flight Center, for observing the solar spectrum from 1 Å to 400 Å. Data for three EUV emission lines obtained with a grating spectrometer are shown in Figure 2. Results concerning active regions appear to be confirmed by these new data. New spectra obtained by OSO-III at shorter wavelengths are shown in Figures 3 and 4. These data were taken during a flare of importance 3 on March 22, 1967. Tentative identifications of strong emission lines of iron have been indicated. A detailed discussion of these identifications has been published elsewhere (Neupert *et al.*, 1967). Of interest in the current discussion is the variation with time of these prominent emission lines. In general we find that the highest stages of ionization rise to maximum the most rapidly and show the greatest amount of variability. Preliminary

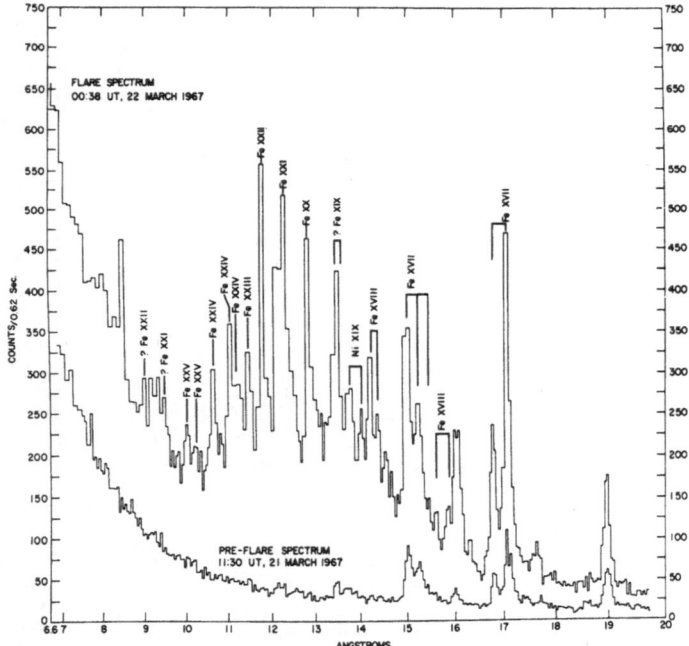

Fig. 4. *Comparison of the solar spectrum between 6·3 Å and 20·0 Å obtained during a flare on March 22, 1957, with a spectrum obtained on the previous day when no flares were in progress. Tentative identification for new transition arrays of Fe xxv–Fe xx are indicated. Spectral resolution is insufficient to allow resolution of lines within each array. Emission lines of Ni xix observed in laboratory spectra by Feldman et al. (1967) are also indicated.*

results for the flare of March 22 are given in Figure 5, which shows the strong line at 1·87 Å (Fexxv) reaching maximum intensity earlier than the lines of lower stages of ionization. Thus, at least for some flares, we observe the same characteristic as noted for active regions; that is, that the highest stages of ionization observable tend to increase most rapidly, with lower stages appearing later (compare Figure 1 with Figure 5). Higher stages of ionization are emitted during X-ray bursts than at other times, however.

More detailed studies of intensity with time for these lines, and comparison with the emission in the line of HeII at 304 Å, are currently underway.

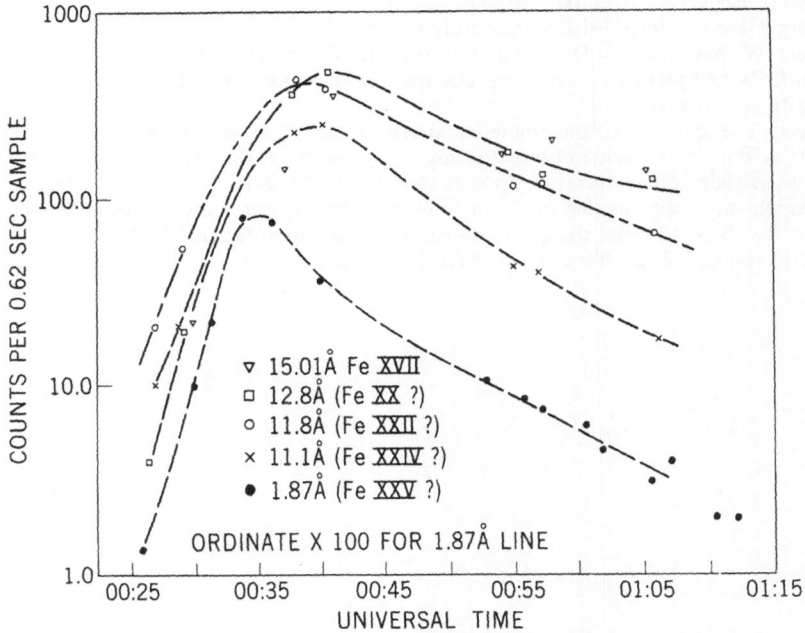

FIG. 5. *Preliminary OSO-III observations on the variation of X-ray line emission with time for a flare of importance 3 on March 22, 1967. Note that peak intensity is reached earlier for the higher stages of ionization than for lower stages.*

## References

Behring, W.E., Neupert, W.M., Nichols, W.A. (1962)     *J. Opt. Soc. Am.*, **S2 (5)**, 597.
Behring, W.E., Neupert, W.M., Lindsay, J.C. (1963)     in *Space Research*, III, Ed. by W. Priester, North-Holland Publishing Co., Amsterdam, p. 814.
Feldman, U., Cohen, L., Swartz, M. (1967)     *Astrophys. J.*, **148**, 585.
Neupert, W.M. (1965)     *Ann. Astrophys.*, **28**, 446.
Neupert, W.M., Gates, W., Swartz, M., Young, R. (1967)     *Ap. J. Letters*, **149**, No. 2, Part 2, L79.
Tousey, R. (1965)     *Ann. Astrophys.*, **28**, 755.
Vauquois, B. (1959)     *Paris Symposium on Radio Astronomy*, ed. by R. N. Bracewell, Stanford University Press, Stanford, Calif., p. 143.

# DISCUSSION

*Noyes:* In the slide showing the variations of 304, 335, 284 Å lines from OSO-III, is the apparent lesser increase of the 284 line during the flare, compared with the increases of the other two, statistically significant?

*Neupert:* Each point plotted represents an average of observations made over a period of 1 hour, so that these data do not give the peak emission reached in each line during the flare. The smaller increase in Fexv compared to Fexvi was also found for several flares observed in 1962 by the OSO-I satellite.

*Krat:* In what way have you done your identification of new coronal lines?

*Neupert:* The first suggestion that iron emission lines might be present was made by Kawabata (*Rept. Ion. Space Research Japan,* **14**, 1960, 405). We have calculated the positions of possible lines (centres of gravity of multiplets) using screening corrections to hydrogenic energy parameters given by Froese (*Canadian J. Phys.,* **41**, 1963, 50).

*Severny:* Have you recorded simultaneously Lα-intensity?

*Neupert:* We have not, but Dr. Hinteregger recorded the region in question.

*Newkirk:* Would you care to mention electron densities and temperatures which would be consistent with your observations?

*Neupert:* The bulk of corona material above an active region reaches temperatures up to $4.0 \times 10^6\,°K$ during the period of most intense activity of the region. On the other hand, when the region has subsided the temperature appears to be about $2.5-3.0 \times 10^6\,°K$. Assuming that the area of the region is the same as that of the underlying plage, the electron densities in the active region appear to be about 10 times the quiet corona. These densities decrease to 2–3 times quiet corona values over residual plage after sunspots have disappeared.

# ON SOME ASPECTS OF XUV SPECTROHELIOGRAMS*

R. TOUSEY, G. D. SANDLIN, and J. D. PURCELL

*(E.O. Hulburt Center for Space Research, U.S. Naval Research Laboratory, Washington, D.C., U.S.A.)*

ABSTRACT

A comparison between XUV and Ca-K spectroheliograms for 9 dates from 1963 to 1967 showed an excellent correlation between plage intensities in Ca-K and He II 304 Å, except for plages near the limb and a few others. Around the limb all but the highest ionization XUV emission lines form a bright ring, usually weaker over the poles. This is an unresolved combination of the limb-brightened emission from the quiet corona and high chromosphere, and emission extending into the corona above plages located as much as several days from limb passage.

In Fe xv and xvi only the localized coronal emissions are observed; these vary in form and intensity from line to line. The 171–500 Å and white-light coronas, recorded on November 12, 1966, correlate well at low altitudes, but beyond 3′ the XUV corona becomes diffuse and without structure.

A comparison has been made between the extreme ultraviolet (XUV) solar images obtained since 1963 by the Naval Research Laboratory (NRL), and the corresponding Ca-K spectroheliograms photographed by the McMath-Hulbert Observatory, giving special attention to the centres of activity and the corona. One rocket-borne instrument is a spectroheliograph, that produces a spectrum of monochromatic solar images, covering $\lambda$ 171–650 Å. Although there is much overlapping, the images in certain important lines are sufficiently separated to permit distinguishing between centres of activity. These lines are He II 304 Å, O IV 554 Å, O V 630 Å, Ne VII 465 Å, Mg IX 368 Å, Fe XV 284 Å, and Fe XVI 335, 361 Å; arranged in this order they come from higher and higher levels in the solar atmosphere; therefore, viewed in these lines, the change in form and intensity of a centre of activity can be followed through the chromosphere into the corona. The resonance lines of Fe IX–XIV between 171 Å and 274 Å are also present, with great intensity; but they lie so close together that the solar images are not well separated.

A second type of instrument, called an XUV heliograph, was first flown on July 27, 1966. This uses a paraboloidal mirror and three aluminum filters in series to form a broadband XUV solar image, including all emissions from 171 Å to about 500 Å. The image has great intensity; with 30-sec exposures XUV coronal emission extending to 3 $R_\odot$ has been recorded (Purcell *et al.*, 1967).

Table 1 lists the dates on which spectroheliograms and heliograms have been ob-

---

* Presented by R. Tousey.

**Table 1**

| Date | Instrument | Image Size | Spatial Resolution | Solar Conditions |
|---|---|---|---|---|
| May 10, 1963 | Spectroheliograph 170–400 Å | 1·8 mm | 1' | Numerous old cycle regions of moderate activity |
| June 28, 1963 | Spectroheliograph 170–400 Å | 1·8 mm | 1' | The data covered zone A. The principal centres of activity were on the back side |
| Sept. 20, 1963 | Spectroheliograph 170–400 Å | 1·8 mm | 40" | The flare-rich proton region was near the 'centre', but the flight occurred 8 hrs before the great proton flare |
| Oct. 20, 1965 | Spectroheliograph 170–700 Å | 4·6 mm | 20" | Only small plages present on front side. The big spot and radio region of Oct. was on the back side |
| Febr. 1, 1966 | Spectroheliograph 170–700 Å | 4·6 mm | 1' | An 'empty centre'. The major regions (January cluster) on the back side marked the first distinctive rise in activity in the new cycle. The features near the East limb were ancestors of the great April region |
| April 28, 1966 | Spectroheliograph 170–700 Å | 4·6 mm | 10" | Region #8279 (NE) was the return of the great April region. The back side was the site of the great proton region of March |
| July 27, 1966 | Heliograph 171–400 Å | 4·6 mm | 1'·5 | Zones A and B were on the limbs. The North member of the double group in the East was the return of the July 7 proton flare and was the site of a flare on July 28. The region just South and to the East of it was the site of proton flares on Aug. 28 and Sept. 2 |
| Nov. 12, 1966 | Heliograph 171–500 Å | 2·3 mm | 30" | Plage #8573 dominated the 'centre' and contained a spot of long duration (six rotations) that formed in Aug., and radiated strongly at radio frequencies. The strong region in the NW was new |
| May 9, 1967 | Spectroheliograph 170–370 Å; 550–700 Å | 4·6 mm | 15" | The only big region on the Sun was on the opposite hemisphere. Zones A and B were again at the limb |
| | Heliograph 171–500 Å | 2·3 mm | 30" | |

tained by NRL, and describes briefly the solar activity and the nature of the photographic results. For the purpose of making intensity correlations between the XUV and Ca-K plages, an arbitrary intensity scale was set up, similar to the one which is used for Ca-K plages at the McMath-Hulbert Observatory. Plages were selected no farther than approximately 70° from the central longitude. The photographic density of each active region in He II 304 Å was estimated relative to the average disk background on a scale rated: 1 = just detectable, to 5 = extremely bright. This was also done for the XUV heliograms.

When the He II 304 Å intensity values were plotted against the McMath-Hulbert values for the same regions in Ca-K, a linear relationship was observed with a scatter of ± half a scale unit, excluding the exceptional cases to be discussed. The intensity estimates of the plages, as recorded by the broad-band XUV heliograph, also correlated well with Ca-K. The plages in He II 304 Å are more intense, relative to the quiet background than in Ca-K, as was found for H-Lyman-α (Purcell and Tousey, 1961). For the higher ionization lines such as Mg IX 368 Å, Fe XV 284 Å, and Fe XVI 335, 361 Å there was no background against which to make a comparison; however, with a few exceptions, the plage intensities in the different lines increased or decreased together.

There were a few cases of plages not close to the limb that were clearly different in the XUV than in Ca-K. The most notable example is shown in Figure 1, where two of the best He II 304 Å images obtained as yet, with 10″ spatial resolution, are compared with images in Ca-K and Hα. The activity on April 28, 1966 lay in a belt in the North. The plage at 9°E, 28°N is strong in Ca-K but is just detectable in He II; in Hα this feature is small and weak, and is surrounded by a ray-like absorption structure. This plage received no McMath-Hulbert number, but was the Western extension of plage # 8278. Its position was about halfway between the two longitude zones of maximum activity.

Great differences between the XUV and the Ca-K images are observed for regions near and on the limb. For example, in Figure 2 two exposures with the XUV heliograph on July 27, 1966 are compared with Ca-K and Hα. Over most of the disk the same plages are present, with similar shapes and intensities. Within 1 or 2 min of the limb and beyond the limb, however, the images show much more emission in the XUV than in Ca-K. The intense XUV emission in the NW is the most striking example; it is not present at all in Ca-K and Hα. This activity is the coronal extension above a plage complex that was 2–3 days over the limb.

The bright, partial ring present in Figure 2 is a characteristic of all XUV heliograms and is much the same in images produced by X-rays of $\lambda < 100$ Å. It is composed of two parts, which cannot be fully separated without greater spatial resolution. One is just the corona above the quiet regions; the emitting layer is optically thin and is much more intense when viewed edge-on. The second contribution is from the corona above the active regions on or within a few days of limb passage. The ring is especially

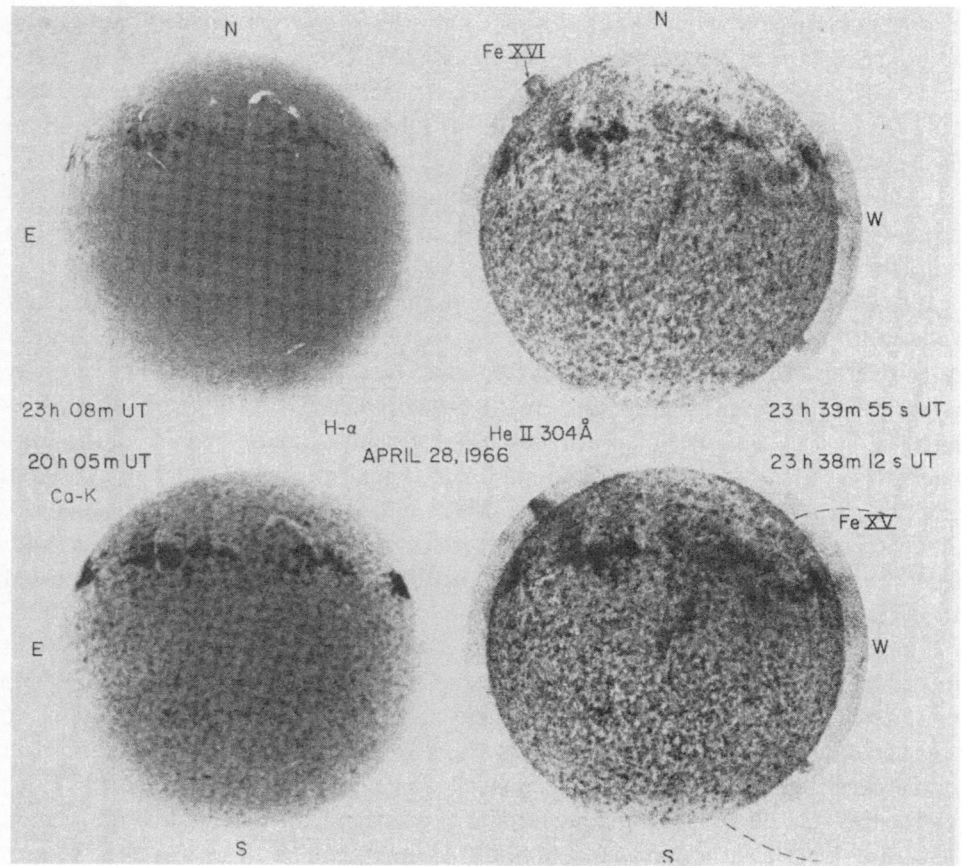

FIG. 1.  *Two HeII 304 Å spectroheliograms obtained on April 28, 1966, compared with Ca-K from Mt. Wilson and Hα from Sacramento Peak. The FeXV 284 Å image, whose position is shown, produces the plages that appear at the centre of the HeII image, and others that are in the West; several prominences are present in HeII, but FeXVI 335 Å accounts for the large limb feature in the NE.*

pronounced in all of the XUV images obtained by NRL because, for every flight date except September 20, 1963, both zones A and B of maximum solar activity (Dodson-Prince and Hedeman, 1967) lay close to the limb, and the disk as a whole showed a relatively 'empty centre'. Therefore, enhancements in the emission ring were prominent; for each enhancement an active centre that appeared to account for it was found either on the limb or within 1–5 days of limb passage.

   The high-spatial resolution spectroheliograms of April 28, 1966 show several interesting limb features. In Figure 3 the range 240–340 Å is reproduced, and in Figure 4 four particular images are brought together for comparison.

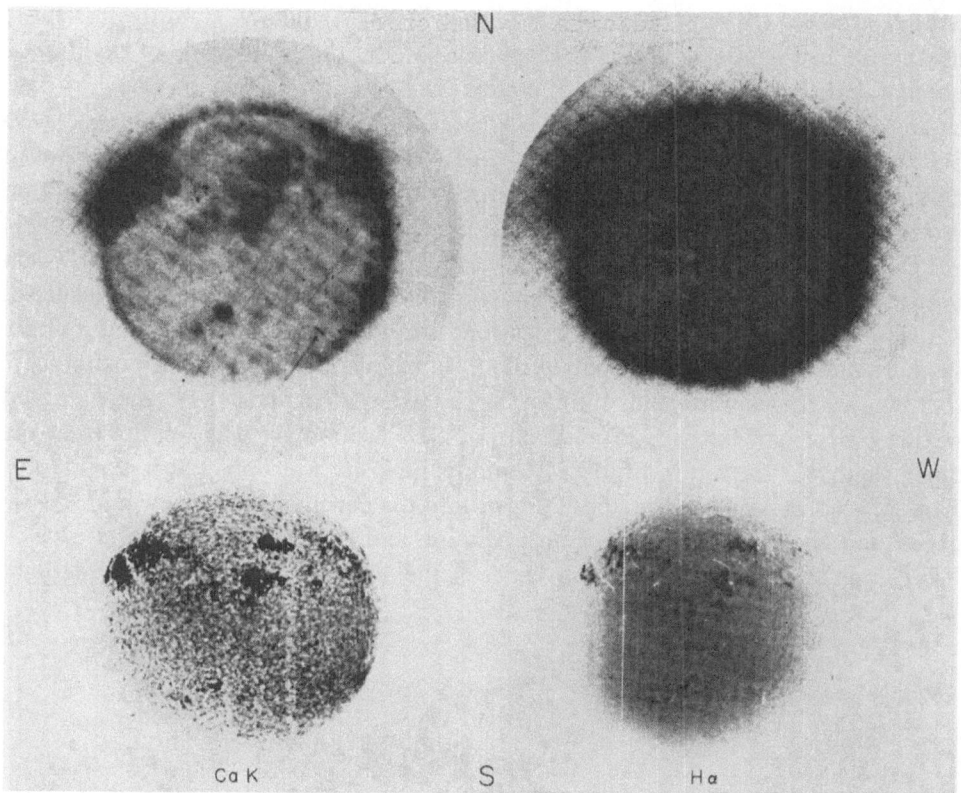

FIG. 2. *Medium and long exposures with the XUV heliograph (171–500 Å) on July 27, 1966, compared with Ca-K and Hα from the Mt. Wilson Observatory. The grid pattern was caused by the mesh on which one of the aluminum filters was supported.*

In Fexv 284 Å and Fexvi 335 Å there is coronal emission in the NW that extends to about 65° apparent latitude and reaches 5′ beyond the limb. This emission is present also in Heii 304 Å, but in this case there is a possibility that it may be caused by Sixi 304 Å instead of Heii (Smith, 1967). This coronal emission is believed to originate from a large active complex between latitudes 15° and 35°, and lying 30°–60°, heliocentric, over the limb; this was the great particle-emitting region of March 1966. The complex was assigned plage numbers 8262 and 8275 and had total area 7100 and intensity 3·5 during the preceding half-period; it was given numbers 8294, 8299, 8300 during the succeeding half-period. Therefore the XUV emission region, seen in projection above the NW limb, was at a height above the photosphere of at least $3·5 \times 10^5$ km ($0·5\ R_{\odot}$) and may have been observed to $10^6$ km ($1·5\ R_{\odot}$).

Close to East-limb passage on April 28, 1966 there were two centres of activity;

plage # 8284 at 24°N latitude crossed the limb one day earlier; plage # 8285 at 18°N latitude crossed two days later. The difference in XUV emission from them is easily observable in Figures 1, 3, and 4. Plage # 8284, on the visible side of the Sun, is observable in all lines, including, of course Ca-K, Hα, and HeII. In Ov, however, its area is very small and its intensity is great. Coronal emission above plage # 8285 on the back side cannot be seen in Hα, Ca-K, HeII, OIV, Ov, and probably not in NeVII. It is conspicuous, however, in MgIX, SiXII, FeXV, and FeXVI. It is most diffuse in MgIX, and appears to have the steepest gradient in FeXVI. Because plage # 8284 lies closer to the limb than # 8285, and has twice the area (3900 vs. 1900), it would be expected to show as much or more emission above the limb, but none was detected. A possible explanation is that the extensions into the corona are not radial, but are strongly tilted under the influence of the local magnetic field. Another possibility is that the two plages were really different in character. Actually # 8285 was new, and was recorded with intensity 3, and sunspot classification A; plage # 8284, on the other hand, was the return of # 8240, and its intensity was only 2·5; it was class J. This suggests that emission extends higher into the corona above a new and intense plage than above one that is old, more diffuse, and less intense.

In order to make a comparison between the XUV images and the white-light

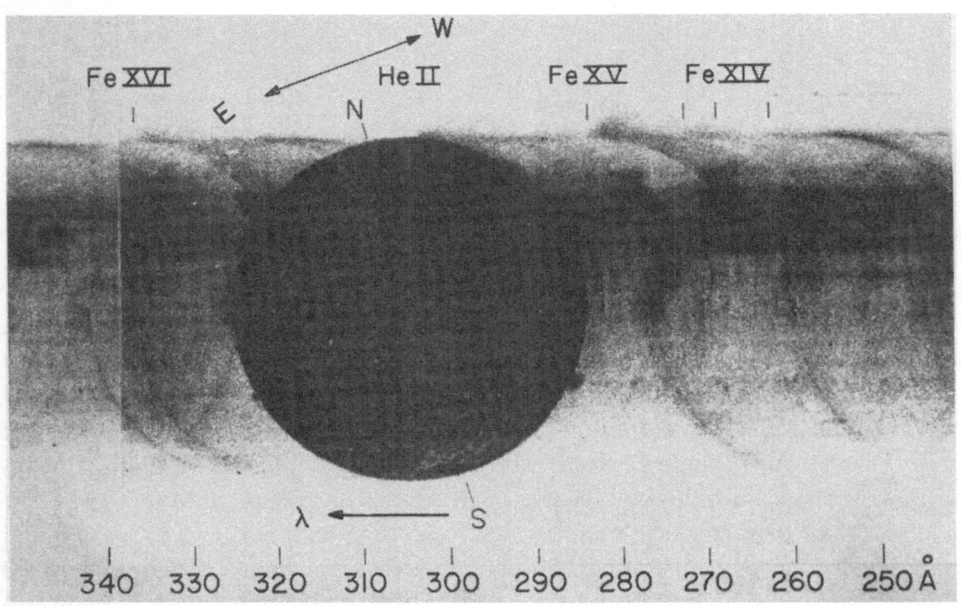

FIG. 3.    *The section, 250–340 Å, of the spectroheliogram of longest exposure on April 28, 1966, printed so as to reproduce the coronal emission in the NW, associated with FeXVI 335 Å, FeXV 284 Å, HeII and SiXI 304 Å, and faintly with FeXIV.*

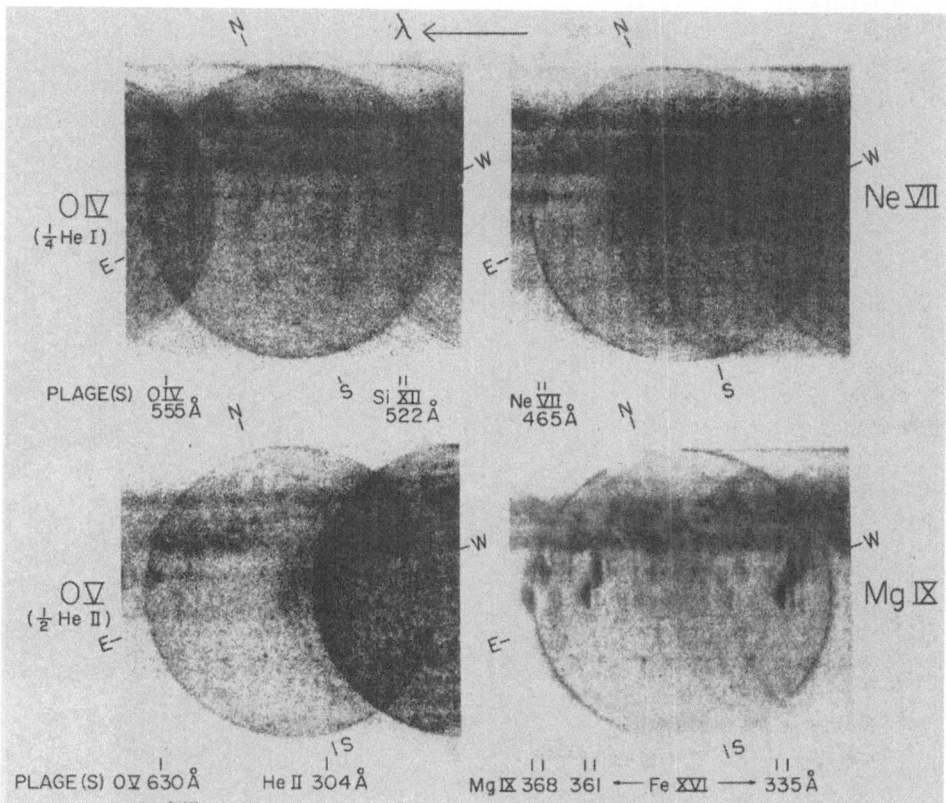

FIG. 4. *Sections of the April 28, 1966 spectroheliogram covering O IV 554 Å, O V 630 Å, Ne VII 465 Å, and Mg IX 368 Å. Also present are a small part of the He I 584 Å image, and half of the second-order image of He II 304 Å. The East-limb features discussed in the test are indicated.*

corona, a flight was scheduled for November 12, 1966, the day of the total eclipse in South America. Figure 5 shows Newkirk's (1967) beautiful eclipse photograph, but with the centre removed and an XUV image introduced to scale. Below, is another XUV image that was exposed for a longer time. Close to the limb the XUV emission matches very well the detail in the white-light corona. This is to be expected, because collisional excitation of the XUV resonance lines follows $n_e^2$, and optical excitation follows $n_e$. Farther out, however, the two types of corona are less similar; the long XUV exposure shows no structure or streamers, but only a diffuse emission, correlating in an approximate fashion with the white-light corona.

The XUV spectroheliograms and heliograms contain a great deal of information about the structure of the chromosphere and corona. But to make use of them a day to day series is required. The spatial resolution should be the greatest that can be

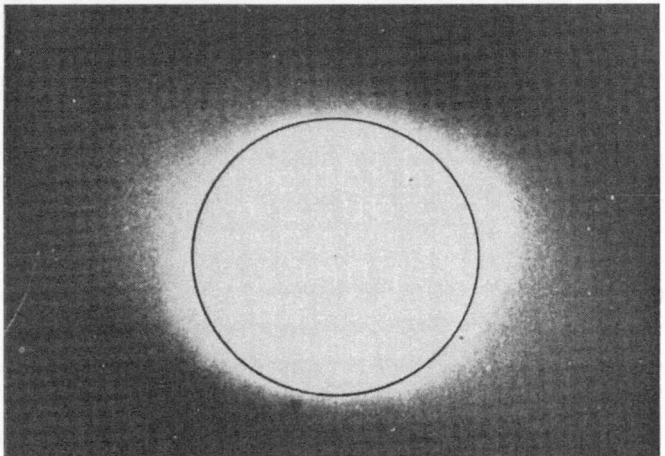

FIG. 5.   *The white-light corona, photographed during the November 12, 1966 eclipse by the Bolivian Expedition of the High Altitude Observatory and The Johns Hopkins University, with the centre replaced by an XUV heliogram obtained on the same date by the Naval Research Laboratory. The second heliogram, of long exposure, shows the extension of XUV emission into the corona. North is at the top and East to the left.*

attained. An instrument to accomplish this with about 1′ resolution is nearing completion at NRL, and will be flown in an Orbiting Solar Observatory (OSO-F) in 1968. This will be followed, 1 to 2 years later, by a photographic spectroheliograph with very high spatial resolution, that will be included in the Astronomical Telescope Mount of the Apollo Applications Program, and will operate over one to two solar rotations.

## Acknowledgments

We wish to express our sincere appreciation to Dr. Helen Dodson-Prince and to Miss Ruth Hedeman of the McMath-Hulbert Observatory for making available their solar data and much helpful advice.

## References

Dodson-Prince, H., Hedeman, E.R. (1967)      *Annals of the IQSY.*
Newkirk, G.Jr. (1967)      *Sky and Telescope*, **33**, 136.
Purcell, J.D., Tousey, R. (1961)      *Mém. Soc. R. Sc. Liège*, **4**, 274.
Purcell, J.D., Tousey, R., Koomen, M.J. (1967)      in *Space Research*, VIII.
Smith, T.S. (1967)      discussion at the meeting of the Amer. Astron. Soc., Boulder, Colo., Oct. 3, 1966.

## DISCUSSION

*Kiepenheuer:* I would like to congratulate Dr. Tousey on the beautiful angular resolution which he obtained in the EUV from a rocket.

*Tousey:* This was just luck!

# LA COMPOSANTE LENTEMENT VARIABLE DES RAYONS X SOLAIRES EN RELATION AVEC LA STRUCTURE DES CENTRES D'ACTIVITÉ

R. MICHARD et Mme E. RIBES*

*(Observatoire Meudon, France)*

## ABSTRACT

The slowly varying component of solar X-rays has been studied through records of the satellite Explorer-30 instrumented by the U.S. Naval Research Laboratory. Correlations with plage areas, and flux at radiofrequencies have been studied. High flux values in the 8–20 Å and, still more, the 1–8 Å bands are related to specific ARs, also characterized by great flare productivity, high intensities on 3 cm, and anomalous magnetic structure.

Le rayonnement X du soleil peut être approximativement divisé en trois composantes:

(1) une composante stable due au rayonnement thermique de l'ensemble de la couronne,

(2) une composante lentement variable (CLV) due à l'émission des centres d'activité,

(3) les sursauts directement associés aux éruptions chromosphériques.

A l'aide des mesures effectuées par le satellite Solrad-8 = Explorer-30 = 1965-93A de l'USNRL, et reçues par les stations du Centre National d'Etudes Spatiales, nous avons étudié la CLV pour le semestre mars–août 1966. Dans certains cas les mesures du réseau du CNES ont été complétées par celles publiées dans *Solar Geophysical Data*.

Un indice journalier du flux solaire moyen dans les bandes 44–60 Å, 8–20 Å et 1–8 Å a été calculé en utilisant autant que possible exclusivement les observations effectuées en dehors des sursauts. La Figure 1 indique les variations de cet indice pour les deux bandes de plus courtes longueurs d'ondes. On constate la très grande variabilité du flux X, bien que l'élimination des sursauts et le calcul des moyennes journalières réduisent considérablement les variations réelles.

La 'composante stable' est totalement indécelable sur 1–8 Å et presque négligeable dans la bande 8–20 Å. Il n'en est pas de même sur 44–60 Å où un flux important reste présent même durant les minimums d'activité.

On notera aussi sur la Figure 1 d'importantes variations de la 'couleur' du rayonnement X solaire définie par le rapport des flux 1–8 et 8–20 Å. Ce rapport est < 0·01

---

* Presented by P. Simon.

*Kiepenheuer (ed.), Structure and Development of Solar Active Regions*, 420–430. © *I.A.U.*

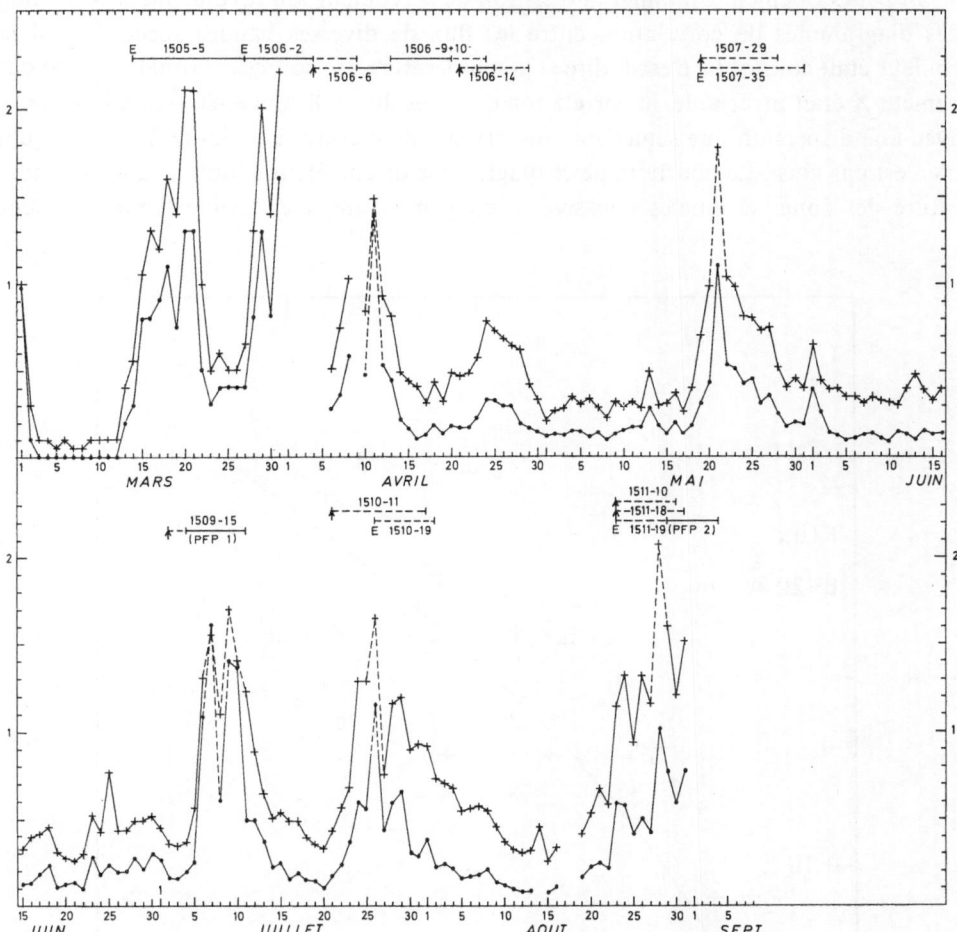

FIG. 1.    *Evolution de la composante lentement variable des flux X solaires de mars à septembre 1966 dans les bandes 8–20 et 1–8 Å. Les points sont les moyennes journalières de mesures effectuées en dehors des sursauts associés aux éruptions, et recues par les stations du Centre National d'Etudes Spatiales. Les données de mars proviennent de l'USNRL (Solar Geophysical Data). En traits interrompus: mesures douteuses.*

*En haut sont indiqués les passages de CA (cités dans le texte et dans l'Appendice) responsables de l'émission au cours de diverses périodes remarquables. En trait plein: périodes où un seul CA produit tout le flux X dans les bandes considérées. E = CA passant au bord est. Flèche = CA naissant sur le disque.*

pour des périodes très calmes (mars 3–13); il peut atteindre des valeurs proches de 0·1 pour des périodes d'intense activité, tandis que sa valeur moyenne est de 0·042.

On peut également examiner la question de la 'couleur' du rayonnement X à l'aide des diagrammes de corrélation entre les flux de diverses bandes spectrales. Si la couleur était constante, c'est-à-dire si la température des sources coronales du rayonnement X était invariable, la corrélation entre les flux 1–8 Å et 8–20 Å serait linéaire, avec une dispersion due seulement aux erreurs de mesure. La Figure 2 montre qu'il n'en est pas ainsi. La courbure de ce diagramme de corrélation indique que la température des zones coronales émissives croît généralement en même temps que leur

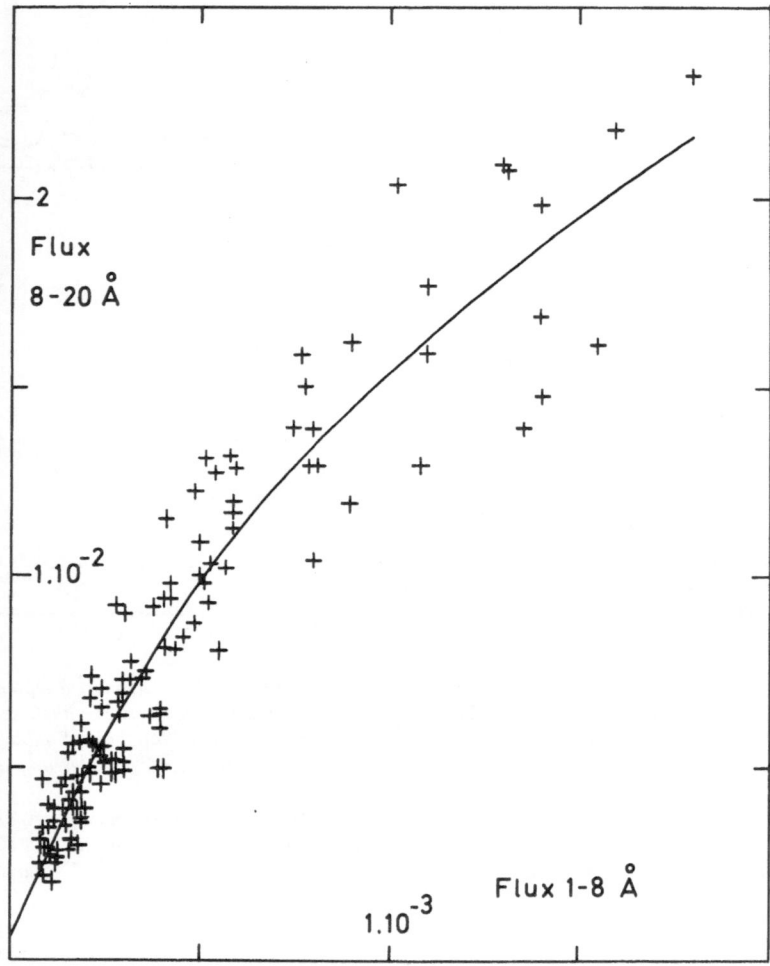

FIG. 2.    *Diagramme de corrélation entre flux 1–8 et 8–20 Å de la composante lentement variable. Les points sont les moyennes journalières de mars à août 1966.*

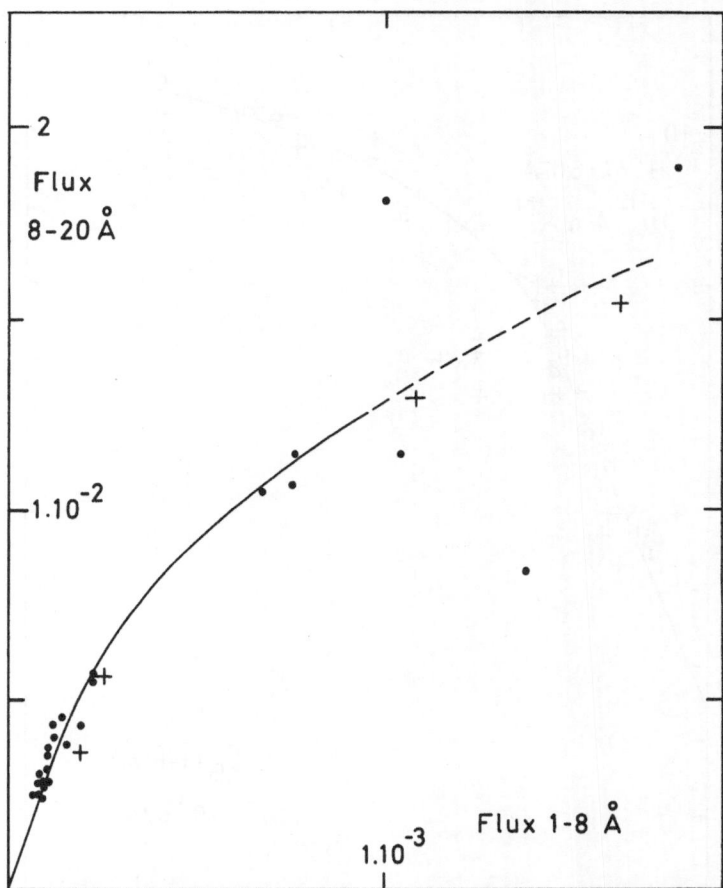

FIG. 3.   *Corrélation entre flux 1–8 et 8–20 Å (hors sursauts) des 3 au 6 juillet 1966. Points = mesures de l'USNRL (Friedman et Kreplin, 1967). Croix = Moyennes journalières de l'Observatoire de Meudon.*

densité, le rayonnement X de la CLV devenant plus dur quand son intensité augmente. Le même phénomène est mis en évidence de manière encore plus sensible sur les diagrammes de corrélation portant sur des époques choisies: la Figure 3 est relative à la période du 3 au 6 juillet, phase de croissance du centre actif 1509–15 (McM 8362) qui produisit l'éruption à protons du 7 juillet (Friedman et Kreplin, 1967). Les diagrammes de corrélation entre flux 1–8 et flux 44–60 Å indiquent également le net 'durcissement' des émissions solaires dues aux sources les plus intenses (Figure 4).

Les variations du flux X avec le cycle solaire (Kreplin et Gregory 1966), les photographies directes (Friedman, 1963; Underwood et Muney, 1967), l'étude des éclipses (Landini *et al.*, 1966) indiquent que la CLV des rayons X provient des centres d'activité (CA). Mais il est évident que l'intensité et le spectre de cette émission sont prodigieusement variables d'un CA à un autre. Nous avons essayé de relier l'émission

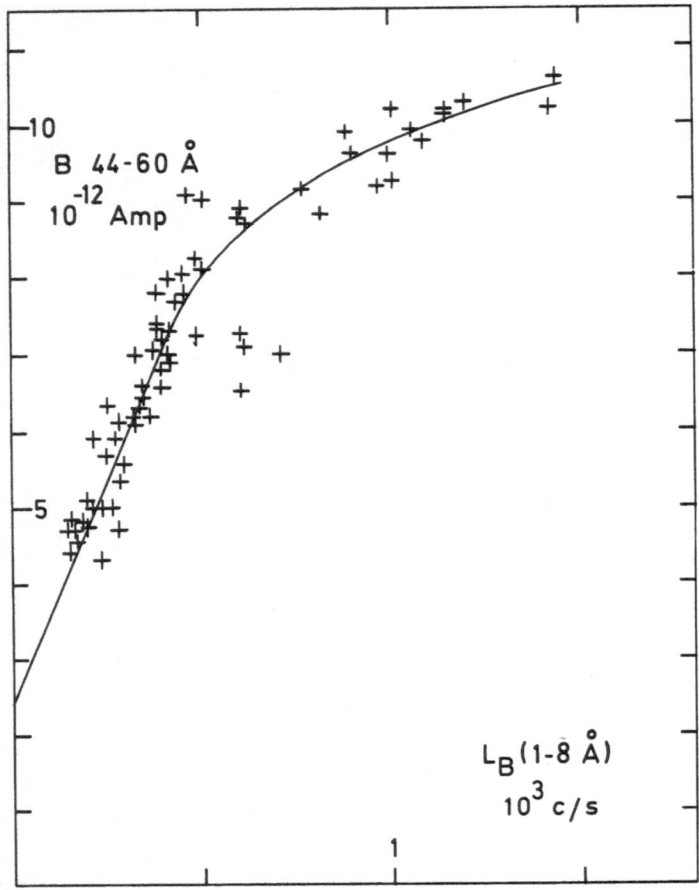

FIG. 4.   *Corrélation entre flux 1–8 et 44–60 Å (hors sursauts) d'après les mesures du CNES pour août 1966.*

X des CA à leurs caractères optiques et radioélectriques. Pour cela on peut faire appel à des techniques de corrélation, on peut aussi examiner directement les cas où un seul CA est responsable de la quasi-totalité du flux observé. Ces cas sont relativement fréquents durant la phase d'activité solaire modérée que nous étudions: les périodes correspondantes ont été notées sur la Figure 1.

La Figure 5 donne le diagramme de corrélation entre le flux 8–16 Å et l'aire des taches, corrigée de la perspective, d'après les mesures de l'Observatoire de Rome-Monte Mario, corrélation que l'on peut qualifier d'assez bonne. Elle est en tout cas sensiblement meilleure que la corrélation entre le même flux et l'aire des plages du calcium pondérée par leur brillance. La Figure 6 nous donne un exemple d'une 'bonne' corrélation entre le flux X 8–16 Å et le flux radioélectrique à 3 cm mesuré par l'Observatoire de Toyokawa.

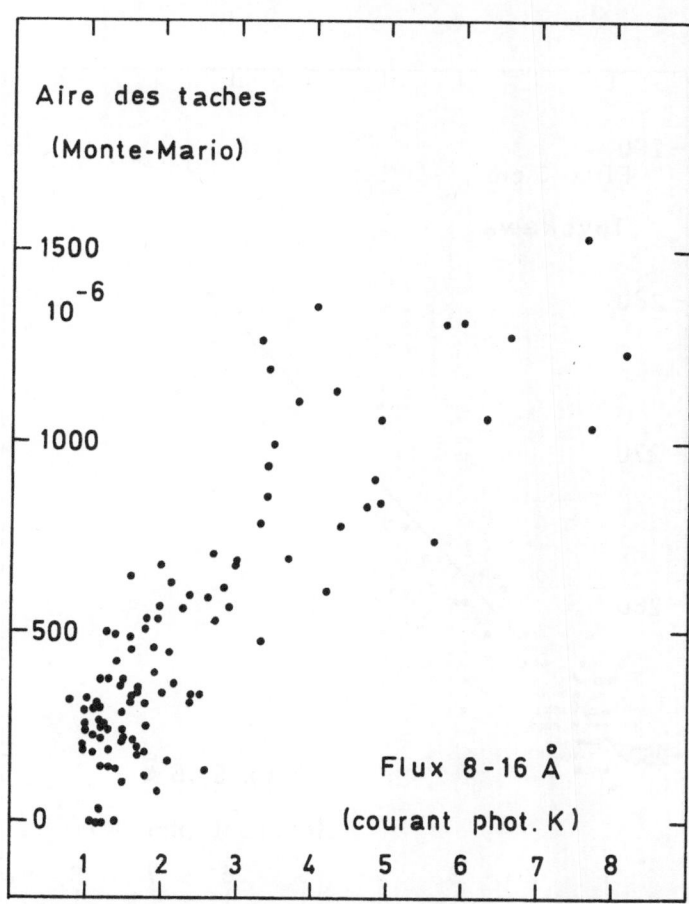

Fig. 5.  *Corrélation entre flux 8–20 Å et aire des taches corrigée de la perspective (mesures de l'Observatoire de Rome).*

Après ces exemples nous pourrons résumer les résultats de l'examen de nombreux diagrammes de corrélation dans le Tableau 1.

Les flux en rayons X mous (44–60 Å) sont très bien corrélés avec les flux radio-électriques aux fréquences voisines de 3000 MHz, assez bien avec l'aire des plages et médiocrement avec l'aire des taches. Quand on passe à la bande 8–16 Å, la corrélation avec l'aire des taches s'améliore beaucoup mais celle liant le flux X à l'aire des plages se dégrade; d'autre part la corrélation avec le flux à 10000 MHz reste excellente tandis qu'elle décroît beaucoup pour les fréquences plus basses. Pour les rayons X durs (1–8 Å), toutes les corrélations deviennent beaucoup moins bonnes, leurs variations selon le paramètre considéré restant qualitativement semblables à celles du flux 8–16 Å.

L'étude de quelques cas particuliers permet d'expliquer le caractère assez vague des corrélations entre flux radioélectriques et rayons X durs. Les 10–12 avril 1966 par

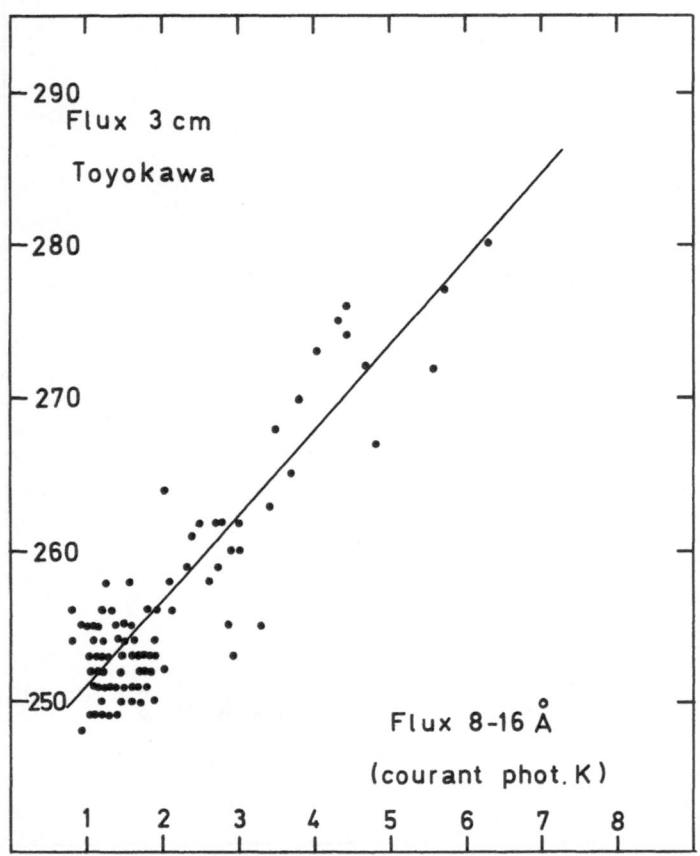

FIG. 6.   *Corrélation entre flux 8–20 Å et flux radioélectrique sur 3 cm (mesures de l'Observatoire de Toyokawa).*

## Tableau 1

### Coefficients de corrélation entre flux X, indices solaires et flux radioélectriques

|                                    | Flux 1–8 | Flux 8–16 | Flux 44–60 |
|------------------------------------|----------|-----------|------------|
| Aire pondérée des plages           | 0·28     | 0·57      | 0·77       |
| Aire corrigée des taches (Rome)    | 0·59     | 0·81      | 0·70       |
| Flux 3 cm (Toyokawa)               | 0·60     | 0·87      | 0·82       |
| Flux 8 cm (Toyokawa)               | 0·58     | 0·84      | 0·88       |
| Flux 10·7 cm (Ottawa)              | 0·49     | 0·77      | 0·88       |
| Flux 14 cm (Toyokawa)              | 0·40     | 0·66      | 0·85       |

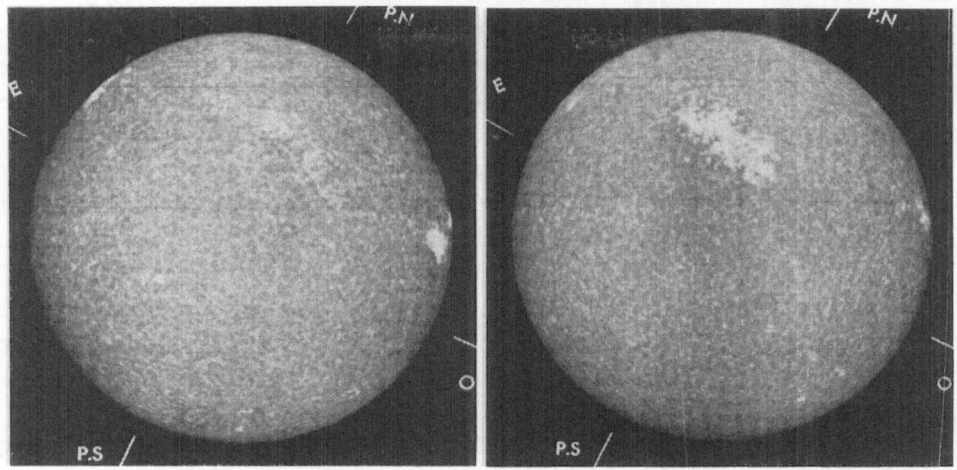

Fig. 7.   *Comparaison des CA visibles les 10 et 17 avril 1966 en liaison avec les variations des flux X et centimétriques entre ces deux dates.*

| 11 avril 1966: | | | 17 avril 1966: | | |
|---|---|---|---|---|---|
| 8–20 | $1{\cdot}48 \times 10^{-2}$ | | 8–20 | $0{\cdot}31 \times 10^{-2}$ | |
| 1–8 | $1{\cdot}40 \times 10^{-3}$ | | 1– 8 | $0{\cdot}12 \times 10^{-3}$ | |
| 2·800 MHz | 94 | Ottawa | 2·800 MHz | 95 | Ottawa |
| 3·750 | 108 | Toyokawa | 3·750 | 102 | Toyokawa |
| 10·000 | 261 | id. | 10·000 | 253 | id. |

exemple (Figure 7) nous observons des flux X intenses dus à la présence du petit CA 1506-6. Les 15–17 avril nous sommes en présence du groupe de CA 1505-9-10: cette région comporte une plage très vaste mais des taches petites, avec une activité éruptive modérée. Le flux X est alors réduit d'un facteur 3 dans la bande 8–16 Å et supérieur à 5 dans la bande 1–8 Å, tandis que les flux centimétriques sont très voisins dans les deux cas. De même lors du passage de la région 1509-15 (3–10 juillet) la croissance des flux X est beaucoup plus spectaculaire que celle observée sur 8 ou 10 cm.

## Discussion

Il est clair que l'émission X 8–16 Å et plus encore 1–8 Å est concentrée en des 'points chauds' du CA, alors que la CLV centimétrique, pratiquement insensible à la température, est plus directement contrôlée par la surface du CA (telle qu'elle apparaît sur les spectrohéliogrammes de la raie K).

Les émissions à 8–10 cm tendent à s'associer au mieux avec l'émission X 44–60 Å correspondant à des températures relativement basses. L'émission à 3 cm paraît plus sensible que celles de plus basse fréquence à des propriétés des CA autres que leurs

surface (Avignon *et al.*, 1966), telle que leur structure magnétique. Aussi ses corrélations avec les émissions X à 8–16 Å et même 1–8 Å sont-elles meilleures. Pour des
longueurs d'ondes radioélectriques plus grandes (14 et 30 cm) les corrélations avec les
flux X se dégradent rapidement. Il serait intéressant d'étendre ces comparaisons à des
émissions radioélectriques de très haute fréquence. A ce propos notons que le CA
1506-6 que nous avons signalé comme ayant un rayonnement X anormalement élevé
par rapport à sa dimension et à son intensité centimétrique, a aussi présenté une
intensité anormalement grande sur 17 GHz d'après les observations de Tsuchiya et
Nagane (1967).

Nous avons identifié et examiné les CA responsables des maximums remarquables
du flux solaire en rayons X durant la période considérée (cf. Appendice). De cet
examen on peut conclure que ces maximums sont dus à la présence *d'au moins un* CA
ayant les caractéristiques suivantes:

(1) comporter un groupe de taches en phase de croissance et d'évolution
rapide;

(2) appartenir à la classe C (complexe) de la classification magnétique des Régions
Actives de l'Observatoire de Meudon (avec une préférence pour les sous-classes $C_p$
et $C\gamma$ (Martres *et al.*, 1966);

(3) appartenir (en général) aux types magnétiques $\beta\gamma$, $\gamma$ et $\delta$ de l'Observatoire du
Mount Wilson;

(4) produire un grand nombre d'éruptions par jour;

(5) présenter une intensité remarquable sur 3 cm d'après les observations interféro-
métriques de la Station de Nançay.

Les corrélations étroites qui existent entre ces divers caractères sont d'ailleurs bien
connues.

Les observations spectroscopiques de l'émission X du soleil dans les bandes 8–20
(Blake *et al.*, 1965; Culhane *et al.*, 1967) et 1–8 Å (Friedman, 1967) étudiées ici, montrent qu'elle est due à des condensations coronales de *température sensiblement
supérieure* à la température moyenne de la couronne. Il est intéressant de noter que
la formation de ces condensations exceptionnelles est conditionnée par la structure
magnétique spéciale du CA: géométrie complexe, forts 'gradients' du champ longitudinal aux changements de polarité, variations rapides. Il est probable que les propriétés magnétiques du CA influent à la fois sur les possibilités de piégeage des ions
coronaux, et sur l'efficacité du processus de chauffage.

### Remerciements

Notre reconnaissance va à nos collègues de l'USNRL qui nous ont fourni généreu-
sement les données nécessaires à la réception des mesures d'Explorer-30, et aux
ingénieurs et techniciens du Centre National d'Etudes Spatiales qui ont assuré cette
réception.

## Appendice: Centres d'activité correspondant aux maximums remarquables du flux X solaire (Mars–Août 1966)

(1) Numéro dans le *Quarterly Bulletin on Solar Activity* Coordonnées de Carrington.
(2) Numéro de l'Observatoire de McMath.
(3) Numéro de l'Observatoire de Mount Wilson.
(4) Dates où le CA contribue au flux solaire.
(5) Dates où le CA est responsable de la quasi-totalité du flux X solaire.
(6) Types magnétiques du groupe de taches selon l'Observatoire de Mount Wilson.
(7) Classe magnétique du CA selon l'Observatoire de Meudon (Martres et al., 1966).
(8) Nombre d'éruptions observées pendant les jours indiqués dans la colonne (4).
(9) Remarques sur l'évolution du CA.

| (1) | (2) | (3) | (4) | (5) | (6) | (7) | (8) | (9) |
|---|---|---|---|---|---|---|---|---|
| 1505-5 146, 18 N | 8207 | 16000 | 13–28 mars | 13–27 mars | $\beta\gamma$ et $\gamma$ | C$\gamma$ du 17 au 22 | 110 | Jeune au bord E; décroissant à l'O. |
| 1506-2 332, 27 N | 8223 | 16004 | 27 mars – 9 avril | 28 mars – 5 avril | $\beta\gamma$; $\gamma$ le 2 | C$\gamma$ à Cp | 34 | Jeune au bord E; décroissant à l'O. |
| 1506-6 288, 22 N | 8240 | 16008 | 6–13 avril | 9–12 avril | $\beta$f puis $\beta\gamma$ | Cc | 30 | Né le 6 dans plage pré-existante. Max. au bord O. |
| 1507-29 68, 21 S | 8302 | 16035 | 19–27 mai | | $\beta$p* | Ca puis Cp | 17 | Né le 19 dans plage pré-existante; décroissant à l'O. |
| 1507-35 0, 14 N | 8310 | 16037 | 19–31 mai | | $\beta$f | Si; Cp le 24 | 7 | Jeune au bord E; très décroissant à l'O. |
| 1509-15 210, 35 N | 8362 | 16067 | 4–11 juillet | 4–10 juillet | $\beta\gamma$ puis $\gamma$ | Ca puis C$\gamma$ | 51 | Né le 3 dans plage pré-existante. Max. à l'O. |
| 1510-11 270, 37 N | 8408 | 16089 | 21 juillet – 1 août | 22–27 juillet? | $\beta$p | Cp les 25–27 | 12 | Né le 21 juillet; décroissant à l'O. |
| 1510-19 180, 24 N | 8414 | 16092 | 26 juillet – 2 août | | $\beta$p; $\delta$ les 1–2 | Si** | 12 | Paraît à l'E. le 26; très décroissant à l'O. |
| 1511-10 248, 7 N | 8454 | 16115 | 23–30 août | | $\beta$p puis $\beta\gamma$ | Si; Cp le 27 | 21 | Né le 22 août; décroissant à l'O. |
| 1511-18 196, 23 N | 8459 | 16111 | 23–31 août | | $\beta$p; $\beta\gamma$ $\beta$f | Ca à Si | 15 | Né le 22 août; très décroissant à l'O. |
| 1511-19 182, 20 N | 8461 | 16114 | 23 août – 4 sept. | 29 août – 4 sept. | $\beta\gamma$ puis $\delta$ | Ca puis C$\gamma$ | 42 | Paraît à l'E. le 22; croissant irrégulièrement; un peu décroissant à l'O. |

* Observations magnétiques très incomplètes.
** CA bipolaire anormalement compact.

## Bibliographie

Avignon, Y., Martres, M-J., Pick, M. (1966)    *Ann. Astrophys.*, **29**, 33.
Blake, R.L., Chubb, T.A., Friedman, H., Unzicker, A.E. (1965)    *Astrophys. J.*, **142**, 1.
Culhane, J.L., Evans, K., Pounds, K.A. (1967)    *Nature*, **214**, 41.
Friedman, H. (1963)    *Ann. Rev. Astr. Astrophys.*, **1**, 59.
Friedman, H. (1967)    communication personelle.
Friedman, H., Kreplin, R.W. (1967)    8th International Space Science Symposium, London.
Kreplin, R.W., Gregory, B.N. (1966)    dans *Space Research*, VI, Spartan Books, Washington, p. 1011.
Landini, M., Russo, D., Tagliafferi, G.L. (1966)    *Nature*, **211**, 393.
Martres, M-J., Michard, R., Soru-Iscovici, I. (1966)    *Ann. Astrophys.*, **29**, 245.
Tsuchiya, Asushi, Nagane, Kiyoshi (1967)    *Solar Phys.*, **1**, 121.
Underwood, J.H., Muney, W.S. (1967)    *Solar Phys.*, **1**, 129.

## DISCUSSION

*Neupert:* Did you use sunspot number or area in your correlation with X-rays?

*P. Simon:* Sunspot area was used.

*Krüger:* How many values are used in your statistics of calculation of correlation coefficients?

*P. Simon:* One value each day.

*Krüger:* Do there exist some comparisons between the gradients of spectra of the s-component of solar radio emission and those of X-ray emission?

*P. Simon:* Until now, there is no direct comparison, but it is obvious, according to the last part of this report, that there is a direct correlation between the X-ray spectrum and the centimetric spectrum.

*Lundbak:* Would you kindly explain how in the scheme presented the relation is established between the indices given and the cm-values in the first column?

*P. Simon:* It is the conventional computation of the correlation function.

# X-RADIATION STUDIES OF THE CORONA

K. A. POUNDS, K. EVANS and P. C. RUSSELL

*(University of Leicester)*

## ABSTRACT

The coronal X-ray emission has been studied in a series of sunpointed Skylark-rocket flights over the period August 1964 to August 1967. Two types of instrument have been used, namely, an array of pinhole cameras, to study the distribution of X-ray sources across the solar disk, and a set of Bragg crystal spectrometers, to examine the detailed spectrum of the emission.

Six rocket flights in all (Skylarks 301, 302, 303, 306, 307, and 406) have provided solar X-ray photographs, covering the wave-band 12–60 Å. The main conclusions of this work may be summarised as follows:

(1) Of over twenty separate calcium-plages areas, all those having an area-intensity product greater than 300 were associated with enhanced coronal X-radiation, this 'threshold' being merely a function of X-ray camera sensitivity.

(2) The plages studied were of from 1st to 4th rotation, indicating the enduring nature of the local coronal X-ray enhancement.

(3) Crude spectral analysis, derived from different filters used in a given camera array, showed an increasing contrast between the active region and disk emissions below 25 Å, indicating the probability of the active region coronal material being rather hotter.

(4) Photographs of regions on the disk and on the limb showed a typical coronal active region to produce more than 50% of its total X-ray flux from a region within 1 arc minute cube. Electron densities up to 10 times the normal coronal values are required by the measured X-ray brightness of the larger regions.

(5) The major part of the emission above 20 Å arises from the (cooler) background corona. Examination of this coronal disk emission revealed strong limb brightening, polar darkening (particularly in the 1964–65 photographs) and a band of slightly enhanced X-radiation near latitude 35 °N. The latter observation was noted on successive flights, in October 1965 and February 1966, and clearly followed the tilt in aspect angle of the solar disk.

Skylarks 304 and 305 were equipped with uncollimated Bragg crystal spectrometers. The former payload, flown in May 1966, revealed strong line emission down to the scan limit at 11 Å. Emission lines of O VII and VIII, Ne IX and X, Fe XVII and XVIII and Ni XIX have been identified. All but the O VII series were found to have dominant components arising from a large active region on the solar disk, and an analysis has shown that the complete active-region spectrum is consistent with an electron temperature of 3 million degrees. Intercomparison of the different emission lines indicated coronal abundances of iron, nickel and neon comparable to those derived earlier by Pottasch from an analysis of the solar UV spectrum.

Provisional analysis of the latest flight (Skylark 305 in August 1967), employing crystals of beryl and EDDT in addition to the KAP flown on Skylark 304, has shown strong emission lines of the helium-like ions Mg XI and Si XIII, with weaker but significant intensity in Lyman-$\alpha$ of Mg XII (8·4 Å) and Si XIV (6·2 Å), the latter emission requiring the presence of active region material at a temperature of near $10^7$ °K.

## DISCUSSION

*Newkirk:* Does your emission measure $N_e^2 V \sim 2 \times 10^{48}$ refer to the high or low ionization lines?

*K. Evans:* High ionization.

*Kiepenheuer (ed.), Structure and Development of Solar Active Regions,* 431. © *I.A.U.*

# OBSERVATIONS OF ENERGETIC X-RAYS
# FROM QUIESCENT SOLAR ACTIVE REGIONS*

Loren W. Acton and Philip C. Fisher**

*(Lockheed Palo Alto Research Laboratory, Palo Alto, Calif., U.S.A.)*

ABSTRACT

A measurement of the solar X-ray spectrum in the 1–30 KeV interval was made by rocket-borne instruments at 0015 UT on April 5, 1967. These measurements show that the solar X-ray spectrum, which falls off steeply with energy below 4 KeV, becomes less steep above this energy. These data are taken to indicate that energetic, non-thermal, electrons are present or continuously produced in quiescent solar active regions. No optical flare activity was reported for several hours around the time of these observations.

## 1. The Experiment

A solar active region represents a remarkable concentration of energetic phenomena exemplified by the tremendous outpouring of energy associated with solar flares which, although long recognized, still remain to be understood. A study of the quiescent state of active regions can advance our knowledge of the physical nature of such regions and hence the sources of flare energy.

An experiment to investigate the quiescent state of solar active regions was carried out by instruments flown on a spin-stabilized rocket, which was launched from White Sands Missile Range, N.M., at 0015 UT on April 5, 1967. The nose of the spinning rocket was pointed approximately at the zenith with a yaw cone having a half angle of about 5°. X-ray detectors were mounted so as to look out the sides of the rocket. The Sun was located in the West at a zenith angle of approximately 75°. The experiment did not provide any information concerning the positions of the X-ray sources on the Sun. However, recent X-ray photographs of the Sun (Giacconi *et al.*, 1965; Pounds and Russell, 1966; Underwood and Muney, 1967; Zhitnik *et al.*, 1967) seem to establish conclusively that the emission of the harder X-rays is restricted to centres of activity.

The object of the experiment was to measure the solar X-ray emission in the 1 KeV (12 Å) to 30 KeV (0·4 Å) region. The radiation sensors were eight gas-filled proportional counters with a total area of 0·34 m². The signals from these proportional

* This work has been supported by the National Aeronautics and Space Administration under contract NASw-1398 and by the Lockheed Independent Research Program.

** Presented by L. W. Acton.

counters were processed by five different pulse-amplitude measuring systems which were equipped with a variety of provisions for reducing the deleterious effects of the cosmic-ray environment. The primary instrument, which yielded the most precise measurement of the solar X-ray spectrum, consisted of four proportional counters of total area 0·23 m². Figure 1 illustrates the computed photon-detection efficiency

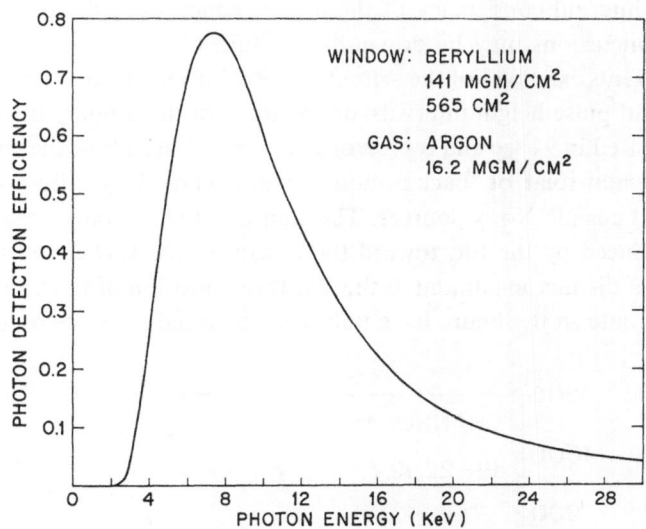

FIG. 1. *Typical photon-detection efficiency of the proportional counters used in the primary system. Four such counters were used to feed the primary pulse-height analyzer. The energy resolution (full width at half maximum) of these counters at 6 KeV was better than 20% over the entire window area.*

of each of the four counters of the primary instrument. These counters were equipped with field-shaping electrodes so that the gain and resolution were uniform to within a few percent over the entire window area. Pulses from these counters were fed into a single analog-to-digital pulse-height analyzer which was developed specifically for this experiment (Reed *et al.*, 1968). The analyzer was designed so that high pulse rates could be handled and a complete 16-channel spectrum was read out every 5 milliseconds, with no dead time introduced by the read-out process. The channels of this analyzer were arranged so that the channel widths in the 3–8·5 KeV interval were approximately 0·5 KeV. The higher energy channels were set wider because of the anticipated lower photon flux at these energies. A very important feature of this X-ray detection system was a pulse rise-time discrimination circuit which rejected a high percentage of the pulses produced in the proportional counters by cosmic rays while accepting the pulses produced by X-rays. This circuit was developed especially for this experiment by Dr. R. C. Catura of this laboratory. The technique was first used by Mathieson and Sanford (1963).

## 2. The Data

It is important to emphasize that the following discussion of the experimental results represents a preliminary analysis of the data from the flight. The possibility exists that systematic errors remain, both in our understanding of the performance of the instruments and in the data-reduction techniques which have been used. However, the internal consistency of the initial reduction of these data is such that preliminary conclusions may be drawn at this time.

Figure 2 presents examples of the azimuthal distribution of detector counts recorded in three different pulse-height intervals, or channels, of the primary instrument. These data were acquired in 94 sec and represent a sum over 176 rolls of the spinning rocket.

Most of the non-solar or 'background' signal is caused by diffuse cosmic X-rays and unresolved cosmic X-ray sources. The sinusoidal variation with azimuth of this signal is introduced by the tilt, toward the North, of the rocket roll axis. Channels 5 and 15 have a distinct maximum in the West, the position of the Sun at the time of the flight. The data in the figure have not been corrected for system dead-time.

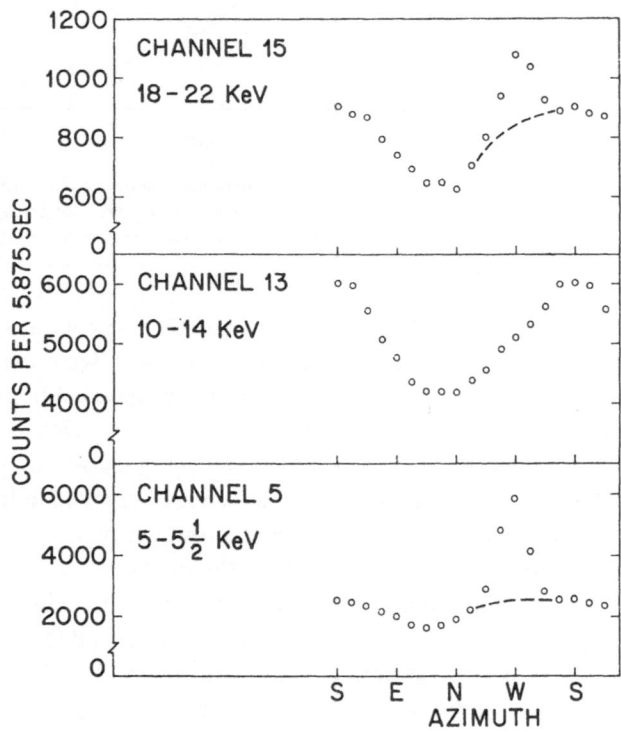

FIG. 2. *Azimuthal distribution of signals from three channels of the primary system summed over 176 rolls of the spinning rocket. The width of the solar peak corresponds to the detector field of view in the direction normal to the spin axis of the rocket.*

The solar signal is not evident in channel 13 where the signal to background ratio is poorer than in the other channels.

A preliminary solar spectrum, derived from a partial reduction of the flight data, is presented in Figure 3. The statistical uncertainties of these data are small, and in most cases are roughly the size of the points which have been plotted. However, with the exception of the measurements below 3 KeV, the data have *not* been corrected

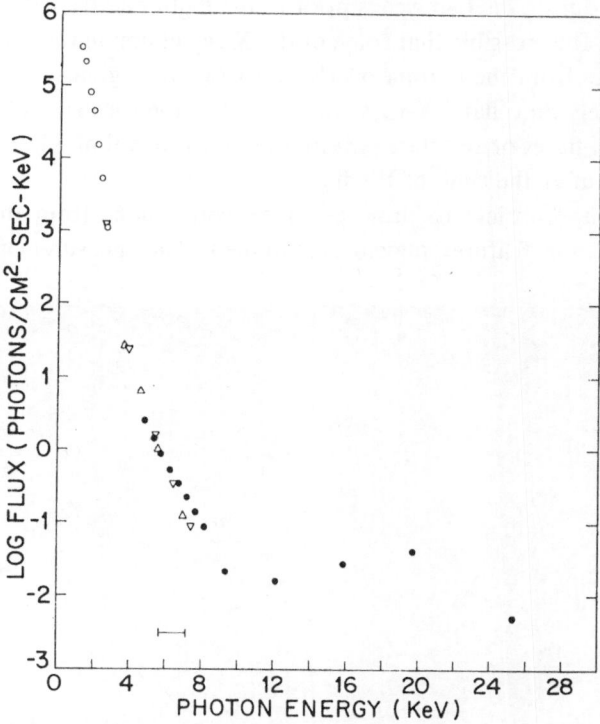

FIG. 3.   *Solar X-ray spectrum derived from a preliminary analysis of a portion of the data from four of the five X-ray detection systems flown on April 5, 1967. The data from each system are indicated by different symbols. Those points from the primary system are represented by filled circles. With the exception of the data indicated by the open circles, none of the data have been corrected for system dead-time. It is anticipated that the properly corrected data will not exhibit a pronounced minimum in the spectrum in the 8–14 KeV interval. The horizontal bar indicates the energy resolution (full width at half maximum) of the primary system at 6·4 KeV.*

for the dead-time of the pulse-amplitude measuring systems. These corrections are in general less than 20%, but they may significantly alter the appearance of the spectrum at the higher energies. We anticipate that the apparent minimum in the spectrum in the 8–14 KeV interval will disappear or become less pronounced when the corrections for dead-time are properly taken into account.

### 3. The Interpretation

Figure 4 is a reproduction of a Sacramento Peak Observatory Hα filtergram taken approximately 9½ hours prior to the time of the flight. At the time of the flight there was a complex of old active regions on the Sun in the Northern hemisphere. There were also four active regions in the Southern hemisphere, two of which produced the only flares within 12 hours of our flight. The two extremely active regions that had crossed the disk during the two weeks prior to the flight had passed the Western limb two days earlier. It is possible that some of the X-ray emission recorded by the rocket instruments came from the corona overlying these two regions.

It seems unlikely that 'flare' X-rays contributed to the spectrum of Figure 3. There were no reported flares or sub-flares within the time interval of 3 hours prior to flight time to 7 hours after the time of the flight.

While it is inappropriate to draw extensive conclusions from these preliminary results the following features may be mentioned. The measurements of the solar

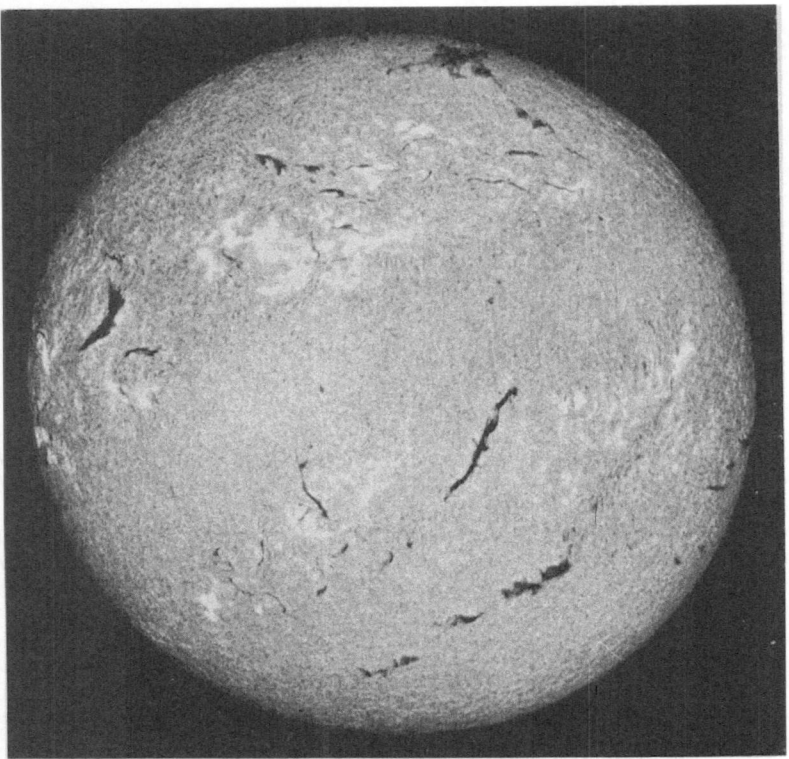

FIG. 4.   *Hα filtergram of the Sun taken at 1448 UT on April 4, 1967. (Courtesy of Sacramento Peak Observatory.)*

spectrum at energies below 4 KeV are consistent with those of other workers (Bowles *et al.*, 1967). The observed X-ray intensity lies between that reported by Chodil *et al.* (1965) and that of Underwood given by Bowles *et al.* (1967). Fritz *et al.* (1967) have shown that the X-ray spectrum at energies below 6 KeV is dominated by emission lines.

The slope of the X-ray spectrum above 4 KeV in energy is markedly less steep than at the lower energies. A second change in the nature of the solar X-ray spectrum appears to occur in the 8–12 KeV interval where the spectrum becomes approximately flat, extending to the upper limit of the most reliable data at 22 KeV. The flux which we observed around 15 KeV is almost two decades higher than the upper limit given by Peterson *et al.* (1966) for February 18, 1966 and about five decades below that reported by Chubb *et al.* (1966) from their pioneering observations made in 1959. However, our data appear to be consistent with the measurements reported by Peterson (1967) for the 7·7–12·5 KeV and 12–22 KeV channels of his OSO-3 instrument.

We believe that the extension of the solar X-ray spectrum above approximately 6 KeV is strongly indicative of non-thermal processes taking place in the active regions on the Sun which were quiescent at the time of our flight. Indeed, the relatively flat spectrum extending to high energies beyond 12 KeV may be bremsstrahlung from relativistic electrons either trapped or continually produced within the centres of activity. Further analysis of these data and additional measurements of the kind made during this flight show promise for helping to understand the physics of solar active regions.

**Note added in proof:** It has not yet been possible to reconcile the solar data of the three large-area systems for photon energies greater than 8 KeV. Therefore we consider it advisable to treat with caution the data above 8 KeV from this experiment until such time as these particular measurements are better understood or until independent confirmation of a flat solar X-ray spectrum in the 8-22 KeV interval is obtained.

## References

Bowles, J. A., Culhane, J. L., Sanford, P. W., Shaw, M. L., Cooke, B. A. (1967) *Planet. Space Sci.*, **15**, 931.

Chodil, G., Jopson, R. C., Mark, Hans, Seward, F. D., Swift, C. D. (1965) *Phys. Rev. Lett.*, **15**, 605.

Chubb, T. A., Kreplin, R. W., Friedman, H. (1966) *J. geophys. Res.*, **71**, 3611.

Fritz, G., Kreplin, R. W., Meekins, J. F., Unzicker, A. E., Friedman, H. (1967) *Astrophys. J.*, **148**, L133.

Giacconi, R., Reidy, W. P., Zehnpfennig, T., Lindsay, J. C., Muney, W. S. (1965) *Astrophys. J.*, **142**, 1274.

Mathieson, E., Sanford, P. W. (1963) in *Proc. Int'l. Symp. on Nuclear Elec.*, p. 65.

Peterson, L. E. (1967) Paper presented at the XIII General Assembly of the IAU, Prague, August 23, 1967.

Peterson, L.E., Schwartz, D.A., Pelling, R.M., McKenzie, D. (1966)      *J. geophys. Res.*, **71**, 5778.

Pounds, K.A., Russell, P.C. (1966)      in *Space Research*, VI, p. 38.

Reed, R.D., Bakke, J.C., Reagan, J.B., Acton, L.W. (1968)      *IEEE Trans. on Nuclear Sci.*, **NS-16** (in press).

Underwood, J.H., Muney, W.S. (1967)      *Solar Phys.*, **1**, 129.

Zhitnik, I.A., Krutov, V.V., Malyavkin, L.P., Mandel'shtam, S.L., Cheremukhin, G.S. (1967) *Kosm. Issled.*, Akad. Nauk SSSR (Moscow), **5**, 276.

# X-RAY PICTURE OF THE SUN TAKEN WITH
## FRESNEL ZONE PLATES

G. ELWERT

*(Lehrstuhl für Theoretische Astrophysik, Universität Tübingen, Germany)*

It is well known that Fresnel zone plates act as a lens with a focal length inversely proportional to the wavelength. At Professor Moellenstedt's Institute of Applied Physics in Tübingen, a technique has been developed to manufacture micro-zone plates electronoptically from Buckbee-Mears zone plates. These micro-zone plates have a diameter of 0·5 mm approximately, 38 zones and a focal length of 30 cm for X-rays of nearly 50 Å. Their resolving power is of the order of a few seconds of arc (Einighammer *et al.*, 1966). One finds, however, experimentally as well as theoretically that the sharply defined image is surrounded by a halo having a diameter of a few minutes of arc for the above-mentioned zone plates. Whereas for a point source the intensity of a halo is poor and therefore unimportant, for extended sources it becomes large as a result of the superposition of the contributions from the individual points of the source. This has been demonstrated by the photometer curves of the images of circular sources by Einighammer (1966) (Figure 1). The halo is not noticeable in case of very narrow sources (curves *a* and *b*). With increasing diameter of the source

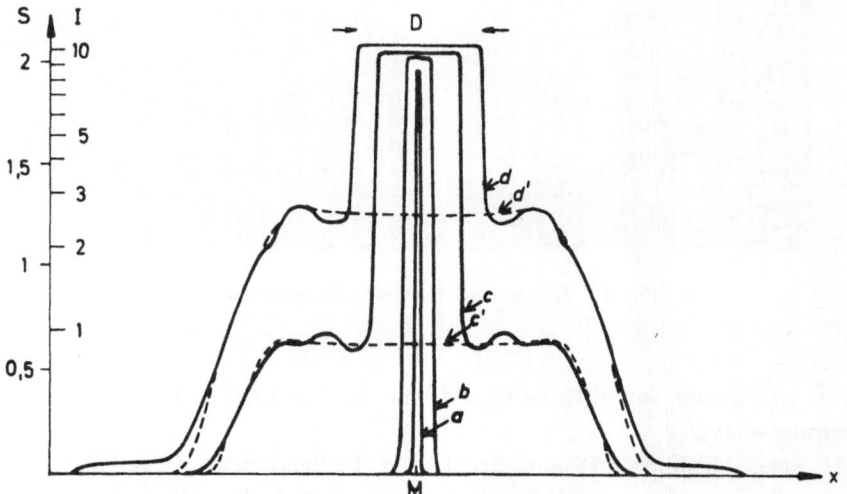

FIG. 1.   *Photometer curves of the images of circular sources (according to Einighammer, 1966).*

*Kiepenheuer (ed.), Structure and Development of Solar Active Regions, 439–443. © I.A.U.*

the intensity of the halo rapidly increases (curves *c* and *d*). For the same source the brightness distribution of the halo agrees practically with the one obtained by a pinhole camera, provided the diameter of the hole and that of the zone plate are equal; these brightness distributions are represented by the dashed lines *c'd'* in Figure 1.

During a previous congress, Dr. Friedman kindly consented to insert some plates in one of his pinhole cameras. The zone plates were built by Von Grothe under the guidance of Professor Moellenstedt for the line of O vi at 34 Å and the line of Si x

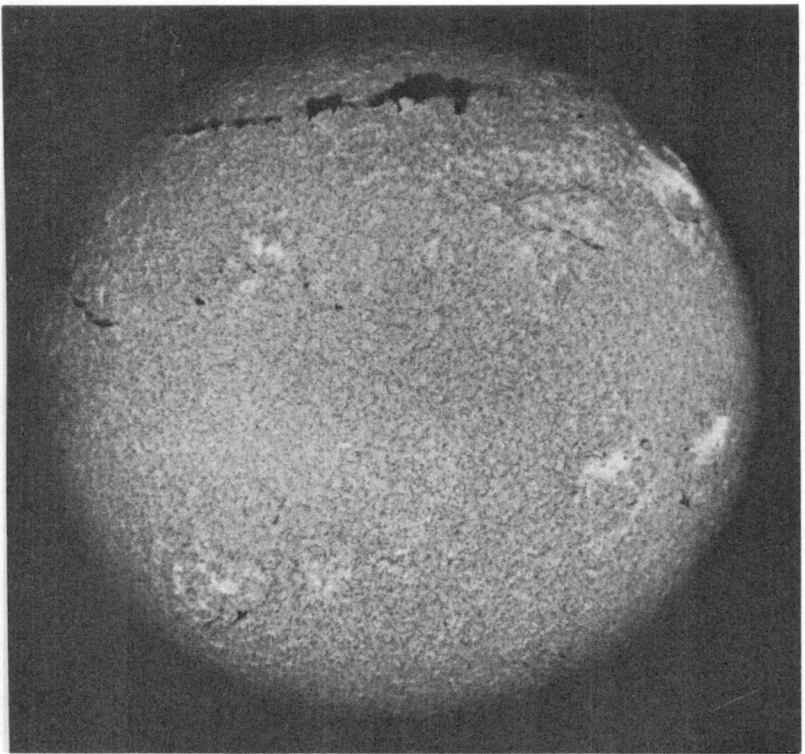

FIG. 2.  *Hα-spectroheliogram taken on October 2.*

at 51 Å, which were found to be strong lines by Tousey. The launching took place on October 4, 1966.

A Hα-spectroheliogram (Figure 2) taken at the Sacramento Peak Observatory on October 2 shows two plages in the neighbourhood of the West limb seen on the right side, one of which disappeared on October 4 (Figure 3). As the condensations are situated above the plages, it is to be expected that on the day of launching both X-ray

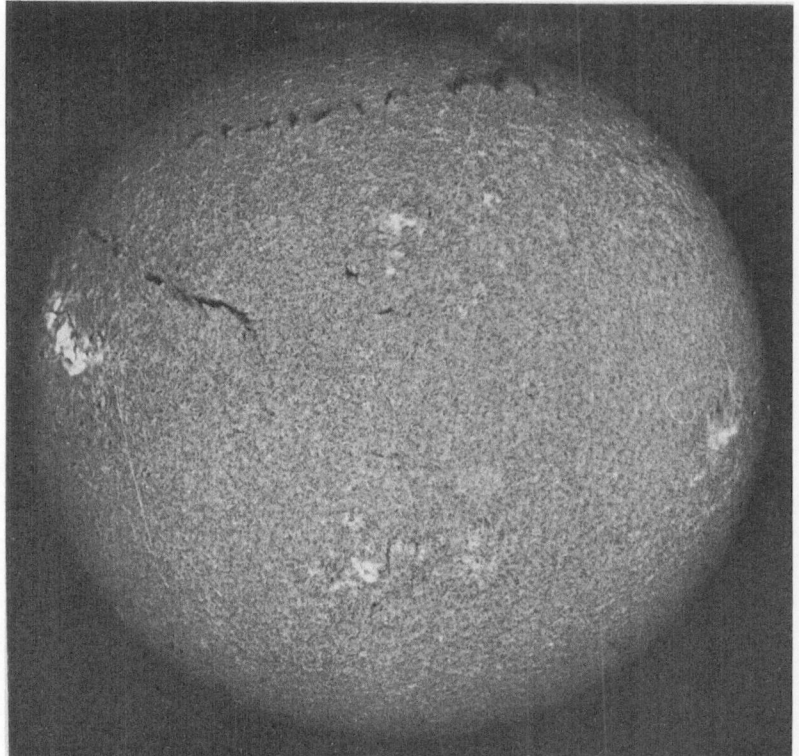

FIG. 3. *Hα-spectroheliogram taken on October 4.*

sources on the limb were visible. This was found to be true. On the East limb, one extended plage appeared on October 4. Unfortunately the attitude control did not work satisfactorily. During the time of exposure the rocket rotated around the direction to the Sun's centre by an angle of 70° nearly, so that the pictures of active regions were smeared.

The photograph taken in the focus of 51 Å (Figure 4) shows smeared arcs with a thickness of approximately 0·2 solar diameter. They are due to the halo, and a small part may also be contributed by the out-of-focussing of the O line of 34 Å. Within these smeared arcs on the West limb, however, narrow arcs of small width can be detected. In addition to that, on the disk and on the East limb very narrow arcs are to be seen. From these sharply defined structures which arise from the central peak of the diffraction pattern, one can conclude that small sources of the Si line at 51 Å exist.

The results of the photograph taken in the focus of 34 Å are somewhat different (Figure 5). On the West limb, a diffused arc, the width of which corresponds to the

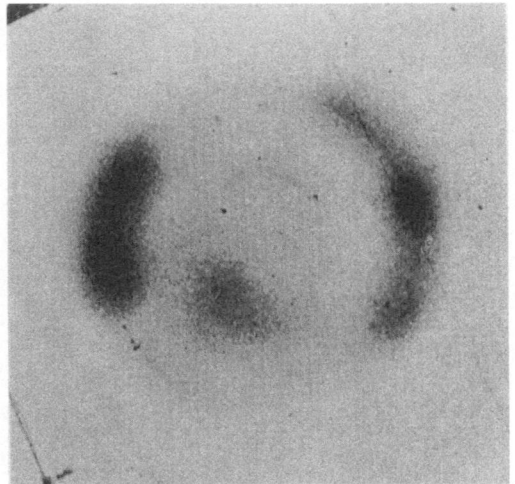

FIG. 4.    *X-ray picture of the Sun taken in the focus of* 51 Å.

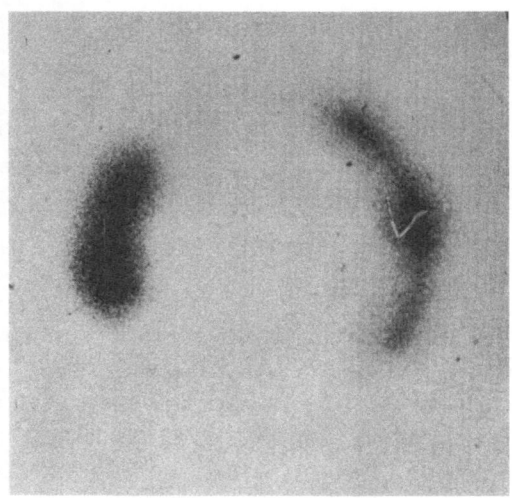

FIG. 5.    *X-ray picture of the Sun taken in the focus of* 34 Å.

halo of the 34 Å radiation, is to be seen. It contains again a relatively small region of emission at 34 Å. However, its extension is somewhat larger than that of the 51 Å radiation. On the East limb and on the disk, on the other hand, no sharply defined structure can be detected.

I have already mentioned that one can show experimentally as well as theoretically that for the same source the brightness distribution of the halo agrees practically with the one obtained by a pinhole camera, provided the diameter of the hole and that of the zone plate are equal. According to Einighammer (1966) it is therefore possible to eliminate the halo experimentally if X-ray pictures of the Sun are taken simultaneously by means of a zone plate and a pinhole camera, as described above. This is supposed to be exploited for later launchings. More details will be published elsewhere.

## Acknowledgements

It is a pleasure to express my gratitude to Professor Moellenstedt and Mr. von Grothe for manufacturing the zone plates and to Dr. Friedman for consenting to launch them. I would also like to thank Dr. Unzicker for arranging the insertion of the zone plates into the rocket. He and Dr. Conrads (Institute for Plasma Physics, Yülich, Germany) kindly procured the filters. Finally I am grateful to Dr. Bruzek for providing the spectroheliograms.

## References

Einighammer, H. (1966)     *Naturwissenschaften*, **53**, 272.
Einighammer, H., Elwert, G., Mayer, U. (1966)     in *Space Research*, VII, p. 1336.

## DISCUSSION

*Burton:* Please would you give details of the types of filters used to cover the zone plates, and of the photographic film used for recording the solar images?

*Elwert:* Filters of Aluminium and Parlodion (Fig. 5) or Macrolon (Fig. 4) have been used. The film used was Kodak SC-5 ultra-fine grain.

# THE SIGNIFICANCE OF THE POLARIZATION OF SOLAR
## SHORT-WAVELENGTH X-RAYS

G. Elwert

*(Lehrstuhl für Theoretische Astrophysik, Universität Tübingen, Germany)*

For the interpretation of the physical events occurring at the beginning of a flare it seems to be significant not only to observe the spectra of the short-wavelength X-ray radiation but also its polarization. Solar X-ray spectra below 10 Å during flares observed by Bowen *et al.* (1964) show in the region of long wavelength a slow fall in their intensity and below 5 Å approximately a rapid fall (Figure 1). This spectral variation cannot be understood on the basis of thermal radiation only. However, it is possible to explain the observed spectra by assuming a superposition of a thermal radiation as well as a non-thermal one. Indeed, one should assume that the fast electrons, during their acceleration, obtain a preferred direction and that only after-

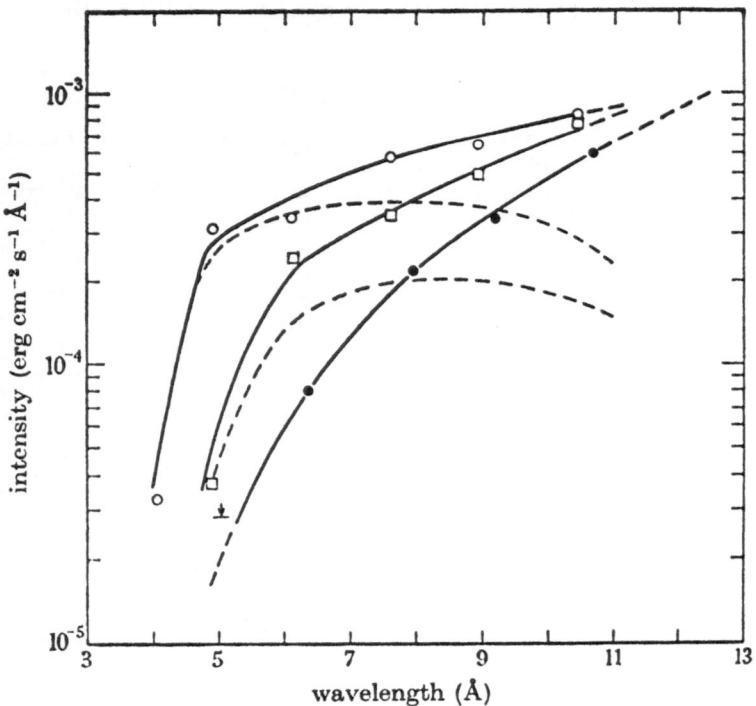

Fig. 1.  *Solar X-ray spectra as observed by Bowen et al. (1964).*

*Kiepenheuer (ed.), Structure and Development of Solar Active Regions,* 444–448. © *I.A.U.*

wards their velocity distribution becomes a Maxwellian one. The assumption of the existence of a non-Maxwellian distribution of electrons in the beginning of a flare is also supported by the fact that the rise in intensity by more than one order of magnitude takes place in a few minutes. This point has already been emphasised by Bowen *et al.*

Accordingly one is led to the interpretation that the non-thermal part is produced by electrons moving in a preferred direction. This interpretation then corresponds to the conjecture of De Jager and Kundu (1963), according to which electrons accelerated in a magnetically neutral zone of a flare and moving towards the photosphere generate X-ray bremsstrahlung in the deeper atmosphere, while the outgoing electrons cause a type-III burst (Figure 2). The model proposed by Takakura and Kai (1966) would also provide electrons moving in a preferred direction. In this case the energetic electrons are trapped in a magnetic tube situated in the lower corona between bipolar sun spots (Figure 3). Now, the point is that for the fast electrons moving in a preferred

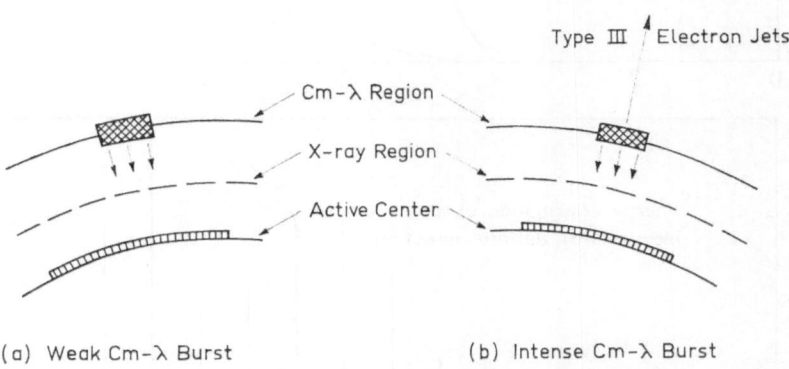

(a) Weak Cm-λ Burst          (b) Intense Cm-λ Burst

FIG. 2.   *Flare model according to De Jager and Kundu (1963).*

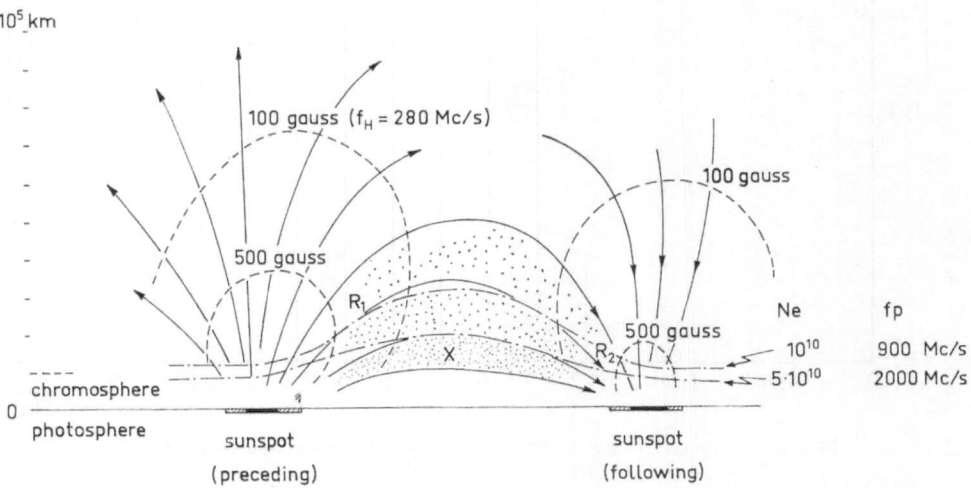

FIG. 3.   *Flare model according to Takakura and Kai (1966).*

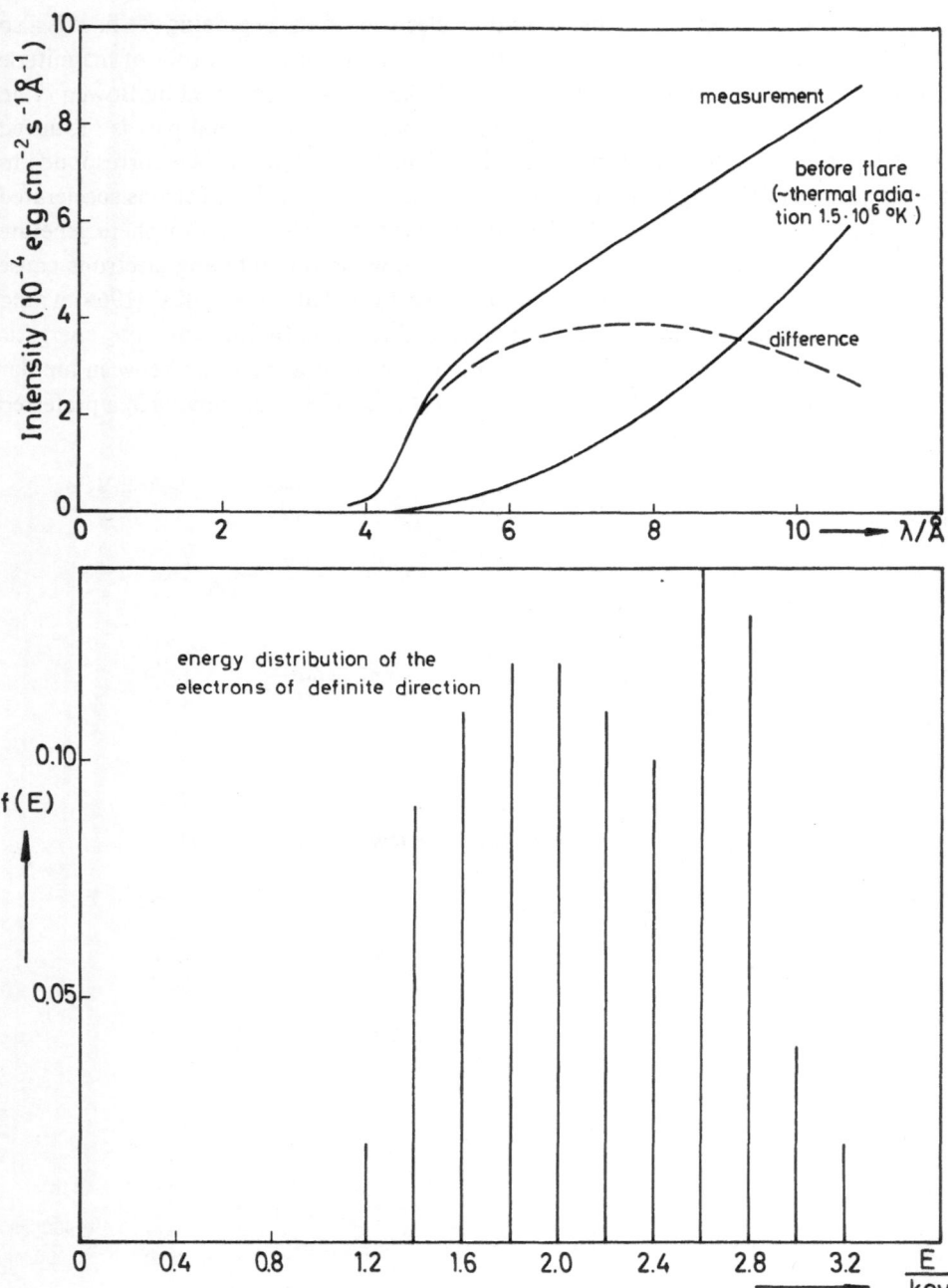

FIG. 4.   *X-ray spectra and distribution function of the electron energies f(E).*

direction the X-ray bremsstrahlung should be appreciably polarized below 8 Å approximately.

To get an idea of the polarization which can be expected, we discuss in the following the uppermost spectrum observed by Bowen *et al.* The intensity of the radiation before the flare corresponding to a temperature of $1 \cdot 5 \times 10^6$ °K is subtracted from the measured one to give the dashed curve. These curves are represented again in Figure 4. The distribution function $f(E)$ of the electron energy, which allows one to represent the difference curve as X-ray bremsstrahlung of directed electrons, was calculated neglecting a possible contribution of recombination radiation. This function $f(E)$ with a maximum of about 2 keV can be approximated by a discrete energy spectrum, which is also shown in Figure 4.

The degree of polarization of the X-radiation was then determined as a function of the angle $\theta$ between the direction of observation and the direction of the electrons (Figure 5). This calculation, as well as the determination of $f(E)$, has been performed by one of my co-workers, Dr. Haug.

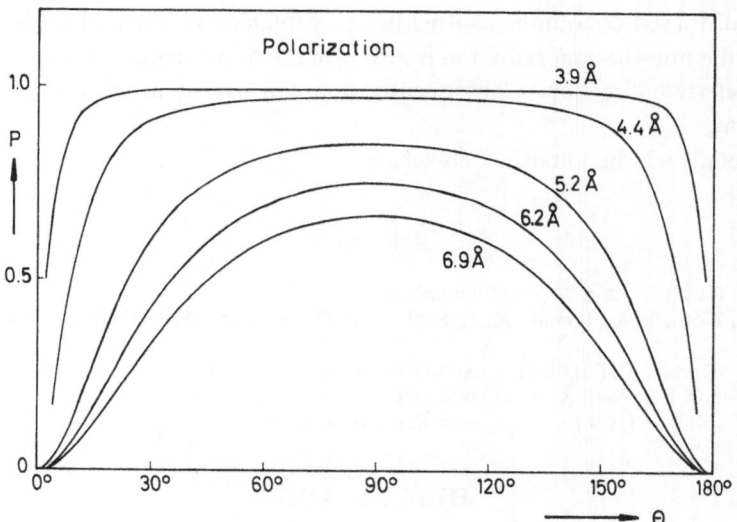

FIG. 5.    *Calculated X-ray polarization P.*

The polarization $P$ attains its maximum value when observed at right angle to the direction of the electron beam ($\theta = 90°$) i.e., in the case of a radial beam corresponding to the model of De Jager and Kundu for a flare at the limb of the Sun. On the other hand, the polarization goes to zero when observed opposite to the direction of the electron beam ($\theta = 180°$), i.e., in the case of electrons moving inward radially for a flare occurring at the centre of the Sun's disk. For the model of Takakura and Kai, according to which the electrons move essentially parallel to the Sun's surface, the maximum value of $P$ would be obtained for a flare at the centre of the Sun's disk.

We see from Figure 5 that the maximum values of polarization are nearly 60% at 7 Å and practically 100% at 4 Å. Observed under 150° with regard to the direction of the electron beam (in case of the model of De Jager and Kundu, i.e., for a flare occurring at a heliocentric angle of 30°), one obtains a polarization of nearly 30% at 7 Å and practically 100% at 4 Å.

The spectrum used in our calculations was observed at the time of maximum intensity of X-ray flares. Thus, our calculations lead to the conclusion that it should be possible to establish during this time an appreciable degree of polarization of short-wave bremsstrahlung X-rays. By measuring the polarization, the models in which the fast electrons have a preferred direction could therefore be tested. The degree of polarization would increase with decreasing wavelength. If no polarization is found, each model in which the electrons have at all a preferred direction would then be ruled out. A proposal to perform these measurements has been accepted by the sun group of ESRO.

As Acton has reported (1968), the short-wavelength X-radiation below 3 Å observed during undisturbed conditions can neither be explained as thermal radiation. Most probably, the non-thermal radiation is also generated as bremsstrahlung by electrons with a preferred direction. Consequently one can expect a substantial degree of polarization.

More details will be published elsewhere.

## References

Acton, L. W. (1968)      in the present volume, p. 432.
Bowen, P. J., Norman, K., Pounds, K. A., Sanford, P. W., Willmore, A. P. (1964)      *Proc. Roy. Soc.*, **281**, 538.
De Jager, C., Kundu, M. R. (1963)      in *Space Research*, III, p. 836.
Elwert, G., Haug, E. (1968)      to be published.
Takakura, T., Kai, K. (1966)      *Publ. astr. Soc. Japan*, **18**, 57.

## DISCUSSION

*Öhman:* Have you also developed some new schemes for measuring polarization?

*Elwert:* The polarization may be measured by standard technique, i.e. Thomson or Barkla effect; one measures the intensity of the scattered radiation as a function of the direction. Another possibility is to use the Borrmann effect. Borrmann has shown that X-rays incident under Braggs angle on a perfect crystal undergo no absorption and that the forward-diffracted beam is polarized orthogonal to the plane of scattering.

Part VI

# TRANSIENT PHENOMENA

# ANALYSIS OF SOME SOLAR FLARES FROM OPTICAL, X-RAY, AND RADIO OBSERVATIONS*

R. Falciani, M. Landini, A. Righini, M. Rigutti

*(Arcetri Observatory, Firenze, Italy)*

### ABSTRACT

Using an improved isodensitometric technique it has been possible to study in great detail the photometric structure and the evolution of eight flares. A comparison has been made between the evolutive curves and the ones obtained from measurements of solar X-rays and radio fluxes at $\lambda\lambda$ 3·2 and 21 cm.

A reduction of the flare areas (and of the emitted energies) before the flash phase and a continuous pulsation of the flare have been observed. Further it seems that the flares associated with radio bursts or X-ray events are those which show regions of a sufficiently high intensity, the emitting areas not being a very important parameter.

The correlation in time between the various examined aspects of the AR seems to indicate that the sequence for the beginning of the different phenomena is, in general: optical flare, X-ray events, radio events.

It has been pointed out several times in the past that flare patrol is handled mostly as if it were to be used only by geophysicists (Severny, 1965a). Nevertheless in the last years much progress has been made, particularly from an observational point of view. The same cannot be said for the analysis of the photographic data. All of us know that flare classification is still left to a more or less personal judgment of the observer, that the area of a flare is very often obtained by empirical and questionable methods, and that there only exist very few and scanty photometric studies of the intensity distribution within the flare region and of the photometric evolution of the flare (Ballario, 1958, 1959; Billings and Roberts, 1953; Dodson *et al.*, 1956; Ellison *et al.*, 1960; Van Gruithuyzen and Houtgast, 1959; Russo and Righini, 1961). Of course this is due to the fact that a detailed photometric analysis of a number of frames for a single flare is a huge and very time-consuming task.

However, at the Prague General Assembly we briefly described a photographic technique to obtain rapidly a network of very reliable isodensity thresholds from photographs of extended sources. We used this technique to get isophotes from corona photographs and the success obtained in this kind of work suggested us to try to use the technique also for the analysis of Hα-flare photographs.

So we analyzed eight solar flares recorded at Arcetri to test the possibilities of our method.

At the same time we used some X (NRL Solrad satellite) and radio ($\lambda\lambda$ 3·2 and 21 cm)

* Presented by M. Rigutti.

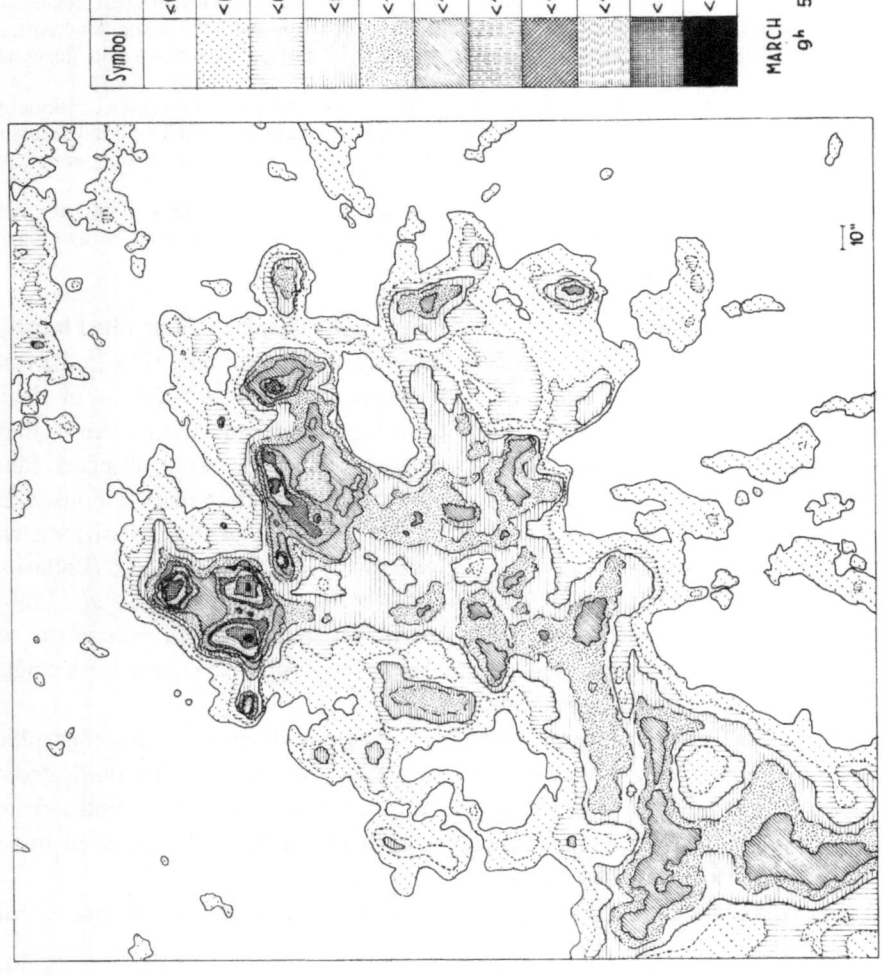

| Symbol | D | log I |
|--------|------|-------|
|  | <0.67 | <0.050 |
|  | <0.73 | <0.035 |
|  | <0.84 | MEAN CHROMOSPR. LEVEL |
|  | <0.96 | <0.035 |
|  | <1.09 | <0.075 |
|  | <1.22 | <0.110 |
|  | <1.41 | <0.170 |
|  | <1.56 | <0.215 |
|  | <1.68 | <0.250 |
|  | <1.76 | <0.280 |
|  | <1.91 | <0.320 |

MARCH 21 - 1966
9ʰ 56ᵐ U.T.

FIG. 1.   Isophotes of the March 21, 1966 flare. D is the photographic density of the isophote on the original filtergram and I is the corresponding intensity in terms of the nearby mean undisturbed chromospheric intensity level.

results of observations performed at the Arcetri Observatory to try to get some correlations between the various phenomena from single flares.

With our technique (De Gregorio *et al.*, 1966, 1967*a, b, c*) we can obtain from a filtergram a high contrast print of the flare region in such a way that the copy shows only two regions: one completely transparent and another completely black. A very reliable isophote can then be obtained. By changing the exposure time in the process it is possible to get a number of isophotes from a single filtergram. The density corresponding to the line separating the transparent and black regions (the isophote) is obtained by photographing a photometric wedge together with the original. Each isophote encloses regions of the flare showing photographic densities larger, or smaller, than the one corresponding to the isophote itself. So, the isophote is a density threshold. By difference of the areas enclosed by two successive isophotes one gets the area of a density step in the flare. This area is transformed into energy through the calibration curve and the photographic density of the nearby undisturbed chromosphere. The energy emitted from a given flare intensity level upward, is obtained by adding the energies of the single steps starting from the considered level.

A picture of the intensity structure of the flare will be obtained by superimposing the different isophotes (Figure 1). Finally, taking into account the foreshortening effect, it is possible to compare the energies of different flares. The measurements of the areas can be performed photoelectrically using the black-and-transparent prints as diaphragms in a suitable instrument. All the numerical reductions can be made by a computer. So, the whole analysis process may easily be done automatically or semi-automatically.

The choice of the flares we took into consideration for this first try has been made on the basis of the coincidence of optical and X-ray or radio observations. Unfortunately, for various reasons these coincidences were not very numerous.

Figures 2–5 are examples of evolutive curves of four of the considered flares (the ordinate scales in Figs. 2–5 are the same if the limb-darkening laws for flares and undisturbed chromosphere coincide).

We would like to say at this point that the technique is very appropriate for analyzing flares as far as sensitivity, reliability, rapidity and cost are concerned, also with a sun image of about 20 mm in diameter. We made the first print using an enlargement of a factor 2, and arranged the successive operations in a semi-automatic way but are now preparing a new instrumentation, basically very simple and inexpensive, to operate almost fully automatically.

It is very difficult to give a complete picture of the results of our analysis. We will here point out the most important features we obtained from our research.

(1) The flash phase of the optical flare is preceded by a more or less large pulsation of the active region; just before the flash a rapid and more or less conspicuous diminution of the emitted energy takes place at all levels of intensity in the flare (i.e., also in the emitting areas).

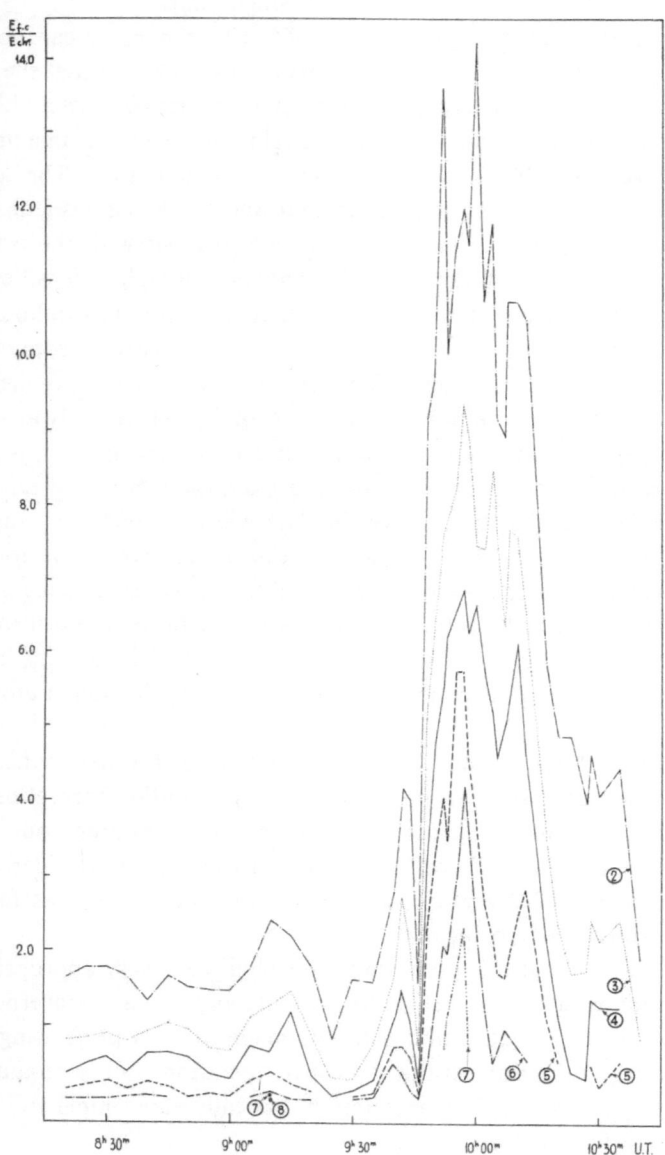

Fig. 2.   *Evolutive curves of the October 4, 1965 flare. In the ordinates are the energies $E_{f.c}$ emitted by the flare from the intensity level $I_f$ (over the nearby mean chromospheric level $I_{ch.}$) up, in terms of the energy $E_{ch.}$ emitted by an area, covering 1 mm² on the original frame, of the nearby undisturbed chromosphere. 1 refers to $I_f = 1\cdot23\ I_{ch}$. 2 refers to $I_f = 1\cdot32\ I_{ch}$. 3 refers to $I_f = 1\cdot42\ I_{ch}$. 4 refers to $I_f = 1\cdot52\ I_{ch}$. 5 refers to $I_f = 1\cdot62\ I_{ch}$. 6 refers to $I_f = 1\cdot74\ I_{ch}$. 7 refers to $I_f = 1\cdot86\ I_{ch}$. 8 refers to $I_f = 2\cdot00\ I_{ch}$. 9 refers to $I_f = 2\cdot14\ I_{ch}$. 10 refers to $I_f = 2\cdot29\ I_{ch}$.*

(2) The pulsation of the flare lasts during the whole life of the flare with several maxima and minima, sometimes rather important.

(3) As it is seen in Figure 1 the region around the flare shows densities smaller than that of the near-undisturbed chromosphere. The presence of this nimbus already observed by other researchers (Ellison *et al.*, 1960, 1961; Severny, 1965*b*), seems to be a constant feature of the flare. On this point we will come back in the very near future.

(4) Sometimes it happens that the optical flare associated with radio bursts does

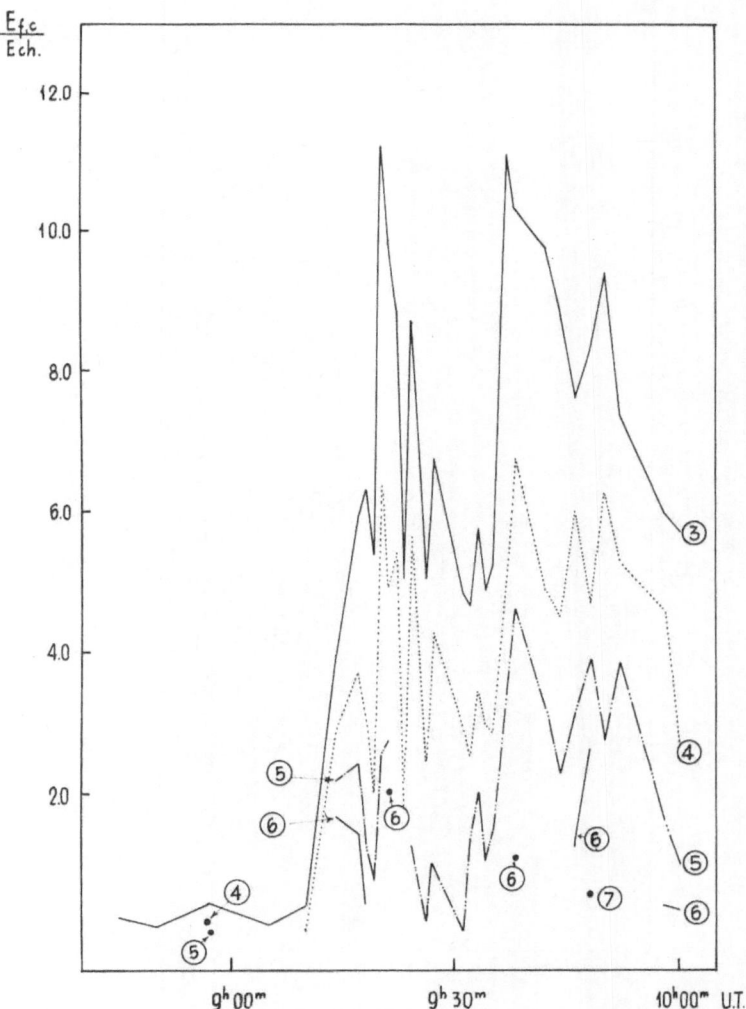

FIG. 3.  *Evolutive curves of the March 16, 1966 flare. For explanations see Figure 2.*

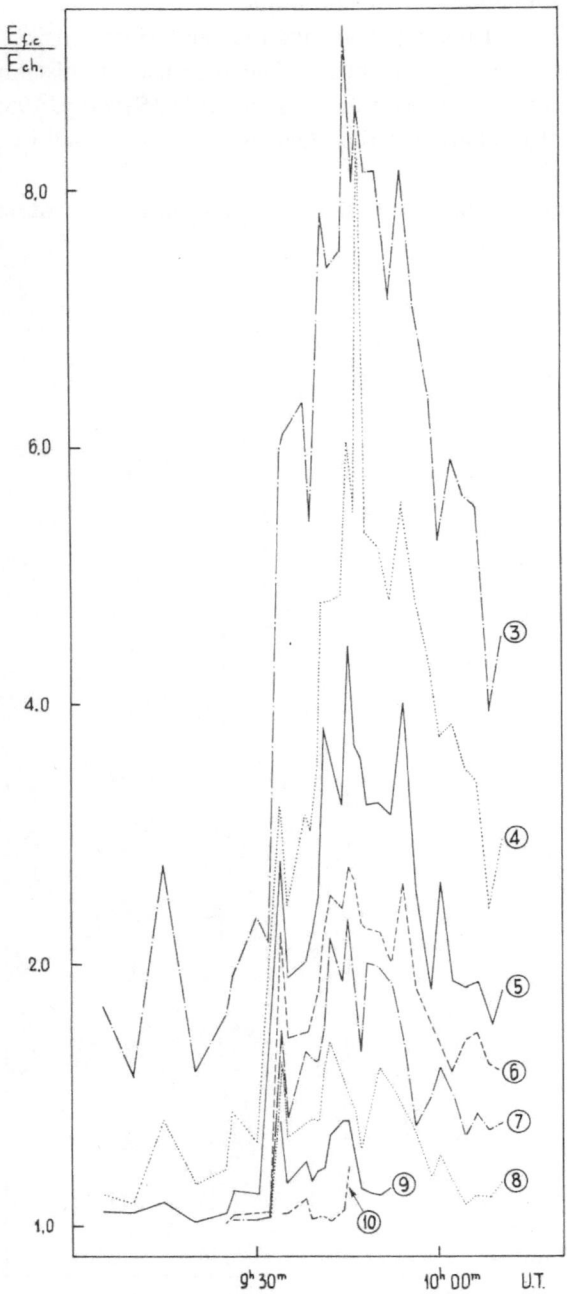

FIG. 4. *Evolutive curves of the March 21, 1966 flare. For explanations see Figure 2.*

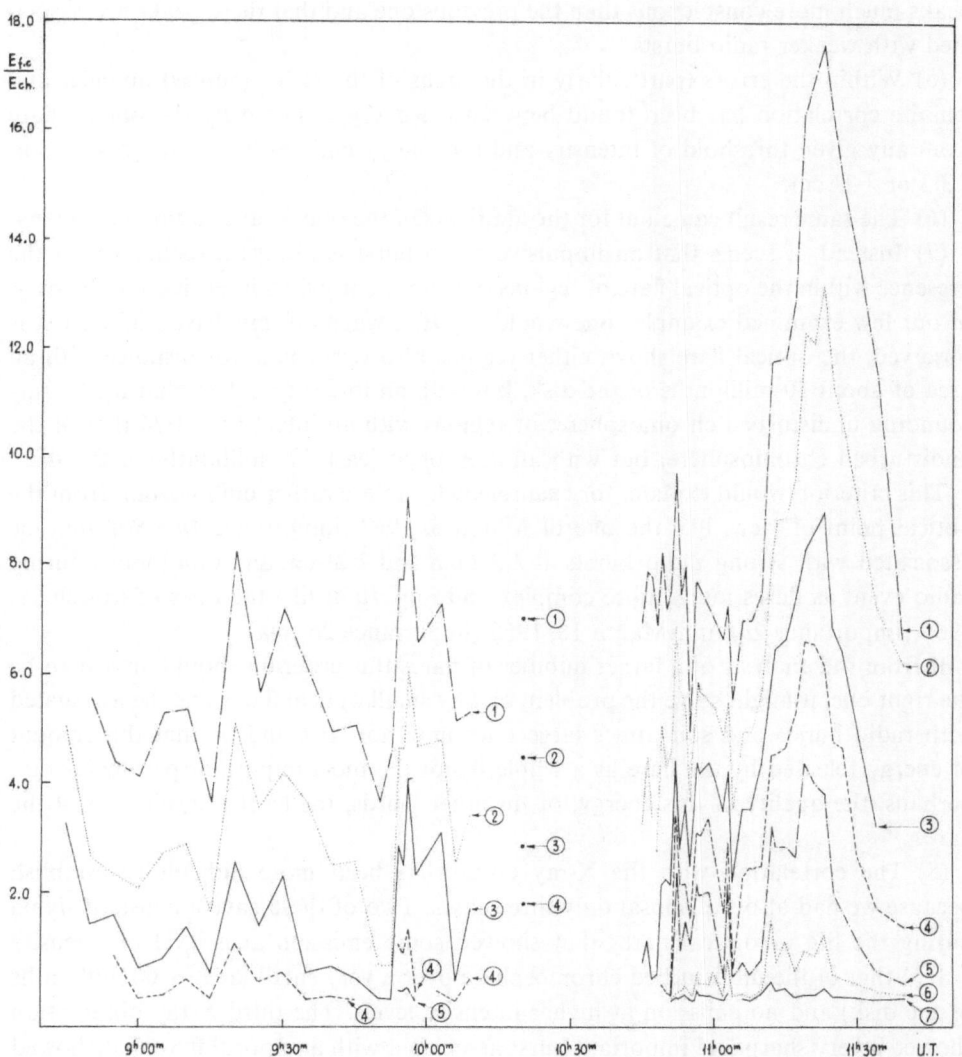

FIG. 5.   *Evolutive curves of the March 29, 1966 flare. For explanations see Figure 2.*

not end its life with the associated radio phenomenon. The optical flare may last much longer and show other conspicuous maxima of energy emission. In this connection it can be noted that (see e.g. the event of March 29, 1966; Figures 5 and 8) after a strong radio burst associated with an optical peak, the flare can show variations with peaks much more conspicuous than the previous one and that these peaks are associated with weaker radio bursts.

(5) Within the errors (particularly in the areas of the radio sources) no clear and definite correlation has been found between the energy emitted by the optical flare from any given threshold of intensity and the energy emitted by the radio bursts at $\lambda$ 3·3 or $\lambda$ 21 cm.

(6) The same result came out for the gradients of the optical and radio phenomena.

(7) Instead, it seems that an impulsive radio burst is always associated with the presence within the optical flare of regions of a sufficiently high intensity. On the basis of our few examined examples one would say that when an impulsive radio burst is observed, the optical flare shows either regions also very small, for instance with an area of about 30 millionths of the disk, but with an intensity $\geqslant 1·86$ that of the surrounding undisturbed chromosphere, or regions with an intensity $\simeq 1·74$ that of the undisturbed chromosphere, but with an area of at least 400 millionths of the disk.

This criterion would explain, for example, why a flare rather unimportant from the optical point of view, like the one of March 3, 1967 (importance $1n + Sn$), may be associated with strong radio bursts at $\lambda$ 3·2 cm and $\lambda$ 21 cm and to a long-enduring radio event as flares much more complex and important like the ones of March 29, 1966 (importance $2b$) and March 16, 1966 (importance $2b + 4n$).

If from the analysis of a larger number of flares this criterion should appear to be the right one, it might solve the problem of why small optical flares may be associated with radio bursts and sometimes large ones may not. It could be that the amount of energy released by the flare as a whole is not the most important parameter but, perhaps, the quality of this energy, or, in other words, the level of excitation of the flare.

(8) The correlation with the X-ray events has been more difficult to establish because we had at our disposal only three cases. Two of these gave a constant signal during the life of optical flares that showed some emission at a level of intensity $= 1·74$ that of the undisturbed chromosphere over a very small area (6–60 millionths of the disk) and no emission at higher intensity levels. The third X-ray observation showed a very sharp and important burst associated with an optical flare that showed emissions from intensity levels up to 2·3 that of the undisturbed chromosphere. At this level – the 'top' of the flare – the emitting area was about 100 millionths of the disk. We may say, very tentatively, having also in mind the hard X-ray event associated with flare-like points studied recently by De Jager (1967), that the X-ray events might be associated with optical flares which show regions of high intensity ($\geqslant 2$ that of the undisturbed chromosphere) with an area also very small (i.e. few millionths of the

disk). In this connection, we can mention a not yet published result by Landini, Noci and Tagliaferri regarding the position and size of one X-ray source observed during the solar eclipse of May 20, 1966. According to these authors the source coincided with the very brightest region of the optical flare. The size of the X-ray source was about 300 millionths of the disk.

(9) We also tried to get some information about the association in time of the various observations. Figures 6–9 show schematically the times of beginning, ending and maxima of the various phenomena. For the optical observations we considered only three intensity steps, the ones from intensity levels 1·42, 1·74, and 1·86 times the value of the undisturbed chromosphere intensity upward.

Although the time resolution of the optical observations is not very high, it seems rather clear that the disturbance begins in general at the lower intensity levels of the optical flare. With a delay ranging from 1 to 5 min, the high-excitation regions in the flare flash up and then, with a delay of about 2 min, the X-ray event, if it occurs, takes place. With respect to the high-excitation regions the radio events begin with a delay of about 2–3 min. However, in one case (flare of March 3, 1967) the radio events preceded the optical flare's beginning.

The conclusion can be that in general the time sequence for the beginning of the various examined phenomena in an active region should be: optical flare, X-ray events, and radio events.

As far as the maxima of energy releases are concerned it seems that, when there is an association, they are more or less simultaneous. Our time resolution of the optical observations is too poor at the moment to have the possibility of saying something more on this point.

These are, at present, our experimental results. Unfortunately we had not enough time to try to get a model of the flares consistent with the observed features. On the other hand, to get a model from an insufficiently large number of phenomena might be a wrong way to go on with the problem or simply a waste of time. So, although certain facts, as e.g. the contraction of the optical flare before the flash phase, seem to be well enough established, we think that the first thing to be done in the near future is to examine much more experimental data than we have done so far. Furthermore, it is evident that the needed time resolution for the optical observation is much better than our present one and that we need as much information as possible about solar X-radiation bursts. It would be very nice to know quantitatively also the evolution of the magnetic fields before and during the flare. Our purpose is to go on working on this line possibly with the cooperation of some other researchers.

## References

Ballario, M. C. (1958)     *Mem. Soc. astr. ital.*, **29**, 89.
Ballario, M. C. (1959)     *Mem. Soc. astr. ital.*, **29**, 439.
Billings, D. E., Roberts, W. O. (1953)     *Astrophys. J.*, **118**, 429.

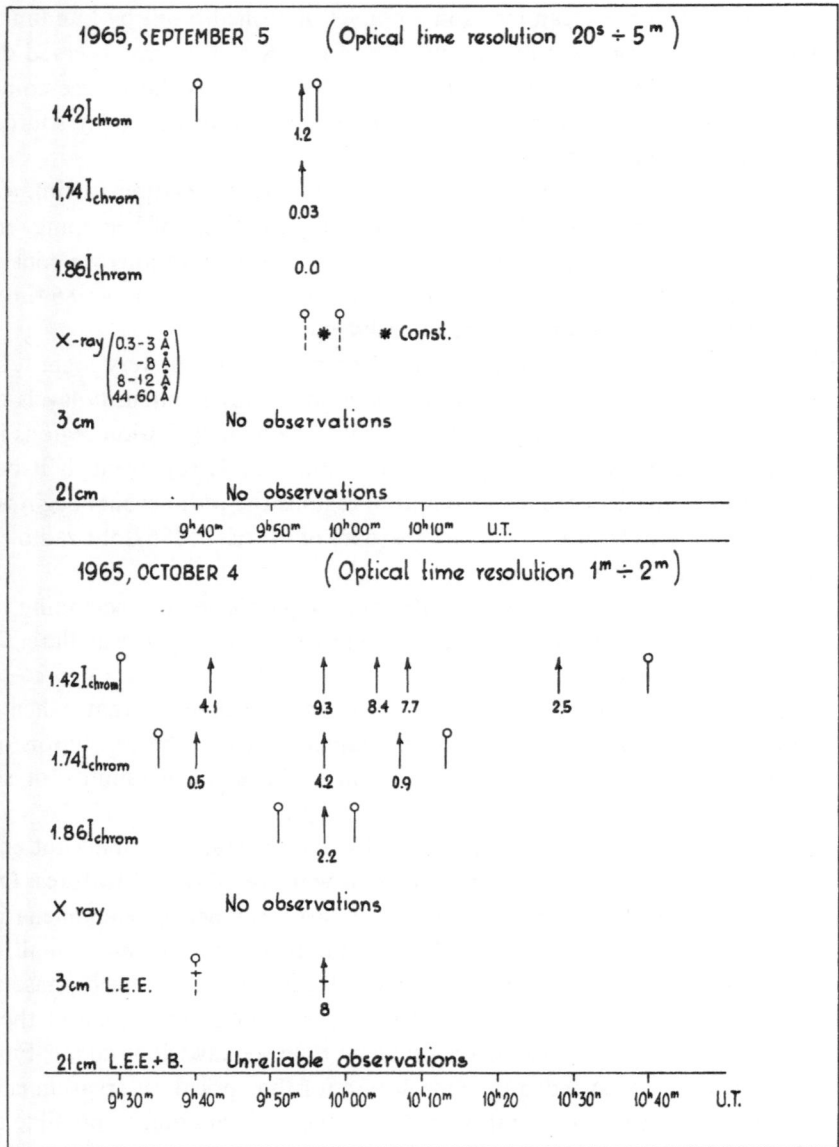

FIG. 6.   *Schematical representation of the association in time of various observations for two AR.*
↑ *indicates the maxima of the considered phenomena and the numbers under the arrows are the energies*
*of these maxima;* ⁝ *indicates the beginning and the ending of the considered phenomenon;* ⁝ *indicates*
*the beginning and the ending of the observations;* ⁝ *indicates the presence of isolated points, not a peak*
*in the evolutive curve; L.E.E. indicates a long-enduring event (♀ in the graphs) and B. indicates a*
*superimposed impulsive burst. As for the symbol 1·42 $I_{chrom}$ and similar ones, see the text. The figures*
*under the arrows describing the optical phenomenon indicate the values $(E_{f.c})/(E_{ch}.)$, while those under*
*the arrows describing the radio phenomenon indicate the ratio of the energy emitted by the radio source*
*to the one emitted by an equivalent area of the mean undisturbed disk.*

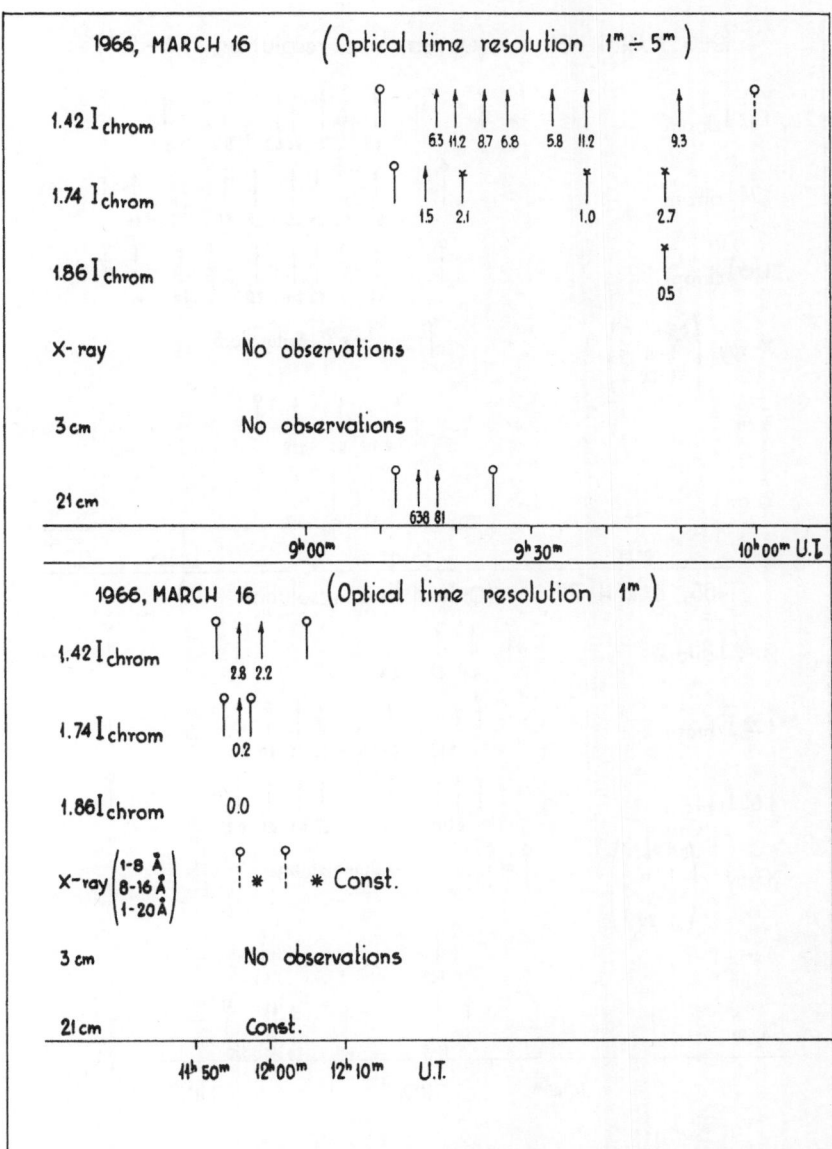

FIG. 7. *Schematical representation of the association in time of various observations for two AR. The notations are the same as in Figure 6.*

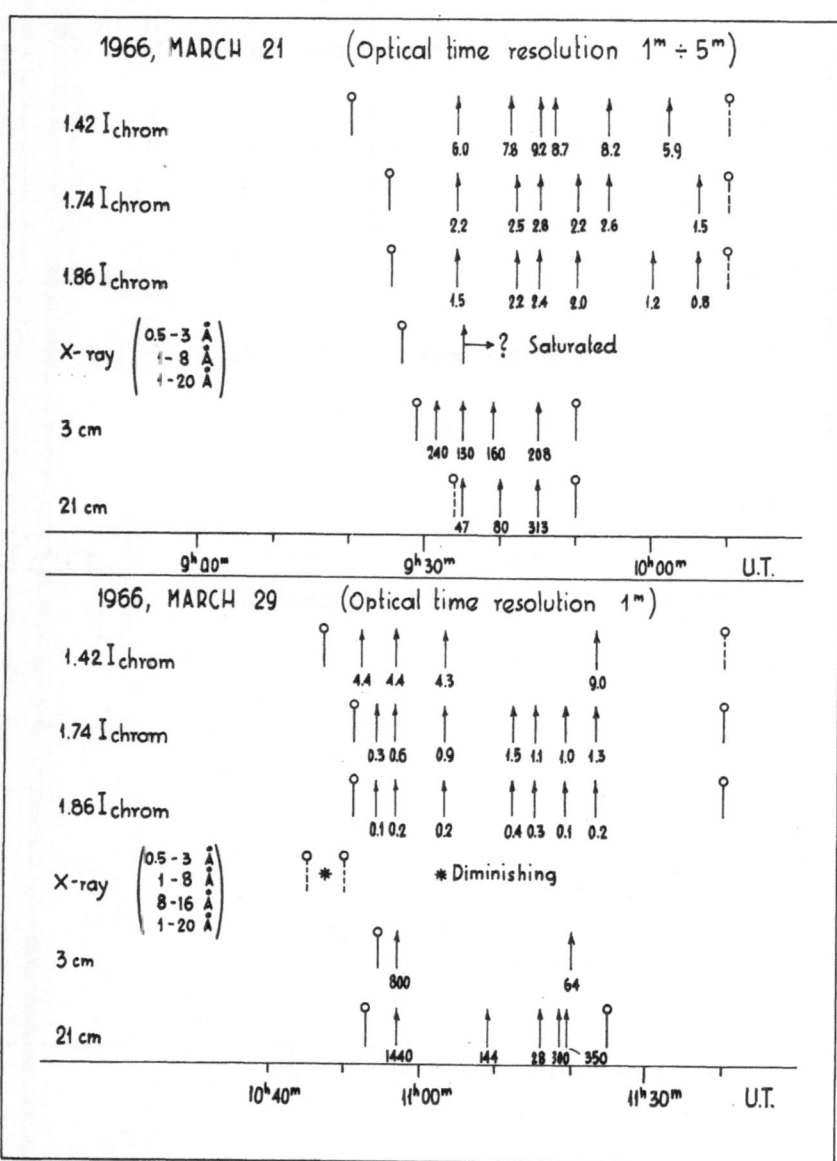

FIG. 8.  *Schematical representation of the association in time of various observations for two AR.*
*The notations are the same as in Figure 6.*

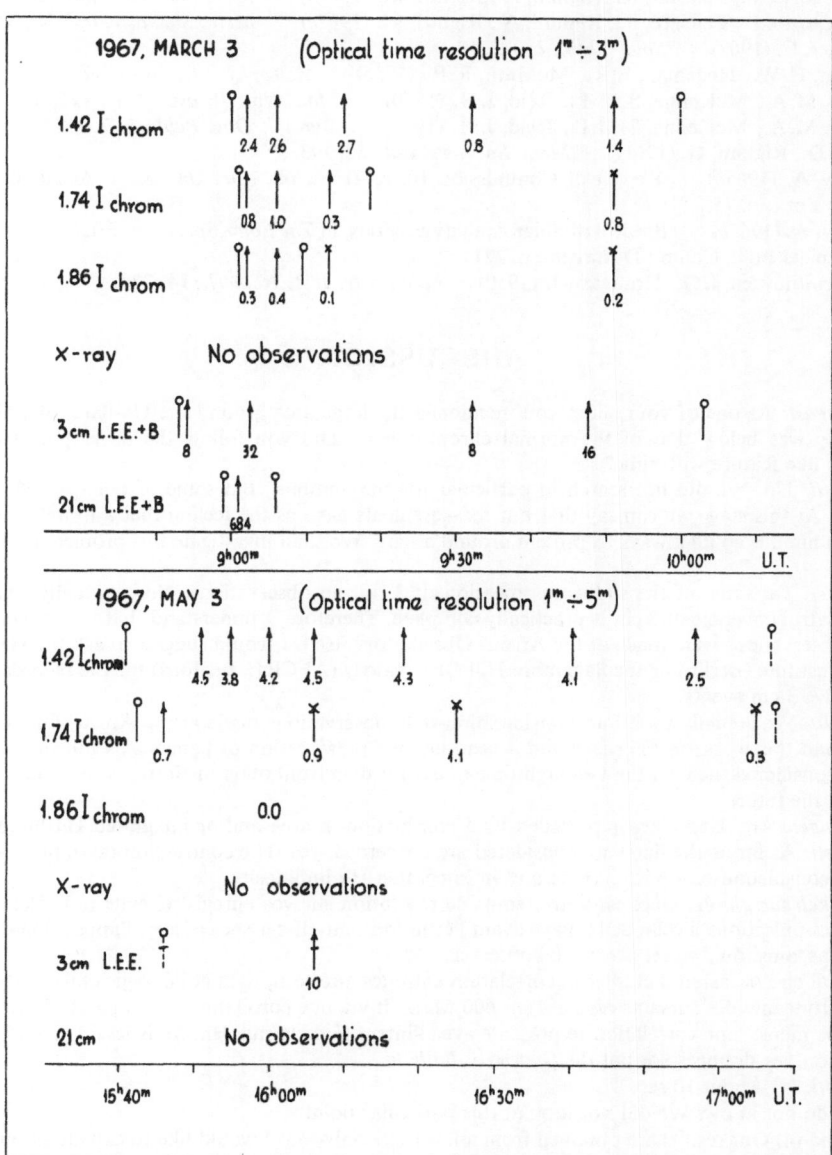

FIG. 9. *Schematical representation of the association in time of various observations for two AR. The notations are the same as in Figure 6.*

De Gregorio, P., Falciani, R., Righini, A., Rigutti, M. (1966)     *Mem. Soc. astr. ital.*, **37**, 807.
De Gregorio, P., Falciani, R., Righini, A., Rigutti, M. (1967*a*)     *Mem. Soc. astr. ital.*, **38**, 33.
De Gregorio, P., Falciani, R., Righini, A., Rigutti, M. (1967*b*)     *Mem. Soc. astr. ital.* **38**, 519.
De Gregorio, P., Falciani, R., Righini, A., Rigutti, M. (1967*c*)     *Mem. Soc. astr. ital.* **38**, 531.
De Jager, C. (1967)     *Solar Phys.*, **2**, 327.
Dodson, H.W., Hedeman, E.R., McMath, R.R. (1956)     *Astrophys. J.*, Suppl., **2**, 241.
Ellison, M.A., McKenna, S.M.P., Reid, J.H. (1960)     *Mon. Not. R. astr. Soc.*, **122**, 491.
Ellison, M.A., McKenna, S.M.P., Reid, J.H. (1961)     *Dunsink Obs. Publ.*, **1**, 2.
Russo, D., Righini, G. (1961)     *Mem. Soc. astr. ital.*, **32**, 193.
Severny, A. (1965*a*)     Report of Commission 10, in *Trans. int. astr. Un.*, XIIa, Academic Press, New York, p. 75.
Severny, A. (1965*b*)     Spectra of Solar Activity Regions, in *The Solar Spectrum*, Ed. by C. de Jager, D. Reidel Publ. Comp., Dordrecht, p. 221.
Van Gruithuyzen, I.G., Houtgast, J. (1959)     *Bull. astr. Inst. Netherl.*, **14**, 279.

# DISCUSSION

*Houtgast:* At one of your slides you mentioned the large area around the Hα-flare, of which the intensity was below that of the normal chromosphere. Did you follow the development of this nimbus-like feature with time?

*Rigutti:* No. We did not search in particular for the 'nimbus', but some of our cases showed it clearly. At this stage we can say that our measurements gave us the feeling that a more or less important nimbus could always be present around a flare. We shall investigate this problem in the near future.

*Fokker:* On some of the slides was mentioned: 3 cm, no observations. Now, actually, the solar radio-patrol coverage at 3 cm is practically complete. Therefore, I understand that in these cases no 3-cm observations were made at the Arcetri Observatory itself. I would suggest to ask the Heinrich-Hertz Institute (Berlin) or the Sagamore Hill Observatory (AFCRL, Bedford) for the records of the respective 3-cm events.

*Rigutti:* We actually took into consideration only observations made at the Arcetri Observatory. We began this work on February and a dead-line for presentation of papers was put on April 15. So we considered there was not enough time to ask for data from other institutions. We will certainly do it in the future.

*De Jager:* Are flares always preceded by a contraction in area and/or integrated Hα luminosity?

*Rigutti:* As far as the flares we considered are concerned, yes. The contraction takes place – more or less conspicuously – both in area and in integrated Hα luminosity.

*Koeckelenbergh:* (1) Quel est votre temps de résolution sur vos enregistrements radioélectriques?

(2) Les pulsations d'éclat de la plage avant l'éruption sont-elles associées avec l'apparition de sous-éruptions, ainsi qu'il m'est arrivé de l'observer?

(3) J'ai eu l'occasion d'étudier la corrélation entre les aires éruptives et l'énergie émise pendant la phase croissante des sursauts associés sur 600 MHz. Il y a une corrélation qui apparaît dans ce cas.

(4) De même, une corrélation se présente avec l'intensité de la raie Hα. Mais ici il y a une grande dispersion, les données sortant du *Quarterly Bulletin*.

*Rigutti:* (1) About 10 sec.

(2) I do not know. We did not look at this particular point.

(3) and (4) Our results are obtained from a few cases only, but I would like to call attention to the fact that the flare areas given in the *Quarterly Bulletin* come from eye estimates, which are very rough and questionable.

# INTERACTION OF MAGNETIC FIELDS AND
# THE ORIGIN OF PROTON FLARES

L. KŘIVSKÝ

*(Astronomical Institute, Ondřejov, Czechoslovakia)*

ABSTRACT

The magnetic model of a proton flare is discussed, in agreement with the known characteristic features of proton-flare development, on the basis of the interaction of an emerging magnetic channel from the photosphere, and its being wedged into the field of an already existing interaction of two systems of magnetic fields of a complex active region with spots. This explains the origin of the two emission filaments following the splitting and of the 'annular' emission filament, entering the corona, connected by a fine loop structure with two split filaments in the chromosphere. The acceleration of particles and their ejection will take place in a number of created 'zero line' spaces and in spaces where magnetic lines of force are reconnected.

An attempt is made to explain some Hα phenomena in connection with proton flares, by means of interactions of magnetic fields in the chromosphere and in the lower corona.

## 1. Characteristic Features of Proton Flares

We know the following typical characteristic features of proton-flares development:

(1) Proton flares occur in groups of certain types of spots along a 'magnetic axis' on the boundary of the magnetic field of spots polarities (Bumba, 1958; Ellison *et al.*, 1961) in complex multi-pole magnetic systems, especially in the case of an interaction of two or more groups of spots and their magnetic fields (Antalová, 1967).

(2) A transient occurrence of a characteristic Y-type phase in the period of an expanding flare phase (Křivský, 1963*a*, *b*) and the appearance of a 'channel' from two or more roughly parallel emission filaments (Ellison *et al.*, 1961; Křivský, 1963*b*).

(3) One of the flare filaments is usually a stationary one, the second filament shows, with flares on the disc, a tangential component of movement, and, in reality, ascends at a slanting angle (Křivský, 1963*b*, 1964).

(4) The moving emission filament can conceal for a transient period by its movement even large spots in the group, attains in the initial stage, that is during the disruption of the Y-phase, a velocity of 10–20 km/sec, its velocity then falls down to a few km/sec, and finally stops altogether (Křivský, 1963*b*).

(5) The emerging emission filament in Hα has an annular structure, is noticeable even in projections onto the disk, and is the first type of a fine loop structure. When emerging above the limb the filament acts as an intensive ring-shaped coronal condensation,

evident in Hα and in the Fe xiv 5303 Å and Ca xv 5694 Å lines (Slonim, 1963; Waldmeier, 1960; Křivský, 1968).

(6) From the opening mouth of the flare 'channel' it is possible to notice the propagation of a 'disturbance' in the form of absorption or alternately also emission traces over large distances, or an absorption filament emerges from the canal. Sometimes it is 'loose', sometimes it is anchored in the shape of an arc (Křivský, 1963b; Křivský and Nešpor, 1967).

(7) The developing flare channel shows, especially in the later stages of development, a filamentary structure, joining both the flare-emission filaments moving apart from each other (second type of loop structure) (Křivský, 1968; Bruzek, 1964).

(8) In the declining stage of the flare (following the maximum brightness) large arcs and loops, noticeable on the disk in the Hα absorption, and above the limb as large loop-type prominences either appear or renew their activity (Kleczek and Křivský, 1960). This third type of the loop structure is spatially the largest and it appears, that the anchoring of the system of loops need not lead to large spots in groups on either side.

(9) Further characters typical for flares with a powerful ejection of fast particles were compiled by Švestka (1966). These are especially: flares with ejection of cosmic or even subcosmic rays were observed in an integral light (McCracken, 1959), some parts of the flare-emission filaments touch with the umbra of large spots (Dodson and

---

FIG. 1.    (a) A new system of magnetic channel-tube with a considerable density of plasma energy $E_3$ and magnetic energy $B_3$, identical with the flare-emission filament in Hα, emerges and is pushed between two large magnetic systems with an interaction region (with zero point $N_0$). It is a system with lines of force running in opposite direction as regards the higher placed, already created system. In the contact space of the opposite-directed magnetic fields there occurs the condensation and acceleration of plasma, and as a result, also an emission. Such spaces are dotted on the cross-section. The interconnection of magnetic lines will create roughly two systems of interaction.

(b) A more detailed picture of a further stage of development in the space of a complex interaction. On the sides of the emerging canal two new spaces of interaction, $N_1$ and $N_2$ are created (in the cross-section plane they are seen as points, in fact they are lines). This is a stage of splitting the filament of the flare in Hα into two filaments and at the same time the rise of an emerging annular system (rotor). The condensation and the acceleration of the plasma already takes place in the spaces of newly reconnected lines of force (Syrovatskij, 1966). These spaces can lie very close to the spots, which are identical with the contact of the emission filaments of the flare in Hα and with some spots' umbrae. The arrows mark the directions of the principal plasma pressures and drifts.

(c) If one of the original macrosystems, e.g. the one seen in the picture on the right-hand side, will comply with the inequality of $E_2 + B_2 < E_3 + B_3$, then the zero point $N_2$ and the corresponding regions of reconnecting lines will be moved at a slanting angle into space above and behind the spot $S_2$. This shifting is identical with the moving flare emission filament after this is separated. Moreover, a new emerging system of a sort of magnetic rotor is created in space with condensed plasma, which in some places can produce instabilities of various kinds connected with an exceptional acceleration of particles and their penetration into coronal and interplanetary space. The large arrows drawn in interrupted lines demonstrate the predominant resulting movement.

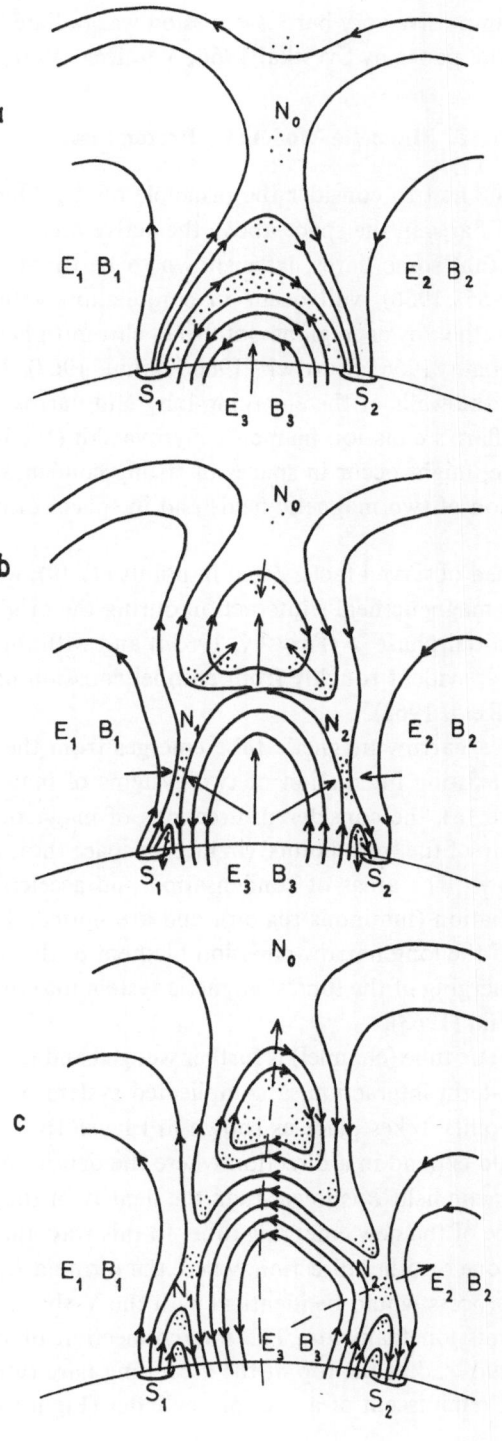

Hedeman, 1960), the impulsive very hard X-emission was radiated in the period of the Y-phase (Křivský in the paper by Švestka, 1966; Valníček, 1967).

## 2. Magnetic Model of a Proton Flare

Sweet (1956) was the first to consider the principle of the interaction of magnetic fields for the origin of flares in the space above the active areas of groups. In further works it was proved that some flares, later shown to be the flares with ejection of particles (Křivský, 1965, 1966), will occur in conjunction with the emergence of magnetic tubes from the lower regions into the chromosphere and the corona (Slonim, 1965; Kawabata, 1966; Sturrock, 1966; Banin, 1967). It was stated earlier (Křivský, 1963b) that the walls of the emerging long and narrow tube (channel) will be identical with the flare's emission filaments. Syrovatskij (1966) asserts that radiation, i.e. a visible flare, might occur in spaces of strong condensation of plasma in a system of an interaction of two magnetic fields and in spaces of reconnected lines of force.

Due to the mentioned observed facts, listed in points (1)–(9), we submit for discussion the model of the magnetic field's interaction during the origin and development of proton flare with initial phase Y (Křivský, 1963a) and with the following phase in the shape of a channel, evident roughly from parallel emission filaments of the flare (Ellison et al., 1961; Reid, 1965).

The new system of a narrow magnetic tube emerges from the lower levels into a region of an already existing interaction of two systems of magnetic fields (vertical cross-section in Figure 1a), the considered directions of magnetic lines of forces are important for the origin of the 'zero points' (in fact, in space these are 'zero lines') and are marked with arrows. The areas of condensation and acceleration of plasma are identical with the radiation (luminous region), and are dotted. This process is identical with the origin of the long narrow-emission filament of the flare. Essentially, we are considering the emerging of the tube's magnetic system into the centre of a system of a shape used by Wild (1963).

The emerging magnetic tube (channel) is further wedged and pushed into the system of a preceding longer-term interaction, a complicated system of interaction and the appearance of 'zero points' takes place as shown in Figure 1b.

The tube walls would expand in a direction where the density of the kinetic energy of plasma $E$ and magnetic field $B$ is lower than the density of the plasma energy and of the field in the space of the new emerging tube. In this way, the emission filaments would move either in one or other direction, where the formula $E_{1,2} + B_{1,2} < E_3 + B_3$ would be valid. The process would be identical with the Y-shape phase and with the appearance of two emission filaments. The interconnection of magnetic fields and their annihilation would lead at the top of the ascending flare tube to the origin of a new magnetic system, reminiscent of a rotatory cylinder (Figure 1b, c). This process would be identical with the appearance and emergence of the peak annular emission filament, as seen by the author himself many times with these flares.

The transport of material in Hα would be very complicated and would be subject to the directions of the interacting magnetic fields, and its vertical cross-section would resemble the shape 8 . The possibility of this shape was considered, but for other reasons, by Banin (1966) and, in a more detailed model without taking into account the interaction of fields, by Kawabata (1966). The appearance of the top annular rotor (which is identical with the emerging emission filament and with condensation up to high altitudes) was explained by Kawabata as a result of a close, indefined, turbulent process.

During the later disintegration of this system, as a result of the conversion of a magnetic energy into an energy of an ejection of particles, and the restoration of the original shape of the fields in the higher chromosphere and the lower corona, a system of large spatial loops would be created (on the disk of the absorption arcs and loops emerging from the 'flare channel'). The process of contraction and control of dispersed particles by renewed magnetic lines of a large spatial character would already be asserted (Křivský and Nešpor, 1967). This process, occurring after the expanding phase of the flare and covering the time period of the interflare stages agree with a model sketched by Carmichael (1964, Figs. 54-4 and 54-5).

The acceleration of particles to high velocity (cosmic and sub-cosmic rays) would take place primarily in the 'zero lines' $N$ spaces (Dungey, 1956) and further, in the spaces of reconnecting lines of force (Sturrock, 1966; Syrovatskij, 1966). Their escape into space would occur at the beginning of the stage when the new tube would be wedged and pushed into the preceding system and to the creation of an ascending annular rising system (rotor). This phase of development is identical with the phase of a Y-shaped flare and the splitting of the channel into the shape of an orifice in the period of the first maximum burst on the radio waves (Křivský, 1963b), and can sometimes be accompanied by effects in Hα and in the radio spectra, which indicate the evolution and the very fast propagation of shock waves (Wild, 1963). The mechanism of Syrovatskij (1966) could be considered as an acceleration and escape mechanism for the particles.

The model presented here is based on the assumption that two main neutral regions, $N_1$ and $N_2$, have been formed. In the case when a newly created region of interaction $N_3$ occurs near the top of an emerging magnetic tube, two (instead of one) closed magnetic cylinders are formed. These cylinders have to be considered regions where extraordinary dense clouds of particles (eventually relativistic) together with HM-waves can be created, the waves collide under various angles. Relativistic particles should reach their largest energies at the beginning of the quickly evolving 'tearing' process mentioned above, when the gradients of magnetic fields on the regions of interaction are enormous and when the total magnetic energy involved in the process of conversion to the kinetic energy of particles is still very large.

## References

Antalová, H. (1967)     *Bull. astr. Inst. Csl.*, **17**, 61.
Banin, V.G. (1966)     *Izv. Krym. astrofiz. Obs.*, **35**, 190.
Banin, V.G. (1967)     *Izv. Sib. Izmiran*, **2**.
Bruzek, A. (1964)     *Astrophys. J.*, **140**, 745.
Bumba, V. (1958)     *Izv. Krym. astrofiz. Obs.*, **19**, 105.
Carmichael, H. (1964)     AAS-NASA Symp. Phys. of Solar Flares, Washington, 451.
Dodson, H.W., Hedeman, E.R. (1960)     *Astr. J.*, **65**, 51.
Dungey, J.W. (1956)     *Sixth I.A.U. Symp.*, Stockholm, Cambridge, 1958, p. 135.
Ellison, M.A., McKenna, S., Reid, J.H. (1961)     *Dunsink Obs. Publ.*, **1**, 53.
Kawabata, K. (1966)     *Rep. Ionosph. Space Res. Japan.*, **20**, 107.
Kleczek, J., Křivský, L. (1960)     *Bull. astr. Inst. Csl.*, **11**, 165.
Křivský, L. (1963a)     *Nuovo Cim.*, X-27, 1017.
Křivský, L. (1963b)     *Bull. astr. Inst. Csl.*, **14**, 77.
Křivský, L. (1964)     *Bull. astr. Inst. Csl.*, **15**, 75.
Křivský, L. (1965)     *Bull. astr. Inst. Csl.*, **16**, 27.
Křivský, L. (1966)     *Bull. astr. Inst. Csl.*, **17**, 141.
Křivský, L. (1968)     *Bull. astr. Inst. Csl.*, **20** (in print).
Křivský, L., Nespor, Ju.I. (1967)     *Izv. Krym. astrofiz. Obs.*, **36**, 98.
McCracken, K.G. (1959)     *Nuovo Cim.*, X-13, 1081.
Reid, J.H. (1965)     NATO Adv. Study Inst. (Planetary and Stellar Magn).
Slonim, Ju.M. (1963)     *Soln. Dann. No. 4*, 67.
Slonim, Ju.M. (1965)     *Soln. Dann. No. 1*, 51.
Sturrock, P.A. (1966)     *Nature*, **211**, 695.
Švestka, Z. (1966)     *Space Sci. Rev.*, **5**, 388.
Sweet, P.A. (1956)     *Sixth Int. Astr. Union Symp.*, Stockholm, Cambridge 1958, p. 123.
Syrovatskij, S.I. (1966)     *Astr. Zu.*, **43**, 340.
Valníček, B. (1967)     *Bull. astr. Inst. Csl.*, **18**, 249.
Waldmeier, M. (1960)     *Z. Astrophys.*, **51**, 1.
Wild, J.P. (1963)     *The Solar Corona* (Proc. Intern. Astr. Union Symp. No. 16) New York, p. 115.

# A MODEL OF SOLAR FLARES

P. A. STURROCK

*(Institute for Plasma Research, Stanford University, Stanford, Calif., U.S.A.)*

ABSTRACT

A model of solar flares is proposed in which the preflare state comprises a bipolar magnetic-field structure associated with a bipolar photospheric magnetic region. At low heights, the magnetic-field lines are closed but, at sufficiently great heights, the lines are drawn out into an open structure comprising a bipolar flux tube containing a 'neutral sheet' or 'sheet pinch'. Such a sheet pinch is probably related to a coronal streamer. The energy stored in the closed-field region is derived from photospheric motion whereas energy stored in the open-field region is derived from the non-thermal energy flux which heats the corona and drives the solar wind.

The flare itself is identified with reconnection of magnetic field by the tearing-mode resistive instability. If the thickness of the sheet pinch is determined by resistive diffusion and a growth time of the bipolar region of order 1 day, the transverse dimension will be about $10^4$ cm. The rise time of the tearing-mode instability is then a few seconds, compatible with the characteristic time of Type-III radio bursts. One can understand that the time-scale of the reconnection process is of order $10^2$–$10^3$ sec if reconnection proceeds by the Petscheck mechanism, with the modification that resistive diffusion is replaced by the more rapid Bohm diffusion.

The evolution of a flare, according to this model, appears to fit a number of the observational characteristics of flares.

It seems unlikely that it will ever be possible to deduce the mechanism of solar flares directly and unambiguously from observations. Progress in understanding this complex phenomenon can probably come only by a long process of trial and error. In attempting to understand such a phenomenon as it is explained to him by observers, a theorist will sketch an interpretation which may develop into a full-fledged theory. Such a theory would be subjected to two types of tests: that of internal physical and mathematical consistency, and that of agreement between components of the theory and observational facts. Since the development of a detailed theory of such a phenomenon could well use up many man-years of theoretical labor, it is sensible to first examine a theory in outline form. This is roughly what is meant by constructing a 'model' of the phenomenon. My purpose here is to present a model of solar flares and I hope to learn from your response the extent to which characteristics of this model fit the characteristics of the real phenomenon.

A solar flare is an explosion which, as far as we can tell, usually occurs spontaneously. This implies that the basic mechanism of a solar flare is some kind of instability. It has been agreed for some time that a solar flare is an electromagnetic process (Cowling, 1953), and this implies that the instability is a plasma instability. Furthermore, it is generally believed that the energy released in the flare is stored in magnetic form (Gold and Hoyle, 1960). Since a flare seems to occur in the chromosphere and corona, and has no apparent influence upon the photosphere, the magnetic field of a preflare state can not be in its lowest-energy state, consistent with the boundary conditions of given normal magnetic field at the photosphere. This means that there must be currents in the preflare magnetic-field pattern. However, since the corona has a low gas density and pressure, one might imagine that the magnetic field must be virtually force-free, and Schmidt (1968), following a similar suggestion by Gold (1964) has proposed that the preflare energy is stored in a force-free magnetic field.

One objection to this proposal is that such a force-free field will be composed of closed magnetic-field lines. It is then hard to understand how electrons accelerated at the very beginning of a flare can travel almost radially outward and produce a Type-III radio burst (Wild *et al.*, 1963). The idea proposed by Schmidt and Gold also traces the source of flare energy to movement of magnetic-field lines at the photosphere. But this aspect of the idea is hard to reconcile with the occurrence of homologous flares (Ellison, 1963) which in their idealized form repeat exactly, without significant change in the photospheric magnetic field. For these reasons, one is persuaded to look for an alternative method for storing magnetic energy and an alternative method for providing this energy.

It is at this point relevant to direct one's attention to the solar wind (Dessler, 1967), the existence of which insures that some fraction of the magnetic-field lines emerging from the photosphere have an open configuration. If we now consider a simple bipolar magnetic region and construct a field pattern which has both closed and open magnetic-field lines, we arrive at a pattern of the type shown in Figure 1. This pattern contains a Y-type neutral point (or, more correctly, a 'neutral line'). The configuration of the magnetic field in the neighborhood of the neutral point is quite sensitive to the

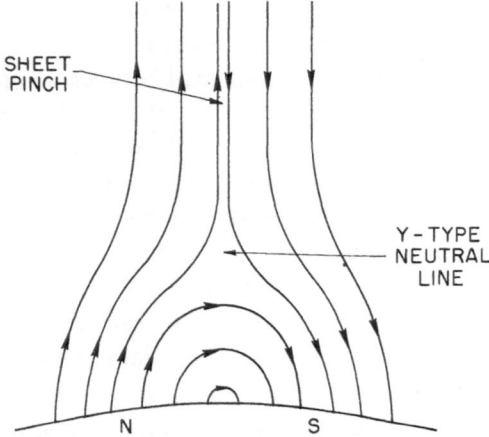

FIG. 1.    *Schematic representation of magnetic-field pattern above a bipolar magnetic region, showing the transition from closed field lines at low heights to open field lines at great heights, involving Y-type neutral line and sheet pinch.*

equality or inequality of gas pressures in the closed-field region and the open-field region. The diagram is drawn for the case that the gas pressure in the closed-field region is greater than that in the adjacent open-field region. The magnetic-field lines then take the form of a cusp, which is strongly reminiscent of the helmet shape often seen in coronal structures (Billings, 1966). If the gas pressure in the closed-field region were less than that in the adjacent open-field region, the field configuration in the neighborhood of the neutral point would be that of an inverted 'T'. If there were a filament (quiescent prominence) in this structure, it would be comparatively low down above the line dividing regions of opposite magnetic polarity at the photosphere.

The important point about this configuration is the existence of a 'neutral sheet' or, as it is called in plasma physics, a 'sheet pinch' (Furth *et al.*, 1963), in which there is a thin region of high-density plasma, the pressure of which balances the adjacent magnetic pressure. (Such a structure exists also in the magnetic tail of the earth (Ness, 1965).) A sheet pinch appears to be metastable, yet the associated magnetic field certainly contains 'free energy', which can be released if the magnetic field can be 'reconnected'. In the present model, such energy can be released without changing the connection of magnetic-field lines at the photosphere. The shape and location of the sheet pinch above a bipolar magnetic region are similar to that of a coronal streamer when a streamer is associated with a bipolar magnetic region. This suggests that a coronal streamer is in fact the visible manifestation of such a sheet pinch. Dr. Sheldon Smith of the NASA-Ames Research Laboratory has a very fine eclipse photograph which bears out this suggestion, and an article on this observation and interpretation will be published in the near future.

With this model of the preflare state we may now inquire into the development of the flare itself. Energy may be released by reconnection of the magnetic field, and such reconnection may be effected by the tearing-mode resistive instability analyzed in the linear approximation by Furth *et al.* (1963). The growth rate of this instability depends critically on the thickness of the sheet pinch, a dimension of which we have no direct knowledge. However, we can probably assign a lower limit to the thickness of the sheet pinch by considering the lifetime of the configuration and the relative diffusion rate of plasma and magnetic field. Spitzer (1962) gives a convenient formula for relating the diffusion time $\tau_D$, the temperature $T$ of a fully ionized plasma, and the characteristic length scale $L$:

$$\tau_D \approx 10^{-12.7} T^{3/2} L^2. \tag{1}$$

If the temperature of the corona is taken to be $10^{6.4}\,^\circ\text{K}$, and if $\tau_D$ is identified with the lifetime of a coronal streamer which is perhaps of order 1 day in an active region, i.e. $10^5$ sec, we find that $L$, which measures the thickness of the sheet pinch, is of order $10^4$ cm, i.e. 0·1 km. This estimate of the thickness is much smaller than one might have supposed without considering the high conductivity and small diffusivity of the coronal plasma.

The time-scale $\tau_I$ of the most rapidly growing wave of the tearing-mode instability is given (Furth *et al.*, 1963) by

$$\tau_I = \tau_D^{\frac{1}{2}}\tau_A^{\frac{1}{2}} \tag{2}$$

where

$$\tau_A = Lv_A^{-1}, \tag{3}$$

$v_A$ being the Alfvén velocity given by

$$v_A = B(4\,\pi\rho)^{-\frac{1}{2}}, \tag{4}$$

where $B$ is the mean magnetic-field strength and $\rho$ the mean gas density. Since the magnetic pressure inside the sheet is being balanced by gas pressure on the outside, the Alfvén velocity will in fact be determined by the plasma temperature and be comparable with the speed of sound. For $T = 10^{6\cdot4}\,°\mathrm{K}$, we find that $v_A = 10^{7\cdot5}$ cm sec$^{-1}$. This means that, for the sheet we are considering, $\tau_A = 10^{-3\cdot5}$ sec, so that $\tau_I = 10^{-7}$ sec.

The above time-scale of 5 sec for the growth rate of the tearing-mode instability may seem to be too short to be relevant to solar flares. However, this is probably not the case. Type-III radio bursts frequently occur at the beginning of a solar flare. The fine structure of a Type-III radio burst can have a time-scale of about 1 sec. Hence, the growth rate which we have computed is none too short. Indeed, we should possibly conclude that the growth rate should be even more rapid so that the thickness of the sheet pinch is typically even less than we have assumed.

The preceding considerations indicate that the onset of instability is very rapid, and further calculations of this stage of the flare process indicate that sufficient electrons will be accelerated to high enough energy to explain the principal characteristics of Type-III radio bursts. However, the main energy release of a flare must be effected by a process with very different parameters. The time-scale for release of the bulk of the energy of a very large flare (Smith and Smith, 1963) is of order $10^3$ sec. Since the transverse dimensions of a very large flare are typically $10^{10}$ cm, this implies that the mean relative diffusion velocity of magnetic field and plasma is about $10^7$ cm sec$^{-1}$. Petschek (1964) has considered the possibility that this very rapid reconnection takes place in a very thin layer (the magnetohydrodynamic treatment of his mechanism leads to a sheet thickness of order $10^{-4}$ cm). However, the magnetohydrodynamic equations may not be applied to the coronal plasma when the relevant length-scale is so very small.

If the Type-III phenomenon represents the initial build-up of the tearing-mode instability, then the 'flash phase' of a flare presumably represents a process which takes place after the instability has developed to the highly non-linear level of amplitude at which the plasma is best described as 'turbulent'. If this is so, the rapid reconnection of the magnetic-field lines may be ascribed to anomalous diffusion in a turbulent plasma. In order to determine whether the reconnection rate can be sufficiently rapid, we follow Spitzer (1962) in adopting the Bohm diffusion coefficient as that which characterizes a highly turbulent plasma:

$$D_B = \frac{1}{16} \frac{ckT}{eB} \approx 10^{2\cdot6} B^{-1} T. \tag{5}$$

This means that the relative diffusion velocity is given by

$$v_D \approx Db^{-1} \approx 10^{2\cdot6} B^{-1} b^{-1} T, \tag{6}$$

where $b$ is the length-scale characterizing magnetic and plasma fluctuations transverse to the magnetic field. If $b$ is identified with the small-scale structure developed by the tearing-mode instability, we find from the article by Furth et al. (1963) that we should adopt

$$b \approx (\tau_D/\tau_A)^{-\frac{1}{4}} L. \tag{7}$$

For the particular values we adopted, this gives $b \approx 10^2$ cm. Since we are considering the state resulting from the onset of instability, we need to evaluate the plasma temperature after the occurrence of instability rather than before it. This temperature may be estimated by equating the magnetic-energy density $(1/8\pi) B^2$ before the flare to the plasma energy density $2 \times \frac{3}{2} nkT$ after the flare (where $n$ is the number density of electrons and of ions outside the initial sheet pinch). If we take $B \approx 10^2$ gauss and $n \approx 10^{10}$ cm$^{-3}$, we find that the plasma is heated to a temperature of $10^8$ °K. When these values are substituted into Equation (6), we find the anomalous diffusion velocity to be $10^{6\cdot6}$ cm sec$^{-1}$, which is very close to the required value of $10^7$ cm sec$^{-1}$ quoted above. It seems, therefore, that the very rapid rise-time of the light curve of a flare may be ascribed to anomalous diffusion, of the Bohm type, which follows the development of very small-scale turbulence resulting from the tearing-mode instability.

The above considerations may perhaps best be interpreted as a modification of the Petschek (1964) model of the reconnection of the magnetic-field lines in a solar flare, in which the resistive layer is replaced by a turbulent layer, and resistive diffusion by Bohm diffusion. The thickness of the layer is then estimated to be $10^2$ cm (which is in fact the transverse length-scale of the most rapidly growing wave of the tearing-mode instability, in the model discussed earlier) rather than the $10^{-4}$ cm, as estimated by Petschek.

The decay time of the light curve of a flare is typically about ten times the rise time (Smith and Smith, 1963), and this part of the curve has an exponential form. It appears, therefore, that the slow decay is likely to be caused by the slow diffusion of heat and energetic particles out of the instability region down to the chromosphere, where the Hα and other prominent emission lines are produced.

The gross characteristics of the evolution of a flare according to the above model is shown schematically in Figure 2. We see that the stream of heat and particles, as guided along magnetic-field lines to the chromosphere, will give rise to a pair of bright filaments, as is typically observed (Ellison, 1963). As the process continues, the two regions at which heating occurs will move progressively away from the line dividing

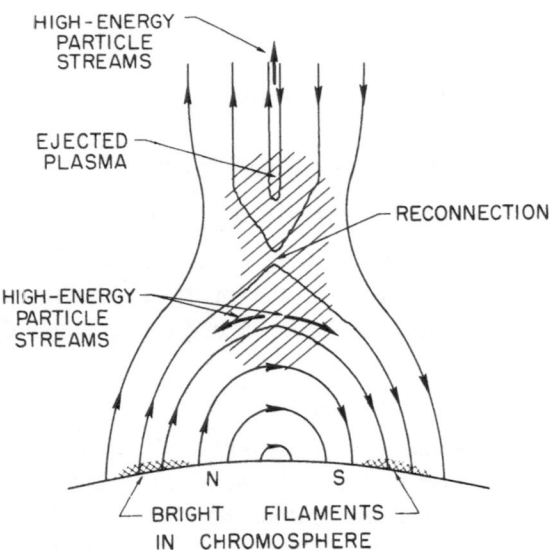

FIG. 2.   *Schematic representation of reconnection of magnetic-field lines during a flare, showing ejection of plasma and heating of chromosphere by particle streams to form two bright filaments.*

opposite magnetic polarities, at the photosphere, giving the appearance that the two bright filaments are moving slowly away from each other. Since the plasma effected by the reconnection process is believed to be highly turbulent, there will be stochastic acceleration (Sturrock, 1966a; Hall and Sturrock, 1968) of both electrons and protons, but these will escape only slowly from the 'magnetic bag' formed by the disordered magnetic field. Such a moving cloud of electrons might give rise to what is called a moving Type-IV radio burst (Wild *et al.*, 1963). Furthermore, the collisionless shock wave which probably forms ahead of the outward-moving plasma cloud may be the origin of a Type-II radio burst (Wild *et al.*, 1963). One may also note that this magnetic cloud will ultimately be ejected completely from the Sun. On arrival in the vicinity of the Earth, this cloud would give rise to a geomagnetic storm.

The paper presented at this meeting by Rust (1968) ascribes surges and Ellerman bombs to small magnetic regions surrounded by larger regions of opposite magnetic polarity, such as 'satellite sunspots'. It appears that the flare process described in this article may be applied also to a ring-type sheet pinch, such as may form above the boundary of a satellite sunspot. If the Y-type neutral line of this configuration is low in the atmosphere, perhaps at chromospheric heights, the emission line may well be absorbed at the centre of the line, as is characteristic of the 'moustache' shape of the emission features of Ellerman bombs (Severny, 1964). The ejected plasma would then be much more dense than is typical of flares of the type we have discussed earlier, which lead to the ejection of coronal plasma. The ejection of a column of chromospheric plasma may well be the proper interpretation of surges.

In conclusion, we shall inquire briefly into the origin of the energy which is stored in the open magnetic-field pattern shown in Figure 1. This is believed to be derived from the non-thermal energy flux which goes to heat the corona and drive the solar wind. This, according to Osterbrock (1961) is typically $10^7$ erg cm$^{-2}$ sec$^{-1}$, but may be higher by a factor of 10 in active regions. Since the area covered by a large flare is $10^{20}$ cm$^2$ and the recovery time between a series of intense homologous flares (Ellison, 1963) is about 1 day ($10^5$ sec), we see that the total non-thermal energy ejected in this time, into this region, is $10^{33}$ erg. Hence we can explain the energy released during a flare if only a few percent of this energy is converted into the 'free energy' associated with the open magnetic-field structure of Figure 1.

Since accounts of this model were first published (Sturrock, 1966b, 1968), it has come to the author's attention that very similar ideas have previously been expressed by Carmichael (1964).

## Acknowledgements

This work was supported by the Air Force Office of Scientific Research, Office of Aerospace Research, United States Air Force, under Contract AF 49(638)1321.

## References

Billings, D.E. (1966)   *A Guide to the Solar Corona*, Academic Press, New York, p. 65.
Carmichael, H. (1964)   in *Proc. AAS-NASA Symp. on the Physics of Solar Flares, NASA SP-50*, National Aeronautics and Space Administration, Washington, D.C., p. 451.
Cowling, T.G. (1953)   in *The Sun*, Ed. by G. P. Kuiper, University of Chicago Press, Chicago, p. 583.
Dessler, A.J. (1967)   *Rev. Geophys.*, **5**, 1.
Ellison, M.A. (1963)   *Q.J.R. astr. Soc.*, **4**, 62.
Furth, H.P., Killeen, J., Rosenbluth, M.N. (1963)   *Phys. Fluids*, **6**, 459.
Gold, T. (1964)   in *Proc. AAS-NASA Symp. on the Physics of Solar Flares, NASA SP-50*, National Aeronautics and Space Administration, Washington, D.C., p. 389.
Gold, T., Hoyle, F. (1960)   *Mon. Not. R. astr. Soc.*, **120**, 89.
Hall, D.E., Sturrock, P.A. (1968)   *Phys. Fluids* **10**, 2620.
Ness, N.F. (1965)   *J. Geophys. Res.*, **70**, 2989.
Osterbrock, D.E. (1961)   *Astrophys. J.*, **134**, 347.
Petschek, H.E. (1964)   in *Proc. AAS-NASA Symp. on the Physics of Solar Flares, NASA SP-50*, National Aeronautics and Space Administration, Washington, D.C., p. 425.
Rust, D.M. (1968)   in the present volume, p. 77.
Schmidt, H.U. (1968)   in the present volume, p. 95.
Severny, A.B. (1964)   *A. Rev. Astr. Astrophys.*, **2**, 363.
Smith, H.J., Smith, E. v. P. (1963)   *Solar Flares*, MacMillan, New York, p. 102.
Spitzer, L. (1962)   *Physics of Fully Ionized Gases*, Interscience, New York, 2nd ed., pp. 42, 47.
Sturrock, P.A. (1966a)   *Phys. Rev.*, **141**, 186.
Sturrock, P.A. (1966b)   *Nature*, **211**, 695.
Sturrock, P.A. (1968)   Solar Flares, in *Proc. Enrico Fermi int. Summer School of Physics*, 39th Course: *Plasma Astrophysics, Varenna, Italy, 1966*, p. 168.
Wild, J.P., Smerd, S.F., Weiss, A.A. (1963)   *A. Rev. Astr. Astrophys.*, **1**, 305.

# DISCUSSION

*De Jager:* Any theory of flares should explain that the greater part of the energy of the larger types of flares is contained in particles accelerated to energies of the order of 10–100 KeV. In Sturrock's flare theory the *optical flare* is explained as caused by the collisions of chromospheric matter with downward-moving energetic particles. With a view to the fairly great energy which these particles presumably have it looks very difficult to produce anything like the optical flare by this interaction. The energy will mainly be emitted at shorter wavelength, not in Hα and other lines.

*Sturrock:* The primary energy-release mechanism is the tearing-mode instability which gives rise to strong electric fields associated with intense filamentary currents. When such an electric field develops in an open-field region in a low-density part of the corona, one expects that a large part of the energy released by the instability will appear as high-energy electrons and protons. However, the electric field associated with the closed-field region is developed at smaller heights and therefore in regions of higher density. The main result of the electric field is likely to be a conduction current rather than a runaway current, so that the resulting heating of the chromosphere is better interpreted as being due to thermal conduction from coronal heights, rather than a stream of high-energy particles.

*De Feiter:* An additional difficulty is that the downward motion of these energetic particles should last as long as the optical flare (hence 10 min to 1 or more hours). Since a plasma excited to high energies (10–100 KeV) will lose its energy in a few minutes, Sturrock's theory makes necessary the assumption of a continuous acceleration of energetic particles, during a time as long as the period of visibility of the optical flare.

*Sturrock:* Suppose a large flare releases $10^{32}$ erg in a closed-field region of height $10^{10}$ cm and area $10^{20}$ cm$^2$. If this energy is radiated in $10^3$ sec, the downward heat flux will be $10^9$ erg cm$^{-2}$ sec$^{-1}$, compatible with a maximum temperature of $10^{7.3}$ °K. To meet the total energy requirement, the mean density must be $10^{10}$ cm$^{-3}$. The time-scale for cooling of this hot gas by bremsstrahlung radiation is in excess of $10^5$ sec, much longer than the time-scale for cooling by conduction. The time-scale for cooling by gyro-radiation is about $10^6$ sec. Hence the present theory does not require acceleration after the flash phase of a flare.

*Newkirk:* Your model requires the existence of open magnetic-field lines above the active region. How do these lines become opened initially?

*Sturrock:* There have been open field lines as long as there has been a solar wind. The boundary between closed and open field lines is sensitive to the pressure differential between the plasma in these two regions. Since plasma in the closed field region cannot escape, it tends to have a higher temperature, density and pressure than that in the adjacent open field region. It seems that this allows a progressive ejection of closed field lines along the sheet pinch so that the bipolar flux tube grows from the inside out.

*Wilcox:* When energetic solar flare particles are detected by spacecraft near the Earth, does your model give any prediction for the configuration to be expected in the interplanetary magnetic field?

*Sturrock:* I have not investigated this question. My initial reaction is that the first high-energy particles would arrive along field lines with the normal spiral angle, but the main cloud of plasma would arrive with a magnetic field oriented transverse to the Sun–Earth line.

*Davis:* I have several questions on your interesting model; but most of them are related to the following feature. The tearing instability that produces the energy results in an increase in the flux of the arch region, i.e., the flux of the tubes both of whose ends pass into the photosphere. There is a corresponding decrease in the flux of the tubes that lead from the solar surface out into interplanetary space. After the flare is over and the energy supply is to be replenished, the original flux configuration must be restored. I do not understand how this can be done by heating the gas in the arches and pushing them up higher. Instead, it will be necessary to cut and reconnect lines of force.

*Sturrock:* The process which I just mentioned, in my reply to Dr. Newkirk, results in the ejection of closed field lines out along the sheet pinch into interplanetary space. As far as the flare region is concerned, these field lines are then effectively open. There is therefore a slow conversion of closed to open field lines without a *bona fide* reconnection by a resistive process.

*Winckler:* We note that the direct verification that Type-III bursts are actually caused by streams of relativistic electrons, by space measurements or X-ray bursts, is lacking at present. One should be cautious in using this idea to support a flare theory in a critical way.

*Sturrock:* There seems no doubt that Type-III bursts comprise electromagnetic radiation from a moving 'wake' of plasma oscillations in the corona. The only effective mechanism for exciting plasma oscillations is by an electron stream, and this supposition is supported by the high brightness temperatures of Type-III bursts which may be attributed to the two-stream instability. The additional fact that the radial velocity of the moving source is compatible with that of an electron stream results in a very strong case for the view that Type-III bursts are caused by electron streams.

# THE HIGH-ENERGY FLARE PLASMA

C. DE JAGER

*(University Observatory 'Sonnenborgh' and Space Research Laboratory,
Utrecht, The Netherlands)*

## 1. Acceleration of Particles in Flares

A solar flare has various aspects: the *optical flare* is often associated with emissions in the microwave or X-ray regions: this indicates the occurrence of a highly excited plasma, which we call the *high-energy flare plasma*. The existence of the high-energy flare plasma was first shown by radio observations in the microwave regions (Hachenberg) and later confirmed by X-ray observations in the energy range $10^2$–$10^6$ eV.

The picture that is gradually emerging from the various kinds of observations of flares and flare-associated phenomena is that during the very first phase of solar flares particles are accelerated to energies of the order $10^3$–$10^6$ eV; observations pointing in that direction are the occurrence of X-ray bursts covering that energy range, the occurrence of microwave radiobursts in the first minutes of solar flares, and the emission of type-III radio bursts, also mostly occurring in the very first phases. The velocity of the type-III excitation sources, which ranges between $c/6$ to $0.9\,c$ with maximum occurrence at $c/2$ points to electron energies in the range 7–700 keV with a maximum around 80 keV. This is in agreement with X-ray burst observations in the deka-keV range.

Other phenomena, like the emission of solar protons and the occurrence of metric type-IV bursts, are rather related to later phases of the solar flare event, and show that after some time further acceleration occurs to energies of the order $10^7$–$10^9$ eV. Hence, secondary electromagnetic particle acceleration with a factor 1000 occurs in a second phase of solar flares.

## 2. The Origin of the Flare Energy and the Acceleration Process

According to Kiepenheuer and Bruzek the energy output in the form of *radiation* is (for strong flares) of the order of $10^{32}$ ergs, while in the acceleration of *particles* an energy of $10^{32.5}$ erg is involved. Assuming for the high-energy flare a volume of $10^{26}$ cm$^3$, it is clear that this energy could be obtained if in this volume the magnetic field would be reduced from e.g. 500 to 400 gauss. No other energy sources seem likely to be sufficient for producing the flare energy. Hence, the energy of solar flares could arise from a partial annihilation of the magnetic field. This, in turn, means the establishment of a large electric field and consequent current flow, since the essential aspect of this annihilation process is the consumption of the magnetic energy of the current by a discharge.

Whereas it is difficult but not impossible to construct a model of a solar flare that might explain magnetic-field annihilation by a discharge (see e.g. Alfvén and Carlqvist, 1967), the real problems are those of the primary origin of this energy and of its accumulation in a magnetic field. Since flares may recur at intervals of the order of a day in the same active region, it is clear that the flare process is due to a storage of energy and a subsequent release. The time-scale for the establishment of the field may be of the order of a day.

The only source that seems capable of producing sufficient energy is the subphotospheric convective energy flux, which we know to be suppressed in the region below sunspots. The missing energy flux in an average sunspot is of the order of $4 \times 10^{29}$ erg sec.

If the mechanism which transforms the missing convective energy flux into magnetic energy had an efficiency of 100%, the energy necessary to produce a large flare would be provided in about 1000 sec. If the efficiency was only 0·01, one flare could be produced in 24 hours.

## 3. Conclusions

The purpose of this note was to draw attention to certain aspects and problems related to the origin of the flare energy, its transformation into magnetic energy, the acceleration of particles during the first phase of the flare, and the secondary acceleration in a later phase.

(a) The energy emitted in solar flares is so large, and flares occur so often that the energy can only be provided by a mechanism transforming the missing subphotospheric convective energy flux in a sunspot into magnetic energy.

(b) This magnetic energy should be released, perhaps through the establishment of a large electric field and a subsequent discharge.

Occasionally, this may produce the high-energy flare, where in a magnetic bottle about $10^{22}$–$10^{29}$ electrons are contained with energies of $10^3$–$10^6$ eV. This means that the electric voltage over the discharge need not necessarily exceed $10^6$ volts, which is smaller than assumed in most current flare theories.

(c) In a later phase of the flare, maybe 15–60 min later, secondary acceleration of the electrons may occur; this may lead to an average increase of the energy by a factor of about 1000. These accelerated particles may produce the metric type-IV radio bursts (electrons), and the high-energy particle fluxes observed near the Earth as subrelativistic or relativistic particle events.

### Reference

Alfvén, H., Carlqvist, P. (1967)    *Solar Phys.*, **1**, 220.

# DISCUSSION

*Krüger:* You mentioned the picture of the two steps of acceleration process of particles within a flare. Do you assume the action of different acceleration mechanisms or more likely of a cyclic acceleration process?

*De Jager:* Little is known, if anything, of the acceleration mechanism. Yet I believe the two acceleration mechanisms to differ. For the first phase I would think of a mechanism like the one proposed by Alfvén and Carlqvist (*Solar Phys.*, **1**, 1967, 220); the second phase might rather be a Fermi-type acceleration.

*Krüger:* In some cases there seems to be a relation between the duration of cm-type IV bursts or their rising time and the energy of emitted particles.

*Schmidt:* I wonder whether it might be conceivable that the comparatively young and closed field-lines of a flare-producing active centre intermingle on a scale much smaller than the flare dimension with the rather old fieldlines originating in the weak and extended field patches and stretching radially outward into the sector structure of the solar wind. It seems to me that such a concept would be also consistent with the observed formation of new photospheric flux occurring without much interference with the surrounding older flux as well as with the results of Mrs. Pick concerning the large solid angles filled by energetic protons.

# THE OCCURRENCE AND POSSIBLE MEANING OF THE 'NIMBUS'

J. HOUTGAST

*(Sterrewacht 'Sonnenborgh', Utrecht, The Netherlands)*

There are few observations of cases in which a nimbus has developed around a solar flare of major importance. Speculations have been made about the nature of this obscuring cloud. We shall give here some consequences of the assumption that we have to do with scattering by a cloud of electrons, ejected together with the protons during the formation of the flare.

Since the electron receives radiation from a solid angle $2\pi$, half the scattering cross-section should be taken, that is $\frac{1}{2} \times 0.66 \times 10^{-24}$ cm$^{-1}$.

If we start from a darkening in a nimbus of 3% (Houtgast, 1962), we get $\frac{1}{2} \times 0.66 \times 10^{-24} N_{tot.} = 0.03$, which yields for $N_{tot.} = 10^{23}$.

For a height of the cloud, assumed equal to its cross radius, of about 100000 km $= 10^{10}$ cm, $\bar{n}_e = 10^{13}$ per cm$^3$.

For comparison: in a prominence $n = 10^{10}$–$10^{11}$ cm$^{-3}$, at the base of the chromosphere the total number of particles is $10^{16}$ cm$^{-3}$, for a height of 1000 km it is $10^{13}$ cm$^{-3}$.

Since the scattering of electrons holds for all wavelengths, the observation in continuous light is to be preferred for avoiding the influence of chromospheric structure as seen in Hα and other lines.

The gas of protons and electrons in a nimbus would cause a series of absorptions due to free-free and bound-free transitions. Of these the Balmer continuum is the most important. It can compete with electron scattering for $T_e$ of the order of 10000° and $n = 10^{10}$–$10^{11}$ cm$^{-3}$, increasing with $n$. So it would be interesting to observe a nimbus in a wavelength region below the Balmer discontinuity, say at $\lambda$ 3600 Å.

Other explanations of the nimbus phenomenon are not excluded, but we wish to stress the need for observations which could give more information. I should very much like to receive available, unpublished, and future observational data.

## References

Houtgast, J. (1962)    *Bull. astr. Inst. Netherl.*, **17**, 56.
    (1963)    in *The Solar Corona, Symposium No. 16 of I.A.U.*, Ed. by John W. Evans, p. 231.

## DISCUSSION

*Kiepenheuer:* The question of whether the nimbus effect could be explained by electron scattering had been discussed already at the Cloudcroft Symposium (J. W. Evans, *Proc. IAU Symposium No. 16:*

*The Solar Corona*, New York 1963, p. 233). The main difficulty involved seems to me the very large total number of electrons and consequently ions required to produce the observed effect, which is of the order of $10^{43}$ electrons respectively ions. It is equivalent to about $10^5$ times the electron respectively ion number of a well-developed prominence or of a flare surge. I can not quite see, how such a large amount of material can be injected without observable effects.

*Severny:* The picture shows the nimbus for a flare just at the limb and it gives an idea about the maximal height of the layer producing the 'nimbus'. The height of the layer responsible for nimbus cannot be larger than $10^4$ km ($= 10^9$ cm).

*De Jager:* Since the nimbus is a H$\alpha$ phenomenon, its explanation should be looked for in the H$\alpha$-source function. If it were due to electron scattering it should as well be visible in the continuous radiation.

*Beckers:* If the nimbus were due to electron scattering I would expect it to be bright when seen on the disk in the centre of H$\alpha$. This because of the large thermal motions of the electrons which cause Doppler shifts of about 10 Å r.m.s. for the incident radiation at a temperature of 10000 °K. So the electrons see radiation coming from the continuum near the H$\alpha$ line causing a strong increase in brightness (factor 5). Another factor to take into account is the solar limb darkening causing the electron cloud to go into emission near the limb for wavelengths where the limb darkening is large. At H$\alpha$ this may only be a small effect, however.

*Švestka:* An electron density of $10^{13}$ cm$^{-3}$ exists only in narrow filaments of the flare, the volume of which is much smaller than you have considered. I also suspect very strongly that with the number of free electrons you give, we should have to observe continuous emission behind the Balmer limit, which never has been the case. And as for the limb event shown by Professor Severny, we should observe some emission hill on the limb, since the light scattered by the electrons to all directions must become visible on the limb. But we do not observe anything like that.

*Houtgast:* The foregoing remarks are very much to the point and are welcome reasonings regarding the nimbus problem. Once more they stress the need for a quantitative treatment, as well theoretically as from the observational side.

# FLARE-PRODUCED CORONAL WAVES

Friedrich Meyer

(*Max-Planck-Institut für Physik und Astrophysik, München, Germany*)

ABSTRACT

We present a coronal propagation model for fast chromospheric disturbances following flares that were first observed by Moreton and Ramsey, and deduce estimates of average coronal magnetic fields (6 gauss) and disturbance energies ($\leqslant 10^{29}$ erg).

Flare-produced hydromagnetic waves were first observed by Moreton (1960, 1965) and Ramsey as chromospheric disturbances on the wings of the Hα line. They travel across the solar disk with speeds of the order of 1000 km/sec, can excite filaments to oscillations and trigger other flares.

In the event of September 20, 1963 observed by Moreton and Ramsey (Moreton, 1965; cf. also Zirin, 1966, p. 400ff.), such a wave apparently was caused by the sudden brightening of the flare. It propagated away from its origin with a nearly constant speed of 750 km/sec into an angular sector of about 60° covering a distance of several hundred thousand kilometers. The widths of the disturbance was about 100000 km, and the Hα Doppler shift indicated a material motion of about 10 km/sec away from the observer followed by a recovery in the opposite direction.

The high constant propagation speed and the low wave amplitude of this event are incompatible with a chromospheric propagation. For a temperature of $1·6 \times 10^4$ °K the chromospheric sound velocity is 18 km/sec and for a density of $3 \times 10^{10}$/cm$^3$ and a 5 gauss magnetic field the chromospheric Alfvén velocity is about 55 km/sec. The 750 km/sec wave then would have moved with a Mach number of the order of 10. A shock wave of such strength would have shown large wave amplitudes (of the order of the propagation speed) and strong dissipation, i.e. slowing down. Both such effects are not observed.

These difficulties disappear if a coronal propagation is considered.* In the corona the sound and Alfvén speeds both are about a factor of 10 higher than in the chromosphere. This results from a coronal temperature roughly 100 times larger and a coronal density 100 times smaller than their corresponding chromospheric values.

In the corona the large ratio of the magnetic to the internal gas-energy density clearly separates the 3 hydromagnetic modes. This ratio is 10 at the bottom of the corona and increases upwards (for a magnetic field $B=5$ gauss, a temperature of

---

\* The coronal propagation of such disturbances was independently suggested by H. U. Schmidt (private communication).

1·6 million degrees and a particle density of $3 \times 10^8/\text{cm}^3$). The slow mode is reduced to a sound-wave traveling along the guiding magnetic field. The incompressible Alfvén mode moves with Alfvén speed $V_A = B/(4\pi\rho)^{1/2}$ (where $\rho$ is the mass density) but is also confined to propagation along the fieldlines. The fast hydromagnetic mode moves also at Alfvén speed (slightly modified by the sound velocity) but can propagate in all directions with respect to the magnetic field. It is this latter mode that has to be made responsible for the fast disturbances spreading across the solar disk.

The propagation of this mode in the corona is strongly influenced by downward refraction in the coronal density gradient and upward reflection at the chromospheric interface. The refraction is caused by the density-dependent increase of the Alfvén speed with height and prevents the energy from escaping upwards. The sudden density increase between corona and chromosphere produces the almost complete reflection. Only a small fraction of the energy penetrates into the chromosphere and causes the observed Hα-motions. The propagation is further influenced by dispersion which must finally disintegrate any disturbance into wavetrains of different speed. Waves of long horizontal extent reach higher up where the Alfvén speed is larger. They therefore travel faster than small-scale waves.

For a quantitative comparison we simplify the corona to an isothermal gas of 1·6 million degrees lying above a chromosphere a hundred times cooler and denser. For simplicity we first investigate the case when the magnetic field is vertical and constant (Figure 1). The first approximation consists in neglecting the small chromospheric penetration altogether and assuming a perfectly reflecting interface.

As is suggested by the dispersion we consider the individual harmonic waves which together make up a wave packet.

For each value of the horizontal wave number $k$ there exists a discrete set of height-dependent eigenfunctions with $n = 0, 1, 2, \ldots$ nodes (the vertical set is discrete rather than continuous because of the exponential increase of the signal velocity $V_A$ with

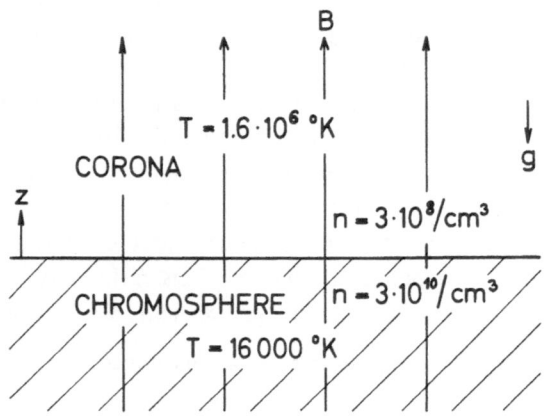

FIG. 1.   *The corona – chromosphere model.*

height). To each wave $(k, n)$ belongs an eigenfrequency $\omega = \omega_n(k)$. From this one determines the phase velocity $V_{\mathrm{ph}} = \omega/k$ and the group velocity $V_{\mathrm{gr}} = d\omega/dk$. Figure 2 gives the result of such an evaluation in form of a dispersion diagram. It shows the nine lowest eigenfrequencies as a function of $k$. The six horizontal lines belong to Alfvén modes. Their group velocity is zero since they cannot propagate across the magnetic field. The three sloping lines belong to the fast mode with a remarkably

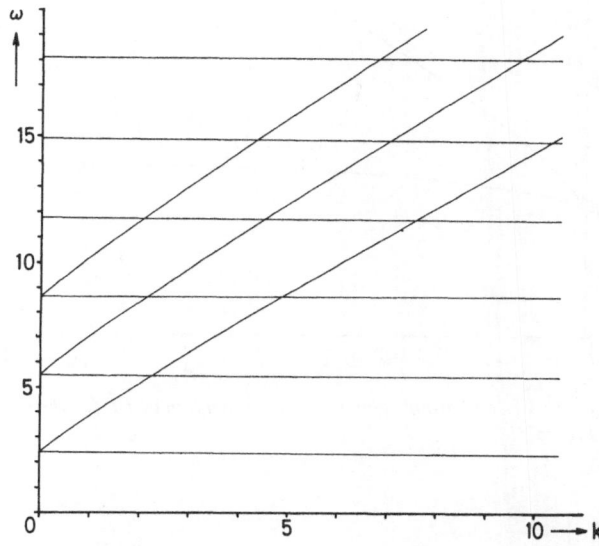

FIG. 2. *Angular frequency $\omega$ over horizontal wave number $k$ in the corona. The magnetic field is vertical. Unit of length is twice the coronal density-scale height. Unit of time is the unit of length divided by the Alfvén speed at the bottom of the corona.*

constant group velocity. In Figure 3 we plot phase and group velocity as a function of $1/k$. The small variation of the group velocity in the range of the observed dimensions ($\approx 3 \leqslant k \leqslant 10$) will account for the long distance across the Sun over which the disturbance on the September 20, 1963 event could be followed.

Equating the observed velocity of this disturbance to the group velocity for the characteristic wave number ($k \approx 7$) one can determine the physical value of the unit velocity, i.e. the Alfvén speed at the bottom of the corona. With a density of $3 \times 10^8/\mathrm{cm}^3$ this leads to a coronal magnetic field of 6 gauss. It is an average field over the dimension of the wave. The value depends only on the square root of the assumed coronal density. Though earlier determinations of undisturbed photospheric fields suggested lower values of the coronal field, the 6 gauss may be reasonable in view of recent observations. They found strong photospheric flux concentrations that should at least partially spread out into the corona.

The influence of an inclined magnetic field on the propagation of such coronal

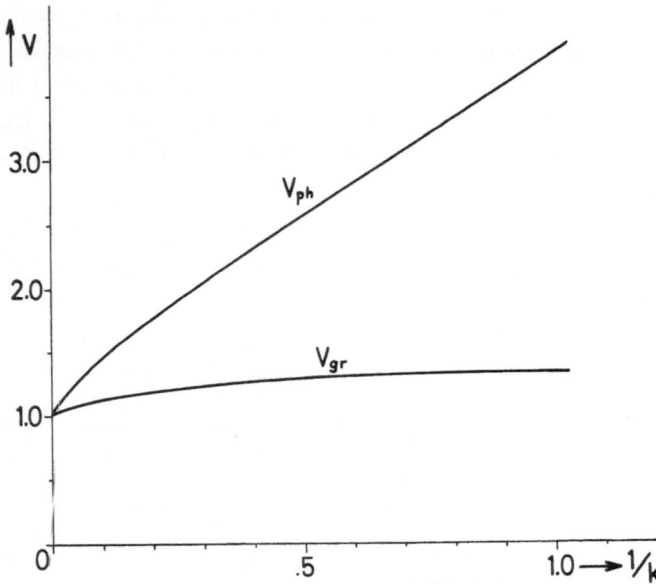

FIG. 3. *Phase velocity* $V_{ph}$ *and group velocity* $V_{gr}$ *as a function of* $1/k$ *for the lowest order fast mode* $(n = 0)$. *Units as in Figure 2.*

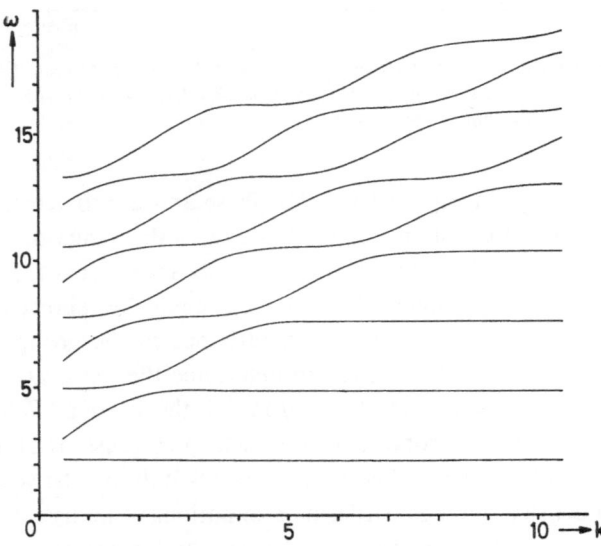

FIG. 4. *Dispersion diagram for a magnetic field inclined 30° to the vertical. Direction of horizontal wave vector at 45° to the horizontal magnetic-field component. Units as in Figure 2.*

disturbances was investigated together with E. Meyer-Hofmeister on a recent visit to the Sacramento Peak Observatory. We found that fast modes are not changed if they propagate in the direction of the horizontal field component. Alfvén modes for this case have lowered eigenvalues. This is due to an increased 'optical path' along the inclined fieldlines.

For other directions of this horizontal propagation fast and Alfvén modes are superposed to give the eigenfunctions. This is caused by the lower boundary at which fast modes on reflection also excite Alfvén modes, and vice versa. The resulting mixture of fast and Alfvén modes in each eigenfunction changes with the wave number $k$. This leads to the characteristic frequency variation for the eigenmodes in Figure 4. That dispersion diagram is an example for general propagation in an inclined magnetic field. The change of character expresses itself in the group velocity which for a particular eigenvalue varies from near zero values of the Alfvén mode to high values of the fast mode. Thus in inclined fields the dispersion becomes very strong for all directions except those close to the direction of inclination. From this effect we expect an apparent concentration of observed disturbances along that preferred direction. For the September 20, 1963 event such a relation is indicated. A surge preceding the wave phenomenon pointed in projection against the disk in a direction which coincided with that of the later wave propagation. Since the surge had to follow the magnetic field we conclude that it also was tilted in the same direction.

One can determine amplitudes of the coronal waves from the observed chromospheric motions. To do this one has to relax the condition of perfect reflection at the interface and to supplement the coronal pattern by a chromospheric part. This changes the coronal propagation very little. A small imaginary part of the eigenvalues of about 1% is introduced by the damping which is caused by the energy loss into the chromosphere. The amplitudes of the corona are now connected with those in the chromosphere. For the case of a vertical magnetic field this next order approximation has been performed. For the September 20, 1963 event we infer a coronal amplitude maximum of about 200 km/sec. It is about $\frac{1}{4}$ of the local Alfvén speed. A check on this estimate comes from filament oscillations caused by such coronal waves. The theoretical prediction agrees in the order of magnitude with the observed oscillation amplitudes. The energy contained in the observed disturbance is estimated from this amplitude as $\leqslant 10^{29}$ erg which is a small fraction of the total flare energy.

## References

Moreton, G. E., (1960)    *Astronom. J.*, **65**, 494.
Moreton, G. E., (1965)    in *Stellar and Solar Magnetic Fields. I.A.U. Symp. 22*, North-Holland Publishing Co., Amsterdam, p. 371.
Zirin, H. (1966)    *The Solar Atmosphere*, Blaisdell Publishing Co., Waltham, Mass., p. 400ff, Fig. 13.11 and 13.12.

# THE OBSERVATION OF 10-50 KeV SOLAR FLARE X-RAYS
# BY THE OGO SATELLITES AND THEIR CORRELATION
# WITH SOLAR RADIO AND ENERGETIC PARTICLE EMISSION

R. L. ARNOLDY, S. R. KANE and J. R. WINCKLER

*(School of Physics and Astronomy, University of Minnesota,
Minneapolis, Minn., U.S.A.)*

## ABSTRACT

More than 70 cases have been observed of energetic solar flare X-ray bursts by large ionization chambers on the OGO satellites in space. The ionization chambers have an energy range between 10 and 50 KeV for X-rays and are also sensitive to solar protons and electrons. A study has been made of the X-ray microwave relationship, and it is found that the total energy released in the form of X-rays between 10 and 50 KeV is approximately proportional to the peak or total energy simultaneously released in the form of microwave emission. For a given burst the rise time, decay time and total duration are similar for the 10–50 KeV X-rays and the 3 to 10 cm radio emission. Roughly exponential decay phases are observed for both emissions with time constants between 1 and 10 min. All 3 or 10 cm radio bursts with peak intensity greater than 80 solar flux units are accompanied by an X-ray burst greater than $3 \times 10^{-7}$ ergs cm$^{-2}$ sec$^{-1}$ peak intensity. The probability of detecting such X-ray events is low unless the radio spectrum extends into the centimetric range of wavelengths. The best correlation between cm-$\lambda$ and energetic X-rays is observed for the first event in a flare. Subsequent structure and second bursts may not correspond even when the radio emission is rich in the microwave component. The mechanism for the energetic X-rays is shown to be bremsstrahlung probably of fast electrons on a cooler plasma. If the radio emission is assumed to be synchrotron radiation then a relationship is developed between density and magnetic field which meets the observed quantitative results. One finds, on the average, that $5 \times 10^{-54}$ joules m$^{-2}$ (CPS)$^{-1}$ of microwave energy at the Earth are required per electron at the Sun to provide the radio emission for the various events.

A strong correlation between interplanetary solar flare electrons observed by satellite and X-ray bursts is shown to exist. This correlation is weak for solar proton events. One may infer a strong propagation asymmetry for solar flare electrons along the spiral interplanetary magnetic field.

## 1. Introduction

Since September 1964 ionization chambers flown in space have detected numerous solar flare X-ray bursts in the energy range from 10 to 50 keV, originating in solar flare disturbances. In the present paper we shall present the most recent summary of these observed events and discuss their relationship with the microwave solar radio bursts and with the observations of energetic electrons and protons ejected into space by the same flares. The central object of such a study is to reach a better understanding of the solar flare processes and the nature of the instability which generates flares but, in particular, to understand the processes which gave rise to suprathermal particles so frequently observed from flares.

*Kiepenheuer (ed.), Structure and Development of Solar Active Regions*, 490–509. © *I.A.U.*

The X-rays are detected by ionization chambers carried for long periods of time outside the magnetosphere by the OGO-I and OGO-III satellites. Details of the instrumentation and previous work on this program have been summarized in several publications (Kane *et al.*, 1966; Arnoldy *et al.*, 1967*a, b*). The range of energies covered by the present experiments is similar to, but in general somewhat less than, the X-ray events detected previously by balloons flown near the top of the atmosphere (Peterson and Winckler, 1959; Winckler *et al.*, 1961; Vette and Casal, 1961; Anderson and Winckler, 1962; Hofmann and Winckler, 1963). Several summaries of this very energetic bremsstrahlung emission from flares and their relationship to the radio and optical features are available, based on the older results (Winckler, 1963; Friedman, 1964; Kundu, 1965). In a general sense the energetic flare X-rays of energy above 10 KeV appear as bursts of duration between 1 and 20 min in very good time simultaneity with the 'explosive' phase of flares (Moreton, 1964). Previously, the observations of these energetic X-rays were made by chance on high-altitude balloons carrying ion chambers or scintillation counters. These early experiments detected only the most energetic quanta from the flare due to the atmospheric absorption above the balloons. These rather exceptional events have given rise to considerable speculation about processes which could produce energetic quanta, such as the inverse Compton process (Acton, 1964; Shklovsky, 1964, 1965; Zheleznyakov, 1965), the synchrotron process (Stein and Ney, 1963), or nuclear processes giving gamma-rays. The results of the present study show that flares of all sizes from Class 1S to Class 3B emit such energetic X-rays and that their origin is probably bremsstrahlung following the suprathermal heating of electrons in the magnetic-plasma medium in the solar active region. Our recent investigation (Arnoldy *et al.*, 1967*b*) has shown that the X-ray bursts are well-correlated with the direct observations immediately afterward in space of energetic electrons greater than 40 KeV energy which may well come from the same source (Lin and Anderson, 1967). In this paper and the previous related accounts we also have found evidence that the acceleration of solar flare protons which are now widely observed in interplanetary space or at the Earth may arise from a process disjoint from that responsible for the X-ray electron emission.

## 2. Discussion of the OGO Experiments

The X-rays were detected with an 18-cm diameter aluminum wall ionization chamber filled with argon gas at 3·5 atm pressure. Identical instruments were provided for the OGO-I (launched September, 1964) and the OGO-III (launched June, 1966) satellites, both of which continue to give data. Details of the instrumentation are given in our previous publications (Kane *et al.*, 1966; Arnoldy *et al.*, 1967*b*). The ion-chamber response has a lower limit of 10 KeV, has a maximum response between 20 and 30 KeV and an effective upper limit between 50 and 100 KeV depending on the type of spectrum characteristic of the X-rays. In this paper the ion-chamber rates

R. L. ARNOLDY ET AL.

## Table 1

### Ion chamber response

**Primary X-Ray Spectrum** $dN/dE = e^{-E/E_0}$ photons cm$^{-2}$ sec$^{-1}$ KeV$^{-1}$

| $E_0$ (KeV) | Chamber Rate ($N$ pulses sec$^{-1} \times 10^3$) | | | | Incident Energy Flux $> 10$ KeV ergs cm$^{-2}$ sec$^{-1}$ | Conversion Factor ergs cm$^{-2}$ sec$^{-1}$ ($N$ pulses sec$^{-1} \times 10^3$) |
| --- | --- | --- | --- | --- | --- | --- |
| | 10–16 KeV | 16–106 KeV | 106–150 KeV | Total | | |
| 7 | 0·026 | 0·116 | $5 \cdot 1 \times 10^{-8}$ | 0·142 | $1 \cdot 7 \times 10^{-8}$ | $1 \cdot 2 \times 10^{-7}$ |
| 20 | 0·106 | 1·16 | $1 \cdot 8 \times 10^{-3}$ | 1·27 | $4 \cdot 7 \times 10^{-7}$ | $3 \cdot 7 \times 10^{-7}$ |
| 50 | 0·165 | 2·61 | 0·07 | 2·85 | $3 \cdot 8 \times 10^{-6}$ | $1 \cdot 3 \times 10^{-6}$ |

are given in the various figures in terms of a standardized arbitrary rate designated as normalized pulses per $\sec \times 10^3$ (NPPS $\times 10^3$). The response characteristics and factors for converting the chamber rate to absolute energy flux are given in Table 1. For the analysis in this paper, we have assumed a rather steep exponential type X-ray spectrum with $E_0 = 7$ KeV. The threshold sensitivity of the chamber for X-rays is $3 \times 10^{-7}$ ergs cm$^{-2}$ sec$^{-1}$. The chamber responds also to protons above 12 MeV, to electrons above 700 KeV, and frequently detects particle events in space closely following the X-ray bursts from the same flare.

Recently some solar X-ray events have been detected simultaneously by both

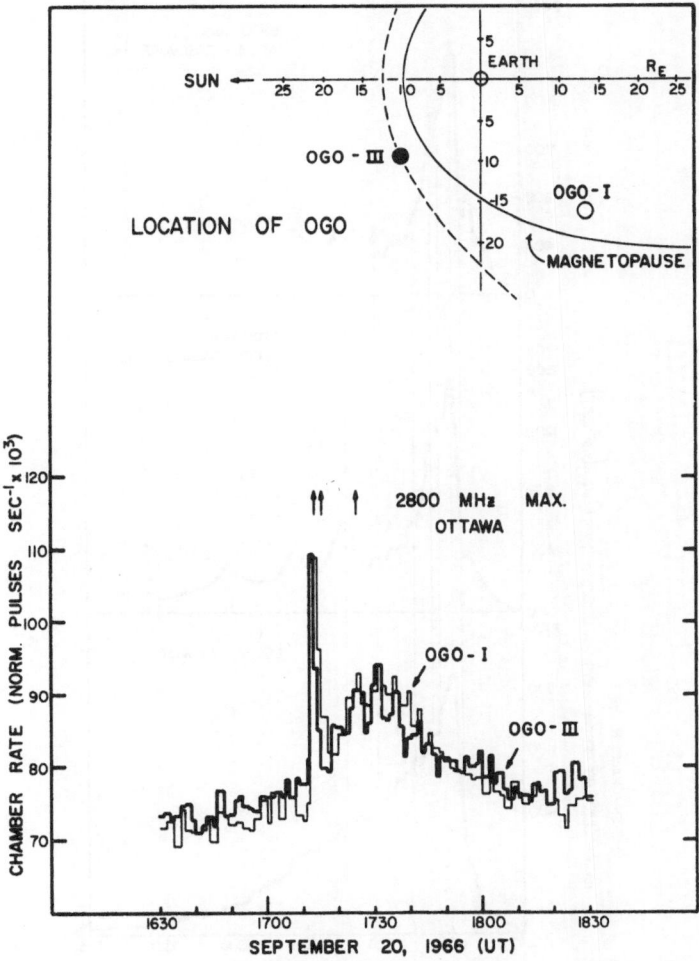

FIG. 1.  *A solar flare X-ray increase observed simultaneously by the ion chamber on OGO-I and OGO-III satellites. This X-ray burst has a sharp preliminary peak at 1712 UT and a broad maximum centered at 1730. A satisfactory 10-cm radio correlation exists.*

OGO-I and III. Figure 1 shows such an example and also delineates the orbital positions of the two satellites with respect to the magnetospheric structure. Such simultaneous events make the identification of the X-ray increase completely certain and help to distinguish spurious cases due to electron bursts associated with the magnetosphere. The example shown in Figure 1 is also interesting because of the sharp burst at 1712 UT and the rather smooth maximum at 1730. The close agreement of the two ion chambers measuring the same event shows that the calibrations used were correct and that no drift has occurred in the calibration of the OGO-I instrument over a period of $2\frac{1}{2}$ years.

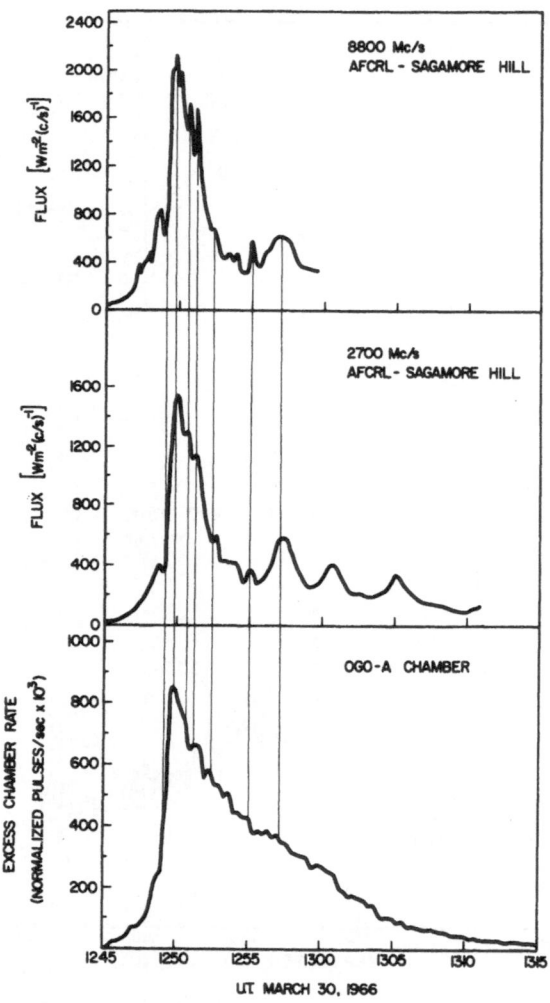

FIG. 2. *Comparison of X-ray intensities (lower) and 3 and 10-cm radio emission (above) for a large complex flare event.*

### 3. Correlation of X-Ray and Radio Bursts

In Figure 2 we have a more or less typical large flare event producing a considerable intensity of energetic X-rays and radio emission in the microwave range. We note a close similarity in the structure and rather good correlation of the start and time of maximum intensity. However, the fluctuations in the radio emission are much larger and do not always correspond to fluctuations in the X-ray intensity, which is observed in general to be much smoother. The decay of the March 30 event shown in Figure 2 is given in Figure 3, and one notes that it is roughly exponential but with the previously noted fluctuations in the radio emission. Approximately 70 events have now been analyzed constituting most of the cases observed up to December 31, 1966. The

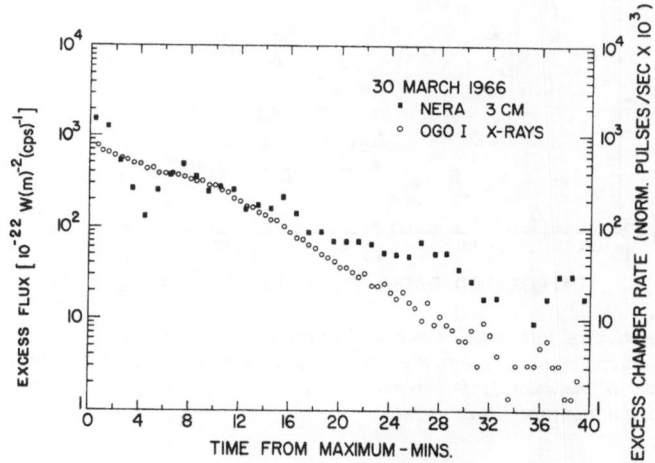

FIG. 3.    *The decay of a complex event showing similar trends for X-rays and 3-cm radio emission.*

correlation plot of the integrated radio and X-ray flux is shown in Figure 4. The data are separated into two periods representing events prior to June 30, 1966 and events from July 1 to December 31, 1966, although a reasonable line may be constructed for the earlier period. With the advance of the solar cycle, there seems to be a systematic effect such that relatively more X-ray emission occurs with the same radio emission. Also, the scatter increases considerably especially for lower values of the fluxes. The solar cycle trend shown in this graph may possibly be associated with the appearance in the second half of 1966 of many centers of activity and sunspot zones in both solar hemispheres.

### 4. Interpretive Remarks about the X-Ray Radio Relationship

Because of the very close morphological relation between the X-ray production and

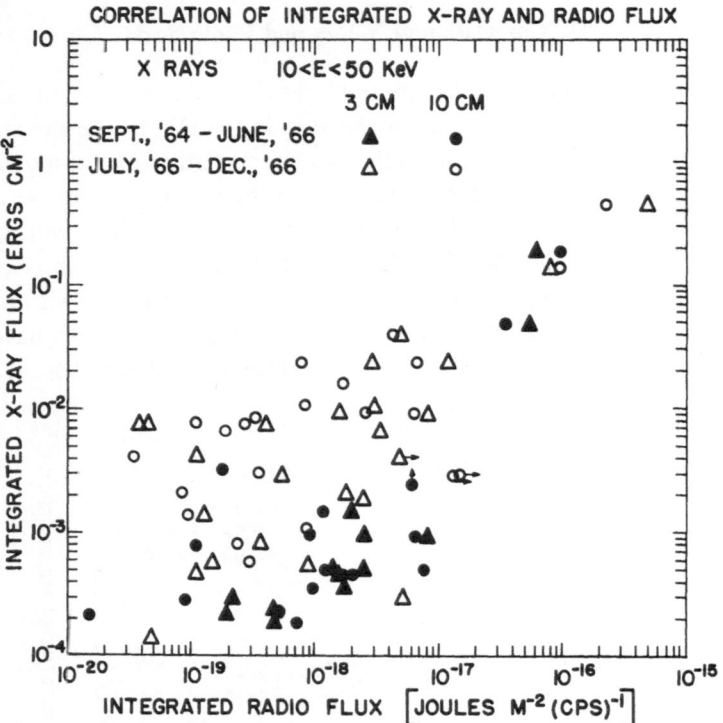

FIG. 4. *Correlation of integrated X-ray and integrated radio flux for the observed events. The integral has been carried out over the first event in a flare occurrence. A roughly linear relation exists, but the events later in the solar cycle between July 1966 and December 1966 appear relatively more rich in X-ray emission than the earlier events.*

the centrimetric range of radio emission it is very plausible to search for a model or to propose a situation in which both types of electromagnetic emission come from a common source. In the paper of Peterson and Winckler (1959) the source of the X-rays was assumed to be bremsstrahlung from energetic electrons in an energy range around 500 KeV. The radio emission was then assumed to be by the synchrotron process from the same electrons. This lead to the difficulty that about $10^4$ times too much radio emission was expected compared to the observed. Takakura (1963, 1966) examined the situation and proposed that the region of emission was different for the X-rays and the radio bursts and was able to adjust the radio power at the same time retaining the concept that the same energy region of electrons was responsible for both emissions.

The examples given in this paper have been presented purposely to show the complex character of the situation. Early in the flare event there appears to be a very close relationship between X-rays and microwaves and as time progresses the radio emission assumes a time structure frequently not closely related to the smoothly disappearing

X-ray burst. It will thus be difficult or impossible to form a single model applicable to all events. However, for the initial part of the events where the correlation is very strong between X-rays and microwaves, it may be possible. In our previous paper (Arnoldy *et al.*, 1967*b*) we attempted to visualize the simplest possible situation that was consistent with the experimental facts and did not assume special kinds of processes not specifically dictated by the observations. We recall the approximate proportionality between the X-ray and radio emission as shown above and the similarity in duration and decay rate of the two types of emission. Two possible cases are considered: (a) that the characteristic time constant is determined entirely by the time variation of the basic energy source itself with all other time constants associated with specific processes (for example, synchrotron emission) being shorter than this. In case (b) the source is impulsive but a single dominant electron-decay process determines the time constant for both the X-ray and radio emission. This time constant could be that required for the suprathermal electrons responsible for the bremsstrahlung to disappear by collision loss in the plasma. We consider that the same plasma region and probably the same electrons are producing both the X-ray and radio emission. The plasma is entrained in a magnetic field above the solar active centre. These fields must play a major role in the acceleration of particles and in the emission and propagation of radio-frequency energy. For any quantitative calculations the emission process must be made specific. Our model assumes that very hot energetic or suprathermal electrons lose energy predominantly by collisions with a much cooler plasma. This point of view has also been suggested by Elwert (1961). Neither the present measurements, nor the original measurements of Peterson and Winckler (1959) can provide exact energy discrimination. Thus a re-interpretation of the March 20, 1958 event of Peterson and Winckler, made by Chubb *et al.* (1966) in terms of an exponential spectrum with $E_0$ about 60 KeV is probably acceptable. However, balloon scintillation counter measurements by Anderson and Winckler (1962) showed directly the presence of photons $> 150$ KeV energy, and Cline *et al.* (1967) have shown flare bremsstrahlung spectra at 100 KeV. We do not consider plausible the concept that there exists a complete plasma at an enormously elevated temperature and that one is justified in using the thermal bremsstrahlung approach for computing the X-ray emission power for such events. The thermal bremsstrahlung approach with very high temperatures has been consistently proposed by the NRL group (see e.g. Chubb *et al.*, 1966). It is true that the concept of temperature is often applied to one component of a medium, e.g. electron energies are frequently given in terms of electron temperature if the energy distribution appears to be exponential. However, one might expect that use of the concept of temperature would imply that the medium is close to a thermodynamic equilibrium condition and that one is therefore justified in applying this same temperature to compute many types of processes such as, e.g., the distribution of ionization states, spectral emission, etc. There appears to be no evidence at all that the temperatures deduced from other means in flare regions reach the enormous values of $10^{8\circ}$ which one is forced to assume in order to produce the hard X-rays by thermal bremsstrahlung. Optical measurements (De Jager, 1959) frequently show $10000-20000°$ as typical. We therefore favor the point of view that the electron heating in the flare region is highly non-equilibrium and may be associated with such phenomena as magneto-hydrodynamic waves and that there exists an energy distribution characteristic of a much lower temperature with a large suprathermal tail.

We now consider a quantitative estimate of the bremsstrahlung emission. In case (a), where the time behavior of the event is determined entirely by the energy source, the collision lifetime is very short. One can use the thick target bresmsstrahlung equation given by Koch and Motz (1959) for non-relativistic electrons

$$\varepsilon = 5 \times 10^{-4} Z \frac{E_K}{m_0 c}, \tag{1}$$

where $\varepsilon$ is the efficiency defined as (total energy radiated)/(total beam energy), $E_K$ the kinetic energy, and $Z$ the atomic number of the target. Considering 100-KeV electrons for a large event such as March 30, 1966 or July 7, 1966 a beam energy of $3 \times 10^{30}$ ergs is required, which is equivalent to $2 \times 10^{32}$ electrons. If we consider case (b), where an impulsive injection of 100-KeV electrons is assumed the collision loss determines the decay of the event (from 1 to 10 min) and gives density estimates of $3 \times 10^{10}$ atoms cm$^{-3}$ to $3 \times 10^9$ atoms cm$^{-3}$ for the lifetime range of 1 to 10 min respectively. The bremsstrahlung power is calculated also from the results of Koch and Motz (1959) valid for electron energies of 10–100 KeV incident on neutral hydrogen. Quantitatively, the relationship reduces to the following, where $P$ is the number of photons between 10 and 50 KeV per cm$^2$ per sec measured at the Earth, and $N_H$ and $N_e$ are respectively the densities of hydrogen and energetic electrons situated in the volume $V$.

$$P = 3 \times 10^{-43} N_H N_e V \text{ photons cm}^{-2} \text{ sec}^{-1}. \tag{2}$$

If $N_H$ can be estimated from observed event lifetimes, and if $P$ is measured, then Equation (2) may be used to compute the total number of electrons, $N_e V$. Again for a large X-ray event the value of $P$ is $1 \cdot 5 \times 10^4$ and from the observed mean lifetime of 300 sec $N_H$ is estimated to be $5 \times 10^9$ cm$^{-3}$. For this density $10^{37}$ electrons are required in the impulsive injection process, i.e. essentially the same number as calculated under assumption (a).

It is necessary to point out that in the thick target case (a) we begin with 100 KeV electrons but the energy is degraded to zero in the thick target as they are brought to rest. Thus the emission occurs for all energies between 0 and 100 KeV. A portion of this emitted spectrum is not detected by the ionization chamber which results in an error of probably not more than a factor of 2 in applying the Koch and Motz thick target equation. In case (a) the integrated bremsstrahlung energy over the entire observed X-ray burst is used for the comparison. In case (b) we utilize the Koch and Motz thin target approximation for electrons of the unique energy of 100 KeV equal to the assumed injection energy. For comparison with the theory therefore the measured peak X-ray burst intensity is used rather than the integrated intensity.

Considering the radio emission this is often attributed to synchrotron radiation. Although this is certainly a plausible mechanism the detailed measurements of polarization and other factors do not exclude a thermal source accompanied by propagation effects which produce polarization. In fact, Kundu (1965) suggests that the microwave-burst events may frequently be a mixture of thermal and synchrotron emission. Considering a large event such as March 30, 1966 and July 7, 1966 the microwave

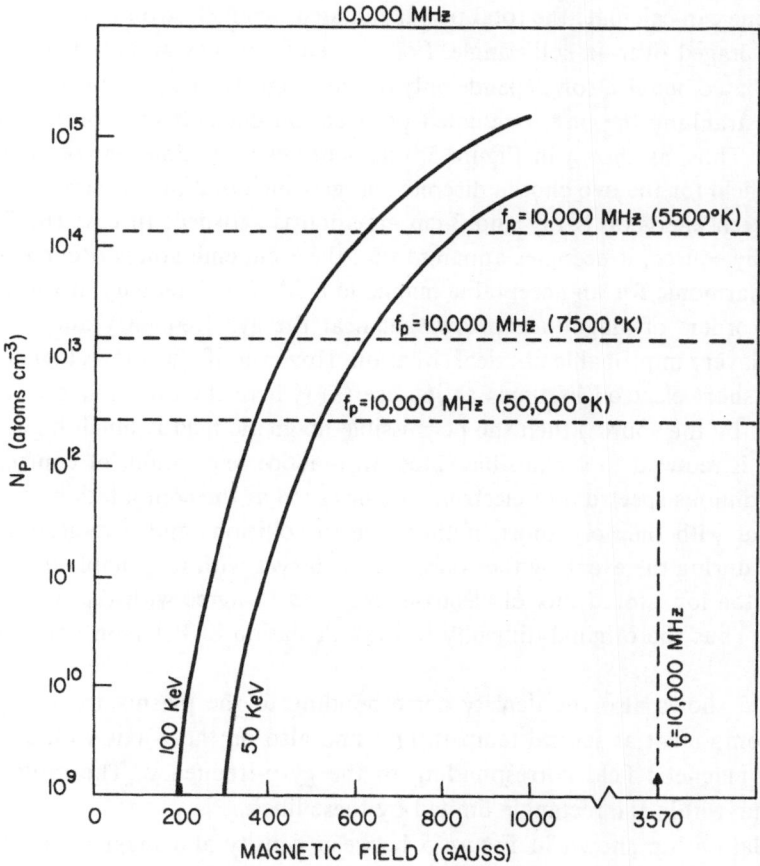

FIG. 5. *The density required to give the observed bremsstrahlung power per electron expressed as a function of the magnetic field required to give the observed radio emission at 10000 MHz by the synchrotron process. The relation is shown for discrete energies of 50 KeV and 100 KeV. Any point along the curves will satisfy the observed proportionality between the microwave and X-ray emission under the assumptions presented in the text. The densities at which the plasma frequency of the medium is equal to 10000 MHz is indicated for various temperatures. The magnetic-field strength which gives an electron gyro-frequency of 10000 MHz is also shown.*

energy received at earth at 3 cm was approximately $6 \times 10^{-17}$ joules m$^{-2}$ (CPS)$^{-1}$. Following our basic assumption that the same electrons are responsible for both processes, about $5 \times 10^{-54}$ joules m$^{-2}$ (CPS)$^{-1}$ of microwave energy at the Earth are required per electron at the Sun on the average. This number will be approximately the same for all events as long as the proportionality between X-ray and microwave emission is valid. The large difference in size from event to event is thus principally due to the difference in number of electrons in the source.

For the purposes of discussion we have assumed that the microwave energy is from the synchrotron emission of electrons of intermediate energy. Following the work of Takakura (1960) for electrons of intermediate energy and assuming 50 and 100 KeV energies one can calculate the total power radiated by an electron in uniform circular motion averaged over $4\pi$ solid angle. For a given frequency and electron energy the power radiated per electron depends only on the magnetic field $B$. On the other hand, for bremsstrahlung the power radiated per electron depends only on the density of hydrogen. Thus, as shown in Figure 5, one achieves a relation between density and magnetic field for the two chosen discrete energies such that the observed proportionality between energetic X-ray and 3-cm emission is satisfied. In case (b) (impulsive injection by source) it becomes apparent that the 3-cm emission is occurring close to the 20th harmonic for an acceptable magnetic field. The emissivity at this harmonic is several orders of magnitude below that near the gyrofrequency and we consider this to be a very improbable physical situation. However, if one allows higher densities and very short electron lifetimes as in case (a) (where the duration of the event is controlled by the source) then the permissible magnetic field is much higher and the harmonic is reduced to a plausible value. In practice one would, of course, wish to use a continuous spectrum of electron energies such as the computation of Takakura (1966). But with the very short lifetime due to collisions and a re-generation continuously during the event by the source one achieves with reasonable magnetic field strengths the low stored flux of electrons required to agree with the observed radio emission. Thus the original difficulty in the calculation of Peterson and Winckler is resolved.

Figure 5 shows also the density corresponding to the plasma frequency for the ionized component at several temperatures and also on the $x$-axis with an extended scale the magnetic field corresponding to the gyro-frequency. The radio emission must occur within the rectangle limited by these lines.

The relationship shown in Figure 5 between density and magnetic-field strength corresponding to the discrete energies 50 KeV and 100 KeV may possibly define some direct physical situation in the flare region. We show in Figure 6 a mean curve for 75 KeV and a family of curves for different values of the ratio, $\beta$, of thermal energy to magnetic energy for a chosen temperature of 7500 °K. One might suppose that during an active time the magnetic field above the sunspot regions would be carrying a large plasma density and that possibly the $\beta$-value might tend to be constant in different portions of the field. However, the empirical curve for 75 KeV spans several orders of magnitude variation in $\beta$, as shown in Figure 6. If, indeed, there is any direct physical significance to this curve then it implies that the magnetic field is more heavily loaded with plasma for high field strengths and very lightly loaded in the upper corona where the field strengths are small.

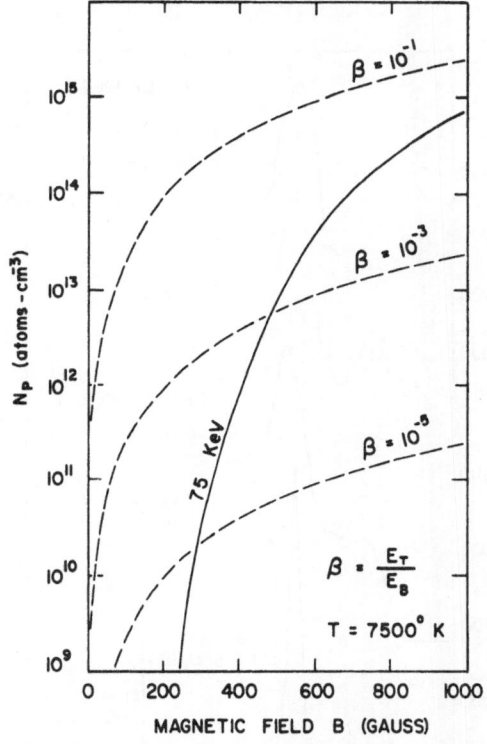

FIG. 6.    *An examination of the hypothesis that the magnetic-field density curve (for mean energy 75 KeV) physically represents the thermal to magnetic-energy relationship in the plasma.*

The discussion above may be considered applicable to the 'first' event in a flare where a good correlation between X-ray and radio emission is observed. The complex character of the X-ray radio relationship is shown by many events. Figure 7 shows data from various sources relating to the large flare on July 7, 1966. Besides the radio emission and the OGO ionization chamber record, a high-energy scintillation counter also on the OGO satellite (Cline *et al.*, 1967) is plotted. We note that the 25 KeV range of X-rays again is similar in its structure and time duration to the radio emission. The >80 KeV X-rays, however, show a much shorter time constant reminiscent of the older events observed by balloons. However, the maximum intensity in all ranges seems to occur at closely the same time.

Another event representative of many already observed (Figure 8) shows that the energetic X-ray emission is correlated closely with centimetric range of frequencies and that the appearance later in the event of a spectrum rich in the metric region does not correlate with X-ray emission. This may be interpreted to show that the production region of the X-rays is deep in the chromosphere and that the electrons responsible for the metric wave emission are at greater heights in the corona.

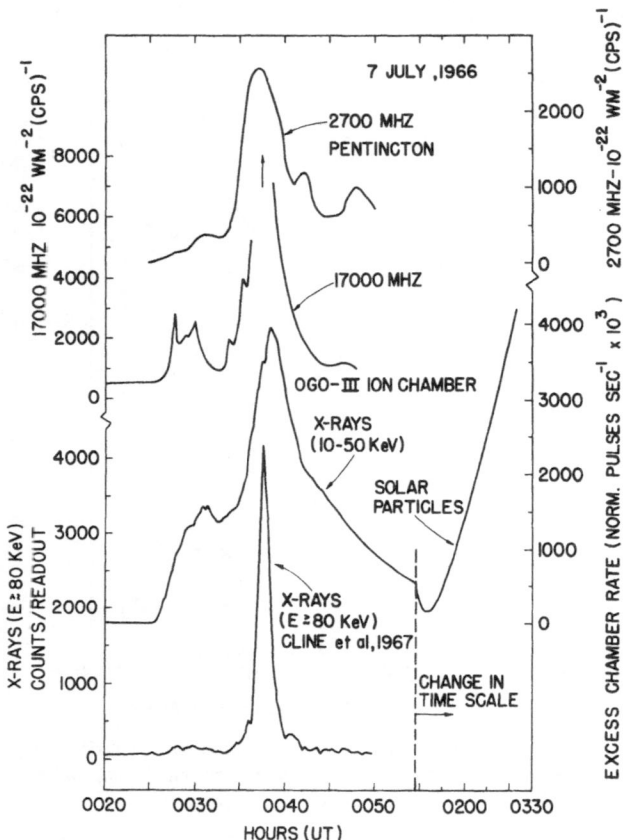

Fig. 7.    *Radio and hard X-ray emission for the flare of July 7, 1966. Note that while the 25 KeV X-ray range shows a time structure similar in duration to the microwave emission, the > 80 KeV X-ray emission is very short, like cases observed earlier by balloon equipment.*

In the inverse sense sometimes double X-ray bursts occur for which the second is not well correlated with a maximum in the radio emission. Such a case is shown in Figure 9. In Figure 10 we see again an event observed earlier by balloon apparatus (Hofmann and Winckler, 1963) in which there is a large double microwave event but for which only the first part produces a detectable X-ray increase. This particular event on July 20, 1961 has been widely studied (Bruzek, 1964; Ellison *et al.*, 1961) and occurred on the solar limb.

## 5. Energetic Particle Relationships

The OGO ionization chamber can detect solar particle events as well as X-ray bursts and frequently shows intensity increases following a flare which are clearly due

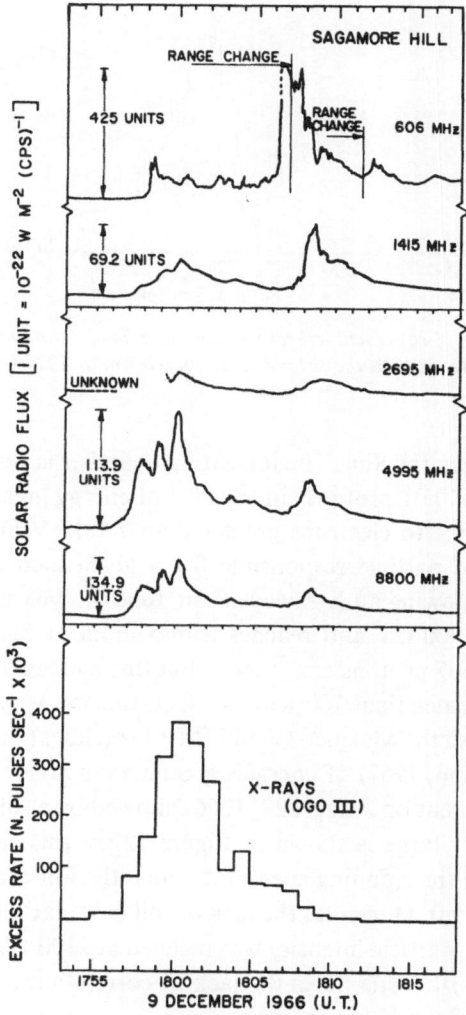

FIG. 8.    *An example of a good X-ray radio correspondence for the first event in the flare beginning at 1757 but showing the lack of observable X-rays corresponding to the predominantly metric event beginning at 1806.*

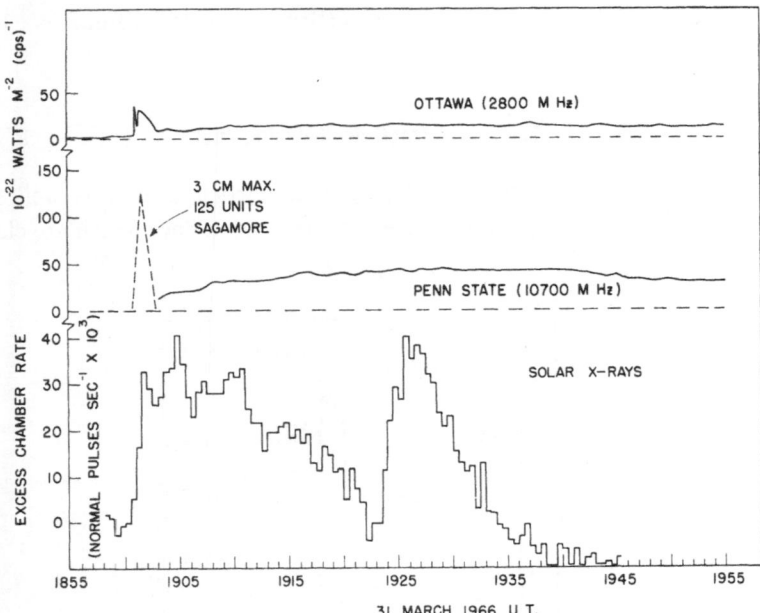

FIG. 9. *A large double X-ray event corresponding to a long-duration microwave increase. No detectable radio event corresponds to the second X-ray maximum at 1927 UT.*

to charged particles from the Sun. The ionization chamber is very sensitive to protons and can detect a flux of 0·01 protons $cm^{-2}$ $sec^{-1}$ of energy greater than 12 MeV. The chamber also is sensitive to electrons greater than 700 KeV but in principle cannot distinguish the type of particle responsible for a given increase. An example of a small particle event following an X-ray event on June 5, 1965 is shown in Figure 11. The event begins at 1900 UT and reaches a maximum at 2100 UT. Interpreted as protons, this implies 0·07 protons $cm^{-2}$ $sec^{-1}$ but this increase could also presumably be due to electrons greater than 700 KeV. In fact, this event produced an identified-electron increase in both the Mariner-IV and IMP-I satellites (Van Allen and Krimigis, 1965; Lin and Anderson, 1967) of energies greater than 40 KeV.

A very large X-ray event on August 28, 1966 followed by a solar proton event which eventually became very large is shown in Figure 12. In this case the X-ray increase was roll modulated by the spinning spacecraft when the ion chamber was eclipsed by the body of the spacecraft. One notes the lack of roll modulation for the particle event. The maximum in solar particle intensity was reached at 2100 UT with an ion-chamber rate of 15000 NPPS × $10^3$. One notes the lack of correlation between the large radio maximum at 1605 and the X-ray intensity.

The release into space of electrons above 40 KeV energy associated with solar flares has now been observed on many occasions by the IMP spacecraft (Lin and Anderson,

1967). It is possible to use IMP data to determine which OGO particle events did not contain electrons and presumably were due to solar protons. Also both the IMP electron events and the proton events can be correlated with the occurrence of energetic X-ray bursts and microwave emission. A complete summary with tables for the period September, 1964 through June, 1966 has been given in our previous paper (Arnoldy *et al.*, 1967*b*). These correlations may be summarized as follows:

(a) Of 8 proton events observed by the ionization chamber but not detected as electron events on IMP satellite, 7 were not associated with solar flare energetic

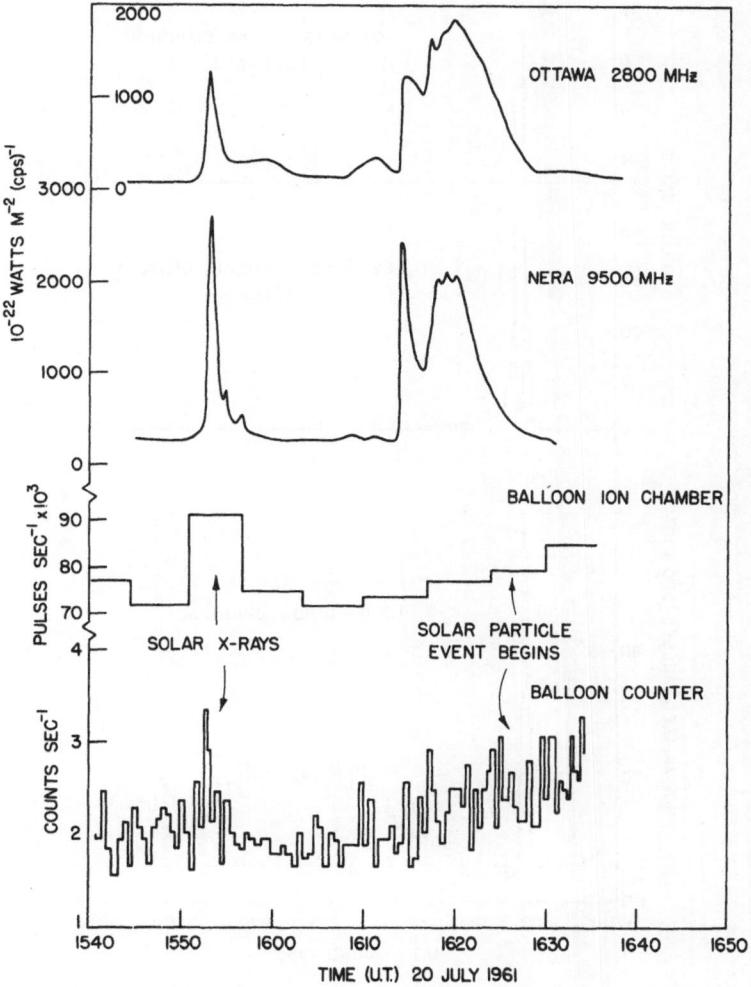

FIG. 10.  *An example of a limb event observed by balloon equipment in which the first increase produces a solar X-ray event but the second large microwave increase has no detectable X-rays. A solar proton increase from this West limb flare has begun by 1612. (See Hofmann and Winckler, 1963.)*

X-rays. An example of such an event was given as Figure 20 of our previous paper (Arnoldy *et al.*, 1967*b*). It appears, therefore, that the production of solar protons is not necessarily closely correlated with the processes producing electrons and energetic bremsstrahlung.

(b) Eight large electron events detected when the OGO and IMP spacecraft were simultaneously operating in interplanetary space had associated X-ray bursts. The remaining 4 electron events were very small and were not recorded by OGO as X-rays

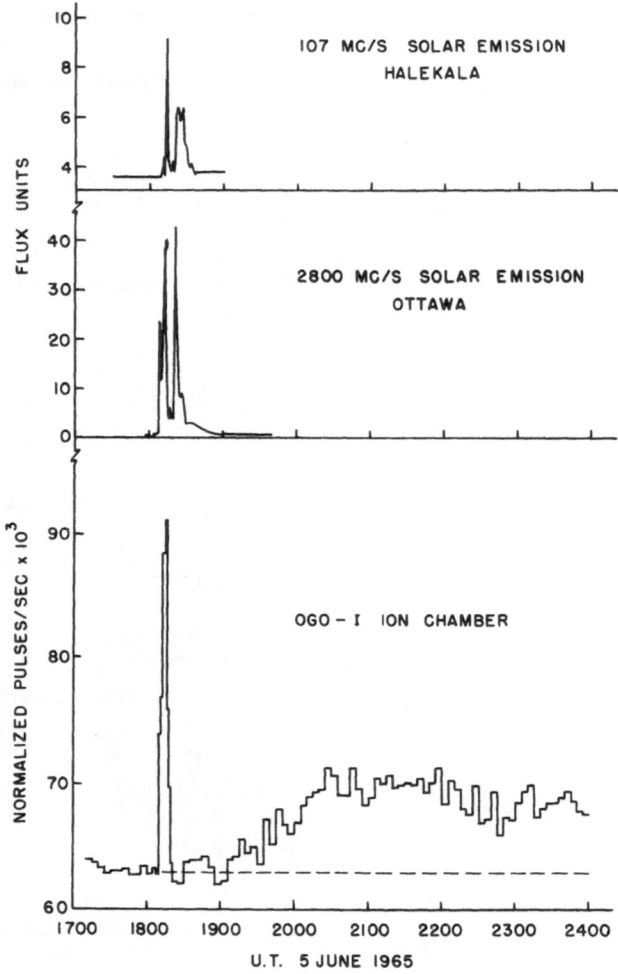

FIG. 11. *A small X-ray event showing the ensuing solar particle event observed by the ionization chamber. These particles are not specifically identified by the OGO ion chamber but may be solar protons. Solar electrons were observed by several experiments (see text) and correspond to this particle increase.*

nor was there microwave emission. This rather good correlation of X-ray bursts with interplanetary electron events suggests that the flare electrons that leak out into interplanetary space might be from the same suprathermal source as those responsible for the energetic X-rays. For the June 5, 1965 event the observed number of electrons measured directly in space above 40 KeV can be provided by the leakage into interplanetary space from the flare region of about 0·1% of the number of electrons required to produce the corresponding X-ray event.

(c) The reverse correlation where one begins with known OGO X-ray bursts and compares IMP data for interplanetary electrons shows 12 events with no interplanetary electrons. It is striking that all 12 were produced by flares near or East of central meridian. It is very reasonable to assume that the absence of electrons in these cases is due to a propagation asymmetry between the Sun and the Earth similar

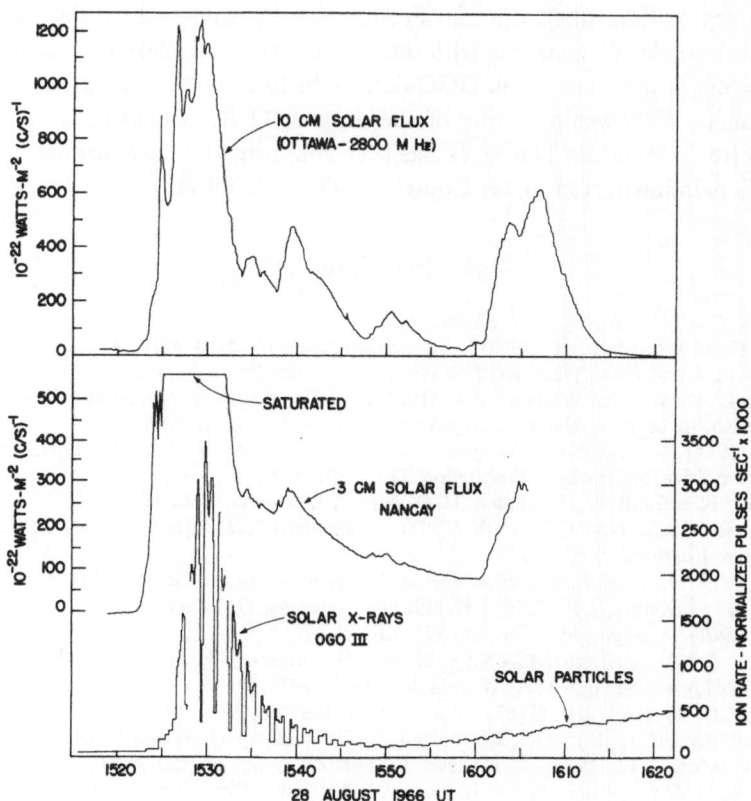

FIG. 12.   *The great event on August 28, 1966. Note the almost complete roll modulation of the X-ray burst with the 96-sec roll period of OGO-III. This roll modulation is not present in the solar particle increase beginning at 1550 UT. Only the initial features of the complex radio burst are reflected in the X-ray profile.*

to that previously observed for high-energy solar protons caused by the spiral inter-planetary magnetic field.

(d) It is generally found that very large flares such as July 7, 1966 or August 28, 1966 produce all types of energetic solar phenomena simultaneously. X-rays, micro-wave emission, interplanetary electrons and solar protons are all observed.

## Acknowledgments

We are indebted to the radio observatories of Penn State, Nancay, Nera, Toyokawa and Ottawa who have furnished many original records for this study and to Dorothy Trotter, World Data Center A for Solar Activity, for lists of outstanding occurrences. Preliminary use has been made of optical data prepared for us by Helen W. Dodson-Prince and E. Ruth Hedeman of McMath-Hulbert Observatory. We are also grateful to Dr. M. Pick of Meudon Observatory for important assistance with solar radio and optical interpretation, to R. Lin and Professor K. A. Anderson of the University of California, Berkeley for assisting with the IMP electron correlations, and to Mr. Karl Pfitzer for major assistance with OGO-data reduction. Data analysis was assisted by a Guggenheim Fellowship to one of the authors (J. R. Winckler) for study at the Observatoire de Meudon. This work has been supported by the National Aeronautics and Space Administration under Contract NO. NAS5-2071.

## References

Acton, L. W. (1964)      *Nature*, **204**, 64–65.
Anderson, K. A., Winckler, J. R. (1962)      *J. geophys. Res.*, **67**, 4103–4117.
Arnoldy, R. L., Kane, S. R., Winckler, J. R. (1967a)      *Solar Phys.*, **2**, 171.
Arnoldy, R. L., Kane, S. R., Winckler, J. R. (1967b)      *University of Minnesota Technical Report CR-97*. Also, to be published in *Astrophysical Journal*, February, 1968.
Bruzek, A. (1964)      *AAS-NASA Symposium on the Physics of Solar Flares*, Scientific and Technical Information Division, NASA, Washington, D.C., 301–322.
Chubb, T. A., Kreplin, R. W., Friedman, H. (1966)      *J. geophys. Res.*, **71**, 3611–3622.
Cline, T. L., Holt, S. S., Hones, Jr., E. W. (1967)      *Goddard Tech. Report X-611-67-348*, 1965 Cospar Symposium, London.
De Jager, C. (1959)      in *Handbuch der Physik*, Springer-Verlag, Heidelberg, *LII*, p. 205.
Ellison, M. A., McKenna, S. P., Reid, J. H. (1961)      *Dunsink Obs. Publ.*, **1**, 75.
Elwert, G. (1961)      *J. geophys. Res.*, **66**, 391–401.
Friedman, H. (1964)      *AAS-NASA Symposium on the Physics of Solar Flares*, Scientific and Technical Information Division, NASA, Washington, D.C., 147–157.
Hofmann, D. J., Winckler, J. R. (1963)      *J. geophys. Res.*, **68**, 2067–2099.
Kane, S. R., Pfitzer, K. A., Winckler, J. R. (1966)      *University of Minnesota Technical Report CR-87*.
Koch, H. W., Motz, J. W. (1959)      Reviews of *Modern Physics*, **31**, 920–955.
Kundu, M. K. (1965)      *Solar Radio Astronomy*, Interscience Publishers, New York, pp. 530–534.
Lin, R. P., Anderson, K. A. (1967)      *Solar Phys.*, **1**, 446–464.
Moreton, G. E. (1964)      *AAS-NASA Symposium on the Physics of Solar Flares*, Scientific and Technical Information Division, NASA, Washington, D.C., 209–212.
Peterson, L. E., Winckler, J. R. (1959)      *J. geophys. Res.*, **64**, 697–701.
Shklovsky, J. (1964)      *Nature*, **202**, 275.

Shklovsky, I.S. (1965)          *Soviet Astr. – AJ*, **8**, 538–543.
Stein, W.A., Ney, E.P. (1963)          *J. geophys. Res.*, **68**, 65–81.
Takakura, T. (1960)          *Publ. astr. Soc. Japan*, **12**, 326–346.
Takakura, T. (1963)          private communication.
Takakura, T., Kai, K. (1966)          *Publ. astr. Soc. Japan*, **18**, 57–76.
Van Allen, J.A., Krimigis, S.M. (1965)          *J. geophys. Res.*, **70**, 5737–5751.
Vette, J.I., Casal, F.G. (1961)          *Phys. Rev. Lett.*, **6**, 334–336.
Winckler, J.R. (1963)          *AAS-NASA Symposium on the Physics of Solar Flares*, Scientific and
  Technical Information Division, NASA, Washington, D.C., 117–127.
Winckler, J.R., May, T.C., Masley, A.J. (1961)          *J. geophys. Res.*, **66**, 316.
Zheleznyakov, V.V. (1965)          *Soviet Physics – Astronomy*, **9**, 73–76.

Part VII

# PROTON FLARE PROJECT (PFP)

PART VII

PROTOPLAST PRODUCTION

# PROTON FLARE PROJECT

*Introduction and Summary*

*reported by*

Z. ŠVESTKA

*(Astronomical Institute, Ondřejov, Czechoslovakia)*

All contributions presented at the Friday morning session concerned results of the Proton Flare Project (PFP), which was organized by the IAU Commission 10 under the sponsorship of the IQSY Committee, from May 1 to September 30, 1966. As explained by Z. Švestka in his introductory talk, this project had four main aims:

(1) To observe proton flares shortly after the minimum of the solar activity, when, on the rising part of the solar cycle, some proton flares already appear, but not too many of them, so that the individual proton flare phenomena are fairly isolated. This makes it easy to study all the effects of such a proton flare in the interplanetary space and in the Earth's surroundings, and particularly, it allows a detailed study of the isolated active region, in which the proton flare appears.

(2) To get some practical experience in the forecasts of proton flares and verify the reliability and the practical use of them.

(3) With the aid of these forecasts, to give to the solar and geophysical observatories and to the launching sites the possibility to get prepared for the coming event. It was hoped that in this way one might get very detailed observations of the proton-flare active region, of the proton flare itself, and of its effects in the interplanetary space and in the Earth's magnetosphere and ionosphere.

And finally, it has been intended to publish all the results of such a study in one homogeneous series of publications, so that the final result would be a fairly complete picture of the whole proton-flare event, including not only the proton flare itself and its effects, but also a study of the birth and development of the active region in which the proton flare appeared.

The main burden of the organization of this project was carried by Dr. Simon and his co-workers at the Meudon Observatory in France. Dr. Simon served as the chief coordinator of the project and he was also responsible for the forecasts of proton flares and for the collection of the results.

The first PFP proton-flare alert was announced on July 5, for a new solar region, which developed very fast in the Northern solar hemisphere. A SPARMO balloon was launched in evening hours on July 6, and 4 hours later, shortly after midnight, a proton flare actually appeared in the suspected active region. This was a very favourable event from the point of view of the PFP programme, because the proton flare

*Kiepenheuer (ed.), Structure and Development of Solar Active Regions*, 513–535. © *I.A.U.*

occurred in a quiet period as a completely isolated event, and the active region in which this flare appeared was born a few days before in the visible hemisphere of the solar disk. The proton flare was also associated with a small but distinct GLE. Therefore, this event has been selected for a detailed study.

The results of this study will be published in Volume 3 of the *IQSY Annals*, which also will contain, in other two volumes, the Proceedings of the IQSY Symposium, which was held in July in London. Following the wish of many participating scientists, however, two meetings on the PFP results were organized before this publication, the first one during the COSPAR meeting in London, on July 27 and 28, 1967, and the second one in Budapest, during this IAU Symposium, on September 8. The London meeting was mainly concerned with the effects of the proton flare in the solar corona, interplanetary space, and the Earth's surroundings, as measured by the space techniques. In Budapest, on the contrary, the preference has been offered to ground-base observations of the solar active region, in which the proton flare formed.

Since all PFP papers will be published in the *IQSY Annals*, including those presented in Budapest, we do not consider it useful to publish these contributions here in full. Instead, we have decided to publish in these Proceedings only a brief summary of the PFP session in Budapest, to give to the readers general information on the results reported here, and anybody who is interested in them can find the full text of all the contributions in the *IQSY Annals*, Volume 3 (published under the auspices of the IQSY Committee by the MIT Press in 1968).

The programme of the Friday morning session on PFP results concerning the structure and development of the active region which produced the proton flare of July 7, 1966 was as follows (in the following Summary these papers are referred to by their series numbers):

1. A. B. Severny (Crimea): Magnetic fields and proton flares (The evolution of the magnetic field).
2. G. Brückner (Göttingen) and M. Waldmeier (Zürich): Distribution of magnetic fields in photospheric and chromospheric layers and its correlation to the flare event of July 8, $12^h53^m-13^h40^m$ UT. (Presented by E. v. P. Smith.)
3. P. S. McIntosh (Boulder): Birth, evolution and fine structure of proton-flare associated sunspots.
4. G. Newkirk, R. T. Hansen, and S. Hansen (Boulder): Development of the white-light corona in the proton region.
5. A. Krüger (Berlin): Remarks on the S-component of the radio emission.
6. M. J. Martres and M. Pick (Meudon): Summary on the development of the active region, based on papers 1–5, and on the following contributions, which were not presented verbally in Budapest:
   6a. C. Popovici and A. Dimitriu (Bucharest): The H-alpha plage.
   6b. T. Fortini and M. Torrelli (Rome): The calcium plage.

6c. J. L. Leroy (Pic-du-Midi): Photométrie des raies coronales 5303 Å and 6374 Å.

6d. M. N. Gnevyshev (Kislovodsk): Coronal observations.

6e. H. Friedman and R. W. Kreplin (Washington): The slowly varying component of X-ray emission.

6f. H. Tanaka, T. Kakinuma, and S. Enome (Nagoya): The S-component of the radio emission.

7. A. Bruzek (Freiburg): Flares in the active regions.

8. L. Křivský (Ondřejov): Complex study of energy loss of the active region.

9. V. A. Banin (Irkutsk), L. D. de Feiter, and A. D. Fokker (Utrecht): Summary on the activity of the active region, based on papers 7 and 8, and on the following contributions, which were not presented verbally in Budapest:

9a. B. Valníček (Ondřejov), G. Godoli (Catania), and F. Mazzucconi (Arcetri): The West-limb activity in the H-alpha line.

9b. J. L. Leroy (Pic-du-Midi): Photographie en H-alpha, H-beta et $D_3$ de la protubérance active du 9 juillet 1966.

9c. G. Stiber (Saltsjöbaden): Polarization measurements of the July 11th event.

9d. E. Hurtovenko, N. Morozhenko, and A. Rachubovsky (Kiev): Active prominences on July 9 and 11, 1966.

9e. K. Kai (Mitaka) and O. Yudin (Gorky): Radio bursts.

9f. H. Friedman and R. W. Kreplin (Washington): The X-ray emission events preceding the flares.

9g. H. W. Dodson-Prince (Michigan): The behaviour of the active region prior to the proton flare based on $\lambda$ sweep records.

9h. R. R. Fisher and G. R. Mann (Haleakala): Variations in the active region.

9i. A. Caldwell (Culgoora) and M. McCabe (Haleakala): Optical observations of the proton flare.

10. L. B. Demkina, B. A. Ioshpa, E. I. Mogilevsky, and V. N. Obridko (Moscow): Local magnetic field decay.

11. H. W. Dodson-Prince and E. R. Hedeman (Michigan): Late activity of the active region.

Apart from these 11 contributions two more papers were presented during the Friday morning session, by C. Sawyer and J. H. Kinsey. These papers were partly related to the PFP July 7 event, but they discussed other flares as well, and therefore they are published separately from the following summary.

## Summary

On June 25 and 26 at 33 °N a large region, which covered almost 30° in longitude, was passing the East limb of the Sun. It was an old expanding region and the McMath active region No. 8362, which later on produced the proton flare on July 7, was born on June 28 on the Eastern periphery of this area (6b). This is in agreement with the general conclusions by Bumba and Howard (1965) that new active regions form inside or very close to old expanding magnetic fields.

The development of this new active region in the following days also closely resembled the scheme described by Bumba and Howard. The region expanded along the borders of three supergranular cells (6b) and the first spots also formed along

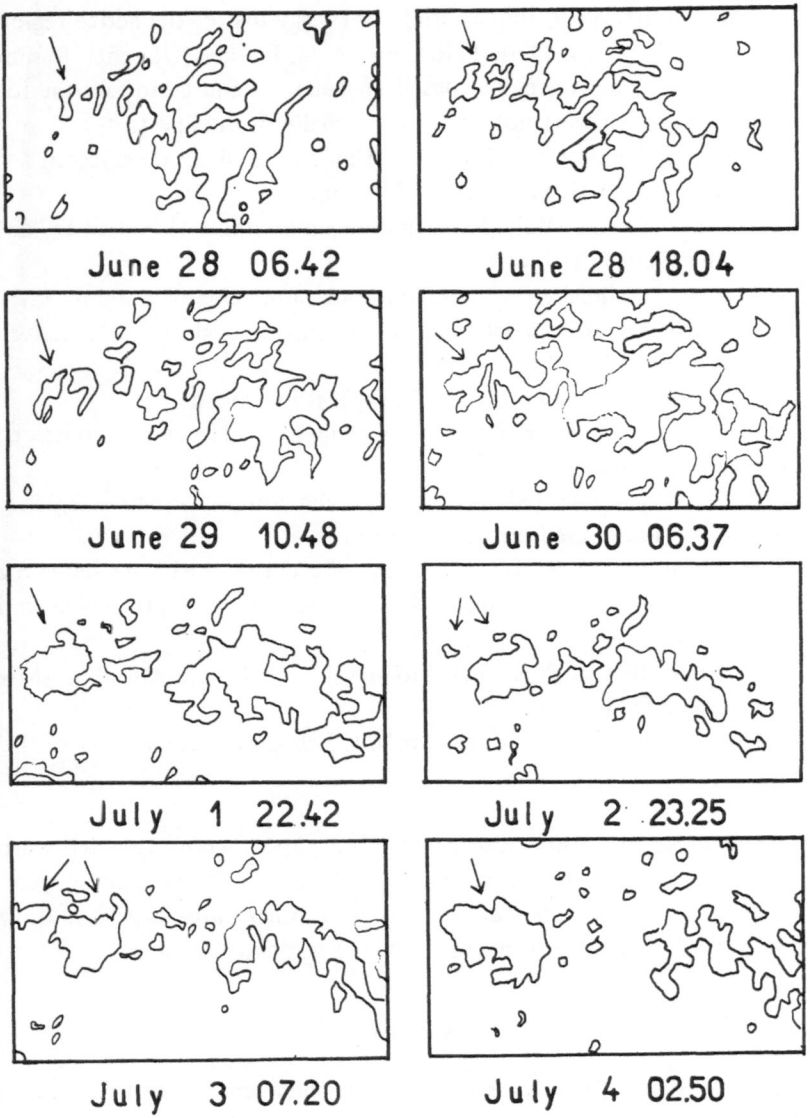

FIG. 1. *Development of the calcium plage during the first days of life of the active region. The newly born active region No. 8362 is marked by an arrow (Fortini and Torrelli, 6b).*

supergranular boundaries (3). The development of the calcium plage is drawn in Figure 1, according to photographs (6b) which the reader can find in the *IQSY Annals*. The increase of the active region was fairly slow until a new luminous grain appeared on July 2, 2° East of the older plage and merged with it on July 4. At about the same time, on July 3 afternoon, the Hα plage also took a compact and elongated form, which developed to its maximum brightness on July 6 (6a).

The first spots (3) were observed on June 30 and all possessed the magnetic polarity of following spots in the Northern hemisphere. The evolution of the sunspot group proceeded with the successive development of two bipolar groupings of spots East of the original spots, and the follower polarity was the first to appear in each of these pairs. McIntosh emphasizes that in both pairs the leader spots were North of the followers, contrary to the normal occurrence. The appearance of the second bipolar pair coincided with the commencement of rapid growth of the entire region on July 3.

The respective positions of these groups were quite remarkable (6). Since their appearance, the sunspots constituted two ranges of inverse polarity, with the neutral line roughly parallel to the equator (Figure 2). Between $14^h$ on July 4 and $8^h$ on July 5, the $\delta$ configuration was built up, characterized by spots of inverse polarity within the same penumbra. Finally, between July 5 and 6, the proton-flare active A configuration (Avignon *et al.*, 1963) was definitely built up. After about $12^h$ UT on July 6 everything seemed to be prepared for a proton-flare occurrence: The sunspot group was formed by two parallel rows of spots of opposite magnetic polarity embedded in a common penumbra, and the Hα plage (6a) began to exhibit two bright ribbons which had a symmetrical disposition with respect to the axis of the group and just covered the penumbra of the two ranges of spots (Figure 3).

These twelve hours preceding the proton flare (which appeared at $0^h26^m$ on July 7) are quite interesting from several points of view. First, it is of interest that during this time the Hα plage already possessed a shape very similar to the shape of the proton flare itself (6a). The penumbra between the large umbrae in the central part of the group developed exceptionally dark and thick filaments parallel to the rows of umbrae (3). Eighteen $\lambda$-sweep records in the Hα line made during this period at McMath-Hulbert Observatory (9g) showed heavy absorption in the active region and in its surroundings and motions of absorbing material closer and closer to the large spot of Northern polarity. There was a continuous production of small flares with increasing area from one subflare to the following one, but no larger flare appeared during this period. Another peculiar characteristic of this phase of development was the occurrence of small bursts at around 10 000 Mc/s, partly associated with the sub-flares, which had no counterpart at lower frequencies, below 8000 Mc/s (9, 9e, 9g). It is noteworthy that the production of these high-frequency bursts was not in the least disturbed by the occurrence of the proton flare itself. A somewhat similar burst of activity was observed in the proton-flare active region which passed the central meridian on July 13·8, 1961, but we hardly know any other previous centres of

Z. ŠVESTKA

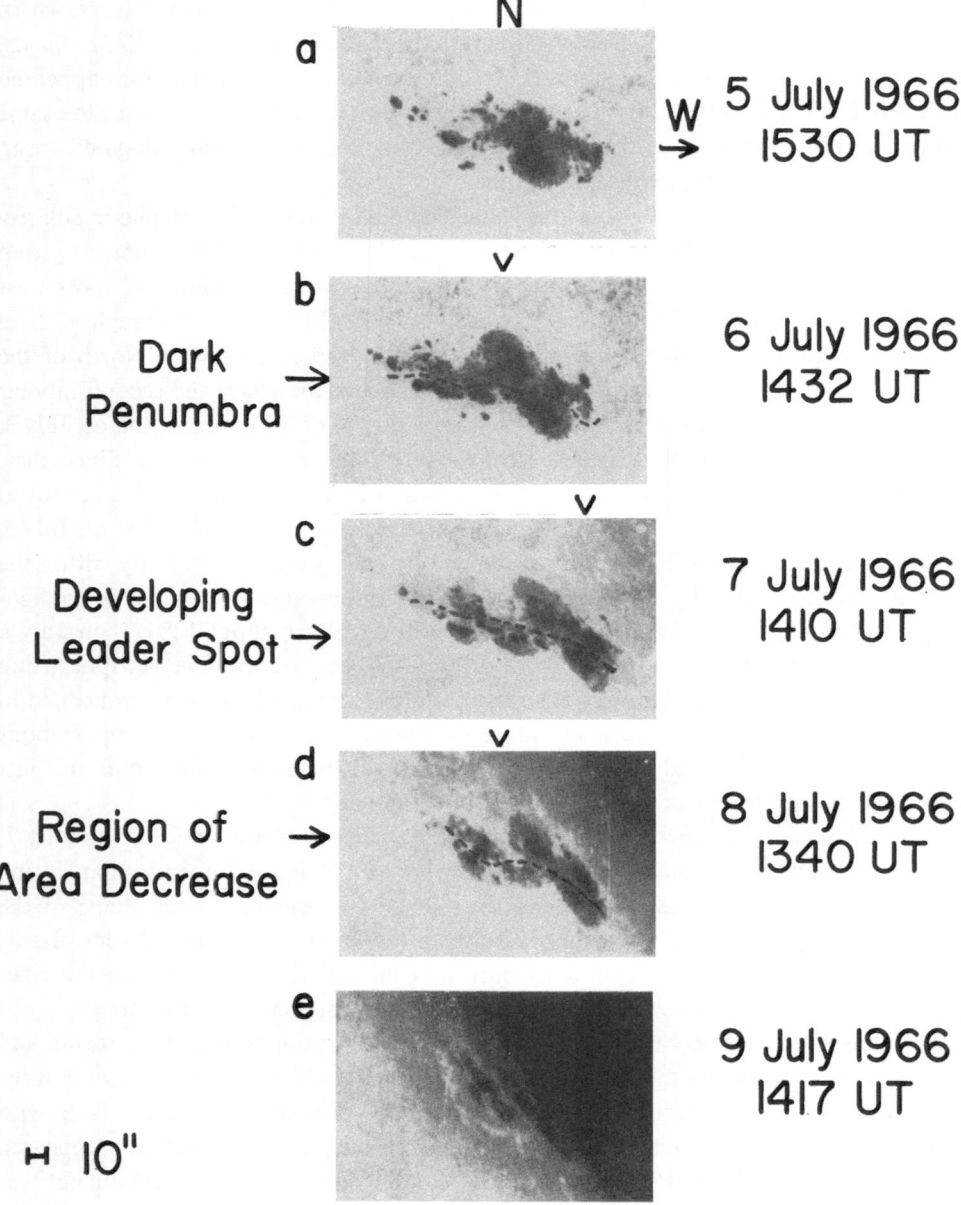

FIG. 2. *Development of the sunspot group, with neutral line of the magnetic field drawn inside the group (McIntosh, 3; Martres and Pick, 6).*

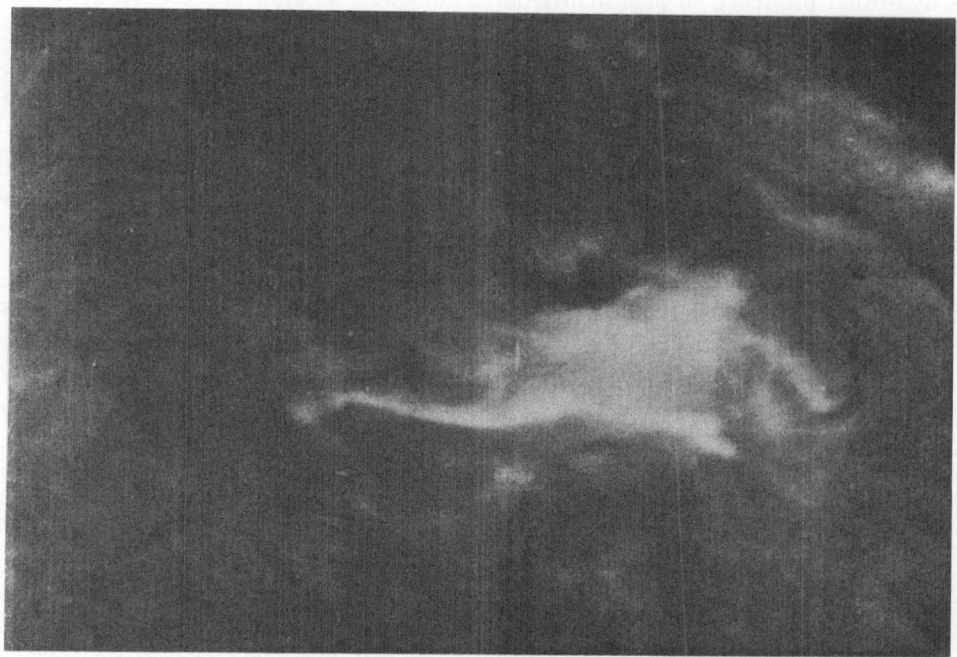

FIG. 3.   *The Hα plage on July 7, at 0ʰ18ᵐ UT, a few minutes before the proton-flare appearance (Banin et al., 9).*

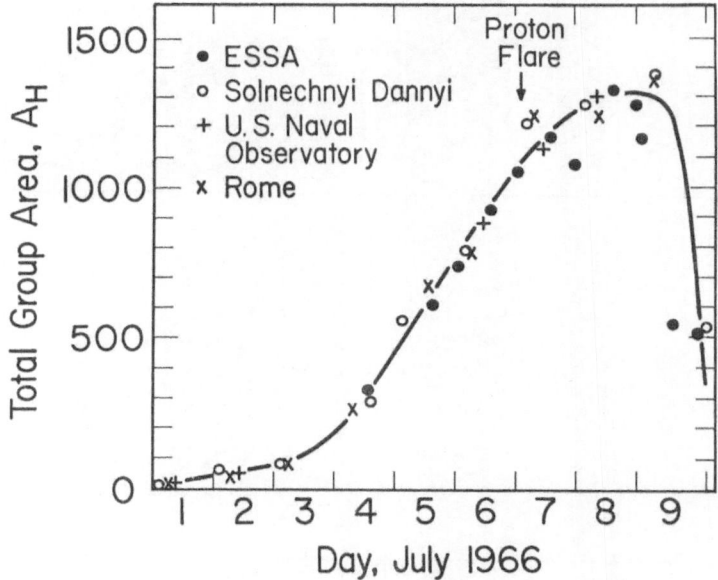

FIG. 4.   *Time variation of the sunspot-group area (McIntosh, 3).*

activity that produced so persistently bursts that were so systematically restricted to very high frequencies (9).

The total area of the sunspot group (3) began to increase at a fast rate since July 3, as one can see from Figure 4. Contrary to some previous observations of proton-flare regions, this increase continued for at least 2 days following the proton flare. However, the large umbrae near the centre of the group, which underlaid the brightest parts of the proton flare, began to decay within half-a-day of the time of the flare, which might have been related to its occurrence (3; Sawyer, 1968).

From July 4 to July 7, 27 records of the magnetic field of the active region were obtained at the Crimean Observatory (1). The comparison of longitudinal-field maps for different days shows that they are very similar (an example is shown in Figure 5): on each map we can identify the same three strong magnetic 'hills', As, Bs, Cs of Southern polarity in the Northern part of the map, and 2–3 $A_N$, $B_N$, $C_N$ not so strong

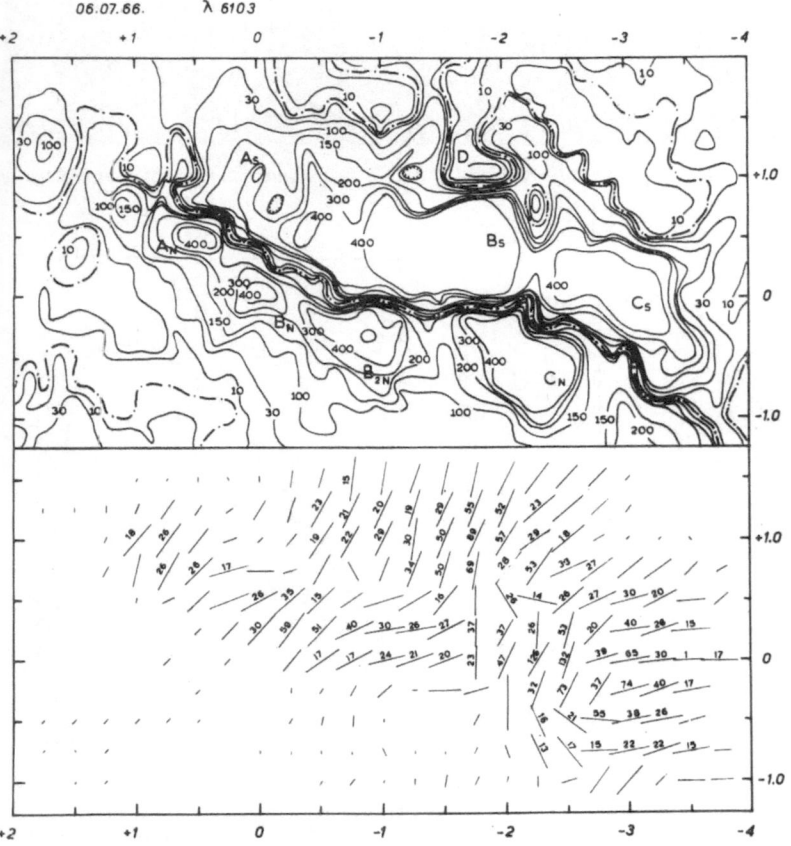

FIG. 5.    *Magnetic map of the longitudinal (above) and transversal field (below) on July 6 (Severny, 1).*

'hills' of Northern polarity located to the South just opposite to the preceding ones.

The total magnetic energy of the active region was increasing from $1-2 \times 10^{32}$ ergs on July 4·3 up to $20 \times 10^{32}$ ergs on July 6·4, and decreased back to about the initial value after the proton-flare occurrence (Figure 6). Also the gradients of the longitudinal magnetic field along the straight lines joining the magnetic hills A, B, C, show the same behaviour, increasing, on an average, from initial values of about 0·1 gauss/km to the peak value of $\sim 1$ gauss/km on July 6·2 and decreasing again to 0·1–0·2 gauss-km on July 7·2. When comparing measurements in $\lambda 5250$ and $\lambda 6103$ lines Severny found the magnetic flux as well as the gradient and the relative increase of these quantities larger in the deeper layer ($\lambda 5250$ Å). The general character of the magnetic field inside the sunspot group was very similar to that one observed in the group which produced the proton flare of July 16, 1959 (Howard and Severny, 1963). Due to the gap in

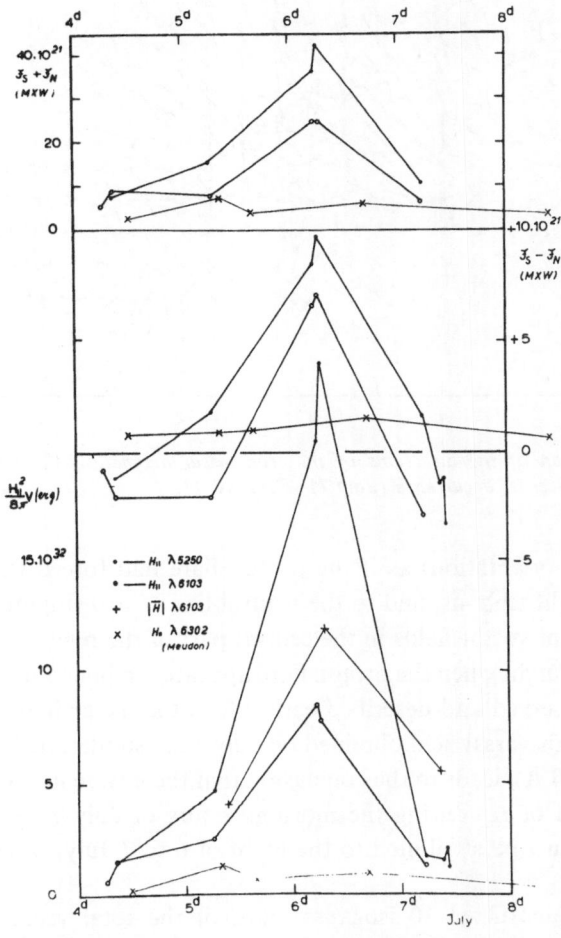

FIG. 6.  *Time variation of the magnetic flux and total magnetic energy in the active region (Severny, 1).*

observations between July 6·4 and 7·2, however, one cannot say whether the decrease of the magnetic flux and of the gradients closely preceded or followed the proton-flare appearance.

Figure 5 also shows a representative map of the transversal field as it looked according to the Crimean measurements, before the proton-flare appearance. Appreciable difference in directions of the transverse field first appeared on the maps obtained on July 7, 5 hours after the flare onset. While the directions had formed roughly something like a cross in the middle of the map before the flare appearance (Figure 5), on July 7 Severny found instead of it a stream of purely horizontal di-

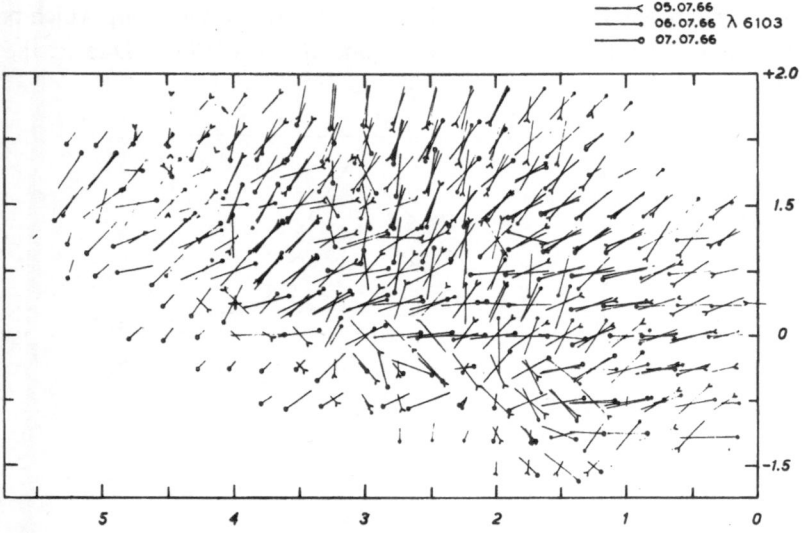

FIG. 7.   *Comparison of the directions of the transversal magnetic field before the proton flare (July 5 and 6) and after its occurrence (July 7) (Severny, 1).*

rections (in E–W orientation) as if the proton flare had forced the directions to be parallel to its bright ribbons, and to the neutral line $H_{\parallel} = 0$ (Figure 7). Thus, we find a rotation by 90° of vector fields in the central part of the region between July 6 and July 7, during the night when the proton flare appeared, a phenomenon which Severny (1964) already observed and described earlier for other solar flares. The fact that all six maps of the transversal field obtained on July 7 are similar, including those for the lowest level ($\lambda 4808$ Å), leads to the conclusion that there were no appreciable changes in the vector field of $H_{\perp}$ during the morning hours of July 7, so that all observed changes indeed must be attributed to the night of 6 to 7 July, when the proton flare appeared.

Severny also constructed 10 isogauss maps of the total vector of the magnetic field $|\mathbf{H}|$ based on observed maps of $H_{\parallel}$ and $H_{\perp}$. Examples of these maps are on

Figure 8 and show the main process of the fission of large magnetic tubes of force into small pieces, the process observed recently by Gopasyuk (1967) for decaying groups. One also can see from this figure that isogauss contours (for a given strength) are broader in $\lambda 4808$ than in $\lambda 5250$ indicating that the magnetic flux and energy at the lower level was higher than at the upper one. Severny concludes from it that the magnetic field of the active region was concentrated at deep layers of the solar atmosphere.

The slowly varying component of the radio emission associated with the proton-flare active region (5, 6f) began to increase on July 3, and the general evolution of the radio flux was quite similar to the development of the sunspot area: The flux was

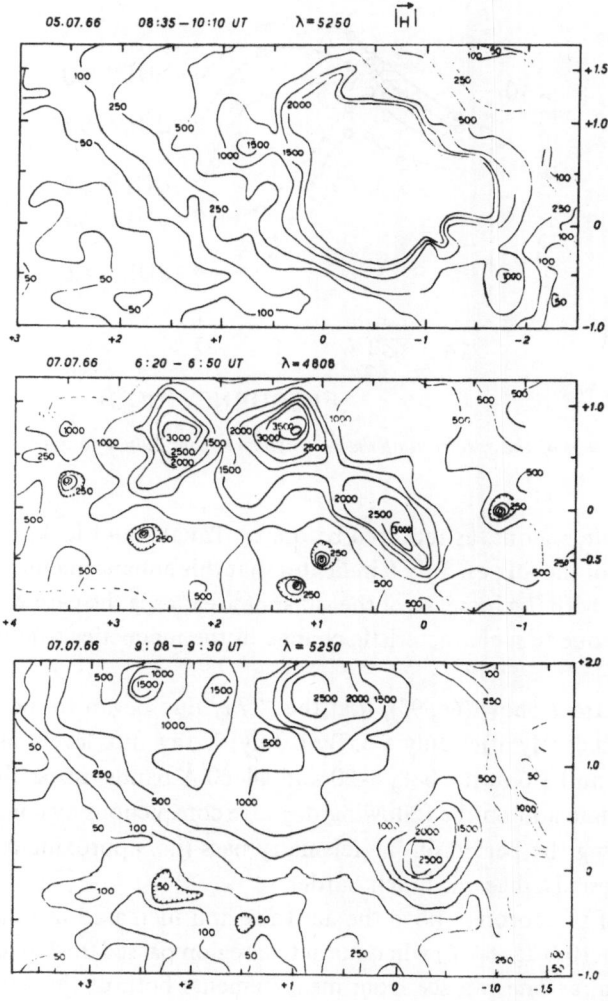

FIG. 8. *Total vector magnetic maps of July 5 and 7 (Severny, 1).*

increasing up to July 8 and rapidly decreased on the following days. During this period, however, the spectral distribution of the radio intensity changed quite substantially, as one can see from Figure 9. Until July 4·5, the flux density decreased with the increasing frequency between 4 and 9·4 GHz, as commonly observed for the majority of the active centres (Swarup *et al.*, 1963). Between July 4·5 and 5·0, however, the flux at 9·4 GHz increased substantially so that the spectrum became relatively flat in the 4–9·4 GHz frequency interval for all the remaining days until July 8. Such a behaviour is in agreement with the generally observed spectral characteristics of

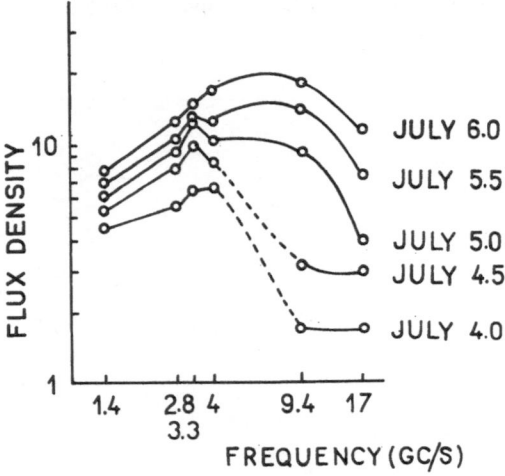

FIG. 9.   *Time variation of the spectrum of the s-component from July 4 to July 8 (Tanaka et al., 6f).*

proton-flare active regions, as reported before by Tanaka and Kakinuma (1964). The sudden increase of the flux on July 5 indicates that this enhancement of 3 cm radiation is not associated with the increase of the sunspot area, as is the case with longer wavelengths, but it is due to a characteristic change of the magnetic structure of the active region (6).

X-ray observations show (6e, 9f), that the X-ray flux began to increase slightly on July 4, and significantly after July 5·5. By 6 July, X-ray flux levels had increased by factors of 15, 5, and 1·6 in the 0–8, 8–20 and 44–60 Å bands, respectively. The X-ray records on this date and on the following days are conspicuous by considerable fluctuations even during the period of one telemetry pass (i.e. approximately 10 min). The emission spectrum also became much harder.

The density of the corona above the active region increased dramatically between June 27 and July 10 (Figure 10) when the active region passed the Eastern and Western limbs of the Sun, as one can see from measurements both of the white-light corona (4) and of coronal line intensities (6c, 6d). Both Gnevyshev and Newkirk *et al.* confirm

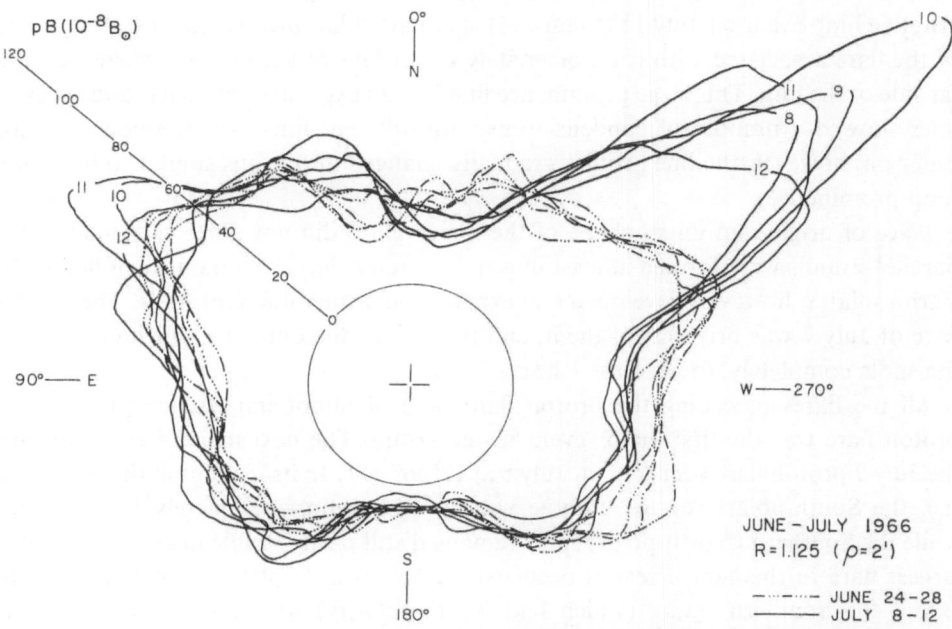

FIG. 10. *A superposition of radial plots of K-coronameter data for the limb passages of the active region on June 24–28 and July 8–12 for a scan height of R = 1·125 R⊙. Dates for the individual traces in July are indicated (Newkirk et al., 4).*

that the maximum of coronal intensity did not coincide exactly with the position of the proton flare, but it was shifted a few degrees to the North.

The deduced coronal electron density on July 10 (i.e., 3 days after the proton event) was only slightly larger than over other active regions within the range of heights above 0·3 solar radii, but it significantly exceeded the density in other active regions in the low coronal layers (4). This observation of a unique, low-elevation coronal condensation (also found for the limb passage on September 5, shortly after the two other PFP proton-flare occurrences) suggests that proton flares eject material into the corona. Newkirk *et al.* think that the expanding series of loop prominences and the expanding condensation represent different aspects of the same phenomenon brought on by the emergence of a magnetic dome from the lower atmosphere.

Flare activity (7) began on July 3 and increased parallel to the growth of the active region. On July 6 through July 9 the flare activity (including subflares) took place during about 50% of the time, but it was relatively low as well as for the importance of the largest flares as for the number of flares other than subflares. No really large event (importance-3 or 4 flares) took place in the visible hemisphere. Three class-2b flares only occurred on the disk, the proton flare of July 7 being the most important of them. A total of 74 subflares, 34 importance-1 flares and three importance-2b flares were observed in the active region from July 3 through July 10. A conspicuous eruptive limb event on July 11 (Figure 11) is of particular interest, but the importance of the flare associated with it unfortunately is not known because it occurred on the far side of the Sun. This large prominence first had an explosive character, and spectral lines showed a number of condensations with different line-of-sight velocities (9d). Later on, however, the line profiles gradually changed into forms similar to those for loop prominences.

Place of origin and initial shape of the major flares did not differ very much, and parallel strands were formed at least in a rudimentary shape in a number of flares (7). Various flares, however, developed and expanded in rather different ways. The proton flare of July 7 was brightest of them, and it also was the only one which covered all the spots completely, for at least 1 hour.

All the flares preceding the proton flare were of minor importance, so that the proton flare was the first major event in the group. The next major flare following the July 7 proton flare occurred on July 8 at $12^h36^m$ UT. In its maximum the Northern (i.e. the South-polar) row of umbrae was covered almost completely by the flare, while the Southern (North-polar) spots remained still partly visible in Hα. The second largest flare in this active region occurred on July 9 at $03^h05^m$ UT. All large spots except the Southern leader (which had North-polarity) were covered by the flare within a few minutes, and since $03^h29^m$ one could observe loop prominences developing from the flare region (see an example in Švestka, 1968).

Comparing the 'non-proton' July 9 flare with the July 7 proton flare we find an appreciable difference in shape and development, which may partly be due to different arrangement of the spots. Bruzek emphasizes that the July 9 flare, surprisingly, looked much more like the proton flares as they were observed in past years than the July 7 flare did. The areas of the two flares were about the same size at maximum brightness. The July 7 flare, however, was much brighter and it expanded much more and for a longer period finally covering all spots in the group (7).

It is well known that loop-prominence systems are closely associated with proton flares (Bruzek, 1964). It is of interest that in this case loop prominences did not accompany the proton flare itself, but another large flare produced by the same active region 2 days later. This seems to confirm Švestka's (1968) conclusion that loop-prominence systems are associated with proton-flare active regions and not directly with proton flares themselves.

The great eruptive prominence of July 11 (Figure 11) was an outstanding phenomenon and its form seems to demonstrate the existence of a complex magnetic-field structure above the active region (9a). At the beginning of its development the prominence was divided in two parts which developed separately. One part developed approximately in a direction parallel to the limb and towards the observer. The other part, which was the main one, developed in a direction inclined at 40° to the limb and its velocity exceeded 300 km/s. The main part of the prominence, however, as

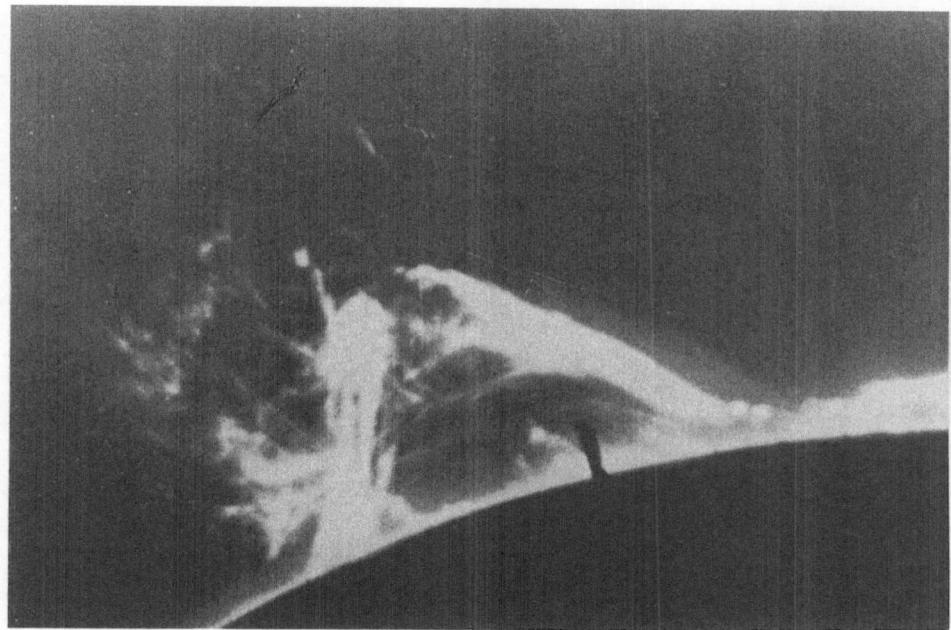

FIG. 11. *The eruptive prominence of July 11, photographed in the Hα line at 9ʰ52ᵐ UT (Valníček et al., 9a).*

well as the hypothetic large flare below it, remained behind the solar limb. Valníček and his co-authors come to the conclusion that we meet here with a twisted prominence, which started in a direction parallel to the plane of the disk and after completing a twist returned to the chromosphere in a direction nearly perpendicular to the initial direction.

In the radio-frequency range, until July 5 only occasional bursts with small intensity were observed (9e). Since 12ʰ UT on July 5 bursts began to occur more frequently and the number of bursts had the maximum at 15ʰ–24ʰ on July 6. As we mentioned above, however, most of them were observed at 3 cm only, without any counterpart at longer wavelengths, though observations were carried out continuously at many frequencies. The number of flares does not show such a maximum. High radio-emission activity continued after the proton-flare appearance and its fall was slower than its rise which, in fact, took 1 day only.

The July 7 proton-flare event was associated with an outstanding type-IV burst. Apart from it, other four significant bursts were produced by the proton-flare region: The next greatest one to the burst associated with the proton event occurred on July 9 at 02ʰ30ᵐ UT and accompanied the large flare which was already described above. The spectral diagram of this burst could also be classified as a type-IV event, its flux

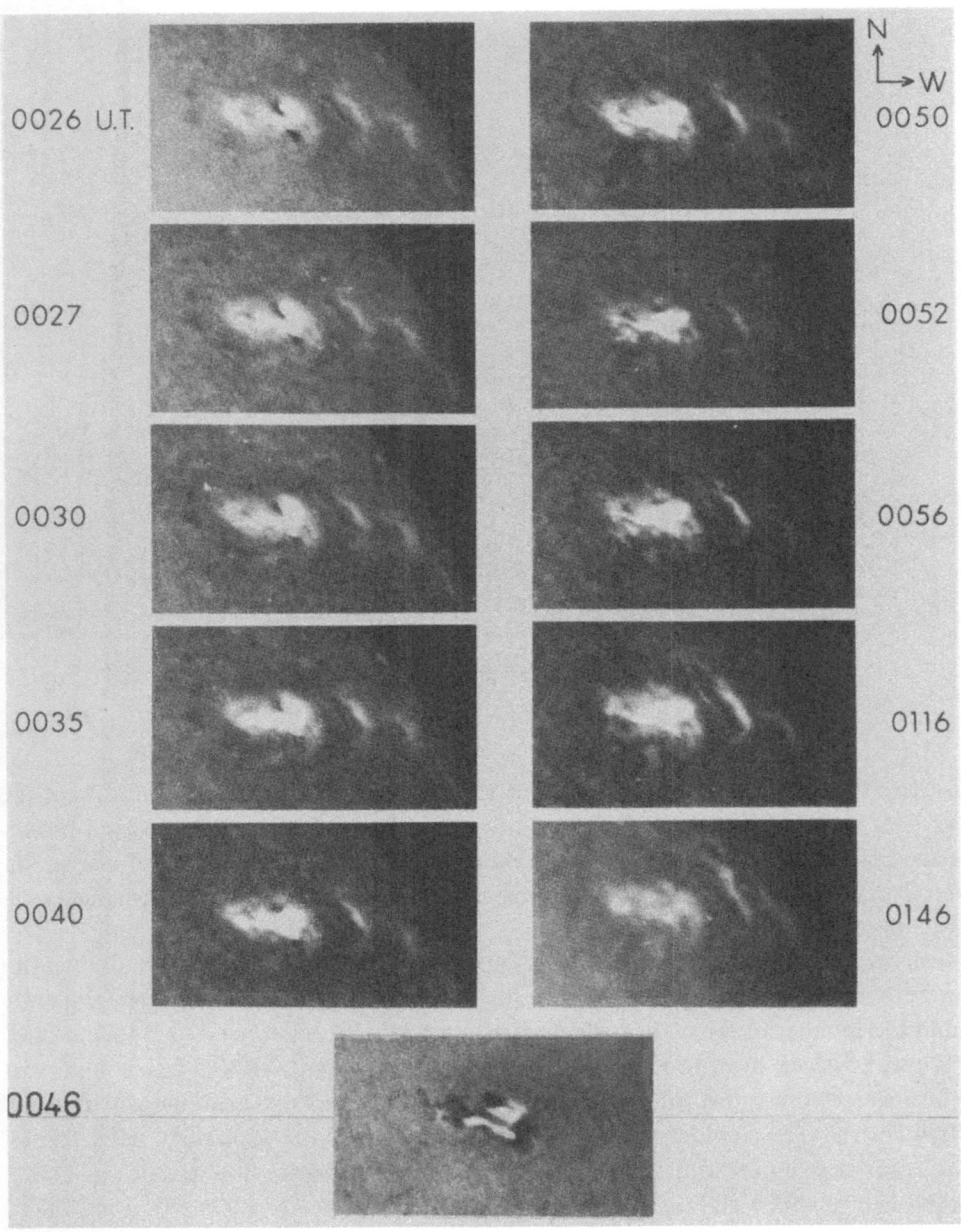

FIG. 12. *Time development of the proton flare in the Hα line (Caldwell and McCabe, 9i). The off-band picture at the bottom taken at $0^h46^m$ UT shows the position of the two bright ribbons amongst the spots (Dodson-Prince, 9g).*

density, however, was much smaller than on July 7. The other greater bursts were associated with two flares of importance 1B and 2B on July 8, and with the large eruptive prominence on July 11.

X-ray bursts also occurred very frequently on July 6. On ten telemetry passes (9f) the X-ray intensity was observed to change within a period of 5–15 min, in some cases dramatically, and in most cases these rapid variations were clearly associated with solar flares from the proton-flare active region. During the proton flare all of the photometer amplifiers were heavily saturated until $01^h57^m$ and then the flare X-ray emission decayed slowly until $9^h$ UT on July 7. During this decay-time no variability of the X-ray flux was recorded, but it started again in the UT afternoon on July 7.

Křivský (8) constructed a summation curve of all SID effects produced by the proton-flare region from values $I \times D$ (where $I$ is the SID-event importance and $D$ its duration) and demonstrated the slope of this summation curve as a characteristic of the time distribution in the 'energy loss' of the active region. The slope started to increase on July 3 and changed remarkably in the UT morning hours of July 6. Since that time the slope remained fairly constant until July 10. This again confirms that the character of the activity in the region changed substantially on July 6 after the A-configuration of the sunspots had been built. Martres and Pick (6) distinguish two phases in the development of the active region before the proton-flare occurrence: The first phase leads to the formation of the appropriate structure, whereas the second phase, which begins somewhere between the 5th and 6th of July is the elaboration of the proton flare itself.

The proton flare itself (Figure 12) occurred as a flare event of importance 2B at $0^h26^m$ UT on July 7, when the active region was about 50° West from the central meridian. The heliographic coordinates of the flare were 35N 47W. It commenced as two small bright areas adjacent to the larger spots and the areas expanded rapidly within the plage (9i). Finally, the shape of the flare kept the form of two, parallel, bright filaments stretched along the spot rows, i.e. the typical proton-flare formation discovered by Ellison et al. (1961). The Southern flare filament was longer than the Northern one and emission covered almost completely the umbrae (6, 6i). The structure of the flare and its relationship to the sunspots is seen more clearly in sketches made from off-band pictures (Figure 13). No flare nimbus nor loop prominences could be detected in association with the proton flare (6i).

During the lifetime of the flare an activation of filaments and of sympathetic flares was observed (6, 6i). Two filaments to the East of the active region disappeared abruptly and the large dark filament in this region disappeared gradually. In the same active region or close to it, two flares appeared during the life of the flare, one at $0^h31^m$ (position 35N 62W) and the other one at $0^h50^m$ (position 35N 56W).

Figure 14 shows a superposition of the area occupied by the proton flare, on the combined magnetic map containing the main 'hills' of the longitudinal field and the directions of the transversal one as recorded by Severny (1) on July 7 between $5^h$ and

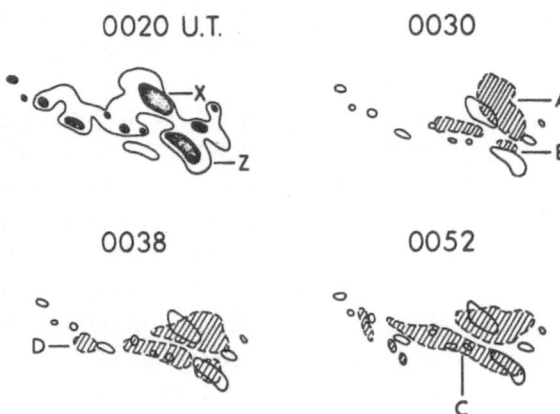

FIG. 13.  *Sketches of the flare made from Hα off-band pictures (Caldwell and McCabe, 9i).*

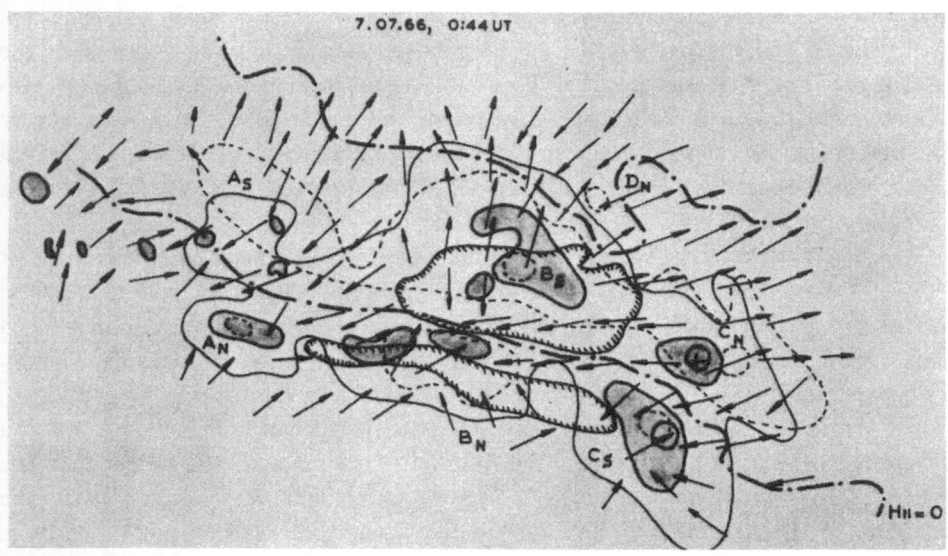

FIG. 14.  *Superposition of the proton flare (contoured area) on the maps of the longitudinal and transversal magnetic field. $H_\parallel = 0$ is the neutral line and the dark areas represent sunspots (Severny, 1).*

$6^h$ UT. One can clearly see that the two bright ribbons of the proton flare appeared simultaneously in regions of opposite magnetic polarity. One of the flare areas is just to the North of the neutral line $H_\parallel = 0$ and in contact with it, and the other ribbon is about 8″ to the South from the neutral line. This distribution of the flaring areas is in full agreement with the recent results obtained by Moreton and Severny (1968) for the very active group of September 17–26, 1963.

Both ribbons were parallel to the neutral line. The flare as a whole appeared in the region of crossing or bifurcation of directions of the transverse field, which again is in agreement with earlier results of Moreton and Severny (1968) and Severny (1964). Severny points out that particular interest is given by the position of the flare on a map showing distribution of vertical electric currents $j_z$ calculated with the aid of observed data on $H_{\parallel}$ and $H_{\perp}$ from the relation

$$\mathbf{j} = \frac{c}{4\pi} \cdot \operatorname{rot} \mathbf{H}.$$

Examples of these maps for July 6 and 7 are presented in Figure 15. The demarcation line between positive and negative currents is parallel to the line $H_{\parallel} = 0$ in the middle of the region and sometimes coincides with this line, so that, taking also into account the directions of $H_{\perp}$, it is not excluded that electric currents are connecting the magnetic regions of opposite polarity and form a pattern similar to the pattern of magnetic lines of force. Severny's maps show clearly that both areas of the flare were just above places with the strongest electric currents, in accordance with the recent results of Moreton and Severny (1968). Severny thinks that this gives support to the Alfvén and Carlqvist (1967) theory of flares as interruptions in electric-current filaments.

The distribution of magnetic fields in photospheric and chromospheric layers after the proton-flare appearance was also studied by Brückner and Waldmeier (2). They used $\lambda 5250$ and H$\alpha$ lines, and their results are generally in agreement with Severny's conclusions. Their measurement was carried out on July 7, at $12^h00^m$ UT, and they found tremendous differences of the photospheric and chromospheric field strength. The H$\alpha$ fields strike only 80 gauss, while the photospheric fields go up to 2000 gauss. Some parts of the region, particularly in the centre of the group and in parts of the two large preceding spots, also showed opposite polarity of the H$\alpha$ and $\lambda 5250$ fields. In the centre of the group, the third polarity had a tendency to join the magnetic-field regions of the same polarity lacing the opposite polarity region. The authors tried to superpose the flare of July 8 on their magnetic map. Even when the flare occurred only 1 day later, the flare bright knots could be identified with the largest field gradients in the neighbourhood of the large spots.

The active region that produced the proton flare of July 7, continued through at least two subsequent rotations (11). In the first of these, in late July, spot area and radio emission were greatly diminished but the calcium plage had increased in area by 50%. Flares continued to occur in the region and the major flare of July 28, at $22^h16^m$ UT is of special interest. Its importance was 2B or greater, and again, as the major flares of July 7, 8, and 9, it consisted primarily of two bright ribbons. It differed, however, from these flares in this respect that the H$\alpha$-flare emission was far from all spots. Nevertheless, the flare was associated with an enhancement of radio emission for more than 2 hours, most intense at lower frequencies, and it also produced a

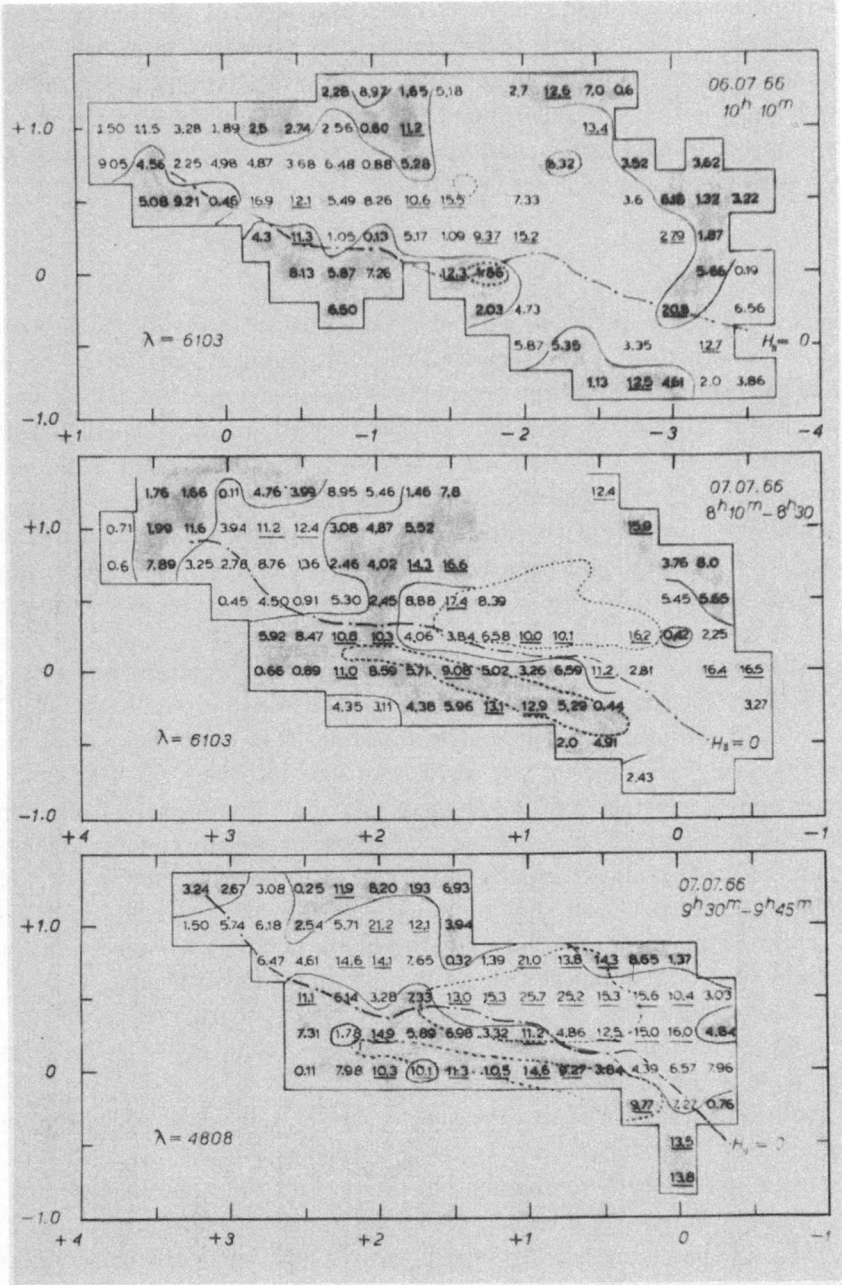

FIG. 15.   *Maps showing the distribution of vertical electric currents $j_z$. Black and white areas are regions of oppositely directed current density (Severny, 1).*

strong X-ray emission, of about one half of the intensity of the X-ray enhancement associated with the July 7 proton-flare event. It is clear that even in late July, the region was still capable of producing a major flare.

In the August rotation, as Dodson and Hedeman point out, the post-proton flare region of July, at latitude N33 through differential rotation became colongitudinal at 182°, with a previously following region in latitude N22. This region at lower latitude produced major proton flares on August 28 and September 2, during the second active PFP period.

Demkina *et al.* (10) investigated the magnetic-field decay of the active region after the proton-flare appearance. While before the July 7 flare an approximate equality of magnetic fluxes had been conserved in the sunspot group, after the proton flare a sudden growth of the Southern-polarity magnetic flux and considerable decrease of the Northern-polarity flux was observed. After that the active region came to the opposite invisible side of the solar disk, but one can suppose that this run of development continued, since in the next rotation in late July the group looked like a relatively stable unipolar spot of Southern polarity, with the magnetic class $\alpha p$. It may be of interest to note that the area of the remaining spot was near to the area of a supergranule at this time (10). On 6 days during this late July transit, small ephemeral spots of the opposite polarity were observed following the large spot making the spot group on those days of magnetic class $\beta p$ (11). Magnetographic measurements at Mount Wilson showed that the extensive and relatively bright plage associated with the sunspot group was bipolar. On the third rotation in late August only a small $\alpha$-type spot without penumbra was remaining in the active region, and the calcium plage, though greatly fragmented and reduced in intensity, was still a detectable feature (11).

At this time, however, the activity already was shifted to the second proton-flare region, located at lower latitude in the close vicinity of the active region discussed in this summary. Obviously, both these regions appeared in one complex of activity, which dominated solar activity during the entire second half of 1966 (11). Already in the late July rotation, the activity of the studied region was observed in its tail part situated close to this neighbouring active region, which fully developed only during the next rotation in August (10). Demkina *et al.* suggest that this subsequent rapid development of this neighbouring group, which was a fairly inactive small group during the three previous rotations, might have been stimulated by the magnetic field of the decaying high-latitudinal group which produced the proton flare on July 7.

## References

Alfven, H., Carlqvist, P. (1967)  *Solar Phys.*, **1**, 220.
Avignon, Y., Martres, M.J., Pick, M. (1963)  *C.R. Ac. Sc.*, **256**, 2112.
Bruzek, A. (1964)  *Astrophys. J.*, **140**, 746.
Bumba, V., Howard, R. (1965)  *Astrophys. J.*, **141**, 1492.
Ellison, M.A., McKenna, M.P., Reid, J.H. (1961)  *Publ. Dunsink Obs.*, **1**, 53.

Gopasyuk, S. (1967)        *Izv. Krym. astrofiz. Obs.*, **36**, 56.
Howard, R., Severny, A.B. (1963)        *Astrophys. J.*, **137**, 1242.
Moreton, G., Severny, A.B. (1968)        *Solar Phys.*, **3**, 282.
Sawyer, C. (1968)        in the present volume, p. 543.
Severny, A.B. (1964)        *Izv. Krym. astrofiz. Obs.*, **31**, 159.
Švestka, Z. (1968)        in the present volume, p. 287.
Swarup, G. *et al.* (1963)        *Astrophys. J.*, **137**, 1251.
Tanaka, T., Kakinuma, T. (1964)        *Rep. Ionosph. Space Res. Japan*, **18**, 32.

# DISCUSSION

*Fokker:* For centres of activity far from the centre of the solar disk (i.e. relatively close to the limb) the neutral line, as it is observed, does not, in principle, correspond with the line at which the magnetic field is parallel to the solar surface. I should like to ask, how large the difference in position between the observed neutral line and the true line of horizontal magnetic field can be. At what distance to the centre of the disk does the difference become important?

*Severny:* My experience shows that the best is simply to avoid recording the regions very near to the limb. But practically we should not have essential differences due to projection effect if region is not more far than 60°–70°. It can be checked by comparing Hα m.f. records with those in λ5250 to show whether we have effects of such a kind or not.

*McIntosh:* Did I interpret your slide correctly that the dissolution of the magnetic field into smaller parts occurred only after the proton flare?

*Severny:* Observations were made a day apart and were not near time of proton flare. I cannot say exactly when the dissolution occurred.

*Švestka:* I would like to point out that there were made only two measurements, widely apart, on July 6 and shortly after the flare on July 7. The magnetic energy and the gradients were high on July 6 and much decreased on July 7. But one cannot decide whether this decrease occurred before or after the proton-flare occurrence.

*Severny:* There are examples in the past, according to the Crimean measurements, that the decrease already occurred before the proton-flare appearance.

*Krat:* At what place in the profile of Hα were set the slits of the magnetograph when Brückner and Waldmeier measured the magnetic field?

*Wiehr:* The Hα-magnetograms were taken with the magnetograph similar to that designed by Babcock. The two exit slits covered the region from the line centre to ±0·8 Å.

*Krat:* Then in fact the magnetograms were not measured in the chromosphere but in the photosphere.

*Smith:* I do not agree that even that part of the Hα-line profile is formed in the photosphere. It is almost entirely from the chromosphere.

*Krat:* I think that due to superposition of emission and absorption in the central part of Hα at every place on the solar disk and especially in flares no reliable values of the magnetic-field strength can be obtained in this way.

*Severny:* As far as I can guess from the private talk with Dr. Brückner our results relating to the magnetic field in the proton-flare region of July 7 are in general agreement. But frankly speaking I do not believe that measurements of magnetic fields in Hα we are doing in Crimea as well as that of Dr. Brückner are reliable because of the very strong influence of emission on measurements of magnetic fields in the case of active regions filled up with the emission from plages and flares.

*Newkirk:* The features which McIntosh mentions as occurring in the proton-flare regions may well be characteristic of *every* region of intense activity. Have you any evidence that such features are unique to proton-flare regions?

*McIntosh:* I have not examined enough material to give a good answer to your question. I have looked at two other proton regions photographed at Mt. Wilson and they also show similar features. A number of sunspot drawings also indicate similar features. I do not recall ever seeing such features in non-proton regions.

*Neupert:* Could the observations of an electron-density component low in the corona mentioned by Newkirk be associated with the possible existence of a permanent or sporadic condensation?

*Newkirk:* The resolution of the K coronameter (1 min of arc) is not sufficient to define the size of this component accurately. However, it appears to be larger than the typical permanent condensation as defined by Waldmeier and is comparable in size to what is considered to be the 'active region enhancement'.

*Sawyer:* Examining the sunspot drawings with magnetic-field measurements made at various observatories, I was struck by a single spot umbra with rather strong field with both polarities present. This umbra was on the equator side of the group, just East of the large leading spot. On July 6, several different observatories recorded both polarities in this umbra, and I expected to hear some discussion of this situation at these meetings.

*Smith:* Did I understand Dr. Severny to say that longitudinal neutral lines through the umbrae of sunspots are often observed, even for non-proton flare regions?

*Severny:* Oh yes, there are many such cases. I remember the maps for sunspot groups in September 1961, May 1962, September 1963 and others having this feature. It is quite a common occurrence. But note that this only refers to the longitudinal fields, the transverse fields may be strong and complex and the occurrence of line $H_{\parallel} = 0$ inside umbrae can simply mean that we have inclusion of strong transverse field there.

*Öhman:* With reference to our paper describing the limb event of July 11, 1966, I want to add, that Stiber has completed now his discussion of the material selected for the purpose of measuring the polarization of the continuous spectrum. The polarization is found to be somewhat smaller than that predicted by the theory of electron scattering, particularly near the limb. This suggests that also other mechanisms produce the continuous spectrum. From a discussion of his intensity and polarization measurements Stiber finds an electron density of $3 \times 10^{11}$ per cm$^3$.

*Severny:* We should be extremely careful with conclusions obtained with a magnetograph for July 8 and the following days, because the active region considered was very near to the limb, and effect of projection could possibly produce the apparent decay. The conclusion of decay contradicts also the observation that area proceeded to increase after proton flare, and moreover an important flare appeared also on July 11 at the very limb.

# OBSERVATIONS OF THE SOLAR PROTON EVENT OF AUGUST 28, 1966*

J. H. KINSEY** and F. B. McDONALD

*(NASA/Goddard Space Flight Center, Greenbelt, Md., U.S.A.)*

### ABSTRACT

The CsI scintillator telescope aboard IMP-III indicated a peak intensity of 6·25 proton/sec cm² ster for a SPE (Solar Proton Event) commencing during the period 1600 to 1700 UT on August 28, 1966. This SPE appears to have been associated with the importance 2–3 flare occurring at 22N/06S at 1522 UT on this date. The comparison of this event to other large proton events occurring during the period May 25, 1965 to September 30, 1966, is made.

## 1. Experiment

The data for this analysis was supplied by the GSFC medium-energy scintillator cosmic-ray telescope, aboard IMP-III. This device, shown schematically in Figure 1, was also flown on IMP-I, IMP-II, and the recently launched IMP-IV, thus providing the long-term monitoring of H and He in the range of 16–80 MeV/nucleon. A brief description of this telescope is given by Balasubrahmanyan *et al.* (1966) and Bryant *et al.* (1962).

Briefly, the telescope consists of a 1-mm CsI $\Delta E$ detector 5 cm in diameter, $A$, separated by 10 cm from a 2-cm CsI E-$\Delta E$ detector $B$ of the same diameter. The latter is surrounded by an anti-coincidence counter of Pilot-B scintillator plastic, $C$. Three photomultiplier tubes look at the light output of each of these detectors. The thin detector measures the energy loss $\Delta E$ of a particle completely penetrating it, while the thick detector has a light output proportional to the amount of energy E-$\Delta E$ deposited by a particle stopping in it after passing through the thin detector. It is readily seen that the desired coincidence for determining the charge-mass ratio of the incident particle is $AB\bar{C}$, i.e., the particle must penetrate $A$ and $B$ but not pass into C.

The total counts obeying this coincidence are sampled for 40 sec every 400 sec data cycle and stored in an accumulator. Six times during this data cycle the energy of a particle satisfying the $AB\bar{C}$ coincidence is sampled and analyzed by two 512-channel differential pulse-height analysers, one each for the $\Delta E$ and E-$\Delta E$ crystals. A count from these two PHA's may be considered to occupy a discrete position on a $\Delta E$ versus E-$\Delta E$ plot, as shown in Figure 2. Because of their unique charge-to-mass ratio,

---

* Presented by J. H. Kinsey.
** Grad. Res. Asst. University of Maryland.

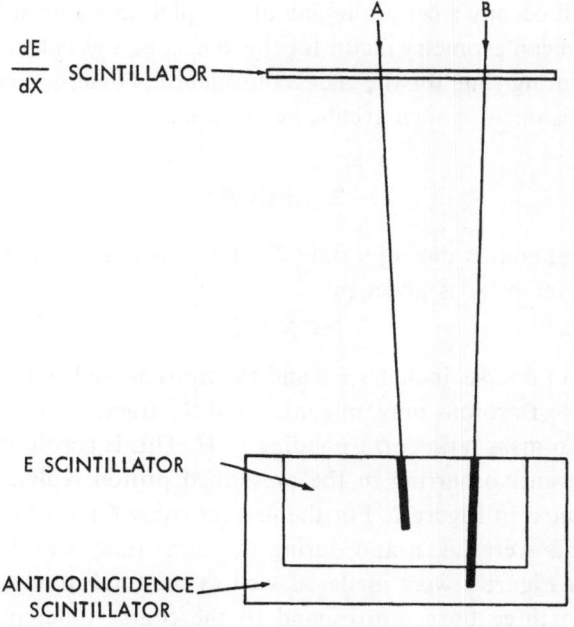

FIG. 1.    *Schematic of IMP-III medium-energy cosmic-ray telescope.*

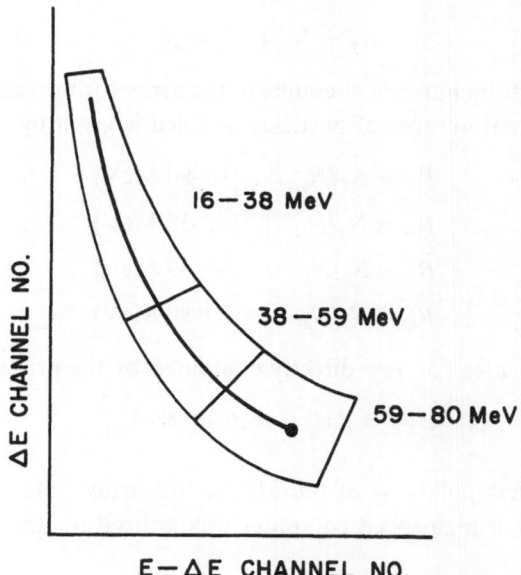

FIG. 2.    *Schematic of ΔE vs. E–ΔE array with proton curve and differential energy-counting regions shown.*

proton counts will occupy a certain region of this plot, as signified by the solid curve in Figure 2. The mean geometry factor for this telescope was calculated to be 3·4 cm² ster, and the counting time for the $AB\bar{C}$ coincidence is determined by the time that the accumulator is open to such events, i.e. ∼40 sec.

## 2. Analysis

If $N_c$ is the total counts during a time $T$, and $G$ is the geometry factor then the incident particle intensity is given by

$$I = N_c/TG. \tag{1}$$

This intensity, of course, includes He and electrons as well as H. In order to determine the intensity of protons only, one must find the fraction of the total counts that have the charge-to-mass ratio corresponding to H. This is accomplished by counting the number of events occurring in the prescribed proton region in the $\Delta E$ versus $E-\Delta E$ array depicted in Figure 2. For the present study 6 hour averages of the $AB\bar{C}$ coincidence counts were taken and during the same time period the counts in the 'boxes' shown in Figure 2 were made, as well as the total number of counts in the entire array. The three boxes correspond to the coarse differential energy ranges 16–38 MeV, 38–59 MeV, and 59–80 MeV.

If $N_A$, $N_B$ and $N_C$ are the number of counts in each of the boxes in ascending energy, respectively, then the total number of protons counted between 16 and 80 MeV is

$$N_T = N_A + N_B + N_C. \tag{2}$$

Now if $N_M$ is the total number of counts in the array for the same period, the ratios of the protons to total number of particles sampled is given by

$$R_T = N_T/N_M \quad \text{(16–80 MeV)} \tag{3}$$

$$R_A = N_A/N_M \quad \text{(16–38 MeV)} \tag{4}$$

$$R_B = N_B/N_M \quad \text{(38–59 MeV)} \tag{5}$$

$$R_C = N_C/N_M \quad \text{(59–80 MeV)}. \tag{6}$$

The proton intensities are now directly computed by the products,

$$J_T = R_T I \quad \text{(16–80 MeV)}, \tag{7}$$

etc.

Because of the manipulations of the 512 by 512 arrays and the large number of related calculations, a high-speed computer was utilized for these computations.

## Results

The solar proton event (SPE) of August 28, 1966 appears from the above calculations to have a double-peaked structure, the two peaks occurring at 2100 on August 28 and 0300 on August 30, as shown in Figure 3. Recalling that the time resolution of this analysis is 6 hours, the commencement times of these two events are at 1500 (August 28) and 2100 (August 29). The peak-proton intensities for $16 \leqslant E \leqslant 80$ MeV are 6·25 and 4·2 protons/sec cm² ster, respectively. It is noted (ESSA, 1966) that there was a 3B-importance flare at 1522 on August 28 and a 1N importance flare at 2114 on August 29, as well as the observation of type-IV radio emission at 1547 and 2032 on these two dates, respectively, both of intensity 2+. There is not enough information at present to determine whether the second peak is the result of another flare, or if it represents the arrival of energetic storm particles.

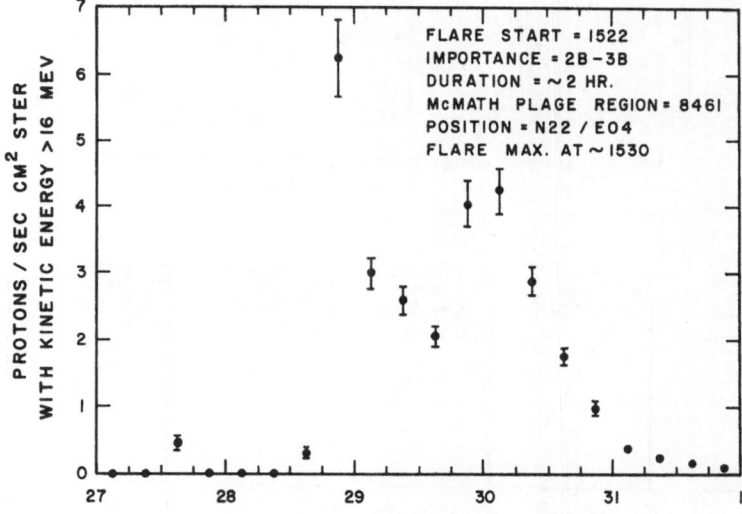

FIG. 3.  *Time history of the SPE of August 28, 1966.*

In Figure 4 are shown linear plots of the intensity versus energy for the three differential energy ranges, giving a crude but simple picture of the spectrum of these particles. From the steep slope of these curves, one can conclude that to within a few percent the peak intensities can be interpreted as being correct for all energies $\geqslant 16$ MeV for the solar protons.

In Figure 5 we show the relative differential intensities as a semilog plot of the exponential rigidity as given by

$$\frac{dJ}{dR} = \left(\frac{dJ}{dR}\right)_0 \exp(-R/R_0). \tag{8}$$

The characteristic differential intensity $(dJ/dR)_0$ and characteristic rigidity $R_0$ as yet have no precisely defined physical meaning, but serve as a basis for comparing different events as suggested by Rinehart (1967). For this particular event the mean

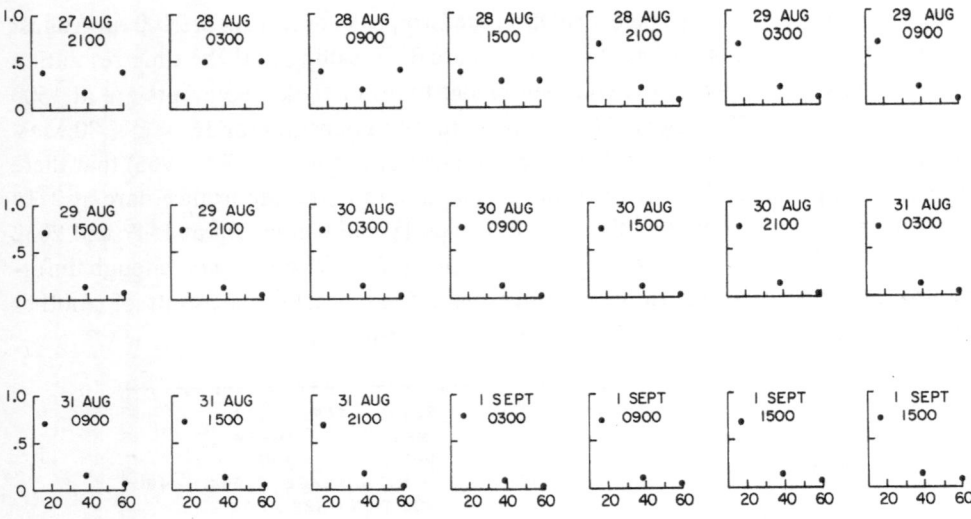

FIG. 4.    *Six-hour energy spectra of the August 28, 1966 SPE.*

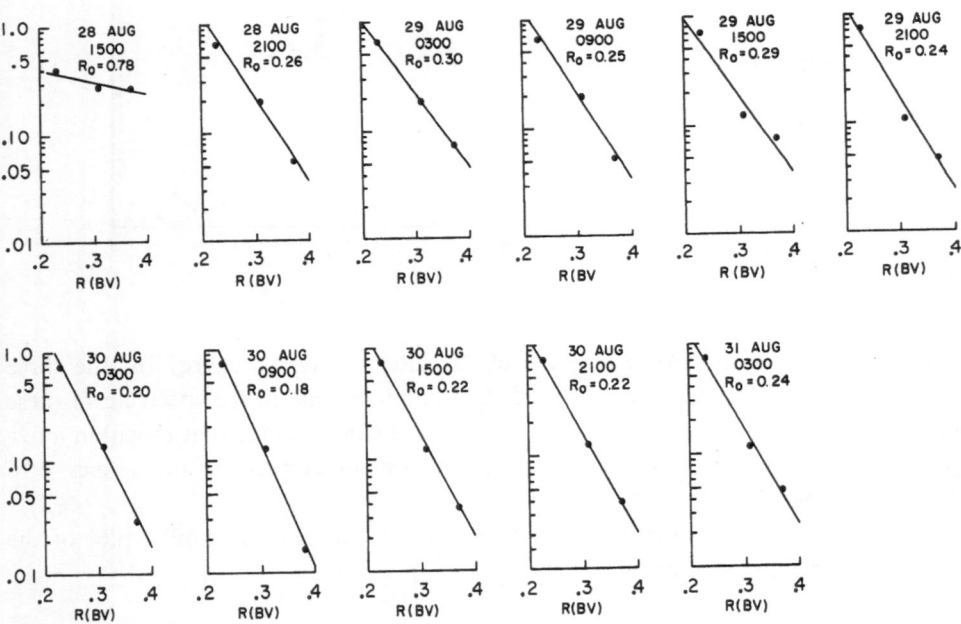

FIG. 5.    *Six-hour exponential rigidity spectra of the August 28, 1966 SPE.*

value of $R_0$ is $\sim 0.25$. This value is comparable to values Rinehart has computed for SPE's of several years ago.

### 3. Correlation with other SPE's and Related Phenomena

Fichtel and McDonald (1967) have suggested that an arbitrary criterion of 0·5 protons/sec cm² ster is a good value for the lower limit of the peak intensity of these proton events for correlation with optical flares and type-IV radio emission. This seems to be generally the case for the events presented in Table 1. Herein are tabulated all events appearing above the mean-background level of 0·007 protons/sec cm² ster. Except for the events of October 4, 1965, March 24, 1966, and the complex event starting on September 2, 1966, this does seem to be the case. To our knowledge there are some geographic gaps in the radio spectrographic average of the Sun during these particular periods, or at least we were unable to find records of type-IV emission at those times.

This survey of IMP medium-energy detector data is being extended to give a catalog of proton events during most of the period of minimum solar activity to be published.

### Table 1

**Summary of abrupt prolonged proton-intensity increases above mean background[a] for energies > 16 MeV for IMP-C, May 29, 1965 to September 30, 1966**

| Date | Time[b] | Peak Intensity (proton/sec cm²ster) | Optical Flare Time | Impor. | Location | Type IV Time | Int. |
|------|---------|-------------------------------------|------|--------|----------|------|------|
| 4 Oct. 1965 | 0900 UT | 2·6 | 0935 | 2 | S21/W31 | | |
| 19 Jan. 1966 | 0900 | 0·015 | | | | | |
| 24 Mar. 1966 | 0300 | 4·4 | 0225 | 3B | N18/W37 | | |
| 29 Apr. 1966 | 0900 | 0·055 | | | | | |
| 2 May. 1966 | 2100 | 0·15 | 0808 | 2N | N16/E53 | 1216 | 1 − |
| 25 Jun. 1966 | 1500 | 0·11 | 1523 | 1B | S25/W09 | 1607 | 2 |
| 4 Jul. 1966 | 2100 | 0·045 | | | | | |
| 7 Jul. 1966 | 0300 | 23·0 | 0022 | 2B | N35/W45 | 0053 | 3 |
| 13 Jul. 1966 | 1500 | 0·12 | 1625 | 1N | N22/E90 | | |
| 16 Jul. 1966 | 2100 | 0·21 | | | | | |
| 30 Jul. 1966 | 2100 | 0·028 | | | | 2330 28 Jul. | 3 |
| 28 Aug. 1966 | 1500 | 6·25 | 1522 | 3B | N23/E04 | 1547 | 2 + |
| 29 Aug. 1966 | 2100 | 4·2 | 2114 | 1N | N07/W71 | 2032 | 2 + |
| 2 Sept. 1966 | 0900 | > 38·0[c] | 0541 | 2B | N22/W57 | | |
| 14 Sept. 1966 | 0300 | 0·88 | | | | | |
| 20 Sept. 1966 | 1500 | 0·087 | 1738 | 2B | N03/W15 | | |
| 26 Sept. 1966 | 0300 | 0·053 | | | | 1312 | 1 − |
| 27 Sept. 1966 | 1500 | 0·14 | | | | 1313 | 2 |

[a] Mean background level ∼0·007 proton/sec cm²ster.
[b] Count averaging time is 6 hours, hence time given represents period from 3 hours before to 3 hours after.
[c] This event was so large that the detector and electronics saturated for ∼36 hours. C.E. Fichtel by private communication has indicated that emulsion measurements show a peak intensity of the order of 500 protons/sec cm² ster.

## References

Balasubrahmanyan, F.K., Hagge, D.E., Ludwig, G.H., McDonald, F.B. (1966)    *J. geophys. Res.*,
   **71**, 1771.
Bryant, D.A., Ludwig, G.H., McDonald, F.B. (1962)    *IRE Trans. nucl. Sci.*, **9**, 376.
ESSA (1966)    *Solar-Geophysical Data*, CRPL-FB-265.
Fichtel, C.E., McDonald, F.B. (1967)    *Ann. Rev. of Astr. Astrophys.*, **5**, 351.
Rinehart, M.C. (1967)    *J. geophys. Res.*, **72**, 3459.

# SUNSPOT CHANGES FOLLOWING PROTON FLARES

CONSTANCE SAWYER

*(Institute for Telecommunication Sciences and Aeronomy, Environmental Science
Services Administration, Boulder, Colo., U.S.A.)*

ABSTRACT

Although the area of some sunspot groups declines suddenly after a flare that produces energetic
protons, other groups show a delayed or gradual decline, or continue to grow for several days after
the flare. The mean area of sunspot groups that produced flares with polar-cap absorption declines
gradually and continuously, beginning within a day after the flare.

This paper describes a search for a characteristic change in sunspot area at the time
of a major flare in the group. The total energy released by such a flare, in the form
of particles and electromagnetic radiation, sometimes amounts to a noticeable
fraction of the total magnetic energy in the sunspot group, and no other energy
supply is sufficient. Cases in which the total magnetic flux and the field gradient
declined measurably after a particularly energetic flare have been described (Gopasyuk
*et al.*, 1963; Howard and Severny, 1963), and Howard (1963) noted that the total area
of the sunspot group decreased after several flares that produced an increase in cosmic-
ray flux at ground level (GLE). On the other hand, Bruzek (1960) found that both
sunspots and their magnetic fields developed smoothly around the time of major
flares, and Newton and Howe (1952) found no unusual area change, on the average,
at the time of flares of importance 3 and 3+. We ask here whether a measurable
decrease in sunspot-group area is usual after flares that produce protons of relativistic
(GLE) or sub-relativistic (PCE) energy.

First, the area and appearance of the sunspot group at the time of 3 proton flares
were examined in some detail. The first of these flares occurred early on July 7, 1966,
and was accompanied by a ground-level cosmic-ray increase and by rather weak
polar-cap absorption. Routine daily measurements of the total area (penumbra and
umbra) of the sunspot group made at different observatories each showed that the
group continued to grow after the flare (Table 1). The growth was rapid enough that
even the projected area increased although the group was already West of central
meridian. Figure 1 shows the total corrected area of the sunspot group through its
passage. A more detailed look at the group's development convinces us, however,
that the total area is not relevant. Figure 2 shows the spot group before the flare,
with the flare position indicated. Although both the leading and following parts of

*Kiepenheuer (ed.), Structure and Development of Solar Active Regions*, 543–550. © *I.A.U.*

## Table 1

### Sunspot-group area measured before and after proton flares

$A_H$ (mill. hem.) $= A_D$ (mill. disk) $\times \frac{1}{2} \times$ secant of central angle

| Flare time | Time of measurement | Projected area, $A_D$ (mill. disk) | Corrected area, $A_H$ (mill. hem.) | Source |
|---|---|---|---|---|
| 1966 July 07·0 | July 06·2 | 1134 | 797 | *Solnechnye Dannye* |
| | 07·2 | 1282 | 1235 | |
| | July 06·3 | 1082 | 780 | Rome |
| | 07·3 | 1400 | 1255 | |
| | July 06·4 | 1151 | 887 | U.S. Naval Observatory |
| | 07·4 | 1175 | 1140 | |
| 1966 Aug 28·6 | Aug 28·2 | 814 | 425 | *Solnechnye Dannye* |
| | 29·2 | 1564 | 810 | |
| | Aug 28·2 | 738 | 387 | Rome |
| | 29·2 | 859 | 448 | |
| | Aug 28·5 | 241 | 125 | U.S. Naval Observatory |
| | 29·5 | 799 | 416 | |
| 1966 Sept 02·2 | Sept 02·2 | 1048 | 904 | *Solnechnye Dannye* |
| | 03·2 | 712 | 934 | |
| | Sept 01·3 | 1273 | 917 | |
| | 02·3 | 986 | 967 | Rome |
| | 03·3 | 620 | 989 | |
| | Sept 01·4 | 872 | 636 | U.S. Naval Observatory |
| | 02·6 | 872 | 939 | |

the group grew after the flare, the umbrae in the central portion that lay directly under the flare did decay. Figure 3 shows the area development of two separate umbrae: the growing Northern umbra in the leading part of the group, and the Northernmost central umbra, which lay under intense flare emission and which decayed after the flare. At the same time, the penumbra in the central region, which was particularly dark before the flare, became weaker and broken with bright patches. These changes began, as closely as can be determined (within several hours), at the time of the flare. The close spatial and temporal association leads us to consider this group as a case in which significant decay of the sunspot-group area followed a major flare. The magnetic field of this group varies quite differently than its area (Severny, 1967). Integrated energy, net flux, and gradient of the longitudinal field all reach a maximum a day before the flare, and decline thereafter.

Now we turn to the sunspot group that produced on August 28, 1966 a flare with moderate polar-cap absorption and on September 2, a still more energetic flare with fairly large PCA. Table 1 and the corrected-area curve sketched in Figure 4 show that

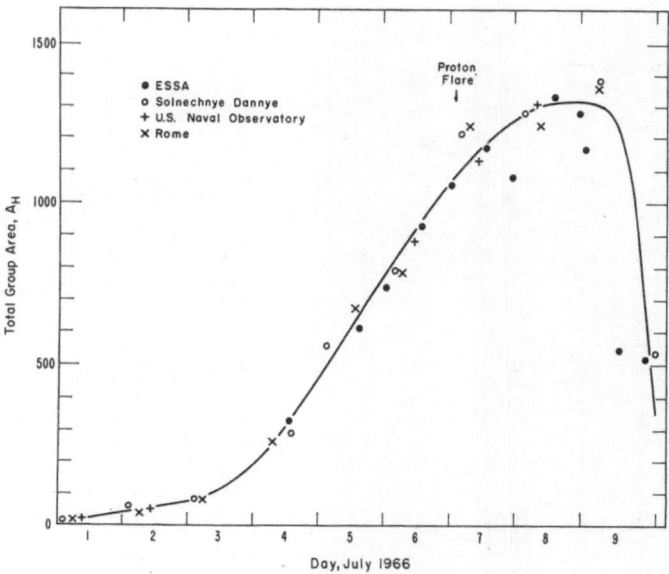

FIG. 1. *Total area, corrected for foreshortening, of the sunspot group in which occurred the proton flare of July 7, 1966.*

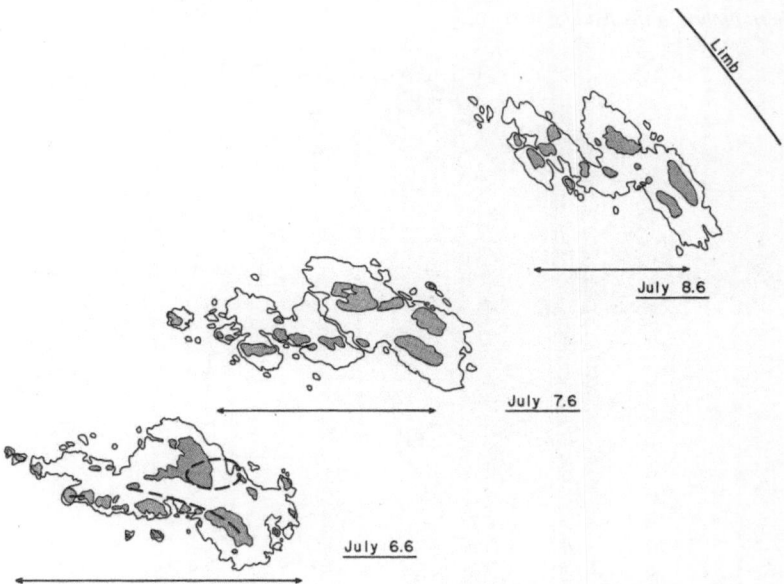

FIG. 2. *Sketches of the spot group before and after the proton flare of July 7, 1966. The position of the flare is indicated in the earliest sketch. The straight line under each sketch, proportional to the cosine of the central angle of the sunspot group, helps to estimate the effect on the apparent area of foreshortening. Note the decay of the two umbrae that were largest before the flare and that lay directly under the flare and note also the break-up of the penumbra in the flare region.*

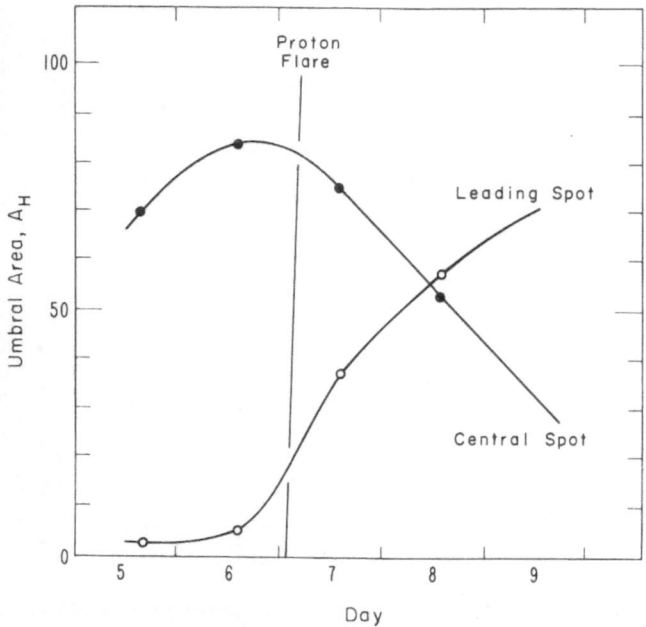

FIG. 3. *Corrected area of the Northernmost umbra in the central portion of the spot group and of the Northern umbra in the leading part.*

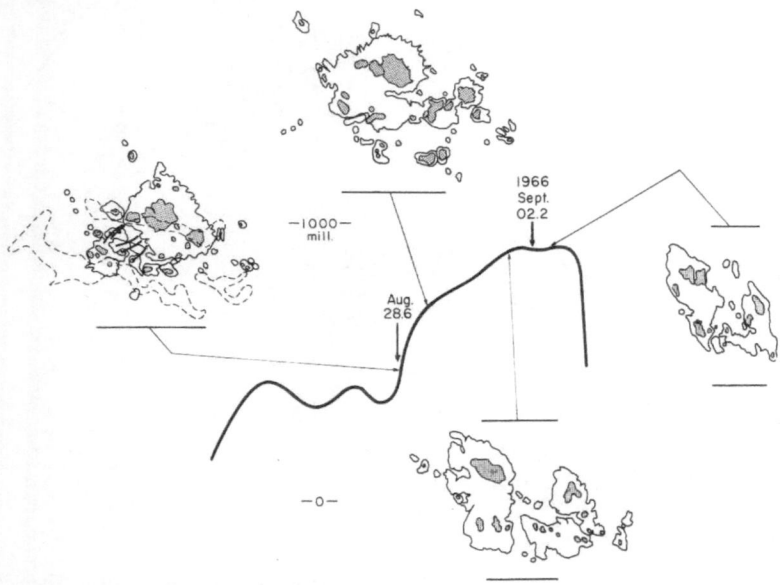

FIG. 4. *The curve of corrected area of the sunspot group from August 22 to September 4, 1966, with sketches of the group before (or during) and after the proton flares of August 28·6 and September 02·2. The length of the horizontal lines with each sketch is proportional to the cosine of the angular distance of the group from disk center. The position of the August 28 flare is indicated by dashed lines. In this case, parts of the group that were directly under the flare of August 28 – for example, the leading Northern umbra, continued to grow after the flare.*

the total area of this group, too, continued to increase during the several days between the two flares, and for about a day after the second flare. In this case, we are unable to find a part of the group that was especially close to the flare and that clearly decayed after the flare. The obvious changes in the group are the separation of the leading portion and the coalescence of the growing, following part with the central portion; these seem to fit the picture proposed by Gopasyuk *et al.* (1963) of the approach of spots of opposite polarity, and ejection of a secondary spot of the same polarity as the principal umbra.

Descriptions found in the literature of spot groups with great flares present an equally ambiguous picture. A flare on February 28, 1942 produced the first recognized cosmic-ray increase and the first recorded solar radio burst. The Greenwich observers note of the parent spot group that "... a marked decline sets in after February 28". Nevertheless, a second cosmic-ray increase was observed a week later, when this region, with area only half its maximum value, was near West limb. On the other hand, the group that produced the GLE flare of July 25, 1946 is described as "remarkably stable, undergoing very little radical change throughout its transit". We may also reread with interest Carrington's (1859) description of the sunspot group in which he observed a white-light flare: "It was impossible, on first witnessing an appearance so similar to a sudden conflagration, not to expect a considerable result in the way of alteration of the details of the group in which it occurred; and I was certainly surprised, on referring to the sketch which I had carefully and satisfactorily (and I may add fortunately) finished before the occurrence, at finding myself unable to recognize any change whatever as having taken place."

The contradictory evidence presented by individual cases leads us to examine the Greenwich measures of sunspot-group areas at the times of a larger number of flares. Figure 5 shows the corrected area of the sunspot group measured before and after 14 GLE flares. The tail of each arrow is at the relative area and central meridian distance of the sunspot group at the time of the daily measurement made before the flare, and the arrowhead at the point representing the measurement made after the flare. The area does indeed decrease in the majority of cases, but there are clear exceptions. Furthermore, since most of these flares occurred when the group was West of central meridian, we should expect in any case that the area would decrease as the group approached the limb.

The shaded curve shows the mean corrected area of large ($A_H \geqslant 400$ mill.) sunspot groups that were observed through a complete or nearly complete disk passage. The curve is very similar to that for a large sample of sunspot groups that produced PCE flares, and although we do not understand the large asymmetry, we believe it is an appropriate estimate of the way a sunspot group's area varies, regardless of individual flares. Most of the area changes at the time of GLE flares are similar to the expected change during normal development and rotation of such a group. Perhaps it is significant that of three groups that showed an anomalous area *increase*, two produced a second GLE flare (connected by broken lines). We should also note that the sunspot group that produced the flare of July 16, 1959 (after PCE flares on July 10 and 14) did reach its maximum area within a day after the GLE flare, declining thereafter.

Finally, we looked for evidence of a general decrease in sunspot-group area follow-

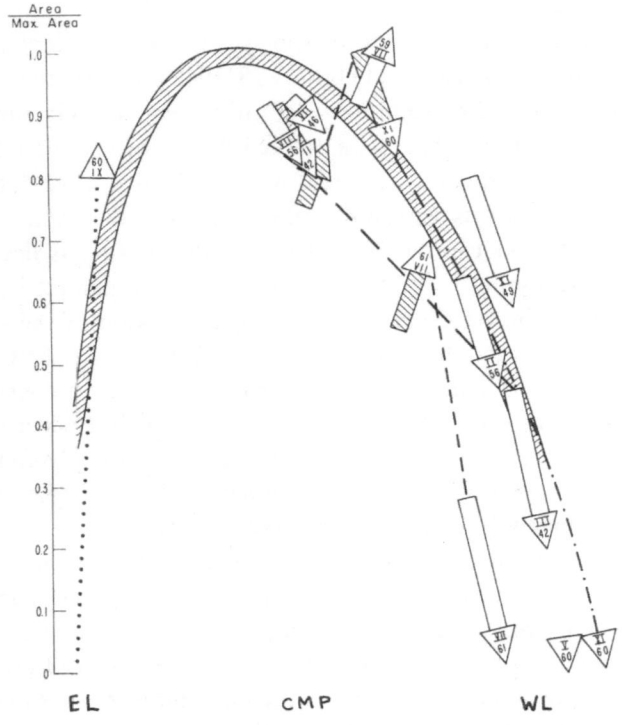

FIG. 5.   *The arrows show the relative corrected area and the central meridian distance of the sunspot group before and after each of 14 GLE flares. The shaded curve shows the expected area change during development and rotation of large, long-lived spot groups. Dotted lines connect GLE flares that occurred in the same group, and shafts of arrows representing the early flares are shaded.*

ing a proton flare. Sixty-three PCE flares that occurred between 1956 and 1963 in sunspot groups with area measured within one day both before and after the flare formed the observational sample. The lower, solid curve in Figure 6 shows the mean value of the sunspot-group area on successive days around the flare day. The day of the flare marks the beginning of a consistent decline in area. The area change due to rotation and development of the group was assumed to follow the curve shown in Figure 5, and each area was divided by the relative area appropriate to its central distance, to give the corrected area plotted as the upper curve in Figure 6. The run of the corrected areas is generally similar to that of the uncorrected mean areas, except that the correction pushes the maximum ahead to the first measurement after the flare.

We conclude that, in general, sunspot-group area declines after a proton flare (or that the flare occurs near maximum development of the sunspot group). There are clear exceptions to this rule, when the group continues to grow after the flare. Many

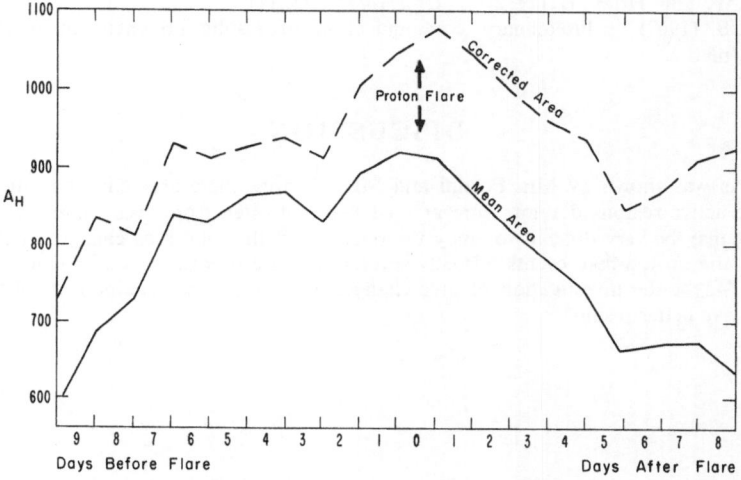

FIG. 6. *Mean sunspot-group area (lower, solid curve) around the time of 63 PCE flares. Areas shown in the upper, broken curve are corrected for effects depending on the central distance (see text). The standard error of the mean corrected area on day 0 is 79 millionths of the solar hemisphere.*

times, the spots show no sign of decay until hours after the flare, and then the area decreases gradually and smoothly. Sunspot area would seem to respond much less sensitively to the occurrence of an energetic flare than do magnetic parameters, and to vary somewhat independently of the field. Isolated cases do exist when the decline of sunspot area sets in suddenly and immediately after the flare, and strongly suggests a physical connection.

### Acknowledgements

Sacramento Peak Observatory loaned the films of flares and sunspots that formed the basis of the first part of this study. We thank Patrick S. McIntosh for generously sharing the copies he made of selected frames from these films. We are grateful to Fred Ward for sending copies of magnetic tapes of Greenwich sunspot data prepared under his direction. Carolyn McLellan and Dennis Schatz carried out much of the statistical analysis reported in the last section.

This work was partially supported by NASA Contract R-102.

CONSTANCE SAWYER

## References

Bruzek, A. (1960) *Z. Astrophys.*, **50**, 110.

Carrington, R.C. (1859)     *Mon Not. R. astr. Soc.*, **20**, 13.

Gopasyuk, S.I., Ogir, M.B., Severny, A.B., Shaposhnikova, E.F. (1963)     *Izv. Krym. astrofiz. Obs.*, **29**, 15.

*Greenwich Photoheliographic Results 1942 and 1946*, Her Majesty's Stationery Office, London.

Howard, R. (1963)     *Astrophys. J.*, **138**, 1312.

Howard, R., Severny, A.B. (1963)     *Astrophys. J.*, **137**, 1242.

Newton, H.W. and Howe, H. (1952)     *Observatory*, **72**, 111.

Severny, A.B. (1967)     Preliminary communication of results presented at IQSY Meeting, London, July.

## DISCUSSION

*Bumba:* As was shown by Mrs. Fortini and Mrs. Martres, there are some indications that the proton-flare active regions develop from at least two and often from even more sunspot groups. Therefore it may be very difficult to study the relations of the total area changes of this complex situation to the proton-flare events. Usually several days are needed for the individual spotgroup to interact. Maybe the investigation of area changes of the individual components of the complex group may give better results.

Part VIII

# RADIO STRUCTURE OF AN ACTIVE REGION

# HOMOLOGY OF SOLAR RADIO EVENTS

A. D. FOKKER

*(Sterrewacht 'Sonnenborgh', Utrecht, The Netherlands)*

ABSTRACT

Successive flares within the same centre of activity sometimes produce radio events that are remarkably similar. The occurrence of such homologous radio events is commonly restricted to periods of less than 48 hours.

Typical for certain centres of activity is their tendency to give rise to impulsive microwave bursts. Such bursts may occur repeatedly during a period of more than a week. The distributions of the lengths of periods during which series of homologous bursts and of impulsive microwave bursts occur are compared.

Ellison *et al.* (1960) called attention to the occurrence of homologous flares. These are flares which occur successively in the same active region, correspond in position relative to the local sunspots, and show a common pattern of structure and development.

The concept of homology can be extended to the flare-associated radio events, since occasionally two or more flares in the same centre of activity give rise to radio events that are remarkably similar in one or more of the main frequency bands: centimetric, decimetric and metric. Sometimes the intensity curves of two radio events show detailed similarity over a wide range of frequencies. But even if the intensity curves do not resemble each other in detail, two radio events may be similar in that their responses in the various frequency bands have comparable durations and intensities. On the basis of the correspondence of intensity and duration it is possible to select pairs of homologous radio events in the tabulations of distinctive events in the *Quarterly Bulletin on Solar Activity*. On applying a certain system of rating the degree of correspondence of duration and intensity in the various frequency bands, it is possible to derive a quantitative measure for the extent to which two radio events are homologous. We determined the measure of homology for several pairs of radio events that produced responses in each of the cm, dm, and m frequency bands. Two events were considered as homologous if this measure surpassed a certain value.

Among 69 pairs of events that occurred in the same centre of activity during the period 1959–61, there were 14 homologous pairs, or 20%; among 83 pairs of events taken arbitrarily from different centres of activity, only 4, or 5% were homologous.

We thus conclude that there is a significant tendency for homologous radio events to occur within one centre of activity.

*Kiepenheuer (ed.), Structure and Development of Solar Active Regions*, 553–555. © *I.A.U.*

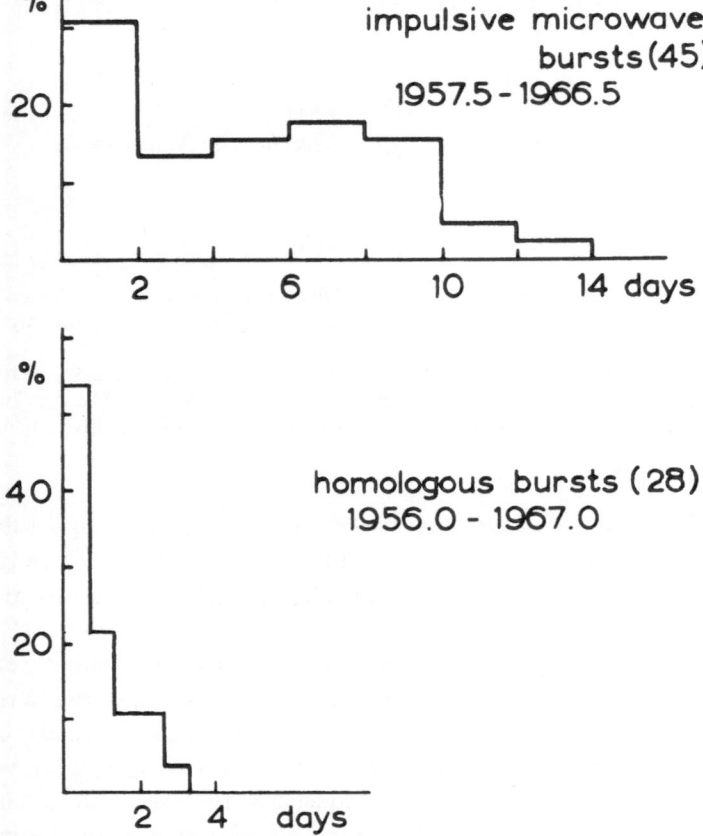

FIG. 1. *Histograms of time intervals between the first and the last events that belong to a series of impulsive microwave bursts and of homologous radio events, respectively.*

In a centre of activity homologous events succeed each other at time intervals which seldom surpass 48 hours. A histogram of such time intervals is given in the lower part of Figure 1.

It is interesting to draw into comparison the occurrence, within one centre of activity, of the well-defined class of impulsive microwave bursts. The impulsiveness of a burst can be determined quantitatively as the steepness of the rise of intensity. If this figure surpasses a certain value, the microwave burst may be considered an impulsive one.

Impulsive microwave bursts need not at all be homologous: their intensity, duration and detailed intensity curves may be quite different. Fokker and Roosen (1961) showed that the impulsiveness of microwave bursts is a very characteristic feature of bursts that occur in specific centres of activity. Contrary to the homology, the impulsiveness of microwave bursts may be preserved during several days in succession.

A histogram of time intervals during which impulsive microwave bursts occurred associated with flares in one and the same centre of activity is given in the upper part of Figure 1.

There is an outstanding difference between the two histograms in Figure 1. The type of condition which gives rise to the occurrence of homologous radio events apparently does not persist for much longer than 2 days. On the other hand, the impulsiveness of microwave bursts is governed by situations that may persist for longer than a week.

We can only guess what the factors may be that govern the occurrence of a particular type of radio event. The impulsiveness may, for instance, be related to the existence of strong gradients in the magnetic field near the place of origin of the flare. Homology perhaps occurs only if the detailed structure of the magnetic field remains unchanged and if the acceleration of particles takes place in exactly the same location.

There is a suggestion that the histogram for the impulsive microwave bursts consists of two components. It will be of interest to see whether this feature is preserved as the available material increases. If genuine, the effect would seem to imply that there exist two conditions of a different nature, and both may give rise to an impulsive microwave burst.

### References

Ellison, M. A., McKenna, S. M. P., Reid, J. H. (1960)    *Dunsink Obs. Publ.*, **1**, No. 1.
Fokker, A. D., Roosen, J. (1961)    *Bull. astron. Inst. Netherl.*, **16**, 83.

## DISCUSSION

*De Jager:* Your histogram of impulsive microwave bursts extends over 14 days, which is the maximum observable time of an active region. Did you correct your observations for the fact that at least part of the observations extended over a time interval < 14 days?

*Fokker:* No corrections were made.

# SOME RESULTS ON SOLAR ACTIVITY AT 408 MHz

B. CLAVELIER

*(Observatoire de Meudon (S. et O.), France)*

ABSTRACT

We present the first results obtained with the 408 MHz interferometer of Nançay. The typical activity observed at 408 MHz is the 'noise storm'. Undoubtedly with this apparatus we could distinguish multiple centres, generally double.

The comparison with corresponding optical observations and with Nançay observations at frequency 169 MHz allowed us to associate the centres of noise storm at 408 MHz with eruptive centres lying in 'Anomalous Active Regions' and to describe the schematical structure of active zones.

The results presented here have been obtained with the 408 MHz interferometer that we have built at Nançay (Clavelier, 1967).

The 1·5 km base line gives 1ʹ7 resolving power with an interlobe of 25ʹ. This interferometer allows us to measure the East–West dimension of solar sources greater

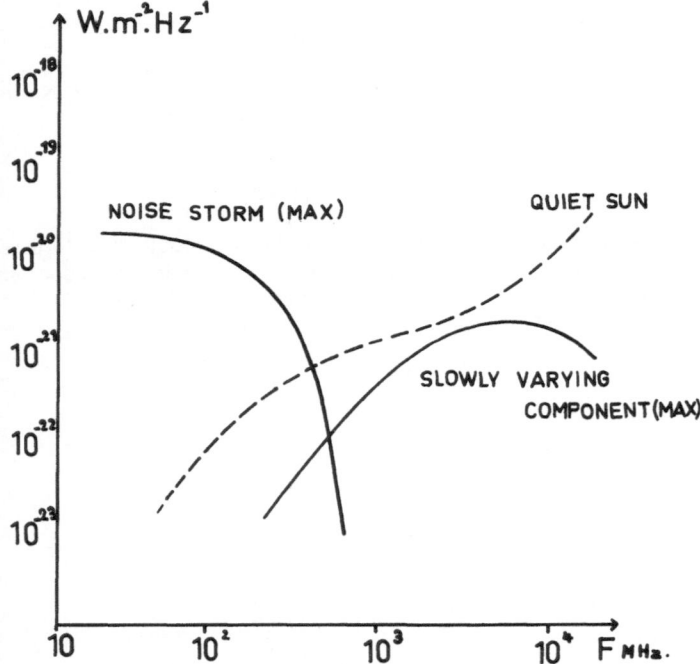

FIG. 1.   *Spectra of noise storms and slowly varying component.*

*Kiepenheuer (ed.), Structure and Development of Solar Active Regions, 556–564. © I.A.U.*

than 1' of arc, and to separate two sources 1.7 distant. The position of an emitting centre on the solar disk is determined with an accuracy of 20″ (of arc).

With this instrument we can analyse centres with a flux greater than $10^{-23}$ W.m$^{-2}$. Hz$^{-1}$.

The following results are derived from observations carried out during the period from October 1, 1965 to July 14, 1966.

The spectrum of the slowly varying component shows us that the fluxes at 408 MHz are weaker than those at higher frequencies. Likewise, fluxes from noise storms must be weaker, at 408 MHz than for metric wavelengths (Figure 1).

Our observations confirm the weakness of the mean flux (Figure 2): $2 \times 10^{-22}$ W.m$^{-2}$.Hz$^{-1}$ for 90% of the centres.

On the other hand, whereas one thought of meeting mainly the slowly varying component at 408 MHz, we found that the noise storm activity was the most important. Finally, besides the bursts that we will not discuss here, the two most durable types of emission are given in Table 1.

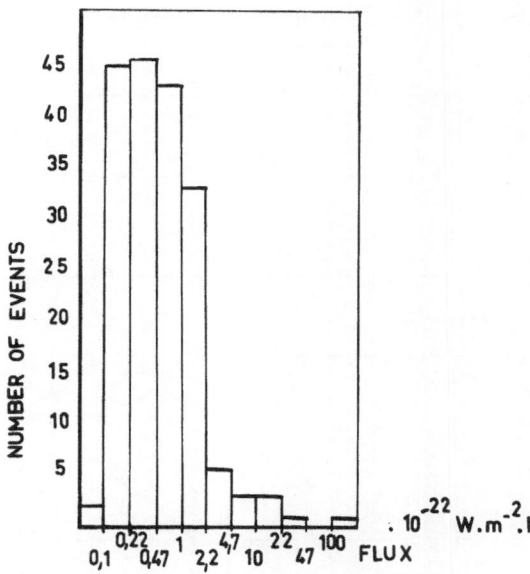

FIG. 2.   *Histogram of flux densities.*

**Table 1**

| *Slowly Varying Component* | *Noise storms* |
|---|---|
| Large $> 2'.5$ | Narrow $< 2'$ |
| Brightness temperature $T_b < 5 \times 10^6$ °K | $T_b > 5 \times 10^6$ °K |
| No storms at 169 MHz | Storm simultaneously at 169 MHz |
| Not variable | Variable ($+$ some types 1) |
| Associated with faculae | Associated with active centres (with spots) |

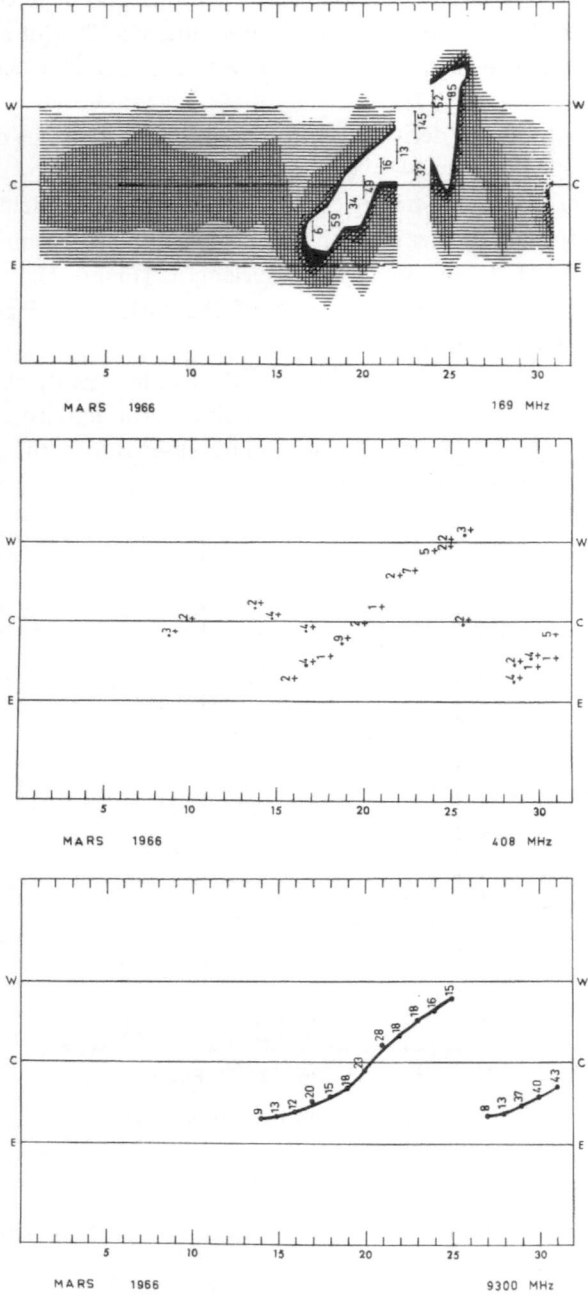

FIG. 3.    *Maps of solar activity at 9300, 408 and 169 MHz (March 1966).*

In order to study certain properties of the radio centres, we have been led to associate the radioelectric centres with optical centres. For that purpose, we made use of the solar photographs of the Meudon-Observatory and the lists of centres found in the *Quarterly Bulletin on Solar Activity*.

Among 175 centres observed at 408 MHz, 123 were associated with centres of activity where spots were visible, 46 were associated with diffuse faculae without spot; they are coronal condensations of low brightness temperature and of large diameter, and 6 were ambiguous.

With the three interferometers (at 169 MHz, 408 MHz, and 9300 MHz) of the radio observatory at Nançay we establish a monthly map of solar radioelectric activity giving position and flux of the radio centres. These maps (Figure 3) allow us to follow the evolution of solar activity at 408 MHz and to compare it with corre-

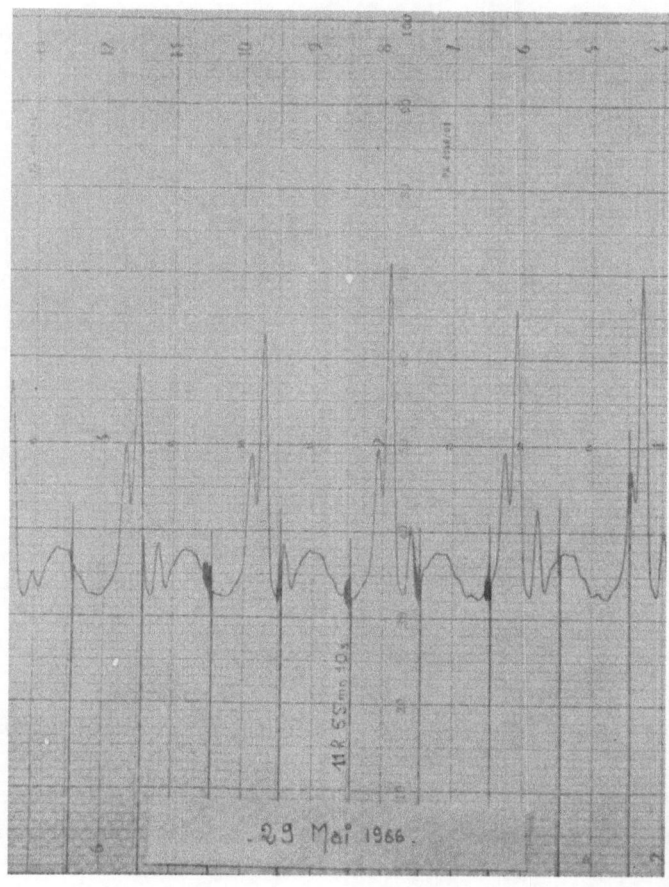

FIG. 4.   *Record of May 29, 1966, showing a double centre.*

sponding activities at 169 and 9300 MHz. From these maps we conclude that the centres at 408 MHz are more stable than those at 169 MHz but less stable than centres at 9300.

On certain days we can observe two centres close enough to be associated with the same active centre. We are dealing here with multiple centres (generally double) as confirmed by the associated optical centres (Figure 4).

This had not yet been firmly established for this type of activity (Le Squeren had thought of associating two centres at 169 MHz, with the same active region in only a few cases (Le Squeren, 1963)). For all these double centres at 408 MHz the corresponding optical centres appear as two distinct regions with spots. We shall see that these two regions correspond to two different eruptive zones each associated with a radio component.

Two methods for measuring altitude give an average height between 70000 and 80000 km.

On the diagram (Figure 5) we note $d$ and $\delta$: $d$=radio centre to solar meridian distance, $\delta$=optical centre to solar meridian distance.

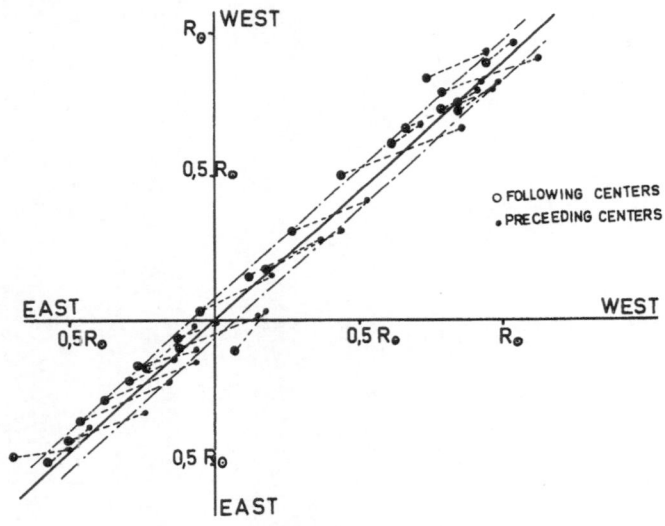

FIG. 5. *Height (double centres).*

Preceding centres are noted by a dot and following centres by a cross. We found that:

(1) The average slope of the lines joining preceding and following centres shows that the radio centres are more distant (on the average) than the optical centres by a factor of 2 (the mean distance between radio centres being 3' of arc).

(2) Preceding centres are found on a parallel to $D$ (70000 km of altitude) shifted to the West and following centres to the East (average altitude: 70000 km).

Let us explain now, how we have associated the noise storms centres at 408 MHz with the eruptive activity. Although the activity at 408 MHz notably differs from that observed at lower frequencies, we may think that it shows a certain analogy.

In particular, former studies have shown (Le Squeren, 1963) that the centres of activity in the metric range appear following flares.

In the following we will use a very simplified classification of the active regions derived from Martres *et al.* (1966):

(1) The simple ARs with lines of polarity inversion slightly inclined on the meridian.

(2) The complex ARs for which the opposite polarities are irregularly located with respect to several lines of inversion. In the simple Active Regions they distinguish between:

(a) The simple and normal ARs that present no anomaly as to magnetic structure and aspect of the group of spots, and

(b) The simple but anomalous ARs having either an anomalous inclination of the group of spots on the parallels or a following spot greater than the preceding spot.

Thereafter we will only retain the following classes:

(1) Normal ARs corresponding to simple and normal ARs defined above (not eruptive).

(2) Anomalous ARs corresponding to simple and anomalous ARs and to complex ARs (eruptive).

From the optical observations of 'Meudon' we were able to determine which class of AR our centres could be associated with.

We have found that, with the exception of three centres belonging to a normal AR, all other centres (simple and double) are associated with an anomalous AR.

Inversely most of the anomalous AR are accompanied by 408 MHz storm activity. It is of interest to mention that the two events of July 5, 1966 and September 3, 1966 are two important exceptions: these cases show an obvious anomalous configuration with numerous eruptions; however, no activity at 408 MHz was detected. Those events were accompanied with proton emissions.

It is possible to describe further the association between radio centre and AR, by studying the eruptive zones that appeared in these regions during the days of observations.

In the case of double radio-electric centres, despite the great distance between chromospheric AR and radio centres (70000 km) we have tried to associate them with two distinct eruptive centres in the same active region. For that purpose we have considered as distinct two groups of eruptive centres more than 1.7 apart (resolving power of our instrument). We found that out of 17 cases of double radio centres, 10 correspond to two eruptive zones located in the same AR and more than 1.7 apart.

The conclusion of this comparative study with optical observations is that the noise-storm type activity at 408 MHz can, beyond doubt, be associated with the

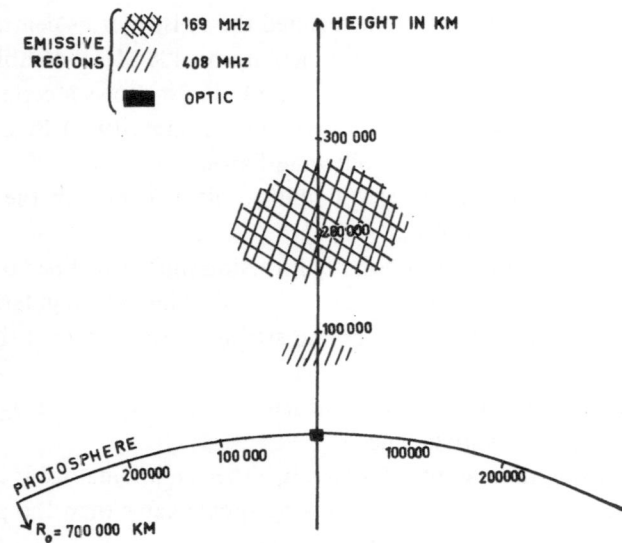

FIG. 6.   *Schematic structure above an active region (simple radio centre).*

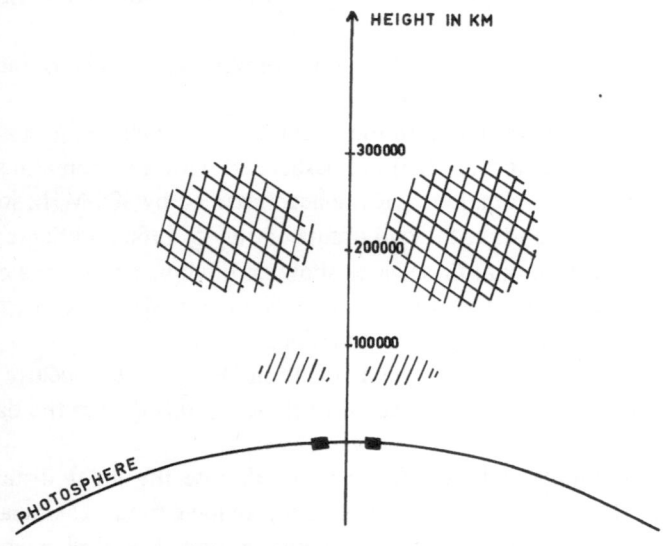

FIG. 7.   *Schematic structure for a double centre.*

existence of an anomalous AR. Furthermore, one could put forward the theory that any radio centre is associated with a distinct eruptive zone.

   Results obtained at 169 and 408 MHz allow us to draw a schematic picture (Figures 6 and 7) of the structure of the active zones. The diameter increases with altitude (but nothing can be said about the extension in latitude). We cannot exclude the possibility that the simple centre might have a multiple structure not resolved by our instruments.

As to double centres, one would be tempted to identify them with a bipolar structure. However, since each of the individual centres seems associated with an eruptive region with its bipolar structure (as linked to an inversion line), one would rather think that the radio components of these centres are independent.

Another fact can confirm either one hypothesis or the other, that is the relation between the fluxes of the two radio centres. We have studied the comparative evolution of head and tail fluxes in the same double centres.

On November 8, 1965 and March 31, 1966, when the components were clearly visible for more than 1 hour, we have not found any correlation between fluxes of the two centres.

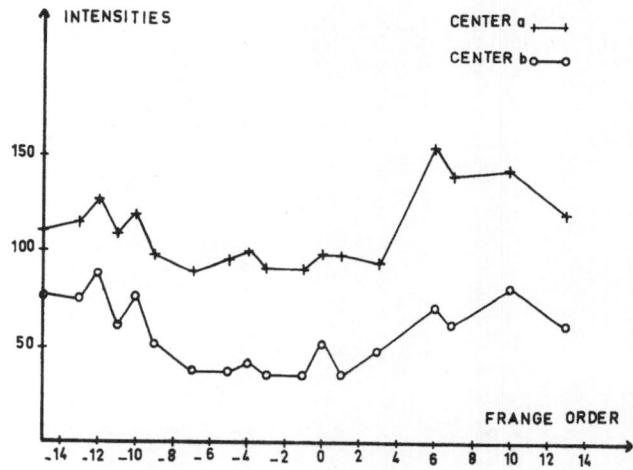

FIG. 8.    *Evolution of the components a and b of the double centre on October 4, 1965.*

Had this been confirmed later on, the second interpretation would be more valid.

Let us note, however, that on October 4, 1965 a strong correlation was found between the two fluxes (Figure 8), but it was a type-IV burst.

The study of the polarization of emissive centres at 408 MHz with a better resolving power will permit us to add further details on the schematic structure described above.

## References

Clavelier, B. (1967)      *Ann. Astrophys.*, **30**, 895.
Le Squeren, A. M. (1963)      *Ann. Astrophys.*, **26**, 97.
Martres, M.J., Michard, R., Soru-Iscovici, J. (1966)      *Ann. Astrophys.*, **29**, 246.

# DISCUSSION

*Koeckelenbergh:* Je n'ai pas bien compris quelle était la résolution angulaire de votre interferomètre?

*Clavelier:* 1 ʹ. 7 arc.

*Koeckelenbergh:* Est-ce un interferomètre lineaire?

*Clavelier:* Oui, 16 miroirs.

*Elgaróy:* When you observe a noise-storm source with double structure, is there any connection between changes in the leading and the preceding part of the source? (See the present paper)

# CONDITIONS OF ACCELERATION OF SOLAR ELECTRONS, AND DETERMINATION OF THE MAGNETIC FIELD IN THE HIGH CORONA FROM THE CHARACTERISTICS OF A TYPE-IV BURST

A. Boischot and B. Clavelier

*(Observatoire de Paris-Meudon, France)*

### ABSTRACT

The study of a moving type-IV burst shows that acceleration of electrons can happen in the high corona, and a mechanism is proposed for such an acceleration. The low-frequency cut-off of radio spectrum is interpreted as due to the effect of the coronal plasma upon the synchrotron emission, and this leads to an accurate determination of the intensity of the magnetic field at an altitude of $1 R_\odot$. Then it is possible to compute the energy and the density of the relativistic electrons.

On September 14, 1966, a flare was observed on the West limb of the Sun, more precisely on an active center located 10° behind the limb. For this event several types of solar radio bursts were detected, mainly a type II beginning at 10·17 UT, a first type IV at 10·43 UT, and a second one at 11·00 UT. The latter has been studied at Nançay on 408 and 169 MHz and may be described as follows:

(1) The emission begins at exactly the same time on both frequencies, and has the same profile (Figure 1). This is also true for all other frequencies on which the burst has been observed.

(2) The position of the centres of emission determined on 169 and 408 MHz shows three remarkable characteristics (Figure 2):

(a) The emissions on 169 and 408 MHz come from the same source;

(b) When the burst starts, this source is already located at an altitude of one solar radius (this is a minimum value, assuming a radial position relative to the optical flare.

(c) The source moves upward with a constant velocity of 530 km/s and was followed up to an altitude of two solar radii (in the radial hypothesis).

(3) The size of the source has the same value on 169 and 408 MHz and is constant during the main part of the burst. This source size, $80\,000 \pm 10\,000$ km is much smaller than the displacement of the source during its ascension through the corona. Only near the end of the burst the source increases rapidly in size.

These characteristics are those of a moving metric type IV (Kundu, 1965). It is generally accepted that this type of burst is due to synchrotron emission by relativistic

*Kiepenheuer (ed.), Structure and Development of Solar Active Regions, 565–569. © I.A.U.*

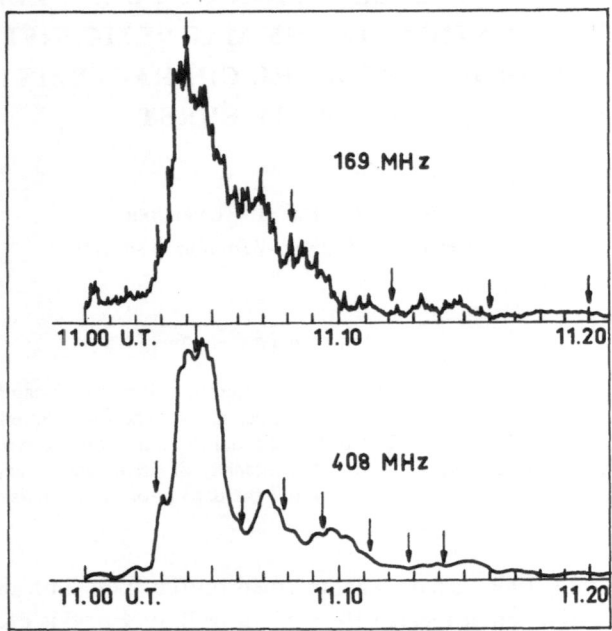

FIG. 1. *The second type IV of the September 14 event, observed on 169 MHz and 408 MHz. The arrows show the times at which the position of the centre of emission has been determined.*

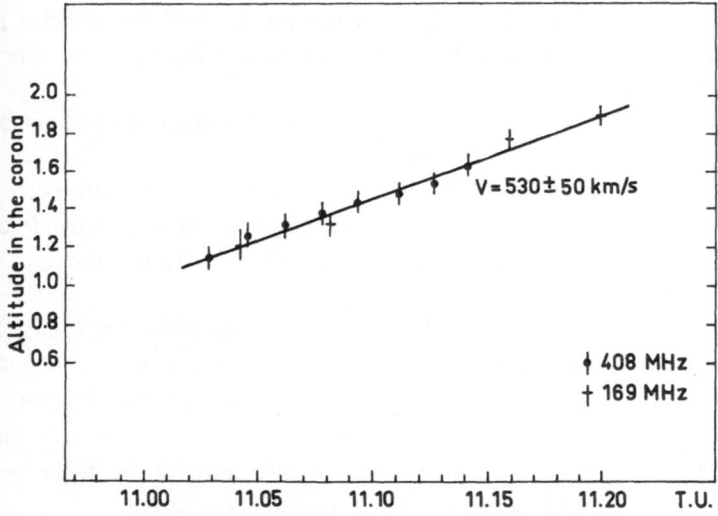

FIG. 2. *Position of the centre of emission on 169 MHz and 408 MHz. Altitudes and velocity are given assuming a radial outward motion.*

electrons accelerated during the flare. But up to now it was difficult to study the properties of these electrons because of the lack of information about the source of the
emission. The present observations allow us to make a finer analysis of this problem.

## 1. Mechanism of Acceleration of the Electrons

When the burst starts, the source is already at an altitude of 700000 km in the
corona. This is much higher than the critical heights where the plasma frequency is
equal either to 169 MHz or 408 MHz. Then we are sure that the waves suffer no
perturbation, cut-off or refraction, from the coronal plasma and that we are observing
the source exactly as it is. The fact that the emission starts at exactly the same time
on each frequency supports this conclusion, for any effect of the plasma would be
chromatic.

As the emission begins only at $1 R_\odot$, we have to conclude that the relativistic
electrons do not exist lower, or in other words that the acceleration takes place at this
altitude, where nothing noticeable is seen in the visible range. Moreover, the burst
shows a rapid increase (1 min) followed by a much longer quasi-exponential decrease,
and this means that the acceleration takes place in a relatively narrow layer of
60000 km at the most.

Mangeney and Lacombe proposed a mechanism to explain this acceleration by the
action of perturbation coming from the chromosphere and ascending in the corona
with the measured speed of 530 km/s. (Boischot *et al.*, 1967).

When the Alfvén-Mach number is greater than 1, the perturbation propagates as
a soliton, and an instability occurs for a critical Mach number equal to

$$M^* = 1 + \frac{3}{8}\left(\frac{8\pi N_e kT}{H^2}\right).$$

This instability leads to plasma oscillations and turbulence in which the electrons
are accelerated by the mechanism described by Tsytovich. The acceleration then takes
place at the altitude where the perturbation attains this critical Mach number.

For a velocity of 500 km/s, and $T = 2 \times 10^6\,°K$, this happens when $N_e = 10^7\,\mathrm{cm}^{-3}$
and $H = 0.5$ gauss, i.e., at an altitude of $1 R_\odot$ as we shall determine later on.

## 2. Influence of the Coronal Plasma upon the Synchrotron Radiation (Razin effect)

The burst spectrum is given in Figure 3. It shows sharp cut-off toward high and
low frequencies, and cannot be interpreted by any energy spectrum of the electrons
in classical synchrotron theory.

The low-frequency cut-off has a slope at about +4, and cannot be explained either
by a low-energy cut-off in the energy spectrum, or by free-free or synchrotron absorption.

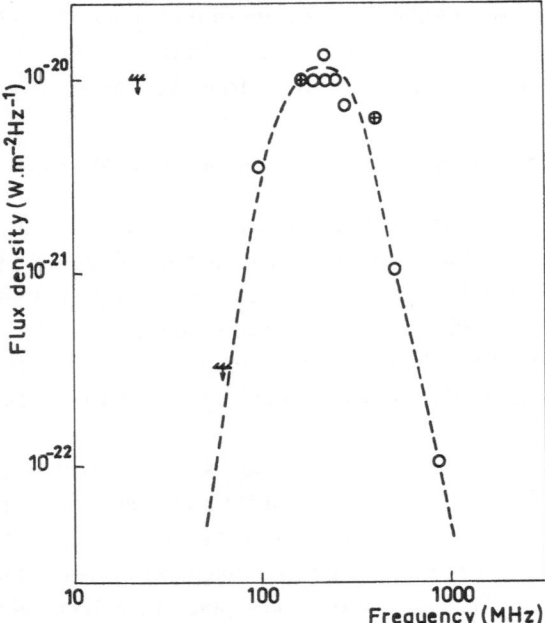

FIG. 3.  *Spectrum of the moving type IV of September 14, 1966.*

This low-frequency cut-off is due to the effect of the coronal plasma on the synchrotron emission. The synchrotron effect is generally computed by assuming that the electrons radiate in free space, with a refractive index equal to unity. But it is known that if we take account of a departure of $n$ from unity, the emission on low frequencies is much less intense, giving a quasi-exponential cut-off (Ginzburg and Syrovatskii, 1965).

It is shown that this effect is noticeable at frequencies smaller than a critical frequency given by

$$v_c = 20 \frac{N_e}{H},$$

where $v_c$ is in Hertz, $N_e$ the density of the thermal electrons in cm$^{-3}$ and $H$ the magnetic field in gauss.

In our case, $v_c = 400$ MHz, and we can take $N_e = 10^7$ cm$^{-3}$ at an altitude of 1 $R_\odot$. This leads to $H = 0·5$ gauss and gives one of the more accurate determinations of the magnetic field in the high corona.

## 3. Energy and Density of the Relativistic Electrons

The high-frequency part of the radio spectrum has a very steep slope, of spectral index $-4$ if we assume a power law. This would correspond to an energy spectrum of the electrons of the same shape with a spectral index of $+9$. But this high-frequency slope is more likely due to a high-energy cut-off of the electrons.

On the other side, the effect of the coronal plasma upon the synchrotron mechanism prevents a good determination of the low-energy part of the electron spectrum because the corresponding electrons radiate very little.

Then, in a first approximation, it is possible to explain the observed spectrum by quasi-monoenergetic electrons, and, in this case, to compute their energy and density.

The maximum of the spectrum may be estimated, in absence of Razin effect, to be around 25 MHz, and the maximum intensity $10^{-20}\mathrm{W.m^{-2}Hz^{-1}}$.

From the frequency of maximum and the value of magnetic field which has been determined above, it is possible to derive the energy of the electrons by the classical formulae:

$$v_m = 4.6 \times 10^{-6} H_\perp E_{ev}^2 .$$

We find an energy of 3 MeV. The density of the relativistic electrons is then given by

$$N = 7.2 \times 10^{22} \frac{I_v}{LH} ,$$

where $I_v$ is the brightness of the source and $L$ its linear dimension.

Finally, assuming that the source has in the three dimensions the same size that has been measured radially, we find for the density of the relativistic electrons: $N = 6000$ el/cm$^3$.

## References

Boischot, A., Clavelier, B. (1967)     *Astrophys. Lett.*, **1**, 7.
Boischot, A., Clavelier, B., Lacombe, C., Mangeney, A. (1967)     *C.R. Acad. Sci.*, **265**, 1151.
Ginzburg, V.L., Syrovatskii, S.I. (1965)     *A. Rev. Astr. Astrophys.*, **3**, 297.
Kundu, M.R. (1965)     *Solar Radio Astronomy*, Interscience Publishers, New York, p. 383.

## DISCUSSION

*Krüger:* In principle, spectra with a small frequency extent could be expected also when there are particles of a very limited energy range, but this can be excluded in your case?

*Boischot:* In the classical theory of synchrotron emission it is shown that if there is a low-energy cut-off in the spectrum of electrons, or even in the case of monoenergetic electrons, the spectral index of the low-energy part of the radio emission cannot be larger than 1/3. This leads to much wider spectrum than observed.

*Elske Smith:* I think you have to be cautious in using the electron density in the corona in your relation to find the magnetic field. If there is an inhomogeneity in the corona because of the association with an active region, there may be an increase in the electron density by a factor 3, perhaps even 10, leading to a corresponding decrease in the magnetic field.

*Boischot:* I think that the coronal density over an active centre is known within a factor of 2 or 3. Now, the main inaccuracy comes probably from the determination of $v_c$ from the observed spectrum, but more detailed calculations for different energy spectra of electrons must improve this point.

# TENDENCIES TO REPEATING OF TYPE-IVm BURSTS AND THEIR RELATIONS TO THE STAGE OF DEVELOPMENT OF THE SUNSPOT GROUP

A. BÖHME*

*(Heinrich-Hertz-Institute of Solar Terrestrial Physics, Berlin-Adlershof, D.D.R.)*

ABSTRACT

It has been shown that moving type-IV bursts at decameter waves are infrequent events compared with the moving type-IV bursts at meter waves. They belong preferably to sunspot groups of relatively homogeneous magnetic structure. Moving type-IV bursts at decameter waves have a very low tendency to repeat. This statement indicates that the magnetic structures producing them have only a short lifetime. The short duration of these bursts cannot be interpreted by radiation damping of high energetic electrons. Furthermore it can be shown that the storm continuum at decameter waves is related preferably to sunspot groups of complex magnetic structure. Probably stationary type-IV bursts at decameter waves are more representative for the acceleration of high-energetic particles than the moving type-IV bursts at decameter waves.

In a recent study Antalova (1967) has shown that type-IV bursts originate only during certain stages of development of the associated sunspot groups. She classified seven different spot-group configurations in the photosphere, which can produce type-IV bursts. Now it was our aim to find out whether some of the sub-types of the type-IV burst are related to special configurations of the sunspot groups. In the present investigation the type-IV bursts were taken from the catalogues of Pick-Gutmann, Švestka-Olmr and Fokker, the *Quarterly Bulletin*, and from the measurements of the Heinrich-Hertz-Institute. The classification of the sunspot groups was based upon the Antalova scheme for which drawings were used obtained by the Potsdam Solar Observatory. The distribution of the type-IV bursts on the various configurations obtained by us agrees well with the distribution given by Antalova.

The first sub-type under consideration was the moving type-IV burst at 40 MHz and lower frequencies. From interferometric observations made by Weiss (1963) it can be seen that at 40 MHz a moving type-IV burst is a rather infrequent event compared with the frequency range between 170 and 100 MHz. There is also indirect evidence for this from the observations illustrated by Figure 1, where the number of the type-IV bursts we have measured at 111 and 40 MHz, together with measurements of the National Bureau of Standards (Boulder) at 108 and 18 MHz, are plotted

---

* Presented by A. Krüger.

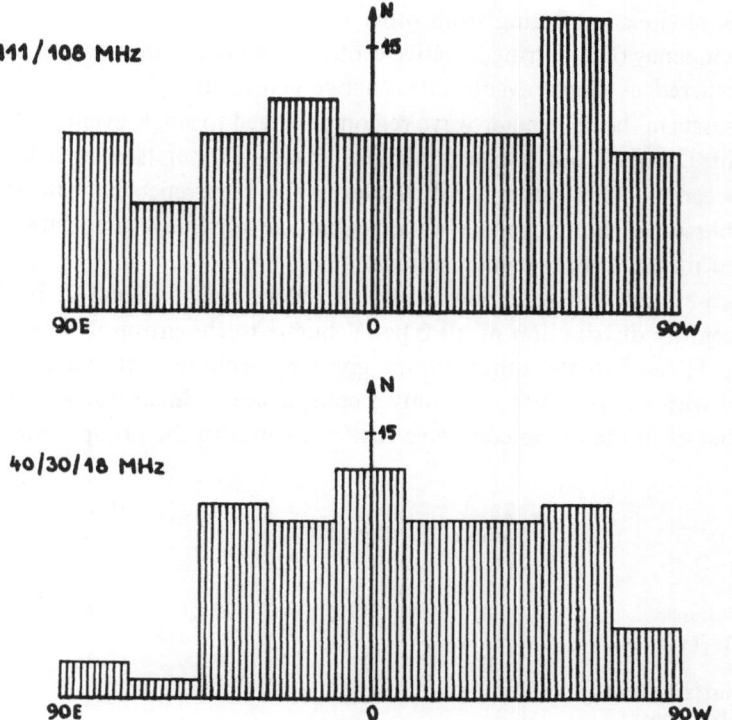

FIG. 1. *Center-limb variation of the number of type-IV bursts at 111/108 and 40/30/18 MHz.*

in dependence of the distance from the central meridian on the Sun. While there are at 111/108 MHz hardly any center-limb variations, the number of bursts in the decameter wave region decreases remarkably towards the limb, which suggests the absence of a moving component.

Though we had no interferometric measurements available, the attempt was made to find out whether or not the type-IV bursts observed at our institute have a moving component at 40, 30, and 23 MHz. Fortunately, since 1961 polarization observations in the decameter-wave region were available and we assigned a moving component to those bursts which exhibited either a reversal of the sense of the polarization between a first part and a second part, or which – with the flare near to the limb – showed, in accordance with the leading-spot hypothesis, an extraordinary polarization. Based on these experiences we have assigned a moving component also to the type-IV bursts measured prior to 1961, if at the beginning there was a continuum with a frequency drift of the order of some minutes, a duration greater than 5 and smaller than 25 min, corresponding to the flash phase of the centimeter bursts.

Together with the cases reported by Weiss and Philip (1964) material on 33 moving type-IV bursts in the decameter-wave region was available. These bursts have some properties obviously differing from other type-IV bursts. In general, type-IV bursts have the tendency of occurring in active centres repeatedly. Only 26 % of all the type-IV bursts occurred as single events in an active centre. But 21 out of the 33 moving type-IV bursts in the decameter-wave region belonged to single events in the respective active centre. Only 2 out of the remaining 12 type-IV bursts occurred repeatedly in the same centre. This rather weak tendency for recurrence indicates that moving type-IV bursts in the decameter-wave region are obviously associated with rather short-lived magnetic-field configurations.

This assumption is also supported by consideration of Figure 2. In the first line can be seen the distribution of all type-IV bursts to the group named 'complex', to the group H' and to the other groups given by Antalova. It can be seen that, as compared with all type-IV bursts, only a conspicuously small fraction of the moving type-IV bursts in the decameter-wave region belongs to the group named 'complex'.

| | Total number of bursts | Complex | H' | HE, E', R, F', 2F, D | Unclassified |
|---|---|---|---|---|---|
| All type-IV bursts | 198 | 43% | 27% | 18% | 12% |
| Moving type-IV bursts at 40 MHz | 33 | 27 | 37 | 36 | – |
| Type-IV bursts without stationary component at 40 MHz | 33 | 42 | 33 | 25 | – |
| Weak stationary type-IV bursts at 40 MHz | 27 | 41 | 37 | 15 | 7 |
| Storm continua at 40 MHz and lower frequencies | 22 | 86 | 5 | 9 | – |

FIG. 2.    *Distribution of type-IV bursts at the different stages of development of the related spot groups.*

On the other hand, 70 % of the type-IV bursts which have no moving component in the decameter-wave region belong to the group named 'complex'. This observation suggests that obviously the ejection of a plasma cloud with the great speed observed for moving type-IV bursts in the decameter-wave region occurs mainly in not very inhomogeneous magnetic fields.

After Warwick and Wood Haurwitz (1962) type-IV bursts in the decameter-wave region are especially typical for the production of high-energy particles in the active centre. In order to check if the short duration observed by Weiss and Philip of the moving type-IV bursts in the decameter-wave region could be explained by the radiation damping of high-energy electrons, we compared the maximum intensities at 3000 MHz of the type-IV bursts with and without a moving component in the decameter-wave region, because the maximum intensity in the centimeter region can also be regarded as an indication for the existence of high-energy particles. It was then, however, found that the intensity of type-IV bursts with a moving component in the decameter-wave region at 3000 MHz, on the average, is less than that of the type-IV bursts without a moving component. This suggests that the short duration of the moving type-IV bursts cannot be explained by radiation damping and that, further, the high ejection speeds are caused by magnetic-field configurations rather than by high-energy production in the flare.

In addition to the moving type-IV bursts there has been investigated also the stationary type-IV burst in the decameter-wave region. As is recognizable from Figure 2, the storm continua with a component at 40 MHz and less are associated

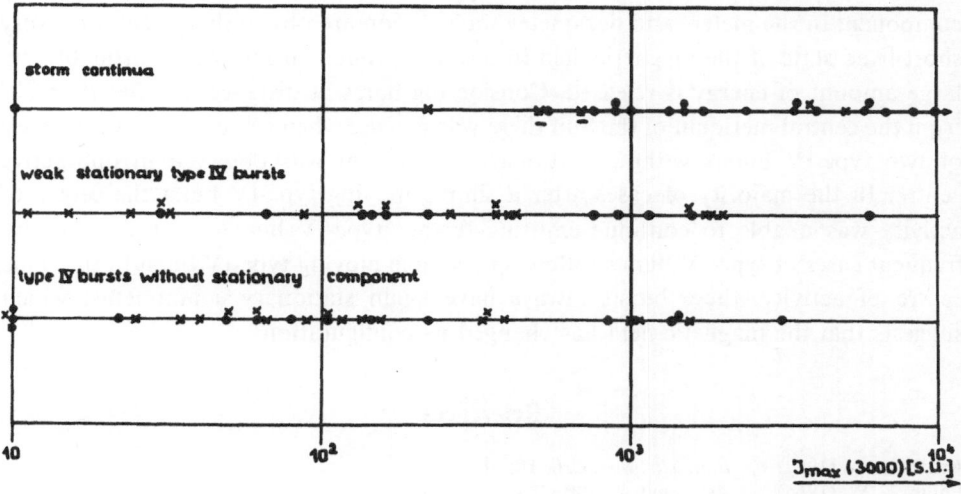

FIG. 3. *The distribution of $I_{max}(3000)$ for type-IV bursts with and without stationary component at 40 MHz.*

in 86% of the cases with spots of the class 'complex'. This value is far in excess of the average of the total of the type-IV bursts. Obviously, the configurations producing storm continua are very stable. Nearly all the active centres, which for one or two solar rotations have produced type-IV bursts, have emitted during their formation one or more storm continua with a component in the decameter-wave region. It is, further, of some interest that the only two storm continua which had no component in the decameter-wave region belonged to the spot groups of class H' or R, respectively.

Only 41% of the centres which only produced stationary type-IV bursts of less than 2 hours' duration and low intensity belonged to the class named 'complex'. As can be seen from Figure 3 a distinction between storm continua and stationary type-IV bursts of low intensity in the decameter-wave region is indicated by the intensity of the component in the cm-wave region. The figure contains the distribution of $I_{max}(3000)$ for storm continua, weak stationary bursts, and type-IV bursts without a stationary component in the decameter-wave region. It can be seen that storm continua in the decameter-wave region preferably have intense components at 3000 MHz. Hence, it follows that *just* the strong stationary bursts and *not* the moving type-IV bursts in the decameter-wave region are the ones typical for the production of high-energy particles, in spite of the fact that stationary bursts are likely to be interpreted as Čerenkov radiation and, hence, not expected to be produced by particles with energies in the MeV region.

In conclusion it can be said that the type-IV bursts, which have no stationary component in the meter- and decameter-wave region are obviously typical for a very short-lived state of the magnetic field in the spot groups during which, probably, no large amount of energy is released. Considering flares at distances smaller than 60° from the central meridian of the Sun there was no case when a subsequent occurrence of two type-IV bursts without stationary component was observed in one active centre. In the majority of cases after a simple moving type-IV burst the centre of activity was unable to continue emitting further type-IV bursts. In the rather in-frequent cases of type-IV bursts following a simple moving type-IV burst in the same centre of activity, these bursts always have again stationary components, which suggests that the magnetic field has changed its configuration.

## References

Antalova, A. (1967)      *Bull. astr. Inst. Csl.*, **18**, 61.
Philip, K.W. (1964)      *Astrophys. J.*, **139**, 723.
Warwick, C.S., Wood Haurwitz, M.J. (1962)      *J. geophys. Res.*, **67**, 1317.
Weiss, A.A. (1963)      *Austr. J. Phys.*, **16**, 526.

# PROPERTIES OF SOURCES OF THE SLOWLY VARYING COMPONENT OF 2 cm SOLAR RADIO EMISSION

V. G. NAGNIBEDA

(*Astronomical Observatory of Leningrad University, Leningrad V-178, U.S.S.R.*)

To study the nature of the local sources of solar radio emission connected with active regions, it is important to investigate the structure of such sources and their emission spectra. These problems are being investigated in detail by a group of workers of the Radio Astronomy Department of the Pulkovo Observatory led by G. B. Gelfreikh. The Pulkovo large radio telescope used for the observations allows them to investigate the solar radio sources at the whole cm-wavelength range with a high resolution reaching 40 sec of arc at the 2-cm wave. Observations are taken at 2·0, 3·2, 4·4, 6·6, and 9-cm waves. The author observes at the 2-cm wave.

The results of the observations at the 2-cm wave with a high resolution permit, in the first place, to draw the reliable conclusion that the radio source over the active region is not a homogeneous formation. It turned out that the structure of the radio source depends on the structure of the magnetic field of the active region and optically observed structures – spot groups and calcium plages connected with the magnetic field. Up to now a clear indication on the existence of radio sources over separate spots in one group was made by Pulkovo radio astronomers only by making use of the data obtained in the polarized emission during the Sun eclipse on April 20, 1958 (Korolkov *et al.*, 1960). The same result for an unpolarized emission was obtained by a number of observers during the Sun eclipse on May 20, 1966.

In particular, G. Apushkinsky and V. Nagnibeda got this result at 0·8 cm, 3·3 and 10-cm wavelengths from the record of the eclipse of the source connected with the spot group no. 57 *(Solnechnye dannye)* in the Southern hemisphere of the Sun. The one-dimensional brightness distributions were taken from the record of the opening and the closing of the radio source (Apushkinsky *et al.*). Having one-dimensional distributions in two directions corresponding to the opening and the closing of the radio source, it is possible to draw two-dimensional images of the radio source (Figure 1). It is seen from the picture that the radio source consists of two main details corresponding to two groups of spots of different polarity. Some bright small detail stands out on a less bright background. (The brightness temperatures of a bright detail are equal $4 \times 10^6$, $0·6 \times 10^6$, and $0·08 \times 10^6$ °K at the waves 10, 3·3 and 0·8 cm respectively.)

Our observations with the Pulkovo large radio telescope show a complicated structure of an enhanced radio emission over active regions too. A comparison of

*Kiepenheuer (ed.), Structure and Development of Solar Active Regions*, 575–580. © *I.A.U.*

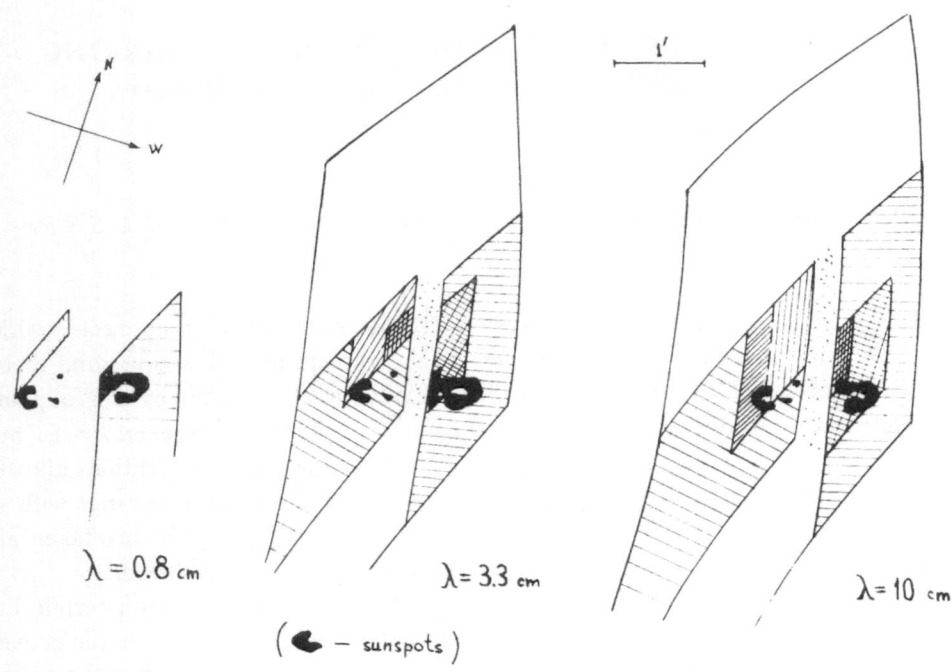

FIG. 1. *Two-dimensional images of the radio source, May 20, 1966.*

records of enhanced radio emission and curves of calcium-plages brightness distribution provides some interesting results. Such distributions were obtained from calcium spectroheliograms on the microphotometer with a split, corresponding to the antenna pattern. The radio-emission intensity over the calcium plages repeats in many respects the plage curves, increasing in the region of the spot groups. But over the spots themselves, in the place where the magnetic field of the spot penetrates into the Sun's atmosphere, the radio-emission intensity increases sharply, forming a bright radio source. In Figure 2 are shown radio sources associated with spot groups no. 144 and no. 145 (using the numeration of the bulletin *Solnechnye dannye*) observed in September 1966. Plage-brightness distributions are also given there.

First of all, it is necessary to point out that the radio sources over the spots have a small width, not more than 0".7. Separate narrow sources are clearly seen over the main and tail spots of the bipolar group no. 145. It all means that the radio source probably is not high over the photosphere as, according to the recent paper of Livshitz *et al.* (1966), the magnetic field of the spot disperses strongly already on the height of 2000–3000 km. On the other hand, this fact points to the high brightness temperature of the emission of such sources, which differs sharply from the brightness temperature of the surrounding background. Indeed, the brightness temperature of

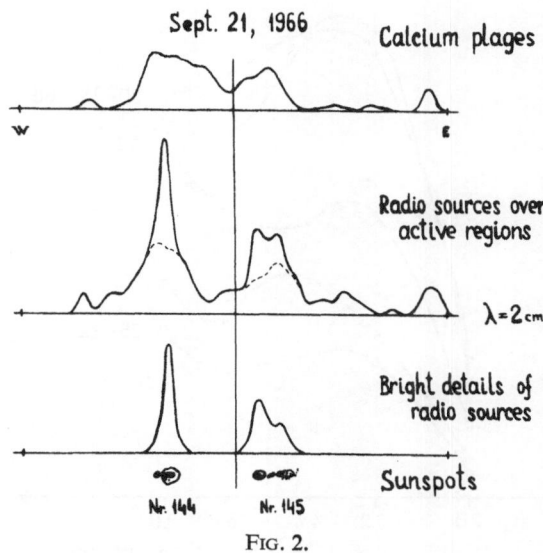

FIG. 2.

the bright detail of the radio source over the group no. 144 reached $2-2\cdot5 \times 10^5\,°K$, whereas the background temperature is scores of thousands of degrees. It means that the mechanism of radio emission of the bright part of the source and the background are different. To assume the thermal nature of the radio emission of the local sources it is necessary to explain the existence of strongly heated regions at a small height over the photosphere, which is probably rather difficult to do.

It is interesting to note that the flux from the nucleus of the source over the group no. 144 falls when the source moves to the limb of the Sun, but the size of the source decreases. Thus the brightness temperature remains constant all the time. This fact can be explained by assuming that we deal with a flat source of a considerable optical thickness.

In solving the problem of the radio-emission mechanism an important role is played by the spectrum of this emission, especially at short cm wavelengths. It follows from the observations in Pulkovo that the solar local radio sources have various spectra (Akhmedov *et al.*, 1966), though most of them possess a marked decrease of the flux in the 3·2–2·0 cm wavelengths. Radio sources over the groups no. 144 and no. 145 have such a spectrum (Figure 3). It is seen from the figure that in the 3–9 cm wavelengths the spectrum is approximately flat, but at the 3–2 cm wavelengths the flux decreases sharply. In our opinion, this fact excludes the presence in the radio emission of a source with an essential contribution of the bremsstrahlung of the thermal electrons demanding for its explanation a gyro-radiation of the thermal electrons at the gyro-frequency and its harmonics. This mechanism also explains the decrease of the radio-emission flux at all wavelengths when the source of the group no. 144 moves to the limb of the Sun.

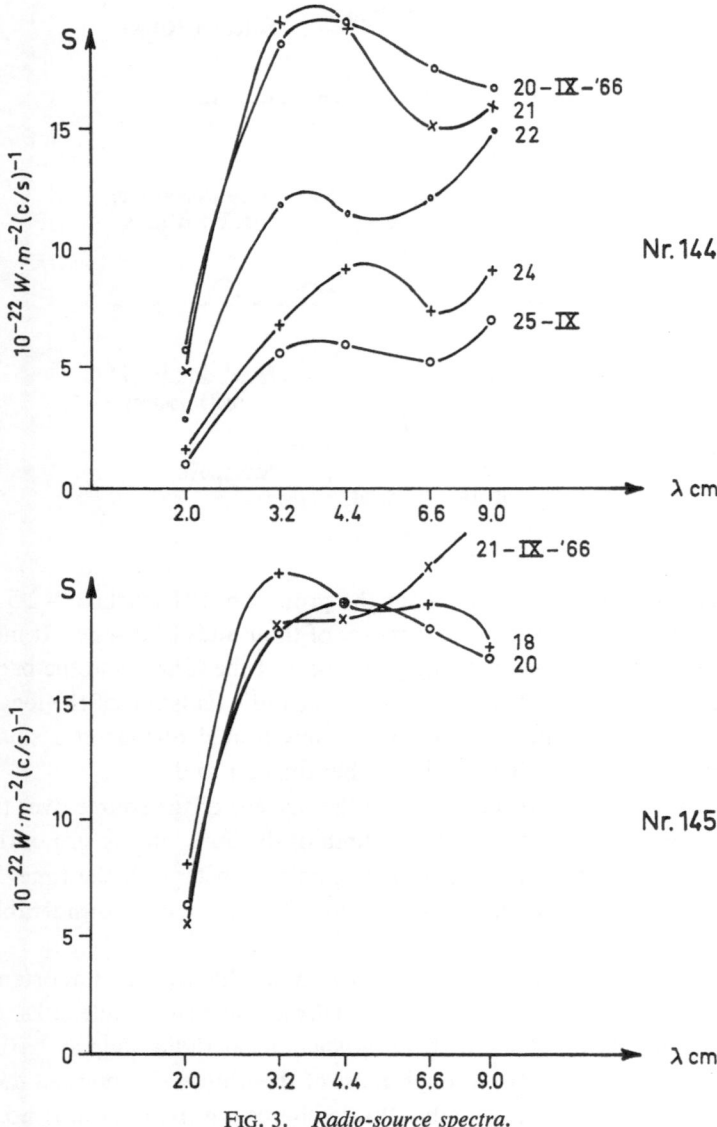

FIG. 3. *Radio-source spectra.*

It follows from our observations that at the 2-cm wavelengths the share of the radio emission connected with the calcium plage in the total flux of the radio source is rather great, falling markedly with the increase of the wavelength (in the cm range). On the other hand, only in high-resolution observations in unpolarized emission it is possible to separate the bright nucleus of the source connected with the spot. This can be seen from the record obtained on September 21, 1966, with 0'.7 and 4' width beam (the latter being obtained by smoothing) (Figure 4). Therefore the spectrum of

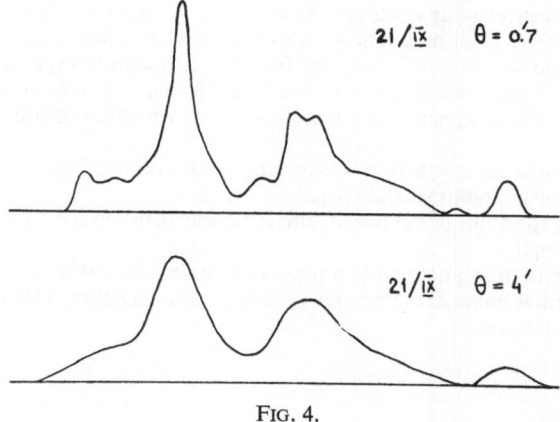

FIG. 4.

such a source plotted on the basis of observations with insufficiently high resolution, will have an extra short-wave part. This probably was the case in the paper of Tsuchiya and Nagane (1967).

The following conclusions can be drawn from the above:

(1) In the distribution of the brightness of the radio source the complicated structure of the active region is well seen: (a) it is connected with the spot component which for bipolar groups of great longitude gives two sources; (b) the component is connected with calcium plages.

(2) The small size of bright details of the radio source implies that their height over the photosphere cannot be large, and the high brightness temperature demands the existence of strong heated regions at this height.

(3) The observed radio-emission spectrum of bright details of the radio source at the cm wavelengths (3–10 cm) requires for its explanation a gyro-radiation of the thermal electrons and excludes the essential influence of the bremsstrahlung.

### References

Akhmedov, Sh., Borovik, V., Korzhavin, A., Nagnibeda, V., Peterova, N., Spitkovsky, V. (1966) *Soln. Dann. Bjull.*, **2**.
Apushkinsky G., Grebinsky, A., Enikeev, R., Levtchenko, M., Nagnibeda, V.        (to be published).
Korolkov, D. V., Soboleva, N. S., Gelfreikh, G. B. (1960)    *Izv. glav. astr. Obs. Pulkove*, **21**, 81.
Livshitz, M., Obridko, V., Pikel'ner, S. (1966)    *Astr. Zh.*, **43**, 1135.
Tsuchiya, A., Nagane, K. (1967)    *Solar Phys.*, **1**, 121.

## DISCUSSION

*De Jager:* The very steep decrease of spectral intensities for $\lambda \lesssim 2$ cm is surprising. What is the absorption mechanism?

*Nagnibeda:* The steep decrease of intensity for $\lambda < 3$ cm is the observed fact. Such a spectrum of the slowly varying component of solar emission was explained by Zhelezniakov ('*The Radio Emission of the Sun and the Planets*', 1964) on the basis of gyro-resonance radiation of the thermal electrons. He showed that the steep decrease of the intensity for $\lambda < 3$ cm might be explained by the decrease of the height of emission layers with frequencies $\omega \simeq 2\omega_x$, $3\omega_x$ ($\omega_x =$ gyro-frequency) and its passage from the corona to the chromosphere and the decrease of the $\tau$ because of insufficient magnetic-field strength.

*Krüger:* Have you made any distinction between the spectra originating in the bright cores of the sources of the s-component and its surroundings?

*Nagnibeda:* Yes, the spectrum of the bright core is steeper than the spectrum of its surroundings at the 2–3·2-cm wavelengths.

*Krüger:* I should like to draw attention to a paper of Zlotnik, who calculated spectra of the s-component basing on thermal bremsstrahlung and gyro-resonance emission. This paper will appear in the *Soviet Astronomical Journal*.

# SOME PROPERTIES OF THE SOURCES OF SLOWLY VARYING COMPONENT AND OF BURSTS AT 612 Mc/s*

G. Swarup, M. R. Kundu, V. K. Kapahi and J. D. Isloor

*(Tata Institute of Fundamental Research, Bombay 5, India)*

A high-resolution 24-element, E–W interferometer situated at Kalyan near Bombay has been used to study the sources of slowly varying component and of bursts at a frequency of 612 Mc/s. This interferometer has a half-power beamwidth of 2·8 min of arc, and the fan beams are located about a degree apart in the sky. Thus, a strip scan is obtained roughly every 4 min as the Sun drifts through the beams. About 20 slowly varying sources and 11 bursts were observed during the period June 1965 to February 1967 at 612 Mc/s.

## 1. Slowly Varying Component

The half-power widths of the sources of S-component determined after removing the estimated quiet Sun level from the recorded strip scans, were found to vary from about 7 min of arc to about 13 min of arc. Figure 1a shows a histogram of the angular size of the sources.

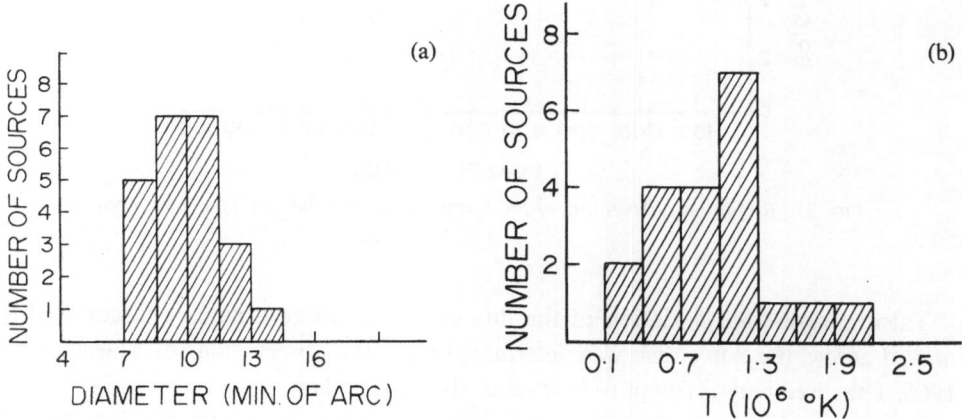

Fig. 1. *The distribution of (a) the angular size and (b) the brightness temperature of the sources of slowly varying component.*

* Presented by M. R. Kundu.

*Kiepenheuer (ed.), Structure and Development of Solar Active Regions, 581–584. © I.A.U.*

In the absence of data on daily flux values of the Sun around 600 Mc/s for the entire period, flux values at 1000 Mc/s from Toyokawa were used to normalize the observed scans. A multiplication factor of 0·765 was used to reduce the Toyokawa daily flux values at 1000 Mc/s to those at 606 Mc/s. This factor was determined by comparing the flux values of Sagamore Hill at 606 Mc/s when they became available with those of Toyokawa. This method seems to be justified since there is a good correlation between the Toyokawa and Sagamore Hill flux values. The resulting brightness temperatures of the sources are shown in the form of a histogram in Figure 1b. As shown in Figure 2, the brightness temperature is roughly proportional to the sunspot area.

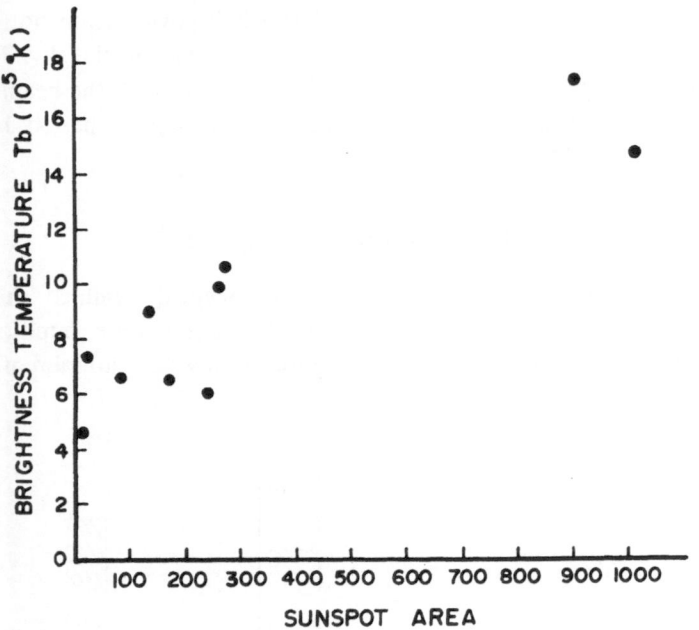

FIG. 2.    *Brightness temperature of the S-component as a function of sunspot area.*

The centre-to-limb variation of the flux density of three isolated sources as they moved across the Sun's disk was determined from the observations of March–May, 1966. The directivity is found to be steeper than cosine law.

In order to determine the height of a source, we have plotted the displacement of the source from the axis of rotation against the sine of the angle of rotation of the Sun. Comparing the rate of rotation of the source with that of the most probable optical source, the height was determined. The heights of three sources having CMP on March 20, April 17, and May 15, 1966 were found to lie between 40000 and 65000 km above the photosphere.

## 2. Bursts

Most of the 11 bursts recorded during the period March 1966 to March 1967 lasted for less than about 10 min, appearing only on a few successive scans (Figure 3). Only one long-duration burst lasting over 4 hours was recorded on January 18, 1967.

The angular width of the burst sources was generally found to be about 3 min of

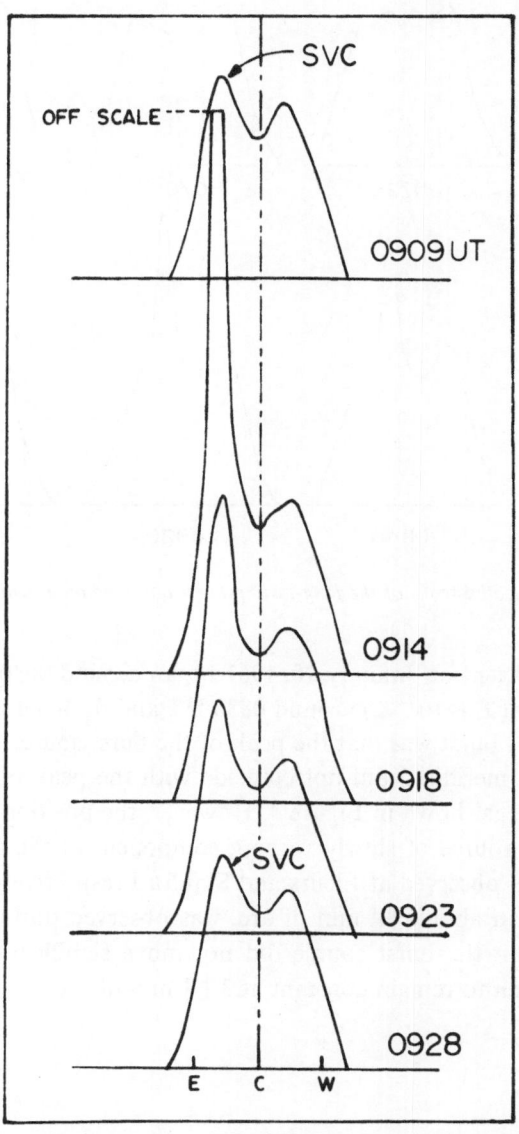

FIG. 3.    *Successive strip scans during a burst recorded on March 16, 1966.*

arc; three sources, however, being less than 1·5 min of arc wide. The maximum size recorded was only 4·9 min of arc. The total Sun flux values of Toyokawa at 1000 Mc/s were used (with a multiplication factor of 0·765) to normalize the scans and to determine the total flux for each burst. The brightness temperatures computed by assuming the sources to be circular vary between $10^6$ °K and $10^8$ °K. No burst was observed near the limb to enable a determination of the height of the burst source.

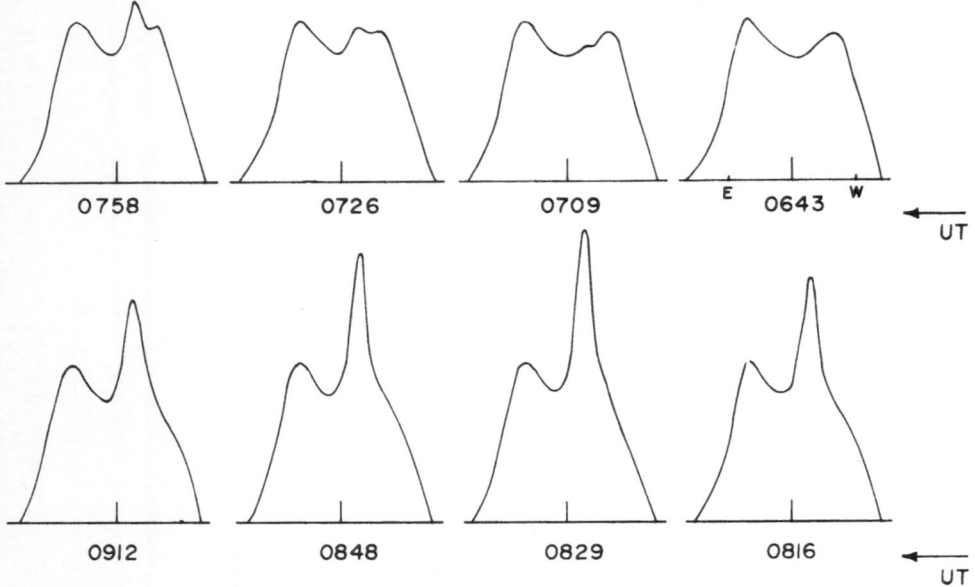

FIG. 4.  *Selected scans of the long-lasting burst observed on January 18, 1967.*

The long-lasting event on January 18, 1967 began around 0630 UT, gradually rose to its peak intensity ($T_b \sim 10^7$ °K) around 0830 UT and declined thereafter. An interesting feature of this burst was that the peak of the burst source (situated 4·5 min of arc West of central meridian) did not coincide with the peak of a source of slowly varying component, as shown in Figure 4. However, the position of the burst agreed well with that of a source of slowly varying component at the shorter wavelengths of 21 cm and 9·1 cm observed at Fleurs and Stanford respectively. No change in the position, greater than about $\pm \frac{1}{2}$ min of arc, was observed during the period of the burst, indicating that the burst source did not move significantly. The size of the burst source was seen to remain constant at $3 \pm \frac{1}{2}$ min of arc.

# SATELLITE OBSERVATIONS OF SOLAR RADIO BURSTS*

R. G. Stone, H. H. Malitson, J. K. Alexander, and C. R. Somerlock

*(NASA Goddard Space Flight Center, Greenbelt, Md., U.S.A.)*

The second Advanced Technology Satellite (ATS-II) carries a radio-astronomy experiment designed to perform radio-noise measurements in the range 0·5 to 3·0 MHz from above the terrestrial ionosphere. The spacecraft was launched into an 11 000-km apogee, 180-km perigee orbit on April 6, 1967, and since there was considerable solar activity in the ensuing months it is possible to present a preliminary description of the kind of solar observations this experiment will provide.

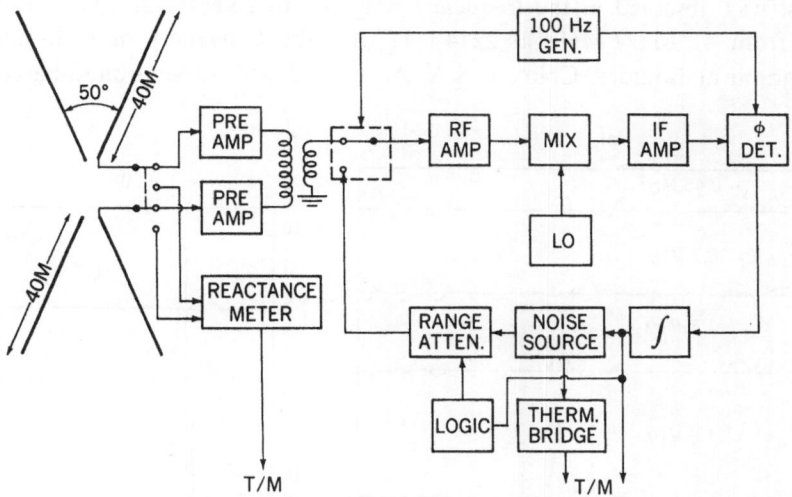

Fig. 1.    *Block diagram of ATS-II radio-astronomy experiment.*

The essential features of the radio-astronomy experiment are shown in block diagram form in Figure 1. The satellite carries four 40-m booms in an X-configuration for a gravity-gradient stabilization experiment. One pair of these booms is used to form a dipole antenna which is a half-wavelength at about 1·8 MHz. The dipole is connected through a pair of high impedance pre-amplifiers to a Ryle-Vonberg type radiometer. The radiometer consists of a receiver which is switched between the

---

* Presented by M. R. Kundu.

*Kiepenheuer (ed.), Structure and Development of Solar Active Regions*, 585–587. © *I.A.U.*

antenna and a reference-noise source at 100 Hz. The difference between the antenna signal and the noise-source signal is synchronously detected to provide an error signal which adjusts the noise-source output to match the antenna signal. The noise-source output, which is then equal to the noise power from the antenna, is measured in a precision thermistor bridge and telemetered to the ground. The system has an effective bandwidth of 40 kHz and an integration time constant of about 1 sec. The radiometer steps through seven frequencies – 0·45, 0·7, 0·9, 1·1, 1·6, 2·2, and 3·0 MHz – in 40 sec. Once during each 40-sec cycle the antenna is connected to a reactance meter to provide a measurement of the antenna capacitance. Since the antenna is shared with a second experiment on the satellite, radio-noise measurements are obtained for a period of 10 min, and then the experiment is off during alternate 10-min periods.

Numerous solar radio bursts have been observed with the ATS-II experiment, and although data analysis is still in progress we can describe phenomena being seen in a preliminary fashion. Figure 2 shows two type-III bursts observed on May 20, 1967. The first was observed in the frequency range 41 to 20 MHz at 2028 UT, and the second from 41 to 27 MHz at 2214 UT, with the University of Colorado radio spectrograph at Boulder, Colo., U.S.A. Associated with these ground-based obser-

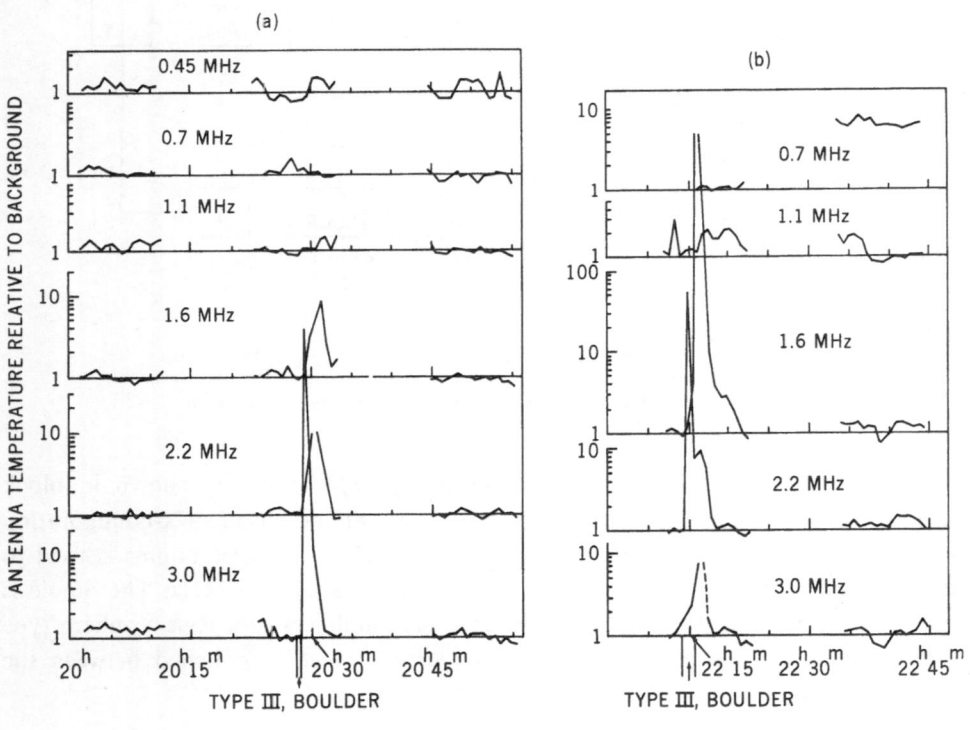

FIG. 2.  *Examples of Type-III solar radio bursts observed at low frequencies.*

vations, one can see abrupt enhancements in the low-frequency noise levels down to 1 MHz. If the type-III bursts are generated by moving disturbances which trigger radiation at the local electron-plasma frequency, then we are observing phenomena out to the order of 30 solar radii. By measuring the time of occurrence of the burst at each frequency and noting the time delay for the burst to occur at lower frequencies, one can deduce the velocity of the disturbance in the corona. For a reasonable model of the coronal electron-density distribution we get velocities which are typically 0·1–0·3 c. From the burst durations at each frequency we can estimate the temperature distribution in the outer corona. If one assumes that electron collisions are the principal mechanism for the decay, we find temperatures the order of $2 \times 10^5$ °K at 15 $R_\odot$. Of course, these numbers are derived from a very small sample of measurements; when analysis of a larger data sample is completed we can expect a much more reliable picture of the interaction of energetic particle streams with the outer corona.

On several occasions we have observed intense noise bursts associated with complex radio events including bursts of types III, II and IV observed on the ground. In such cases the noise level has been found to exceed the background level by at least 3 orders of magnitude from 3 MHz down to at least 0·7 MHz. If one assumes an apparent source diameter of 1° at 1 MHz, then this implies an equivalent brightness temperature in excess of $10^{15}$ °K. Such intense non-thermal radio bursts appear to be associated with the initial phases of complex, energetic, flare-associated events.

# COMPARISON OF 8-mm SOLAR RADIO FEATURES WITH LOCAL MAGNETIC FIELDS AND CHROMOSPHERIC FEATURES

V. Efanov, I. Moiseev, and A. Severny

*(Crimean Astrophysical Observatory, Nauchny, Crimea, U.S.S.R.)*

The new big 22-m radio-telescope of Crimean Observatory shown on Figure 1 appeared to be of very good quality: for radio emission at $\lambda = 8$ mm the antenna beam is $1.6$, the efficiency $\cong 0.4$, and the error of automatic pointing (guiding) does not exceed $\pm 20''$. A Dicke type radiometer was used as receiver possessing a sensitivity $\cong 2°K$ at the time constant $\tau = 1$ sec, and pass-band $\sim 30$ MHz. The error in the

Fig. 1.

*Kiepenheuer (ed.), Structure and Development of Solar Active Regions, 588–593. © I.A.U.*

contrast determination does not exceed $\pm 0.5\%$ of the level nearly corresponding to that of the quiet Sun. The scanning of the solar image is made in right ascension with velocity 0·5 sec in sec of time, the distance between successive scans in $\delta$ being 1.'5. An example of a single scan is shown in Figure 2, where the scan through Jupiter (for the determination of antenna beam) is also presented (for description see Moiseev, 1968).

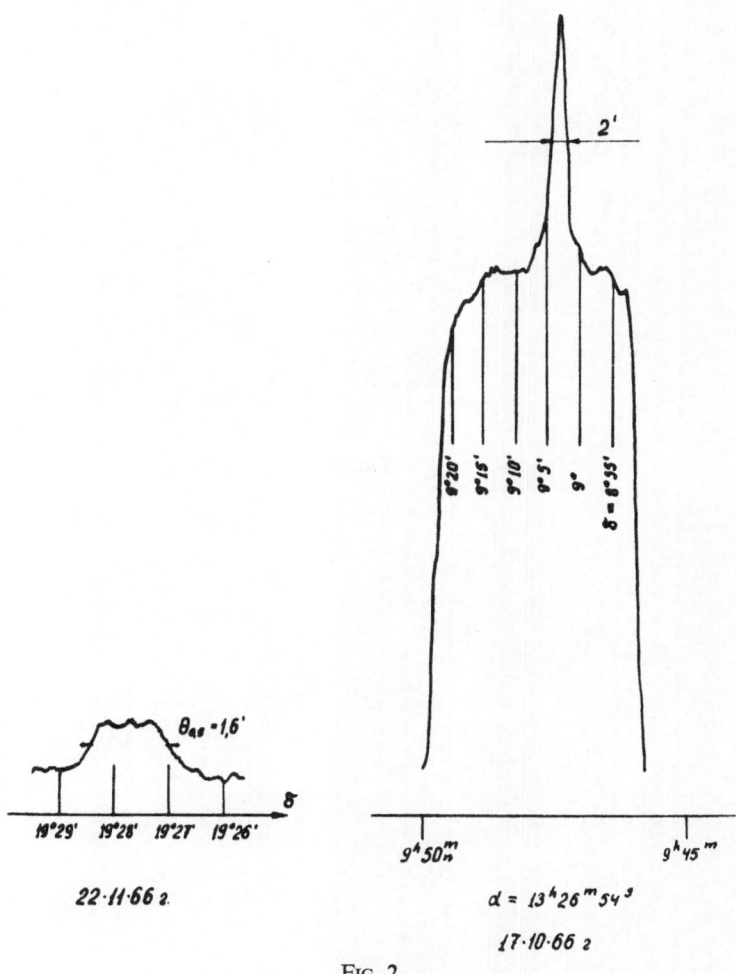

FIG. 2.

The maps of radio-isophotes (in % of undisturbed background) were obtained for October 24, 25, 26, 1966, simultaneously with the maps of isogauss of longitudinal magnetic fields over the whole disk recorded with the aid of solar magnetograph of Crimean Solar tower with the resolution $2''5 \times 27''$ and sensitivity $\simeq 2$ gs. These overlapping maps are shown in Figures 3–5 (full lines are radio-isophotes and thin lines

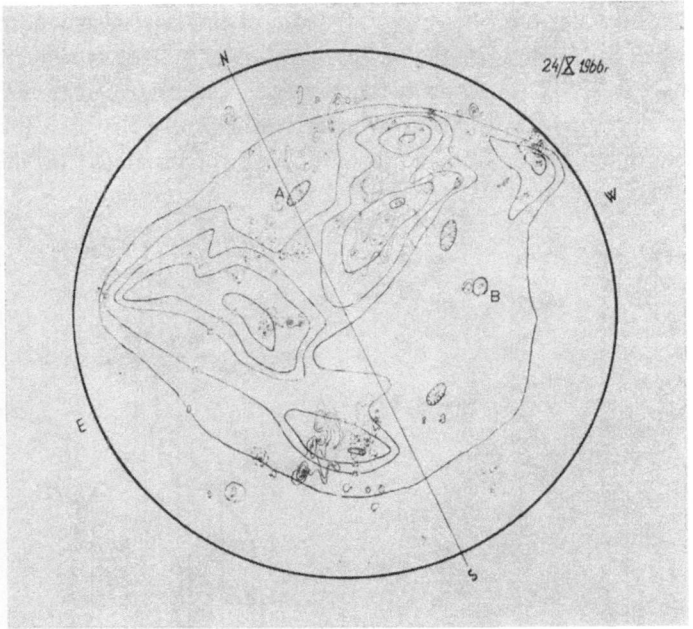

Fig. 3.

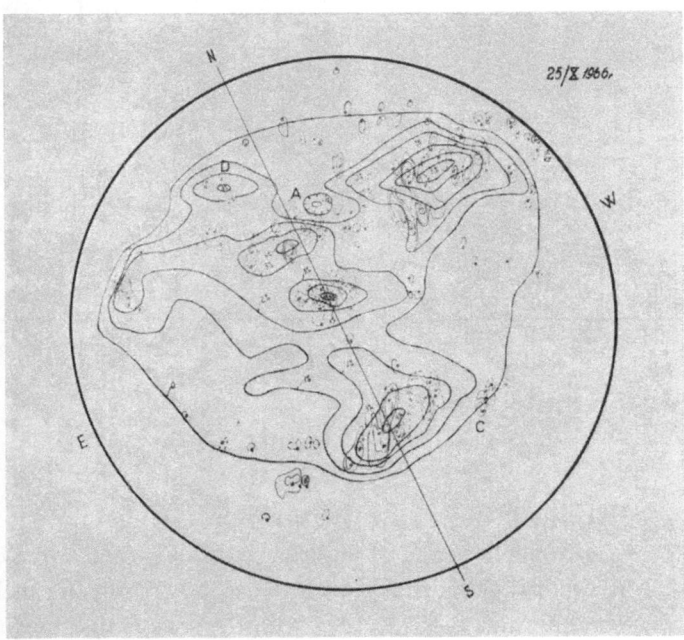

Fig. 4.

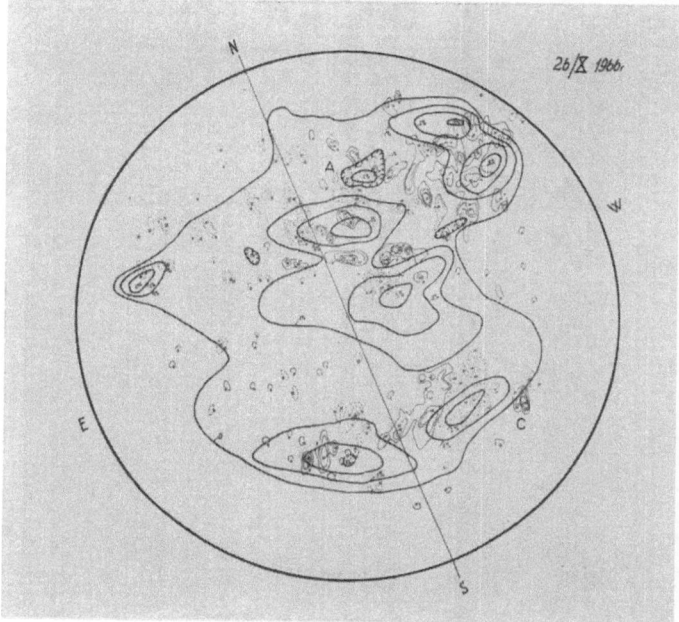

FIG. 5.

show magnetic-field strength starting with 5 or 10 gs). The main results of comparison between these two maps as well as between these maps and K and Hα-spectrohelio-grams obtained for the same day at double spectroheliograph of the Crimean tower (see example in Figure 6, Oct. 25, 1966) follow:

(1) All features of enhanced 8-mm radio emission are found in the regions of local magnetic fields. The reverse is not true – appreciable local magnetic fields are observed without enhanced 8-mm radio emission.

(2) In several cases the local magnetic fields are not accompanied by enhanced 8-mm emission when a dark filament is observed in the area in question. (Some examples are presented in Figures 3–5). Sometimes the presence of a high contrast dark filament leads to the depression in the level of enhanced radio emission. (See examples: region A on the map 24·10 and 26·10.)

(3) In some places (the regions A for the map 24·10 and 26·10) we observe the depression of 8-mm radio emission below the undisturbed level of the quiet Sun, and these regions are the regions without local magnetic fields but they are occupied by dark filaments (regions dashed from inside). The depression of radio emission in mm region associated with dark filaments was first observed by Khangildin (1964).

(4) One compact active region with developed local bipolar magnetic field and sunspots and plages did not show enhanced radio emission in 8 mm (region C in

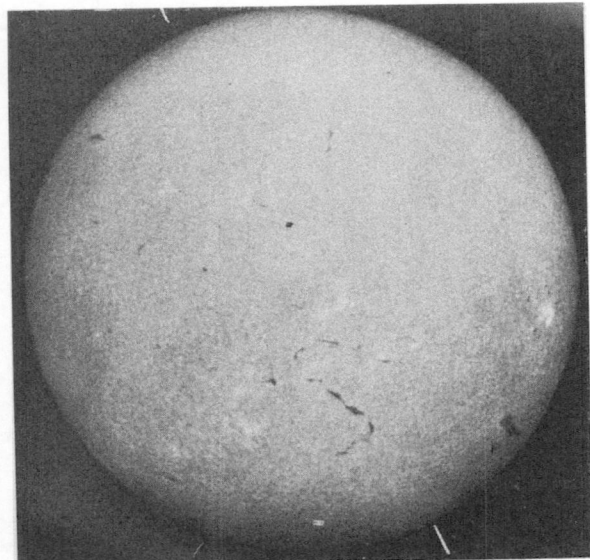

FIG. 6a.

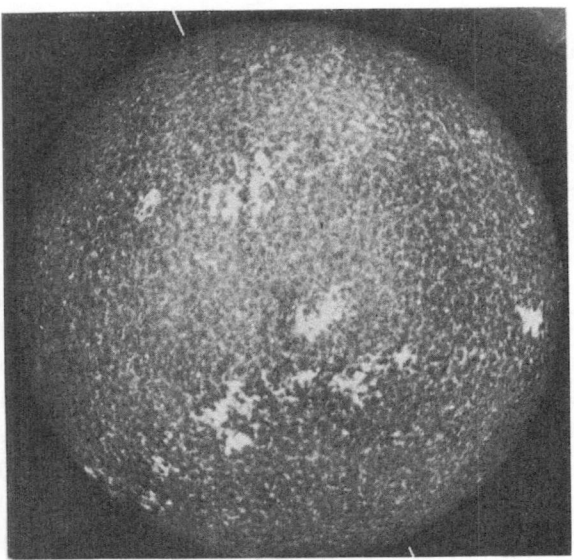

FIG. 6b.

Figure 4 and 5). This case can eventually be connected with the influence of the limited resolving power of the radio-telescope.

(5) The most interesting are two regions where no appreciable features on K and Hα-spectroheliograms can be found but they are occupied by local magnetic fields and enhanced 8-mm radio emission as well (the region B on the map 24·10 and D on the map 25·10). This could probably be compared with the recent comparison by Tousey of the regions enhanced in X-rays with the regions of cm enhanced radio emission having no counterpart in K-spectroheliograms. Possibly these peculiar regions are some condensations confined by magnetic fields having their 'roots' deep in the photosphere but elevated *above* the level of chromosphere responsible for K-line emission. If charged particles are trapped in these regions and magnetic fields are pulsating, as we illustrated in an earlier talk (Severny, 1967) they could be able to produce X-rays due to accelerations exerted in the varying magnetic fields.

## References

Khangildin, U. (1964)     *Astr. Zu.*, **41**, 302.
Moiseev, I. (1968)     *Izv. Krym. astrofiz. Obs.*, **38** (in press).
Severny, A. (1967)     *Astr. Zu.*, **44**, 481, see also page 233 of this volume.

## DISCUSSION

*De Jager:* Could you detect limb brightening (or darkening) at 8 mm in the quiet Sun?

*Severny:* It is difficult to say definitely. Sometimes the records across the disk look like a 'table', i.e. they show abrupt fall of intensity which remains about the same level inside the disk.

*De Jager:* Were the regions of enhanced 8-mm emission without associated optical features perhaps related to optical magnetic features that appeared later, or that has disappeared already?

*Severny:* We have records of magnetic fields and radio emission only for 3 days, so the data are not sufficient to have a definite answer, but magnetic features, as a rule, appear earlier and disappear later than optical features and this is why we can expect the regions of enhanced radio emission being connected with these magnetic features.

# RADIO EMISSION OF SPOTGROUPS

J. Kleczek, J. Olmr        and        A. Krüger
*(Astronomical Institute,*        *(Heinrich-Hertz-Institut,*
*Ondřejov, C.S.R.)*        *Berlin-Adlershof, D.D.R.)*

## ABSTRACT

The slowly varying component (SVC) of solar radio emission and burst activity associated with different types of spotgroups from the past solar minimum have been investigated statistically.

The solar minimum is an appropriate period for studying some aspects of solar activity. The isolation of events in space and time facilitates, among other things, the study of the relation between optical and radio phenomena. The days with no or only one spot group seem convenient for studying the slowly varying component and burst activity of individual types of spotgroups. As the sequence of Zürich types approximately corresponds to the evolution of active centres, one can get statistically an idea of the evolution of active regions in radio waves, which is otherwise possible only by interferometric methods.

In the last solar minimum there were 251 days without a spotgroup and 521 days with one group only, which challenge our attention. Fraunhofer maps were used as photospheric data.

We made the assumption that the photospheric situation does not change during a whole calendar day. It is not possible to verify how much our results are influenced by this assumption. The radio data used are spectral types of bursts and 1420 MHz radioheliograms published in the *Quarterly Bulletin on Solar Activity* and single frequency records from the Heinrich-Hertz Institute as well as from some other observatories. We assumed that the recorded radio events were produced by the one spotgroup present on the Sun on the selected days. This assumption does not distort our results, because the burst activity associated with plages is relatively low. The full text of the paper will be published in the *Bull. Astr. Inst. Czechoslovakia*, and only some of its main results will be mentioned here:

(1) The slowly varying component is composed of two parts: One connected with plages and the other with active centres containing sunspots. Their mean values are represented in Figure 1 which, at the same time, gives an impression of the average development of SVC of a single centre of activity. The SVC from plages seems to vary with the phase of the solar cycle, as may be seen from Figure 2. However, during the minimum period no clear correlation could be found between the SVC flux and any measure of plage activity (i.e. area, corrected area, Arcetri index, intensity of the

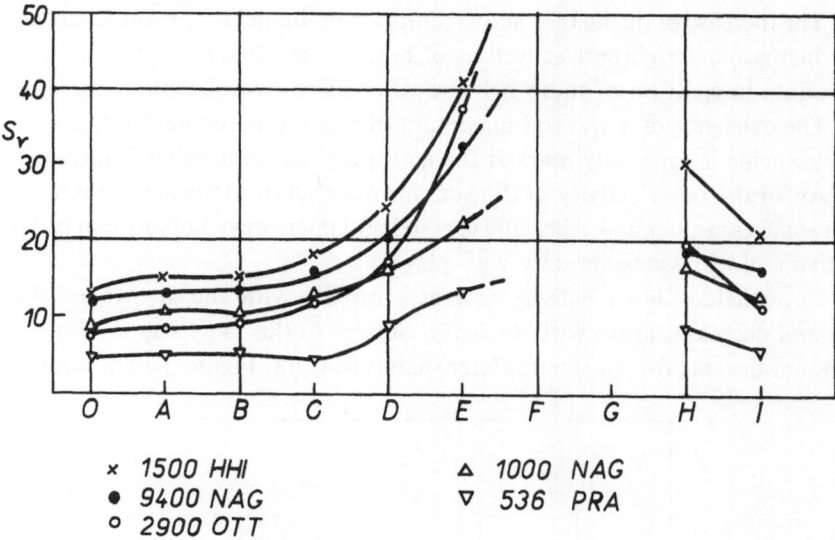

FIG. 1. *Mean values of flux of radiation for days with no spotgroup (O) or only one spotgroup (Zürich types A, B, ... I). According to this figure, the flux of an active region becomes significant after it has reached the type C.*

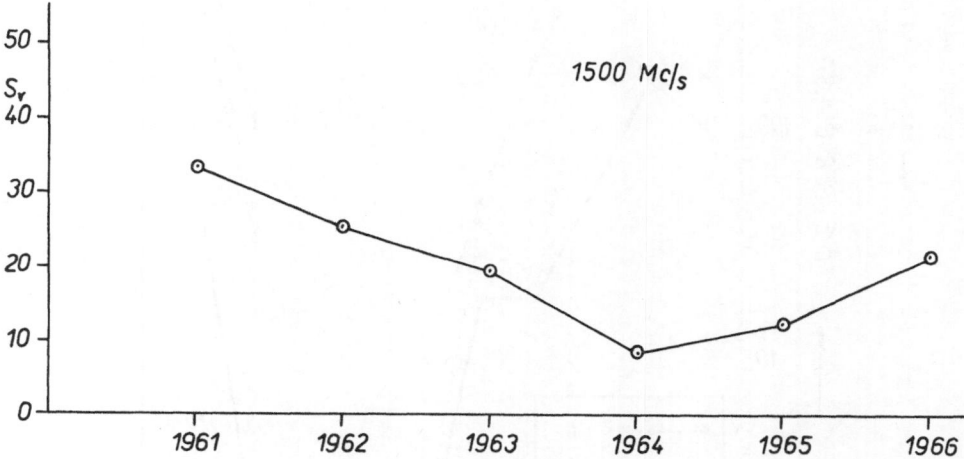

FIG. 2. *Mean yearly values of flux for days without spotgroups seem to depend on the phase of the cycle.*

brightest plages, and combinations of them). In this connection the role of a basic component is not yet clarified.

(2) The contribution of active centres to the total SVC flux depends upon the type of the spotgroup, as may be seen from Figure 1. It becomes noticeable for type C or later.

(3) The increase of the flux for active centres containing D, E, F and G groups is due to the increase of brightness as well as of bright area. This result has been obtained by isophote integration of radio heliograms at 21-cm wavelength.

(4) The existence of a spectral maximum due to a predominant influence of local magnetic fields is especially marked for active regions containing E groups.

(5) As for the burst activity of the Sun, the material from the past minimum shows, that in some cases, bursts – especially type-III and microwave bursts – can be produced by active regions associated only with plages.

(6) Nevertheless, burst activity increases rapidly with the growth of the active centre and decreases again with its decay. Similar to the SVC, also the burst activity becomes noticeable for spotgroups later than type C (cf. Figure 3). The connection to

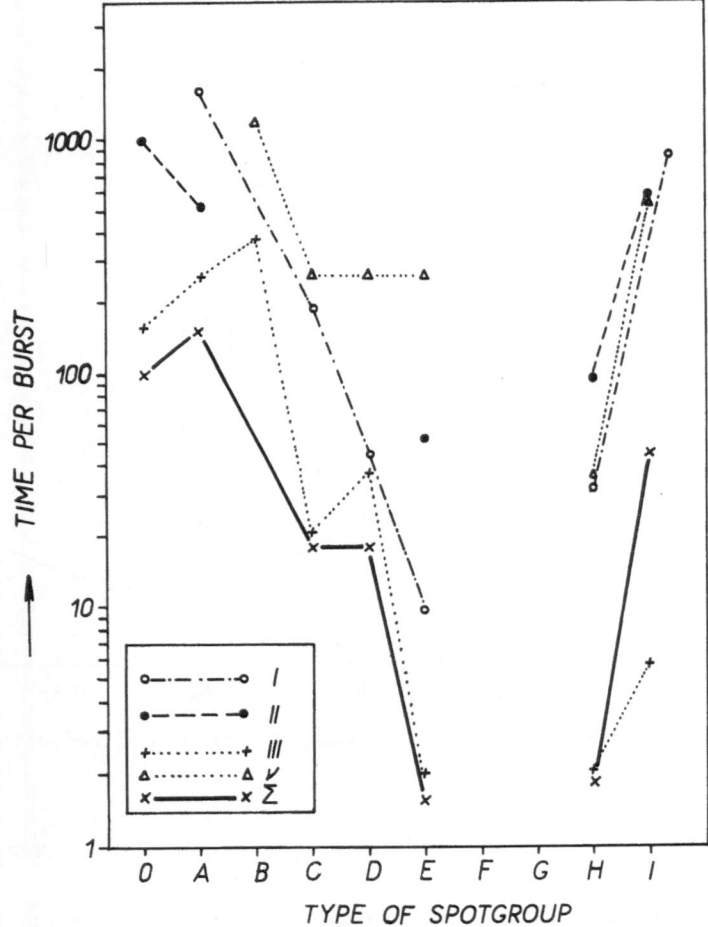

FIG. 3.    *Mean time interval between two following bursts depends on the type of burst and on the producing spotgroup.*

the type of related spotgroup is closest for microwave bursts and becomes looser with decreasing radio frequency of the burst events.

(7) In general, the burst activity of a given type of spotgroup seems also to depend upon the phase of the solar cycle. It has been considerably higher for the 1966 groups than for the previous years of the minimum.

## Conclusions

The results are not surprising but verify the close connection between the stage of spot development and the associated radio emission, both for the SVC and for different types of bursts. Because of the special role of spotgroups of D and E type shown here, this investigation underlines the importance of stronger complex magnetic fields for an increase of electron density and temperature as indicated by the SVC and resulting instabilities leading to different types of radio bursts and flares.

# QUELQUES RELATIONS ENTRE SURSAUTS RADIOÉLECTRIQUES SOLAIRES SUR ONDES DÉCIMÉTRIQUES ET CARACTÈRES MORPHOLOGIQUES DES ÉRUPTIONS CHROMOSPHÉRIQUES ASSOCIÉES

A. KOECKELENBERGH

*(Observatoire Royal de Belgique, Bruxelles, Belgium)*

### ABSTRACT

This work analyses some correlations or associations between certain features of solar radio bursts at 600 Mc/s and of associated chromospheric flares respectively. Subject of the analysis are 149 radio events observed at the Humain station and 125 associated flares observed at Uccle during the years 1957–61. It is found that the radioelectric energy of the bursts is associated with the eruptive area and the Hα intensity of the flares, whereas the duration of the bursts is associated with the Hα broadening. For bursts connected with a second phase, the eruptive area and the importance of the magnetic field have a bearing on the characteristics of the radio event.

Ce travail analyse les corrélations ou associations entre facteurs descriptifs des sursauts radioélectriques solaires sur 600 MHz et des éruptions associées. Le matériel d'observation consiste en 241 événements radioélectriques, observés sur 600 MHz à la station de Humain durant la période de juillet 1957 à décembre 1961 pendant les heures de surveillance effective au filtre de Lyot de l'Observatoire Royal de Belgique à Uccle. L'analyse porte sur 149 de ces événements et sur 125 éruptions associées.

Chaque événement radioélectrique et chaque éruption a été caractérisé par un certain nombre de facteurs. Seuls les facteurs qui donnent lieu à une association significative entre eux ont été retenus. L'étude conduit aux conclusions suivantes:

(1) La quantité d'énergie radioélectrique émise au cours d'un sursaut (c'est-à-dire le produit du flux par la durée) est corrélée à l'aire éruptive.

(2) L'intensité Hα des éruptions à leur maximum est corrélée à l'énergie rayonnée pendant la phase croissante du sursaut, alors que l'élargissement de la raie Hα est associé à la durée totale du sursaut. Le nombre d'atomes excités au niveau chromosphérique est lié à l'énergie rayonnée sur ondes décimétriques. L'agitation des atomes est pour sa part associée à la durée du sursaut. Ces constatations confirment le caractère d'onde mécanique dans le plasma qu'on s'accorde à attribuer à la perturbation.

(3) En ce qui concerne la seconde phase des sursauts, il s'agit sans doute de type IVB: radiation de synchrotron émise assez bas dans la couronne. L'aire éruptive et

l'importance du champ magnétique (en grandeur et en dimensions) déterminent la réserve énergétique susceptible d'être émise par rayonnement.

Une étude plus complète sera publiée ultérieurement.

## Remerciements

L'auteur adresse ses vifs remerciements à Monsieur le Professeur R. Coutrez qui lui a indiqué l'intérêt de ces recherches ainsi qu'au Service de Calcul Numérique de l'Université Libre de Bruxelles, au Comité National Belge pour la Coopération Géophysique et au Fonds National de la Recherche Scientifique.

# THE EFFECT OF COMPRESSION AND EXPANSION OF PLASMA ON THE GENERATION OF SYNCHROTRON RADIATION

S. J. Gopasyuk, N. N. Erushev, and Y. I. Neshpor

*(Crimean Astrophysical Observatory, Nauchny, Crimea, U.S.S.R.)*

### ABSTRACT

We consider the variation of the synchrotron-flux density of relativistic electrons with the power law spectrum when the region of generation of this emission initially experiences a homogeneous compression and then an expansion. Ionization losses have been taken into account. The velocities of compression and expansion have been taken as constant. It is shown that in the cases of compression or expansion the flux density at a given frequency changes as

$$S(t) \sim S_0 K^{\gamma}(t)$$

where $S_0$ = flux density before the compression, $\gamma$ = index of the power law spectrum, $K(t) = (H(t))/(H_0(t=0))$, and $H$ is the magnetic-field strength. In the case of compression $K(t) > 1 \cdot 0$ and in the case of expansion $K(t) < 1 \cdot 0$.

The results obtained are applied to an explanation of the increasing and decreasing parts of impulsive bursts of the centimeter range. Such a description of the impulsive bursts has allowed us to estimate both the parameters of the radiating region and the parameters of the differential energetic spectrum of relativistic electrons.

# THE GREAT BURST OF MAY 23, 1967*

JOHN P. CASTELLI, JULES AARONS     and     GLEN A. MICHAEL
*(Air Force Cambridge Research*     *(Air Weather Service, Ent AFB,*
*Laboratories, Bedford, Mass., U.S.A.)*          *Colo., U.S.A.)*

## ABSTRACT

One of the largest radio bursts on record took place on May 23, 1967. Peak flux densities ranged between 23 000 flux units ($10^{-22}$ wm$^{-2}$ Hz$^{-1}$) at 8800 MHz to about 370 000 units at 606 MHz. In addition to the high-accuracy measurements of the peak flux densities at 606, 1415, 2695, 4995, and 8800 MHz, sweep frequency observations from 19–39 MHz show Type-IV emission with Type-II bursts occurring during the Type-IV continuum. The associated flare was clearly visible in white light. In reviewing and compiling microwave-radio data recorded during earlier white-light flares, it was found that all but one of eight events listed by Svestka (1966) had high microwave flux densities associated with them. In comparing radio-burst intensity with optical flare importance for the series of three flares between 1809 and 2150 UT on May 23, there is only moderate agreement. The first radio burst was small; the third was by far the largest, while the second flare had the highest optical classification. The flux densities of the third burst may have been the highest ever recorded in the decimeter portion of the radio spectrum and amongst the largest four in the 8800 MHz region. It is suggested that the details of the development of the flare might be followed on radio flux-density plots to determine detailed correlation with particle events in space and with terrestrial effects.

* Presented by J. P. Castelli.

# THE DEVELOPMENT AND STRUCTURE OF AN ACTIVE REGION*

C. DE JAGER

(University Observatory and Space Research Laboratory, Utrecht, The Netherlands)

## 1. Introduction

The concept *active region*, previously called 'centre of activity', was introduced by D'Azambuja as "the totality of all visible phenomena accompanying the birth of sunspots". This idea was first placed in the centre of attention during the Convegno Volta in 1952 when Kiepenheuer put forward his paper: 'Was ist ein Aktivitätszentrum auf der Sonne?'. At that time our knowledge of an active region could still be condensed in a five-page paper. During the present symposium we have listened to 89 communications, all dealing with certain aspects of the active region. Particular attention was given to the role played by supergranulation and by detailed magnetic fields; this was possible since we now have precise observations with good spatial and time resolution of the magnetic field, and very detailed information on the related optical structures.

The present review is called 'Development and Structure of an Active Region', as it is the development that is essential; structures will be dealt with only in as far as they are important for our understanding of the way an active region develops. The mechanism behind the development of an active region may now be defined, with Kiepenheuer, as "the interplay of solar convection, differential rotation, and magnetic fields on the Sun".

A general remark that should be made after this Symposium refers to the difference between the resolution attained optically on one side, and in X-ray and radio observations on the other hand. With optical methods a spatial resolution of the order of $1''$ can occasionally be reached; sometimes even better observations are made. Radio and X-ray observations are still in the range of tens of seconds of arc or even minutes of arc. This hinders the intercomparison of coronal observations (which rely on radio and X-ray observations) with photospheric and chromospheric data. One of the most important and imperative developments in radio and X-ray astronomy will be to improve the resolution to something around $1''$.

---

* Invited summarizing review of the Budapest Symposium.

*Kiepenheuer (ed.), Structure and Development of Solar Active Regions*, 602–608. © *I.A.U.*

## 2. Large-Scale Development of the Active Region

As shown in Kiepenheuer's introductory paper the broad development of an active region can best be understood on the basis of Babcock's theory, but certain problems remain. The first is how to neutralise the poleward moving fields: it appears that the magnetic flux that moves poleward is about ten times larger than the flux eventually arriving. An answer to this problem has been given during the Budapest Symposium by Stenflo: when the weaker fields in and around an active region are taken into account there appears to be a better flux balance in the two magnetically oppositely polarized parts of active regions, than has been assumed hitherto. Beckers and Schröter found that the fields of the magnetic knots partly or wholly balance those of the spots.

A harder problem, not compatible with Babcock's theory, is that apparently new active regions or even new-cycle regions may appear to originate in old ones, or they may occur at the places of old ones. It remains to be examined whether this observation would break down Babcock's theory or not.

A further general aspect concerns the origin of solar active regions. There are clear indications that certain longitudes are favoured. They seem to be situated about 180° apart; both Mrs. Dodson-Prince and Švestka produced data pointing in that direction. However, as can be seen from a comparison of Figure 1 of Mrs. Dodson-Prince with figure 3 of Švestka, it is not clear whether the two favourite longitudes rotate with the normal photospheric speed (Carrington) or whether they find their basis in sub-photospheric regions rotating at a much faster rate.

In any case it seems that the notion *plage family* may be justified and may merit further attention.

## 3. The First Phase of Development of an Active Region

There are indications that the basic difference between the active and the quiet parts of the Sun is a quantitative rather than a qualitative one. The basic magnetic structures, also visible in the quiet Sun, appear in larger quantities in the active region. We should particularly stress the importance of the supergranulation for the development of solar activity.

In the quiet Sun some essential scales of length may be defined (see also the paper by Simon and Weiss), as given in Table 1.

The magnetic field is concentrated in the supergranulation boundaries where it is directed radially; there is no component at right angles.

Next we describe the various phases of development of an active region. The first phase is the arrival of magnetic flux at the surface. This is a rapid event, occurring during about 1 day; it soon leads to the development of a *bipolar magnetic region*. Here the question may be asked whether real bipolar magnetic regions do exist or

## Table 1

| object | life-time | characteristic size | size related to |
|---|---|---|---|
| granulation; spicules | 10 min | 1000 km | scale height in sub-photospheric region |
| mottles | 5–10 min | 1000 km | ? |
| rosettes | 1 day | 5000–10000 km | |
| supergranulation | some days | 30000 km | scale height somewhere deep in the convective region |
| large network | ? | 300000 km | giant cells in the deep convective zone |

whether refined magnetic measurements will eventually show that only complex magnetic regions occur on the disk.

Prior to the outbreak of spots the brightness difference in photospheric levels increases in the sense that the intergranular space becomes darker. Slightly later, pores originate; a fine example, from a movie-picture, was given by Rösch. The intensity in the pore is about 0·6 times the continuum intensity. It is very important to realise that the pores and spots never originate *inside* the supergranulation, but always at the *boundaries*, apparently due to a concentration of the magnetic flux by the supergranular motion field.

Still obscure is the part played in this connection by the objects called 'magnetic knots', 'invisible spots', or 'satellite spots'. These three terms refer to the same object! The first seems to be the best, since upon more detailed examination the 'invisible' spots nearly always appear to become visible, though they are less dark than the pores; the intensity depression is about 0·1 times the continuum intensity. They have magnetic fields of the order of 1000 gauss (up to 1400 gauss) and diameters between 1″ and 12″. An important and still open question: what is the role of the magnetic knots in the formation of spots?

During the development of spotgroups, individual spots may show strong motions with respect to each other. After the formation of spots the facular region extends in area: the K emission appears to spread over the whole supergranular region. At that time the first phase of development of an active region may be considered as ended; the main phase of development sets in; it is characterized by intense chromospheric and coronal activity and in particular by flares. We will discuss the main phase of development of the photospheric and chromospheric parts of the active region in Section 4, the coronal and interplanetary parts in Section 5; flares will be discussed in Section 6.

## 4. The Main Phase of Development; (a) Photosphere and Chromosphere

The areal development of the photospheric and chromospheric facular regions was investigated particularly well during the CSSAR period (Cooperative Study of Solar

Active Region). It appears that the plages reach their maturity in less than 1 month and decline slowly in activity in a period of the order of 3–5 months. It is known, of course, that occasionally plages may live much longer.

The relation of the facular fine structure and network to the supergranulation and the magnetic field should be mentioned. Plages, in particular those that are not yet too bright, show in chromospheric images a kind of network structure which seems to extend into the quiet chromospheric network. However, it should be made clear that the identity of the network in active regions and a possible active-region super-granulation has not yet been proved. The concept supergranulation is defined by the photospheric-chromospheric motion field; supergranulation is a large structure in which material is moving towards the borders where the material apparently moves downward. At the borderline magnetic fields occur. In the quiet Sun the supergranu-lation appears to be identical with the chromospheric network; the same need not apply to the network in the active region.

Oscillatory motions have been discovered both in the quiet and in the active regions. Their periods, being of the order of 5 min in the undisturbed photosphere increases to 45 min in the disturbed region; the amplitude seems to decrease with increasing field. These motions are correlated to brightness variations; not to variations of the magnetic field.

An important problem with regard to the sunspot is that of the energy balance in spots. This problem, in turn, is directly connected to that of the 'missing energy flux': where does the sub-photospheric convective energy flux go, that does not escape in the form of radiation in the region of a sunspot? This energy flux is large and, if used 100% efficiently, would provide the energy for a large flare in about 1000 sec. Danielson argued that part of this energy flux may escape sidewards in sub-photos-pheric regions in the form of gravity waves. It is still unclear what fraction of the energy moves up vertically in the form of hydromagnetic waves.

The diameter of the umbral granules may be at most 500 km; but they may even be as small as 160 km. In the latter case the umbral granules would have the same surface brightness as the normal photosphere, which is not impossible; they have the photospheric colour temperature too.

An interesting feature in the chromosphere above active regions is formed by the 'bright points', seen in the Hα wings (moustaches, bombs). They have diameters of 1″ to 5″ and lifetimes of the order of 25 min. Bruzek drew attention to the close relation between these features and his newly discovered arch-shaped filaments. These filaments occur between small areas of opposite polarity in active regions. They cross the neutral line and nearly always connect bright points situated at their bases. Material motions with velocities of the order of some tens of km per second occur along the filaments which tend to expand.

There is a clear difference between the above-mentioned arched filaments and the other filamentary structures in or above the solar chromosphere: the fibrilles and the classical filaments, which are nothing but the prominences projected on the disk. High-resolution pictures show a relation between filaments and the fibrilles: the first may be the result of some shearing motion. It is not unlikely that these fibrilles are related to a sheared magnetic-field structure carrying the filaments. It may be significant that fibrilles situated beyond the ends of filaments are directed more or less perpendicularly to the filamentary direction, whereas the fibrilles at the sides of the filament tend to be inclined at only small angles to it.

## 5. Main Phase of Development of an Active Region; (b) the Coronal and Interplanetary Part

By means of observations of forbidden lines in the visible spectral region the solar corona is observed at the limb; disk observations are made by means of X-ray and radio measurements. In all cases the resolution is bad: at the limb because we integrate over a long line of sight, and on the disk because the observing techniques are still rather primitive.

It appears that the ratio between the intensities of the red and the green coronal lines vary systematically during the development of an active region. A similar behaviour has been found for various ion lines in the EUV. It appears very hard to explain this variation without the introduction of inhomogeneities in the coronal extention of the active region; in particular this seems to be necessary in the beginning phase of the development of the coronal active region; the inhomogeneities may disappear later.

Attempts have been made to compute the coronal magnetic fields from the given photospheric boundary values, assuming a potential field. The results of these computations appear to be in conflict with other observations. This clearly indicates that the coronal parts of active regions are not current-free, at least not at the 'small' scale of the corona. However, it may be that on a larger, interplanetary scale the field is indeed current-free.

It is also clear that the structures *above* the coronal active regions are for the greater part defined by the magnetic field, which, in turn, is defined by three different quantities: the photospheric field, coronal electrical currents, and the solar wind. It is a general rule that above *all* active chromospheric regions dense coronal regions occur. The temperature of these dense regions is of the order of $2 \cdot 5 \times 10^6 \, ^\circ\text{K}$. Recent measurements of the electron densities in coronal active regions show these to be about 2–5 times the quiet coronal density. However, the *shape* of the active region depends on the magnetic structure and its state of evolution. The well-known coronal bushes and helmets may correspond to closed field lines and occur in the early phase of the life history of the coronal active regions. In later phases one observes the streamers, characterized by their pinching tendency, and the rays, with their sharp structures extending radially from the Sun. The rays have been seen with rocket-borne coronameters, and appear to stretch out to distances of many solar radii. They do not show any pinching tendency and apparently correspond to a structure with completely open field lines.

In the interplanetary medium the sector structure, discovered by Wilcox and Ness, is related to plages in the preceding parts of each of the sectors. It is not impossible that these structures are current-free and correspond to closed field lines.

## 6. Flares

The Proton Flare Project (PFP) organized by Švestka and Simon has appeared to be extremely useful for increasing our knowledge about the very energetic events related to solar flares. *Where* and *when* do flares occur?

New additional information has been advanced for showing that flares occur preferably in magnetically complex active regions; in particular near regions where the gradient of the longitudinal field is strong. Flares tend to favour regions where new plages or spotgroups are born, in particular where they are included in old ones. Dr. S. Smith and Howard have mentioned the importance of the reversed polarity regions, which produce more flares than simple bipolar magnetic regions. In particular islands of opposite polarity in large magnetic regions are places favoured by flares. The Ellerman bombs seem to favour the regions where $|B| = 0$.

Flares seem to occur in the above-mentioned regions at moments when the magnetic energy of the active region has greatly increased. Some cases have been mentioned by Severny. In one case it appeared that the magnetic energy already started to decrease, a few hours before the occurrence of the flare.

After the occurrence of the flare the transverse field component may appear to have changed its direction by 90° and tends to be oriented more parallel to the 'neutral' line.

Flares are known to show a repetitive tendency; for radio flares it has been shown that the period during which homologous flares can occur in one active region is not longer than about 1–2 days.

It seems a fairly general rule for plages to show a small decrease in brightness just prior to the brightness increase of the flare; later the integrated Hα brightness may show pulsations.

Optical flares are often accompanied by surges and similar phenomena, indicating the acceleration of a low-temperature plasma. These accelerations seem to be related to the propagation of the fast hydromagnetic mode.

The *optical flare plasma* finds its counterpart in the *high-energy flare plasma*. It is a plasma, containing particles accelerated to energies of the order of 10 to a few hundreds of keV's. The lower the average energy of the plasma the larger the amount of accelerated electrons; very roughly the product of the average particle energy with the total number of accelerated electrons is constant. The acceleration to these energies takes place in the very first phase of the flare process as shown by X-ray and microwave type-IV radio emission, as well as by type-III radiobursts. A second acceleration, whereby the energy is multiplied by another factor, of the order of $10^3$, seems to occur in a later phase of the flare; the metric type-IV emission suggests that it occurs perhaps about a quarter to a full hour later.

Since this later acceleration may be due to a kind of Fermi process, flare theories basically need only explain the first acceleration process. Three different sources for the flare energy have been mentioned. The first is the shearing energy of photospheric motions, a process advocated by Sturrock. Another source may be the 'missing' convective energy flux from below the sunspots which would produce sufficient energy for a large flare in about 15 min (De Jager). A third possible energy source was mentioned by Danielson and is the turbulent energy flux inside sunspots. However, it is not sufficient simply to mention the energy source; further unknown factors are (a) how to *store* the energy: perhaps in the form of magnetic energy; and (b) how to *free* it. As a possible triggering mechanism a current discharge as propagated in studies by Severny and Alfvén might be mentioned.

Little has been said about coronal events occurring in connection with or after flares; radio and direct optical measurements indicate that the electron density in the low corona seems to increase after a flare.

## 7. The Last Phases of the Active Region

With regard to the last phases we only want to make brief mention of a few problems; no clear solutions have as yet been given.

What is the character and the scope of the poleward motion of filaments and the coronal activity; how does it relate to the problem of the dissolution and the partial neutralisation of the magnetic field? What happens to the field lines in the low solar corona in the last phases of the active region; are they neutralised and how does this happen? What is the significance of the ghost fields in relation to the evolution of active regions? Future research may answer these questions.